Springer Optimization and Its Applications

Volume 172

Aims and Scope

Optimization has continued to expand in all directions at an astonishing rate. New algorithmic and theoretical techniques are continually developing and the diffusion into other disciplines is proceeding at a rapid pace, with a spot light on machine learning, artificial intelligence, and quantum computing. Our knowledge of all aspects of the field has grown even more profound. At the same time, one of the most striking trends in optimization is the constantly increasing emphasis on the interdisciplinary nature of the field. Optimization has been a basic tool in areas not limited to applied mathematics, engineering, medicine, economics, computer science, operations research, and other sciences.

The series **Springer Optimization and Its Applications (SOIA)** aims to publish state-of-the-art expository works (monographs, contributed volumes, textbooks, handbooks) that focus on theory, methods, and applications of optimization. Topics covered include, but are not limited to, nonlinear optimization, combinatorial optimization, continuous optimization, stochastic optimization, Bayesian optimization, optimal control, discrete optimization, multi-objective optimization, and more. New to the series portfolio include Works at the intersection of optimization and machine learning, artificial intelligence, and quantum computing.

Volumes from this series are indexed by Web of Science, zbMATH, Mathematical Reviews, and SCOPUS.

More information about this series at http://www.springer.com/series/7393

Roman A. Polyak

Introduction to Continuous Optimization

Springer

Roman A. Polyak
School of Mathematical Sciences
Technion, Haifa, Israel

ISSN 1931-6828 ISSN 1931-6836 (electronic)
Springer Optimization and Its Applications
ISBN 978-3-030-68711-3 ISBN 978-3-030-68713-7 (eBook)
https://doi.org/10.1007/978-3-030-68713-7

This Springer imprint is published by the registered company Springer Nature Switzerland AG
The registered company address is: Gewerbestrasse 11, 6330 Cham, Switzerland

In memory of our beloved Lubochka

"Nothing in the World takes place without Optimization, and there is no doubt that all aspects of the World, that have a rational basis, can be explained by Optimization Methods."

1744 Leonard Euler

Preface

For decades, substantial effort had been made to transform a constrained optimization problem into a sequence of unconstrained problems. The fundamental progress in unconstrained optimization stimulated such effort.

The results of this effort were summarized in the remarkable classic Sequential Unconstrained Minimization Technique (SUMT) by Anthony Fiacco and Garth McCormick.

Any constrained optimization problem can be replaced by an equivalent unconstrained, but a non-smooth optimization problem. The function of the equivalent problem is non-smooth exactly at the minima, no matter how smooth the data is for the original problem.

Any SUMT method replaces the non-smooth function by its smooth approximation, which is scaled by a positive scaling parameter. The most popular approximations are based on the log-barrier and the interior distance functions. Such approximations however, are singular at the solution. Therefore, the quality of these approximations at the solution and its neighborhood is very poor.

To find an accurate solution for the initial problem, one has to find the minima of such a smooth approximation under a very small scaling parameter. This can be an impossible task due to the vanishing condition number of the corresponding Hessian. Therefore, in the late 1970s and early 1980s, SUMT got less popular.

At this point, it was clear to the author that the efficiency of a SUMT type technique can be substantially improved by using the Lagrange multipliers explicitly. It led to the nonlinear rescaling theory and exterior point methods.

The basics of the nonlinear rescaling theory and the exterior point methods have been developed in the early 1980s, but, unfortunately, the results were published only 10 years later.

Meanwhile, in 1984, N. Karmarkar published his projective transformation method for LP calculation. Soon after, P. Gill et al. found similarity between N. Karmarkar's method and Newton log-barrier method for LP calculation. It reignited interest in the classical log-barrier and interior distance functions.

The SUMT techniques were replaced by the path-following methods, which do not find at each step a primal minimizer.

Instead, it alternates one Newton step toward the minimizer with a careful scaling parameter update. In other words, from the so-called "warm" start on, the path-following methods are moving from one "warm" start to another "warm" start, reducing the objective function value by a factor, which depends only on the size of the problem.

The path-following trajectory is contained in the interior of the feasible set, therefore such methods were called Interior Point Methods.

Using path-following technique, C. Gonzaga and J. Renegar not only established polynomial complexity, but substantially improved N. Karmarkar's complexity bound for LP.

Most importantly, in contrast to the ellipsoid method, the theoretical complexity of the interior point methods turned out to be much more consistent with their practical performance.

From this point on, for many years, the interior point methods became the mainstream in Modern Optimization.

In the yearly 1990s, G. Dantzig's simplex method, one of the ten best algorithms of the twentieth century, lost it dominance, in particular, when it comes to the large-scale LP.

In the mid-1980s, an important simplification of N. Karmarkar's method, the affine scaling algorithm for LP had been found independently by E. Barnes and R. Vanderbei et al. The affine scaling algorithm played an important role in understanding the numerical efficiency of interior point methods.

Soon after, the Optimization community learned that the affine scaling algorithm, in fact, had been introduced in 1967 by Ilya Dikin, a student of Leonid Vitalievich Kantorovich.

Similar to L.V. Kantorovich's work on LP before WW2, the affine scaling algorithm waited 20 years before the Optimization community became aware of its existence.

Meanwhile, a few thousand papers were published on interior point methods. There was a need to understand the roots and to have a unique and general point of view on these developments.

The fundamental change in the political system of the former Soviet Union, called "Perestroika," made it possible to leave the Soviet Union and travel freely abroad.

Therefore, in the late 1980s, the Optimization community became aware of the remarkable self-concordance theory, soon after it was developed by Yurii Nesterov and Arkady Nemirovski. The self-concordance theory explains the interior point methods complexity results from a unique and general point of view.

The authors also extended the results on some Nonlinear Optimization problems and Semidefinite Optimization.

At the heart of the self-concordance theory, there are two notions: the self-concordant functions and the self-concordant barrier.

It turned out that a strictly convex and three time differentiable function is self-concordant, if its Legendre-Fenchel invariant is bounded.

Both the Legendre-Fenchel invariant and the Legendre-Fenchel identity are critical throughout the book, because the duality is the main theme of the book.

Duality is the key ingredient of both the nonlinear rescaling and the Lagrange transformation theories and correspondent exterior point methods. Duality is used for proving convergence, establishing convergence rate, finding complexity bounds, and making exterior point methods efficient and robust.

The exterior point methods are based on ideas fundamentally different from both SUMT and IPMs and can be used for constrained optimization problems, which can't be equipped with self-concordant barrier.

Over the last 30 years, exterior point methods were proven to be efficient tools, in particular, when it comes to large-scale and difficult constrained optimization problems, which require accurate solutions.

One of the main purposes of the book is to show what makes the exterior point methods numerically attractive and why.

The nonlinear rescaling theory is the foundation for one of the best constrained optimization solver PENNON.

The book has five parts. The first part contains basics of calculus, convex analysis, elements of unconstrained optimization, as well as classical results of Linear and Convex Optimization.

The second part contains the basics of self-concordance theory and interior point methods, including complexity results for LP, QP, and QP with quadratic constraint, semidefinite and conic programming.

In the third part, we describe the nonlinear rescaling and Larangian transformation theories and exterior point methods. It includes the theories of modified barrier and exterior distance functions and correspondent exterior point methods.

We consider the primal, the dual, and the primal–dual aspects of the exterior point methods.

In the fourth part, we consider three important problems of finding equilibrium: optimal resource allocation - an alternative for LP, nonlinear input-output equilibrium - an alternative for the classical Wassily Leontief's input-output model, and J. Nash's Equilibrium of n persons concave game.

For all these problems, we use first-order projection methods because projection on their feasible sets is a low-cost operation.

The methods, in fact, are pricing mechanisms for establishing equilibrium. Their convergence, convergence rate, and complexity were established under natural assumptions on the input data.

In the final part of the book, we consider several important applications arising in economics, structural optimization, medicine, and statistical learning theory, just to mention a few.

We also provide numerical results obtained by solving a number of both real life and test problems. The results strongly justify the theoretical findings. In particular, the numerical results confirmed, typical for the exterior point methods, the so-called "hot" start phenomenon, which leads to substantial convergence acceleration in the final phase of the computational process.

A large part of the book contains results that have never been covered in a systematic way in optimization literature.

The book can be used in a two-semester course in continuous optimization for master's students in mathematics. It can also be used by graduate students in applied mathematics, computer science, and economics as well as by researchers working in optimization and those applying optimization methods for solving real-life problems.

There has been a profound "Perestroika" in the field of Continuous Optimization in the last 30 years. We tried to cover only some of the basic ideas and results, which, as it seems to us, transformed the field.

The turning point in Optimization coincided with the turning point in the author's life.

In February 1988, after almost 9 years of my tenure as a "refuznik," I finally left the Soviet Union and at the end of June 1988 arrived in Boston. Two days later, I took the liberty to participate (without an invitation) in an unforgettable event: the Interior Point Methods Workshop at the Bowdoin College in Maine.

This was my first interaction with Western colleagues. I was very impressed not only with the quality of the presentations, but also with the very warm and friendly environment at the Workshop.

I was also pleasantly surprised that David Shanno and Garth McCormick were aware of my Modified Barrier Functions manuscript, which I had smuggled to the West long before my departure from the Soviet Union. Later, I learned that David handled my manuscript as an AE of "Mathematical Programming" and Garth was one of the referees.

Soon after, I was offered a visiting scientist position at the Mathematical Sciences Department, IBM T.J. Watson Research Center. The original 9-month appointment lasted 4 years. For me it was a transition from Hell to Heaven.

Then, the George Mason University, where I was a Math. and OR Professor for 21 years. Currently I am a visiting professor in the Department of Mathematics at the Technion.

Monsummano, Italy
Haifa, Israel

Roman A. Polyak

Contents

Chapter 1
Introduction

The first steps in optimization go back to ancient times, when several isoperimetric problems were solved.

The first optimization tools appeared together with calculus in the seventeenth century. Pierre de Fermat introduced in 1629 the necessary condition for an unconstrained optimum. Later Isaac Newton and Gottfried Wilhelm Leibniz formulated the second-order optimality conditions for an unconstrained optimum in terms of second derivatives. Isaac Newton and Joseph Raphson developed one of the most celebrated methods in mathematics, the Newton–Raphson method for solving nonlinear equations and finding unconstrained optima.

At the end of the eighteenth century, Joseph–Louis Lagrange introduced the necessary condition for optimization problems with equality constraints. Later Adrien–Marie Legendre introduced his transformation, which became an important instrument in modern optimization.

In the nineteenth century, Augustin–Louis Cauchy introduced his gradient method, one of the basic tools in numerical optimization.

The first steps in modern optimization were made in the mid-1930s of the last century, when a number of real life problems, dealing with optimal allocation of limited resources, attracted the attention of Leonid Vitalievich Kantorovich, then a young and brilliant mathematician.

L.V. Kantorovich found that all these problems are about finding the minimum or the maximum of a linear function under linear equality and inequality constraints. Obviously, L.V. Kantorovich recognized that one can find the solution of such problem among vertices of the correspondent polyhedron. Then, L.V. Kantorovich established optimality criteria for such vertex by introducing Lagrange multipliers, which are now known as shadow prices.

His booklet *Mathematical Methods in the Organization and Planning of Production* has been recognized in the optimization community as the first step in the creation of a new mathematical discipline – linear programming (LP).

The 1930s was not the best time in the Soviet Union for discussions of pricing issues in a socialist economy.

R. A. Polyak, *Introduction to Continuous Optimization*, Springer Optimization and Its Applications 172, https://doi.org/10.1007/978-3-030-68713-7_1

L.V. Kantorovich called the Lagrange multipliers “objectively conditioned estimates.” Still, his book *The Best Use of Economic Resources* was published with a 20 years’ delay.

In 1975 Leonid Kantorovich shared with Tjalling Koopmans the Nobel Prize in Economics “for their contributions to the theory of optimal allocation of resources.”

Wide areas of LP applications arising in economics, industrial and military fields, as well as the specific mathematical structure of LP, attracted mathematicians after WW2.

In 1947 George B. Dantzig introduced the simplex method for solving LP problems. For several decades simplex method was the method of choice for solving LP. Its efficiency was widely recognized; it was one of ten best algorithms in the twentieth century.

G. Dantzig’s contributions to both the theory and computational aspects of LP were fundamental. Together with his colleagues in the 1950s and early 1960s, he transformed LP into a new discipline with its own theory and methods. Progress in implementation of the simplex method together with much improved computational capabilities allowed solving real life problems from wide areas of application.

George B. Dantzig summarized the results in his book *Linear Programming and Extensions*, published in 1963.

Substantial progress in understanding the mathematical aspects of LP along with impressive real-world applications lead to forming optimization as a mathematical discipline that considerably differs from other areas of mathematics. It stimulated correspondent research around the world, in particular in the former Soviet Union, where, in fact, LP was born.

Almost at the same time John von Neumann introduced the LP duality and pointed out to the connection between matrix games and dual pairs of LP.

The Western optimization community was very active in the 1950s, much progress was made in understanding LP problems from both theoretical and numerical stand point.

The results were summarized in the volume *Linear Inequalities and Related Systems* edited by Harold Kuhn and Albert Tucker. L.V. Kantorovich initiated translation into Russian, and the book became the main reference in optimization in the Soviet Union at the end of the 1950s.

The first steps in modern nonlinear optimization were made in the 1950s.

In 1951 H. Kuhn and A. Tucker found necessary and sufficient optimality condition for convex optimization. As it turned out W. Karush proved the correspondent theorem in his Master thesis in 1939.

In 1955 K. Frisch introduced the log-barrier function for constrained optimization, which was the first step in developing later by A. Fiacco and G. McCormick the SUMT.

In 1956 M. Frank and P. Wolfe introduced the conditional gradient method for quadratic programming (QP).

The 1960s and the 1970s were the “golden age” for optimization in both Western and Eastern optimization communities.

In the Soviet Union in the early 1960s, a few centers with a number of talented young mathematicians got involved in optimization.

General necessary optimality conditions, which can be traced to L.V. Kantorovich's paper (1940), have been developed in the 1960s for wide classes of constrained optimization problems in both finite and infinite dimensional spaces (Dubovitskii–Miljutin, Girsanov, Pshenichnyj, E. Golshtein, Ioffe, Tikhomirov, Demyanov, Rubinov, Halkin, Neustadt, Mangasarian, Fromovitz).

A very influential book *Convex Analysis* by R.T. Rockafellar was written in 1967 and published in 1970.

A number of now classical methods for convex optimization have been developed: the feasible direction methods in finite and Hilbert spaces (Zoutendijk, Zuchovitsky, R. Polyak, Primak, later Demyanov, Rubinov, E. Polak, Ben-Tal); the method for simultaneous solution of the primal and the dual convex optimization problems (R. Polyak); non-smooth optimization (Shor, Ermoliev, B. Polyak, Eremin, later Wolfe, Lemarechal, Goffin, Mordukhovich); SUMT (Fiacco, McCormick later Grossman, Kaplan, Auslender, Cominetti, Haddou); the projected gradient method (Rosen, A. Goldstein, Levitin, B. Polyak); the conditional gradient method for the convex optimization (Demyanov, Rubinov, Levitin, B. Polyak); the minimization methods for a class of concave functions under linear constraints (Zuchovitsky, R. Polyak, Primak); the center methods (Huard, Newman, A. Levin later E. Polak, Mifflin, Grossman, Kaplan, Sonnevend, Primak); the augmented Lagrangians, multipliers, quadratic proximal point methods (Hestenes, Powell, Rockafellar, Bertsekas, B. Polyak, E. Golshtein, Tretyakov, Moreau, Martinet, Yosida); and the convergence acceleration in convex optimization (R. Polyak).

In their classical paper "Minimization methods under constraints" (1966), E. Levitin and B. Polyak established new standards: along with the convergence, the convergence rate was estimated for a number of constrained optimization methods.

In 1967 L.V. Kantorovich's student Ilya Dikin introduced the affine scaling (AS) method for LP. Unfortunately, Dikin's result was practically unknown for about 20 years. The AS method has been rediscovered by Barnes (1986) and independently by Vanderbei et al. (1986) as a simplification of N. Karmarkar's method.

It becomes evident that progress in LP is very much dependent on continuous optimization. Such a point of view was justified in the 1970s, when independently N. Shor and A. Nemirovski–D. Yudin developed the ellipsoid method – another product of continuous optimization.

Leonid Khachiyan, using the ellipsoid method for LP, proved its polynomial complexity. The result led to substantial progress in the complexity theory and got an indisputable theoretical value.

Unfortunately, numerically the ellipsoid method for LP was a disappointment: the exponential simplex method almost always outperformed the polynomial ellipsoid method.

Measuring properly efficiency of optimization methods became one of the main trends in optimization in the early 1980s. The book *Problem Complexity and Method Efficiency in Optimization* by Nemirovski and Yudin (1983) turned the efficiency of optimization methods and the complexity of classes of optimization problems into one of the main issues in optimization.

The basic results of extensive developments in linear and nonlinear optimization in the 1960s and 1970s have been summarized in a number of books, in particular in a very informative book *Introduction to Optimization* by B. Polyak (1987).

The 1980s and the 1990s were decades of IPMs. All started with Karmarkar's (1984) projective transformation method for LP.

Then, the connection of N. Karmarkar's method with the Newton log-barrier method for LP calculation established by Gill et al. (1986), much stronger complexity results found by Gonzaga (1989) and Renegar (1988), and the SC theory by Nesterov and Nemirovski (1994), as well as much improved consistency of the theoretical complexity bounds and the practical performance, lead to an unprecedented activity in IPMs.

In the early 1990s, simplex method lost its dominance after Lustig et al. (1991, 1992, 1994) developed their packages based on IPMs. This was the turning point in optimization.

For a number of years, when IPMs was the "hottest" topic in optimization, the author together with some of his colleagues and students promoted EPMs.

The EPMs attracted substantially less attention than IPMs, mainly because of the lack of polynomial complexity.

On the other hand, EPMs are based on ideas, which are fundamentally different from those of both SUMT and IPMs. Over the last three decades, EPMs prove to be robust and efficient in finding solutions with high accuracy for large-scale and difficult constrained optimization problems.

The simplest NR scheme employs smooth, monotone increasing, strictly concave functions with particular properties to transform the constraints of a given constrained optimization problem into an equivalent set of constraints. The constraint transformation is scaled by a positive scaling (penalty) parameter.

The Lagrangian for the equivalent problem (LEP) is the main NR instrument. The NR method alternates finding an approximation for the primal unconstrained LEP's minimizer with Lagrange multipliers update, while the scaling parameter can be fixed or updated from step to step.

The shifted log-barrier transformation, used in the framework of NR, leads to the LEP called modified barrier function (MBF).

The MBF is free from the main shortcomings of both the classical Lagrangian for the original problem and log-barrier function and at the same time combines their best features.

Under the standard second-order optimality condition and a fixed, but large enough, scaling parameter, the correspondent EPM converges to the primal–dual solution with Q —linear rate, and the ratio is inversely proportional to the penalty parameter.

Also the MBF is smooth and strongly convex at the primal solution under optimal Lagrange multipliers. Moreover, MBF keeps these very important properties in the neighborhood of the primal minimizer under any Lagrange multiplier vector from the neighborhood of the dual solution.

It contributes to numerical stability, substantially speeds up convergence in the final phase and allows finding solutions with high accuracy.

The interior and exterior point methods and renewed interest to the first-order optimization technique, in particular to the fast gradient methods, as well as new and very important applications, fundamentally reshaped the field of continuous optimization.

The book is about the most important, as it seems to us, results and ideas that transformed continuous optimization in the last 30 years.

To make the book self-contained, we provide in Chapter 2 basic facts from calculus and convex analysis. Particular attention is given to the Legendre–Fenchel transform. It leads to two important notions, Legendre–Fenchel identity and Legendre–Fenchel invariant, which are critical throughout the book.

In Chapter 3 we concentrate on two unconstrained optimization methods: the gradient and the Newton–Raphson. Efficient unconstrained optimization techniques are critical for constrained optimization, because, as we will see later, there are a number of ways a constrained optimization problem can be reduced to a sequence of unconstrained problems.

Along with the classical gradient method (1847), which goes back to Augustin–Louis Cauchy (1789–1857), we consider the subgradient method (1962) for convex, but non-smooth functions by N. Shor, as well as the fast gradient method (1983) by Yu. Nesterov, for convex functions with a Lipschitz gradient.

The rate of convergence and the complexity of gradient-type methods, under various assumptions on convexity and smoothness of the function, are our main focus.

For Newton's method, along with the classical local quadratic convergence, we consider its complexity, that is, the number of Newton steps required for finding an ε- approximation to the solution.

For strictly convex function, Newton's method, generally speaking, converges only locally, near the solution.

Therefore, we introduce the regularized Newton method, which is free from this limitation. We took the parameter of the quadratic regularization equal to the Euclidean norm of the gradient at the point.

Due to the vanishing regularization parameter, the regularized Newton method converges globally and keeps asymptotic quadratic convergence rate for any strictly convex and smooth enough function, for which the minimizer exists. The area of quadratic convergence is characterized through basic smoothness and convexity parameters of the given function.

In Chapter 4 we consider optimality conditions for optimization problems with both inequality constraints and equations. On the top of the standard first- and second-order necessary and sufficient conditions, we proved the Eigenvalues Theorem. The Theorem selects the essential parameters and shows how their interaction with optimal Lagrange multipliers leads to the second-order sufficient optimality conditions.

In the second part of the chapter, we consider some classical methods for equality constrained optimization with emphasis not only on the primal but also on the dual and the primal–dual aspects.

In particular, the classical Courant's penalty method (1943) turns out to be equivalent to the classical Tikhonov's regularization method (1963) for the dual problem.

Along with the standard Augmented Lagrangian (AL) results, we introduce the primal–dual AL method.

Instead of finding approximation for the primal unconstrained optimizer following by the Lagrange multipliers update, the primal–dual AL at each step solves a linear system of equations and, instead of linear convergence rate, converges locally with quadratic rate.

The key idea: the penalty parameter is update inversely proportional to the vanishing merit function. For the primal–dual setting, the condition number of the AL Hessian is irrelevant; therefore, a drastic increase of the penalty parameter does not compromise computations and allows to prove the local quadratic convergence rate.

In Chapter 5 we consider basic facts of LP and convex optimization theory, as well as a few classical methods, sometimes using a non-traditional approach.

In particular, SUMT methods are viewed as an interior regularization for the dual problem. It allows to develop a unified approach for convergence analysis and establish the error bounds for a wide class of SUMT methods. The Legendre–Fenchel identity is the key instrument.

Along with the classical simplex method, we consider three IPMs for LP calculations: the Newton log-barrier, the primal–dual, and the affine scaling.

After N. Karmarkar published his projective transformation method, thousands of papers were published in a relatively short time. What was lacking is a general viewpoint on all these developments and a better understanding of the IPM's roots.

In the late 1980s, Yurii Nesterov and Arkadi Nemirovski published their SC theory. There are two critical notions on which the SC theory is based: the SC function and the SC barrier.

We recall a strictly convex and three times differentiable function is self-concordant if its Legendre–Fenchel invariant is bounded.

The boundedness of the Legendre–Fenchel invariant leads to an important differential inequality. By integrating it sequentially four times, one obtains the basic properties of SC functions. All these results comprise the first part of Chapter 6.

In the second part, the SC properties are used for establishing complexity bounds for LP. The final part contains the primal–dual predictor–corrector method. Also the complexity results were extended for QP, as well as QP with quadratic constraints and semidefinite and conic programming.

The EPMs are products of the NR and the LT approaches to constrained optimization. Duality is critical for developing EPMs, their convergence analysis, and establishing convergence rate, as well as for making EPMs fast and numerically stable.

The LT approach is also used for establishing connection between interior and exterior point methods. In the next three chapters, we discuss the NR and LT theory and correspondent EPMs.

In Chapter 7 we describe the basic facts of the NR theory and EPMs. The first NR results, in particular, the MBF theory and methods, were developed in the early 1980s.

The purpose of the research was to find an alternative to SUMT that is free from its most essential deficiencies. Unfortunately, the corresponding publication appeared in "Mathematical Programming" more than 10 years later, in 1992.

Roughly speaking, the NR is to SUMT for inequality constrained optimization as AL is to Courant's penalty method for equality constrained optimization (ECO).

Numerical realization of NR leads to the Newton NR method. Newton's method finds an approximation for LEP's primal minimizer, which is used for the Lagrange multipliers update.

The NR method is equivalent to the interior proximal point method for the dual problem. The entropy – like distance of the dual prox – is based on the kernel, that is, the negative of the Legendre–Fenchel transform of the constraints transformation; see Polyak and Teboulle (1997).

The fundamental difference between SUMT and NR is the latter has no limitation for constraints violation, while the dual vector is always positive due to the way the Lagrange multipliers updated and the properties of the constraint transformations.

Thus, NR methods are primal exterior and dual interior. The primal feasibility is not due to an unbounded increase of the scaling parameter, but rather due to the "soft power" of the LM update.

In fact, the LM update can be viewed as a pricing mechanism for establishing equilibrium between the objective function reduction and the penalty for constraints violation. The equilibrium is given by the Karush–Kuhn–Tucker's Theorem.

We consider three aspects of NR methods: the primal, the dual, and the primal–dual.

Under the standard second-order sufficient optimality condition; the NR converges with Q-linear rate for any fixed, but large enough scaling parameter, no matter the original problem is convex or not. Moreover, the primal–dual NR, under such assumptions and smooth enough input data, converges locally with quadratic rate.

In Chapter 8 we consider three important NR realizations. The first is MBF, an alternative to K. Frisch's log-barrier function; the second is exterior distance function (EDF), an alternative to P. Huard's interior distance function (IDF); and the third is log-sigmoid (LS) Lagrangian, an alternative to the smoothing technique, introduced by Chen and Mangasarian (1995).

The primal MBF is equivalent to the dual proximal point method with Kullback–Leibler's entropy distance function.

The dual prox is closely related (see Eggermont (1990)) to the well-known multiplicative EM algorithm by Shepp and Vardi (1982) (see also Vardi et al. (1985)).

For LP calculation Powell (1995) proved: the primal MBF sequence converges to the Chebyshev center of the optimal face under any fixed positive scaling parameter.

In contrast to K. Frisch's log-barrier function, the MBF is not singular at the primal solution. Moreover, if the input data is smooth enough, then for any nondegenerate constrained optimization problem and a fixed, but large enough scaling parameter, the MBF is strongly convex at the primal solution under optimal Lagrange multipliers. It keeps this property in the neighborhood of any primal minimizer for any Lagrange multipliers vector from the neighborhood of the dual solution.

Together with the Q—linear convergence rate, it leads to the "hot" start phenomenon: there exists a point on the MBF primal trajectory that after each LM update the primal iterate remains in the Newton area for the minimizer of the updated MBF. In other words, from the "hot" start on, the MBF moves from one "hot" start to another "hot" start.

It reduces dramatically the number of Newton steps per Lagrange multipliers update, which improves substantially both the convergence rate and the complexity bound.

The neighborhood of the primal–dual solution, where the "hot" start occurs, is characterized through the basic parameters of a non-degenerate convex optimization problem. For the non-degenerate QP, much improved complexity bounds were established by Melman and Polyak (1996).

Due to the "hot" start, for all NR realizations obtaining an extra digit of accuracy is much easier toward the end of the process, which is a fundamental departure from SUMT.

After a few updates, the Lagrange multipliers, which correspond to the passive constraints, become negligibly small. This leads to a substantial reduction of the number of constraints.

Finally, the shifted log-barrier function is a SC barrier, so for a number of constrained optimization problems, including LP and QP, one can use IPM far from the solution and MBF near the solution. It improves substantially the standard complexity bounds for wide classes of non-degenerate LP, QP, and QP with quadratic constraints.

The IDF is the second important IPM tool. The minimizer of IDF is the "center" of the relaxation feasible set, which is the intersection of the objective function level set at the attained level and the feasible set. The relaxation feasible set is updated at each step, using the objective function value at the new "center." The obtained "centers" define the central path.

Using the central path, Renegar (1988) substantially improved Karmarkar's complexity bound for LP calculation.

However, application of the center method for constrained optimization problems, for which IDF is not a SC barrier, leads to substantial difficulties due to IDF singularity at the primal solution.

Therefore, we introduce the exterior distance function (EDF) and develop the corresponding theory.

The log-barrier and the shifted log-barrier functions are used to transform the objective function and the constraints of a given constrained optimization problem into an equivalent one. Then, the correspondent LEP is used in the NR framework.

The EDF is free from the main IDF deficiencies and possesses all the best MBF properties. On top of it, the EDF has an extra tool, the center, which can be used for improving the convergence rate.

The third NR realization leads to an alternative to the smoothing technique by C. Chen and O. Mangasarian.

The classical LS is modified and used in the NR framework with "dynamic" scaling parameters, that is, each constraint has its own scaling parameter, which is updated, at each step, inversely proportional to the corresponding current Lagrange multiplier.

Such scaling parameters update was suggested by Tseng and Bertsekas (1993) for exponential transformation.

First, we show that the correspondent NR method is equivalent to the interior proximal point method with second-order Fermi–Dirac entropy distance.

Second, the interior prox, in turn, is equivalent to the dual interior ellipsoid method in the rescaled from step to step dual space. This is the basis for convergence analysis, which produced a number of new convergence results.

In Chapter 9 we consider a new class of multipliers methods based on LT. The basic LT scheme transforms terms of the classical Lagrangian for the original problem, associated with constraints. The transformation is rescaled by a positive scaling parameter. The corresponding multipliers method finds an approximation for the primal LT's minimizer and then uses it for the Lagrange multipliers update, while the scaling parameter can be fixed or updated from step to step.

The primal exterior LT is equivalent to the dual interior prox with Bregman-type distance. The distance function is induced by a kernel, which is the negative of the Legendre–Fenchel transform of the original transformation. The prox method with Bregman-type distance, in turn, is equivalent to the interior ellipsoid method for the dual problem.

The LT with truncated MBF transformation was studied by Matioli and Gonzaga (2008); they called it MBF2 method. They show that MBF2 is equivalent to the dual interior proximal point method with Bregman's distance induced by the standard log-barrier function.

The correspondent kernel, which is the negative of the Legendre–Fenchel transform of the shifted log-barrier function, is a SC function. Therefore, the dual prox is equivalent to the dual interior ellipsoid method with Dikin's ellipsoids.

Application of the LT method with truncated MBF transformation to LP leads to affine scaling-type method for the dual LP.

In Chapter 10 we consider a general problem of finding a nonlinear equilibrium (NE) and two first-order methods: pseudo-gradient projection (PGP) and extra pseudo-gradient (EPG) for finding NE.

Particular emphasis is given to finding NE for the optimal resource allocation, as an alternative to LP, finding nonlinear input–output equilibrium as an alternative to the classical input–output model by Wassily Leontief and finding John Nash's equilibrium in n persons concave game.

Finding NE for each of these problems is equivalent to solving particular variation inequality (VI). Each VI is defined by an operator with specific properties and a very simple feasible set, projection on which is a low-cost operation.

The PGP and the EPG are, in fact, pricing mechanisms for finding NE in optimal resources allocation and for establishing nonlinear input–output equilibrium.

What is most important: for each of these methods, matrix by vector multiplication is the main computational operation per step, whereas application of the IPMs even for LP calculations requires solving a linear system of equations at each step.

Application of PGP and EPG for finding J. Nash equilibrium leads at each step to projection on the probability simplexes, which is also a low-cost operation.

For both PGP and EPG methods and all models, under consideration, the convergence is proven, the convergence rate is estimated, and complexity bounds are established under natural assumptions on the input data.

In Chapter 11 we consider a number of important applications, arising in structural optimization, health care, data analysis, and economics, just to mention a few. The applications required solving large-scaled constrained optimization problems or finding NE. The numerical results obtained, justified the theoretical findings.

In particular, for the EPMs we systematically observed the "hot" start.

Based on NR, the PENNON package, designed by Kočvara and Stingl (2005, 2007, 2015), has been widely used for solving structural design, free material optimization, optimal control, and chemical engineering problems, just to mention a few.

PENNON has proven to be an efficient tool for both constrained and semidefinite optimization problems.

One of the main focus of the book is the duality, which has been systematically used for analyzing old methods and designing new ones, improving their robustness, convergence rate, and complexity bounds. Duality is also used for identifying pricing mechanisms for establishing equilibrium in optimization, game theory, and economics.

Substantial part of the book has never been covered in a systematic way.

The book can be used for a two-semester course in "Continuous Optimization" for master students in mathematics with Chapters 2–6 covered in the first and Chapters 7–11 in the second semester. It can also be used by graduate students in apply mathematics, computer science, and economics.

Chapter 11 and Section 5.11 were written together with Dr. Igor Griva, who was instrumental in developing primal–dual interior and exterior point methods and correspondent software, as well as in solving hundreds of both real life and test constrained optimization problems.

It gives me a pleasure to thank Sofya Kamenkovich–Kolodizner for her excellent work in preparing the manuscript for publication.

Writing the book was a long and a difficult process, which would be impossible without excellent conditions and stimulating environment I found at the Mathematical Sciences Department, IBM T.J. Watson Research Center (1988–1993), Mathematical Sciences and SEOR Departments at George Mason University (1993–2013), and the Department of Mathematics at the Technion (2013–2020).

Chapter 2
Elements of Calculus and Convex Analysis

2.0 Introduction

Along with elements of calculus, we consider some basic facts of convex analysis, including convex functions, convex set, and their most important properties.

Particular attention is given to Legendre–Fenchel (LF) transform, which is a product of a general duality principle.

Application of the LF transform to a strictly convex and smooth function leads to LF identity—the basic tool for establishing dual equivalents for a number of constrained optimization methods. Dual equivalence is the key element of the convergence analysis. It is also used for establishing error bounds.

For strictly convex and three times differentiable functions, the LF transform leads to LF invariant.

It turns out that boundedness of the LF invariant leads to important class of self-concordant (SC) functions – the centerpiece of the interior point methods.

2.1 Elements of Calculus

2.1.1 Differentiation of Scalar Functions

A scalar function $f : \mathbb{R}^n \to \mathbb{R}$ is differentiable at $x \in \mathbb{R}^n$ if there is a vector $a \in \mathbb{R}^n$, such that for any $y \in \mathbb{R}^n$ we have

$$f(y) = f(x) + \langle a, y - x \rangle + o(||y - x||),$$

where $\langle u, \mathrm{v} \rangle = \sum_{i=1}^{n} u_i \mathrm{v}_i$ and $||u|| = \langle u, u \rangle^{\frac{1}{2}}$ is Euclidean norm in $\mathbb{R}^n$. Vector a is called gradient of f at x and denoted $\nabla f(x)$, so that

R. A. Polyak, *Introduction to Continuous Optimization*, Springer Optimization and Its Applications 172, https://doi.org/10.1007/978-3-030-68713-7_2

$$f(y) = f(x) + \langle \nabla f(x), y - x \rangle + o(||y - x||). \tag{2.1}$$

In other words, a function f is differentiable at x if one can find a linear approximation $\widetilde{f}(y) = f(x) + \langle \nabla f(x), y - x \rangle$ such that $|f(y) - \widetilde{f}(y)| = o(||y - x||)$.

Exercise 2.1. Show that $\nabla f(x) = \left(\frac{\partial f}{\partial x_1}, \ldots, \frac{\partial f}{\partial x_n} \right)$.

For example, let $f : \mathbb{R}^n \to \mathbb{R}$ be a quadratic function $f(x) = \langle Ax, x \rangle / 2 - \langle b, x \rangle$, where $A \in \mathbb{R}^{n \times n}$ is a symmetric matrix and $b \in \mathbb{R}^n$. Let $\Delta = y - x$, then

$$f(y) = \frac{1}{2}\langle A(x + \Delta), x + \Delta \rangle - \langle b, x + \Delta \rangle = f(x) + \langle Ax - b, \Delta \rangle + \frac{1}{2}\langle A\Delta, \Delta \rangle,$$

since $\langle Ax, y \rangle = \langle Ay, x \rangle$.

From $|\langle A\Delta, \Delta \rangle| \leq \|A\| \|\Delta\|^2$ with $\|A\| = \max_{\|x\|=1} \|Ax\|$ follows $|\langle A\Delta, \Delta \rangle| = o(||\Delta||)$. Thus, $f(x)$ is differentiable at x and

$$\nabla f(x) = Ax - b.$$

Function f is differentiable on a set $\Omega \subset \mathbb{R}^n$ if it is differentiable at all points of Ω. Function f is differentiable if it is differentiable on the entire space $\mathbb{R}^n$.

Let $f : \mathbb{R}^n \to \mathbb{R}$ be differentiable on a segment $[x, y]$ and

$$x_\tau = x + \tau(y - x), \quad 0 \leq \tau \leq 1.$$

We consider the restriction $\varphi(\tau) = f(x_\tau)$ of f on $[x, y]$. Using (2.1) we obtain

$$\begin{aligned} \varphi'(\tau) &= \lim_{\Delta\tau \to 0} \frac{\varphi(\tau + \Delta\tau) - \varphi(\tau)}{\Delta\tau} = \lim_{\Delta\tau \to 0} \frac{f(x_\tau + \Delta\tau(y - x)) - f(x_\tau)}{\Delta\tau} \\ &= \lim_{\Delta\tau \to 0} \frac{\langle \nabla f(x_\tau), \Delta\tau(y - x) \rangle + o(\Delta\tau)}{\Delta\tau} = \langle \nabla f(x_\tau), y - x \rangle . \end{aligned}$$

For $\tau = 0$ we have the directional derivative of f at the point x in the direction $d = y - x$:

$$\nabla f(x; d) = \langle \nabla f(x), d \rangle = \langle \nabla f(x), y - x \rangle. \tag{2.2}$$

Using the Newton–Leibniz formula

$$\varphi(1) - \varphi(0) = \int_0^1 \varphi'(\tau) d\tau$$

we obtain $f(y) - f(x) = \int_0^1 \langle \nabla f(x_\tau), y - x \rangle d\tau$. Therefore,

$$f(y) = f(x) + \langle \nabla f(x), y - x \rangle + \int_0^1 \langle \nabla f(x_\tau) - \nabla f(x), y - x \rangle d\tau. \tag{2.3}$$

We say that the gradient ∇f is Lipschitz continuous in $\mathbb{R}^n$ if for any $x, y \in \mathbb{R}^n$ there exists $L > 0$ such that

$$\|\nabla f(x) - \nabla f(y)\| \leq L\|x - y\|. \tag{2.4}$$

The class of functions $f : \mathbb{R}^n \to \mathbb{R}$, which have a Lipschitz continuous gradient, plays an important role in optimization. For such functions (2.3) yields

$$|f(y) - f(x) - \langle \nabla f(x), y - x \rangle| \leq \frac{L}{2}\|y - x\|^2. \tag{2.5}$$

From (2.5) follows

$$\begin{aligned} f(x_0) + \langle \nabla f(x_0), x - x_0 \rangle - \frac{L}{2}\|x - x_0\|^2 &\leq f(x) \\ &\leq f(x_0) + \langle \nabla f(x_0), x - x_0 \rangle + \frac{L}{2}\|x - x_0\|^2. \end{aligned}$$

In other words, the graph of f lies between the graphs of two quadratic functions.

From the mean value theorem follows existence $0 \leq \theta \leq 1$, that $\varphi(1) - \varphi(0) = \varphi'(\theta)$, which leads to Lagrange formula

$$f(y) = f(x) + \langle \nabla f(x_\theta), y - x \rangle. \tag{2.6}$$

Finally, let us consider the following two important properties of the gradient.

First, from the Cauchy–Schwarz inequality, we have

$$-\|\nabla f(x)\|\|d\| \leq \langle \nabla f(x), d \rangle \leq \|\nabla f(x)\|\|d\|,$$

which leads to $\langle \nabla f(x), d \rangle \geq -\|\nabla f(x)\|$ for any d: $\|d\| = 1$.
By taking $\bar{d} = -\nabla f(x)/\|\nabla f(x)\|$ we obtain

$$-\|\nabla f(x)\| = \langle \nabla f(x), \bar{d} \rangle = \min\left\{\langle(\nabla f(x), d \rangle : \|d\| = 1\right\}.$$

In other words, $\bar{d}$ is the direction of the fastest local decrease of f or the steepest descent direction.

Second, let $\partial \mathscr{L}_f(r) = \{x \in \mathbb{R}^n : f(x) = r\}$ be the boundary of the sublevel set $\mathscr{L}_f(r) = \{x \in \mathbb{R}^n : f(x) \leq r\}$. Consider the set $T(x)$ of directions d tangent at a point x. Then for any $d \in T(x)$, we have

$$\langle \nabla f(x), d \rangle = 0.$$

Indeed, for any $d \in T(x)$, there exists a sequence $\{x_s\}_{s \in N} \subset \partial \mathscr{L}_f(r)$ such that $d = \lim_{x_s \to x} \frac{x_s - x}{\|x_s - x\|}$. Now from

$$f(x_s) = f(x) + \langle \nabla f(x), x_s - x \rangle + o\left(\|x_s - x\|\right) = f(x)$$

follows $\langle \nabla f(x), x_s - x \rangle + o\left(\|x_s - x\|\right) = 0$, or $\lim_{x_s \to x} \left[\left\langle \nabla f(x), \frac{x_s - x}{\|x_s - x\|}\right\rangle + \frac{o(\|x_s - x\|)}{\|x_s - x\|}\right]$ $= \langle \nabla f(x), d \rangle = 0$.

2.1.2 Differentiation of Vector Functions

A vector function $g : \mathbb{R}^n \to \mathbb{R}^m$ is said to be differentiable at $x \in \mathbb{R}^n$ if there is a matrix $A \in \mathbb{R}^{m\times n}$ such that for any $y \in \mathbb{R}^n$ we have

$$g(y) = g(x) + A(y-x) + o(||y-x||).$$

Matrix A is called Jacobian of the vector function g at $x \in \mathbb{R}^n$ and denoted $\nabla g(x)$. Thus,

$$g(y) = g(x) + \nabla g(x)(y-x) + o(||y-x||).$$

In other words, a vector function is differentiable at x if it admits a linear approximation at x, that is,

$$\|g(y) - g(x) - \nabla g(x)(y-x)\| \le o(||y-x||).$$

If $g : \mathbb{R}^n \to \mathbb{R}^m$ is differentiable at every point of set Ω, we say that g is differentiable on Ω. Vector function $g : \mathbb{R}^n \to \mathbb{R}^m$ is differentiable if it is differentiable at all points of $\mathbb{R}^n$.

Exercise 2.2. Show that for a differentiable vector function $g(x) = (g_1(x), \ldots, g_m(x))^T$ the elements of its Jacobian $\nabla g(x)$ are given by $\nabla g(x)_{ij} = \partial g_i(x)/\partial x_j$.

Let $g : \mathbb{R}^n \to \mathbb{R}^m$ be differentiable at x and $h : \mathbb{R}^m \to \mathbb{R}^s$ be differentiable at $g(x)$. Then the following chain rule for differentiation of a composite function $h(g(x))$

$$\nabla[h(g(x))] = \nabla h(g(x))\nabla g(x)$$

is valid, where the right-hand side contains the product of two matrices ∇h and ∇g.

The mean value theorem does not hold for vector functions. So, generally speaking, there is no $0 < \theta < 1$ such that

$$g(y) = g(x) + \nabla g(x_\theta)(y-x).$$

On the other hand, if $g(x)$ is differentiable on $[x,y]$, then it follows from (2.6) that for a given vector $\mathrm{v} \in \mathbb{R}^m$ there is $0 \le \theta(\mathrm{v}) \le 1$ such that

$$\langle g(y), \mathrm{v}\rangle - \langle g(x), \mathrm{v}\rangle = \langle \nabla g(x_{\theta(\mathrm{v})})(y-x), \mathrm{v}\rangle. \tag{2.7}$$

Besides,

$$g(y) = g(x) + \int_0^1 \nabla g(x_\tau)(y-x)d\tau, \tag{2.8}$$

or

$$g(y) = g(x) + \nabla g(x)(y-x) + \int_0^1 (\nabla g(x_\tau) - \nabla g(x))(y-x)d\tau. \tag{2.9}$$

If $\nabla g(x)$ satisfies the Lipschitz condition on $[x,y]$, then from (2.9) we obtain

$$\|g(y) - g(x) - \nabla g(x)(y-x)\| \leq \frac{L}{2}\|y-x\|^2, \tag{2.10}$$

similarly to (2.5)

2.1.3 Second Derivatives

A scalar function $f : \mathbb{R}^n \to \mathbb{R}$ is said to be twice differentiable at $x \in \mathbb{R}^n$ if there is a symmetric matrix $H \in \mathbb{R}^{n\times n}$ such that for any $y \in \mathbb{R}^n$ we have

$$f(y) = f(x) + \langle \nabla f(x), y-x\rangle + \frac{1}{2}\langle H(y-x), y-x\rangle + o(\|y-x\|^2).$$

The matrix H is called Hessian of f at $x \in \mathbb{R}^n$ and denoted $\nabla^2 f(x)$. In other words, a function f is twice differentiable at $x \in \mathbb{R}^n$ if it admits a quadratic approximation

$$\widetilde{f}(y) = f(x) + \langle \nabla f(x), y-x)\rangle + \frac{1}{2}\langle \nabla^2 f(x)(y-x), y-x\rangle,$$

such that $|f(y) - \widetilde{f}(y)| = o(\|y-x\|^2)$.

Let C^2 be a class of twice continuously differentiable functions $f : \mathbb{R}^n \to \mathbb{R}$. We can sharpen the estimates obtained earlier for functions in C^2. Consider the scalar function $\varphi(\tau) = f(x_\tau)$. For a twice differentiable f, we have $\varphi'(\tau) = \langle \nabla f(x_\tau), y - x\rangle$ and $\varphi''(\tau) = \langle \nabla^2 f(x_\tau)(y-x), y-x\rangle$. Then Taylor's formula with the integral reminder

$$\varphi(1) = \varphi(0) + \varphi'(0) + \int_0^1 \left(\int_0^t \varphi''(\tau)d\tau\right) dt$$

yields

$$f(y) = f(x) + \langle \nabla f(x), y-x\rangle + \int_0^1 \left(\int_0^t \langle \nabla^2 f(x_\tau)(y-x), y-x\rangle d\tau\right) dt. \tag{2.11}$$

If Hessian $\nabla^2 f$ satisfies Lipschitz condition

$$\|\nabla^2 f(x) - \nabla^2 f(y)\| \leq M\|x-y\|, \tag{2.12}$$

then from (2.11) follows

$$\begin{aligned}
&|f(y) - f(x) - \langle \nabla f(x), y-x\rangle - \frac{1}{2}\langle \nabla^2 f(x)(y-x), y-x\rangle| \\
&\leq \int_0^1 \left(\int_0^t \|(\nabla^2 f(x_\tau) - \nabla^2 f(x))\|\|y-x\|^2 d\tau\right) dt \\
&\leq \int_0^1 M\frac{t^2}{2}\|y-x\|^3 dt = \frac{M}{6}\|y-x\|^3.
\end{aligned}$$

From Taylor's formula with Lagrange reminder follows existence $0 \leq \theta \leq 1$ that

$$\varphi(1) = \varphi(0) + \varphi'(0) + \frac{1}{2}\varphi''(\theta).$$

Therefore,

$$f(y) = f(x) + \langle \nabla f(x), y - x \rangle + \frac{1}{2}\langle \nabla^2 f(x_\theta)(y-x), y-x \rangle.$$

The following Lemma establishes the necessary and sufficient condition on $f \in C^2$ that guarantees Lipschitz condition (2.4) on ∇f.

Lemma 2.1. *A function $f \in C^2$ satisfies (2.4) if and only if*

$$\|\nabla^2 f(x)\| \leq L \tag{2.13}$$

for any $x \in \mathbb{R}^n$.

Proof. Indeed, for any x and $y \in \mathbb{R}^n$, from (2.8) with $g(x) = \nabla f(x)$ follows

$$\nabla f(y) = \nabla f(x) + \int_0^1 \nabla^2 f(x_\tau)(y-x)d\tau. \tag{2.14}$$

From (2.14) we have

$$\|\nabla f(y) - \nabla f(x)\| = \left\| \int_0^1 \nabla^2 f(x_\tau)(y-x)d\tau \right\| \leq \|y-x\| \int_0^1 \|\nabla^2 f(x_\tau)\| d\tau.$$

Using (2.13) we obtain (2.4).

On the other hand, for $f \in C^2$, $u \in \mathbb{R}^n$ and $t > 0$, we have

$$\|\nabla f(x+tu) - \nabla f(x)\| = \left\| \int_0^t \nabla^2 f(x+\tau u)u d\tau) \right\| \leq tL\|u\|.$$

Therefore,

$$\|\nabla^2 f(x)u\| = \lim_{t\to 0} \left\| \frac{\nabla f(x+tu) - \nabla f(x)}{t} \right\| \leq L\|u\|,$$

and (2.13) holds. □

Exercise 2.3. Show that

1. $\nabla^2[\langle Ax, x\rangle/2 - \langle b, x\rangle] = A$, where A is a symmetric $n \times n$ matrix, $b \in \mathbb{R}^n$;
2. $\nabla^2\|x\| = I\|x\|^{-1} - xx^T\|x\|^{-3}$, $x \neq 0$, where I is the identical $n \times n$ matrix ;
3. $\nabla^2\langle c, x\rangle^2 = 2cc^T$, where $c \in \mathbb{R}^n$.

We use notation $A \succeq 0$ for symmetric positive semidefinite matrices, that is, matrices satisfying $\langle Ax, x\rangle \geq 0$ for any $x \in \mathbb{R}^n$. Notation $A \succ 0$ means that A is positive definite, that is, $\langle Ax, x\rangle > 0$ for any $x \neq 0$. Notation $A \succeq B$ ($A \succ B$) means that $A - B \succeq 0$ ($A - B \succ 0$, respectively).

Let $0 \leq m < M < \infty$ is min and max eigenvalue of A, then the condition number of A is

$$\varkappa = \text{cond}(A) = mM^{-1}. \tag{2.15}$$

Corollary 2.1. *It follows from (2.12) that*

$$-M\|x-y\|I \preceq \nabla^2 f(x) - \nabla^2 f(y) \preceq M\|x-y\|I$$

or

$$\nabla^2 f(y) - M\|x-y\|I \preceq \nabla^2 f(x) \preceq \nabla^2 f(y) + M\|x-y\|I. \tag{2.16}$$

In the future we will use the following condition number

$$\varkappa(x) = \text{cond}(\nabla^2 f(x)) = m(x)M^{-1}(x) \tag{2.17}$$

of Hessian $\nabla^2 f$ at $x \in \mathbb{R}^n$.

2.1.4 Convex Functions in $\mathbb{R}^n$

Function $f : \mathbb{R}^n \to \mathbb{R}$ is said to be convex if

$$f(x_\lambda) = f((1-\lambda)x+\lambda y) \leq (1-\lambda)f(x)+\lambda f(y) = \lambda(f(y)-f(x))+f(x) \tag{2.18}$$

holds for any $x, y \in \mathbb{R}^n$ and any $0 \leq \lambda \leq 1$.

The definition has an intuitive geometric interpretation: the graph of f lies below the chord, which joins points $(x, f(x))$ and $(y, f(y))$. A function $f : \mathbb{R}^n \to \mathbb{R}$ is concave if $-f$ is convex.

Exercise 2.4. Show that if f_i, $i = 1, \ldots, m$, are convex then both $F_1 = \sum_{i=1}^m \gamma_i f_i$, $\gamma_i \geq 0$, and $F_2 = \max_{1 \leq i \leq m} f_i$ are also convex.

Exercise 2.5. Let $f : \mathbb{R}^n \to \mathbb{R}$ be convex. Show that for any $x_1, \ldots, x_k$ and $\lambda_i \geq 0$, $\sum_{i=1}^k \lambda_i = 1$, the following Jensen's inequality holds:

$$f(\lambda_1 x_1 + \cdots + \lambda_k x_k) \leq \lambda_1 f(x_1) + \cdots + \lambda_k f(x_k).$$

Hint: Use induction on k.

Exercise 2.6. Function $f : \mathbb{R}^n \to \mathbb{R}$ is convex if and only if for any x and $d \in \mathbb{R}^n$ the function $\varphi(t) = f(x+td)$ is convex.

The behavior of convex functions is rather remarkable. They are not only continuous at any interior point, but Lipschitz continuous; moreover, they are differentiable at any given direction.

Lemma 2.2. *For any convex combination* $x = \sum_{i=1}^k \lambda_i x_i$, $\lambda_i \geq 0$, $\sum_{i=1}^k \lambda_i = 1$, *of vectors* $x_1, \ldots, x_k \in \mathbb{R}^n$, *we have*

$$f(x) \leq \max_{1 \leq i \leq k} f(x_i).$$

Proof. The proof follows immediately from the Jensen inequality. In fact,

$$f(x) = f\left(\sum_{i=1}^k \lambda_i x_i\right) \leq \sum_{i=1}^k \lambda_i f(x_i) \leq \max_{1 \leq i \leq k} f(x_i) \sum_{i=1}^k \lambda_i = \max_{1 \leq i \leq k} f(x_i).$$

□

Theorem 2.1. *A convex function* $f : \mathbb{R}^n \to \mathbb{R}$ *is continuous at any point* $x_0 \in \mathbb{R}^n$.

Proof. Without loss of generality, we assume $x_0 = 0$. Let us consider a converging sequence $\{x_s\}_{s \in N}$ such that $\lim_{s \to \infty} x_s = x_\infty = 0$. Using the convexity of f, we get

$$f(x_s) \leq (1 - \|x_s\|) f(0) + \|x_s\| f(x_s / \|x_s\|).$$

Observe that for any point $x_s / \|x_s\|$, all its coordinates lie within the closed segment $[-1, 1]$. Therefore, by Lemma 2.2, $f(x_s / \|x_s\|) \leq \max_{1 \leq i \leq n} f(\pm e_i) = M$, where $e_i = (0, \ldots, 0, 1, 0, \ldots, 0)$ with the only nonzero coordinate at the i-th place. Consequently,

$$\limsup_{s \to \infty} f(x_s) \leq (1 - \|x_\infty\|) f(0) + \|x_\infty\| M = f(0).$$

On the other hand,

$$f(0) \leq \frac{\|x_s\|}{1 + \|x_s\|} f(-x_s / \|x_s\|) + \frac{1}{1 + \|x_s\|} f(x_s).$$

Using again the same reasoning, we obtain $f(0) \leq \liminf_{s \to \infty} f(x_s)$, and hence $\lim_{s \to \infty} f(x_s) = f(0)$. Therefore, f is continuous at $x_0 = 0$. □

Theorem 2.2. *A continuously differentiable function* $f : \mathbb{R}^n \to \mathbb{R}$ *is convex if and only if for any two vectors* x *and* y *from* $\mathbb{R}^n$ *we have*

$$f(y) \geq f(x) + \langle \nabla f(x), y - x \rangle. \tag{2.19}$$

Proof. From (2.18) follows

$$f(x_\lambda) - f(x) \leq \lambda (f(y) - f(x)),$$

therefore

$$\lim_{\lambda \to 0} \frac{f(x_\lambda) - f(x)}{\lambda} = \lim_{\lambda \to 0} \frac{(f(x) + \lambda(y - x)) - f(x)}{\lambda} = \langle \nabla f(x), y - x \rangle \leq f(y) - f(x).$$

Conversely, write (2.19) for the pair x, x_λ and for the pair y, x_λ. Multiplying the first inequality by $(1 - \lambda)$ and the second by λ and adding them, we obtain (2.18).

□

Theorem 2.3. *A continuously differentiable function $f : \mathbb{R}^n \to \mathbb{R}$ is convex if and only if for any two vectors x and y from $\mathbb{R}^n$ the gradient monotonicity condition*

$$\langle \nabla f(x) - \nabla f(y), x - y \rangle \geq 0 \tag{2.20}$$

holds

Proof. From (2.19) follows

$$f(x) \geq f(y) + \langle \nabla f(y), x - y \rangle, \quad f(y) \geq f(x) + \langle \nabla f(x), y - x \rangle.$$

By adding these inequalities, we obtain (2.20).
Conversely, (2.20) yields

$$\begin{aligned} f(y) &= f(x) + \int_0^1 \langle \nabla f(x_\tau), y - x \rangle d\tau \\ &= f(x) + \langle \nabla f(x), y - x \rangle + \int_0^1 \langle \nabla f(x_\tau) - \nabla f(x), y - x \rangle d\tau \\ &= f(x) + \langle \nabla f(x), y - x \rangle + \int_0^1 \frac{1}{\tau} \langle \nabla f(x_\tau) - \nabla f(x), x_\tau - x \rangle d\tau \\ &\geq f(x) + \langle \nabla f(x), y - x \rangle, \end{aligned}$$

and the convexity of f follows from Theorem 2.2. □

Inequality (2.19) is called the first-order convexity condition. If f is twice continuously differentiable, then the following second-order convexity condition holds.

Theorem 2.4. *A twice continuously differentiable function $f : \mathbb{R}^n \to \mathbb{R}$ is convex if and only if for any $x \in \mathbb{R}^n$ Hessian $\nabla^2 f(x)$ is positive semidefinite, that is,*

$$\nabla^2 f(x) \succeq 0. \tag{2.21}$$

Proof. Let $f \in C^2$ be convex, $d \in \mathbb{R}^n$ be an arbitrary direction, and $x_\tau = x + \tau d$, $\tau > 0$. Then in view of (2.7) and (2.20), we obtain

$$0 \leq \tau^{-1} \langle \nabla f(x_\tau) - \nabla f(x), x_\tau - x \rangle = \langle \nabla f(x_\tau) - \nabla f(x), d \rangle = \int_0^\tau \langle \nabla^2 f(x_\theta) d, d \rangle d\theta.$$

We get (2.21) by letting $\tau \to 0$.
Conversely, let (2.21) holds for all $x \in \mathbb{R}^n$, then from (2.11), we have

$$\begin{aligned} f(y) &= f(x) + \langle \nabla f(x), y - x \rangle + \int_0^1 \left(\int_0^\tau \langle \nabla^2 f(x_\lambda)(y - x), y - x \rangle d\lambda \right) d\tau \\ &\geq f(x) + \langle \nabla f(x), y - x \rangle, \end{aligned}$$

and the convexity of f follows from Theorem 2.2. □

Let us consider some examples of differentiable convex functions.

1. A linear function $f(x) = \alpha + \langle a, x\rangle$ is simultaneously convex and concave.
2. Let A be symmetric and positive semidefinite, then the quadratic function

$$f(x) = \alpha + \langle a, x\rangle + \frac{1}{2}\langle Ax, x\rangle$$

   is convex (since $\nabla^2 f(x) = A \succeq 0$).
3. The following functions of one variable are convex:

$$\begin{aligned} f(x) &= -\ln x, \quad x > 0, \\ f(x) &= |x|^p, \quad p > 1, \\ f(x) &= x\ln x, \quad x > 0. \end{aligned}$$

Exercise 2.7. Show the convexity of the functions

$$f(x) = \sum_{i=1}^{m} e^{\alpha_i + \langle a_i, x\rangle},$$

arising in *Geometric Programming* and

$$f(x) = \sum_{i=1}^{m} |\langle a_i, x\rangle - b_i|^p, \; p > 1$$

arising in *L_p-approximation* problem.

Exercise 2.8. Show that for a convex function f with Lipschitz continuous gradient ∇f, the following inequalities hold for any pair $x, y \in \mathbb{R}^n$:

1.
$$0 \leq f(y) - f(x) - \langle \nabla f(x), y - x\rangle \leq \frac{L}{2}\|x - y\|^2; \tag{2.22}$$

2.
$$f(y) \geq f(x) + \langle \nabla f(x), y - x\rangle + \frac{1}{2L}\|\nabla f(x) - \nabla f(y)\|^2; \tag{2.23}$$

3.
$$\langle \nabla f(x) - \nabla f(y), x - y\rangle \geq \frac{1}{L}\|\nabla f(x) - \nabla f(y)\|^2. \tag{2.24}$$

Hints:

1. Inequality (2.22) follows from (2.5).
2. For a fixed $x \in \mathbb{R}^n$, consider $\varphi : \mathbb{R}^n \to \mathbb{R}$ given by $\varphi(y) = f(y) - \langle \nabla f(x), y\rangle$. It is convex in $y \in \mathbb{R}^n$ with Lipschitz continuous gradient $\nabla\varphi(y) = \nabla f(y) - \nabla f(x)$ and $\nabla\varphi(x) = 0$; therefore, $\varphi(x) \leq \varphi(y - L^{-1}\nabla\varphi(y))$. Applying (2.22) to the right-hand side of the last inequality, one gets (2.23).
3. Consider (2.23) with x and y interchanged.

Inequality (2.24) is known as the co-coercive of ∇f with parameter L. It follows from Theorem 2.3 that (2.23) implies convexity of f. Applying the Cauchy–Schwarz inequality to (2.24), we restore the Lipschitz condition (2.4).

If $f:\mathbb{R}^n \to \mathbb{R}$ has a minimizer, that is, if there exists $x^* \in \mathbb{R}^n$ such that $\nabla f(x^*) = 0$, then from (2.22) and (2.23) follows

$$\frac{1}{2L}\|\nabla f(x)\|^2 \leq f(x) - f(x^*) \leq \frac{L}{2}\|x - x^*\|^2. \tag{2.25}$$

Exercise 2.9. Show that from Lipschitz continuity of ∇f follows:

1. Convexity of $\varphi(x) = \frac{L}{2}\|x\|^2 - f(x)$.
2. For a twice differentiable f and any $x \in \mathbb{R}^n$, we have $LI \succeq \nabla^2 f(x) \succeq 0$.

2.1.5 Strictly and Strongly Convex Functions in $\mathbb{R}^n$

Function $f:\mathbb{R}^n \to \mathbb{R}$ is strictly convex if (2.19)–(2.21) are satisfied as strict inequalities.

Exercise 2.10. Show that from $f(y) > f(x) + \langle \nabla f(x), y - x\rangle$ follows (2.20) as a strict inequality.

A continuously differentiable function $f:\mathbb{R}^n \to \mathbb{R}$ is strongly convex on $\mathbb{R}^n$ if there exists a constant $m > 0$ such that for any x and $y \in \mathbb{R}^n$ the following inequality holds

$$f(y) \geq f(x) + \langle \nabla f(x), y - x\rangle + \frac{1}{2}m\|y - x\|^2. \tag{2.26}$$

The constant $m > 0$ is called the convexity modulus of f.

For any strongly convex function, a unique minimizer x^* exists, and the following bound holds for any $x \in \mathbb{R}^n$:

$$f(x) \geq f(x^*) + \langle \nabla f(x^*), x - x^*\rangle + \frac{m}{2}\|x - x^*\|^2 = f(x^*) + \frac{m}{2}\|x - x^*\|^2.$$

Combining (2.25) and (2.26), we obtain the following bounds:

$$\frac{m}{2}\|x - x^*\|^2 \leq f(x) - f(x^*) \leq \frac{L}{2}\|x - x^*\|^2.$$

Exercise 2.11. If f_1 is strongly convex with the convexity modulus m_1 and f_2 is strongly convex with the convexity modulus m_2 and $\alpha \geq 0$, $\beta \geq 0$, then $f = \alpha f_1 + \beta f_2$ is strongly convex with the convexity modulus $\alpha m_1 + \beta m_2$.

Exercise 2.12. For a strongly convex f, the following bounds hold for any x and $y \in \mathbb{R}^n$:

1.
$$\langle \nabla f(x) - \nabla f(y), x - y \rangle \geq m\|x - y\|^2. \tag{2.27}$$
Hint: Consider (2.26) with x and y interchanged.

2.
$$(1-\alpha)f(x) + \alpha f(y) \geq f((1-\alpha)x + \alpha y) + \alpha(1-\alpha)\frac{m}{2}\|x-y\|^2$$
for any $0 \leq \alpha \leq 1$.
Hint: Apply (2.26) twice for $x_\alpha = (1-\alpha)x + \alpha y$ and x and for x_α and y .

3.
$$f(y) \leq f(x) + \langle \nabla f(x), y - x \rangle + \frac{1}{2m}\|\nabla f(x) - \nabla f(y)\|^2. \tag{2.28}$$
Hint: Consider $\varphi(y) = f(y) - \langle \nabla f(x), y \rangle$, then $\varphi(y)$ is strongly convex and $\nabla \varphi(x) = 0$. From (2.26) follows
$$\begin{aligned} \varphi(x) &= \min_u \varphi(u) \geq \min_u \{\varphi(y) + \langle \nabla \varphi(y), u - y \rangle + \frac{1}{2}m\|u-y\|^2\} \\ &= \varphi(y) - \frac{1}{2m}\|\nabla \varphi(y)\|^2. \end{aligned}$$

4.
$$\langle \nabla f(x) - \nabla f(y), x - y \rangle \geq \frac{1}{m}\|\nabla f(x) - \nabla f(y)\|^2.$$
Hint: Apply (2.28) with x and y interchanged.

Exercise 2.13. Show that twice continuously differentiable function $f : \mathbb{R}^n \to \mathbb{R}$ is strongly convex in $\mathbb{R}^n$ if and only if
$$\nabla^2 f(x) \succeq mI. \tag{2.29}$$
Hint: Apply (2.27).

Exercise 2.14. Show that for a strongly convex function the following holds:

1. $f(x) - f(x^*) \geq \frac{m}{2}\|x - x^*\|^2$.
2. $\langle \nabla f(x), x - x^* \rangle \geq m\|x - x^*\|^2$.
3. $\|\nabla f(x)\| \geq m\|x - x^*\|$.

Theorem 2.5. *If $f : \mathbb{R}^n \to \mathbb{R}$ is strongly convex with convexity modulus $m > 0$ and the gradient ∇f is Lipschitz continuous with the Lipschitz constant $L > m$, then for any $x, y \in \mathbb{R}^n$*
$$\langle \nabla f(x) - \nabla f(y), x - y \rangle \geq \frac{mL}{m+L}\|x-y\|^2 + \frac{1}{m+L}\|\nabla f(x) - \nabla f(y)\|^2. \tag{2.30}$$

Proof. From the strong convexity of f with the convexity modulus $m > 0$ follows the convexity of $\varphi(x) = f(x) - \frac{m}{2}\|x\|^2$. Besides,
$$\|\nabla \varphi(x) - \nabla \varphi(y)\| \leq (L - m)\|x - y\|.$$

Application of the co-coercivity property (2.24) to φ leads to the inequality

$$\langle \nabla f(x) - \nabla f(y) - m(x-y), x-y \rangle \geq \frac{1}{L-m} \|\nabla f(x) - \nabla f(x) - m(x-y)\|^2,$$

or

$$\langle \nabla f(x) - \nabla f(y), x-y \rangle \geq \frac{Lm}{L-m} \|x-y\|^2$$
$$+ \frac{1}{L-m} \|\nabla f(x) - \nabla f(y)\|^2 - \frac{2m}{L-m} \langle \nabla f(x) - \nabla f(y), x-y \rangle,$$

which gives

$$\frac{L+m}{L-m} \langle \nabla f(x) - \nabla f(y), x-y \rangle \geq \frac{Lm}{L-m} \|x-y\|^2 + \frac{1}{L-m} \|\nabla f(x) - \nabla f(y)\|.$$

The bound (2.30) follows directly from the last inequality. □

Exercise 2.15. Show that for any $x, y \in \mathbb{R}^n$, $\tau \in [0,1]$ and convex $f : \mathbb{R}^n \to \mathbb{R}$ with Lipschitz continuous gradient ∇f, the following inequalities hold:

$$\frac{\tau(1-\tau)}{2L} \|\nabla f(x) - \nabla f(y)\|^2 \leq (1-\tau) f(x) + \tau f(y) - f(x_\tau) \leq \tau(1-\tau) \frac{L}{2} \|x-y\|^2.$$

Hint: Use (2.23) for x_τ and x and for x_τ and y. Multiply the corresponding inequalities by $1-\tau$ and τ, respectively, and use the following inequality (check it first):

$$\tau \|a_1 - u\|^2 + (1-\tau) \|a_2 - u\|^2 \geq \tau(1-\tau) \|a_1 - a_2\|^2$$

for any $a_1, a_2, u \in \mathbb{R}^n$.

The gradient ∇f, which satisfies (2.27), is a strongly monotone operator. The value $\varkappa = \frac{m}{L}$ is called the condition number of the strongly monotone operator $\nabla f : \mathbb{R}^n \to \mathbb{R}^n$.

2.2 Convex Sets

2.2.1 Open and Closed Sets

Let $\Omega \subset \mathbb{R}^n$, a vector $x \in \Omega$ is an interior point in Ω if there exists $\varepsilon > 0$ such that $\{y \in \mathbb{R}^n : \|y - x\| \leq \varepsilon\} \subset \Omega$. The set of interior points in Ω is called the interior of Ω and is denoted $\operatorname{int}\Omega$. A set Ω is called open if $\operatorname{int}\Omega = \Omega$, that is, if every point in Ω is an interior point.

A set Ω is closed if its complement $\mathbb{R}^n \setminus \Omega$ is an open set. The closure of a set Ω is $\operatorname{cl}\Omega = \mathbb{R}^n \setminus \operatorname{int}(\mathbb{R}^n \setminus \Omega)$.

There is another way to describe closed sets. A set Ω is closed if it contains all its limit points, that is, for any converging sequence $\{x_s\}_{s\in N} \subset \Omega$, we have $\lim_{s\to\infty} x_s = x \in \Omega$. The boundary of Ω is $\partial\Omega = \Omega \setminus \operatorname{int}\Omega$. A point $x \in \Omega$ belongs to the boundary if for any $\varepsilon > 0$ there exist $u \in \Omega$ and $\mathrm{v} \notin \Omega$ such that $\|x-u\| \le \varepsilon$ and $\|x-\mathrm{v}\| \le \varepsilon$ or if there exist points in Ω and in $\mathbb{R}^n \setminus \Omega$ arbitrary close to x.

A set $\Omega \subset \mathbb{R}^n$ is a compact if it is closed and bounded in $\mathbb{R}^n$.

Exercise 2.16. Show the following:

1. $\operatorname{cl}\Omega = \Omega \cup \partial\Omega$.
2. A set Ω is closed if $\partial\Omega \in \Omega$ and open if $\partial\Omega \cap \Omega = \emptyset$.

2.2.2 Convex Sets

A set Ω is convex if along with any pair $x_1, x_2 \in \Omega \subset \mathbb{R}^n$ it contains the entire segment $[x_1, x_2] = \{x : x = (1-t)x_1 + tx_2,\ 0 \le t \le 1\}$. Any point of the segment is a convex combination of x_1 and x_2. A point $x \in \Omega$ is called an extreme point if it is not an interior point of any segment that belongs to Ω. For $x_1, \dots, x_m \in \Omega$ a point

$$x = \sum_{i=1}^{m} t_i x_i, \quad t_i \ge 0, \quad \sum_{i=1}^{m} t_i = 1$$

is a convex combination of $x_1, \dots, x_m$. A set Ω is called strictly convex if $tx_1 + (1-t)x_2 \in \operatorname{int}\Omega$ for any $0 < t < 1$.

There are a number of connections between convex functions and convex sets. Convex functions are a source of convex sets. On the other hand, convex sets can be a source of convex functions, for example, support functions, which we consider later.

Exercise 2.17. Show that the following sets are convex, provided Ω_1 and Ω_2 are convex.

1. Intersection $\Omega_1 \cap \Omega_2$.
2. Sum $\Omega_1 + \Omega_2 = \{x = x_1 + x_2 :\ x_1 \in \Omega_1,\ x_2 \in \Omega_2\}$.
3. Cartesian product $\Omega_1 \times \Omega_2 = \{(x_1, x_2) : x_1 \in \Omega_1, x_2 \in \Omega\}$.

Proposition 2.1.

(a) Consider a mapping $\psi : \Omega \to \mathbb{R}^m$ defined by the formula $\psi(x) = Ax + b$, where A is an $n \times m$ matrix and $b \in \mathbb{R}^m$. If $\Omega \subset \mathbb{R}^n$ is a convex set, then $\psi(\Omega) = \{y = Ax + b :\ x \in \Omega\}$ is a convex set.

(b) The inverse affine image

$$\psi^{-1}(\Omega') = \{x \in \mathbb{R}^n : \psi(x) \in \Omega'\}$$

of a convex set $\Omega' \subseteq \psi(\Omega)$ is convex.

Proof.

(a) If $y_1 \in \psi(\Omega)$ and $y_2 \in \psi(\Omega)$, then there exist x_1 and $x_2 \in \Omega : y_1 = \psi(x_1) = Ax_1 + b$ and $y_2 = \psi(x_2) = Ax_2 + b$. Let us consider

$$y(t) = (1-t)y_1 + ty_2 = (1-t)(Ax_1+b) + t(Ax_2+b) = A((1-t)x_1 + tx_2) + b.$$

From the convexity of Ω follows that $x = (1-t)x_1 + tx_2 \in \Omega$; therefore, $y(t) \in \psi(\Omega)$, that is, $\psi(\Omega)$ is convex.

(b) Let $x_1, x_2 \in \psi^{-1}(\Omega')$ then $Ax_1 + b = y_1$, $Ax_2 + b = y_2$ for some y_1 and $y_2 \in \Omega'$. Consider $x(t) = (1-t)x_1 + tx_2$, then $Ax(t) + b = (1-t)(Ax_1+b) + t(Ax_2+b) \in \Omega'$. Therefore $x(t) \in \psi^{-1}(\Omega')$. □

Let $\Omega \subset \mathbb{R}^n$, the convex hull of Ω is the set

$$\text{chull}\,\Omega = \{x = \sum_{i=1}^{r} t_i x_i : t_i \geq 0, \sum_{i=1}^{r} t_1 = 1\},$$

where $(x_1, x_2, \ldots, x_r)$ is any set of points from Ω.

A simplex in $\mathbb{R}^n$ is the convex hull of $e_1, \ldots, e_n$, where $e_i = (0, \ldots, 1, \ldots, 0)$ that is $S_n = \{x \in \mathbb{R}^n : x = \sum_{i=1}^{n} x_i e_i, x_i \geq 0, i = 1, \ldots, n,\ \sum_{i=1}^{n} x_i = 1\}$.

Theorem 2.6 (Caratheodory). *Let $\Omega \subset \mathbb{R}^n$, then every element of* chull Ω *can be represented as a convex combination of at most n+1 elements of Ω.*

Below we consider several examples of convex sets.

The set

$$\mathbb{S}^n = \{A \in \mathbb{R}^{n \times n} : A^T = A\}$$

of symmetric matrices is a $\frac{n(n+1)}{2}$ dimensional vector space. We use $\mathbb{S}^n_+$ to denote the set of symmetric positive semidifine matrices

$$\mathbb{S}^n_+ = \{A \in \mathbb{S}^n : A \succeq 0\}.$$

Finally, let us introduce the set

$$\mathbb{S}^n_{++} = \{A \in \mathbb{S}^n : A \succ 0\}$$

of symmetric, positive definite matrices, that is, $A = A^T$ and $(Ax, x) > 0$ for $x \neq 0$.

Exercise 2.18. Show that $\mathbb{S}^n_+$ and $\mathbb{S}^n_{++}$ are convex sets.

Exercise 2.19. Show that the solution set

$$\{x \in \mathbb{R}^n : x_1 A_1 + \ldots + x_n A_n \preceq A_0\}$$

of a linear matrix inequality (LMI), where $A_0, A_1, \ldots, A_n \in \mathbb{S}^m$, is a convex set in $\mathbb{R}^n$.

For a given vector x_0 and $r > 0$, the Euclidian ball with a center x_0 and radius r is given by

$$B(x_0,r) = \{x : \|x - x_0\| \leq r\} = \{x : \langle x - x_0, x - x_0 \rangle \leq r^2\} = \{x_0 + ru : \|u\| \leq 1\}.$$

Exercise 2.20. Prove that $B(x_0,A)$ is a closed convex set.

Theorem 2.7 (Perturbations of Compact Convex Sets). *Let c_i:$\mathbb{R}^n \to \mathbb{R}$, $i = 0$, $1,\ldots,m$, be concave functions and*

$$\Omega = \{x \in \mathbb{R}^n : c_i(x) \geq 0,\ i = 0, 1, \ldots, m\}$$

be a nonempty and bounded set, then for any $\varepsilon = (\varepsilon_0, \varepsilon_1, \ldots, \varepsilon_m) \in \mathbb{R}^{m+1}_+$, the set

$$\Omega_\varepsilon = \{x \in \mathbb{R}^n : c_i(x) \geq -\varepsilon_i,\ i = 0, 1, \ldots, m\}$$

is bounded.

Proof. It suffices to prove the statement in the case when all ε_i except for one (say, ε_0) equal 0. So, we have to prove that $\Omega_{\varepsilon_0} = \{x \in \mathbb{R}^n : c_0(x) \geq -\varepsilon_0,\ c_i(x) \geq 0, i = 1, \ldots, m\}$ is bounded. Let us assume the contrary, then we can take $x_1 \in \Omega$ and find a ray through x_1 that intersects the boundary of Ω, but not of Ω_{ε_0}. Take x_2 on this ray so that $c_0(x_2) = -\delta < 0$ and $c_i(x_2) \geq 0$, $i = 1, \ldots, m$. Using concavity of $c_0(x)$, we obtain

$$-\delta = c_0(x_2) \geq \lambda c_0(x_1 + \frac{1}{\lambda}(x_2 - x_1)) + (1 - \lambda)c_0(x_1)$$

for $0 < \lambda < 1$. Keeping in mind that $c_0(x_1) \geq 0$, we have

$$c_0(x_1 + \frac{1}{\lambda}(x_2 - x_1)) \leq \frac{1}{\lambda}(-(1 - \lambda)c_0(x_1) - \delta) \leq -\frac{\delta}{\lambda}.$$

By taking $\lambda < \frac{\delta}{\varepsilon_0}$, we can find a point $\bar{x} = x_1 + \frac{1}{\lambda}(x_2 - x_1)$ on the ray such that $c_0(\bar{x}) < -\varepsilon_0$, that is, $\bar{x} \notin \Omega_{\varepsilon_0}$, a contradiction. □

Corollary 2.2. *If f is convex, all c_i are concave and $X^* = \text{Argmin}\{f(x) : c_i(x) \geq 0, i = 1, \ldots, m\}$=$\{x \in \mathbb{R}^n : f(x) = f(x^*), c_i(x) \geq 0, i = 1, \ldots, m\}$ is nonempty and bounded; then for any $\varepsilon = (\varepsilon_0, \varepsilon_1, \ldots, \varepsilon_m) \in \mathbb{R}^{m+1}_+$ and $x^* \in X^*$, the set*

$$X^*_\varepsilon = \{x \in \mathbb{R}^n : f(x) - f(x^*) \leq \varepsilon_0, c_i(x) \geq -\varepsilon_i, i = 1, \ldots, m\}$$

is bounded.

Corollary 2.2 follows directly from Theorem 2.7 for $c_0(x) = f(x^*) - f(x)$.

Exercise 2.21. Show that strictly convex function on closed, convex, and bounded set is strongly convex.

2.2.3 Affine Sets

A set Ω is affine if for any $x_1, x_2 \in \Omega$ and any $t \in \mathbb{R}$ we have $(1-t)x_1 + tx_2 \in \Omega$. Let $x_1, \ldots, x_r \in \Omega$; a vector $x = t_1x_1 + \ldots + t_rx_r$ with $\sum_{i=1}^r t_i = 1$ is an affine combination of $x_1, \ldots, x_r$. An affine set Ω contains affine combinations of any set of points from Ω. Obviously, any affine set is convex, but not every convex set is affine.

Example 2.1. The solution set $X^* = \{x \in \mathbb{R}^n : Ax = b\}$ of a system of linear equations is an affine set, because for $x_1 \in X^*$ and $x_2 \in X^*$ we have $Ax_1 = b, Ax_2 = b$; therefore, for any affine combination $x = t_1x_1 + t_2x_2$, $t_1 + t_2 = 1$, we have $Ax = t_1Ax_1 + t_2Ax_2 = (t_1 + t_2)b = b$.

A linear manifold in $\mathbb{R}^n$ is a translated subspace, that is, a set of vectors

$$y + V = \{y + x : x \in V\},$$

where V is a subspace in $\mathbb{R}^n$.

For a given vector $a \in \mathbb{R}^n$, $a \neq 0$, and a number $\alpha \in \mathbb{R}$, a hyperplane

$$H = \{x \in \mathbb{R}^n : \langle a, x\rangle = \alpha\}$$

is an affine set.

The set $\{x \in \mathbb{R}^n : \langle a, x\rangle \leq \alpha\}, a \neq 0$, is a closed halfspace. A halfspace is a convex set, but not an affine set. Each hyperplane divides $\mathbb{R}^n$ into two halfspaces.

The affine hull of Ω

$$\operatorname{aff}\Omega = \{x = \sum_{i=1}^r t_ix_i \ : \ x_1, \ldots, x_r \in \Omega, \sum_{i=1}^r t_i = 1\}$$

is the set of all affine combinations of points in Ω. It is the intersection of all linear manifolds containing Ω, that is, the smallest affine set that contains Ω.

The affine dimension of Ω is the dimension of its affine hull. Such definition is used in convex analysis, but it is not always consistent with other definitions of dimension. For example, a unit circle $S = \{x \in \mathbb{R}^2 : x_1^2 + x_2^2 = 1\}$ has dimension one, while the affine dimension of S is two, because two is the dimension of the affine hull

$$\operatorname{aff} S = \{x \in \mathbb{R}^2 : x = t_1x_1 + t_2x_2, \ \ x_1 \in S, x_2 \in S, \ \ t_1 + t_2 = 1\} = \mathbb{R}^2.$$

If the affine dimension of $\Omega \subset \mathbb{R}^n$ is less than n, then Ω lies in the affine set that does not coincide with $\mathbb{R}^n$. The relative interior of the set Ω is the interior relative to $\operatorname{aff}\Omega$:

$$\operatorname{relint}\Omega = \{x \in \Omega : B(x,r) \cap \operatorname{aff}\Omega \subset \Omega \text{ for some } r > 0\}.$$

Let $a_1, \ldots, a_n \in \mathbb{R}^n$ and $\alpha_1, \ldots, \alpha_m \in \mathbb{R}$

$$\operatorname{LIN}\{a_1, \ldots, a_m\} = \{a \in \mathbb{R}^n : a = \alpha_1a_1 + \ldots + \alpha_ma_m\}.$$

Exercise 2.22. An ellipsoid with a center $x_0 \in \mathbb{R}^n$ is defined as follows:

$$E(x_0, A) = \{x_0 + Au : \|u\| \leq 1\},$$

where $A \in \mathbb{S}^n_+$. If A is symmetric and positive semidefinite, but singular, then the set $E(x_0, A)$ is called a degenerate ellipsoid. Show that its affine dimension is equal to the rank of A.

2.2.4 Cones

A set C is a cone if for every $x \in C$ and $t \geq 0$ we have $tx \in C$. A set C is a convex cone if it is convex and is a cone. A point $x = t_1x_1 + \ldots + t_kx_k$ is called a conic combination of $x_1, \ldots, x_k$ if all $t_i \geq 0$. Clearly, a set C is a convex cone if and only if it contains all conic combinations of its elements.

For a given convex cone C, the conjugate cone is defined as

$$C^* = \{u \in \mathbb{R}^n : \langle u, x \rangle \leq 0 \text{ for all } x \in C\}.$$

Example 2.2. Consider a halfspace $C = \{x \in \mathbb{R}^n : \langle a, x \rangle \leq 0\}$, $a \neq 0$. Clearly, C is a convex cone. Its conjugate cone is given by

$$C^* = \{u \in \mathbb{R}^n : u = \lambda a, \lambda \geq 0\}.$$

The conic hull of a set Ω is the set of all conic combinations of points in Ω, that is,

$$\text{chull}\,\Omega = \{x = t_1x_1 + \ldots. + t_rx_r : x_i \in \Omega, t_i \geq 0,\ i = 1, \ldots. r\};$$

it is the smallest cone that contains Ω.

Let us consider a polyhedral cone which is defined by the following homogeneous system of linear inequalities:

$$C = \{x \in \mathbb{R}^n : \langle a_i, x \rangle \leq 0,\ i = 1, \ldots, m\}, \tag{2.31}$$

then the following Lemma holds.

Lemma 2.3 (Farkas). *Let $C_i = \{x \in \mathbb{R}^n : \langle a_i, x \rangle \leq 0\}$, $i = 1, \ldots, m$, and $C = C_1 \cap C_2 \cap \ldots \cap C_m \neq \emptyset$, then*

$$C^* = \text{chull}(C_1^* \cup C_2^* \cup \ldots \cup C_m^*) = \{u = \sum_{i=1}^m \lambda_i a_i,\ \lambda_i \geq 0\}.$$

Corollary 2.3. *If for a polyhedral cone C given by (2.31) and a vector $a \in \mathbb{R}^n$, $a \neq 0$, we have $\langle a, x \rangle \leq 0$ for any $x \in C$ then*

$$a = \sum_{i=1}^{m} \lambda_i a_i,\ \lambda_i \ge 0,\ i = 1,\dots,m.$$

We will prove this important Lemma later.

2.2.5 Recession Cones

Let consider a convex set $\Omega \subset \mathbb{R}^n$. A direction d is called a recession direction for Ω if for any point $x \in \Omega$ and any $t > 0$ we have $x + td \subset \Omega$. The recession cone of the set Ω is

$$RC(\Omega) = \{d \in \mathbb{R}^n : x + td \in \Omega \text{ for any } x \in \Omega \text{ and } t \ge 0\}.$$

Example 2.3.

1. For $\Omega_1 = \{(x_1,x_2) : x_1 > 0, x_2 \ge 1/x_1\}$, one has $RC(\Omega_1) = \{(d_1,d_2) : d_1 \ge 0, d_2 \ge 0\}$.
2. For $\Omega_2 = \{(x_1,x_2) : x_2 \ge x_1^2\}$, one has $RC(\Omega_2) = \{(d_1,d_2) : d_1 = 0, d_2 \ge 0\}$.
3. For $\Omega_3 = \{(x_1,x_2) : x_1^2 + x_2^2 \le 1\}$, one has $RC(\Omega_3) = \{(d_1,d_2) : d_1 = d_2 = 0\}$.
4. For $\Omega_4 = \{(x_1,x_2) : x_1 > 0, x_2 > 0\}$, one has $RC(\Omega_4) = \{(d_1,d_2) : d_1 \ge 0, d_2 \ge 0\}$.
5. For $\Omega_5 = \{x \in \mathbb{R}^n : Ax - b \ge 0\}$, one has $RC(\Omega_5) = \{d \in \mathbb{R}^n : Ad \ge 0\}$.

Exercise 2.23. Show that $RC(\Omega_1 \cap \Omega_2) = RC(\Omega_1) \cap RC(\Omega_2)$ for any two convex sets Ω_1, Ω_2.

We will say that $RC(\Omega)$ is trivial if it does not contain any vector $d \neq 0$.

If follows from the definition of the recession cone that a convex set Ω has a trivial recession cone if and only if Ω is bounded.

Let us consider a convex function $f : \mathbb{R}^n \to \mathbb{R}$, a number $\alpha \in \mathbb{R}$, and the sublevel set

$$\mathcal{L}_f(\alpha) = \{x \in \mathbb{R}^n : f(x) \le \alpha\}.$$

It follows from Theorem 2.7 that if $\mathcal{L}_f(\alpha)$ is unbounded for some α_0, it will be unbounded or empty for any $\alpha \in \mathbb{R}$.

Let $f : \mathbb{R}^n \to \mathbb{R}$ be a convex function and $\Omega \subset \mathbb{R}^n$ be a convex set. Due to the continuity of a convex function on $\mathbb{R}^n$ (see Theorem 2.2), the set $\Omega(\alpha) = \Omega \cap \mathcal{L}_f(\alpha)$ is closed and convex for any α. Let us consider the following convex optimization problem:

$$f(x^*) = \min\{f(x) | \, x \in \Omega\}.$$

If the optimal set $X^* = \{x \in \mathbb{R}^n : f(x) = f(x^*)\}$ is not empty, then for any $x^* \in X^*$ and $\alpha^* = f(x^*)$, we obtain $X^* = \Omega(\alpha^*)$. Therefore, the optimal set X^* is closed and convex.

The following Lemma establishes condition, under which the optimal set X^* is not empty and bounded.

Lemma 2.4. *If $f : \mathbb{R}^n \to \mathbb{R}$ is a convex function and Ω is a convex set, then X^* is nonempty and bounded if and only if there exists α such that $\mathscr{L}_f(\alpha) \neq \emptyset$ and*

$$RC(\Omega(\alpha)) = \{0\}. \tag{2.32}$$

Proof. If (2.32) is true for some α, then $\Omega(\alpha)$ is bounded, and $x^* \in \Omega$ exists due to the Weierstrass Theorem.

Conversely, if X^* is nonempty and bounded, then using Theorem 2.7 we see that $\Omega(\alpha)$ is bounded for any α, and hence (2.32) follows. □

Let us consider set $\Omega = \{x \in \mathbb{R}^n : c_i(x) \geq 0,\ i = 1,\ldots,m\}$, where $c_i : \mathbb{R}^n \to \mathbb{R}$ are concave. If X^* is not empty and bounded, then from Theorem 2.7 follows boundedness $\Omega_N = \{x \in \mathbb{R}^n : c_i(x) \geq 0,\ i = 1,\ldots,m;\ c_0(x) = N - f(x) \geq 0\}$ for any $N > f(x^*)$.

Corollary 2.4. *Let f be convex, all c_i $i = 1,\ldots,m$ be concave, and X^* is not empty and bounded set, then by adding one extra constraint $c_0(x) = N - f(x) \geq 0$ with $N > 0$ large enough, we obtain a convex optimization problem equivalent to* (2.32) *with a bounded feasible set $\Omega := \{x \in \Omega : c_0(x) \geq 0\}$.*

For N large enough, adding an extra constraint $c_0(x) = N - f(x) \geq 0$ to the set of constraints that defines Ω cannot affect the solution. Therefore, for the cost of one extra constraint, we can assume that the initial set Ω is bounded, provided that X^* is not empty and bounded.

2.2.6 *Polyhedrons and Polytopes*

A polyhedron is defined as a solution set of a finite number of linear equations and inequalities:

$$\Pi = \{x \in \mathbb{R}^n : \langle a_j,x\rangle \leq \alpha_j,\ j = 1,\ldots,m; \langle c_i,x\rangle = \beta_i, i = 1,\ldots,r\}.$$

So, a polyhedron is an intersection of a finite number of halfspaces and hyperplanes. Subspaces, hyperplanes, rays, and halfspaces are all polyhedrons. The nonnegative orthant

$$\mathbb{R}^n_+ = \{x \in \mathbb{R}^n : x_i \geq 0, i = 1,\ldots,n\}$$

is a polyhedral cone, because it is a polyhedron and a cone.

A bounded polyhedron is called a polytope. For example, the $(n-1)$- dimensional simplex

$$S_n = \{x \in \mathbb{R}^n_+ : \sum_{i=1}^{n} x_i = 1\}$$

is a polytope.

Let $\Pi = \{x \in \mathbb{R}^n : \langle a_i,x\rangle \leq b_i, i = 1,..,q\}$, $q \geq n$ and $I(x) = \{i : \langle a_i,x\rangle = b_i\}$. A point $\bar{x} \in \Pi$ is a vertex (extreme point) if and only if there exist n linearly independent vectors in the set $\{a_i : i \in I(\bar{x})\}$.

In fact, if this set has less than n linearly independent vectors, then the system $\langle a_i, u\rangle = 0$, $i \in I(\bar{x})$ has a solution $\bar{u} \neq 0$. Therefore, for small $t > 0$ points $x_1 = \bar{x} + t\bar{u}$ and $x_2 = \bar{x} - t\bar{u}$ are two different points in Π, and $\bar{x} = \frac{1}{2}(x_1 + x_2)$, which means that $\bar{x}$ is not an extreme point.

On the other hand, if among a_i, $i \in I(\bar{x})$, there are n linearly independent, then $\bar{x}$ is an extreme point. Indeed, assuming the opposite we can find $x_1 \in \Pi$ and $x_2 \in \Pi$ such that $\bar{x} = \frac{1}{2}(x_1 + x_2) \in \Pi$, that is,

$$b_i = \langle a_i, \bar{x}\rangle = \frac{1}{2}\langle a_i, x_1\rangle + \frac{1}{2}\langle a_i, x_2\rangle \leq b_i,$$

for any $i \in I(\bar{x})$. Therefore, $\langle a_i, x_1\rangle = b_i$ and $\langle a_i, x_2\rangle = b_i$, $i \in I(\bar{x})$. However, the system $\langle a_i, x\rangle = b_i$, $i \in I(\bar{x})$ cannot have two different solutions, because among vectors a_i there are n linear independent vectors.

A polyhedron might not have vertices, for example, if it is a halfspace or a subspace. For a polytope the vertices not just exist, they generate the polytope in the following sense.

Proposition 2.2. *A polytope is the convex hull of its vertices.*

Proposition 2.2 is a particular case of the well-known **Krein–Milman Theorem** in a finite dimensional space: if $\Omega \subseteq \mathbb{R}^n$ is a convex compact set, then Ω is a convex hull of its extreme points.

2.3 Closed Convex Functions

So far we considered convex functions $f : \mathbb{R}^n \to \mathbb{R}$. Sometimes, however, we have to deal with convex functions $f : \Omega \to \mathbb{R}$, where Ω is a convex set in $\mathbb{R}^n$. To handle this situation, it will be convenient to assume that the convex function f is defined on $\mathbb{R}^n$ and takes values in $\bar{\mathbb{R}} = \mathbb{R} \cup \{\infty\}$. Then $f : \Omega \to \mathbb{R}$ can be redefined as follows:

$$f(x) := \begin{cases} f(x), \; x \in \Omega \\ +\infty, \;\; x \notin \Omega. \end{cases}$$

We will assume the following standard rules of arithmetic operations: for $a < \infty$, $\infty \pm a = \infty$, $a \cdot \infty = \infty$ $(a > 0)$, $\infty + \infty = \infty$, $\max\{a, \infty\} = \infty$, $0 \cdot \infty = \infty \cdot 0 = 0$, $\inf\{x : x \in \emptyset\} = +\infty$. It is easy to see that under these rules, $f : \mathbb{R}^n \to \bar{\mathbb{R}}$ is convex.

In the future we will deal only with convex functions $f : \mathbb{R}^n \to \bar{\mathbb{R}}$. We do not add value $-\infty$ to $\bar{\mathbb{R}}$, so convex functions, such as

$$f(x) = \begin{cases} -\infty, & |x| < 1, \\ 0, & |x| = 1, \\ \infty, & |x| > 1, \end{cases}$$

are excluded from our considerations.

The set

$$\operatorname{dom} f = \{x \in \mathbb{R}^n : f(x) < \infty\}$$

is called the domain of f. A convex function $f : \mathbb{R}^n \to \bar{\mathbb{R}}$ is called proper if $\operatorname{dom} f \neq \emptyset$.

Lemma 2.5. *A function $f : \mathbb{R}^n \to \bar{\mathbb{R}}$ is convex if and only if for any pair $x, y \in \operatorname{dom} f$ and $\alpha \geq 0$ such that $y + \alpha(y - x) \in \operatorname{dom} f$ we have*

$$f(y + \alpha(y - x)) \geq f(y) + \alpha(f(y) - f(x)). \tag{2.33}$$

Proof. Let $t = \frac{\alpha}{1+\alpha}$ and $\mathrm{v} = y + \alpha(y - x)$, that is, $y = (1 - t)\mathrm{v} + tx$, then (2.33) follows from the convexity of f. Conversely, for $x, y \in \operatorname{dom} f$ and $t \in (0, 1]$ consider $u = tx + (1 - t)y$, then

$$x = \frac{1}{t}u - \frac{(1-t)}{t}y = u + \alpha(u - y)$$

for $\alpha = \frac{1-t}{t}$ and the convexity of f follows from (2.33). □

Lemma 2.6. *If $f : \mathbb{R}^n \to \bar{\mathbb{R}}$ is a convex function, then for any $\alpha \in \mathbb{R}$ the sublevel set $\mathscr{L}_f(\alpha)$ is either convex or empty.*

Proof. Let $\mathscr{L}_f(\alpha) \neq \emptyset$ and $x_1, x_2 \in \mathscr{L}_f(\alpha)$, then the convexity of f implies

$$f((1 - t)x_1 + tx_2) \leq (1 - t)f(x_1) + tf(x_2) \leq (1 - t)\alpha + t\alpha = \alpha;$$

therefore $x = (1 - t)x_1 + tx_2 \in \mathscr{L}_f(\alpha)$. □

Convexity of sublevel sets is only a necessary condition for f to be convex; not every function with a convex sublevel set is convex.

Exercise 2.24. Find an example of a nonconvex function with convex sublevel sets.

Lemma 2.7. *A function $f : \mathbb{R}^n \to \bar{\mathbb{R}}$ is convex if and only if its epigraph*

$$\operatorname{epi} f = \{(x, \alpha) \in \operatorname{dom} f \times \mathbb{R} : f(x) \leq \alpha\}$$

is a convex set.

Proof. The convexity of $\operatorname{epi} f$ implies

$$(x(t)), \alpha(t)) = ((1 - t)x_1 + tx_2; (1 - t)\alpha_1 + t\alpha_2) \in \operatorname{epi} f$$

for any (x_1, α_1), $(x_2, \alpha_2) \in \operatorname{epi} f$ and any $0 < t < 1$. Therefore, $f((1 - t)x_1 + tx_2) \leq (1 - t)\alpha_1 + t\alpha_2$, for any $\alpha_1 \geq f(x_1)$ and $\alpha_2 \geq f(x_2)$, so

$$f((1 - t)x_1 + tx_2) \leq (1 - t)f(x_1) + tf(x_2),$$

and hence f is convex.

Conversely, the convexity of f implies

$$f((1-t)x_1+tx_2) \leq (1-t)f(x_1)+tf(x_2) \leq (1-t)\alpha_1+t\alpha_2$$

for any $\alpha_1 \geq f(x_1), \alpha_2 \geq f(x_2)$, that is, $((1-t)x_1+tx_2,\ (1-t)\alpha_1+t\alpha_2) \in \operatorname{epi} f$. □

The behavior of a convex function on the boundary of $\operatorname{dom} f$ can be rather strange. Let us consider the following example:

$$f(x,y) = \begin{cases} 0, & (x,y): x^2+y^2 < 1, \\ x^2+\ln(2+y), & (x,y): x^2+y^2 = 1, \\ \infty, & (x,y): x^2+y^2 > 1. \end{cases}$$

The domain $\operatorname{dom} f = \{(x,y): x^2+y^2 \leq 1\}$ is convex, the function f is convex, but its epigraph is not a closed set. Moreover, one can take instead of $x^2+\ln(2+y)$ any positive function. To avoid such a behavior, we consider closed convex functions.

A convex function $f: \mathbb{R}^n \to \bar{\mathbb{R}}$ is closed if its epigraph is a closed set.

Let us consider few examples.

Example 2.4.

1. The function

$$f(x,y) = \begin{cases} 0, & x^2+y^2 \leq 1, \\ \infty, & x^2+y^2 > 1, \end{cases}$$

is closed because f is continuous on a closed $\operatorname{dom} f$.

2. The function

$$f(x,y) = \begin{cases} 0, & x^2+y^2 < 1, \\ 1, & x^2+y^2 = 1, \\ \infty, & x^2+y^2 > 1, \end{cases}$$

is not closed because its epigraph is not closed.

3. The function

$$f(x,y) = \begin{cases} \frac{1}{1-x^2-y^2}, & x^2+y^2 < 1, \\ \infty, & x^2+y^2 \geq 1, \end{cases}$$

is closed.

Any convex function with a closed domain that is continuous on $\operatorname{dom} f$ is a closed function. If $f: \mathbb{R}^n \to \bar{\mathbb{R}}$ is convex and $\operatorname{dom} f$ is open, then f is closed if and only if $f(x_s) \to \infty$ for every sequence $\{x_s\}_{x\in N}$, such that $\lim_{s\to\infty} x_s = \bar{x} \in \partial(\operatorname{dom} f)$.

Example 2.5.

(a) The function $f(x) = x\ln x$ for $x > 0$ is convex with $\operatorname{dom} f = \mathbb{R}_{++}$, which is not closed, and $\lim_{x\to 0} x\ln x \neq \infty$; therefore, f is not closed.
(b) The function

$$f(x) = \begin{cases} x\ln x, & x > 0, \\ 0, & x = 0, \end{cases}$$

is convex with the closed $\operatorname{dom} f = \mathbb{R}_{+}$ and is continuous on $\operatorname{dom} f$; therefore, it is a closed convex function.
(c) The function $f(x) = -\ln x$ for $x > 0$ is convex and closed, because $\operatorname{dom} f = \mathbb{R}_{++}$ is open and for any $\{x_s\}_{s\in N} : x_s \to 0$ we have $f(x_s) \to \infty$.

Exercise 2.25. Among functions given below find those which are closed:

1. $f(x) = |x_1 + x_2|$.
2. $f(x) = \|x\| = \langle x, x\rangle^{\frac{1}{2}}$.
3.

$$f(x) = \begin{cases} x^p, & x > 0, 1 \le p < \infty, \\ 1, & x = 0, \\ \infty, & x < 0. \end{cases}$$

4.

$$f(x) = \begin{cases} (a^2 - x^2)^{-\frac{1}{2}}, & |x| < a, \\ \infty, & |x| \ge a. \end{cases}$$

5.

$$f(x,y) = \begin{cases} 0, & \frac{x^2}{a^2} + \frac{y^2}{b^2} < 1, \\ \text{any positive } \psi(x,y), & \frac{x^2}{a^2} + \frac{y^2}{b^2} = 1. \end{cases}$$

Theorem 2.8. *The following statements are equivalent:*

(1) a function f is closed;
(2) sublevel sets $\mathcal{L}_f(\alpha)$ are closed;
(3) f is semicontinuous from below, that is, for any sequence $\{x_s\}_{s\in N}$, such that $\lim_{s\to\infty} x_s = x_0$, one has

$$\liminf_{s\to\infty} f(x_s) \ge f(x_0). \tag{2.34}$$

Proof. Let us show that (1) implies (2). This follows immediately from Lemma 2.6 and the fact that $\mathcal{L}_f(\alpha)$ is an intersection of two closed sets $\operatorname{epi} f$ and $\{(x,t) : t = \alpha\}$. Let (2) hold, then from $x_s \to x_0$, $f(x_s) \to \alpha$ follows $f(x_s) \le \alpha + \varepsilon$ for any $\varepsilon > 0$; therefore, $f(x_0) \le \alpha + \varepsilon$, because $\mathcal{L}_f(\alpha + \varepsilon)$ is closed.

Keeping in mind that $\varepsilon > 0$ can be as small as one wants, we obtain $f(x_0) \le \alpha$, and hence (2.34) is true, which means that f is semicontinuous from below.

Finally, let us show that (1) follows from (3). In fact, if $\mu_s \geq f(x_s)$ and $(\mu_s, x_s) \to (\mu_0, x_0)$, then from (2.34) follows

$$\mu_0 = \liminf_{s\to\infty} f(x_k) \geq f(x_0),$$

which means $(\mu_0, x_0) \in \operatorname{epi} f$; hence, $\operatorname{epi} f$ is a closed set. □

2.3.1 Operations on Closed Convex Functions

We consider several operations on closed convex functions, which retain these properties.

The simplest one is the multiplication by a positive number: $f(x) = tf_1(x), t > 0$. In this case $\operatorname{dom} f = \operatorname{dom} f_1$. It is obvious that if f_1 is closed and convex, the same is true for f.

Proposition 2.3. *Let f_i, $i = 1, \ldots, r$ be closed convex functions, then*

$$f(x) = \sum_{i=1}^{r} f_i(x)$$

is closed and convex on $\operatorname{dom} f = \bigcap_{i=1}^{r} \operatorname{dom} f_i$.

Proof. The convexity of f is obvious. To prove that f is closed, we consider a sequence $\{(x_s, t_s)\}_{s\in N} \subset \operatorname{epi} f$ such that $t_s \geq \sum_{i=1}^{r} f_i(x_s)$, $\lim_{s\to\infty} x_s = \bar{x} \in \operatorname{dom} f$, $\lim_{s\to\infty} t_s = \bar{t}$. For a closed f_i, Theorem 2.8 implies

$$\liminf_{s\to\infty} f_i(x_s) \geq f_i(\bar{x}), \ i = 1, \ldots, r.$$

Therefore, for $f(\bar{x}) = \sum_{i=1}^{r} f_i(\bar{x})$ and $\bar{t}$, we have

$$\bar{t} = \lim_{s\to\infty} t_s \geq \liminf_{s\to\infty} \sum_{i=1}^{r} f_i(x_s) \geq \sum_{i=1}^{r} f_i(\bar{x}) = f(\bar{x}).$$

Thus, $(\bar{x}; \bar{t}) \in \operatorname{epi} f$. □

Proposition 2.4. *Let f_i, $i = 1, \ldots, r$, be closed convex functions, then $f(x) = \max\{f_i(x): i = 1, \ldots, r\}$ is closed and convex on* $\operatorname{dom} f(x) = \bigcap_{i=1}^{r} \operatorname{dom} f_i$.

Proof. Let us consider

$$\operatorname{epi} f = \{(x,t) : t \geq f_i(x), \ i = 1, \ldots, r, \ x \in \bigcap_{i=1}^{r} \operatorname{dom} f_i\} = \operatorname{epi} f_1 \bigcap \cdots \bigcap \operatorname{epi} f_r.$$

It is closed and convex as the intersection of closed and convex sets; therefore, f is closed and convex. □

Lemma 2.8. *Let $\Phi(x,y)$ be a closed convex function in x for any given $y \in Q$, where Q is a given set. Then*

$$f(x) = \sup\{\Phi(x,y) : y \in Q\}$$

is a closed convex function with the domain

$$\operatorname{dom} f = \{x \in \bigcap_{y \in Q} \operatorname{dom} \Phi(\cdot,y) : \Phi(x,y) \leq t \text{ for some } t \in \mathbb{R} \text{ and any } y \in Q\}. \quad (2.35)$$

Proof. For any x that belongs to the right-hand side of (2.35), we have $f(x) < \infty$; therefore, $x \in \operatorname{dom} f$. If $\bar{x}$ does not belong to the right-hand side, then there is $\{y_s\}_{s \in N}$ such that $\Phi(\bar{x}, y_s) \to \infty$; therefore $\bar{x} \notin \operatorname{dom} f$. Besides, $(x,t) \in \operatorname{epi} f$ if and only if for all $y \in Q$ we have $x \in \operatorname{dom} \Phi(\cdot,y)$, $t \geq \Phi(x,y)$, so $\operatorname{epi} f = \bigcap_{y \in Q} \operatorname{epi} \Phi(\cdot,y)$. Hence, f is convex and closed because $\operatorname{epi} \Phi(\cdot,y)$ is convex and closed for each $y \in Q$. □

Example 2.6. Let f_i, $i = 1,\ldots,m$, be closed convex functions and S_m be the $(m-1)$—dimensional simplex, then it follows from Proposition 2.3 that

$$F_\mu(x) = \sum_{i=1}^{m} \mu_i f_i(x)$$

is closed and convex for any $\mu = (\mu_1,\ldots,\mu_m) \in S_m$. Therefore, it follows from Lemma 2.8 that $f(x) = \sup_{\mu \in S_m} F_\mu(x)$ is closed and convex.

2.3.2 Projection on a Closed Convex Set

Projection on a closed convex set is an important operation, which is used in a number of methods for constrained optimization, as well as in convex analysis. In particular, it is used for proving separation theorems. Let $x_0 \in \mathbb{R}^n$ and $\Omega \subset \mathbb{R}^n$ be a closed convex set. The projection of x_0 on Ω is the closest to x_0 point in Ω:

$$P_\Omega(x_0) = \operatorname{argmin}\{\frac{1}{2}\|x - x_0\|^2 : x \in \Omega\}.$$

First of all, $P_\Omega(x_0)$ always exists and is unique for any $x_0 \in \mathbb{R}^n$. In fact, by taking any $y \in \Omega$, we will find $P_\Omega(x_0)$ if we replace Ω

$$\Omega_y = \{x \in \Omega : \|x - x_0\| \leq \|y - x_0\|\}.$$

The set Ω_y is closed, convex, and bounded because the set $\{x \in \mathbb{R}^n : \|x - x_0\| \leq \|y - x_0\|\}$ is closed, convex, and bounded, while Ω is closed and convex. The function $f(x) = \frac{1}{2}\|x - x_0\|^2$ is continuous in x, so due to Weirstrass theorem it has a minimizer on Ω_y. Also, f is strongly convex; therefore, the minimizer $P_\Omega(x_0)$ is unique. If $x_0 \in \Omega$ then $P_\Omega(x_0) = x_0$.

Lemma 2.9. *Let Ω be a closed convex set and $x_0 \notin \Omega$, then*

$$\langle P_\Omega(x_0) - x_0, x - P_\Omega(x_0)\rangle \geq 0 \tag{2.36}$$

for any $x \in \Omega$.

Proof. Let $f(x) = \frac{1}{2}\|x - x_0\|^2$ and $P_\Omega(x_0) = \operatorname{argmin}\{f(x) : x \in \Omega\}$, then a descent direction in Ω for f at $P_\Omega(x_0)$ does not exist, that is, $\langle \nabla f(P_\Omega(x_0)), x - P_\Omega(x_0)\rangle \geq 0, \forall x \in \Omega$.

From the last inequality and $\nabla f(x) = x - x_0$ follows (2.36) □

Exercise 2.26. Show that both closeness and convexity Ω cannot be dropped in Lemma 2.9.

Lemma 2.10. *Let $x_0 \notin \Omega$, then for any $x \in \Omega$, we have the following triangle inequality:*

$$\|x - P_\Omega(x_0)\|^2 + \|P_\Omega(x_0) - x_0\|^2 \leq \|x - x_0\|^2. \tag{2.37}$$

Proof. Keeping in mind (2.36), we obtain

$$\|x - P_\Omega(x_0)\|^2 - \|x - x_0\|^2 = \langle x_0 - P_\Omega(x_0), 2x - P_\Omega(x_0) - x_0\rangle =$$
$$= -2\langle P_\Omega(x_0) - x_0, x - P_\Omega(x_0)\rangle + \langle x_0 - P_\Omega(x_0), P_\Omega(x_0) - x_0\rangle \leq -\|x_0 - P_\Omega(x_0)\|^2,$$

that is, (2.37) holds. □

Let us consider the mapping $\psi : \mathbb{R}^n \to \Omega$, defined by the formula $\psi(x) = P_\Omega(x)$.

Lemma 2.11. *The mapping ψ is continuous and non-expansive, that is,*

$$\|P_\Omega(x) - P_\Omega(y)\| \leq \|x - y\| \tag{2.38}$$

for any $x, y \in \mathbb{R}^n$.

Proof. From Lemma 2.9 follows $\langle x - P_\Omega(x), z - P_\Omega(x)\rangle \leq 0$, for any $z \in \Omega$. In particular, for $z = P_\Omega(y)$ we get $\langle x - P_\Omega(x), P_\Omega(y) - P_\Omega(x)\rangle \leq 0$. Similarly, $\langle y - P_\Omega(y), P_\Omega(x) - P_\Omega(y)\rangle \leq 0$.

By adding the last two inequalities, we obtain

$$\langle P_\Omega(y) - P_\Omega(x), x - P_\Omega(x) + P_\Omega(y) - y\rangle =$$
$$= \|P_\Omega(y) - P_\Omega(x)\|^2 - \langle P_\Omega(y) - P_\Omega(x), y - x\rangle \leq 0.$$

Using the Cauchy–Schwarz inequality, we get

$$\|P_\Omega(y) - P_\Omega(x)\|^2 \leq \langle y - x, P_\Omega(y) - P_\Omega(x)\rangle \leq \|y - x\|\|P_\Omega(y) - P_\Omega(x)\|,$$

that is, (2.38) holds. The continuity of ψ follows from (2.38). □

Exercise 2.27.

1. Prove that for a closed Ω the function

$$d_\Omega(x) = \min\{\|x-u\| : u \in \Omega\} = \|x - P_\Omega(x)\|$$

is continuous.
2. Show that for a closed convex set Ω the function d_Ω is convex.
3. Show that for a closed convex Ω the function $\varphi(x) = \frac{1}{2}d^2_\Omega(x)$ is convex and differentiable.

2.3.3 Separation Theorems

The projection operation naturally leads to separation theorems.

Two sets Ω_1 and Ω_2 are separable if there exists a hyperplane separating them, that is, there is a number α and vector $a \in \mathbb{R}^n$, $a \neq 0$, such that $\langle a,x\rangle \geq \alpha$, for any $x \in \Omega_1$ and $\langle a,x\rangle \leq \alpha$ for any $x \in \Omega_2$. Two sets are strictly separable if there exists $a \in \mathbb{R}^n$, $a \neq 0$, and $\alpha_1 > \alpha_2$ such that $\langle a,x\rangle \geq \alpha_1$ for any $x \in \Omega_1$ and $\langle a,x\rangle \leq \alpha_2$, for any $x \in \Omega_2$.

Theorem 2.9. *Let Ω_1 and Ω_2 be closed convex disjoint sets in $\mathbb{R}^n$ and Ω_2 be bounded, then Ω_1 and Ω_2 are strictly separable.*

Proof. By Exercise 2.27.1, $d(x) = \|x - P_{\Omega_1}(x)\|$ is continuous. Therefore, on a closed and bounded set Ω_2, it attains a minimum. We consider $a_2 = \operatorname{argmin}\{d(x) : x \in \Omega_2\}$, $a_1 = P_{\Omega_1}(a_2)$. Since $\Omega_1 \cap \Omega_2 = \emptyset$, we have $a_1 \neq a_2$. Hence,

$$\|a_1 - a_2\| = \min\{\|x-y\| : x \in \Omega_1, y \in \Omega_2\} > 0$$

and $a_2 = P_{\Omega_2}(a_1)$. It follows from (2.36) that $\langle a_1 - a_2, x\rangle \geq \langle a_1 - a_2, a_1\rangle = \alpha_1$ for any $x \in \Omega_1$ and $\langle a_1 - a_2, x\rangle \leq \langle a_1 - a_2, a_2\rangle = \alpha_2$ for any $x \in \Omega_2$, hence $\alpha_1 - \alpha_2 = \|a_1 - a_2\|^2 > 0$. Thus Ω_1 and Ω_2 are strictly separable. □

Remark 2.1. The boundedness of Ω_2 is essential, for example, two unbounded sets $\Omega_1 = \{x \in \mathbb{R}^2 : x_2 \leq 0\}$ and $\Omega_2 = \{x \in \mathbb{R}^2 : x_2 \geq x_1^{-1}, x_1 > 0\}$ are not strictly separable.

Let $x_0 \notin \Omega$ and Ω be a closed convex set. We say that a hyperplane $H(a,\alpha) = \{x : \langle a,x\rangle = \alpha\}$ separates x_0 from Ω if $\langle a,x\rangle \leq \alpha$, for any $x \in \Omega$ and $\langle a,x_0\rangle > \alpha$.

Theorem 2.10. *Let Ω be a closed convex set and $x_0 \notin \Omega$, then there exists a hyperplane which separates x_0 from Ω.*

Proof. By (2.36),

$$\langle x_0 - P_\Omega(x_0), x\rangle \leq \langle x_0 - P_\Omega(x_0), P_\Omega(x_0)\rangle = \langle x_0 - P_\Omega(x_0), x_0\rangle - \|x_0 - P_\Omega(x_0)\|^2.$$

Since $x_0 \notin \Omega$, we have $\|x_0 - P_\Omega(x_0)\| > 0$; therefore, $\langle x_0 - P_\Omega(x_0), x\rangle < \langle x_0 - P_\Omega(x_0), x_0\rangle$ for any $x \in \Omega$. So the hyperplane $H := \{x \in \mathbb{R}^n : \langle x_0 - P_\Omega(x_0), x - P_\Omega(x_0)\rangle = 0\}$ separates the set Ω from x_0. □

The hyperplane $H(a,\alpha)$ is called supporting for the set Ω at the point x_0 if $x_0 \in H(a,\alpha)$, and $\langle a,x\rangle \le \alpha$, for any $x \in \Omega$.

The following is the **Hahn–Banach Theorem** in a finite dimensional space.

Theorem 2.11. *Let Ω be closed convex set and $x_0 \in \partial\Omega$, then there exists a hyperplane supporting to Ω at x_0.*

Proof. Let $\{x_k\}_{k=1}^\infty$ be a sequence converging to x_0 such that $x_k \notin \Omega, k \ge 1$, then each x_k can be separated from Ω. Let $a_k = \frac{x_k - P_\Omega(x_k)}{\|x_k - P_\Omega(x_k)\|}$ and $\alpha_k = \langle a_k, P_\Omega(x_k)\rangle$. For each $k > 0$, the hyperplane $H(a_k,\alpha_k)$ separates x_k from Ω, that is,

$$\langle a_k, x\rangle \le \alpha_k \le \langle a_k, x_k\rangle \text{ for any } x \in \Omega. \tag{2.39}$$

Since $\|a_k\| = 1$ and $\{\alpha_k\}_{k\in N}$ is bounded, we can assume without loss of generality that $a_0 = \lim_{k\to\infty} a_k$ and $\alpha_0 = \lim_{k\to\infty}\alpha_k$; therefore, taking the limit in (2.39), we obtain

$$\langle a_0, x\rangle \le \alpha_0 = \langle a_0, x_0\rangle \text{ for any } x \in \Omega,$$

and $x_0 \in \partial\Omega$. □

2.3.4 Some Properties of Convex Functions

2.3.4.1 Continuity of Convex Functions

In spite of rather unpredictable behavior of convex functions at the boundary, their behavior at the interior point is rather remarkable. First, similarly to Theorem 2.1, a convex function is continuous at any point in the interior of its domain. Moreover, a convex function f is locally Lipschitz continuous at any point of $\operatorname{int}\operatorname{dom} f$. □

Lemma 2.12. *Let f be convex and $x_0 \in \operatorname{int}\operatorname{dom} f$, then there exists $L > 0$ such that for a small enough $\varepsilon > 0$ and any $x \in B(x_0,\varepsilon)$ we have*

$$|f(x) - f(x_0)| \le L\|x - x_0\|. \tag{2.40}$$

Proof. Chose ε so that $B(x_0,\varepsilon) \subset \operatorname{intdom} f$. Then f is continuous on $B(x_0,\varepsilon)$, and hence $M = \max\{f(x) : x \in B(x_0,\varepsilon)\} < \infty$. Take $x \in B(x_0,\varepsilon)$, $x \ne x_0$, and consider $\alpha = \|x - x_0\|\varepsilon^{-1}$; $0 < \alpha \le 1$. Let $u = x_0 + \alpha^{-1}(x - x_0)$, then

$$\|u - x_0\| = \alpha^{-1}\|x - x_0\| = \varepsilon,$$

that is, $u \in B(x_0,\varepsilon)$. For $x = \alpha u + (1-\alpha)x_0$, we have

$$f(x) \leq \alpha f(u) + (1-\alpha) f(x_0) = f(x_0) + \alpha(f(u) - f(x_0)) \leq f(x_0) + \alpha(M - f(x_0)),$$

or

$$f(x) - f(x_0) \leq \alpha(M - f(x_0)) = \frac{M - f(x_0)}{\varepsilon} \|x - x_0\|.$$

On the other hand, let us consider $\mathrm{v} = x_0 + \frac{1}{\alpha}(x_0 - x)$, then $\|\mathrm{v} - x_0\| = \frac{1}{\alpha}\|x_0 - x\| = \varepsilon$ and $x = x_0 + \alpha(x_0 - \mathrm{v})$. Using Lemma 2.5, we obtain

$$f(x) \geq f(x_0) + \alpha(f(x_0) - f(\mathrm{v})) \geq f(x_0) - \alpha(M - f(x_0)),$$

or

$$f(x) - f(x_0) \geq -\frac{M - f(x_0)}{\varepsilon} \|x - x_0\|,$$

that is,

$$|f(x) - f(x_0)| \leq \frac{M - f(x_0)}{\varepsilon} \|x - x_0\|.$$

Therefore, for any $x \in B(x_0, \varepsilon)$ and $L = \frac{M - f(x_0)}{\varepsilon}$, we obtain (2.40). □

2.3.4.2 Differentiability of Convex Functions

We have seen already that a convex function is continuous and satisfies the Lipschitz condition at any $x_0 \in \operatorname{int} \operatorname{dom} f$.

Let $x \in \operatorname{intdom} f$, then the function f differentiable in direction d at x if the following limit

$$\nabla f(x; d) = \lim_{t > 0} t^{-1} [f(x + td) - f(x)]$$

exists. The value $\nabla f(x; d)$ is called directional derivative of f at x in the direction d.

We will show now that a convex function has directional derivatives at any $x_0 \in \operatorname{intdom} f$.

Theorem 2.12. *A convex function f has a directional derivative in any direction $d \in \mathbb{R}^n$at any $x_0 \in \operatorname{int} \operatorname{dom} f$.*

Proof. Let $x_0 \in \operatorname{int} \operatorname{dom} f$, $d \in \mathbb{R}^n$, $0 < \alpha \leq 1$ and $\varepsilon > 0$ be such that $x_0 + td \in \operatorname{dom} f$ for $0 < t \leq \varepsilon$. The convexity of f implies

$$\varphi(t) = \frac{1}{t}[f(x_0 + td) - f(x_0)] \geq \frac{1}{\alpha t}[f(x_0 + \alpha t d) - f(x_0)] = \varphi(\alpha t),$$

which means that $\varphi(t)$ is monotone decreasing when $t \to 0$. Also for any $t_0 > 0$ such that $x_0 - t_0 d \in \operatorname{dom} f$, we have $\varphi(t) \geq \frac{1}{t_0}[f(x_0) - f(x_0 - t_0 d)]$. Therefore, the limit

$$\lim_{t \to 0} t^{-1} [f(x_0 + td) - f(x_0)] = \nabla f(x_0; d),\ t > 0$$

exists and so is the directional derivative of f in the direction d at x_0. □

Theorem 2.13. *Let f be convex and* $x_0 \in \operatorname{intdom} f$*, then* $\nabla f(x_0;d)$ *is homogeneous of degree one, convex in* $d \in \mathbb{R}^n$*, and*

$$f(x) \geq f(x_0) + \nabla f(x_0; x - x_0) \tag{2.41}$$

holds for any $x \in \operatorname{dom} f$.

Proof. For a given $t > 0$, we consider the derivative at $x_0 \in \operatorname{intdom} f$ in the direction $td \in \mathbb{R}^n$:

$$\begin{aligned}\nabla f(x;td) &= \lim_{\tau \to 0} \tau^{-1}[f(x_0 + \tau t d) - f(x_0)] = t \lim_{\tau \to 0} \frac{1}{\tau t}[f(x_0 + \tau t d) - f(x_0)] \\ &= t \nabla f(x_0; d).\end{aligned}$$

We proved the first part of the theorem. To show that $\nabla f(x_0;d)$ is convex in d, let us consider $d_1, d_2 \in \mathbb{R}^n$; then

$$\begin{aligned}\nabla f(x_0; \lambda d_1 + (1-\lambda)d_2) &\leq \lim \tau^{-1}[f(x + \tau(\lambda d_1 + (1-\lambda)d_2)) - f(x_0)] \\ \leq \lim_{\tau \to 0} \frac{1}{\tau}\{\lambda[f(x_0 + \tau d_1) - f(x_0)] &+ (1-\lambda)[f(x_0 + \tau d_2) - f(x)]\} \\ &= \lambda \nabla f(x_0; d_1) + (1-\lambda)\nabla f(x_0; d_2).\end{aligned}$$

Hence, $\nabla f(x_0;d)$ is convex in d. Finally, let $0 < t \leq 1$, $y \in \operatorname{intdom} f$ and $y_t = x_0 + t(y - x_0)$. Then, from Lemma 2.5 follows

$$f(y) = f(y_t + t^{-1}(1-t)(y_t - x_0)) \geq f(y_t) + t^{-1}(1-t)[f(y_t) - f(x_0)].$$

By taking the limit in $t \to 0$, we obtain (2.41). □

2.3.5 Subgradients

Convex non-smooth functions play an important role in optimization. In particular, the Lagrangian duality is a reach source of convex (concave), but usually non-smooth functions. It turns out that each convex function has a subgradient at each $x \in \operatorname{dom} f$. For non-smooth functions the subgradient is not unique, and there is a reach theory of non-smooth convex functions and non-smooth optimization. If a convex function is smooth, then the gradient ∇f is the only subgradient at each point $x \in \operatorname{dom} f$.

A vector g is called a subgradient of a convex function f at a point $x_0 \in \operatorname{int dom} f$ if for any $x \in \operatorname{dom} f$ we have

$$f(x) \geq f(x_0) + \langle g, x - x_0 \rangle. \tag{2.42}$$

The set of subgradients g of f at x_0 is called the subdifferential $\partial f(x_0)$ of f:

$$\partial f(x_0) = \{g : f(x) \geq f(x_0) + \langle g, x - x_0 \rangle \text{ for any } x \in \operatorname{dom} f\}.$$

We will see later that at any $x_0 \in \operatorname{int}\operatorname{dom} f$, the subdifferential is nonempty, convex, and bounded set. This fundamental property of convex functions we will call subdiferentiability.

Example 2.7. Let

$$f(x) = \begin{cases} x^2, & x \leq 0, \\ x, & x \geq 0. \end{cases}$$

At $x = 0$ the right derivative is equal to 1, while the left derivative is equal to 0. So $\partial f(0) = [0,1]$, $\partial f(-1) = \{-2\}$, $\partial f(1) = \{1\}$.

Exercise 2.28. Show that for $f(x) = |x|$ one has $\partial f(0) = [-1,1]$.

One can view (2.42) as a set of linear constraints defining vectors $g \in \partial f(x_0)$. Therefore, the set $\partial f(x_0)$ is a closed convex set for any $x_0 \in \operatorname{int}\operatorname{dom} f$.

Lemma 2.13. *If for any $x_0 \in \operatorname{dom} f$ the set $\partial f(x_0)$ is nonempty, then f is a convex function.*

Proof. Let $x, y \in \operatorname{dom} f$ and $\alpha \in [0,1]$. Assuming that for $y_\alpha = x + \alpha(y - x)$ the subdifferential $\partial f(y_\alpha)$ is nonempty, we can find $g \in \partial f(y_\alpha)$, so that

$$f(y) \geq f(y_\alpha) + \langle g, y - y_\alpha \rangle = f(y_\alpha) + (1-\alpha)\langle g, y - x \rangle,$$
$$f(x) \geq f(y_\alpha) + \langle g, x - y_\alpha \rangle = f(y_\alpha) - \alpha \langle g, y - x \rangle.$$

Multiplying the inequalities by α and $(1-\alpha)$ and adding them, we obtain

$$\alpha f(y) + (1-\alpha) f(x) \geq f(y_\alpha),$$

and hence f is convex. □

The converse turns out to be true as well.

Theorem 2.14. *Let f be a closed convex function and $x_0 \in \operatorname{intdom} f$. Then $\partial f(x_0)$ is a nonempty, convex, and bounded set.*

Proof. Let us consider the epigraph $\operatorname{epi} f$; it is a convex closed set. At $(f(x_0), x_0) \in \partial(\operatorname{epi} f)$, there exists a supporting hyperplane to the convex set $\operatorname{epi} f$, that is, there exists a pair (τ, g) such that $-\tau t + \langle g, x \rangle \leq -\tau f(x_0) + \langle g, x_0 \rangle$ for any $(t, x) \in \operatorname{epi} f$ or

$$\langle g, x - x_0 \rangle \leq \tau(t - f(x_0)) \tag{2.43}$$

for any $(t, x) \in \operatorname{epi} f$. We can normalize the parameters of the hyperplane $H = \{(t, x) : \tau(t - f(x_0)) - \langle g, x - x_0 \rangle = 0\}$, that is, we can assume that

$$\tau^2 + \|g\|^2 = 1. \tag{2.44}$$

Since for all $t \geq f(x_0)$ the point (t, x_0) belongs to $\operatorname{epi} f$, (2.43) implies $\tau \geq 0$.

Using (2.43) and Lemma 2.13, we obtain

$$\langle g, x - x_0 \rangle \le \tau(f(x) - f(x_0)) \le \tau L \|x - x_0\| \tag{2.45}$$

for some $L > 0$. Let $x = x_0 + \varepsilon g \in \operatorname{dom} f$, then from the last inequality $\varepsilon \|g\|^2 \le \tau \varepsilon L \|g\|$, or $\|g\| \le \tau L$. Hence, (2.44) implies $\tau^2 L^2 + \tau^2 \ge 1$ or $\tau \ge \frac{1}{\sqrt{1+L^2}}$. Therefore, from (2.45) we obtain $\langle d, x - x_0 \rangle \le f(x) - f(x_0)$ where $d = g/\tau \in \partial f(x_0)$, or $\partial f(x_0) \ne \emptyset$.

By choosing $x = x_0 + \varepsilon d / \|d\|$, we get

$$L\varepsilon = L\|x - x_0\| \ge f(x) - f(x_0) \ge \langle d, x - x_0 \rangle = \varepsilon \|d\|.$$

Thus, $\partial f(x_0)$ is bounded.

To show that $\partial f(x_0)$ is a convex set, let us consider $g_1, g_2 \in \partial f(x_0)$. We have $f(x) \ge f(x_0) + \langle g_1, x - x_0 \rangle$, $f(x) \ge f(x_0) + \langle g_2, x - x_0 \rangle$. Multiplying the first inequality by $0 < \lambda < 1$ and the second inequality by $0 < 1 - \lambda < 1$ and adding the results, we obtain

$$f(x) \ge f(x_0) + \langle \lambda g_1 + (1 - \lambda) g_2, x - x_0 \rangle;$$

therefore $\lambda g_1 + (1 - \lambda) g_2 \in \partial f(x_0)$. □

There is an important connection between subdifferentials and directional derivatives.

Theorem 2.15. *Let f be a closed convex function, then the following holds:*

$$\nabla f(x_0; d) = \max\{\langle g, d \rangle : g \in \partial f(x_0)\} \tag{2.46}$$

for any $x_0 \in \operatorname{int}\operatorname{dom} f$ and any given direction $d \in \mathbb{R}^n$.

Proof. From the definition of the subgradient, for any $g \in \partial f(x_0)$ we have

$$\nabla f(x_0; d) \ge \langle g, d \rangle. \tag{2.47}$$

Let us consider $\nabla f(x_0; d)$ as a function in d at $d = 0$, then (2.47) implies every $g \in \partial f(x_0)$ is a subgradient of $\nabla f(x_0; 0)$ or $\partial f(x_0) \subseteq \partial_d \nabla f(x_0; 0)$.

From convexity of $\nabla f(x_0; d)$ in d follows

$$\nabla f(x_0; d) = \nabla f(x_0; d) - \nabla f(x_0; 0) \ge (g, d)$$

for any $g \in n\partial_d \nabla f(x_0; 0)$. Let $d = x - x_0$, then from (2.41) follows

$$f(x) \ge f(x_0) + \nabla f(x_0; x - x_0) \ge (g, x - x_0).$$

It means that any subgradient $g \in \partial d \nabla f(x_0, 0)$ is also a subgradient of f.

Therefore, $\partial_d \nabla f(x_0; 0) \subseteq \partial f(x_0)$; hence, $\partial f(x_0) = \partial_d \nabla f(x_0; 0)$. Keeping in mind that $\nabla f(x_0; d)$ is first-degree homogeneous and convex in d (see Theorem 2.12), we have

$$\tau \nabla f(x_0;u) = \nabla f(x_0;\tau u) \geq \nabla f(x_0;d) + \langle g, \tau u - d\rangle \tag{2.48}$$

for a given $g \in \partial_d \nabla f(x_0;d)$, any $u \in \mathbb{R}^n$ and any $\tau > 0$. Taking $\tau \to 0$ in (2.48), we obtain

$$\nabla f(x_0;d) \leq \langle g,d\rangle. \tag{2.49}$$

On the other hand, (2.48) implies

$$\nabla f(x_0;u) \geq \frac{1}{\tau}\nabla f(x_0;d) + \langle g,u\rangle - \frac{1}{\tau}\langle g,d\rangle.$$

By taking $\tau \to \infty$ we obtain $\nabla f(x_0;u) \geq \langle g,u\rangle$ which together with (2.49) leads to (2.46) □

Theorem 2.16. *Vector x^* is a minimizer of a convex function f if and only if $0 \in \partial f(x^*)$.*

Proof. If $0 \in \partial f(x^*)$ then $f(x) \geq f(x^*) + \langle 0, x - x^*\rangle = f(x^*)$ for all $x \in \operatorname{dom} f$. On the other hand, if $f(x) \geq f(x^*)$ for any $x \in \operatorname{dom} f$ then $0 \in \partial f(x^*)$ by the definition of ∂f. □

In the case of a convex and continuously differentiable $f : \mathbb{R}^n \to \mathbb{R}$, it follows from (2.19) that for any $x_0 \in \mathbb{R}^n$ and any $x \in \mathscr{L}_f(\alpha_0)$ with $\alpha_0 = f(x)$ we have $f(x_0) + \langle \nabla f(x_0), x - x_0\rangle \leq f(x) \leq \alpha_0$. Consequently, $\langle \nabla f(x_0), x - x_0\rangle \leq 0$ for any $x \in \mathscr{L}_f(\alpha_0)$, and hence $\nabla f(x_0)$ is a normal vector to the supporting hyperplane for the sublevel set $\mathscr{L}_f(\alpha_0)$ at x_0.

If f is just convex, then a similar statement holds.

Theorem 2.17. *For any $x_0 \in \operatorname{dom} f$ and $\alpha_0 = f(x_0)$ each vector $g \in \partial f(x_0)$ is the normal to a supporting hyperplane to the sublevel set $\mathscr{L}_f(\alpha_0)$ at the point x_0:*

$$\langle g, x - x_0\rangle \leq 0$$

for any $x \in \mathscr{L}_f(\alpha_0)$.

Proof. For any $g \in \partial f(x_0)$ and any $x \in \mathscr{L}_f(\alpha_0)$, we have $f(x_0) + \langle g, x - x_0\rangle \leq f(x) \leq f(x_0)$ for any $x \in \mathscr{L}_f(x_0)$. □

Corollary 2.5. *Let $\Omega \subset \operatorname{dom} f$ be a closed convex set and $x^* = \operatorname{argmin}\{f(x) : x \in \Omega\}$, then for any $x_0 \in \Omega$ and $g \in \partial f(x_0)$ we have $\langle g, x^* - x_0\rangle \leq 0$.*

2.3.6 Support Functions

Let Ω be a convex set and $g \in \mathbb{R}^n$, then $\psi_\Omega(g) = \sup\{\langle g,x\rangle : x \in \Omega\}$ is called support function of Ω. Due to Lemma 2.8, the function ψ_Ω is closed and convex. For $t > 0$ we have $\psi_\Omega(tg) = \sup\{\langle tg,x\rangle : x \in \Omega\} = t\psi_\Omega(g)$, so ψ_Ω is first-degree homogeneous.

Exercise 2.29. Show that if Ω is bounded, then $\operatorname{dom}\psi_\Omega = \mathbb{R}^n$.

Let us consider two closed convex sets $\Omega_1, \Omega_2 \in \mathbb{R}^n$.

Lemma 2.14.

1. *If for any* $g \in \operatorname{dom}\psi_{\Omega_2}$ *we have* $\psi_{\Omega_1}(g) \le \psi_{\Omega_2}(g)$, *then* $\Omega_1 \subseteq \Omega_2$.
2. *Assume that* $\operatorname{dom}\psi_{\Omega_1} = \operatorname{dom}\psi_{\Omega_2}$ *and for any* $g \in \operatorname{dom}\psi_{\Omega_1}$ *we have* $\psi_{\Omega_1}(g) = \psi_{\Omega_2}(g)$, *then* $\Omega_1 = \Omega_2$.

Proof.

1. Assume there exists $x_0 \in \Omega_1$ such that $x_0 \notin \Omega_2$. It follows from Theorem 2.10 that there exists a vector g such that $\langle g, x_0\rangle > \alpha \ge \langle g, x\rangle$ for every $x \in \Omega_2$. So $g \in \operatorname{dom}\psi_{\Omega_2}$ and $\psi_{\Omega_1}(g) > \psi_{\Omega_2}(g)$, which contradicts our assumption.
2. Using the first statement, we have $\Omega_1 \subseteq \Omega_2$ and $\Omega_2 \subseteq \Omega_1$, that is, $\Omega_1 = \Omega_2$.

□

Exercise 2.30. Show that if f is closed, convex, and differentiable function, then for any $x \in \operatorname{int}\operatorname{dom} f$ one has $\partial f(x) = \{\nabla f(x)\}$.

Lemma 2.15. *Let f be closed and convex with* $\operatorname{dom} f \subseteq \mathbb{R}^m$, $A : \mathbb{R}^n \to \mathbb{R}^m, y = Ax + b$; *then,* $\varphi(x) = f(Ax+b)$ *is closed and convex with* $\operatorname{dom}\varphi = \{x : Ax + b \in \operatorname{dom} f\}$, *and for any* $x \in \operatorname{int}\operatorname{dom}\varphi$ *we have*

$$\partial\varphi(x) = A^T \partial f(Ax+b). \tag{2.50}$$

Proof. For x_1 and x_2 in $\operatorname{dom}\varphi$ and $t \in [0,1]$, we have

$$\begin{aligned} \varphi(tx_1 + (1-t)x_2) &= f(A(tx_1 + (1-t)x_2) + tb + (1-t)b) \\ &= f(t(Ax_1 + b) + (1-t)(Ax_2 + b)) \\ &\le t f(Ax_1 + b) + (1-t) f(Ax_2 + b) = t\varphi(x_1) + (1-t)\varphi(x_2), \end{aligned}$$

so φ is convex. The closeness of its epigraph follows from the continuity of the transformation $y = Ax + b$. By Theorem 2.15, for any $d \in \mathbb{R}^n$ and any $x \in \operatorname{int}\operatorname{dom} f$, $\nabla\varphi(x; d)$ is the support function of $\partial f(x)$ and

$$\nabla f(Ax+b; Ad) = \max\{\langle g, Ad\rangle : g \in \partial f(Ax+b)\} = \max\{\langle g, d\rangle : g \in A^T \partial f(Ax+b)\}.$$

Now Lemma 2.14 yields (2.50). □

Lemma 2.16. *Let* f_i, $i = 1, \ldots, r$, *be closed convex functions and* $t_i \ge 0, i = 1, \ldots, r$. *Then the function* $f = \sum_{i=1}^r t_i f_i$ *is closed and convex and*

$$\partial f(x) = \sum_{i=1}^r t_i \partial f_i(x) \tag{2.51}$$

for any x from $\operatorname{int}\operatorname{dom} f = \operatorname{int}\operatorname{dom} f_1 \cap \ldots \cap \operatorname{int}\operatorname{dom} f_r$.

Proof. The first part of statement was proven in Proposition 2.3. To prove (2.51), we consider $x_0 \in \text{int}(\text{dom} f_1 \cap \text{dom} f_r)$; then for any $d \in \mathbb{R}^n$, we have

$$\nabla f(x_0;d) = \sum_{i=1}^{r} t_i \nabla f_i(x_0;d)$$

$$= \max\{\langle g_1, t_1 d\rangle : g_1 \in \partial f_1(x_0)\} + \ldots + \max\{\langle g_r, t_r d\rangle : g_r \in \partial f_r(x_0)\} =$$

$$\max\{\sum_{i=1}^{r} \langle t_i g_i, d\rangle : g_1 \in \partial f_1(x_0), \ldots, g_r \in \partial f_r(x_0)\} = \max\{\langle g, d\rangle : g \in \sum_{i=1}^{r} t_i \partial f_i(x_0)\}.$$

Note that all $\partial f_i(x_0)$ are bounded. Therefore, (2.51) follows from Theorem 2.15 and Lemma 2.3. □

Finally, let us compute the subdifferential for

$$f(x) = \max\{f_i(x) : 1 \le i \le r\}. \tag{2.52}$$

Lemma 2.17. *Let f_i, $i = 1, \ldots, r$ be closed and convex, then f given by (2.52) is closed and convex, and for any $x \in \text{int dom} f = \cap \text{int dom} f_i$ we have*

$$\partial f(x) = \text{conv}\{\partial f_i(x) : i \in I(x)\}, \tag{2.53}$$

where $I(x) = \{i : f_i(x) = f(x)\}$.

Proof. The first part of the Lemma is proven in Proposition 2.4, so we need to justify (2.53).

Let us consider $x \in \cap_{i=1}^{r} \text{int dom} f_i$ and let $I(x) = \{1, \ldots, m\}$. Then,

$$\nabla f(x;d) = \max_{1 \le i \le m} \nabla f_i(x,d) = \max_{1 \le i \le m} \max\{\langle g_i, d\rangle : g_i \in \partial f_i(x)\}.$$

For any set of numbers $a_1, \ldots, a_m$, we have

$$\max_{i \le i \le m} a_i = \max\{\sum_{i=1}^{m} \lambda_i a_i : \lambda \in \Delta_m\},$$

where Δ_m is the m-dimensional simplex. Therefore,

$$\begin{aligned}
\nabla f(x;d) &= \max_{\lambda \in \Delta_m}\{\sum_{i=1}^{m} \lambda_i \max\{\langle g_i, d\rangle : g_i \in \partial f_i(x)\} \\
&= \max\{\langle \sum_{i=1}^{m} \lambda_i g_i, d\rangle : g_i \in \partial f_i(x), \lambda \in \Delta_m\} \\
&= \max\{\langle g, d\rangle : g = \sum_{i=1}^{m} \lambda_i g_i, g_i \in \partial f_i(x), \lambda \in \Delta_m\} \\
&= \max\{\langle g, d\rangle : g = \text{conv}\{\partial f_i(x) : i \in I(x)\}\}.
\end{aligned}$$

Now (2.53) follows from Lemma 2.14. □

Example 2.8.

1. $f(x) = |\langle a,x\rangle - b|$, $x \in \mathbb{R}^n$. For $\langle a,x\rangle - b \neq 0$ we have

$$\partial f(x) = \begin{cases} a, & \langle a,x\rangle - b > 0 \\ -a, & \langle a,x\rangle - b < 0 \end{cases}$$

To find $\partial f(x)$ for $\langle a,x\rangle - b = 0$, we write $f(x) = \max\{\langle a,x\rangle - b, -\langle a,x\rangle + b\}$ and notice that for $\langle a,x\rangle - b = 0$ the two values in the right-hand side coincide. It follows from Lemma 2.7 that $\langle a,x\rangle - b = 0$ implies $\partial f(x) = \text{conv}\{a,-a\} = [-a,a]$. (see also Exercise 2.29).

2. Let $f(x) = \sum_{i=1}^m |\langle a_i,x\rangle - b_i| = \sum_{i=1}^m f_i(x)$. For a given $x \in \mathbb{R}^n$, we consider three sets of indices: $I_-(x) = \{i : \langle a_i,x\rangle - b_i < 0\}$, $I_+ = \{i : \langle a_i,x\rangle - b_i > 0\}$, and $I_0 = \{i : \langle a_i,x\rangle - b_i = 0\}$. It follows from Lemma 2.16 and Lemma 2.17 that

$$\begin{aligned}\partial f(x) &= \sum_{i\in I_+(x)} \nabla f_i(x) - \sum_{i\in I_-(x)} \nabla f_i(x) + \sum_{i\in I_0(x)} \text{conv}\{a_i,-a_i\} \\ &= \sum_{i\in I_+(x)\cup I_-(x)} \text{sign}(\langle a_i,x\rangle - b_i)a_i + \sum_{i\in I_0(x)} [-a_i,a_i].\end{aligned}$$

3. Let $f(x) = \max_{1\le i\le m} |\langle a_i,x\rangle - b_i|$, and $I_+(x) = \{i : \langle a_i,x\rangle - b_i = f(x)\}$; $I_-(x) = \{i : \langle -a_i,x\rangle + b_i = f(x)\}$. It follows from Lemma 2.17 that

$$\partial f(x) = \text{conv}\{a_i, i\in I_+(x), -a_i, i \in I_-(x)\}.$$

We conclude the section by applying Lemma 2.13 to prove Karush–Kuhn–Tucker's Theorem.

Consider the standard convex optimization problem

$$f(x^*) = \{\min f(x) | x \in \Omega\} \tag{2.54}$$

where $\Omega = \{x : c_i(x) \geq 0, i = 1,\dots,m\}$. We assume that f is convex and all c_i, $i = 1,\dots,m$, are concave.

We say that Slater's condition is satisfied if there exists $x_0 \in \Omega$ such that

$$c_i(x_0) > 0,\ i = 1\dots,m. \tag{2.55}$$

The following is the most important theorem in convex optimization

Theorem 2.18 (Karush–Kuhn–Tucker). *Let f and all $-c_i$ be convex and continuously differentiable functions and Slater's condition (2.55) is satisfied, then existence of $\lambda^* = (\lambda_1^*,\dots,\lambda_m^*) \in \mathbb{R}_+^m$:*

$$\nabla_x L(x^*,\lambda^*) = \nabla f(x^*) - \sum_{i=1}^m \lambda_i^* \nabla c_i(x^*) = 0 \tag{2.56}$$

and

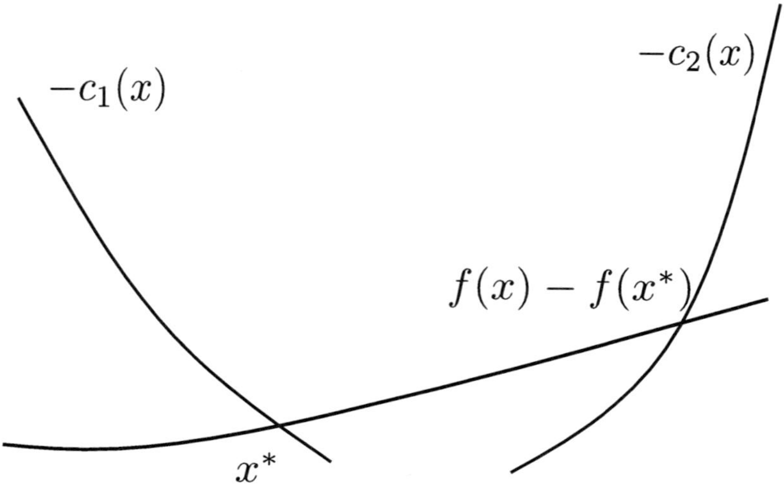

Fig. 2.1 Function $F(x;x^*)$

$$\lambda_i^* c_i(x^*) = 0,\ i = 1,\dots,m, \tag{2.57}$$

is necessary and sufficient condition for $x^* \in \Omega$ to be a solution of (2.54).

Proof. Let x^* be solution of (2.54). Function (see Fig 2.1)

$$F(x;x^*) = \max\{f(x) - f(x^*), -c_i(x),\ i = 1,\dots,m\}$$

is convex in $x \in \mathbb{R}^n$ and

$$\begin{aligned} F(x^*;x^*) &= \min_{x \in \mathbb{R}^n} F(x,x^*) \\ &= \min_{x \in \mathbb{R}^n} \max\{f(x) - f(x^*); -c_i(x), i = 1,\dots,m\} = 0. \end{aligned} \tag{2.58}$$

Indeed, for any $x \in \Omega$ we have $f(x) - f(x^*) \geq 0$ and $c_i(x) \geq 0$, $i = 1,\dots,m$; hence, $F(x;x^*) \geq 0$ for any $x \in \Omega$. For any $x \notin \Omega$ there exists i_0 such that $c_{i_0}(x) < 0$; therefore, $\max\{-c_i(x) : i = 1,\dots,m\} > 0$ and again $F(x;x^*) \geq 0$. In other words, $F(x;x^*) \geq 0$ for any $x \in \mathbb{R}^n$, and $F(x^*;x^*) = 0$; therefore, (2.58) holds.

From Lemma 2.17 follows

$$\partial F(x^*;x^*) = \text{conv}\{\nabla f(x^*), -\nabla c_i(x^*), i \in I(x^*)\},$$

where $I(x^*) = \{i : c_i(x^*) = F(x^*;x^*)\}$.

From Theorem 2.16 follows existence of $\bar{\lambda} = (\bar{\lambda}_0, \bar{\lambda}_i, i \in I(x^*))$:

$$
\begin{aligned}
&\bar{\lambda}_0 \nabla f(x^*) - \sum_{i \in I(x^*)} \bar{\lambda}_i \nabla c_i(x^*) = 0, \\
&\bar{\lambda}_0 + \sum_{i \in I(x^*)} \bar{\lambda}_i = 1, \bar{\lambda}_i \geq 0, \; i \in \{0\} \cup I(x^*).
\end{aligned}
\tag{2.59}
$$

Also $\bar{\lambda}_0 > 0$, because otherwise (2.59) implies $\sum_{i \in I(x^*)} \bar{\lambda}_i \nabla c_i(x^*) = 0$, which contradicts Slater condition (2.55).

Dividing (2.59) by $\bar{\lambda}_0$, we obtain (2.56) with $\lambda_i^* = \bar{\lambda}_i / \bar{\lambda}_0$, $i \in I(x^*)$, and $\lambda_i^* = 0$, $i \bar{\in} I(x^*)$. Besides, for (x^*, λ^*) the complementarity condition (2.57) holds true.

We proved the necessary part of the theorem.

Conversely, let (2.56)–(2.57) holds. For any given $\lambda \in \mathbb{R}^m_+$ Lagrangian $L(x, \lambda) = f(x) - \sum_{i=1}^{n} \lambda_i c_i(x)$ is convex in $x \in \mathbb{R}^n$; therefore, from (2.19) and (2.56) follows

$$
L(x, \lambda^*) - L(x^*, \lambda^*) \geq \langle \nabla_x L(x^*, \lambda^*), x - x^* \rangle = 0.
$$

Keeping in mind complementarity condition (2.57), we obtain $L(x; \lambda^*) \geq f(x^*)$. Hence, $f(x) \geq f(x^*) + \sum \lambda_i^* c_i(x)$. Therefore, $f(x) \geq f(x^*)$ for any $x \in \Omega$. □

2.4 The Legendre–Fenchel Transformation

The LF transform is a product of a general duality principle: any smooth curve $F(x, y) = 0$ is, on the one hand, a locus of pairs $(x; y)$, which satisfy the given equation and, on the other hand , an envelope of a family of its tangent lines.

An application of the LF transform to a strictly convex and smooth function leads to the LF identity. For strictly convex and three times differentiable function, the LF transform leads to the LF invariant.

Although the LF transform has been known for more then 200 years, both the LF identity and the LF invariant are critical in modern optimization.

2.4.1 Basic LF Transformation Property

Consider a convex function $f : \mathbb{R}^n \to \bar{\mathbb{R}}$. The Legendre transformation of f is given by

$$
f^*(s) = \sup_{x \in \mathbb{R}^n} \{\langle s, x \rangle - f(x)\}. \tag{2.60}
$$

The transformation (2.60) is often called Fenchel–Legendre transformation.

The function f^* is called the conjugate to f. The conjugate function is closed and convex regardless of the properties of f, because the value of f^* at s is the supremum of a set of affine in $s \in \mathbb{R}^n$ functions $\langle s, x \rangle - f(x)$. Besides, it follows from (2.60) that for $x \in \operatorname{dom} f$, $s \in \operatorname{dom} f^*$ we have

$$f^*(s) + f(x) \geq \langle s, x \rangle, \tag{2.61}$$

which is called the Fenchel–Young inequality.

Given f^*, one can construct the conjugate to f^*:

$$f^{**}(x) = \sup_{s \in \mathbb{R}^n} \{\langle x, s \rangle - f^*(s)\}.$$

Theorem 2.19.

1. *For all* $x \in \mathbb{R}^n$ *we have*
$$f(x) \geq f^{**}(x). \tag{2.62}$$
2. *If any of* f, f^*, *and* f^{**} *is proper and convex, then the other two possess the same properties.*
3. *If* f *is closed, proper, and convex, then* $f = f^{**}$.

Proof.

1. The Fenchel–Young inequality (2.61) implies
$$f(x) \geq \sup_{s \in \mathbb{R}^n} \{\langle s, x \rangle - f^*(s)\} = f^{**}(x),$$
that is, (2.62) holds.
2. If f is convex and proper, then $\operatorname{epi} f$ is not empty and does not contain vertical lines. Therefore, it is contained in a closed halfspace that corresponds to a nonvertical hyperplane $(s, x) + f(x) = \alpha$. Therefore,
$$f^*(-s) = \sup_{x \in \mathbb{R}^n} \{\langle -s, x \rangle - f(x)\} \leq -\alpha;$$
hence f^* is not identically equal to ∞. Since f is proper, there exists $\bar{x}$ such that $f(\bar{x}) < \infty$. Taking into account (2.61) for an arbitrary $s \in \mathbb{R}^n$ and $x = \bar{x}$, we see that $f^*(s)$ is well defined at any point of $\mathbb{R}^n$. Therefore f^* is proper. From the properness of f^*, using the same argument, we obtain the properness of f^{**}.
Assume now that f^* is convex and proper. Inequality (2.62) implies that $f(x)$ is well defined at any point $x \in \mathbb{R}^n$. Also, f cannot be identically equal to ∞, since then f^* by definition would be not proper. Hence, f is proper. In other words, a convex function f is proper if and only if its conjugate is proper.
Using the same arguments for f^* and f^{**}, we obtain 2.
3. Again from the properness of f it follows that $\operatorname{epi} f$ does not contain a vertical line. Let $x \in \operatorname{dom} f^{**}$ and $y \geq f^{**}(x)$. To arrive to a contradiction, let us assume that (x, y) does not belong to $\operatorname{epi} f$. Since $\operatorname{epi} f$ is closed, Theorem 2.10 implies the existence of a nonvertical hyperplane separating (x, y) and $\operatorname{epi} f$: $\langle s, u \rangle - \mathrm{v} < \beta < \langle s, x \rangle - y$ for any $(u, \mathrm{v}) \in \operatorname{epi} f$. Since $y \geq f^{**}(x)$ and $(u, f(u)) \in \operatorname{epi} f$, we get
$$\langle s, u \rangle - f(u) < \beta < \langle s, x \rangle - f^{**}(x).$$

Therefore,

$$\sup_{u \in \mathbb{R}^n} \{\langle s,u\rangle - f(u)\} \le \beta < \langle s,x\rangle - f^{**}(x)$$

or $f^*(s) < \langle s,x\rangle - f^{**}(x)$, that is, $f^{**}(x) < \langle s,x\rangle - f^*(s)$. But the last inequality contradicts the definition of f^{**}; therefore, $f(x) \le f^{**}(x)$, for any $x \in \mathbb{R}^n$, which together with (2.62) leads to $f = f^{**}$. □

2.4.2 The LF Identity and the LF Invariant

Let $F : \mathbb{R}^2 \to \mathbb{R}$, be smooth, then the corresponding smooth curve can be viewed as a locus

$$L_F = \{(x;y) \in \mathbb{R}^2 : F(x,y) = 0\}$$

of pairs $(x;y)$ that satisfy the given equation and, at the same time, as an envelope

$$E_F = \{(X;Y) \in \mathbb{R}^2 : F_x^{'}(x;y)(X-x) + F_y^{'}(x;y)(Y-y) = 0, (x;y) \in L_F\}$$

of its tangent lines.

Application of this duality principle to $F(x,y) = y - f(x)$ leads to the LF transform (2.60).

The LF transform for strictly convex function f leads to the LF identity

$$f^{*'}(s) \equiv f^{'-1}(s) \tag{2.63}$$

and the LF invariant

$$LFINV(f) = \left| -\frac{d^3 f^*}{ds^3}\left(\frac{d^2 f^*}{ds^2}\right)^{-3/2} \right| = \left| \frac{d^3 f}{dx^3}\left(\frac{d^2 f}{dx^2}\right)^{-3/2} \right|. \tag{2.64}$$

We will see later that both LF identity and LF invariant are critical in modern optimization. In this section we derive (2.63) and (2.64) using the LF transform.

Let $f : \mathbb{R} \to \mathbb{R}$ (see Fig. 2.2) be smooth and strictly convex. For a given $s = tg\varphi$, let us consider $l = \{(x,y) : y = sx\}$. The corresponding tangent to the curve L_f with the same slope is defined as follows:

$$T(x) = \{(X,Y) \in \mathbb{R}^2 : Y - f(x) = f^{'}(x)(X-x) = s(X-x)\}.$$

In other words $T(x)$ is a tangent to the curve $L_f = \{(x,y) : y = f(x)\}$ at the point x such that $f^{'}(x) = s$. For $X = 0$, we have $Y = f(x) - sx$. The conjugate function f^* at the point s is defined as: $f^*(s) = -Y = -f(x) + sx$. Therefore, (see Fig 2.2)

$$f^*(s) + f(x) = sx. \tag{2.65}$$

More often f^* is defined as follows

$$f^*(s) = \max_{x \in \mathbb{R}}\{sx - f(x)\}. \tag{2.66}$$

Keeping in mind that $T(x)$ is the supporting hyperplane to the epi $f = \{(y,x) : y \geq f(x)\}$, the maximum in (2.66) is reached at x, such that $f^{'}(x) = s$; therefore, we can rewrite (2.65) as follows:

$$f^*(f^{'}(x)) + f(x) \equiv f^{'}(x)x. \tag{2.67}$$

For a strictly convex f, we have $f^{''}(x) > 0$. Therefore, due to the inverse function theorem, the equation $f^{'}(x) = s$ can be solved for x, that is,

$$x(s) = f^{'-1}(s). \tag{2.68}$$

Using (2.68), we obtain the dual representation of (2.67)

$$f^*(s) + f(x(s)) \equiv sx(s). \tag{2.69}$$

Also, it follows from $f^{''}(x) > 0$ that $x(s)$ in (2.66) is unique, so f^* is as smooth as f, that is, if $f \in C^1$ then $f^* \in C^1$. Variables x and s are not independent, they are linked through equation $s = f^{'}(x)$.

By differentiating (2.69) we obtain

$$f^{*'}(s) + f^{'}(x(s))x^{'}(s) \equiv x(s) + sx^{'}(s). \tag{2.70}$$

Keeping in mind $f^{'}(x(s)) = s$, from (2.70) we obtain the identity

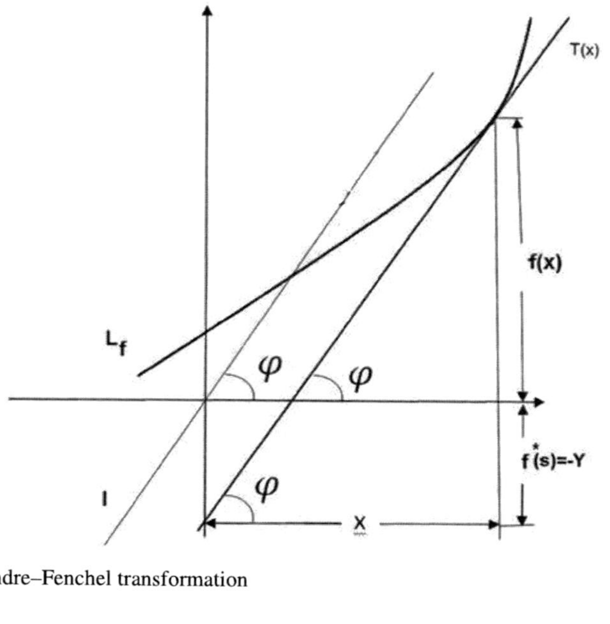

Fig. 2.2 Legendre–Fenchel transformation

$$f^{*'}(s) \equiv f^{'-1}(s), \tag{2.71}$$

which is called LF identity.

Keeping in mind that $s = f^{'}(x)$, from (2.68) and (2.71) we obtain

$$\frac{df^*(s)}{ds} = x. \tag{2.72}$$

On the other hand, we have

$$\frac{df(x)}{dx} = s. \tag{2.73}$$

If $f \in C^2$ then $f^* \in C^2$ and from (2.72) and (2.73) follows

$$a)\frac{d^2 f^*(s)}{ds^2} = \frac{dx}{ds} \text{ and } b)\frac{d^2 f(x)}{dx^2} = \frac{ds}{dx}. \tag{2.74}$$

From

$$\frac{dx}{ds} \cdot \frac{ds}{dx} = 1$$

and (2.74) we get

$$\frac{d^2 f^*}{ds^2} \cdot \frac{d^2 f}{dx^2} = 1. \tag{2.75}$$

The following theorem establishes the relation between third derivatives of f and f^*, which leads to the notion of the LF invariant.

Theorem 2.20. *If f is strictly convex and $f \in C^3$, then*

$$\frac{d^3 f^*}{ds^3} \cdot \left(\frac{d^2 f^*}{ds^2}\right)^{-3/2} + \frac{d^3 f}{dx^3} \cdot \left(\frac{d^2 f}{dx^2}\right)^{-3/2} = 0.$$

Proof. By differentiating (2.75) in x and keeping in mind that $s = f^{'}(x)$, we obtain

$$\frac{d^3 f^*}{ds^3} \cdot \frac{ds}{dx} \cdot \frac{d^2 f}{dx^2} + \frac{d^2 f^*}{ds^2} \cdot \frac{d^3 f}{dx^3} = 0.$$

In view of (2.74b), we have

$$\frac{d^3 f^*}{ds^3} \cdot \left(\frac{d^2 f}{dx^2}\right)^2 + \frac{d^2 f^*}{ds^2} \cdot \frac{d^3 f}{dx^3} = 0. \tag{2.76}$$

By differentiating (2.75) in s and keeping in mind (2.74a), we obtain

$$\frac{d^3 f^*}{ds^3} \frac{d^2 f}{dx^2} + \left(\frac{d^2 f^*}{ds^2}\right)^2 \frac{d^3 f}{dx^3} = 0. \tag{2.77}$$

From (2.75) and (2.76) follows

$$\frac{d^3 f^*}{ds^3} \cdot \left(\frac{d^2 f}{dx^2}\right)^2 + \frac{1}{\frac{d^2 f}{dx^2}} \frac{d^3 f}{dx^3} = 0$$

or

$$\frac{d^3 f^*}{ds^3} \left(\frac{d^2 f}{dx^2}\right)^3 + \frac{d^3 f}{dx^3} = 0.$$

Using (2.75), from (2.77) we obtain

$$\frac{d^3 f^*}{ds^3} \cdot \frac{1}{\frac{d^2 f^*}{ds^2}} + \left(\frac{d^2 f^*}{ds^2}\right)^2 \frac{d^3 f}{dx^3} = 0.$$

Therefore,

$$\frac{d^3 f^*}{ds^3} \left(\frac{d^2 f^*}{ds^2}\right)^{-3/2} + \frac{d^3 f}{dx^3} \left(\frac{d^2 f^*}{ds^2}\right)^{3/2} = 0$$

and, using (2.75) again, we obtain

$$\frac{d^3 f^*}{ds^3} \left(\frac{d^2 f^*}{ds^2}\right)^{-3/2} + \frac{d^3 f}{dx^3} \left(\frac{d^2 f}{dx^2}\right)^{-3/2} = 0. \tag{2.78}$$

□

Corollary 2.6. *From (2.78) follows*

$$-\frac{d^3 f^*}{ds^3} \left(\frac{d^2 f^*}{ds^2}\right)^{-3/2} = \frac{d^3 f}{dx^3} \left(\frac{d^2 f}{dx^2}\right)^{-3/2}$$

or

$$LFINV(f) = \left| -\frac{d^3 f^*}{ds^3} \left(\frac{d^2 f^*}{ds^2}\right)^{-3/2} \right| = \left| \frac{d^3 f}{dx^3} \left(\frac{d^2 f}{dx^2}\right)^{-3/2} \right| \tag{2.79}$$

For a strictly convex and three time differentiable f boundedness of the $LFINV(f)$ defines the class of SC functions, introduced by Yu. Nesterov and A. Nemirovski in the late 1980s.

They developed the SC theory – the centerpiece of the interior point methods (IPMs). The SC theory and the role of the $LFINV(f)$ will be discussed in Chapter 6.

Notes

2.1 For elements of Calculus see Ch. 1 in B. Polyak (1987), for classes of differentiable functions see 1.2.2 in Nesterov (2004). For convex function, see Ch. 3 in Boyd and Vanderberghe (2004), Ch. 1 in B. Polyak (1987), Ch. 1 in Rockafellar (1970), Ch. 2 Rockafellar and Wets (2009). For strongly convex functions see Ch. 9 in Boyd and Vanderberghe (2004), Ch. 2 in Nesterov (2004), Ch. 1 in B. Polyak (1987).

2.2 For convex sets, cones, recession cones, polehedra see Ch. 1 and 2 in Auslender and Teboulle (2003), Ch. 3 in Ioffe and Tihomirov (2009), Ch. 9 and 10 in Polyak (1987), Ch. 1 and 2 in Rockafellar (1970), Ch.2 and 3 in Rockafellar and Wets (2009).

2.3 For closed convex functions and operations on it see, Ch. 3 in Boyd and Vanderberghe (2004); Ioffe and Tichomirov (1968), Ch. 3 and 4 in Ioffe and Tihomirov (2009), Ch. 1 in Mordukhovich (2018), Ch. 3 in Nesterov (2004), Ch.5 in Rockafellar (1970), Ch 8 in Rockafellar and Wets (2009).

2.4 For the LF transform see Ioffe and Tichomirov (1968), Ch. 3 in Ioffe and Tihomirov (2009), Ch.5 in Rockafellar (1970), Ch. 11 and 13 in Rockafellar and Wets (2009). For the LF invariant and its connection with SC functions see Polyak (2016).

Chapter 3
Few Topics in Unconstrained Optimization

3.0 Introduction

Unconstrained optimization is, on the one hand, a classical topic in optimization, which goes back to the seventeenth century. On the other hand, it is a modern one, which undergone substantial transformation in the last few decades.

In this chapter our main focus is the two basic methods for unconstrained optimization: Newton–Raphson's method (1669), introduced by Newton (1642–1726) and Raphson (1648–1715), and gradient method (1847), introduced by Cauchy (1789–1857).

We start with Fermat's (1607–1665) first-order necessary condition and then consider the second-order necessary and sufficient condition – foundation for unconstrained optimization.

Along with classical gradient method, we consider subgradient method (1962) introduced by N. Shor (1937–2006) for nondifferentiable convex unconstrained optimization and Yu. Nesterov's fast gradient method (1983) for convex functions with Lipschitz gradient.

We conclude the chapter with Newton's method.

First, we consider local convergence of Newton's method and global convergence of a particular damped Newton method (DNM) for strongly convex function. The size of Newton area is characterized, and the total number of Newton steps required for finding ε - approximation to the solution is estimated.

It is well known that for strictly convex functions, Newton's method, generally speaking, does not converge from any starting point.

We introduced regularized Newton method (RNM), which guarantees convergence from any starting point with asymptotic quadratic rate. It became possible due to the special regularization at each point with a vanishing regularization parameter equal to the Euclidean norm of the gradient.

The area of quadratic convergence of RNM is characterized, and the complexity of damped RNM is estimated.

R. A. Polyak, *Introduction to Continuous Optimization*, Springer Optimization and Its Applications 172, https://doi.org/10.1007/978-3-030-68713-7_3

3.1 Optimality Conditions

We start with optimality condition for an unconstrained minimizer. Let $f : \mathbb{R}^n \to \mathbb{R}$ be convex. In this chapter we are concerned with the following problem:

$$f(x^*) = \{\min f(x) | x \in \mathbb{R}^n\}. \tag{3.1}$$

We assume throughout this chapter that the optimal set $X^* = \{x \in \mathbb{R}^n : f(x) = f(x^*)\}$ is not empty and bounded. For convex f, it guarantees by the boundedness of the sublevel set $\mathscr{L}_f(\alpha)$ for some α (see Lemma 2.4).

3.1.1 First-Order Necessary Condition

A point x^* is called a local minimizer of $f : \mathbb{R}^n \to \mathbb{R}$ if one can find $\varepsilon > 0$ such that for all $x \in B(x^*, \varepsilon)$ we have $f(x) \geq f(x^*)$. A point x^* is called a global minimizer if $f(x) \geq f(x^*)$ for all $x \in \mathbb{R}^n$.

The following theorem establishes the necessary condition for a local minimizer.

Theorem 3.1 (Fermat). *Let x^* be a local minimizer of $f : B(x^*, \varepsilon) \to \mathbb{R}$ and f be differentiable at x^*, then*

$$\nabla f(x^*) = 0. \tag{3.2}$$

Proof. Let x^* be a local minimizer, then $f(x) \geq f(x^*)$ for any $x \in B(x^*, \varepsilon)$. The differentiability of f at x^* implies

$$f(x) = f(x^*) + \langle \nabla f(x^*), x - x^* \rangle + o(\|x - x^*\|);$$

therefore,

$$\langle \nabla f(x^*), x - x^* \rangle + o(\|x - x^*\|) \geq 0 \text{ for any } x \in B(x^*, \varepsilon). \tag{3.3}$$

Let $x \in B(x^*, \varepsilon)$ and $x \neq x^*$, then for $d = (x - x^*)\|x - x^*\|^{-1}$ from (3.3) follows $\langle \nabla f(x^*), d \rangle \geq 0$. Also from (3.3) follows $\langle \nabla f(x^*), -d \rangle \geq 0$, so $\langle \nabla f(x^*), d \rangle = 0$. By setting $d = e_i = (0, \ldots, 1, 0, \ldots 0)$, $i = 1, \ldots, n$, we obtain (3.2). □

Equation (3.2) is only necessary condition for x^* to be a local minimizer. The solution of (3.2) can be a local maximum of f or just a saddle point (consider $f(x) = x^3$ at $x = 0$).

There is, however, an important class of functions, for which the necessary condition (3.2) is sufficient as well. This is the class of convex functions considered in Chapter 1.

Theorem 3.2. *Let $f : \mathbb{R}^n \to \mathbb{R}$ be a convex function differentiable at x^* and $\nabla f(x^*) = 0$, then x^* is a global minimizer of f on $\mathbb{R}^n$.*

Proof. For a convex and differentiable at x^* function f, we have

$$f(x) \geq f(x^*) + \langle \nabla f(x^*), x - x^* \rangle, \forall x \in \mathbb{R}^n;$$

therefore, x^* is a global minimizer. □

Thus, for convex functions the necessary optimality condition is also sufficient. Later we will see that this is true for convex constrained optimization problems as well.

A vector x^* is an isolated local minimizer if $f(x^*) < f(x)$ for any $x \in B(x^*, \varepsilon), x \neq x^*$. A vector x^* is an isolated global minimizer if $f(x^*) < f(x)$ for any $x \in \mathbb{R}^n, x \neq x^*$.

3.1.2 Second-Order Necessary Condition

Theorem 3.3. *Let x^* be a minimizer of $f : B(x^*, \varepsilon) \to \mathbb{R}$ and f be twice continuous differentiable at x^*, then*

$$\nabla^2 f(x^*) \succeq 0. \tag{3.4}$$

Proof. Since $\nabla f(x^*) = 0$ for a small enough $\varepsilon > 0$, we have

$$f(x) = f(x^*) + \frac{1}{2}\langle \nabla^2 f(x^*)(x - x^*), x - x^* \rangle + o(||x - x^*||^2) \geq f(x^*)$$

for any $x \in B(x^*, \varepsilon)$; therefore, for $d = (x - x^*)\,||x - x^*||^{-1}$, we have $\langle \nabla^2 f(x^*)d, d \rangle + o(||d||^2) \geq 0$, which means $\langle \nabla^2 f(x^*)d, d \rangle \geq 0$ for any d: $||d|| = 1$; therefore, (3.4) holds true. □

3.1.3 Second-Order Sufficient Condition

The second-order sufficient condition is given by the following theorem.

Theorem 3.4. *Let $f : B(x^*, \varepsilon) \to \mathbb{R}$ be twice continuous differentiable at x^*, $\nabla f(x^*) = 0$ and $\nabla^2 f(x^*) \succ 0$, then x^* is an isolated local minimizer.*

Proof. From twice differentiability of f and $\nabla f(x^*) = 0$ at x^* follows

$$f(x) = f(x^*) + \frac{1}{2}\langle \nabla^2 f(x^*)(x - x^*), x - x^* \rangle + o(\|x - x^*\|^2). \tag{3.5}$$

Since $\nabla^2 f(x^*) \succ 0$ its minimum eigenvalue $m > 0$ and condition

$$\nabla^2 f(x^*) \succeq mI \tag{3.6}$$

holds.

Then, since $\frac{o(r)}{r} \to 0$, there exists $\bar{r} > 0$ such that for all $r \in (0, \bar{r}]$ we have $o(r) \leq \frac{1}{4}rm$. Hence, for $r > 0$ small enough and any $x \in B(x^*, r) = \{x : \|x - x^*\| \leq r\}$, we have

$$|o(\|x - x^*\|^2)| \leq \frac{1}{4}m\|x - x^*\|^2.$$

Therefore, from (3.5) follows

$$\begin{aligned} f(x) &= f(x^*) + \frac{1}{2}\langle \nabla^2 f(x^*)(x - x^*), x - x^* \rangle + o(\|x - x^*\|^2) \\ &\geq f(x^*) + \frac{1}{2}m\|x - x^*\|^2 + o(\|x - x^*\|^2) \\ &\geq f(x^*) + \frac{1}{4}m\|x - x^*\|^2 > f(x^*). \end{aligned}$$

□

3.2 Nondifferentiable Unconstrained Minimization

We will see later that both gradient and Newton's methods are based on the first- and the second-order smooth approximation at a point. Such approximation allows finding a descent direction at each step. This is the main idea in convergence analysis.

It is not the case in non-smooth optimization. However, a convex function at each point $x_0 \in \mathbb{R}^n$ has a subdifferential $\partial f(x_0)$, which according to Theorem 2.12 is not empty, closed, convex, and bounded set. For each subgradient $g \in \partial f(x_0)$ and any $x \in \mathbb{R}^n$, we have $f(x) - f(x_0) \geq \langle g, x - x_0 \rangle$. Therefore, for any $x^* \in X^*$ and $x_0 \notin X^*$, we obtain

$$\langle g, x^* - x_0 \rangle \leq f(x^*) - f(x_0) < 0. \tag{3.7}$$

In other words, for any subgradient $g \in \partial f(x_0)$ and any $x^* \in X^*$, the direction $-g$ makes an acute angle with the direction $d_0 = x^* - x_0$. Although $-g$ is not necessarily a decent direction for f, it is "consistent" with d_0. Therefore, by moving along $-g$, one reduces the distance to $x^* \in X^*$.

3.2.1 Subgradient Method

Let us first consider the subgradient method with a fixed step length $t > 0$. Starting from $x_0 \in \mathbb{R}^n$, the subgradient method generates a sequence $\{x_s\}_{s \in N}$ by the following formula:

$$x_{s+1} = x_s - t\frac{g_s}{\|g_s\|}, \tag{3.8}$$

where $g_s \in \partial f(x_s)$ and $\|g_s\| \neq 0$. If $\|g_s\| = 0$, then $x^* = x_s$. Let us fix an arbitrary $x^* \in X^*$. It turns out that by using any $g_s \in \partial f(x_s)$ method (3.8) generates a sequence $\{x_s\}_{s\in N}$ that the boundary $\partial \mathcal{L}_f(\alpha_s) = \{x \in \mathbb{R}^n : f(x) = f(x_s) = \alpha_s\}$ of the correspondent level set $\{x \in \mathbb{R}^n : f(x) \leq f(x_s) = \alpha_s\}$ contains points close to x^* for $t > 0$ small enough.

Let $d_s(x^*) = \min\{\|u - x^*\| : u \in \partial \mathcal{L}_f(\alpha_s)\}$ be the distance from x^* to $\partial \mathcal{L}_f(\alpha_s)$.

Theorem 3.5. *If $f : \mathbb{R}^n \to \mathbb{R}$ is convex, then for a fixed $t > 0$, any given $\varepsilon > 0$ and a given $x^* \in X^*$, there exists s_ε such that*

$$d_{s_\varepsilon}(x^*) \leq \frac{t(1+\varepsilon)}{2}. \tag{3.9}$$

Proof. Let us assume that (3.9) is not true, then for any $s \geq 0$ we have

$$d_s(x^*) > \frac{t(1+\varepsilon)}{2}. \tag{3.10}$$

From (3.8) we obtain

$$\|x_{s+1} - x^*\|^2 = \|x_s - x^*\|^2 - 2t\langle x_s - x^*, \frac{g_s}{\|g_s\|}\rangle + t^2. \tag{3.11}$$

Let $l = \{x \in \mathbb{R}^n : \left\langle x - x_s, \frac{g_s}{\|g_s\|}\right\rangle = 0\}$ and $d(x^*; l) = \min\{\|x^* - x\| : x \in l\} = \|z_s - x^*\|$, then $y_s = [z_s, x^*] \cap \partial \mathcal{L}_f(\alpha_s)$ is such that $d_l(x^*) = \min\{\|z - x^*\| : z \in l\} = \|z_s - x^*\| \geq \|y_s - x^*\| \geq d_s(x^*)$. (see Fig. 3.1)

From (3.11) we have

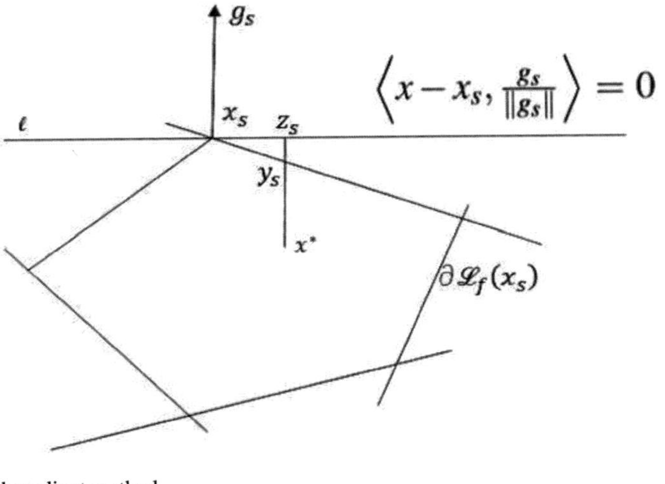

Fig. 3.1 Subgradient method

$$\|x_{s+1}-x^*\|^2 \le \|x_s-x^*\|^2 - 2t(d_s(x^*) - \frac{t}{2}). \tag{3.12}$$

If (3.10) holds for all $s \ge 0$ and a given $\varepsilon > 0$, then (3.11) and (3.12) yield $\|x_{s+1} - x^*\|^2 \le \|x_s - x^*\| - t^2\varepsilon$ for all $s \ge 0$, or $\|x_{N+1} - x^*\|^2 \le \|x_0 - x^*\|^2 - Nt^2\varepsilon$, which is impossible for N large enough. Therefore, there exists s_ε such that (3.9) holds. □

Corollary 3.1. *For any given $\delta > 0$, there exist $t > 0$ and s_δ such that*

$$0 < f(x_{s_\delta}) - f(x^*) \le \delta. \tag{3.13}$$

Proof. The proof follows from Lemma 2.12 and Theorem 3.5.

In other words, for a given $t > 0$ there exist ρ_t and a large enough s such that $x_s \in B(x^*, \rho_t)$ and $\rho_t \to 0$ with $t \to 0$.

This simple, but not trivial, result was proven first in 1961 by N. Z. Shor (1937–2006), who made fundamental contributions in non-smooth optimization theory and methods.

To guarantee convergence of the subgradient method to X^*, one has to decrease the step length from step to step. Let us consider the following subgradient method:

$$x_{s+1} = x_s - t_s \frac{g_s}{\|g_s\|}, \tag{3.14}$$

where $t_s > 0$, $t_s \to 0$, $\sum_{s=1}^{\infty} t_s = \infty$ and $\|g_s\| \ne 0$. Examples of $\{t_s\}_{s\in N}$ are: (a) $t_s = (s+1)^{-1}$; (b) $t_s = \frac{\alpha}{s^\beta}$, $0 < \beta \le 1$; (c) $t_s = \frac{1}{(s+1)\ln(s+1)}$.

Let $\varphi_k = \min\{f(x_s) : 1 \le s \le k\}$ be the record value of f in k steps.

Theorem 3.6. *For any convex $f : \mathbb{R}^n \to \mathbb{R}$, the subgradient method (3.14) converges in record value φ_k, that is,*

$$\lim_{k\to\infty} \varphi_k = f(x^*). \tag{3.15}$$

Proof. By assuming the opposite, we have

$$f(x_s) \ge \bar{f} > f^* \text{ for any } s \ge 0.$$

Let us consider $\bar{x}$ such that $f(\bar{x}) < \bar{f}$. It follows from the continuity of f existence $\tau > 0$ such that for $x \in B(\bar{x}, \tau)$ we have $f(x) \le \bar{f}$, then

$$x_s(\tau) = \bar{x} + \tau \frac{g_s}{\|g_s\|} \in B(\bar{x}, \tau).$$

On the other hand,

$$f(x_s(\tau)) \ge f(x_s) + \langle g_s, x_s(\tau) - x_s\rangle \ge \bar{f} + \langle g_s, \bar{x} - x_s\rangle + \langle g_s, x_s(\tau) - \bar{x}\rangle =$$

$$= \bar{f} + \langle g_s, \bar{x} - x_s\rangle + \tau\|g_s\|,$$

that is,

$$\langle g_s, \bar{x} - x_s\rangle + \tau\|g_s\| \le 0,$$

or

$$\langle g_s, x_s - \bar{x} \rangle \geq \tau \|g_s\|. \tag{3.16}$$

Let us estimate the distance from x_{s+1} to $\bar{x}$. Using (3.11) with x^* replaced by $\bar{x}$ and taking into account (3.16), we obtain

$$\|x_{s+1} - \bar{x}\|^2 \leq \|x_s - \bar{x}\|^2 - 2t_s\tau + t_s^2.$$

Since $t_s \to 0$, it follows

$$\|x_{s+1} - \bar{x}\|^2 \leq \|x_s - \bar{x}\|^2 - t_s\tau,$$

for $s \geq s_0$. Summing up, we get $\|x_{s_0} - \bar{x}\|^2 \geq \tau \sum_{s \geq s_0} t_s = +\infty$, which is impossible. Therefore, (3.15) holds. □

Let X^* be not empty and bounded, then it follows from Theorem 3.6 that $\min_{1 \leq s \leq k} \rho(x_s, X^*)$ converges to 0 as $k \to \infty$, where $\rho(x_s, X^*) = \min\{\|x_s - x^*\| : x \in X^*\}$.

From (3.14) follows

$$\|x_{s+1} - x^*\| = \|x_s - x^* + t_s \frac{g_s}{\|g_s\|}\| \geq \|x_s - x^*\| - t_s. \tag{3.17}$$

Keeping in mind that t_s converges to zero very slowly, it follows from (3.17) that we cannot expect fast convergence of subgradient method.

Therefore, substantial effort was made in the 1960s–1970s to improve convergence of subgradient method. In particular, subgradient method with space dilation in the direction of subgradient leads to ellipsoid method, which was independently discovered by N. Shor and A. Nemirovski and D. Yudin in the mid-1970s.

In 1979, Leonid Khachiyan (1952–2005) have shown that ellipsoid method for LP calculations has polynomial complexity.

3.3 Gradient Methods

The ability to solve efficiently unconstrained optimization problems is important also for constrained optimization, because there are a number of ways to reduce a constrained optimization problem into a sequence of unconstrained optimization problems.

In this section we concentrate on gradient methods. Their convergence, rate of convergence, and complexity obviously depend on the smoothness and convexity properties of f.

3.3.1 Gradient Method

Let us consider a convex and continuously differentiable function $f : \mathbb{R}^n \to \mathbb{R}$. For its linear approximation at $x_0 \in \mathbb{R}^n$, we have

$$\tilde{f}(x) = f(x_0) + \langle \nabla f(x_0), x - x_0 \rangle . \tag{3.18}$$

The best descent direction at $x_0 \in \mathbb{R}^n$ is the steepest descent direction

$$d(x_0) = -\nabla f(x_0) \|\nabla f(x_0)\|^{-1} = \text{argmin}\{f(x_0) + \langle \nabla f(x_0), d\rangle : \|d\| = 1\}. \tag{3.19}$$

One step of the steepest descent method consists of finding

$$\hat{x} = x_0 - t_0 \nabla f(x_0) \|\nabla f(x_0)\|^{-1}, \tag{3.20}$$

where $t_0 = \text{argmin}\{f(x_0 + td(x_0)) : t > 0\}$.

Finding t_0 is, generally speaking, an infinite procedure. We need a better way for finding $t > 0$ to guarantee convergence of the generic gradient method

$$\hat{x} = x - t\nabla f(x). \tag{3.21}$$

If gradient ∇f satisfies Lipschitz condition, then from (2.5) we have

$$f(x) \leq f(x_0) + \langle \nabla f(x_0), x - x_0 \rangle + \frac{L}{2}\|x - x_0\|^2.$$

Using (3.21) we obtain

$$\begin{aligned} f(x) &\leq f(x_0) - t\|\nabla f(x_0)\|^2 + \frac{L}{2}t^2 \|\nabla f(x_0)\|^2 \\ &= f(x_0) - t\left(1 - \frac{tL}{2}\right) \|\nabla f(x_0)\|^2. \end{aligned} \tag{3.22}$$

The natural choice of t is

$$\hat{t} = \text{argmax}\left\{t\left(1 - \frac{tL}{2}\right) : t > 0\right\} = \frac{1}{L}.$$

By reiterating (3.21) with $t = \frac{1}{L}$ from (3.22), we obtain the following sequence

$$\{x_s\}_{s \in N} : x_{s+1} = x_s - \frac{1}{L}\nabla f(x_s),$$

for which we have

$$f(x_{s+1}) \leq f(x_s) - \frac{1}{2L}\|\nabla f(x_s)\|^2. \tag{3.23}$$

Summing up (3.23) from $s = 0$ to $s = k$, we obtain

$$\frac{1}{2L}\sum_{s=0}^{k}\|\nabla f(x_s)\|^2 \le f(x_0)-f(x_{k+1}) \le f(x_0)-f(x^*). \tag{3.24}$$

In view of boundedness of X^*, we have $f(x^*) > -\infty$, so (3.24) implies

$$\lim_{s\to\infty}\|\nabla f(x_s)\| = 0.$$

Let us consider

$$\|\nabla f_k^*\| = \min_{0\le s\le k}\|\nabla f(x_s)\|, \tag{3.25}$$

then from (3.24) and (3.25) we obtain

$$\|\nabla f_k^*\| \le \sqrt{k^{-1}\left[2L\left(f(x_0)-f(x^*)\right)\right]}. \tag{3.26}$$

Bound (3.26) has two basic weaknesses: the value $\|\nabla f_k^*\|$ does not always characterize either $f(x_k)-f(x^*)$ or $\|x_k-x^*\|$. More importantly, error bound $O(k^{-0.5})$ is rather weak.

Bound (3.26) can be substantially improved by using new arguments as well as some extra properties of f.

Let us consider the following quadratic approximation of f at the point $x\in\mathbb{R}^n$:

$$\psi(x,X) = f(x) + \langle\nabla f(x), X-x\rangle + \frac{L}{2}\|X-x\|^2,$$

and

$$x_L = \mathrm{argmin}_{X\in\mathbb{R}^n}\,\psi(x,X) = x - \frac{1}{L}\nabla f(x). \tag{3.27}$$

It follows from (3.27) that if $x=x_s$, then $x_L = x_{s+1}$.

Beck–Teboulle's Lemma will be used in convergence analysis of gradient methods with Lipschitz gradient condition.

Lemma 3.1. *Let $f:\mathbb{R}^n\to\mathbb{R}$ be convex and Lipschitz gradient condition (2.4) holds, then for any given $x\in\mathbb{R}^n$ such that*

$$f(x_L) \le \psi(x,x_L) \tag{3.28}$$

the following inequality holds for any $X\in\mathbb{R}^n$:

$$f(X)-f(x_L) \ge \frac{L}{2}\|x_L-x\|^2 + L\langle x_L-x, x-X\rangle. \tag{3.29}$$

Proof. From (3.28) and the convexity of f, we have

$$
\begin{aligned}
f(X) - f(x_L) &\geq f(X) - \psi(x, x_L) \\
&= f(X) - f(x) - \langle \nabla f(x), x_L - x \rangle - \frac{L}{2}\|x_L - x\|^2 \\
&\geq \langle \nabla f(x), X - x \rangle - \langle \nabla f(x), x_L - x \rangle - \frac{L}{2}\|x_L - x\|^2 \\
&= \langle \nabla f(x), X - x_L \rangle + \frac{L}{2}\|x_L - x\|^2 - L\|x_L - x\|^2.
\end{aligned}
\tag{3.30}
$$

From the optimality condition for x_L follows

$$\nabla_X \psi(x, x_L) = \nabla f(x) + L(x_L - x) = 0;$$

therefore,

$$\langle \nabla f(x) + L(x_L - x), X - x_L \rangle = 0,$$

or

$$\langle \nabla f(x), X - x_L \rangle = -L \langle x_L - x, X - x_L \rangle \text{ for any } X \in \mathbb{R}^n.$$

From (3.30) we obtain

$$
\begin{aligned}
f(X) - f(x_L) &\geq -L \langle x_L - x, X - x_L \rangle + \frac{L}{2}\|x_L - x\|^2 - L \langle x_L - x, x_L - x \rangle \\
&= \frac{L}{2}\|x_L - x\|^2 + L \langle x_L - x, x - X \rangle .
\end{aligned}
$$

□

Theorem 3.7. *Let $f : \mathbb{R}^n \to \mathbb{R}$ be convex and gradient Lipschitz condition (2.4) is satisfied, then for the sequence $\{x_s\}_{s \in N}$ generated by (3.23) the following bound holds*

$$\Delta_k = f(x_k) - f(x^*) \leq \frac{L}{2k}\|x_0 - x^*\|^2 \tag{3.31}$$

for any $x^* \in X^*$.

Proof. From (3.29) with $X = x^*$, $x = x_s$, $x_L = x_{s+1}$ follows

$$
\begin{aligned}
\frac{2}{L}(f(x^*) - f(x_{s+1})) &\geq \|x_{s+1} - x_s\|^2 + 2\langle x_s - x^*, x_{s+1} - x_s \rangle \\
&= \langle x_{s+1}, x_{s+1} \rangle + \langle x_s, x_s \rangle - 2\langle x^*, x_{s+1} \rangle - 2\langle x_s, x_s \rangle \\
&\quad +2\langle x^*, x_s \rangle \\
&= \|x_{s+1} - x^*\|^2 - \|x_s - x^*\|^2.
\end{aligned}
$$

Summing up the last inequality from $s = 0$ to $s = k - 1$, we obtain

$$\frac{2}{L}\left(kf(x^*) - \sum_{s=0}^{k-1} f(x_{s+1})\right) \geq \|x_s - x^*\|^2 - \|x_0 - x^*\|^2. \tag{3.32}$$

Using (3.29) with $X = x = x_s$, $x_L = x_{s+1}$, we obtain

$$\frac{2}{L}(f(x_s) - f(x_{s+1})) \geq \|x_{s+1} - x_s\|^2,$$

or

$$sf(x_s) - sf(x_{s+1}) \geq \frac{L}{2}s\|x_{s+1} - x_s\|^2.$$

Therefore,

$$sf(x_s) - (s+1)f(x_{s+1}) + f(x_{s+1}) \geq \frac{L}{2}s\|x_{s+1} - x_s\|^2.$$

Summing up the last inequality from $s = 0$ to $s = k-1$, we obtain

$$-kf(x_k) + \sum_{s=0}^{k-1} f(x_{s+1}) \geq \frac{L}{2}\sum_{s=0}^{k-1} s\|x_{s+1} - x_s\|^2.$$

From (3.32) we have

$$kf(x^*) - \sum_{s=0}^{k-1} f(x_{s+1}) \geq \frac{L}{2}\left[\|x^* - x_k\| - \|x^* - x_0\|\right].$$

By adding the last two inequalities, we get

$$k(f(x^*) - f(x_k)) \geq \frac{L}{2}\left[\sum_{s=0}^{k-1} s\|x_{s+1} - x_s\|^2 + \|x^* - x_k\| - \|x^* - x_0\|^2\right],$$

which leads to bound (3.31). □

Bound (3.31) is stronger and more informative than (3.26). Still it can be substantially improved without any extra assumptions on the input data.

This can be done in the framework of the fast gradient (FG) method by Nesterov (1983) with important modifications by A. Beck and M. Teboulle made in their FISTA algorithm (2009).

3.3.2 *Fast Gradient Method*

Let $X_0 = X_1 \in \mathbb{R}^n$, the upper bound for the Lipschitz constant L in (2.4) is given and $t_1 = 1$. We assume that X_k has been found already.

A step of the FG method consists of three parts:

(a) find step size $t_{k+1} = \frac{1+\sqrt{1+4t_k^2}}{2}$,
(b) find the predictor $x_{k+1} = X_k + \frac{t_k - 1}{t_{k+1}}(X_k - X_{k-1})$,
(c) find new approximation

$$X_{k+1} = \operatorname{argmin}\left\{\langle X - x_{k+1}, \nabla f(x_{k+1})\rangle + \frac{L}{2}\|X - x_{k+1}\|^2 : X \in \mathbb{R}^n\right\}$$
$$= x_{k+1} - \frac{1}{L}\nabla f(x_{k+1}). \tag{3.33}$$

First of all, it follows from (b) that

$$t_k \geq \frac{1}{2}(k+1). \tag{3.34}$$

Indeed, for $k = 1$ we have $t_2 = \frac{1+\sqrt{1+4t^2}}{2} = \frac{1+\sqrt{5}}{2} > \frac{3}{2}$.

Assuming that inequality (3.34) is true, then from (b) we obtain

$$t_{k+1} = \frac{1}{2}\left(1 + \sqrt{1 + 4t_k^2}\right) \geq \frac{1}{2}\left(1 + \sqrt{1 + (k+1)^2}\right) \geq \frac{1}{2}(k+2).$$

Let $\Delta_k = f(X_k) - f(x^*)$ and $\mathrm{v}_k = t_k X_k + (t_k - 1)X_{k-1} - x^*$.

Theorem 3.8. *Let $f : \mathbb{R}^n_+ \to \mathbb{R}$ be convex with Lipschitz gradient, then the sequence $\{X_k\}_{k\in N}$ generated by the FG method (a)–(c) converges in value and the following bound holds:*

$$\Delta_k \leq \frac{2L\|x^* - x_0\|^2}{(k+2)^2}. \tag{3.35}$$

Proof. We first establish the following inequality

$$\frac{2}{L}\left(t_k^2\Delta_k - t_{k+1}^2\Delta_{k+1}\right) \geq \|\mathrm{v}_{k+1}\|^2 - \|\mathrm{v}_k\|^2. \tag{3.36}$$

From the basic inequality (3.29) for $X = x^*$, $x = x_{k+1}$, and $x_L = X_{k+1}$ follows

$$d(x^*) - d(X_{k+1}) \geq \frac{L}{2}\|X_{k+1} - x_{k+1}\|^2 + L\left\langle x_{k+1} - x^*, X_{k+1} - x_{k+1}\right\rangle. \tag{3.37}$$

By taking $X = X_k$, $x = x_{k+1}$, and $x_L = X_{k+1}$ from (3.29), we obtain

$$d(X_k) - d(X_{k+1}) \geq \frac{L}{2}\|X_{k+1} - x_{k+1}\|^2 + L\left\langle x_{k+1} - X_k, X_{k+1} - x_{k+1}\right\rangle \tag{3.38}$$

or

$$\frac{2}{L}(\Delta_k - \Delta_{k+1}) = \frac{2}{L}\left(d(X_k) - d(x^*) - (d(X_{k+1}) - d(x^*))\right)$$
$$\geq \|X_{k+1} - x_{k+1}\|^2 + 2\left\langle x_{k+1} - X_k, X_{k+1} - x_{k+1}\right\rangle. \tag{3.39}$$

From (3.37) we have

$$-\frac{2}{L}\Delta_{k+1} \geq \|X_{k+1} - x_{k+1}\|^2 + 2\left\langle x_{k+1} - x^*, X_{k+1} - x_{k+1}\right\rangle. \tag{3.40}$$

Multiplying (3.39) by $(t_{k+1}-1)>0$ and adding the obtained inequality to (3.40), we get

$$\frac{2}{L}\left[(t_{k+1}-1)\Delta_k - t_{k+1}\Delta_{k+1}\right] \geq$$

$$t_{k+1}\|X_{k+1}-x_{k+1}\|^2 + 2\langle X_{k+1}-x_{k+1},(t_{k+1}-1)(x_{k+1}-X_k)+x_{k+1}-x^*\rangle$$
$$= t_{k+1}\|X_{k+1}-x_{k+1}\|^2 + 2\langle X_{k+1}-x_{k+1},t_{k+1}(x_{k+1}-X_k+X_k-x^*\rangle$$
$$= t_{k+1}\|X_{k+1}-x_{k+1}\|^2+$$
$$2\langle X_{k+1}-x_{k+1},t_{k+1}x_{k+1}-(t_{k+1}-1)X_k-x^*\rangle. \tag{3.41}$$

From (b) follows $2t_{k+1}-1=\sqrt{1+4t_k^2}$ or

$$t_k^2 = t_{k+1}^2 - t_{k+1} = t_{k+1}(t_{k+1}-1). \tag{3.42}$$

Multiplying (3.41) by t_{k+1} and keeping in mind (3.42), we obtain

$$\frac{2}{L}\left[(t_{k+1}-1)t_{k+1}\Delta_k - t_{k+1}^2\Delta_{k+1}\right] = \frac{2}{L}\left(t_k^2\Delta_k - t_{k+1}^2\Delta_{k+1}\right) \geq$$
$$\|t_{k+1}(X_{k+1}-x_{k+1})\|^2 + 2t_{k+1}\langle X_{k+1}-x_{k+1},t_{k+1}x_{k+1}-(t_k-1)X_k-x^*\rangle. \tag{3.43}$$

Using the three vectors' identity

$$\|b-a\|^2 + 2\langle b-a,a-c\rangle = \|b-c\|^2 - \|a-c\|^2$$

with $a=t_{k+1}x_{k+1}$, $b=t_{k+1}X_{k+1}$, and $c=(t_{k+1}-1)X_k+x^*$ from (3.43), we obtain

$$\frac{2}{L}(t_k^2\Delta_k - t_{k+1}^2\Delta_{k+1}) \geq$$
$$\|t_{k+1}X_{k+1}-(t_{k+1}-1)X_k-x^*\|^2 - \|t_{k+1}x_{k+1}-(t_{k+1}-1)X_k-x^*\|^2. \tag{3.44}$$

Using (c) we obtain

$$t_{k+1}x_{k+1} = t_{k+1}X_k + (t_k-1)(X_k-X_{k-1}). \tag{3.45}$$

Keeping in mind that $\mathrm{v}_k = t_kX_k + (t_k-1)X_{k-1} - x^*$, from (3.44) and (3.45) we obtain

$$\frac{2}{L}(t_k^2\Delta_k \; - \; t_{k+1}^2\Delta_{k+1}) \geq$$
$$\|\mathrm{v}_{k+1}\|^2 - \|t_{k+1}X_k \; + \; (t_k-1)(X_k-X_{k-1}) - (t_{k+1}-1)X_k - x^*\|^2$$
$$= \|\mathrm{v}_{k+1}\|^2 \; - \; \|t_kX_k - (t_k-1)X_{k-1} - x^*\|^2$$
$$= \|\mathrm{v}_{k+1}\|^2 \; - \; \|\mathrm{v}_k\|^2,$$

that is, (3.36) holds.

It follows from (3.36) that

$$t_k^2 \Delta_k - t_{k+1}^2 \Delta_{k+1} \geq \frac{L}{2}\left[\|v_{k+1}\|^2 - \|v_k\|^2\right];$$

therefore,

$$\begin{aligned} t_{k+1}^2 \Delta_{k+1} + \frac{L}{2}\|v_{k+1}\|^2 &\leq t_k^2 \Delta_k + \frac{L}{2}\|v_k\|^2 \\ &\leq t_{k-1}^2 \Delta_{k-1} + \frac{L}{2}\|v_{k-1}\|^2 \\ &\vdots \\ &\leq t_1^2 \Delta_1 + \frac{L}{2}\|v_1\|^2. \end{aligned}$$

Hence,

$$\begin{aligned} t_{k+1}^2 \Delta_{k+1} &\leq t_1^2 \Delta_1 + \frac{L}{2}\|v_1\|^2 - \frac{L}{2}\|v_{k+1}\|^2 \\ &\leq (f(X_1) - f(x^*)) + \frac{L}{2}\|X_1 - x^*\|^2. \end{aligned} \tag{3.46}$$

For $X = x^*$, $x_L = X_1$, and $x = x_0$, it follows from (3.29) that

$$\begin{aligned} f(x^*) - f(X_1) &\geq \frac{L}{2}\|X_1 - x_0\|^2 + L\langle x_0 - x^*, X_1 - x_0\rangle \\ &= \frac{L}{2}\left[\|X_1 - x^*\|^2 - \|x_0 - x^*\|^2\right]. \end{aligned}$$

Therefore,

$$f(X_1) - f(x^*) \leq \frac{L}{2}\left[\|x_0 - x^*\|^2 - \|X_1 - x^*\|^2\right]. \tag{3.47}$$

From (3.46) and (3.47), we have

$$t_{k+1}^2 \Delta_{k+1} \leq \frac{L}{2}\|x_0 - x^*\|^2.$$

Keeping in mind that $t_{k+1} \geq \frac{1}{2}(k+2)$, we obtain (3.35).

□

So the FG method requires similar to gradient method effort per step, but the convergence rate is much better.

It follows from (3.35) that for a given accuracy $\varepsilon > 0$, it takes

$$k = \frac{\sqrt{2L}\|x_0 - x^*\|}{\sqrt{\varepsilon}}$$

steps for FG method to get $\Delta_k \leq \varepsilon$.

The FG method is optimal in the class of gradient-type methods for unconstrained minimization of convex functions with Lipschitz gradient. To formulate the exact result, we need the following assumption.

Assumption 3.1 *The first-order method M generates a sequence of test points* $\{x_s\}_{s\in N}$ *such that*

$$x_k \in x_0 + \mathrm{LIN}\{\nabla f(x_0), \dots, \nabla f(x_{k-1})\},\ k \geq 1.$$

The main complexity result about such methods is stated in the following Theorem by Yu. Nesterov.

Theorem 3.9. *For any* $1 \leq k \leq \frac{1}{2}(n-1)$ *and any* $x_0 \in \mathbb{R}^n$*, there exists a* C^∞ *function* f*, which gradient satisfy* (2.4) *that for any first-order methods M satisfying Assumption 3.1 the following bounds*

$$f(x_k) - f(x^*) \geq \frac{3L\|x_0 - x^*\|}{32(k+1)^2} \text{ and } \|x_k - x^*\|^2 \geq \frac{1}{8}\|x_0 - x^*\|^2$$

hold.

3.3.3 Gradient Method for Strongly Convex Functions

We consider a class of convex function $f : \mathbb{R}^n \to \mathbb{R}$ that the strong monotonicity (2.27) and Lipschitz condition (2.4) for the gradient are satisfied.

Let us first consider the generic gradient method (3.21). Keeping in mind that $\nabla f(x^*) = 0$, we obtain

$$\hat{x} - x^* = x - x^* - t\,(\nabla f(x) - \nabla f(x^*))\,.$$

Therefore,

$$\|\hat{x} - x^*\|^2 = \|x - x^*\|^2 - 2t\,\langle x - x^*, \nabla f(x) - \nabla f(x^*)\rangle \qquad (3.48)$$
$$+t^2\|\nabla f(x) - \nabla f(x^*)\|^2.$$

Using the strong monotonicity (2.27) and Lipschitz condition (2.4) for ∇f, we obtain

$$\|\hat{x} - x^*\|^2 = \left(1 - 2mt + t^2L^2\right)\|x - x^*\|^2 = q(t)\|x - x^*\|^2.$$

For any $0 < t < 2mL^{-2}$, we have $0 < q(t) < 1$ and $q = \min_{t\geq 0} q(t) = q(mL^{-2}) = 1 - m^2L^{-2} = 1 - \varkappa^2$.

By reiterating (3.21) with $t = \frac{m}{L^2}$, we obtain a sequence $\{x_s\}_{s\in N}$:

$$x_{s+1} = x_s - \frac{m}{L^2}\nabla f(x_s),$$

such that

$$\delta_k = \|x_k - x^*\| \leq \sqrt{1 - \varkappa^2}\|x_{k-1} - x^*\| = q^{0.5k}\delta_0. \qquad (3.49)$$

There are two important advantages of bound (3.49) over the previous bound (3.35).

First, instead of the convergence to the optimal solution in value, we have a convergence of x_k to x^*.

Second, bound (3.49) is asymptotically much stronger than (3.35).

In other words, the strong monotonicity along with the Lipschitz continuity of the gradient ∇f guarantees convergence with a linear rate.

In the following theorem, the ratio and complexity bound for gradient method will be improved without extra assumption.

Theorem 3.10. *If for $f : \mathbb{R}^n \to \mathbb{R}$ conditions (2.4) and (2.27) are satisfied, then:*

(1) for $0 < t < 2/(m+L)$, the following bound holds

$$\|\hat{x} - x^*\|^2 \le (1 - t\frac{2mL}{m+L})\|x - x^*\|^2;$$

(2) for $t = 2/(m+L)$, we have

$$\|\hat{x} - x^*\| \le \left(\frac{1-\varkappa}{1+\varkappa}\right)\|x - x^*\|; \tag{3.50}$$

(3) for a given $\varepsilon > 0$ to obtain $\delta_k = \|x_k - x^*\| \le \varepsilon$, it requires

$$k \ge O(\varkappa^{-1}\ln \varepsilon^{-1}) \tag{3.51}$$

steps.

Proof.

(1) Using Theorem 2.5 and $\nabla f(x^*) = 0$ from (3.21), we obtain

$$\begin{aligned}\|\hat{x} - x^*\|^2 &= \|x - t\nabla f(x) - x^*\|^2 \\ &= \|x - x^*\|^2 - 2t\langle \nabla f(x) - \nabla f(x^*), x - x^*\rangle + t^2\|\nabla f(x)\|^2 \\ &\le \left(1 - t\frac{2mL}{m+L}\right)\|x - x^*\|^2 + t\left(t - \frac{2}{m+L}\right)\|\nabla f(x)\|^2 \\ &\le \left(1 - t\frac{2mL}{m+L}\right)\|x - x^*\|^2.\end{aligned}$$

(2) For $t = \frac{2}{m+L}$ we have

$$\begin{aligned}\|\hat{x} - x^*\|^2 &\le \left(1 - \frac{4mL}{(m+L)^2}\right)\|x - x^*\|^2 \\ &= \frac{(m-L)^2}{(m+L)^2}\|x - x^*\|^2 = \left(\frac{1-\varkappa}{1+\varkappa}\right)^2\|x - x^*\|^2,\end{aligned}$$

that is, (3.50) holds.

(3) By reiterating (3.50) we obtain $\delta_k = \|x_k - x^*\| \le \left(\frac{1-\varkappa}{1+\varkappa}\right)^k \delta_0$. For a given $\varepsilon > 0$ small enough, let us estimate the number of steps k one needs to obtain $\delta_k \le \varepsilon$. Let $\left(\frac{1-\varkappa}{1+\varkappa}\right)^k \delta_0 \le \varepsilon$, then $k \ln \frac{1-\varkappa}{1+\varkappa} \le \ln \frac{\varepsilon}{\delta_0}$. Keeping in mind that $0 < (1-\varkappa)(1+\varkappa)^{-1} < 1$, we obtain

$$k \ge \left(\ln \frac{1+\varkappa}{1-\varkappa}\right)^{-1} \ln \frac{\delta_0}{\varepsilon} = \left(\ln \frac{\delta_0}{\varepsilon}\right)\left(\ln\left(1 + \frac{2\varkappa}{1-\varkappa}\right)\right)^{-1}.$$

Using $\ln(1+x) \le x$, for any $x > -1$ we obtain bound (3.51). □

3.4 Method Regularization

Let $f : \mathbb{R}^n \to \mathbb{R}$ be convex and the optimal set X^* be not empty and bounded. Unconstrained minimization technique, which requires strong convexity of f, cannot be applied for finding $x^* \in X^*$.

In this section we consider A. Tikhonov's (1906–1993) regularization method (1963). It replaces the original problem by a sequence of unconstrained minimization problems with strongly convex functions.

Let $r : \mathbb{R}^n \to \mathbb{R}$ be nonnegative and strongly convex, then for any $\varepsilon > 0$ the function

$$F(x, \varepsilon) = f(x) + \varepsilon r(x)$$

is strongly convex and has a unique minimizer

$$x_\varepsilon = \operatorname{argmin}\{F(x;\varepsilon) : x \in \mathbb{R}^n\}. \tag{3.52}$$

It is intuitively clear that x_ε should converge to $x^* \in X^*$ when $\varepsilon \to 0$.

First, let us consider a monotone decreasing sequence $\{\varepsilon_s\}_{s\in N}$ such that $\lim_{s\to\infty} \varepsilon_s = 0$, which we call a regularization sequence. The corresponding sequence $\{x_s\}_{s\in N}$ of minimizers $x_s = x_{\varepsilon_s}$ is uniquely defined.

The following theorem characterizes the convergence of the sequence $\{x_s\}_{s\in N}$.

Theorem 3.11. *For any convex f, strongly convex r, bounded X^*, and any regularization sequence $\{\varepsilon_s\}_{s\in N}$, the following statements hold true:*

(1) the sequence $\{r(x_s)\}_{s\in N}$ is monotone non-decreasing and $\lim_{s\to\infty} r(x_s) = r(x^*)$;
(2) the sequence $\{f(x_s)\}_{s\in N}$ is monotone non-increasing and $\lim_{s\to\infty} f(x_s) = f(x^*)$;
(3) $\lim_{s\to\infty} x_s = x^* = \operatorname{argmin}_{x\in X^*} r(x)$.

Proof.

(1) From the definition of x_s for any $s \ge 0$, we obtain

$$f(x_s) + \varepsilon_s r(x_s) \le f(x_{s+1}) + \varepsilon_s r(x_{s+1}) \tag{3.53}$$

and

$$f(x_{s+1}) + \varepsilon_{s+1} r(x_{s+1}) \leq f(x_s) + \varepsilon_{s+1} r(x_s). \tag{3.54}$$

Therefore, by adding (3.53) and (3.54), we get

$$(\varepsilon_s - \varepsilon_{s+1})(r(x_s) - r(x_{s+1})) \leq 0.$$

From $\varepsilon_{s+1} < \varepsilon_s$ follows

$$r(x_s) \leq r(x_{s+1}),\ s \geq 0. \tag{3.55}$$

On the other hand, for any $s \geq 0$ we obtain

$$f(x_s) + \varepsilon_s r(x_s) \leq f(x^*) + \varepsilon_s r(x^*); \tag{3.56}$$

hence,

$$\varepsilon_s(r(x_s) - r(x^*)) \leq f(x^*) - f(x_s) \leq 0. \tag{3.57}$$

It means that $\{x_s\}_{s\in N} \subset \mathscr{L}_r(r(x^*))$, where $\mathscr{L}_r(\alpha) = \{x \in \mathbb{R}^n : r(x) \leq \alpha\}$.
Due to the strong convexity of r, the set $\mathscr{L}_r(r(x^*))$ is bounded and so is the sequence $\{x_s\}_{s\in N}$. Therefore, it contains a converging subsequence $\{x_{s_i}\}_{i\in N}$.
Let $\lim_{s_i\to\infty} x_{s_i} = \bar{x}$, then taking the limit in (3.57) for the corresponding subsequence, we obtain $f(\bar{x}) \leq f(x^*)$; therefore, $\bar{x} = x^*$.
Keeping in mind (3.55), we obtain

$$\lim_{s\to\infty} r(x_s) = r(x^*). \tag{3.58}$$

(2) From (3.54) and (3.55), we have

$$f(x_{s+1}) - f(x_s) \leq \varepsilon_{s+1}(r(x_s) - r(x_{s+1})) \leq 0,$$

that is,

$$f(x_{s+1}) - f(x_s) \leq 0,\ s \geq 0.$$

So, the sequence $\{f(x_s)\}_{s=0}^{\infty}$ is monotone non-increasing and bounded from below by $f(x^*)$. Therefore, there exists $\lim_{s\to\infty} f(x_s) = \bar{f} \geq f(x^*)$, but $\bar{f} > f(x^*)$ contradicts $f(\bar{x}) \leq f(x^*)$; therefore, $\lim_{s\to\infty} f(x_s) = f(x^*)$.

(3) For a strongly convex function r, from (3.58) follows $\lim_{s\to\infty} x_s = x^* \in X^*$. Inequality (3.54) holds for any $x^* \in X^*$; therefore, from (3.58) we obtain

$$r(x_s) \leq r(x^*) \text{ for any } x^* \in X^*.$$

Hence, $\lim_{s\to\infty} r(x_s) \leq \min_{u\in X^*} r(u)$ or $\lim_{s\to\infty} x_s = x^* = \operatorname{argmin}_{u\in X^*} r(u)$. □

Exercise 3.1.

1. Construct regularization method for finding $f(x^*) = \min\{f(x) = \frac{1}{2}\langle Ax, x\rangle - \langle b, x\rangle : x \in \mathbb{R}^n\}$ with $r(x) = 1/2\langle B(x-a), x-a\rangle, B > 0$. Find the formula for x_ε.
2. Let $B = I, a = 0$, find x_ε.

3. Construct regularization method for finding $f(x^*) = \min\{f(x) = \|Ax - b\|^2\}$, $r(x) = \frac{1}{2}x^2$.

3.5 Proximal Point Method

Quadratic regularization method consists of finding

$$x_\varepsilon = \operatorname{argmin}\{f(x) + \frac{1}{2}\varepsilon\|x-a\|^2\}.$$

We saw in the previous section that $x_\varepsilon \to x^*$ as $\varepsilon \to 0$, where x^* is the minimizer of $f(x)$, for which

$$x^* = \operatorname{argmin}\{\|x-a\|^2 : x \in X^*\}.$$

For small $\varepsilon > 0$, finding x_ε gets very difficult because the level sets of

$$F_\varepsilon(x) = f(x) + \frac{1}{2}\varepsilon\|x-a\|^2$$

are strongly elongated, that is, it increases sharply in some directions and varies little in other directions.

Let $L_\delta = \{x : F_\varepsilon(x) = F_\varepsilon(x^*) + \delta\}$, then more precisely the ill-conditioning means that the condition number

$$\varkappa = \underline{\lim}_{\delta\to 0}\{\inf_{x\in L_\delta}\|x-x^*\| / \sup_{x\in L_\delta}\|x-x^*\|\}$$

of $x^* \in X^*$ is very small, which slows down substantially the methods for unconstrained minimization. We recall that for smooth and strongly convex functions $\varkappa = mL^{-1}$. It follows from (3.49)–(3.51) that for small $\varkappa > 0$, the gradient methods can be very slow. In case of regularization, the condition number vanishes when $\varepsilon_s \to 0$, which makes it difficult if not impossible to obtain results with high accuracy. The prox method

$$x_{s+1} = \operatorname{argmin}_{u\in\mathbb{R}^n}\{f(u) + \frac{1}{2}\varepsilon\|u-x_s\|^2\} \tag{3.59}$$

is free, to a large extent, from the ill-conditioning phenomenon because $\varepsilon > 0$ is fixed. The function

$$F_\varepsilon(u,x) = f(u) + \frac{1}{2}\varepsilon\|u-x\|^2, \tag{3.60}$$

at each step, has practically the same $\varkappa > 0$ when prox sequence $\{x_s\}_{s\in N}$ approaches the solution set X^*. Let us consider the prox function

$$\psi_\varepsilon(x) = \min_{u\in\mathbb{R}^n} F_\varepsilon(u,x) = f(u(x)) + \frac{1}{2}\varepsilon\|u(x)-x\|^2.$$

Finding $x^* \in X^*$ means finding a fixed point $x^* \in \mathbb{R}^n$ such that $u(x^*) = x^*$. The function $F_\varepsilon(u,x)$ is strongly convex in u for any given $\varepsilon > 0$; therefore, $u(x)$ is the unique solution of the following inclusion:

$$0 \in \partial(f(u) + \frac{1}{2}\varepsilon\|u - x\|^2) = \partial f(u(x)) + \varepsilon(u(x) - x),$$

that is,

$$-u(x) + x \in \varepsilon^{-1}\partial f(u(x)) = Q_\varepsilon(x), \tag{3.61}$$

where ∂f is the subdifferential of a convex function f.
In other words,

$$x = (\varepsilon^{-1}\partial f + I)u,$$

where I is the identical operator in $\mathbb{R}^n$. Therefore,

$$u(x) = (\varepsilon^{-1}\partial f + I)^{-1}x = P_\varepsilon x,$$

where P_ε is the prox operator. We can rewrite prox method (3.59) as follows:

$$x_{s+1} = P_\varepsilon x_s. \tag{3.62}$$

Let us consider some important properties of the prox operator P_ε:

1. Due to the uniqueness of $u(x)$, the prox operator is uniquely defined.
2. For any $g_x = -u(x) + x \in \varepsilon^{-1}\partial f(u(x))$ and any $g_y = -g(y) + y \in \varepsilon^{-1}\partial f(u(y))$,we have

   $$\langle g_x - g_y, u(x) - u(y))\rangle \geq 0.$$

   Therefore,

   $$\langle -u(x) + x + u(y) - y, u(x) - u(y)\rangle \geq 0$$

   or

   $$\langle x - y, u(x) - u(y)\rangle \geq \|u(x) - u(y)\|^2.$$

   Using Cauchy–Schwarz inequality, we obtain

   $$\|x - y\|\|u(x) - u(y)\| \geq \|u(x) - u(y)\|^2,$$

   that is,

   $$\|u(x) - u(y)\| \leq \|x - y\|,$$

   which means that the prox operator P_ε is not expansive.
3. Let us consider the prox function $\psi_\varepsilon(x)$.

Exercise 3.2. Show that $\psi_\varepsilon(x)$ is convex and differentiable.

Lemma 3.2. *The gradient $\nabla\psi_\varepsilon$ satisfies the Lipschitz condition with $L = \varepsilon$.*

Proof. Let us find the gradient $\nabla\psi_\varepsilon(x)$. We have

$$\nabla\psi_\varepsilon(x) = [g(u(x)) + \varepsilon(u(x) - x)]\nabla u(x) - \varepsilon(u(x) - x),$$

where $\nabla u(x)$ is the Jacobian of $u(x)$. Keeping in mind that

$$g(u(x)) + \varepsilon(u(x) - x) = 0$$

and using (3.61)–(3.62), we obtain

$$\nabla \psi_\varepsilon(x) = \varepsilon(x - u(x)) = \varepsilon(I - P_\varepsilon)(x) \in \varepsilon Q_\varepsilon(x) \tag{3.63}$$

or

$$P_\varepsilon(x) + Q_\varepsilon(x) = x. \tag{3.64}$$

Let us show that the gradient $\nabla \psi_\varepsilon(x)$ satisfies the Lipschitz condition with the Lipschitz constant $\varepsilon > 0$. Using the monotonicity of ∂f, from (3.61) we get

$$\begin{aligned} 0 &\leq \langle \partial f(u(x_1)) - \partial f(u(x_2)), u(x_1) - u(x_2) \rangle \\ &= \varepsilon \langle Q_\varepsilon(x_1) - Q_\varepsilon(x_2), P_\varepsilon(x_1) - P_\varepsilon(x_2) \rangle, \end{aligned}$$

or

$$\langle P_\varepsilon(x_1) - P_\varepsilon(x_2), Q_\varepsilon(x_1) - Q_\varepsilon(x_2) \rangle \geq 0. \tag{3.65}$$

From (3.64) we have

$$\|x_1 - x_2\|^2 = \|P_\varepsilon(x_1) - P_\varepsilon(x_2) + Q_\varepsilon(x_1) - Q_\varepsilon(x_2)\|^2,$$

and using (3.65) we obtain

$$\|P_\varepsilon(x_1) - P_\varepsilon(x_2)\|^2 + \|Q_\varepsilon(x_1) - Q_\varepsilon(x_2)\| \leq \|x_1 - x_2\|^2;$$

therefore,

$$\|Q_\varepsilon(x_1) - Q_\varepsilon(x_2)\| \leq \|x_1 - x_2\|.$$

Hence,

$$\|\nabla \psi_\varepsilon(x_1) - \nabla \psi_\varepsilon(x_2)\| \leq \varepsilon \|x_1 - x_2\|,$$

that is, $\nabla \psi_\varepsilon(x)$ satisfies Lipschitz condition with constant $\varepsilon > 0$. □

From (3.63) follows

$$x_{s+1} = x_s - \varepsilon^{-1} \nabla \psi_\varepsilon(x_s). \tag{3.66}$$

The convergence and the convergence rate of $\{x_s\}_{s \in N}$ follows from Theorem 3.7.

3.6 Newton's Method and Regularized Newton Method

3.6.1 Introduction

Newton's method, which has been introduced almost 350 years ago, is still one of the basic tools in numerical analysis, variational and control problems, and optimization both constrained and unconstrained, just to mention a few.

It was used not only as a numerical tool but also as a powerful instrument for proving existence and uniqueness results.

In particular, Newton's method plays a critical role in the classical KAM theory by A. Kolmogorov, V. Arnold, and J. Mozer developed in the 1950s and 1960s; see Arnold (1963).

Newton's method was the main instrument in the interior point methods, which preoccupied the field of optimization for a long time.

Yu. Nesterov and A. Nemirovski showed that a special damped Newton method is particularly efficient for minimization of self-concordant (SC) functions. The results will be covered in Chapter 6.

They showed that from any starting point, a special damped Newton step reduces the SC function value by a constant, which depends only on the Newton decrement. The decrement converges to zero.

By the time it gets small enough, damped Newton method practically turns into Newton's method and generates a sequence, which converges in value with quadratic rate.

They characterized the size of the minimizer's neighborhood, where quadratic rate occurs. It allows establishing complexity of the special damped Newton method for SC function, that is, to find the upper bound for the number of damped Newton step, required for finding an ε- approximation for the solution.

For strictly convex functions, which are not self-concordant, such results, to the best of our knowledge, are unknown.

We consider a new version of damped Newton method (DNM) and introduce damped regularized Newton method (DRNM), which guarantees convergence to the minimizer for any strictly convex function from any starting point. The most important feature of DRNM is regularization at each point. By taking Euclidian norm of the gradient as the regularization parameter, we not only guarantee global convergence of DRNM but also retain local quadratic rate. It became possible due to the vanishing regularization parameter.

The important issue still is the size of the Newton and the regularized Newton areas, where correspondent methods converge with quadratic rate.

In this paragraph we establish complexity bounds for both DNM and DRNM.

First, we estimate the size of Newton areas, where DNM and DRNM converge with quadratic rate.

Then, we estimate the number of steps needed for DNM or DRNM to enter the correspondent Newton areas.

The key ingredients of our analysis are the Newton and the regularized Newton decrements.

The decrements provide the upper bound for the distance from the current approximation to the minimizer. Therefore, they are used in the stopping criteria.

On the other hand, they provide a lower bound for the function reduction at each step at any point out of the Newton or the regularized Newton area.

The bounds are used to estimate the number of DNM or DRNM steps needed to get into the correspondent Newton areas.

3.6.2 Newton's Method

We start with the classical Newton's method for finding a root of a nonlinear equation

$$f(t) = 0,$$

where $f : \mathbb{R} \to \mathbb{R}$ has a smooth derivative f'.

Let us consider $t_0 \in \mathbb{R}$ and the linear approximation

$$\widetilde{f}(t) = f(t_0) + f'(t_0)(t - t_0) = f(t_0) + f'(t_0)\Delta t$$

of f at t_0. We assume that $f'(t_0) \neq 0$.

By replacing f with its linear approximation, we obtain the following equation

$$f(t_0) + f'(t_0)\Delta t = 0$$

for the Newton step Δt.

The next approximation is given by formula

$$t = t_0 + \Delta t = t_0 - (f'(t_0))^{-1} f(t_0). \tag{3.67}$$

By reiterating (3.67) we obtain Newton's method

$$t_{s+1} = t_s - (f'(t_s))^{-1} f(t_s) \tag{3.68}$$

for finding a root of a nonlinear equation $f(t) = 0$.

Let $f \in C^2$ and $f'(t^*) \neq 0$, where t^* is the root, that is, $f(t^*) = 0$. We consider the expansion of f at t_s with the Lagrange remainder

$$0 = f(t^*) = f(t_s) + f'(t_s)(t^* - t_s) + \frac{1}{2} f''(\hat{t}_s)(t^* - t_s)^2, \tag{3.69}$$

where $\hat{t}_s \in [t_s, t^*]$. For t_s close to t^*, we have $f'(t_s) \neq 0$; therefore, from (3.69) follows

$$t^* - t_s + \frac{f(t_s)}{f'(t_s)} = -\frac{1}{2} \frac{f''(\hat{t}_s)}{f'(t_s)}(t^* - t_s)^2.$$

Using (3.68) we get

$$|t^* - t_{s+1}| = \frac{1}{2}\frac{|f''(\hat{t}_s)|}{|f'(t_s)|}|t^* - t_s|^2. \tag{3.70}$$

If $\Delta_s = |t^* - t_s|$ is small, then there exist $a > 0$ and $b > 0$ independent on t_s that $|f''(\hat{t}_s)| \le a$ and $|f'(t_s)| > b$. Therefore, from (3.70) follows

$$\Delta_{s+1} \le c\Delta_s^2, \tag{3.71}$$

where $c = 0.5ab^{-1}$.

This is the key characteristic of Newton's method, which makes the method so important even 350 years after it was originally introduced.

Newton's method has a natural extension for a nonlinear system of equations

$$g(x) = 0, \tag{3.72}$$

where $g : \mathbb{R}^n \to \mathbb{R}^n$ is a vector function with a smooth Jacobian $J(g) = \nabla g : \mathbb{R}^n \to \mathbb{R}^n$. The linear approximation of g at x_0 is given by

$$\widetilde{g}(x) = g(x_0) + \nabla g(x_0)(x - x_0). \tag{3.73}$$

We replace g in (3.72) by its linear approximation (3.73). The Newton step Δx one finds by solving the following linear system:

$$g(x_0) + \nabla g(x_0)\Delta x = 0.$$

Assuming $\det \nabla g(x_0) \neq 0$, we obtain

$$\Delta x = -(\nabla g(x_0))^{-1} g(x_0).$$

The new approximation is given by the following formula:

$$x = x_0 - (\nabla g(x_0))^{-1} g(x_0). \tag{3.74}$$

By reiterating (3.74) we obtain Newton's method

$$x_{s+1} = x_s - (\nabla g(x_s))^{-1} g(x_s) \tag{3.75}$$

for solving a nonlinear system of equations (3.72).

Newton's method for minimization of $f : \mathbb{R}^n \to \mathbb{R}$ follows directly from (3.75) if instead of unconstrained minimization problem

$$f(x^*) = \{\min f(x) | x \in \mathbb{R}^n\} \tag{3.76}$$

we consider the nonlinear system

$$\nabla f(x) = 0, \tag{3.77}$$

which, in case of convex f, is the necessary and sufficient condition for x^* to be the minimizer in (3.76) .

Vector

$$n(x) = -(\nabla^2 f(x))^{-1}\nabla f(x) \tag{3.78}$$

defines the Newton direction at $x \in \mathbb{R}^n$.

Application of Newton's method (3.75) to the system (3.77) leads to Newton's method

$$x_{s+1} = x_s - (\nabla^2 f(x_s))^{-1}\nabla f(x_s) = x_s + n(x_s) \tag{3.79}$$

for solving (3.76).

Method (3.79) has another interpretation. Let $f : \mathbb{R}^n \to \mathbb{R}$ be twice differentiable with a positive definite Hessian $\nabla^2 f$.

The quadratic approximation of f at x_0 is given by the formula

$$\widetilde{f}(x) = f(x_0) + \langle \nabla f(x_0), x - x_0\rangle + \frac{1}{2}\langle \nabla^2 f(x_0)(x - x_0), x - x_0\rangle.$$

Instead of solving (3.76), let us find

$$\bar{x} = \operatorname{argmin}\{\widetilde{f}(x) : x \in \mathbb{R}^n\},$$

which is equivalent to solving the following linear system

$$\nabla^2 f(x_0)\Delta x = -\nabla f(x_0)$$

for $\Delta x = x - x_0$.

We obtain

$$\Delta x = n(x_0),$$

so for the next approximation we have

$$\bar{x} = x_0 - (\nabla^2 f(x_0))^{-1}\nabla f(x_0) = x_0 + n(x_0). \tag{3.80}$$

By reiterating (3.80) we obtain Newton's method (3.79) for solving (3.76).

The local quadratic convergence of both (3.75) and (3.79) is well known.

Away from the neighborhood of x^*, however, both Newton's methods (3:75) and (3.79) can either oscillate or diverge.

Example 3.1. Consider

$$g(t) = \begin{cases} -1, & t \in (-\infty, -1] \\ (t+1)^2 - 1, & t \in [-1, 0] \\ -(t-1)^2 + 1, & t \in [0, 1] \\ 1, & t \in [1, \infty). \end{cases}$$

The function g together with g' is continuous on $(-\infty, \infty)$. Newton's method (3.68) converges to the root $t^* = 0$ from any starting point t: $|t| < \frac{2}{3}$, oscillates between

$t_s = -\frac{2}{3}$ and $t_{s+1} = \frac{2}{3}$, $s = 1, 2, \ldots$, and either diverges or not defined for any t: $|t| > \frac{2}{3}$.

Example 3.2. For $f(t) = \sqrt{1+t^2}$, we have

$$f(t^*) = f(0) = \min\{f(t) : -\infty < t < \infty\}.$$

For the first and second derivative, we have

$$f'(t) = t(1+t^2)^{-\frac{1}{2}}, \; f''(t) = (1+t^2)^{-\frac{3}{2}}.$$

Therefore, Newton's method (3.79) is given by the following formula:

$$t_{s+1} = t_s - (1+t_s^2)^{\frac{3}{2}} t_s (1+t_s^2)^{-\frac{1}{2}} = -t_s^3. \tag{3.81}$$

It follows from (3.81) that Newton's method converges from any $t_0 \in (-1,1)$, oscillates between $t_s = -1$ and $t_{s+1} = 1$, $s = 1, 2, \ldots$, and diverges from any $t_0 \notin [-1,1]$. It also follows from (3.81) that Newton's method converges from any starting point $t_0 \in (-1,1)$ with the cubic rate; however, in both examples the convergence area is negligibly smaller than the area where Newton's method diverges. Note that f is strictly convex in $\mathbb{R}$ and strongly convex in the neighborhood of $t^* = 0$.

Therefore, there are three important issues associated with Newton's method for unconstrained convex optimization.

First, to characterize the neighborhood of the solution, where Newton's method converges with quadratic rate, which is called Newton area.

Second, finding for any strictly convex function such modification of Newton's method that generates convergent sequence from any starting point and retains quadratic convergence rate in the neighborhood of the solution.

Third, to estimate the total number of steps required for finding an ε—approximation for x^*.

3.6.3 Local Quadratic Convergence of Newton's Method

We consider a class of convex functions $f : \mathbb{R}^n \to \mathbb{R}$ that are strongly convex at x^*, that is,

$$\nabla^2 f(x^*) \succeq mI, \tag{3.82}$$

$m > 0$ and their Hessian satisfy Lipschitz condition in the neighborhood of x^*. In other words there is $\delta > 0$, a ball $B(x^*, \delta) = \{x \in \mathbb{R}^n, \|x - x^*\| \leq \delta\}$, and $M > 0$ such that for any x and $y \in B(x^*, \delta)$ we have

$$\|\nabla^2 f(x) - \nabla^2 f(y)\| \leq M\|x - y\|, \tag{3.83}$$

where $\|x\| = \langle x, x \rangle^{\frac{1}{2}}$.

In the following theorem, the Newton area is explicitly characterized through the convexity constant $m > 0$ and Lipschitz constant $M > m$. We will use similar technique later to characterize the regularized Newton area.

Theorem 3.12. *Under conditions* (3.82) *and* (3.83) *for* $\delta = \frac{2m}{3M}$ *and any given* $x_0 \in B(x^*, \delta)$, *the entire sequence* $\{x_s\}_{s \in N}$, *generated by (3.79), belongs to* $B(x^*, \delta)$, *and the following bound*

$$\|x_{s+1} - x^*\| \leq \frac{M}{2(m - M\|x_s - x^*\|)} \|x_s - x^*\|^2, \quad s \geq 1. \tag{3.84}$$

holds.

Proof. From (3.79) and $\nabla f(x^*) = 0$ follows

$$\begin{aligned} x_{s+1} - x^* &= x_s - x^* - [\nabla^2 f(x_s)]^{-1} \nabla f(x_s) = \\ &= x_s - x^* - (\nabla^2 f(x_s))^{-1} (\nabla f(x_s) - \nabla f(x^*)) = \\ &= [\nabla^2 f(x_s)]^{-1} [\nabla^2 f(x_s)(x_s - x^*) - (\nabla f(x_s) - \nabla f(x^*))]. \end{aligned} \tag{3.85}$$

Then we have

$$\nabla f(x_s) - \nabla f(x^*) = \int_0^1 \nabla^2 f(x^* + \tau(x_s - x^*))(x_s - x^*) d\tau.$$

From (3.85) we obtain

$$x_{s+1} - x^* = [\nabla^2 f(x_s)]^{-1} H_s (x_s - x^*), \tag{3.86}$$

where

$$H_s = \int_0^1 [\nabla^2 f(x_s) - \nabla^2 f(x^* + \tau(x_s - x^*))] d\tau.$$

Let $\Delta_s = \|x_s - x^*\|$, then using (3.83) we get

$$\begin{aligned} \|H_s\| &= \| \int_0^1 [\nabla^2 f(x_s) - \nabla^2 f(x^* + \tau(x_s - x^*))] d\tau \| \\ &\leq \int_0^1 \| [\nabla^2 f(x_s) - \nabla^2 f(x^* + \tau(x_s - x^*)) \| d\tau \leq \\ &\leq \int_0^1 M \|x_s - x^* - \tau(x_s - x^*)\| d\tau \leq \\ &\leq \int_0^1 M(1 - \tau) \|x_s - x^*\| d\tau = \frac{M}{2} \Delta_s. \end{aligned}$$

Therefore, from (3.86) and the latter bound, we have

$$\begin{aligned} \Delta_{s+1} &\leq \|(\nabla^2 f(x_s))^{-1}\| \|H_s\| \|x_s - x^*\| \leq \\ &\frac{M}{2} \|(\nabla^2 f(x_s))^{-1}\| \Delta_s^2. \end{aligned} \tag{3.87}$$

From (3.83) follows

$$\|\nabla^2 f(x_s) - \nabla^2 f(x^*)\| \le M\|x_s - x^*\| = M\Delta_s;$$

therefore,

$$\nabla^2 f(x^*) + M\Delta_s I \succeq \nabla^2 f(x_s) \succeq \nabla^2 f(x^*) - M\Delta_s I.$$

From (3.82) follows

$$\nabla^2 f(x_s) \succeq \nabla^2 f(x^*) - M\Delta_s I \succeq (m - M\Delta_s)I.$$

Hence, for any $\Delta_s < mM^{-1}$ the matrix $\nabla^2 f(x_s)$ is positive definite; therefore, the inverse $(\nabla^2 f(x^s))^{-1}$ exists, and the following bound holds:

$$\|(\nabla^2 f(x^s))^{-1}\| \le \frac{1}{m - M\Delta_s}.$$

From (3.87) and the latter bound follows

$$\Delta_{s+1} \le \frac{M}{2(m - M\Delta_s)}\Delta_s^2. \tag{3.88}$$

From (3.88) for $\Delta_s < \frac{2m}{3M}$ follows $\Delta_{s+1} < \Delta_s$, which means that for $\delta = \frac{2m}{3M}$ and any $x_0 \in B(x^*, \delta)$ the entire sequence $\{x_s\}_{s\in N}$ belongs to $B(x^*, \delta)$ and converges to x^* with the quadratic rate (3.88).

The proof is complete □

The neighborhood $B(x^*, \delta)$ with $\delta = \frac{2m}{3M}$ is called Newton area.

In the following section, we consider a new version of damped Newton method, which converges from any starting point and at the same time retains quadratic convergence rate in the Newton area.

3.6.4 Damped Newton Method

To make Newton's method practical, we have to guarantee convergence from any starting point. To this end the step length $t > 0$ is attached to the Newton direction $n(x)$, that is,

$$\hat{x} = x + tn(x), \tag{3.89}$$

where $n(x)$ is the solution of the following system

$$\nabla^2 f(x)n = -\nabla f(x).$$

The step length $t > 0$ has to be adjusted to guarantee a "substantial reduction" of f at each $x \notin B(x^*, \delta)$ and $t = 1$, when $x \in B(x^*, \delta)$.

Method (3.89) is called damped Newton method (DNM).

The following function $\lambda : \mathbb{R}^n \to \mathbb{R}_+$:

$$\lambda(x) = \langle (\nabla^2 f(x))^{-1} \nabla f(x), \nabla f(x) \rangle^{0.5} = [-\langle \nabla f(x), n(x) \rangle]^{0.5}, \tag{3.90}$$

which is called the Newton decrement of f at $x \in \mathbb{R}^n$, plays an important role in DNM.

At this point we assume that $f : \mathbb{R}^n \to \mathbb{R}$ is strongly convex and its Hessian $\nabla^2 f$ is Lipschitz continuous, that is, there exist $\infty > M > m > 0$ that

$$\nabla^2 f(x) \succeq mI \tag{3.91}$$

and

$$\|\nabla^2 f(x) - \nabla^2 f(y)\| \leq M\|x - y\| \tag{3.92}$$

are satisfied for any x and y from $\mathbb{R}^n$.

Let $x_0 \in \mathbb{R}^n$ be a starting point.

Due to (3.91) the sublevel set $\mathscr{L}_0 = \{x \in \mathbb{R}^n : f(x) \leq f(x_0)\}$ is bounded for any given $x_0 \in \mathbb{R}^n$. Therefore, from (3.92) follows existence $L > 0$ that

$$\|\nabla^2 f(x)\| \leq L \tag{3.93}$$

is taking place for any $x \in \mathscr{L}_0$.

We also assume that $\varepsilon > 0$ is small enough; in particular,

$$0 < \varepsilon < m^2 L^{-1} \tag{3.94}$$

holds.

We are ready to describe our version of DNM.

Let $x_0 \in \mathbb{R}^n$ be a starting point and $0 < \varepsilon << \delta$ be the required accuracy. Set $x := x_0$

1. find Newton direction $n(x)$;
2. if the following inequality

$$f(x + n(x)) \leq f(x) + 0.5\langle \nabla f(x), n(x) \rangle \tag{3.95}$$

   holds, then $t(x) := 1$, otherwise $t(x) = m(2L)^{-1}$;
3. set $x := x + t(x)n(x)$;
4. if $\lambda(x) \leq \varepsilon^{1.5}$, then $x^* := x$, otherwise go to 1.

The following theorem proves global convergence of DNM (1)–(4) and establishes the upper bound for the total number of DNM steps required for finding ε-approximation for x^*.

3.6.5 Global Convergence of DNM and Its Complexity

Theorem 3.13. *If $f : \mathbb{R}^n \to \mathbb{R}$ is twice differentiable and conditions (3.91) and (3.92) are satisfied, then for $\delta = \frac{2}{3}\frac{m}{M}$ it takes*

$$N_0 = 9\frac{L^2M^2}{m^5}(f(x_0) - f(x^*)) \tag{3.96}$$

DNM steps to find $x \in B(x^*, \delta)$ by using DNM.

Proof. From (3.93) follows

$$\nabla^2 f(x) \preceq LI. \tag{3.97}$$

On other hand, from (3.91) follows the existence of the inverse $(\nabla^2 f(x))^{-1}$. Therefore, from (3.97) follows

$$(\nabla^2 f(x))^{-1} \succeq L^{-1}I. \tag{3.98}$$

From (3.90) and (3.98), we obtain the following lower bound for the Newton decrement:

$$\lambda(x) = \langle \nabla^2 f(x)^{-1} \nabla f(x), \nabla f(x) \rangle^{0.5} \geq \tag{3.99}$$
$$\geq (L^{-1}\|\nabla f(x)\|^2)^{0.5} = L^{-0.5}\|\nabla f(x)\|.$$

From (3.91) we have

$$\|\nabla f(x)\|\|x - x^*\| \geq \langle \nabla f(x) - \nabla f(x^*), x - x^* \rangle \geq m\|x - x^*\|^2$$

or

$$\|\nabla f(x)\| \geq m\|x - x^*\|. \tag{3.100}$$

From (3.99) and (3.100), we obtain

$$\lambda(x) \geq L^{-0.5} m\|x - x^*\|. \tag{3.101}$$

From (3.94) and the stopping criteria 4 follows

$$(m^2L^{-1})^{0.5}\varepsilon \geq \varepsilon^{1.5} \geq \lambda(x) \geq mL^{-0.5}\|x - x^*\|,$$

or $\|x - x^*\| \leq \varepsilon$, which justifies the stopping criteria 4.

On the other hand, Newton decrement defines the lower bound for the function reduction at each step.

In fact, for Newton directional derivative from (3.78), (3.90), and (3.91) follows

$$\varphi'(0) = \frac{df(x + tn(x))}{dt}|_{t=0} = \langle \nabla f(x), n(x) \rangle =$$

$$- \langle \nabla^2 f(x)n(x), n(x) \rangle \leq -m\|n(x)\|^2. \tag{3.102}$$

Due to the strong convexity of $\varphi(t) = f(x + tn(x))$, the derivative $\varphi'(t) = \langle \nabla f(x + tn(x)), n(x) \rangle$ is monotone increasing in $t > 0$, so there is $t(x) > 0$ such that

$$0 > \langle \nabla f(x + t(x)n(x)), n(x) \rangle \geq \frac{1}{2}\langle \nabla f(x), n(x) \rangle, \tag{3.103}$$

otherwise $\langle \nabla f(x + tn(x)), n(x) \rangle < \frac{1}{2}\langle \nabla f(x), n(x) \rangle \leq -\frac{1}{2}m\|n(x)\|^2, t > 0$ and $\inf f(x) = -\infty$, which is impossible for a strongly convex function f.

It follows from (3.102), (3.103), and monotonicity of $\varphi'(t)$ that for any $t \in [0,t(x)]$ we have

$$\frac{df(x+tn(x))}{dt} = \langle \nabla f(x+tn(x)), n(x) \rangle \le \frac{1}{2}\langle \nabla f(x), n(x) \rangle.$$

Therefore,

$$f(x+t(x)n(x)) \le f(x) + \frac{1}{2}t(x)\langle \nabla f(x), n(x) \rangle.$$

Keeping in mind (3.90), we obtain

$$f(x) - f(x+t(x)n(x)) \ge \frac{1}{2}t(x)\lambda^2(x). \tag{3.104}$$

Combining (3.102) and (3.103), we obtain

$$\langle \nabla f(x+t(x)n(x)) - \nabla f(x), n(x) \rangle \ge \frac{m}{2}\|n(x)\|^2.$$

From the mean value formula applied to $\varphi(t) = \langle \nabla f(x+t(x)n(x)) - \nabla f(x), n(x) \rangle$ follows existence $0 < \theta(x) < 1$ such that

$$t(x)\langle \nabla^2 f(x+\theta(x)t(x)n(x))n(x), n(x) \rangle = t(x)\langle \nabla^2 f(\cdot)n(x), n(x) \rangle \ge \frac{m}{2}\|n(x)\|^2,$$

or

$$t(x)\|\nabla^2 f(\cdot)\|\|n(x)\|^2 \ge \frac{m}{2}\|n(x)\|^2.$$

From (3.91) follows

$$t(x) \ge \frac{m}{2L}, \tag{3.105}$$

which justifies the choice of step length $t(x)$ in DNM 1–4.

Hence, from (3.104) and (3.105) we obtain the following lower bound for the function reduction per step

$$\Delta f(x) = f(x) - f(x+t(x)n(x)) \ge \frac{m}{4L}\lambda^2(x), \tag{3.106}$$

which together with the lower bound (3.101) for the Newton decrement $\lambda(x)$ leads to

$$\Delta f(x) = f(x) - f(x+t(x)n(x)) \ge \frac{m^3}{4L^2}\|x-x^*\|^2. \tag{3.107}$$

It means that for any $x \notin B(x^*, \delta)$, the function reduction at each step is proportional to the square of the distance between current approximation x and the solution x^*.

In other words, "far from" the solution Newton step produces a "substantial" reduction of the function value similar to one of the gradient method.

For $x \notin B(x^*, \delta)$ we have $\|x - x^*\| \ge \frac{2m}{3M}$; therefore, from (3.107) we obtain $\Delta f(x) \ge \frac{1}{9}\frac{m^5}{L^2M^2}$. So it takes at most

$$N_0 = 9\frac{L^2M^2}{m^5}(f(x_0) - f(x^*))$$

Newton steps to obtain $x \in B(x^*, \delta)$ from a given starting point $x_0 \in \mathbb{R}^n$. The proof is complete. □

From Theorem 3.12 follows that $O(\ln\ln\varepsilon^{-1})$ steps are needed to find an ε- approximation to x^* from any $x \in B(x^*, \delta)$, where $0 < \varepsilon << \delta$ is the required accuracy. Therefore, the total number of Newton steps required for finding an ε-approximation to the optimal solution x^* from a starting point $x_0 \in \mathbb{R}^n$ is

$$N = N_0 + O(\ln\ln\varepsilon^{-1}).$$

DNM requires a priori knowledge of two parameters m and L or their corresponding lower and upper bounds.

The following version of DNM is free from this requirement. To adjust the step length $t > 0$, we use the backtracking line search.

The inequality

$$f(x+tn(x)) \leq f(x) + \alpha t\langle\nabla f(x), n(x)\rangle \tag{3.108}$$

with $0 < \alpha \leq 0.5$ is called the Armijo condition.

Let $0 < \rho < 1$, the backtracking line search consists of the following steps:

1. For $t > 0$ check (3.108). If (3.108) holds go to 2. If not set $t := t\rho$ and repeat it until (3.108) holds and then go to 2.
2. Set $t(x) := t$, $x := x + t(x)n(x)$

We are ready to describe another version of DNM, which does not require a priori knowledge of the parameters m and L or their lower and upper bounds.

Let $x_0 \in \mathbb{R}^n$ be a starting point and $0 < \varepsilon << \delta$ be the required accuracy.

1. Compute Newton direction

$$n(x) = -(\nabla^2 f(x))^{-1}\nabla f(x); \tag{3.109}$$

2. set $t := 1$, use the backtracking line search until

$$f(x+tn(x)) \leq f(x) + 0.5t\langle\nabla f(x), n(x)\rangle;$$

3. set $t(x) := t$, $x := x + t(x)n(x)$;
4. if $\lambda(x) \leq \varepsilon^{1.5}$ then $x^* := x$ otherwise go to 1.

The complexity of the DNM with backtracking line search can be established using arguments similar to those in Theorem 3.13

Unfortunately, in the absence of strong convexity of $f : \mathbb{R}^n \to \mathbb{R}$, Newton's method might not converge from any starting point.

In the case of Example 3.2, Newton's method does not converge from any $t \notin (-1, 1)$ in spite of $f(t) = \sqrt{1+t^2}$ being strongly convex and smooth enough in the neighborhood of $t^* = 0$.

In the following section, we consider regularized Newton method (RNM), which eliminates the basic drawback of the classical Newton's method. It generates a converging sequence from any starting point $x_0 \in \mathbb{R}^n$ and retains quadratic convergence rate in the regularized Newton area, which we will characterize later.

3.6.6 *Regularized Newton Method*

Let $f \in C^2$ be a convex function in $\mathbb{R}^n$. We assume that the optimal set $X^* = \text{Argmin}\{f(x) : x \in \mathbb{R}^n\}$ is not empty and bounded.

The corresponding regularized at the point $x \in \mathbb{R}^n$ function $F_x : \mathbb{R}^n \to \mathbb{R}$ is defined by the following formula:

$$F_x(y) = f(y) + \frac{1}{2} \parallel \nabla f(x) \parallel \parallel y - x \parallel^2 . \tag{3.110}$$

For any $x \notin X^*$, we have $||\nabla f(x)|| > 0$; therefore, for any convex function $f : \mathbb{R}^n \to \mathbb{R}$ the regularized function F_x is strongly convex in y for any $x \notin X^*$. If f is strongly convex at x^*, then the regularized function F_x is strongly convex in $\mathbb{R}^n$. The following properties of F_x are direct consequences of the definition (3.110):

1°. $F_x(y)_{|y=x} = f(x)$,

2°. $\nabla_y F_x(y)_{|y=x} = \nabla f(x)$,

3°. $\nabla^2_{yy} F_x(y)_{|y=x} = \nabla^2 f(x) + ||\nabla f(x)|| I = H(x)$,

where I is the identical matrix in $\mathbb{R}^n$.

For any $x \notin X^*$, the inverse $H^{-1}(x)$ exists for any convex $f \in C^2$. Therefore, the regularized Newton step

$$\hat{x} = x + r(x), \tag{3.111}$$

where the regularized Newton direction (RND) $r(x)$ is defined by

$$H(x) r = -\nabla f(x), \tag{3.112}$$

can be performed for any convex $f \in C^2$ from any starting point $x \notin X^*$.

We start by showing that the regularization (3.110) improves the "quality" of the Newton direction as well as the condition number of Hessian $\nabla^2 f(x)$ at any $x \in \mathbb{R}^n$ that $x \notin X^*$.

Let $0 \le m(x) < M(x) < \infty$ be the smallest and largest eigenvalue of the matrix $H(x)$, then

$$m(x)||y||^2 \le \langle \nabla^2 f(x) y, y \rangle \le M(x)||y||^2 \tag{3.113}$$

holds for any $y \in \mathbb{R}^n$.

The condition number of Hessian $\nabla^2 f$ at $x \in \mathbb{R}^n$ is

$$\text{cond}\, \nabla^2 f(x) = m(x)(M(x))^{-1}.$$

Along with the regularized Newton step (3.111), we consider the classical Newton step

$$\hat{x} = x + n(x), \tag{3.114}$$

where $n(x)$ solution of the following system

$$H(x)n = -\nabla f(x).$$

The "quality" of any direction d at $x \in \mathbb{R}^n$ is defined by the following number:

$$0 \leq q(d) = -\frac{\langle \nabla f(x), d \rangle}{\|\nabla f(x)\| \cdot \|d\|} \leq 1.$$

For the steepest descent direction $d(x) = -\nabla f(x) \parallel \nabla f(x) \parallel^{-1}$, we have the best local descent direction, and $q(d(x)) = 1$. The "quality" of the classical Newton direction is defined by the following number:

$$q(n(x)) = -\frac{\langle \nabla f(x), n(x) \rangle}{||\nabla f(x)|| \cdot ||n(x)||}. \tag{3.115}$$

For the RND $r(x)$, we have

$$q(r(x)) = -\frac{\langle \nabla f(x), r(x) \rangle}{||\nabla f(x)|| \cdot ||r(x)||}. \tag{3.116}$$

The following theorem establishes the lower bounds for $q(r(x))$ and $q(n(x))$. It shows that the regularization (3.110) improves the condition number of Hessian $\nabla^2 f$ for all $x \in \mathbb{R}^n, x \notin X^*$.

Theorem 3.14. *Let $f: \mathbb{R}^n \to \mathbb{R}$ be a twice continuous differentiable convex function and bounds (3.113) hold, then:*

1.

$$1 \geq q(r(x)) \geq (m(x) + ||\nabla f(x)||)(M(x) + ||\nabla f(x)||)^{-1}$$
$$= \text{cond } H(x) > 0 \qquad \textit{for any } x \notin X^*.$$

2.

$$1 \geq q(n(x)) \geq m(x)(M(x))^{-1} = \text{cond } \nabla^2 f(x)$$
$$\textit{for any } x \in \mathbb{R}^n.$$

3.

$$\text{cond } H(x) - \text{cond } \nabla^2 f(x) =$$

$$||\nabla f(x)||(1 - \text{cond } \nabla^2 f(x))(M(x) + ||\nabla f(x)||)^{-1} > 0 \tag{3.117}$$

for any $x \notin X^*$, cond $\nabla^2 f(x) < 1$.

Proof.

1. From (3.112), we obtain

$$||\nabla f(x)|| \leq ||H(x)|| \cdot ||r(x)||. \tag{3.118}$$

Using the right inequality (3.113) and 3°, we have

$$||H(x)|| \leq M(x) + ||\nabla f(x)||. \tag{3.119}$$

From (3.118) and (3.119), we obtain

$$||\nabla f(x)|| \leq (M(x) + ||\nabla f(x)||)||r(x)||.$$

From (3.112) the left inequality (3.113) and 3° follows

$$-\langle \nabla f(x), r(x)\rangle = \langle H(x)r(x), r(x)\rangle \geq (m(x) + \|\nabla f(x)\|)\|r(x)\|^2.$$

Therefore, from (3.116) follows

$$q(r(x)) \geq (m(x) + ||\nabla f(x)||)(M(x) + ||\nabla f(x)||)^{-1} = \text{cond } H(x).$$

2. Now let us consider the Newton direction $n(x)$. From (3.114), we have

$$\nabla f(x) = -\nabla^2 f(x) n(x); \tag{3.120}$$

therefore,

$$-\langle \nabla f(x), n(x)\rangle = \langle \nabla^2 f(x) n(x), n(x)\rangle.$$

From (3.120) and the left inequality of (3.113), we obtain

$$q(n(x)) = -\frac{\langle \nabla f(x), n(x)\rangle}{||\nabla f(x)|| \cdot ||n(x)||} \geq m(x)||n(x)|| \cdot ||\nabla f(x)||^{-1}. \tag{3.121}$$

From (3.120) and the right inequality in (3.113) follows

$$||\nabla f(x)|| \leq ||\nabla^2 f(x)|| \cdot ||n(x)|| \leq M(x)||n(x)||. \tag{3.122}$$

Combining (3.121) and (3.122), we have

$$q(n(x)) \geq \frac{m(x)}{M(x)} = \text{cond } \nabla^2 f(x).$$

3 Using the formulas for the condition numbers of $\nabla^2 f(x)$ and $H(x)$, we obtain (3.113) □

Corollary 3.2. *The regularized Newton direction $r(x)$ is a decent direction for any convex $f : \mathbb{R}^n \to \mathbb{R}$, whereas the classical Newton direction $n(x)$ exists, and it is a decent direction only if f is a strongly convex at $x \in \mathbb{R}^n$.*

Under conditions (3.82) and (3.83), RNM retains the local quadratic convergence rate, which is typical for the classical Newton's method.

On the other hand, the regularization (3.110) allows to establish global convergence and estimate complexity of RNM, when the original function is only strongly convex at x^*.

3.6.7 Local Quadratic Convergence Rate of RNM

In this section we consider RNM and determine the neighborhood of the minimizer, where RNM converges with quadratic rate.

Along with assumptions (3.82) and (3.83) for Hessian $\nabla^2 f$, we will use the Lipschitz condition for the gradient ∇f

$$\|\nabla f(x) - \nabla f(y)\| \leq L\|x - y\|, \tag{3.123}$$

which is equivalent to (3.93).

RNM generates a sequence $\{x_s\}_{s \in N}$:

$$x_{s+1} = x_s + r(x_s), \tag{3.124}$$

where $r(x_s)$ is the solution of the following system

$$H(x_s)r = -\nabla f(x_s).$$

The following theorem characterizes the regularized Newton area.

Theorem 3.15. *If (3.82), (3.83), and* (3.123) *hold, then for* $\delta = \frac{2m}{3M+2L}$ *and any* $x_0 \in B(x^*, \delta)$ *as a starting point, the sequence* $\{x_s\}_{s \in N}$ *generated by RNM (3.124) belongs to* $B(x^*, \delta)$*, and the following bound holds:*

$$\Delta_{s+1} = \|x_{s+1} - x^*\| \leq \frac{M + 2L}{2} \cdot \frac{1}{m - M\Delta_s}\Delta_s^2, \; s \geq 1. \tag{3.125}$$

Proof. From (3.124) follows

$$x_{s+1} - x^* = x_s - x^* - \left[\nabla^2 f(x_s) + \|\nabla f(x_s)\| I\right]^{-1} (\nabla f(x_s) - \nabla f(x^*)).$$

Using

$$\nabla f(x_s) - \nabla f(x^*) = \int_0^1 \nabla^2 f(x^* + \tau(x_s - x^*))(x_s - x^*)d\tau,$$

we obtain

$$x_{s+1} - x^* = \left[\nabla^2 f(x_s) + \|\nabla f(x_s)\| I\right]^{-1} H_s(x_s - x^*), \tag{3.126}$$

where

$$H_s = \int_0^1 (\nabla^2 f(x_s) + \|\nabla f(x_s)\| I - \nabla^2 f(x^* + \tau(x_s - x^*)))d\tau.$$

From (3.83) and (3.123) follows

$$\|H_s\| = \| \int_0^1 \left(\nabla^2 f(x_s) + \|\nabla f(x_s)\| I - \nabla^2 f(x^* + \tau(x_s - x^*))\right) d\tau\|$$
$$\leq \| \int_0^1 (\nabla^2 f(x_s) - \nabla^2 f(x^* + \tau(x_s - x^*)))d\tau\| + \int_0^1 \|\nabla f(x_s)\| d\tau$$
$$\leq \int_0^1 \|\nabla^2 f(x_s) - \nabla^2 f(x^* + \tau(x_s - x^*))\| d\tau + \int_0^1 \|\nabla f(x_s) - \nabla f(x^*)\| d\tau$$
$$\leq \int_0^1 M\|x_s - x^* - \tau(x_s - x^*)\| d\tau + \int_0^1 L\|x_s - x^*\| d\tau$$
$$= \int_0^1 (M(1-\tau) + L)\|x_s - x^*\| d\tau = \frac{M+2L}{2}\|x_s - x^*\|. \quad (3.127)$$

From (3.126) and (3.127), we have

$$\Delta_{s+1} = \|x_{s+1} - x^*\| \leq \| \left(\nabla^2 f(x_s) + \|\nabla f(x_s)\| I\right)^{-1} \| \cdot \|H_s\| \cdot \|x_s - x^*\|$$
$$\leq \frac{M+2L}{2}\|(\nabla^2 f(x_s) + \|\nabla f(x_s)\| I)^{-1}\| \Delta_s^2. \quad (3.128)$$

From (3.83) follows

$$\|\nabla^2 f(x_s) - \nabla^2 f(x^*)\| \leq M\|x_s - x^*\| = M\Delta_s; \quad (3.129)$$

therefore, we have

$$\nabla^2 f(x^*) + M\Delta_s I \succeq \nabla^2 f(x_s) \succeq \nabla^2 f(x^*) - M\Delta_s I. \quad (3.130)$$

From (3.82) and (3.130), we obtain

$$\nabla^2 f(x_s) + \|\nabla f(x_s)\| I \succeq (m + \|\nabla f(x_s)\| - M\Delta_s)I.$$

Therefore, for $\Delta_s < \frac{m+\|\Delta f(x_s)\|}{M}$ the matrix $\nabla^2 f(x_s) + \|\nabla f(x_s)\| I$ is positive definite; therefore, its inverse exists, and we have

$$\|(\nabla^2 f(x_s) + \|\nabla f(x_s)\| I)^{-1}\| \leq \frac{1}{m + \|\nabla f(x_s)\| - M\Delta_s} \leq$$
$$\frac{1}{m - M\Delta_s}. \quad (3.131)$$

For $\Delta_s \leq \frac{2m}{3M+2L}$ from (3.128) and (3.131) follows

$$\Delta_{s+1} \leq \frac{M+2L}{2}\frac{1}{m - M\Delta_s}\Delta_s^2. \quad (3.132)$$

Therefore, from (3.132) for $0 < \Delta_s \leq \frac{2m}{3M+2L} < \frac{m+\|\nabla f(x_s)\|}{M}$, we obtain

$$\Delta_{s+1} \leq \frac{3M+2L}{2m}\Delta_s^2 \leq \Delta_s.$$

Hence, for $\delta = \frac{2m}{3M+2L}$ and any $x_0 \in B(x^*, \delta)$ as a starting point, the sequence $\{x_s\}_{s\in N}$ generated by (3.124) belongs to $B(x^*, \delta)$, and bound (3.125) holds.

The proof of Theorem 3.15 is complete. □

Corollary 3.3. *Under conditions of Theorem 3.15 for* $\delta = \frac{2m}{3M+2L}$ *and any* $x \in B(x^*, \delta)$*, Hessian* $\nabla^2 f(x)$ *is positive definite and*

$$\nabla^2 f(x) \succeq m_0 I, \tag{3.133}$$

where $m_0 = m(M+2L)(3M+2L)^{-1}$.

In fact, from (3.92) follows

$$\nabla^2 f(x^*) + M\Delta x I \succeq \nabla^2 f(x) \succeq \nabla^2 f(x^*) - M\Delta x I,$$

so for any $x \in B(x^*, \delta)$ we have

$$\nabla^2 f(x) \succeq \left(m - \frac{2Mm}{3M+2L}\right) I = m(M+2L)(3M+2L)^{-1} I = m_0 I.$$

From the latter inequality follows

$$\|\nabla f(x)\| \|x - x^*\| \geq \langle \nabla f(x) - \nabla f(x^*), x - x^* \rangle \geq m_0 \|x - x^*\|^2,$$

that is, for any $x \in B(x^*, \delta)$ we have

$$\|\nabla f(x)\| \geq m_0 \|x - x^*\|. \tag{3.134}$$

It follows from Theorem 3.15 that $B(x^*, \delta)$ with $\delta = \frac{2m}{3M+2L}$ is the Newton area for RNM.

So it takes $O(\ln \ln \varepsilon^{-1})$ regularized Newton steps to find an ε-approximation for x^* from any $x \in B(x^*, \delta)$ as a starting point.

To make RNM globally convergent, we have to replace RNM by DRNM and adjust the step length. It can be done by backtracking line search, using Armijo condition (3.108) with Newton direction $n(x)$ replaced by regularized Newton direction $r(x)$. In the following section, we introduce another version of DRNM and estimate the number of RNM steps required for finding $x \in \mathbb{B}(x^*, \delta)$ from any given starting point $x_0 \in \mathbb{R}^n$.

3.6.8 Damped Regularized Newton Method

Let us consider the regularized Newton decrement

$$\lambda_r(x) = \langle H^{-1}(x)\nabla f(x), \nabla f(x)\rangle^{\frac{1}{2}} = [-\langle \nabla f(x), r(x)\rangle]^{\frac{1}{2}}. \tag{3.135}$$

We assume that $\varepsilon > 0$ is small enough, in particular,

$$0 < \varepsilon^{0.5} < m_0(L + \|\nabla f(x)\|)^{-0.5}, \tag{3.136}$$

for $\forall x \in \mathscr{L}_0$.

From (3.93) follows

$$(\nabla^2 f(x) + \|\nabla f(x)\| I) \preceq (L + \|\nabla f(x)\|)I. \tag{3.137}$$

On the other hand, for any $x \in B(x^*, \delta)$ from Corollary 3.3, we have

$$\nabla^2 f(x) + \|\nabla f(x)\| I \succeq (m_0 + \|\nabla f(x)\|)I.$$

Therefore, the inverse $(\nabla^2 f(x) + \|\nabla f(x)\| I)^{-1}$ exists, and from (3.137) we obtain

$$H^{-1}(x) = (\nabla^2 f(x) + \|\nabla f(x)\| I)^{-1} \succeq (L + \|\nabla f(x)\|)^{-1} I.$$

Therefore, from (3.135) for any $x \in B(x^*, \delta)$, we have

$$\lambda_{(r)}(x) = \langle H^{-1}(x)\nabla f(x), \nabla f(x)\rangle^{0.5} \geq (L + \|\nabla f(x)\|)^{-0.5}\|\nabla f(x)\|,$$

which together with (3.134) leads to

$$\lambda_{(r)}(x) \geq m_0(L + \|\nabla f(x)\|)^{-0.5}\|x - x^*\|.$$

Then from $\lambda_{(r)}(x) \leq \varepsilon^{1.5}$ and (3.136) follows

$$m_0(L + \|\nabla f(x)\|)^{-0.5}\varepsilon \geq \varepsilon^{1.5} \geq \lambda_{(r)}(x) \geq m_0(L + \|\nabla f(x)\|)^{-0.5}\|x - x^*\|$$

or

$$\|x - x^*\| \leq \varepsilon, \forall x \in B(x^*, \delta).$$

Therefore, $\lambda_{(r)}(x) \leq \varepsilon^{1.5}$ can be used as a stopping criteria.

We are ready to describe DRNM.

Let $x_0 \in \mathbb{R}^n$ be a starting point and $0 < \varepsilon < \delta$ be the required accuracy, set $x := x_0$.

1. Compute regularized Newton direction $r(x)$ by solving the system (3.112);
2. if the following inequality

$$f(x + tr(x)) \leq f(x) + 0.5\langle \nabla f(x), r(x)\rangle \tag{3.138}$$

holds, then set $t(x) := 1$, otherwise set $t(x) := (2L)^{-1}\|\nabla f(x)\|$;

3. $x := x + t(x)r(x)$;
4. if $\lambda_r(x) \leq \varepsilon^{1.5}$, then $x^* := x$, otherwise go to 1.

We consider the global convergence and complexity of DRNM in the following section.

3.6.9 The Complexity of DRNM

We assume that conditions (3.82) and (3.83) are satisfied. Due to (3.82) the solution x^* is unique. Hence, from convexity f follows that for any given starting point $x_0 \in \mathbb{R}^n$ the sublevel set $\mathscr{L}_0$ is bounded; therefore, there is $L > 0$ such that (3.93) holds on $\mathscr{L}_0$.

Let $\mathbb{B}(x^*, r) = \{x \in \mathbb{R}^n : \|x - x^*\| \leq r\}$ be the ball with center x^* and radius $r > 0$ and $r_0 = \min\{r : \mathscr{L}_0 \subset \mathbb{B}(x^*, r)\}$.

Theorem 3.16. *If (3.82) and (3.83) are satisfied and $\delta = \frac{2m}{3M+2L}$, then from any given starting point $x_0 \in \mathscr{L}_0$, it takes*

$$N_0 = \frac{1}{2}\left(\frac{L^2(3M+2L)^3}{(m_0 m)^3}(1+r_0)(f(x_0) - f(x^*))\right) \tag{3.139}$$

DRN steps to get $x \in \mathbb{B}(x^*, \delta)$.

Proof. For the regularized Newton directional derivative, we have

$$\begin{aligned} \frac{df(x+tr(x))}{dt}|_{t=0} &= \langle \nabla f(x), r(x) \rangle = \\ &-\big\langle (\nabla^2 f(x) + \|\nabla f(x)\| I) r(x), r(x) \big\rangle \leq \\ &- (m(x) + \|\nabla f(x)\|)\|r(x)\|^2, \end{aligned} \tag{3.140}$$

where $m(x) \geq 0$ and $\|\nabla f(x)\| > 0$ for any $x \neq x^*$. It means that RND is a decent direction at any $x \in \mathscr{L}_0$ and $x \neq x^*$.

It follows from (3.140) that $\varphi(t) = f(x + tr(x))$ is monotone decreasing for small $t > 0$.

From the convexity of f follows that $\varphi'(t) = \langle \nabla f(x + tr(x)), r(x) \rangle$ is not decreasing in $t > 0$; hence, at some $t = t(x)$ we have

$$\langle \nabla f(x + t(x)r(x)), r(x) \rangle \geq -\frac{1}{2}(m(x) + \|\nabla f(x)\|)\|r(x)\|^2, \tag{3.141}$$

otherwise $\inf f(x) = -\infty$, which is impossible due to the boundedness of $\mathscr{L}_0$.

From (3.140) and (3.141), we have

$$\langle \nabla f(x + t(x)r(x)) - \nabla f(x), r(x) \rangle \geq \frac{m(x) + \|\nabla f(x)\|}{2}\|r(x)\|^2.$$

Therefore, there exists $0 < \theta(x) < 1$ such that

$$t(x)\langle \nabla^2 f(x+\theta(x)t(x)r(x)), r(x)\rangle = t(x)\langle \nabla^2 f(\cdot)r(x), r(x)\rangle$$
$$\geq \frac{m(x)+\|\nabla f(x)\|}{2}\|r(x)\|^2$$

or

$$t(x)\|\nabla^2 f(\cdot)\|\|r(x)\|^2 \geq \frac{m(x)+\|\nabla f(x)\|}{2}\|r(x)\|^2.$$

Keeping in mind $\|\nabla^2 f(\cdot)\| \leq L$, we obtain

$$t(x) \geq \frac{m(x)+\|\nabla f(x)\|}{2L} \geq \frac{\|\nabla f(x)\|}{2L}. \tag{3.142}$$

It means that for $t \leq \frac{\|\nabla f(x)\|}{2L}$, the inequality

$$\frac{df(x+tr(x))}{dt} \leq -\frac{1}{2}\langle \nabla f(x), r(x)\rangle$$

holds; hence,

$$\Delta f(x) = f(x) - f(x+t(x)r(x)) \geq$$
$$\frac{1}{2}t(x)\langle -\nabla f(x), r(x)\rangle = \frac{1}{2}t(x)\lambda_r^2(x). \tag{3.143}$$

Therefore, finding the lower bound for the reduction of f at any $x \in \mathcal{L}_0$ such that $x \notin B(x^*, \delta)$, we have to find the corresponding bound for the regularized Newton decrement.

Now let us consider $x \in B(x^*, \delta)$, then from (3.133) follows

$$\langle \nabla f(x) - \nabla f(x^*), x - x^*\rangle \geq m_0\|x - x^*\|^2 \tag{3.144}$$

for any $x \in B(x^*, \delta)$.

Let $\hat{x} \notin B(x^*, \delta)$, we consider a segment $[x^*, \hat{x}]$. There is $0 < \widetilde{t} < 1$ such that $\widetilde{x} = (1-\widetilde{t})x^* + \widetilde{t}\hat{x} \in \partial B(x^*, \delta)$.

From the convexity f follows

$$\langle \nabla f(x^* + t(\hat{x} - x^*)), \hat{x} - x^*\rangle|_{t=0} \leq \langle \nabla f(x^* + t(\hat{x} - x^*)), \hat{x} - x^*\rangle|_{t=\widetilde{t}} \leq$$
$$\langle \nabla f(x^* + t(\hat{x} - x^*), \hat{x} - x^*\rangle|_{t=1},$$

or

$$0 = \langle \nabla f(x^*), \hat{x} - x^*\rangle \leq \langle \nabla f(\widetilde{x}), \hat{x} - x^*\rangle \leq \langle \nabla f(\hat{x}), \hat{x} - x^*\rangle.$$

The right inequality can be rewritten as follows:

$$\langle \nabla f(\widetilde{x}), \hat{x} - x^*\rangle = \frac{\|\hat{x} - x^*\|}{\delta}\langle \nabla f(\widetilde{x}) - \nabla f(x^*), \widetilde{x} - x^*\rangle \leq \langle \nabla f(\hat{x}), \hat{x} - x^*\rangle.$$

In view of (3.144), we obtain

$$\|\nabla f(\hat{x})\|\|\hat{x}-x^*\| \geq \frac{\|\hat{x}-x^*\|}{\delta}\langle \nabla f(\widetilde{x}) - \nabla f(x^*), \widetilde{x}-x^*\rangle \geq \frac{\|\hat{x}-x^*\|}{\delta} m_0\|\widetilde{x}-x^*\|^2.$$

Keeping in mind that $\widetilde{x} \in \partial B(x^*, \delta)$, we get

$$\|\nabla f(\hat{x})\| \geq m_0\|\widetilde{x}-x^*\| = \frac{2m_0 m}{3M+2L}. \tag{3.145}$$

On the other hand, from (3.123) and $\hat{x} \in \mathscr{L}_0$ follows

$$\|\nabla f(\hat{x})\| = \|\nabla f(\hat{x}) - \nabla f(x^*)\| \leq L\|\hat{x}-x^*\| \leq Lr_0. \tag{3.146}$$

From (3.93) follows

$$\nabla^2 f(x) \preceq LI. \tag{3.147}$$

For any $\hat{x} \notin S(x^*, \delta)$, we have $\|\nabla f(\hat{x})\| > 0$; therefore, $H(\hat{x}) = \nabla^2 f(\hat{x}) + \|\nabla f(\hat{x})\| I$ is positive definite, and system (3.112) has a unique solution

$$r(\hat{x}) = -H^{-1}(\hat{x})\nabla f(\hat{x}).$$

Moreover from (3.147) follows

$$(\nabla^2 f(\hat{x}) + \|\nabla f(\hat{x})\| I) \preceq (L + \|\nabla f(\hat{x})\|)I.$$

Therefore,

$$H^{-1}(\hat{x}) \succeq (L + \|\nabla f(\hat{x})\| I)^{-1} I. \tag{3.148}$$

For the regularized Newton decrement, we obtain

$$\lambda_{(r)}(\hat{x}) = \langle H^{-1}(\hat{x})\nabla f(\hat{x}), \nabla f(\hat{x})\rangle^{0.5} \geq (L + \|\nabla f(\hat{x}\|)^{-0.5}\|\nabla f(\hat{x})\|. \tag{3.149}$$

Keeping in mind

$$\|\nabla f(\hat{x})\| = \|\nabla f(\hat{x}) - \nabla f(x^*)\| \leq L\|\hat{x}-x^*\|$$

from (3.142), (3.146), and (3.149) and definition of r_0, we obtain

$$\Delta f(\hat{x}) \geq \frac{1}{2} t(\hat{x})\lambda_r^2(\hat{x}) \geq \frac{\|\nabla f(\hat{x})\|^3}{4L}(L + \|\nabla f(\hat{x})\|)^{-1} \geq \frac{\|\nabla f(\hat{x})\|^3}{4L^2(1+r_0)}.$$

Using (3.145) we get

$$\Delta f(\hat{x}) \geq \left(\frac{2m_0 m}{3M+2L}\right)^3 \frac{1}{4L^2(1+r_0)}$$

$$= \frac{2(m_0 m)^3}{(3M+2L)^3 L^2}\frac{1}{(1+r_0)}.$$

Therefore, it takes

$$N_0 = (f(x_0) - f(x^*))(\Delta f(\hat{x}))^{-1} = \frac{1}{2}\frac{(3M+2L)^3L^2(1+r_0)}{(m_0 m)^3}(f(x_0) - f(x^*))$$

steps to obtain $x \in \mathbb{B}(x^*, \delta)$ from a given $x_0 \in \mathcal{L}_0$. □

From (3.125) follows that it takes $O(\ln\ln\varepsilon^{-1})$ DRN steps to find an ε-approximation for x^* from any $x \in \mathbb{B}(x^*, \delta)$.

Therefore, the total number of DRN steps required for finding an ε-approximation for x^* from a given starting point $x_0 \in \mathbb{R}^n$ is

$$N = N_0 + O(\ln\ln\varepsilon^{-1}).$$

3.6.10 Newton's Method as an Affine Invariant

Bounds (3.96) and (3.139) depend on the size of Newton and the regularized Newton areas, which, in turn, are defined by convexity constant $m > 0$ and smoothness constants $M > 0$ and $L > 0$. The convexity and smoothness constants are dependent on the given system of coordinate.

Let us consider an affine transformation of the original system given by $x = Ay$, where $A \in \mathbb{R}^{n\times n}$ is a nondegenerate matrix. We obtain $\varphi(y) = f(Ay)$.

Let $\{x_s\}_{s\in N}$ be the sequence generated by Newton's method

$$x_{s+1} = x_s - (\nabla^2 f(x_s))^{-1}\nabla f(x_s).$$

For the correspondent sequence in the transformed space, we obtain

$$y_{s+1} = y_s - (\nabla^2 \varphi(y_s))^{-1}\nabla\varphi(y_s).$$

Let $y_s = A^{-1}x_s$ for some $s \geq 0$, then

$$y_{s+1} = y_s - (\nabla^2\varphi(y_s))^{-1}\nabla\varphi(y_s) = y_s - [A^T\nabla^2 f(Ay_s)A]^{-1}A^T\nabla f(Ay_s) =$$

$$A^{-1}x_s - A^{-1}(\nabla^2 f(x_s))^{-1}\nabla f(x_s) = A^{-1}x_{s+1}.$$

It means that Newton's method is affine invariant with respect to the transformation $x = Ay$. Therefore, the areas of quadratic convergence depends only on the local topology of f.

To get the Newton sequence in the transformed space, one needs to apply A^{-1} to the elements of the Newton original sequence.

Let N such that $x_N : \|x_N - x^*\| \leq \varepsilon$, then

$$\|y_N - y^*\| \leq \|A^{-1}\|\|x_N - x^*\|.$$

From (3.84) follows

$$\|x_{N+1} - x^*\| \le \frac{M}{2(m - M\|x_s - x^*\|)} \|x_N - x^*\|^2.$$

Therefore,

$$\|y_{N+1} - y^*\| \le \|A^{-1}\| \|x_{N+1} - x^*\| \le \frac{1}{2}\|A^{-1}\| \frac{M}{(m - M\varepsilon)} \varepsilon^2.$$

Hence, for

$$\varepsilon \le \frac{m}{M}(1 + 0.5\|A^{-1}\|)^{-1}$$

we have

$$\|y_{N+1} - y^*\| \le \varepsilon.$$

We would like to emphasize that bound (3.139) is global, while conditions (3.82) and (3.83) under which the bound holds are local, at the neighborhood of x^*.

Notes

3.1 For optimality condition, see Section 1.2.1 from Nesterov (2004), Ch. 1.2 from B. Polyak (1987).

3.2 For non-smooth unconstrained optimization, see Ermoliev and Shor (1967), Shor (1998), Goffin (1977), Lemarechal (1980), Mordukhovich (2018), Ch. 3 in Nesterov (2004), Nesterov (1984), B. Polyak (1967, 1987).

3.3 For the gradient method, see Ch. 9 in Boyd and Vanderberghe (2004), Dennis and Schnabel (1996), Hestenes (1969), Pshenichnyj (1994), Kantorovich and Akilow (1964), Ch. 1.2 in Nesterov (2004), B. Polyak (1963), Ch. 1.4 in B. Polyak (1987).

For fast gradient method, see Nesterov (1983) and Nesterov (2004). For FISTA algorithm – important addition to the fast gradient method – see Beck and Teboulle (2009) and Beck and Teboulle (2009a).

3.4 For regularization method see Tikhonov (1963) and Tikhonov and Arsenin (1977).

For quadratic proximal point methods, see Bauschke et al. (2004), Bertsekas (1982), Bertsekas (1999), Goldshtein and Tretiakov (1989), Guler (1991), Guler (1992), Lemarechal (1980), Martinet (1978), Martinet (1970), Moreau (1965), B. Polyak (1987), Reich and Sabach (2010), Rockafellar (1976), Rockafellar (1976a).

For Newton's method see Arnold (1963), Bertsekas (1999), Boyd and Vanderberghe (2004), Dennis and Schnabel (1996), Kantorovich and Akilow (1964), Nesterov (2004), B. Polyak (1987, 2007).

For regularized Newton method, see Polyak (2009) and Polyak (2018).

Chapter 4
Optimization with Equality Constraints

4.0 Introduction

In 1797, Lagrange published in Paris his treatise on the Theory of Analytic Functions. In this fundamental book, he describes his method for solving constrained optimization problems:

"...when a function of several variables has to be minimized or maximized, and there are one or more equations in the variables, it will be sufficient to add the given function and the functions that must equal zero, each one multiplied by an indeterminate quantity, and then to look for the maximum or the minimum as if the variables were independent; the equations that will be found, together with the given equations, will serve for determining all the unknowns."

Obviously, Lagrange multipliers rule is only a necessary condition for equality constrained optimum. The primal–dual vector, which solves Lagrange system, is neither a maximum nor a minimum of the Lagrangian; it is a saddle point. Therefore we have the beautiful duality theory.

Still, for more than 200 years, both Lagrangian and Lagrange multipliers remain the most important tools in constrained optimization.

Lagrangian is the main instrument for establishing optimality conditions, the basic ingredient of the Lagrange duality and the most important tool in numerical constrained optimization. Almost all constrained optimization methods one way or another using Lagrange multipliers.

In the first part of the chapter, we consider the first and second order optimality conditions for constrained optimization problems with equality constraints and with both inequality constraints and equations.

In the second part, we consider penalty, gradient, Newton and Augmented Lagrangian methods for equality-constrained optimization (ECO). We conclude the chapter by considering the primal–dual augmented Lagrangian method for ECO.

R. A. Polyak, *Introduction to Continuous Optimization*, Springer Optimization and Its Applications 172, https://doi.org/10.1007/978-3-030-68713-7_4

4.1 Lagrangian and First-Order Optimality Condition

Let $f : \mathbb{R}^n \to \mathbb{R}$ be continuously differentiable function, C be full rank $m \times n$ matrix $(n > m)$, and vector $a \in \mathbb{R}^m$. We consider the following ECO problem

$$f(u^*) = \min\{f(u)|\, u \in \Omega\}, \tag{4.1}$$

where $\Omega = \{u : c(u) = Cu - a = 0\}$. By splitting matrix $C = [A, B]$ and vector $u = (x; y)$, we can rewrite the constraints as follows

$$c(u) = Cu - a = Ax + By - a = 0, \tag{4.2}$$

where $A : \mathbb{R}^{n-m} \to \mathbb{R}^m, B : \mathbb{R}^m \to \mathbb{R}^m, x \in \mathbb{R}^{n-m}$, *and* $y \in \mathbb{R}^m$.

From rank $B = m$ follows that the system (4.2) can be solved for y

$$y = B^{-1}(-Ax + a).$$

The restriction $\varphi : \mathbb{R}^{n-m} \to \mathbb{R}$, of f on Ω, is given by the following formula $\varphi(x) = f(x, B^{-1}(-Ax + a))$.

Finding $f(u^*)$ is equivalent to finding

$$\varphi(x^*) = \min\{\varphi(x)|\, x \in \mathbb{R}^{n-m}\} \tag{4.3}$$

From Fermat theorem follows that minimizer x^* solves the following system

$$\begin{aligned} \nabla\varphi(x) &= \nabla_x f(x,y) + \nabla_y f(x,y)\nabla_x y(x) \\ &= \nabla_x f(x,y) - \nabla_y f(x,y)B^{-1}A = 0. \end{aligned} \tag{4.4}$$

Let row vector $\lambda = \nabla_y f(x,y)B^{-1}$, then

$$\nabla_y f(x,y) - \lambda B = 0 \tag{4.5}$$

From (4.4) follows

$$\nabla\varphi(x) = \nabla_x f(x,y) - \lambda A = 0,$$

which together with (4.5) leads to

$$\nabla f(u) - \lambda C = 0. \tag{4.6}$$

The system (4.6) together with (4.2) comprises the first-order necessary condition for u^* to be a local minimizer of the problem (4.1).

Let us consider the Lagrangian

$$L(u, \lambda) = f(u) - \sum_{i=1}^{m} \lambda_i c_i(u) = f(u) - \langle \lambda, Cu - a \rangle,$$

for the problem (4.1), and then systems (4.6) and (4.2) can be rewritten as the following Lagrange system of equations

$$\begin{cases} \nabla_u L(u,\lambda) = \nabla f(u) - \lambda C = 0 \\ \nabla_\lambda L(u,\lambda) = -Cu + a = -Ax - By + a = 0, \end{cases} \tag{4.7}$$

which is the first-order necessary condition for u^* to be a local minimizer of ECO (4.1).

The key element of our arguments was the ability to solve the linear system (4.2) for y. For this we need $\operatorname{rank} C = m$. We will use similar arguments for establishing the necessary condition for u^* to be a local minimizer of the following ECO problem with nonlinear constrains, that is

$$f(u^*) = \min\{f(u) |\, u \in \Omega\}, \tag{4.8}$$

where $\Omega = \{u \in \mathbb{R}^n : c(u) = 0\}$ and $c(u) = (c_1(u), \ldots, c_m(u))^T$ $(n > m)$.

We will call u^* regular local minimizer in (4.8) if the Jacobian

$$J(c(u^*)) \equiv \nabla c(u^*) = \begin{pmatrix} \nabla c_1(u^*) \\ \cdots \\ \nabla c_m(u^*) \end{pmatrix} : \mathbb{R}^n \to \mathbb{R}^m$$

is a full rank matrix, that is

$$\operatorname{rank} \nabla c(u^*) = m \tag{4.9}$$

and $\nabla c(u)$ is continuous in $u \in B(u^*, \varepsilon)$.

The following Theorem establishes the first-order necessary condition for u^* to be a regular local minimizer of the ECO (4.8).

Theorem 4.1. *If f and all c_i, $i = 1,..,m$ are continuously differentiable and u^* is a regular local minimizer in (4.8), then there is a vector $\lambda^* = (\lambda_1^*, \ldots, \lambda_m^*)$ that the pair (u^*, λ^*) solves the following Lagrange system*

$$\nabla_u L(u;\lambda) = \nabla f(u) - \sum_{i=1}^{m} \lambda_i \nabla c_i(u) = 0 \tag{4.10}$$

$$\nabla_\lambda L(u;\lambda) = -c(u) = 0. \tag{4.11}$$

Proof. Let us split vector u^* into two vectors $x^* \in \mathbb{R}^{n-m} and y^* \in \mathbb{R}^m$ and the matrix $\nabla c(u^*)$ into two matrices $\nabla_x c(u^*) : \mathbb{R}^{n-m} \to \mathbb{R}^m$ and $\nabla_y c(u^*) : \mathbb{R}^m \to \mathbb{R}^m$. From (4.9) we have

$$\operatorname{rank} \nabla_y c(u^*) = m. \tag{4.12}$$

From (4.12), continuity of ∇c_i, $i = 1, \ldots, m$ and the implicit function theorem (see Appendix) follows that for a small $\varepsilon > 0$ in the neighborhood $B(u^*, \varepsilon)$, the system $c(x,y) = 0$ uniquely defines $y(x) = (y_1(x), \ldots, y_m(x))^T$, that $y(x^*) = y^*$ and

$$c(x, y(x)) \equiv 0, \forall (x, y(x)) \in B(u^*, \varepsilon). \tag{4.13}$$

By differentiating (4.13), we obtain

$$\nabla_x c(x,y(x)) + \nabla_y c(x,y(x))\nabla_x y(x) = \nabla_x c(\cdot) + \nabla_y c(\cdot)\nabla_x y(\cdot) = 0. \tag{4.14}$$

Keeping in mind (4.12) and continuity of ∇c_i, $i = 1,..,m$ for small $\varepsilon > 0$ and any $(x,y(x)) \in B(u^*,\varepsilon)$ from (4.14) follows

$$\nabla_x y(\cdot) = -(\nabla_y c(\cdot))^{-1}\nabla_x c(\cdot). \tag{4.15}$$

For $x = x^*$ we have

$$\nabla_x y(x^*) = -(\nabla_y c(x^*,y^*))^{-1}\nabla_x c(x^*,y^*) = -(\nabla_y c(u^*))^{-1}\nabla_x c(u^*). \tag{4.16}$$

Let us consider the restriction

$$\varphi(x) = f(x,y(x)) \tag{4.17}$$

of f on the nonlinear manifold $Q = \{u \in \mathbb{R}^n : c(u) = 0\}$ at the neighborhood of u^*.

For $u^* = (x^*,y^*)$ to be the local minimizer in (4.8), it is necessary for x^* to be local minimizer of $\varphi(x) = f(x;y(x))$. Keeping in mind (4.15), we obtain

$$\begin{aligned}\nabla\varphi(x^*) &= \nabla_x f(u^*) + \nabla_y f(u^*)\nabla_x y(x^*) \\ &= \nabla_x f(u^*) - \nabla_y f(u^*)(\nabla_y c(u^*))^{-1}\nabla_x c(u^*) = 0.\end{aligned} \tag{4.18}$$

By introducing the LM row vector

$$\lambda^* = \lambda(u^*) = \nabla_y f(u^*)(\nabla_y c(u^*))^{-1} = 0 \tag{4.19}$$

we obtain

$$\nabla_y f(u^*) - \lambda^*\nabla_y c(u^*) = 0. \tag{4.20}$$

From (4.19) follows

$$\nabla\varphi(x^*) = \nabla_x f(u^*) - \lambda^*\nabla_x c(u^*) = 0. \tag{4.21}$$

Combining (4.20) and (4.21), we obtain

$$\nabla_u L(u^*,\lambda^*) = \nabla f(u^*) - \lambda^*(\nabla c(u^*)) = \nabla f(u^*) - \sum_{i=1}^{m}\lambda_i^*\nabla c_i(u^*) = 0, \tag{4.22}$$

which together with

$$\nabla_\lambda L(u^*,\lambda^*) = -c(u^*) = 0 \tag{4.23}$$

comprises the first-order necessary condition for u^* to be a local minimizer in ECO (4.8).

□

Another proof of Theorem 4.1 is based on L. Lusternik's (1899–1981) theorem, which we will also use later to prove second-order necessary and sufficient optimality condition for u^* to be a regular minimizer in (4.8).

Let $\Omega \subset \mathbb{R}^n$ and $u \in \Omega$, then a vector $d \in \mathbb{R}^n$ is a tangent vector to Ω at u if for all small enough $\tau > 0$ one can find a curve

$$u(\tau) \in \Omega : \|u(\tau) - (u + \tau d)\| = o(\tau).$$

If Ω is convex than any feasible direction $d : u + \tau d \in \Omega, \tau > 0$ is a tangent vector to Ω, but not every tangent vector d is feasible. Consider

$$\Omega = \{u : u_1^2 + u_2^2 \leq 1\},$$

then at $u = (\frac{\sqrt{2}}{2}, \frac{\sqrt{2}}{2}) \in \partial\Omega$ any vector $d \in \mathbb{R}^2 : d_1 + d_2 \leq 0$ is tangent to Ω, but only $d \in \mathbb{R}^2 : d_1 + d_2 < 0$ are feasible.

Let $T_\Omega(u^*)$ be the set of vectors tangent to Ω at u^*. The following theorem characterize the set $T_\Omega(u^*)$.

Theorem 4.2 (L. A. Lusternik). *Let $\Omega = \{u \in \mathbb{R}^n : c_i(u) = 0, i = 1, \ldots, m\}$, $u^* \in \Omega$, the gradients $\nabla c_i(u^*)$, $i = 1, \ldots, m$ are linear independent and continuous in $B(x^*, \varepsilon)$, then*

$$T_\Omega(u^*) = \{d \in \mathbb{R}^n : \langle d, \nabla c_i(u^*)\rangle = 0, i = 1, \ldots, m\},$$

that is the tangent vectors to Ω at the point u^* is a subspace orthogonal to the vectors $\nabla c_i(u^*)$, $i = 1, \ldots, m$.

Along with Lusternik's theorem, see Lusternik (1934), we need the following Lemma.

Lemma 4.1. *Let $A \in \mathbb{R}^{n\times m}$, $T_0 = \{u \in \mathbb{R}^n : Au = 0\}$, and $\langle c, u\rangle \geq 0, \forall u \in T_0$, then there is a vector $\mathrm{v} \in \mathbb{R}^m : c = A^T\mathrm{v}$ and $\langle c, u\rangle = 0, \forall u \in T_0$.*

Proof. We consider

$$T = \{u \in \mathbb{R}^n : u = A^T\mathrm{v}, \mathrm{v} \in \mathbb{R}^m\},$$

which is convex and closed as a subspace of $\mathbb{R}^n$. Therefore if $c \bar{\in} T$, then c can be strongly separated from T, i.e., there is $a \in \mathbb{R}^n$ that $\langle a, c\rangle < 0$ and $\langle a, u\rangle \geq 0$, $\forall u \in T$. Then

$$0 \leq \langle a, u\rangle = \langle a, A^T\mathrm{v}\rangle = \langle Aa, \mathrm{v}\rangle, \forall \mathrm{v} \in \mathbb{R}^m,$$

which is possible only if $Aa = 0$, so $a \in T_0$, but it is impossible because $\langle a, c\rangle < 0$.

The obtained controversy shows that $c \in T$; therefore $c = A^T\mathrm{v}$ and $\langle c, u\rangle = \langle A^T\mathrm{v}, u\rangle = \langle \mathrm{v}, Au\rangle = 0, \forall u \in T_0$. □

Now let us prove Theorem 4.1, using Lusternik's theorem and Lemma 4.1.

For any tangent to Ω at the point u^* vector d, there exists curve $u(\tau)$ such that $c_i(u(\tau)) = 0, i = 1, \ldots, m$, and $\|u^* + \tau d - u(\tau)\| = o(\tau)$. Therefore, we have

$$f(u(\tau)) = f(u^* + \tau d + o(\tau)e) = f(u^*) + \tau\langle\nabla f(u^*), d\rangle + o(\tau),$$

where $e = (1,..,1) \in \mathbb{R}^n$. For a local minimizer u^* and small $\tau > 0$, we have $f(u(\tau)) \geq f(u^*)$; therefore

$$\langle \nabla f(u^*), d \rangle \geq 0, \forall d \in T(u^*). \tag{4.24}$$

From Lusternik's theorem follows that the set of tangent vectors to Ω at u^* is given by

$$T_\Omega(u^*) = \{d \in \mathbb{R}^n : \langle \nabla c_i(u^*), d \rangle = 0, i = 1, \ldots, m\}.$$

It means that (4.24) holds for any $d \in T_\Omega(u^*)$; therefore, due to Lemma 4.1, there exists λ^* that

$$\nabla f(u^*) = \nabla c(u^*)^T \lambda^*$$

or

$$\nabla f(u^*) - \sum_{i=1}^{m} \lambda_i^* \nabla c_i(u^*) = 0, \tag{4.25}$$

which along with $c(u^*) = 0$ leads to the Lagrangian system (4.10)–(4.11). We saw that the regularity of u^* is a substantial part of Theorem 4.1.

The following Fritz John's theorem (1949) does not require regularity condition.

Theorem 4.3 (F. John). *If u^* is the local minimizer in (4.1), functions f and c_i continuous differentiable in the neighborhood of u^*, then there exists $\bar{\lambda}_0, \bar{\lambda}_1, \ldots, \bar{\lambda}_m$ not at all equal zero that*

$$\bar{\lambda}_0 \nabla f(u^*) - \sum_{i=1}^{m} \bar{\lambda}_i \nabla c_i(u^*) = 0. \tag{4.26}$$

John's theorem follows immediately from Theorem 4.1.

In fact, if u^* is a regular minimizer, then (4.26) follows from (4.25) with $\bar{\lambda}_0 = 1, \lambda_i = \lambda_i^*$.

If u^* is not a regular minimizer, then $\nabla c_i(u^*)$, $i = 1, \ldots, m$ are linear dependent, that is, there exists vector $\mu = (\mu_1, \ldots, \mu_m)$ that

$$\sum_{i=1}^{m} \mu_i \nabla c_i(u^*) = 0$$

and not all μ_i equal zero. Therefore (4.26) holds with $\bar{\lambda}_0 = 0$, $\lambda_i^* = \mu_i$, $i = 1, \ldots, m$.

The regularity of the minimizer u^* guarantees $\bar{\lambda}_0 \neq 0$, and (4.25) means that $\nabla f(u^*)$ belongs to the subspace orthogonal to the tangent subspace $T_\Omega(u^*)$.

4.2 Second-Order Necessary and Sufficient Optimality Condition

In this paragraph, we consider the second-order optimality conditions for ECO problem (4.1). We start with the following second-order necessary optimality condition.

Theorem 4.4. *Let u^* be the regular minimizer and f and all c_i, $i = 1,\dots,m$ be twice continuously differentiable in $B(u^*,\varepsilon)$, and then there is a vector $\lambda^* = (\lambda_1^*,\dots,\lambda_m^*)^T$ that*

$$\langle \nabla^2_{uu} L(u^*,\lambda^*)d,d\rangle \geq 0$$

for

$$\forall d \in T_\Omega(u^*) = \{d : \langle \nabla c_i(u^*),d\rangle = 0, i = 1,\dots,m\}.$$

Proof. Due to Lusternik's theorem for any $d \in T_\Omega(u^*)$, there is a curve $u(\tau)$, that for small $\tau > 0$, $\|u^* + \tau d - u(\tau)\| = o(\tau)$, and $c_i(u(\tau)) = 0$. Therefore, for the minimizer u^*, we obtain

$$f(u^*) \leq f(u(\tau)) = L(u(\tau),\lambda^*) = L(u^*,\lambda^*) + \langle \nabla_u L(u^*,\lambda^*), u(\tau) - u^*\rangle \quad (4.27)$$
$$+\frac{1}{2}\langle \nabla^2_{uu} L(u^*,\lambda^*)(u(\tau)-u^*), u(\tau)-u^*\rangle + o(\tau^2).$$

From $f(u^*) = L(u^*,\lambda^*)$, (4.25) and (4.27) follow

$$\langle \nabla^2_{uu} L(u^*,\lambda^*)d,d\rangle \geq 0, \forall d \in T_\Omega(x^*). \quad (4.28)$$

□

The second-order sufficient condition for u^* to be an isolated local minimizer in ECO (4.8) follows from the correspondent condition for the unconstrained optimization problem

$$\varphi(x^*) = \min\{\varphi(x) \mid x \in \mathbb{R}^{n-m}\},$$

that is

$$\langle \nabla^2_{xx}\varphi(x^*)\mathrm{v}_x, \mathrm{v}_x\rangle > 0, \forall \mathrm{v}_x \in \mathbb{R}^{n-m}. \quad (4.29)$$

The following theorem establishes such condition.

Theorem 4.5. *Let f and all $c_i, i = 1,\dots,m$ be twice continuously differentiable in $B(u^*,\varepsilon)$, then the following condition*

$$\begin{aligned} &(a)\ \langle \nabla^2_{uu} L(u^*,\lambda^*)\mathrm{v},\mathrm{v}\rangle > 0, \forall \mathrm{v} : \nabla c(u^*)\mathrm{v} = 0 \\ &(b)\ \operatorname{rank} \nabla c(u^*) = m \end{aligned} \quad (4.30)$$

is sufficient for u^* to be an isolated minimizer in ECO problem (4.8).

Proof. From rank $\nabla c(u^*) = m$, smoothness of ∇f, ∇c_i, $i = 1,\ldots,m$, and the implicit function theorem, see Appendix, follows that the system

$$c(x,y) = 0,$$

in the neighborhood $B(u^*,\varepsilon)$ uniquely defines vector—function $y(x) = (y_1(x),\ldots, y_m(x))^T$: $c(x,y(x)) \equiv 0$ and $y^* = y(x^*)$.

For the gradient of the restriction, we obtain

$$\nabla\varphi(x) = \nabla_x f(\cdot) + \nabla_y f(\cdot)\nabla_x y(\cdot),$$

where $\nabla_x y(x) = J(y(x))$ is Jacobian of the vector–function $y(x)$.

Now let us consider the Hessian of the restriction $\varphi(x)$ at $x = x^*$.

We obtain

$$\begin{aligned}\nabla^2_{xx}\varphi(x^*) &= \nabla^2_{xx}\varphi = \nabla^2_{xx}f + \nabla^2_{xy}f\nabla_x y \\ &\quad + \nabla_x y^T(\nabla^2_{yx}f + \nabla^2_{yy}f\nabla_x y) \\ &\quad + \sum_{j=1}^m \frac{\partial f}{\partial y_j}\nabla^2_{xx}y_j.\end{aligned}$$

Let

$$\bar{H}_0 = \nabla^2_{xx}f + \nabla^2_{xy}f\nabla_x y + (\nabla_x y)^T(\nabla^2_{yx}f + \nabla^2_{yy}f\nabla_x y),$$

then

$$\nabla^2_{xx}\varphi(x^*) = \bar{H}_0 + \sum_{j=1}^m \frac{\partial f}{\partial y_j}\nabla^2_{xx}y_j \tag{4.31}$$

To determine Hessians $\nabla^2_{xx}y_j$, $j = 1,\ldots,m$, we use the identities

$$c_i(\cdot) \equiv c_i(x,y(x)) \equiv 0,\ i = 1,\ldots,m. \tag{4.32}$$

By differentiating identities (4.32) twice, we obtain

$$\begin{aligned}0 \equiv \nabla^2_{xx}c_i(x^*) &= \nabla^2_{xx}c_i = \nabla^2_{xx}c_i + \nabla^2_{xy}c_i\nabla_x y + (\nabla_x y)^T(\nabla^2_{yx}c_i + \nabla^2_{yy}c_i\nabla_x y) \\ &\quad + \frac{\partial c_i}{\partial y_1}\nabla^2_{xx}y_1 + \cdots + \frac{\partial c_i}{\partial y_m}\nabla^2_{xx}y_m,\ i = 1,\ldots,m.\end{aligned}$$

Let

$$\bar{H}_i = \nabla^2_{xx}c_i + \nabla^2_{xy}c_i\nabla_x y + (\nabla_x y)^T(\nabla^2_{yx}c_i + \nabla^2_{yy}c_i\nabla_x y).$$

The following system of matrix equations can be solved for $\nabla^2_{xx}y_j$, $j = 1,\ldots,m$ because $\det(\nabla_y c(u^*)) \neq 0$:

$$\sum_{j=1}^m \frac{\partial c_i}{\partial y_j}\nabla^2_{xx}y_j = -\bar{H}_i,\ i = 1,\ldots,m. \tag{4.33}$$

Instead, let us multiply both sides of (4.33) by λ_i^*. After summing up the equations in $1 \leq i \leq m$, we obtain

$$\sum_{j=1}^{m} \lambda_i^* \sum_{j=1}^{m} \frac{\partial c_i}{\partial y_j} \nabla_{xx}^2 y_j = -\sum_{i=1}^{m} \lambda_i^* \bar{H}_i. \tag{4.34}$$

From the regularity of u^* follows the first-order necessary conditions (4.10). Therefore we can rewrite (4.34) as follows

$$\sum_{j=1}^{m} \nabla_{xx}^2 y_j \sum_{i=1}^{m} \lambda_i^* \frac{\partial c_i}{\partial y_j} = \sum_{j=1}^{m} \frac{\partial f}{\partial y_j} \nabla_{xx}^2 y_j = -\sum_{i=1}^{m} \lambda_i^* \bar{H}_i. \tag{4.35}$$

From (4.31) and (4.35), we obtain

$$\nabla_{xx}^2 \varphi(x^*) = \bar{H}_0(u^*) - \sum_{i=1}^{m} \lambda_i^* \bar{H}_i(u^*) \tag{4.36}$$

Consider any $\mathrm{v}_x \in \mathbb{R}^{n-r}$, $\mathrm{v}_y = \nabla_x y(x^*) \mathrm{v}_x$, $\mathrm{v} = (\mathrm{v}_x, \mathrm{v}_y)$. Then

$$\begin{aligned}
\langle \bar{H}_0(u^*)\mathrm{v}_x, \mathrm{v}_x\rangle &= \langle \nabla_{xx}^2 f(u^*)\mathrm{v}_x, \mathrm{v}_x\rangle + \langle \nabla_{xy}^2 f(u^*)\nabla_x y(x^*)\mathrm{v}_x, \mathrm{v}_x\rangle \\
&\quad + \langle \nabla_x y^T(x^*)\nabla_{yx}^2 f(u^*)\mathrm{v}_x, \mathrm{v}_x\rangle + \langle \nabla_x y(x^*)^T \nabla_{yy}^2 f(u^*)\nabla_x y(x^*)\mathrm{v}_x, \mathrm{v}_x\rangle \\
&= \langle \nabla_{xx}^2 f(u^*)\mathrm{v}_x, \mathrm{v}_x\rangle + \langle \nabla_{xy}^2 f(u^*)\mathrm{v}_y, \mathrm{v}_x\rangle \langle \nabla_{xx}^2 f(u^*)\mathrm{v}_x, \mathrm{v}_y\rangle \\
&\quad + \langle \nabla_{yy}^2 f(u^*)\mathrm{v}_y, \mathrm{v}_y\rangle = \langle \nabla^2 f(u^*)\mathrm{v}, \mathrm{v}\rangle
\end{aligned}$$

that is

$$\langle \bar{H}_0(u^*)\mathrm{v}_x, \mathrm{v}_x\rangle = \langle \nabla_{xx}^2 f(u^*)\mathrm{v}, \mathrm{v}\rangle. \tag{4.37}$$

Also for $1 \leq i \leq m$, we have

$$\begin{aligned}
\langle \bar{H}_i(u^*)\mathrm{v}_x, \mathrm{v}_x\rangle &= \langle \nabla_{xx}^2 c_i(u^*)\mathrm{v}_x, \mathrm{v}_x\rangle + \langle \nabla_{xy} c_i(u^*)\nabla_x y(x^*)\mathrm{v}_x, \mathrm{v}_x\rangle \\
&\quad + \langle \nabla_x y^T(x^*)\nabla_{yx}^2 c_i(u^*)\mathrm{v}_x, \mathrm{v}_x\rangle + \langle \nabla_x y^T(x^*)\nabla_{yy}^2 c_i(u^*)\nabla_x y(x^*)\mathrm{v}_x, \mathrm{v}_x\rangle \\
&= \langle \nabla_{xx}^2 c_i(u^*)\mathrm{v}_x, \mathrm{v}_x\rangle + \langle \nabla_{xy}^2 c_i(u^*)\mathrm{v}_y, \mathrm{v}_x\rangle + \langle \nabla_{yx}^2 c_i(u^*)\mathrm{v}_x, \mathrm{v}_y\rangle \\
&\quad + \langle \nabla_{yy}^2 c_i(u^*)\mathrm{v}_y, \mathrm{v}_y\rangle = \langle \nabla_{uu}^2 c_i(u^*)\mathrm{v}, \mathrm{v}\rangle,
\end{aligned} \tag{4.38}$$

From (4.36)–(4.38) follows

$$\begin{aligned}
\langle \nabla_{xx}^2 \varphi(x^*)\mathrm{v}_x, \mathrm{v}_x\rangle &= \langle \bar{H}_0(u^*)\mathrm{v}_x, \mathrm{v}_x\rangle - \sum_{i=1}^{m} \lambda_i^* \langle \bar{H}_i(u^*)\mathrm{v}_x, \mathrm{v}_x\rangle \\
&= \langle (\nabla_{uu} f(u^*) - \sum_{i=1}^{m} \lambda^* \nabla_{uu}^2 c_i(u^*))\mathrm{v}, \mathrm{v}\rangle = \langle \nabla_{uu} L(u^*, \lambda^*)\mathrm{v}, \mathrm{v}\rangle
\end{aligned} \tag{4.39}$$

From (4.32) follows

$$\nabla_x c(\cdot) = \nabla_x c(x, y(x)) = \nabla_x c(\cdot) + \nabla_y c(\cdot)\nabla_x y(\cdot) = 0.$$

For any $v_x \in \mathbb{R}^{n-m}$, we obtain

$$\nabla_x c(x^*, y(x^*))v_x = \nabla_x c(u^*)v_x + \nabla_y c(u^*)v_y = \nabla c(u^*)v = 0. \tag{4.40}$$

From (4.39), (4.40), and (4.30a) follows (4.29). Therefore x^* is an isolated local minimizer of $\varphi(x)$ and u^* is an isolated local minimizer of ECO (4.8). □

Let m_0 and M_0 be the min and max eigenvalues of $\nabla^2_{uu} f(u^*)$, m_i and M_i min and max eigenvalues of $\nabla^2_{uu} c_i(u^*)$, g and G min and max eigenvalues of the Gramm matrix $\nabla_x y(x^*)^T \nabla_x y(x^*)$. We consider two sets of indexes $I_+ = \{i : \lambda_i^* > 0\}$ and $I_- = \{i : \lambda_i^* < 0\}$.

We are ready to prove the following eigenvalues theorem.

Theorem 4.6. *Under conditions of Theorem 4.5 and* $m_0 - \sum_{i \in I_+} \lambda_i^* M_i - \sum_{i \in I_-} \lambda_i^* m_i > 0$, *the following matrix inequalities hold*

$$(1+G)(M_0 - \sum_{i \in I_+} \lambda_i^* m_i - \sum_{i \in I_-} \lambda_i^* M_i) I^{n-m} \succeq \tag{4.41}$$

$$\nabla_{xx}\varphi(x^*) \succeq (1+g)(m_0 - \sum_{i \in I_+} \lambda_i^* M_i - \sum_{i \in I_-} \lambda_i^* m_i) I^{n-m},$$

where I^{n-m}- *identical matrix in* $\mathbb{R}^{n-m}$.

Proof. For any $v \in \mathbb{R}^{n-m}$, we have

$$m_0 \langle v, v \rangle \le \langle \nabla^2_{uu} f(u^*) v, v \rangle \le M_0 \langle v, v \rangle \tag{4.42}$$

and

$$m_i \langle v, v \rangle \le \langle \nabla^2_{uu} c_i(u^*) v, v \rangle \le M_i \langle v, v \rangle, \ 1 \le i \le m. \tag{4.43}$$

From (4.43) for $i \in I_+$ follows

$$\lambda_i^* m_i \langle v, v \rangle \le \langle \lambda_i^* H_i(u^*) v, v \rangle \le \lambda_i^* M_i \langle v, v \rangle$$

or

$$-\lambda_i^* M_i \langle v, v \rangle \le -\langle \lambda_i^* H_i(u^*) v, v \rangle \le -\lambda_i^* m_i \langle v, v \rangle \tag{4.44}$$

For $i \in I_-$, we obtain

$$\lambda_i^* m_i \langle v, v \rangle \ge \langle \lambda_i^* H_i(u^*) v, v \rangle \ge \lambda_i^* M_i \langle v, v \rangle$$

or

$$-\lambda_i^* m_i \langle v, v \rangle \le -\langle \lambda_i^* H_i(u^*) v, v \rangle \le -\lambda_i^* M_i \langle v, v \rangle \tag{4.45}$$

From (4.39), (4.44), and (4.45) follows

$$(m_0 - \sum_{i \in I_+} \lambda_i^* M_i - \sum_{i \in I_-} \lambda_i^* m_i) \langle v, v \rangle \le \langle \nabla^2_{xx}\varphi(x^*) v_x, v_x \rangle \le \tag{4.46}$$

$$(M_0 - \sum_{i \in I_+} \lambda_i^* m_i - \sum_{i \in I_-} \lambda_i^* M_i)\langle \mathrm{v}, \mathrm{v}\rangle$$

Let us compute

$$\langle \mathrm{v}, \mathrm{v}\rangle = \langle \mathrm{v}_x, \mathrm{v}_x\rangle + \langle \nabla_x y(x^*)^T \nabla_x y(x^*)\mathrm{v}_x, \mathrm{v}_x\rangle.$$

We obtain

$$(1+g)\langle \mathrm{v}_x, \mathrm{v}_x\rangle \le \langle \mathrm{v}, \mathrm{v}\rangle \le (1+G)\langle \mathrm{v}_x, \mathrm{v}_x\rangle.$$

Keeping in mind $m_0 - \sum_{i\in I_-} \lambda_i^* m_i - \sum_{i \in I_+} \lambda_i^* M_i > 0$ from (4.46) follows (4.41). □

Corollary 4.1. *The following inequality*

$$m_0 - \sum_{i \in I_-} \lambda_i^* m_i - \sum_{i \in I_+} \lambda_i^* M_i > 0$$

is sufficient for u^* to be an isolated minimizer of f on Ω.

Let us consider instead of (4.8) a convex optimization problem

$$f(u^*) = \min\{f(u) |\ c_i(u) \le 0, i = 1, \ldots, m\}, \tag{4.47}$$

where $f : \mathbb{R}^n \to \mathbb{R}$ and all $c_i : \mathbb{R}^n \to \mathbb{R}$ are convex, also $c_1(u^*) = \ldots c_r(u^*) = 0$, $r < m$ be active constraints, so $\lambda_i^* \ge 0$, $i = 1, \ldots, r$, and $\lambda_i^* = 0$, $i = r+1, \ldots, m$ for the passive constraints. Also due to convexity f and c_i, $i = 1, .., m$, we have $0 \le m_0 \le M_0$ and $0 \le m_i \le M_i$, $i = 1, .., m$.

Corollary 4.2. *From (4.41) follows*

$$(1+g)(m_0 + \sum_{i=1}^{r} \lambda_i^* m_i) I^{n-r} \preceq \nabla^2_{xx}\varphi(x^*) \preceq (1+G)(M_0 + \sum_{i=1}^{r} \lambda_i^* M_i) I^{n-r}.$$

In particular, f can be linear ($m_0 = 0$) while φ be strongly convex if at least one of $c_i(x)$, $i = 1, \ldots, r$ is strongly convex and correspondent $\lambda_i^* > 0$.

□

4.3 Optimality Condition for Constrained Optimization Problems with Both Inequality Constraints and Equations

In this section we consider the first- and second-order optimality conditions for problems with both inequality and equality constraints.

Let $f; c_i, i = 1, \ldots, q; g_j, j = 1, .., r$ be continuously differentiable functions in $\mathbb{R}^n$. We are concerned with the following problem

$$\begin{aligned} f(x^*) &= \min f(x) \\ \text{s. t. } c_i(x) \geq 0, i = 1, .., q; g_j(x) &= 0, j = 1, \ldots, r. \end{aligned} \tag{4.48}$$

We assume that along with equality constraints $g_j(x) = 0, j = 1, \ldots, r$, there are $m < q$ active at x^* constraints $c_i(x^*) = 0$, $i = 1, \ldots, m$. Local minimizer x^* in (4.48) is called regular if

$$\text{rank} \begin{bmatrix} \nabla c_{(m)}(x^*) \\ \nabla g(x^*) \end{bmatrix} = m + r, \tag{4.49}$$

where $\nabla c_{(m)} : \mathbb{R}^n \to \mathbb{R}^m$ and $\nabla g : \mathbb{R}^n \to \mathbb{R}^r$ Jacobians of $c_{(m)}(x) = (c_1(x), \ldots, c_m(x))^T$ and $g(x) = (g_1(x), \ldots, g_r(x))^T$. We assume $r + m < n$.

The correspondent Lagrangian $L : \mathbb{R}^n \times \mathbb{R}^q_+ \times \mathbb{R}^r \to \mathbb{R}$ is defined as follows

$$L(x, \lambda, \mathrm{v}) = f(x) - \sum_{i=1}^{q} \lambda_i c_i(x) + \sum_{j=1}^{r} \mathrm{v}_j g_j(x).$$

The necessary first- and second-order optimality conditions are given in the following theorem.

Theorem 4.7. *Let x^* be a regular local minimizer in (4.48), that is,* (4.49) *holds and f and all g_j, c_i continuously differentiable, then:*

(1) There are two uniquely defined Lagrange multipliers vectors $\lambda^* = (\lambda_1^*, \ldots, \lambda_q^*)^T \in \mathbb{R}^q_+$ and $\mathrm{v}^* = (\mathrm{v}_1^*, \ldots, \mathrm{v}_r^*)^T \in \mathbb{R}^r$ that

$$\nabla_x L(x^*, \lambda^*, \mathrm{v}^*) = 0 \tag{4.50}$$

holds and complementarity condition

$$\lambda_i^* c_i(x^*) = 0, \; i = 1, \ldots, q$$

is satisfied.

(2) If the strong complementarity condition holds, that is, $\lambda^*_{(m)} = (\lambda_1^*, \ldots, \lambda_m^*) \in \mathbb{R}^m_{++}$ and all f, c_i, g_j are twice continuously differentiable, then

$$\langle \nabla^2_{xx} L(x^*, \lambda^*, \mathrm{v}^*) y, y \rangle \geq 0, \; \forall y \in T(x^*), \tag{4.51}$$

where

$$T(x^*) = \{ y \in \mathbb{R}^n : \nabla c_{(m)}(x^*) y = 0, \nabla g(x^*) y = 0 \}.$$

Proof.

(1) We will use the following penalty function $P_k : \mathbb{R}^n \to \mathbb{R}$

$$P_k(x) = f(x) - k^{-1} \sum_{i=1}^{q} \ln(1 + k c_i(x)) + \frac{k}{2} \|g(x)\|^2 + \frac{\alpha}{2} \|x - x^*\|^2,$$

where $k > 0$ and $\alpha > 0$.

For a small enough $\varepsilon > 0$, let us consider the sequence $\{x_k\}_{k\in N}$ of minimizers:

$$P_k(x_k) = \min\{P_k(x)|x \in B(x^*,\varepsilon)\},$$

which exists due to Weierstrass theorem. Then

$$P_k(x_k) = f(x_k) - k^{-1}\sum_{i=1}^{q}\ln(1+kc_i(x_k)) + \frac{k}{2}\|g(x_k)\|^2 + \frac{\alpha}{2}\|x_k - x^*\|^2 \quad (4.52)$$
$$\leq f(x^*) - k^{-1}\sum_{i=1}^{q}\ln(1+kc_i(x^*)) \leq f(x^*).$$

The sequence $\{x_k\}_{k\in N}$ belongs to $B(x^*,\varepsilon)$; therefore, there is a converging subsequence $\{x_{k_s}\}_{s\in N}$ that $\lim_{k_s\to\infty} x_{k_s} = \bar{x} \in B(x^*,\varepsilon)$. For k_s large enough from (4.52) follows

$$-k_s^{-1}\sum_{i=1}^{q}\ln(1+k_sc_i(x_{k_s})) + \frac{k_s}{2}\|g(x_{k_s})\|^2 \leq$$
$$2(f(x^*) - f(\bar{x}) - \frac{\alpha}{2}\|\bar{x} - x^*\|^2). \quad (4.53)$$

From $\|\bar{x} - x^*\| \leq \varepsilon$ for small $\varepsilon > 0$ follows: for passive at x^* constraints, there is $\sigma > 0$ that $\min\{c_i(\bar{x})|m+1 \leq i \leq q\} = \sigma$. Therefore

$$\lim_{k_s\to\infty} k_s^{-1}\ln(1+k_sc_i(x_{k_s})) = 0, m+1 \leq i \leq q.$$

For active at x^* constraints, we have $c_i(x_{k_s}) > -\frac{1}{k_s}$; therefore, $c_i(\bar{x}) \geq 0$, $i = 1,\ldots,m$ and again $\lim_{k_s\to\infty} k_s^{-1}\ln(1+k_sc_i(x_{k_s})) = 0$, $1 \leq i \leq m$.
From (4.53) follows

$$\|g(\bar{x})\|^2 = \lim_{k_s\to\infty}\|g(x_{k_s})\|^2 = \lim \frac{4}{k_s}(f(x^*) - f(\bar{x}) - \frac{\alpha}{2}\|\bar{x} - x^*\|^2) = 0.$$

Then, $\bar{x}$ is feasible in (4.48), $\bar{x} \in B(x^*,\varepsilon)$, and from (4.53) follows

$$f(\bar{x}) + \frac{\alpha}{2}\|\bar{x} - x^*\| \leq f(x^*). \quad (4.54)$$

Keeping in mind feasibility $\bar{x}$, we have $f(x^*) \leq f(\bar{x})$; therefore from (4.54) follows $\bar{x} = x^*$. So for arbitrary subsequence $\{x_{k_s}\}_{s\in N} \subset \{x_k\}_{k\in N}$, we have $\lim_{k_s\to\infty} x_{k_s} = x^*$; hence $\lim_{k\to\infty} x_k = x^*$. Therefore from some point on constraint, $x \in B(x^*,\varepsilon)$ is irrelevant for finding $P_k(x_k) = \min\{P_k(x)|x \in B(x^*,\varepsilon)\}$, and x_k is, in fact, an unconstrained minimizer, that is

$$\nabla P_k(x_k) = \nabla f(x_k) - \sum_{i=1}^{q}(1+kc_i(x_k))^{-1}\nabla c_i(x_k) + (kg(x_k))\nabla g^T(x_k)$$
$$+\alpha(x_k - x^*) = 0. \quad (4.55)$$

Let us introduce two Lagrange multipliers vectors $\lambda_k = (\lambda_{k,i} = (1+kc_i(x_k))^{-1}$, $i = 1,\ldots,q) \in \mathbb{R}^q_{++}$ and $v_k = (v_{kj} = kg_j(x_k), j = 1,\ldots,r) \in \mathbb{R}^r$.
First of all, let us show that both sequences $\{v_k\}_{k\in N}$ and $\{\lambda_k\}_{k\in N}$ are bounded. Assuming the opposite, we obtain that $Q_k = \sum_{i=1}^q \lambda_{k,i} + \sum_{j=1}^r |v_{kj}| \to \infty$. By dividing the left-hand side of (4.55) by Q_k, we obtain bounded sequences $\{\bar{\lambda}_k = \lambda_k Q_k^{-1}\}$ and $\{\bar{v}_k = v_k Q_k^{-1}\}$ and

$$Q_k^{-1}\nabla f(x_k) - \sum_{i=1}^q \bar{\lambda}_{k,i}\nabla c_i(x_k) + \sum_{j=1}^r \bar{v}_{k,j}\nabla g_j(x_k) + Q_k^{-1}\alpha(x_k - x^*) = 0. \quad (4.56)$$

Without losing generality, we can assume $\lim_{k\to\infty}\bar{v}_k \to \bar{v} \in \mathbb{R}^r, \lim_{k\to\infty}\bar{\lambda}_k = \bar{\lambda} \in \mathbb{R}^q_+$. By taking limit in (4.56) when $k\to\infty$, we obtain

$$-\sum_{i=1}^q \bar{\lambda}_i \nabla c_i(x^*) + \sum_{j=1}^r \bar{v}_j \nabla g_j(x^*) = 0.$$

Note that $\lim_{k\to\infty}\bar{\lambda}_{k,i} = 0, i = m+1,\ldots,q$; therefore

$$-\sum_{i=1}^m \bar{\lambda}_i \nabla c_i(x^*) + \sum_{j=1}^r \bar{v}_j \nabla g_j(x^*) = 0$$

and not all $\bar{v}_j, j = 1,\ldots,r$ and $\bar{\lambda}_i, i = 1,..,m$ are equal zero, which is impossible due to (4.49). Therefore, $\{v_k\}_{k\in N}$ and $\{\lambda_k\}_{k\in N}$ are bounded sequences.
By taking the limit in (4.55), we obtain

$$\nabla f(x^*) - \sum_{i=1}^q \bar{\lambda}_i \nabla c_i(x^*) + \sum_{j=1}^r \bar{v}_j \nabla g_j(x^*) = 0 \quad (4.57)$$

and $\bar{\lambda}_i = 0, i = m+1,\ldots,q$. In view of regularity x^*, the Lagrange multipliers in (4.57) are uniquely defined; therefore $\bar{\lambda} = \lambda^* \in \mathbb{R}^q_+$, $\bar{v} = v^* \in \mathbb{R}^r$, and (4.57) can be rewritten as follows

$$\nabla_x L(x^*,\lambda^*,v^*) = \nabla f(x^*) - \sum_{i=1}^m \lambda_i^* \nabla c_i(x^*) + \sum_{j=1}^r v_j^* \nabla g_j(x^*) = 0.$$

Also the complementarity condition

$$\lambda_i^* c_i(x^*) = 0, i = 1,\ldots,q \quad (4.58)$$

is satisfied.
The proof of the part (1) is completed.

(2) By using second-order unconstrained optimality condition, we obtain

$$\nabla^2_{xx}P_k(x_k) = \nabla^2 f(x_k) - \sum_{i=1}^{q} \lambda_{k,i}\nabla^2 c_i(x_k) + \sum_{j=1}^{r} v_{k,j}\nabla^2 g_j(x_k)$$

$$k\nabla c(x_k)^T \hat{\Lambda}_k \nabla c(x_k) + k\nabla g(x_k)^T \nabla g(x_k) + \alpha I \succ 0^{n,n} \quad (4.59)$$

where $\hat{\Lambda}_k = \Lambda_k(I^q + kc(x_k))^{-1}$. $\Lambda_k = \text{diag}(\lambda_{k,j})_{j=1}^q$, $(I^q + kc(x_k)) = \text{diag}((1 + kc_j(x_k))_{j=1}^q$ and I^q- identical matrix in $\mathbb{R}^q$.
Let consider null space of the matrix

$$N(x_k) = \begin{bmatrix} \nabla c_{(m)}(x_k) \\ \nabla g(x_k) \end{bmatrix},$$

that is $T_k = \{w \in \mathbb{R}^n : N(x_k)w = 0\}$. Let us consider $y \in T(x^*)$ and let y_k be the projection of $y \in T(x^*)$ on T_k, that is

$$y_k = y - N^T(x_k)(N(x_k)N^T(x_k))^{-1}N(x_k)y. \quad (4.60)$$

Due to the regularity condition (4.49), the inverse $(N(x^*)N^T(x^*))^{-1}$ exists and so is the inverse $(N(x_k)N^T(x_k))^{-1}$ for $k > 0$ large enough. The Hessian $\nabla^2_{xx}P_k(x_k)$ is a positive definite; therefore for any y_k, we obtain

$$\langle \nabla^2 P_k(x_k)y_k, y_k\rangle = \langle \nabla^2_{xx}L(x_k, \lambda_k, v_k)y_k, y_k\rangle + \quad (4.61)$$

$$+k\|\nabla g(x_k)y_k\|^2 + k\langle \hat{\Lambda}_k \nabla c(x_k)y_k, \nabla c(x_k)y_k\rangle + \alpha\langle y_k, y_k\rangle > 0.$$

From $x_k \to x^*, \lambda_{k,j} \to 0, j = m+1, \ldots, q$ and $\lambda_{k,j} \to \lambda_j^* > 0, j = 1, \ldots, m$, as well as, $N(x_k)y_k \to N(x^*)y$. By taking the limit in (4.61) when $k \to \infty$ and keeping in mind strong complementarity condition and the fact that α can be arbitrary small, we obtain (4.51). □

The sufficient second-order optimality condition is given below.

Theorem 4.8. *Let f and all c_i, $i = 1, \ldots, q$, g_j, $j = 1, \ldots, r$ be twice continuously differentiable, regularity condition (4.49) holds, and*

$$\langle \nabla^2_{xx}L(x^*, \lambda^*, v^*)y, y\rangle > 0, \forall y \in T(x^*) \quad (4.62)$$

is satisfied, then x^* is an isolated local minimizer in (4.48), and there exists $\gamma > 0$ that

$$f(x) \geq f(x^*) + \frac{\gamma}{2}\|x - x^*\|, \forall x \in B(x^*, \varepsilon) : g(x) = 0, c(x) \geq 0 \quad (4.63)$$

holds.

Proof. To show that x^* is an isolated local minimizer, it is enough to prove (4.63).

Let us assume the opposite, and then there is a sequence $\{x_k\}_{k \in N} : c_i(x_k) = 0, i = 1, .., m$; $g_j(x_k) = 0, j = 1, .., r$ that $\lim x_k = x^*$ and

$$f(x_k) \leq f(x^*) + \frac{\gamma}{2}\|x_k - x^*\|^2. \tag{4.64}$$

Let $\Delta_k = (x_k - x^*)$; we consider a sequence $\{y_k = (x_k - x^*)\|\Delta_k\|^{-1}\}_{k \in N}$. The sequence $\{y_k\}_{k \in N}$ is bounded; therefore there exists a converging subsequence. Without loss of generality, we can assume that $y_k \to y$, then $\|y\| = 1$. Also we have

$$0 = \lim_{\Delta_k \to 0} \frac{g_j(x_k) - g_j(x^*)}{\|\Delta_k\|} = \lim_{\Delta_k \to 0} \frac{g_j(x^* + \|\Delta_k\| y_k) - g_j(x^*)}{\|\Delta_k\|} \tag{4.65}$$

$$= \langle \nabla g_j(x^*), y \rangle, j = 1, \dots r$$

and

$$0 = \lim_{\Delta_k \to 0} \frac{c_i(x_k) - c_i(x^*)}{\|\Delta_k\|} = \lim_{\Delta_k \to 0} \frac{c_i(x^* + \|\Delta_k\| y_k) - c_i(x^*)}{\|\Delta_k\|} = \tag{4.66}$$

$$\langle \nabla c_i(x^*), y \rangle, i = 1, \dots, m.$$

From (4.65) and (4.66) follows

$$y \in T(x^*). \tag{4.67}$$

Using the middle point formula (2.6) for $1 \leq j \leq r$, we have

$$0 = g_j(x_k) - g_j(x^*) = \|\Delta_k\| \langle \nabla g_j(x^*), y_k \rangle + \frac{\|\Delta_k\|^2}{2} \langle \nabla^2 g(\cdot) y_k, y_k \rangle \tag{4.68}$$

and again, using the middle point formula (2.6) for $1 \leq i \leq m$, we obtain

$$0 = c_i(x_k) - c_i(x^*) = \|\Delta_k\| \langle \nabla c_i(x^*), y_k \rangle + \frac{\|\Delta_k\|^2}{2} \langle \nabla^2 c_i(\cdot) y_k, y_k \rangle. \tag{4.69}$$

From (4.64) follows

$$\frac{1}{k}\|x_k - x^*\|^2 \geq f(x_k) - f(x^*) = \|\Delta_k\| \langle \nabla f(x^*), y_k \rangle + \frac{\|\Delta_k\|^2}{2} \langle \nabla^2 f(\cdot) y_k, y_k \rangle. \tag{4.70}$$

Each middle point in (4.68)–(4.70) lies in the segment, which connects x_k and x^*; therefore when $x_k \to x^*$, all middle points converge to x^*.

By multiplying (4.68) on v_j^* and (4.69) on λ_i^* and adding (4.68)–(4.70), we obtain

$$\frac{1}{k}\|\Delta_k\|^2 = \frac{1}{k}\|x_k - x^*\|^2 \geq \|\Delta_k\| \langle \nabla_x L(x^*, \lambda^*, v^*), y_k \rangle + \frac{\|\Delta_k\|^2}{2} \langle \nabla_{xx}^2 L(\cdot, \lambda^*, v^*) y_k, y_k \rangle.$$

Keeping in mind $\nabla_x L(x^*, u^*, \lambda^*) = 0$, we obtain

$$\frac{2}{k} > \langle \nabla^2 L_{xx}(\cdot, \lambda^*, v^*) y_k, y_k \rangle. \tag{4.71}$$

By taking limit in (4.71) and keeping in mind (4.67), we obtain

$$\langle \nabla^2 L_{xx}(x^*, \lambda^*, v^*)y, y \rangle \leq 0, \forall y \in T(x^*),$$

which contradicts (4.62); therefore (4.63) holds. □

Let us assume that f is convex and all c_i, $i = 1, \ldots, m$ are concave functions in $\mathbb{R}^n$ and $I(x^*) = \{i : c_i(x^*) = 0\} = \{1, \ldots, m\}$ be the set of active constraints.

Our main focus in the future will be the following convex optimization problem

$$f(x^*) = \min\{f(x) | c_i(x) \geq 0,\ i = 1, \ldots, q\}. \tag{4.72}$$

Then, the correspondent Lagrangian is

$$L(x, \lambda) = f(x) - \sum_{i=1}^{q} \lambda_i c_i(x).$$

The regularity condition is

$$\operatorname{rank} \nabla c_{(m)}(x^*) = m. \tag{4.73}$$

The sufficient optimality condition for x^* to be an isolated minimizer in (4.8) is given by the following theorem.

Theorem 4.9. *If f and all c_i, $i = 1, \ldots, q$ are twice continuously differentiable, regularity condition* (4.73) *holds and there is $\mu > 0$ that the following condition*

$$\langle \nabla^2_{xx} L(x^*, \lambda^*)y, y \rangle \geq \mu \langle y, y \rangle,\ \forall y \in T(x^*) \tag{4.74}$$

is satisfied, where $T(x^*) = \{y \in \mathbb{R}^n : \nabla c_{(m)}(x^*)y = 0\}$, then x^* is an isolated minimizer in the problem (4.8).

The proof of Theorem 4.9 is similar to the proof of Theorem 4.8.

4.4 Duality for Equality-Constrained Optimization

We are back to the ECO problem

$$f(x^*) = \min\{f(x) | c_i(x) = 0, i = 1, \ldots, m\}. \tag{4.75}$$

Let x^* be a regular minimizer, that is

$$\operatorname{rank} \nabla c(x^*) = m, \tag{4.76}$$

then from Theorem 4.1 follows existence $\lambda^* \in \mathbb{R}^m$ that the pair (x^*, λ^*) solves the Lagrange system (4.10)–(4.11).

Let us consider the dual function

$$d(\lambda) = \inf\{L(x,\lambda)|x \in \mathbb{R}^n\}, \tag{4.77}$$

which is due to Lemma 2.8 is closed and concave.

It follows from Theorem 2.14 that at each $\lambda \in \mathbb{R}^m$ from the subdifferential

$$\partial d(\lambda) = \{g : d(\hat{\lambda}) - d(\lambda) \le \langle g, \hat{\lambda} - \lambda\rangle,\ \forall \hat{\lambda} \in \mathbb{R}^m\} \tag{4.78}$$

of d is a non-empty, bounded, and convex set. If minimizer

$$x(\lambda) = \text{argmin}\{L(x,\lambda)|x \in \mathbb{R}^n\}$$

is unique and f, $c_i \in C^1$, $i = 1,\dots,m$, then the dual function

$$d(\lambda) = L(x(\lambda),\lambda) = \min\{L(x,\lambda)|x \in \mathbb{R}^n\}$$

is differentiable and

$$\nabla d(\lambda) = \nabla_x L(x(\lambda),\lambda)\nabla_\lambda x(\lambda) + \nabla_\lambda L(x(\lambda),\lambda).$$

Keeping in mind

$$\nabla_x L(x(\lambda),\lambda) = 0, \tag{4.79}$$

we obtain

$$\nabla d(\lambda) = \nabla_\lambda L(x(\lambda),\lambda) = -c(x(\lambda)), \tag{4.80}$$

where $c(x(\lambda)) = (c_1(x(\lambda)),\dots,c_m(x(\lambda)))^T$ is the residual vector. In case if $x(\lambda) = (x_1(\lambda),\dots,x_n(\lambda))^T$ is not unique, then for any $x(\lambda) \in \text{Argmin}\{L(x,\lambda)|x \in \mathbb{R}^n\}$, we have

$$-c(x(\lambda)) \in \partial d(\lambda).$$

In fact, let

$$\hat{\lambda} : d(\hat{\lambda}) = L(\hat{x},\hat{\lambda}) = \min_{x\in\mathbb{R}^n} L(x,\hat{\lambda}), \tag{4.81}$$

then for any $\hat{\lambda} \in \mathbb{R}^m$, we have

$$\begin{aligned} d(\hat{\lambda}) = \min\{f(x) - \sum_{i=1}^m \hat{\lambda}_i c_i(x)|x \in \mathbb{R}^n\} \le f(x(\lambda)) - \sum_{i=1}^m \hat{\lambda}_i c_i(x(\lambda)) = \\ f(x(\lambda)) - \sum_{i=1}^m \lambda_i c_i(x(\lambda)) - \langle c(x(\lambda)), \hat{\lambda} - \lambda\rangle = \\ d(\lambda) + \langle -c(x(\lambda)), \hat{\lambda} - \lambda\rangle \end{aligned}$$

or

$$d(\hat{\lambda}) - d(\lambda) \le \langle -c(x(\lambda)), \hat{\lambda} - \lambda\rangle, \forall \hat{\lambda} \in \mathbb{R}^m,$$

therefore

$$-c(x(\lambda)) \in \partial f(\lambda). \tag{4.82}$$

The dual problem

$$\begin{aligned} d(\lambda) &\Rightarrow \max \\ \text{s. t. } \lambda &\in \mathbb{R}^m \end{aligned} \tag{4.83}$$

is an unconstrained concave maximization problem.

For regular minimizer x^*, the Jacobian $\nabla c(x^*) : \mathbb{R}^n \to \mathbb{R}^r$ is a full rank matrix, that is, $\operatorname{rank} \nabla c(x^*) = m < n$; therefore for the min and max eigenvalues g and G of the Gramm matrix $\nabla c(x^*) \nabla c(x^*)^T$, we have

$$0 < g \leq G. \tag{4.84}$$

Let

$$(a)\ 0 < m_0 = \operatorname{mineigval} \nabla^2_{xx} L(x^*, \lambda^*);\ (b)\ M_0 = \operatorname{maxeigval} \nabla^2_{xx} L(x^*, \lambda^*). \tag{4.85}$$

Theorem 4.10. *If $f, c_i, i = 1, \ldots, m$ is twice continuously differentiable, x^* is a regular minimizer, and (4.85) holds, then*

$$(1)\ d(\lambda^*) = f(x^*),\ (2)\ \nabla d(\lambda^*) = -c(x^*)$$

and

$$(3)\ -\frac{g}{M_0} I^m \succeq \nabla^2_{\lambda\lambda} d(\lambda^*) \succeq -\frac{G}{m_0} I^m, \tag{4.86}$$

where I^m- is the identical matrix in $\mathbb{R}^m$.

Proof. We have

$$d(\lambda^*) = L(x^*, \lambda^*) = \min\{L(x, \lambda^*) | x \in \mathbb{R}^n\}. \tag{4.87}$$

Keeping in mind $c(x^*) = 0$, we obtain $d(\lambda^*) = f(x^*)$. From (4.82) and (4.85a) follows

$$\nabla d(\lambda^*) = -c(x^*).$$

For the Hessian of the dual function from (4.80) follows

$$\nabla^2_{\lambda\lambda} d(\lambda) = -\nabla c(x(\lambda)) \nabla_\lambda x(\lambda). \tag{4.88}$$

By differentiating (4.80) in λ, we obtain

$$\nabla^2_{xx} L(x(\lambda), \lambda) \cdot \nabla_\lambda x(\lambda) - \nabla_{x\lambda} L(x(\lambda), \lambda) = 0 \tag{4.89}$$

or

$$\nabla^2_{xx} L(x(\cdot), \cdot) \nabla_\lambda x(\cdot) = (\nabla_x c(x(\cdot)))^T.$$

From smoothness f, c_i, $i = 1, \ldots, m$ and (4.85) follows strong convexity of $L(x, \lambda)$ in $x \in B(x^*, \varepsilon)$. Therefore $(\nabla^2_{xx} L(x(\cdot), \cdot))^{-1}$ exists and

$$\nabla_\lambda x(\cdot) = (\nabla^2_{xx} L(x(\cdot),\cdot)^{-1}(\nabla c(x(\cdot)))^T.$$

From (4.88) follows

$$\nabla^2_{\lambda\lambda} d(\lambda) = -\nabla c(x(\cdot))(\nabla^2 L(x(\cdot),\cdot))^{-1}\nabla c(x(\cdot)))^T.$$

Hence, for any $y \in \mathbb{R}^m$ and $\lambda = \lambda^*$, we obtain

$$\langle \nabla^2_{\lambda\lambda} d(\lambda^*)y,y\rangle = -\langle \nabla c(x^*)(\nabla^2_{xx} L(x^*,\lambda^*))^{-1}\nabla c^T(x^*)y,y\rangle$$

or

$$\langle \nabla^2_{\lambda\lambda} d(\lambda^*)y,y\rangle = -\langle (\nabla^2_{xx} L)^{-1}\nabla c^T y, \nabla c^T y\rangle = -\langle (\nabla^2_{xx} L)^{-1}u,u\rangle,$$

where $\nabla c = \nabla c(x^*)$, $\nabla^2_{xx} L = \nabla^2_{xx} L(x^*,\lambda^*)$, and $u = \nabla c^T y \in \mathbb{R}^n$.

From (4.85) follows

$$M_0 I^n \succeq \nabla^2_{xx} L \succeq m_0 I^n.$$

Therefore,

$$M_0^{-1} I^n \preceq \nabla^2_{xx} L^{-1} \preceq m_0^{-1} I^n,$$

so for $\forall u \in \mathbb{R}^n$, we have

$$-M_0^{-1}\langle u,u\rangle \succeq \langle -\nabla^2_{xx} L^{-1}u,u\rangle \succeq -m_0^{-1}\langle u,u\rangle. \tag{4.90}$$

From (4.90) follows

$$-M_0^{-1}\langle u,u\rangle \succeq \langle \nabla^2_{\lambda\lambda} d(\lambda^*)y,y\rangle \succeq -m_0^{-1}\langle u,u\rangle \tag{4.91}$$

For the Gramian matrix $\nabla c \nabla c^T$, we have

$$G\langle y,y\rangle \geq \langle u,u\rangle = \langle \nabla c \nabla c^T y,y\rangle \geq g\langle y,y\rangle. \tag{4.92}$$

Combining (4.91) and (4.92), we obtain (4.86). □

Exercise 4.1.

1. In view of (4.80), the gradient method for the dual problem is given by the following formula
$$\lambda_{s+1} = \lambda_s + t\nabla d(\lambda_s) = \lambda_s - tc(x(\lambda_s)).$$
Find the step length $t > 0$, which guarantees local Q-linear convergence rate with the best ratio $0 < q < 1$.
2. Find the area, where Newton's method for the dual problem converges locally with quadratic rate.

4.5 Courant's Penalty Method as Tikhonov's Regularization for the Dual Problem

Let f and all c_i, $i = 1, \dots, m$ be twice continuously differentiable function in $\mathbb{R}^n$. We consider ECO problem (4.75).

In 1943, Richard Courant (1888–1972) introduced the penalty function. Let $\pi(t) = \frac{1}{2}t^2$, and then Courant's penalty function $C : \mathbb{R}^n \times \mathbb{R}_{++} \to \mathbb{R}$ is defined as follows

$$C(x,k) = f(x) + k^{-1} \sum_{i=1}^{m} \pi(kc_i(x)) = f(x) + \frac{k}{2} \sum_{i=1}^{m} c_i^2(x) = f(x) + \frac{k}{2} \|c(x)\|^2. \quad (4.93)$$

We assume that unconstrained minimizer $x(k)$ of C exists for any fixed $k > 0$ and then Courant's penalty method generates a sequence $\{x(k)\}_{k \in N}$:

$$C(x(k),k) = \min\{C(x,k) | x \in \mathbb{R}^n\}. \quad (4.94)$$

For the unconstrained minimizer $x(k)$, we have

$$\nabla_x C(x(k),k) = \nabla f(x(k)) + \sum_{i=1}^{m} \pi'(kc_i(x(k))) \nabla c_i(x(k)) = 0. \quad (4.95)$$

Let

$$\lambda_i(k) = -\pi'(kc_i(x(k))), \; i = 1, \dots, m, \quad (4.96)$$

then we can rewrite (4.95) as follows

$$\nabla_x C(x(k),k) = \nabla f(x(k)) - \sum_{i=1}^{m} \lambda_i(k) \nabla c_i(x(k)) = \nabla_x L(x(k), \lambda(k)) = 0, \quad (4.97)$$

where L is the classical Lagrangian for (4.75).

If from $\nabla_x L(x(k), \lambda(k)) = 0$ follows $L(x(k), \lambda(k)) = \min_{x \in \mathbb{R}^n} L(x, \lambda(k))$, then

$$d(\lambda(k)) = L(x(k), \lambda(k)),$$

where $d : \mathbb{R}^m \to \mathbb{R}$ is the dual function. From (4.82) we have

$$-c(x(k)) \in \partial d(\lambda(k)), \quad (4.98)$$

where $\partial d(\lambda(k))$—subdifferential of d at $\lambda(k)$. From (4.95) follows

$$\pi'(kc_i(x(k))) = -\lambda_i(k).$$

From $\pi''(t) \neq 0$ follows existence $\pi'^{-1}(-\lambda_i(k)) = kc_i(x(k))$. From LF identity (2.63), we have $\pi'^{-1} = \pi'^{*}$; therefore

$$c_i(x(k)) = k^{-1} \pi'^{-1}(-\lambda_i(k)) = k^{-1} \pi^{*'}(-\lambda_i(k)),$$

where $\pi^*(s) = \sup\{st - \pi(t)\} = \sup_t\{st - \frac{1}{2}t^2\} = \frac{1}{2}s^2$. From (4.98) follows

$$0 \in \partial d(\lambda(k)) + c(x(k)) = \partial d(\lambda(k)) - k^{-1}\lambda(k).$$

It means that $\lambda(k)$ is a solution of the following inclusion

$$0 \in \partial(d(u) - \frac{1}{2k}\|u\|^2), \tag{4.99}$$

which is the optimality criteria for the following unconstrained maximization problem

$$d(\lambda(k)) - \frac{1}{2k}\|\lambda(k)\|^2 = \max\{d(u) - \frac{1}{2k}\|u\|^2 | u \in \mathbb{R}^m\}. \tag{4.100}$$

Thus, the Courant's penalty method (4.93) is equivalent to A. Tikhonov's (1906–1993) regularization method (4.100) introduced in 1963.

The convergence of the regularization method (4.100) follows from Theorem 3.11

Exercise 4.2. It is shown that the primal sequence $\{x(k)\}_{k\in N}$ is monotone not decreasing in value, i.e.

$$f(x(k+1)) \geq f(x(k)),$$

while the residual norm $\|c(x(k))\|$ is not increasing, i.e.,

$$\|c(x(k+1)\| \leq \|c(x(k))\|$$

and for any $f > -\infty$, we have

$$\|c(x(k))\| \leq O(k^{\frac{1}{2}}).$$

4.6 Gradient Methods for ECO

We saw that if x^* is a regular minimizer for the problem (4.75), then from Theorem 4.1 follows existence of $\lambda^* \in \mathbb{R}^m$ that

$$\nabla_x L(x^*, \lambda^*) = \nabla f(x^*) - \sum_{i=1}^{m} \lambda_i^* \nabla c_i(x^*) = 0 \tag{4.101}$$

$$\nabla_\lambda L(x^*, \lambda^*) = -c(x^*) = 0. \tag{4.102}$$

Therefore, for any regular minimizer x^*, there is $\lambda^* \in \mathbb{R}^m$ and a small $\delta > 0$ that

$$L(x^*, \lambda) \leq L(x^*, \lambda^*) \leq L(x, \lambda^*), \tag{4.103}$$

holds for $\forall x \in B(x^*, \delta) = \{x \in \mathbb{R}^n : \|x - x^*\| \geq \delta\}$ and $\lambda \in \mathbb{R}^m$.

It means that solving the system (4.101)–(4.102) is equivalent to finding the saddle point (x^*, λ^*) from (4.103).

The following gradient method

$$x_{s+1} = x_s - \tau \nabla_x L(x_s, \lambda_s) = x_s - \tau(\nabla f(x_s) - \sum_{i=1}^{m} \lambda_{i,s} \nabla c_i(x_s)) \quad (4.104)$$

$$\lambda_{s+1} = \lambda_s + \tau \nabla_\lambda L(x_s, \lambda_s) = \lambda_s - \tau c(x_s) \quad (4.105)$$

for finding the saddle point was introduced by Arrow and Hurwicz in the 1950s; see, for example, Polyak (1987).

Without extra assumption on the Lagrangian, one can't expect convergence of such method. In fact, the minimizer of $L(x, \lambda^*)$ in x may not exist. If the minimizer exists, it may not coincide with x^*.

However, if the matrix $\nabla^2_{xx} L(x^*, \lambda^*)$ is positive definite

$$\langle \nabla^2_{xx} L(x^*, \lambda^*) \mathrm{v}, \mathrm{v} \rangle \geq \mu \langle \mathrm{v}, \mathrm{v} \rangle, \forall \mathrm{v} \in \mathbb{R}^n, \mu > 0 \quad (4.106)$$

and f, c_i, $i = 1, \ldots, m$ are C^2 functions, then there is $\bar{\tau} > 0$ that for any $0 < \tau < \bar{\tau}$, local convergence of Arrow and Hurwicz method is possible.

Theorem 4.11. *If x^* is a regular minimizer, (4.106) holds, and $\nabla^2 f$ and all $\nabla^2 c_i$, $i = 1, \ldots, m$ satisfy Lipschitz condition in the neighborhood of x^*, then there exists $\bar{\tau} > 0$ that for any $0 < \tau < \bar{\tau}$, the sequence generated by (4.104)–(4.105) locally converges to (x^*, λ^*) with geometric rate.*

Proof. Let $u_s = (x_s - x^*, \lambda_s - \lambda^*)$, and then keeping in mind Lipschitz condition and (4.101)–(4.102), we can rewrite (4.104)–(4.105) as follows

$$u_{s+1} = D u_s + o(\|u_s\|) e,$$

where

$$D = I - \tau B, B = \begin{pmatrix} A & C^T \\ -C & 0 \end{pmatrix},$$

$A = \nabla_{xx} L(x^*, \lambda^*)$, $C = \nabla c(x^*)$-Jacobian of $c(x)$ at $x = x^*$,

$$\|u\| = \max\{\max_{1 \leq i \leq n} \{|x_i^s - x_i^*|, \max_{1 \leq j \leq m} |\lambda_j^s - \lambda_j^*|\}\} \text{ and } e = (1, \ldots, 1)^T \in \mathbb{R}^{n+m}.$$

Let us show that the system

$$\dot{u}(t) = -B u(t) \quad (4.107)$$

is stable and has a solution for any starting point $u_0 = u(0)$ and $u(t) \to 0$ as $t \to \infty$.

We can rewrite the system (4.107) as follows

$$\dot{x}(t) = -A x(t) - C^T \lambda(t) \quad (4.108)$$

$$\dot{\lambda}(t) = C x(t). \quad (4.109)$$

By taking $v(t) = 0.5(\|x(t)\|^2 + \|\lambda(t)\|^2)$ and using (4.106), we obtain

$$\dot{v} = \langle \dot{x}, x\rangle + \langle \dot{\lambda}, \lambda\rangle = -\langle Ax, x\rangle + \langle C^T\lambda, x\rangle - \langle Cx, \lambda\rangle = \langle -Ax, x\rangle \leq -\mu\|x\|^2, \quad (4.110)$$

which means that $v(t)$ monotone is decreasing as $\dot{v}(t) \to 0$.

From (4.110) follows $x(t) \to 0$ when $t \to \infty$. For linear system (4.108)–(4.109) from $x(t) \to 0$ follows $\dot{x}(t) \to 0$; therefore from (4.108), we obtain

$$C^T\lambda(t) = -\dot{x}(t) - Ax(t) \to 0$$

as $t \to \infty$.

From $\operatorname{rank} C = m$ follows $\lambda(t) \to 0$. So $u(t) \to 0$; therefore $\max Re\gamma_i = \gamma < 0$, where γ_i is the eigenvalue of B. In view of Theorem 2 from the Appendix, there is $\bar{\tau} > 0$ that for every $0 < \tau < \bar{\tau}$ the matrix $D = I - \tau B$ has a spectral radius $\rho = \rho(D) < 1$.

It follows from the Theorem 3 of the Appendix that for any $0 < \varepsilon < 1 - \rho$, there is $\delta > 0$ small enough, that if $\|u_0 - u^*\| \leq \delta$, then

$$\|u_s - u^*\| \leq cq^s$$

and $c > 0$ is independent on $0 < q = \rho + \varepsilon < 1$. □

The condition (4.106) is crucial for convergence of the Arrow–Hurwicz method (4.104)–(4.105).

We will see later that results of Theorem 4.11 remain true under much weaker assumptions instead of classical Lagrangian $L(x, \lambda)$ one which uses augmented Lagrangian, which we consider in the following section. Instead of gradient step, one can make full primal minimization step following Lagrange multipliers update. In such case, the primal–dual sequence $\{x_s, \lambda_s\}_{s\in N}$ is defined as follows

$$x_{s+1} : \nabla_x L(x_{s+1}, \lambda_s) = 0 \quad (4.111)$$

$$\lambda_{s+1} = \lambda_s - \tau c(x_{s+1}). \quad (4.112)$$

If $L(x, \lambda_s)$ is strongly convex in x, then x_{s+1} is unique. If f and all c_i, $i = 1, \ldots, m$ belong to C^1, then the dual function $d(\lambda) = L(x(\lambda), \lambda))$ is smooth and $\nabla d(\lambda) = -c(x(\lambda))$. Therefore (4.111), (4.112) is a gradient method for maximization $d(\lambda)$. The convergence of the gradient method depends on the smoothness and concavity properties of dual function d.

Exercise 4.3.

1. Establish convergence properties of the dual method (4.111)–(4.112) if ∇d satisfy Lipschitz condition.
2. Characterize the convergence properties of the dual gradient method under condition of the Theorem 4.10.
   Hint: Use the results of Theorem 3.7

4.7 Newton's Method for Nonlinear System of Equations

Let us consider a vector–function $c : \mathbb{R}^n \to \mathbb{R}^n$. We assume that the system of nonlinear equations

$$c_i(x) = 0, i = 1, \dots, n \ (c(x) = 0) \tag{4.113}$$

has a solution $x^* \in \mathbb{R}^n$, i.e., $c(x^*) = 0$.

Newton's method for solving (4.113) is a linearization method, which instead of (4.113) solves a sequence of linear systems of the same size.

Let x_0 be an initial approximation for x^*; we replace (4.113) by following linear system

$$c(x_0) + \langle \nabla c(x_0), x - x_0 \rangle = 0, \tag{4.114}$$

where $\nabla c(x) = J(c(x))$- Jacobian of $c(x)$.

Assuming that inverse $(\nabla c(x_0))^{-1}$ exists, then from (4.114) follows

$$x - x_0 = -(\nabla c(x_0))^{-1} c(x_0),$$

or

$$x_1 = x_0 - (\nabla c(x_0))^{-1} c(x_0). \tag{4.115}$$

By reiterating (4.115), we obtain Newton's method

$$x_{s+1} = x_s - (\nabla c(x_s))^{-1} c(x_s) \tag{4.116}$$

for solving (4.113).

Generally speaking, one can expect only local convergence of Newton's method (see Example 3.1 from Section 3.6.1).

Therefore, it is important to characterize the neighborhood of x^*, where the convergence occurred.

We assume that there is $\delta > 0$ and small enough $\rho > 0$ that in the neighborhood $B(x^*, \delta) = \{x : \|x - x^*\| \leq \delta\}$ of x^* the following bound holds

$$\|(\nabla c(x))^{-1}\| \leq \rho^{-1}. \tag{4.117}$$

We also assume that for Jacobian ∇c Lipschitz condition

$$\|\nabla c(x) - \nabla c(y)\| \leq L\|x - y\| \tag{4.118}$$

holds.

Theorem 4.12. *Let (4.117) and (4.118) be satisfied and then for any starting point $x_0 \in B(x^*, \delta)$ such that*

$$q = \frac{1}{2} L \rho^{-2} \|c(x_0)\| < 1 \tag{4.119}$$

Newton's method (4.116) generates a sequence $\{x_s\}_{s \in N}$ that

$$\dots < \|c(x_s)\| < \|c(x_{s-1})\| < \dots < \|c(x_1)\| < \|c(x_0)\|$$

and the following bound

$$\|c(x_s)\| \leq \frac{2\rho^2}{L} q^{2^s} \tag{4.120}$$

holds.

Proof. From (2.10) and (4.118) follows

$$\|c(x+y) - c(x) - \nabla c(x)y\| \leq \frac{L}{2}\|y\|^2. \tag{4.121}$$

Let $x = x_s, y = -(\nabla c(x_s))^{-1} c(x_s)$, then $x_{s+1} = x + y$ and from (4.121) follows

$$\|c(x_{s+1})\| \leq \frac{L}{2}\|\nabla c(x_s)^{-1} c(x_s)\|^2. \tag{4.122}$$

From (4.117), (4.122), and Cauchy–Schwarz inequality, we obtain

$$\|c(x_{s+1})\| \leq \frac{L}{2\rho^2}\|c(x_s)\|^2$$

or

$$\|c(x_{s+1})\| \leq \frac{2\rho^2}{L}\left(\frac{L}{2\rho^2}\|c(x_s)\|\right)^2. \tag{4.123}$$

Keeping in mind $q = \frac{L}{2\rho^2}\|c(x_0)\| < 1$, we have

$$\|c(x_1)\| \leq \frac{L}{2\rho^2}\|c(x_0)\|^2 < \|c(x_0)\|.$$

Using the same arguments, we obtain

$$\ldots < \|c(x_s)\| < \|c(x_{s-1})\| < \ldots < \|c(x_0)\|.$$

It means that $x_s \in B(x^*, \delta)$ for any $s \geq 0$; therefore for any x_s, the bound (4.123) holds.

Reiterating (4.123), we obtain

$$\|c(x_s)\| \leq \frac{2\rho^2}{L}\left(\frac{L}{2\rho^2}\|c(x_0)\|\right)^{2^s}. \tag{4.124}$$

Therefore, for any x_0, which satisfy (4.119), we have (4.120).

Let us estimate $\delta > 0$. By linearizing $c(x)$ at x^*, we obtain

$$c(x_0) = c(x^*) + \nabla c(x^*)(x_0 - x^*) + o(\|x_0 - x^*\|e),$$

where $e = (1, \ldots, 1) \in \mathbb{R}^n$.

Then

$$(x_0 - x^*) = (\nabla c(x^*))^{-1}(c(x_0) - o(\|x_0 - x^*\|e).$$

From (4.119) follows $\|c(x_0)\| \leq 2L^{-1}\rho^2$; therefore, keeping in mind (4.117), we obtain

$$\delta = \|x_0 - x^*\| \leq 2\|(\nabla c(x^*))^{-1}\|\|c(x_0)\| \leq 4L^{-1}\rho.$$

□

4.8 Newton's Method for ECO

We recall that the necessary condition for $u^* \in \mathbb{R}^n$ to be a regular local minimizer for ECO (4.8) is the existence of $\lambda^* \in \mathbb{R}^m$ that the pair $y^* = (u^*, \lambda^*)$ solves Lagrange system (4.10)–(4.11) (see Theorem 4.1).

The sufficient condition (4.30) (see Theorem 4.5) for $u^* \in \mathbb{R}^n$ to be an isolated minimizer of ECO problem (4.8) guarantees uniqueness of $y^* \in \mathbb{R}^{n+m}$.

To find $y^* \in \mathbb{R}^{n+m}$, one has to solve the following Lagrange system

$$N(y) = \begin{bmatrix} \nabla_u L(u, \lambda) \\ \nabla_\lambda L(u, \lambda) \end{bmatrix} = \begin{bmatrix} \nabla f(u) - \lambda^T \nabla c(u) \\ c(u) \end{bmatrix} = 0. \tag{4.125}$$

The local convergence with quadratic rate and the size of the neighborhood of y^*, where quadratic rate occurs, are our main concern.

Later we will need the following technical Lemma.

Lemma 4.2. *Let $A \in \mathbb{R}^{n,n}$ be a nonsingular matrix and $||A^{-1}|| \leq M$, then there exists $\varepsilon > 0$ small enough that any matrix $B \in \mathbb{R}^{n,n}$: $||A - B|| \leq \varepsilon$ is nonsingular and the following bound holds*

$$(a)\ ||B^{-1}|| \leq 2M, \quad (b)\ ||A^{-1} - B^{-1}|| \leq 2M^2\varepsilon. \tag{4.126}$$

Proof. We have $B = A - (A - B) = A(I - A^{-1}(A - B)) = A(I - C)$, where $C = A^{-1}(A - B)$. Using $\|A^{-1}\| \leq M$ and small enough $\varepsilon > 0$, we obtain

$$\|C\| \leq \|A^{-1}\|\|A - B\| = M\varepsilon = q < 0.5. \tag{4.127}$$

From (4.127) follows

$$\|(I - C)^{-1}\| \leq \|I\| + \|C\| + \|C^2\| \leq 1 + q + \ldots + q^s + \ldots \leq 2$$

Let us consider

$$B^{-1} = (A(I - C))^{-1} = (I - C)^{-1}A^{-1}.$$

We obtain

$$\|B^{-1}\| \leq \|(I - C)^{-1}\|\|A^{-1}\| \leq 2M. \tag{4.128}$$

Then, from (4.128) and

$$(A^{-1} - B^{-1}) = A^{-1}(B - A)B^{-1}$$

follows

$$\|A^{-1} - B^{-1}\| \le \|A^{-1}\|\|B - A\|\|B^{-1}\| \le 2M^2\varepsilon.$$

□

Let us consider Jacobian

$$\nabla N(y) = \begin{bmatrix} \nabla^2_{uu}L(u,\lambda) & -\nabla c^T(u) \\ \nabla c(u) & 0^{m,m} \end{bmatrix},$$

of $N(y)$, where $\nabla c(u)$ is Jacobian of $c(u) = (c_1(u),..,c_m(u))^T$.

Application of Newton's method (4.116) to the system (4.125) leads to the following Newton's method

$$y_{s+1} = y_s - (\nabla N(y_s))^{-1}N(y_s), \tag{4.129}$$

for ECO (4.8).

First of all we show that Newton's method (4.129) is well defined. It means there is $\delta > 0$ such that for any $y_0 \in B(y^*,\delta)$ as a starting point, the sequence $\{y_s\}_{s\in N}$ defined by (4.129) exists and belongs to $B(y^*,\delta)$. For this we have to show that conditions type (4.117) and (4.118) satisfied $N(y)$.

Lemma 4.3. *If the second-order sufficient condition (4.73)–(4.74) for ECO (4.8) is satisfied and Lipchitz condition*

$$\|\nabla^2 f(u) - \nabla^2 f(\mathrm{v})\| \le L_0\|u - \mathrm{v}\|,$$
$$\|\nabla^2 c_i(u) - \nabla^2 c_i(\mathrm{v})\| \le L_i\|u - \mathrm{v}\|, 1 \le i \le m \tag{4.130}$$

holds, then there is $\delta > 0$ small enough, $\rho > 0$ and $L > 0$, that for any $y \in B(y^*,\delta)$ the following bound

$$\|(\nabla N(y))^{-1}\| \le \rho^{-1} \tag{4.131}$$

holds and Lipschitz condition

$$\|\nabla N(y_1) - \nabla N(y_2)\| \le L\|y_1 - y_2\| \tag{4.132}$$

is satisfied for any y_1, $y_2 \in B(y^*,\delta)$

Proof. Lipschitz condition (4.132) directly follows from (4.130). Now let us show that the matrix $\nabla N(y^*)$ is not singular. Let $w = (t,\mathrm{v})$, $t \in \mathbb{R}^n$, and $\mathrm{v} \in \mathbb{R}^m$, and then we can rewrite the system $\nabla N(y^*)w = 0$ as follows

$$\nabla^2 L(u^*,\lambda^*)t - \nabla c^T(u^*)\mathrm{v} = 0 \tag{4.133}$$

$$\nabla c(u^*)t = 0. \tag{4.134}$$

For any $t \in \mathbb{R}^n$, we have

$$\langle \nabla^2_{uu}L(u^*,\lambda^*)t,t\rangle - \langle \nabla c(u^*)t,\mathrm{v}\rangle = 0 \tag{4.135}$$

$$\nabla c(u^*)t = 0. \tag{4.136}$$

From (4.135), (4.136), and (4.30a) for any $t \in \mathbb{R}^n : \nabla c(u^*)t = 0$ follows existence of $\mu > 0$, that

$$0 = \langle \nabla^2_{uu} L(u^*, \lambda^*)t, t\rangle \geq \mu \|t\|^2.$$

Therefore, $t = 0$ and from (4.133) follows

$$\nabla^T c(u^*)\mathrm{v} = 0. \tag{4.137}$$

Due to (4.30b), the system (4.137) is possible only for $\mathrm{v} = 0$.

Thus, we have

$$\nabla N(y^*)w = 0 \Rightarrow w = 0;$$

therefore, the matrix $\nabla N(y^*)$ is not singular, so there exist $\rho_o > 0$ that

$$\|(\nabla N(y^*))^{-1}\| = \rho_0^{-1}. \tag{4.138}$$

For $\delta > 0$ small enough, there exists $0 < \rho < \rho_0$ that bound (4.131) follows from (4.138), Lemma 4.2, and Lipschitz condition (4.130).

Keeping in mind (4.128) and Lipschitz condition (4.130), we conclude that for the Lagrange system (4.125) condition, (4.117) and (4.118) are satisfied. □

Therefore, from Theorem 4.12 follows

Theorem 4.13. *If the second-order sufficient optimality condition (4.73)–(4.74) holds and Lipschitz condition (4.130) satisfied, then*

1. *For any* $y_0 : \frac{1}{2}L\rho^{-2}\|N(y_0)\| \leq q < 1$, *the sequence* $\{y_s\}_{s\in N}$ *generated by Newton's method (4.129) is well defined*
2. *For the sequence* $\{\|N(y_s)\|\}_{s\in N}$, *we have*

$$\ldots \leq \|N(y_s)\| \leq \|N(y_{s-1})\| \leq \ldots \leq \|N(y_0)\|$$

3. *For any* $s \geq 1$, *the following bound*

$$\|N(y_s)\| \leq \frac{2\rho^2}{L} q^{2^s}$$

holds

Later we will need the results of Theorem 4.13 in a slightly different form.

Let us consider the Lagrange system $N(y) = 0$ for problem (4.8).

Newton step with $y \in B(y^*, \delta)$ as a starting point produces the following new approximation

$$\hat{y} = y - (\nabla N(y))^{-1} N(y) = y + \Delta y \tag{4.139}$$

for y^*.

Theorem 4.14. *Under conditions of Theorem 4.13, there exists* $0 < \delta_0 < \delta$ *and a constant* $c > 0$ *independent on* $y \in B(y^*, \delta_0)$ *that the following bound*

$$\|\hat{y} - y^*\| \le c\|y - y^*\|^2 \tag{4.140}$$

holds and $\hat{y} \in B(y^*, \delta_0)$.

Proof. From the standard second-order sufficient optimality condition (4.73)–(4.74) follows (4.131). From (4.130) follows (4.132). Then, from (4.121) follows

$$\|N(\hat{y}) - N(y) - \nabla N(y)\Delta y\| \le \frac{L}{2}\|\Delta y\|^2. \tag{4.141}$$

From $\Delta y = (\nabla N(y))^{-1}N(y)$ and (4.141) follows

$$\|N(\hat{y})\| \le \frac{L}{2}\|(\nabla N(y))^{-1}N(y)\|^2 \le \frac{L}{2}\rho^{-2}\|N(y)\|^2. \tag{4.142}$$

From $N(y^*) = 0$ and Lipschitz condition (4.130) follows existence $L_0 > 0$ that

$$\|N(y)\| = \|N(y) - N(y^*)\| \le L_0\|y - y^*\| \tag{4.143}$$

holds for every $y \in B(y^*, \delta_0)$.

Therefore, from (4.142), we obtain

$$\|N(\hat{y})\| \le \frac{L}{2}\frac{L_0^2}{\rho^2}\|y - y^*\|^2. \tag{4.144}$$

Keeping in mind Lipschitz condition (4.130) and (4.101)–(4.102) by linearizing (4.125) at $y^* = (u^*, \lambda^*)$, we obtain

$$N(\hat{y}) = \nabla N(y^*)(\hat{y} - y^*) + o(\|\hat{y} - y^*\|)\begin{pmatrix} e_n \\ e_m \end{pmatrix}, \tag{4.145}$$

where $e_r = (1, \ldots, 1)^T \in \mathbb{R}^r$.

From $\|\nabla N^{-1}(y^*)\| \le \rho_0^{-1}$ and (4.145) follows

$$\|\hat{y} - y^*\| = \left\|(\nabla N(y^*))^{-1}\left[N(\hat{y}) - o(\|\hat{y} - y^*\|)\begin{pmatrix} e_n \\ e_m \end{pmatrix}\right]\right\| \le 2\rho_0^{-1}\|N(\hat{y})\|.$$

Combining the last bound with (4.144), we obtain

$$\|\hat{y} - y^*\| \le \frac{LL_0^2}{\rho_0\rho^2}\|y - y^*\|^2, \tag{4.146}$$

Therefore, there is $\delta > 0$ small enough that $\hat{y} \in B(y^*, \delta_0)$ and (4.140) holds with $c = LL_0^2\rho_0^{-1}\rho^{-2}$ for any $y \in B(y^*, \delta_0)$.

□

4.9 Augmented Lagrangian

We already know that the classical Lagrangian $L(x,\lambda) = f(x) - \sum_{i=1}^m \lambda_i c_i(x)$ is the most important tool in constrained optimization. Unfortunately, it has some limitations.

First, the unconstrained minimizer of $L(x,\lambda^*)$ in $x \in \mathbb{R}^n$ may not exist, or it may not coincide with x^*.

Second, the dual function $d(\lambda) = \inf_{x\in\mathbb{R}^n} L(x,\lambda)$ may not exist for some $\lambda \in \mathbb{R}^m$, even under the second-order sufficient condition.

Third, generally speaking, the dual function is non-smooth, no matter how smooth are f and all c_i, $i = 1,\dots,m$.

The augmented Lagrangian (AL), to a large extend, eliminates the shortages of the classical Lagrangian.

For optimization problems with equality constraints, the AL has been independently introduced in 1969 by M. Hestenes (1906–1991) and M. Powell (1936–2015).

Let us fix $k > 0$ and consider the following problem

$$\min\left(f(x) + \frac{k}{2}\|c(x)\|^2\right) \quad \text{s. t. } c(x) = 0, \tag{4.147}$$

which is equivalent to the ECO problem (4.75).

The quadratic augmented Lagrangian $\mathscr{L} : \mathbb{R}^n \times \mathbb{R}^m \times \mathbb{R}_{++} \to \mathbb{R}$

$$\mathscr{L}(x,\lambda,k) = f(x) + \sum_{i=1}^m \lambda_i c_i(x) + \frac{k}{2}\sum_{i=1}^m c_i^2(x) \tag{4.148}$$

is, in fact, a classical Lagrangian for the equivalent problem (4.147). The AL combines the best properties of the classical Lagrangian L and penalty function C given by (4.93) and at the same time free from their drawbacks.

For the primal–dual solution (x^*,λ^*) and any $k > 0$ from (4.148) follows

$$\text{(a) } \mathscr{L}(x^*,\lambda^*,k) = f(x^*), \text{ (b) } \nabla_x\mathscr{L}(x^*,\lambda^*,k) = \nabla_x L(x^*,\lambda^*) = 0 \tag{4.149}$$

and

$$\text{(c) } \nabla^2_{xx}\mathscr{L}(x^*,\lambda^*,k) = \nabla_{xx}L(x^*,\lambda^*) + k\nabla c(x^*)^T\nabla c(x^*)\,. \tag{4.150}$$

If the second-order sufficient optimality condition is satisfied, then from (4.150) and Debreu's Lemma follows existence of $\mu > 0$ that

$$\left\langle \nabla^2\mathscr{L}_{xx}(x^*,\lambda^*,k)\mathrm{w},\mathrm{w}\right\rangle \geq \mu\langle \mathrm{w},\mathrm{w}\rangle, \qquad \forall \mathrm{w} \in \mathbb{R}^n \tag{4.151}$$

for any $k \geq k_0$, if $k_0 > 0$ is large enough.

If f all $c_i \in C^2$, then there exists a small enough $\delta > 0$ that inequality type (4.151) holds for any $y \in B(y^*,\delta) = \{y = (x,\lambda) \in \mathbb{R}^{n+m} : \|y - y^*\| \leq \delta\}$.

Without restricting the generality, we can assume that

$$\inf_{x\in\mathbb{R}^n} f(x) \geq 0. \tag{4.152}$$

Otherwise it is sufficient to replace $f(x)$ by

$$f(x) := \ln\left(1+e^{f(x)}\right).$$

It follows from (4.149b) and (4.151) that x^* is a local minimizer of $\mathcal{L}(x,\lambda^*,k)$. In other words, there is $\delta > 0$

$$\mathcal{L}(x^*,\lambda^*,k) \leq \mathcal{L}(x,\lambda^*,k), \quad \forall x \in B(x^*,\delta) = \{x \in \mathbb{R}^n : \|x-x^*\| \leq \delta\}. \tag{4.153}$$

Moreover, for $\forall x \in \Omega = \{x \in \mathbb{R}^n : c(x) = 0\}$, we have $\mathcal{L}(x,\lambda^*,k) = f(x)$; therefore for the minimizer x^*

$$f(x) = \mathcal{L}(x,\lambda^*,k) \geq f(x^*) = \mathcal{L}(x^*,\lambda^*,k).$$

The AL $\mathcal{L}(x,\lambda^*,k)$ can be rewritten as follows

$$\mathcal{L}(x,\lambda^*,k) = f(x) + \frac{k}{2}\sum_{i=1}^m \left(c_i(x) + \frac{\lambda_i^*}{k}\right)^2 - \frac{1}{2k}\sum_{i=1}^m \lambda_i^{*2}.$$

Outside of $B(x^*,\delta)\cap\Omega$, the second term can be made as large as one wants by increasing $k > 0$.

Therefore, keeping in mind (4.152), we can find large enough $k_0 > 0$ that

$$x^* = \arg\min\{\mathcal{L}(x,\lambda^*,k) | x \in \mathbb{R}^n\} \tag{4.154}$$

for any $k \geq k_0$.

Let us consider Arrow–Hurwicz gradient method using AL (4.142).

We obtain

$$x_{s+1} = x_s - \tau\nabla_x\mathcal{L}(x_s,\lambda_s,k) = x_s - \tau(\nabla f(x_s) - \sum_{i=1}^m(\lambda_{i,s} - kc_i(x_s))\nabla c_i(x_s)) \tag{4.155}$$

$$\lambda_{s+1} = \lambda_s + \nabla_\lambda\mathcal{L}(x_s,\lambda_s,k) = \lambda_s - \tau c(x_s) \tag{4.156}$$

Theorem 4.15. *If the second-order sufficient conditions (4.73)–(4.74) hold and Lipschitz condition (4.130) is satisfied, then for any $k > 0$ large enough, there is small enough $\bar\tau > 0$ that for any $0 < \tau < \bar\tau$ method (4.155)–(4.156) converges to $y^* = (x^*;\lambda^*)$ locally with geometric rate, so there is $0 < q = q(\tau) < 1$ that*

$$\|y^s - y^*\| \leq cq^s$$

and $c > 0$ is independent on q.

Proof. The problem (4.8) is equivalent to

$$f(x^*) = \min\{f(x) + \frac{k}{2}\|c(x)\|^2 | c_i(x) = 0, i = 1,\ldots,m\}. \tag{4.157}$$

The classical Lagrangian $\mathscr{L}(x,\lambda,k)$ for the problem (4.148) is, in fact, the AL for the ECO problem (4.8). From (4.73)–(4.74), Debreu's Lemma, and (4.150) for k large enough follows condition (4.106). Moreover, for $\delta > 0$ small enough due to Lipschitz continuity of $\nabla^2 f$ and $\nabla^2 c_i$, $i = 1,\ldots,m$, the inequality (4.106) remains true for $y \in B(y^*,\delta)$.

So the condition of Theorem 4.11 is satisfied; hence Arrow–Hurwicz method (4.155)–(4.156) for problem (4.157) converges locally with geometric rate. □

This is not the only advantage of the AL.

Using AL for ECO problem (4.157), one can get the duality results of Section 4.4 from second-order sufficient optimality condition instead of assuming strong convexity $L(x,\lambda)$ in x for any $\lambda \in \mathbb{R}^m$.

Exercise 4.4. Prove Theorem 4.5 using AL.

4.10 The Multipliers Method and the Dual Quadratic Prox

The AL can be rewritten as follows

$$\mathscr{L}(x,\lambda,k) = f(x) - \sum_{i=1}^m \lambda_i c_i(x) + k^{-1}\sum_{i=1}^m \pi(kc_i(x)),$$

where $\pi(t) = \frac{1}{2}t^2$

We assume that for a given $(\lambda,k) \in \mathbb{R}^m \times \mathbb{R}^1_{++}$, the unconstrained minimizer $\hat{x}$ of $\mathscr{L}(x,\lambda,k)$ exists, then

$$\hat{x} = \hat{x}(\lambda,k) : \nabla_x \mathscr{L}(\hat{x},\lambda,k)$$

$$= \nabla f(\hat{x}) - \sum_{i=1}^m (\lambda_i - \pi'(kc_i(\hat{x})))\nabla c_i(\hat{x}) = \nabla_x L(\hat{x},\hat{\lambda}) = 0, \tag{4.158}$$

where

$$\hat{\lambda}_i = \hat{\lambda}_i(\lambda,k) = \lambda_i - \pi'(kc_i(\hat{x})) = \lambda_i - kc_i(\hat{x}), i = 1,\ldots,m$$

or

$$\hat{\lambda} = \lambda - kc(\hat{x}). \tag{4.159}$$

From (4.158) follows that the necessary condition for $\hat{x}$ to be an unconstrained minimizer of $L(x,\hat{\lambda})$ in x is satisfied. If $L(\hat{x},\hat{\lambda}) = \min\{L(x,\hat{\lambda}) | x \in \mathbb{R}^n\}$, then $d(\hat{\lambda}) = L(\hat{x},\hat{\lambda})$ and

$$-c(\hat{x}) \in \partial d(\hat{\lambda}). \tag{4.160}$$

From (4.159) follows

$$-c(\hat{x}) = \frac{1}{k}\pi'^{-1}(\hat{\lambda} - \lambda).$$

Using LFID and inclusion (4.160), we obtain

$$0 \in \partial d(\hat{\lambda}) - k^{-1}\sum_{i=1}^{m} \pi^{*'}(\hat{\lambda}_i - \lambda)e_i,$$

which is the optimality condition for $\hat{\lambda}$ to be the maximizer in the following unconstrained maximization problem

$$d(\hat{\lambda}) - k^{-1}\sum_{i=1}^{m} \pi^*(\hat{\lambda}_i - \lambda_i) = \max\{d(u) - k^{-1}\sum_{i=1}^{m} \pi^*(u_i - \lambda_i) : u \in \mathbb{R}^n\}.$$

From $\pi^*(s) = \frac{1}{2}s^2$ follows

$$\hat{\lambda} = \operatorname{argmax}\{d(u) - \frac{1}{2k}\|u - \lambda\|^2 : u \in \mathbb{R}^n\}. \tag{4.161}$$

Thus, the multipliers method (4.158)–(4.159) is equivalent to the quadratic proximal point (prox) method (4.161) for the dual problem.

If $\hat{x}$ is a unique solution of system $\nabla_x L(x, \hat{\lambda}) = 0$, then $\nabla d(\hat{\lambda}) = -c(\hat{x})$ and from (4.161) follows

$$\hat{\lambda} = \lambda + k\nabla d(\hat{\lambda}),$$

which is an implicit Euler method for solving the following system of ordinary differential equations

$$\frac{d\lambda}{dt} = k\nabla d(\lambda),\ \lambda(0) = \lambda_0. \tag{4.162}$$

Let us consider the prox–function $p : \mathbb{R}^m \to \mathbb{R}$ defined as follows

$$p(\lambda) = d(u(\lambda)) - \frac{1}{2k}\|u(\lambda) - \lambda\|^2 = D(u(\lambda), \lambda) =$$

$$\max\{d(u) - \frac{1}{2k}\|u - \lambda\|^2 : u \in \mathbb{R}^n\}.$$

The function $D(u, \lambda)$ is strongly concave in $u \in \mathbb{R}^m$; therefore $u(\lambda) = \operatorname{argmax}\{D(u, \lambda) : u \in \mathbb{R}^n\}$ is unique. The prox–function p is concave and differentiable. For its gradient, we have

$$\nabla p(\lambda) = \nabla_u D(u(\lambda), \lambda) \cdot \nabla_\lambda u(\lambda) + \nabla_\lambda D(u, \lambda),$$

where $\nabla_\lambda u(\lambda)$ is the Jacobian of $u(\lambda) = (u_1(\lambda), \ldots, u_m(\lambda))^T$. Keeping in mind $\nabla_u D(u(\lambda), \lambda) = 0$, we obtain

$$\nabla p(\lambda) = \nabla_\lambda D(u, \lambda) = \frac{1}{k}(u(\lambda) - \lambda) = \frac{1}{k}(\hat{\lambda} - \lambda)$$

or

$$\hat{\lambda} = \lambda + k\nabla p(\lambda). \tag{4.163}$$

In other words, the prox-method (4.161) is an explicit Euler method for the following system

$$\frac{d\lambda}{dt} = k\nabla p(\lambda),\ \lambda(0) = \lambda_0.$$

By reiterating (4.163), we obtain the dual sequence $\{\lambda_s\}_{s\in N}$:

$$\lambda_{s+1} = \lambda_s + k\nabla p(\lambda_s), \tag{4.164}$$

generated by the gradient method for maximization prox function p. The gradient ∇p satisfies Lipschitz condition with constant $L = k^{-1}$. Therefore, from Theorem 3.7 follows

$$\Delta p(\lambda_s) = p(\lambda^*) - p(\lambda_s) \leq O(sk)^{-1}.$$

We saw that the dual aspects of the penalty and the multipliers methods are critical for understanding their convergence properties, and LF identity is the universal instrument for obtaining the duality results.

By reiterating (4.158)–(4.159), we obtain the following Lagrange multipliers method (LMM)

$$x_{s+1} :\ \nabla_x \mathscr{L}(x_{s+1}, \lambda_s, k) = 0 \tag{4.165}$$

$$\lambda_{s+1} :\ \lambda_{i,s+1} = \lambda_{i,s} - kc_i(x_{s+1}), \quad i = 1, \ldots, m \tag{4.166}$$

We would like to emphasize that neither the primal sequence generated by (4.165) nor the dual sequence generated by (4.166) provides sufficient information for analysis of the LMM method (4.165)–(4.166).

Such information one gets by considering the primal–dual map $\Phi : \mathbb{R}^n \times \mathbb{R}^{2m} \times \mathbb{R}_{++} \to \mathbb{R}^{n+m}$ given by formula

$$\Phi(x, \hat{\lambda}, t, k) = \begin{pmatrix} \nabla f(x) - \sum_{i=1}^m \hat{\lambda}_i \nabla c_i(x) \\ c(x) + t - k^{-1}(\hat{\lambda} - \lambda^*) \end{pmatrix},$$

where $t = k^{-1}(\lambda - \lambda^*)$. The map is instrumental for establishing the basic AL convergence results.

To formulate the convergence results, we consider the extended dual feasible set

$$\Lambda(\lambda^*, \delta, k_0) = \{(\lambda, k) \in \mathbb{R}^m \times \mathbb{R}^1_{++} : |\lambda_i - \lambda_i^*| \leq \delta k,\ k \geq k_0,\ i = 1, \ldots, m\},$$

where $\delta > 0$ is small enough and $k_0 > 0$ is large enough.

The basic AL results establishes the following theorem.

Theorem 4.16. *Let f and all c_i, $i = 1,..,m$ be twice continuously differentiable and the second-order sufficient optimality condition* (4.73)–(4.74) *satisfied, then for any pair $(\lambda, k) \in \Lambda(\lambda^*, \delta, k_0)$, the following statements are taking place:*

(1) There exist $\hat{x}$ and $\hat{\lambda}$ given by formulas (4.158)–(4.159);

(2) The following bound

$$\max\left\{||\hat{x}-x^*||, ||\hat{\lambda}-\lambda^*||\right\} \leq \frac{c}{k}||\lambda-\lambda^*|| \tag{4.167}$$

holds and $c>0$ is independent of $k \geq k_0$;
(3) AL $\mathscr{L}(x,\lambda,k)$ is strongly convex in x in the neighborhood of $\hat{x}=\hat{x}(\lambda,k)$.

Proof.

(1) Using $t=k^{-1}(\hat{\lambda}-\lambda^*)$, we transform the neighborhood $\Lambda(\lambda^*,\delta,k_0)$ of the dual solution λ^* into the neighborhood $\Lambda(0,\delta,k_0)=\{(t,k)\in\mathbb{R}^m\times\mathbb{R}_{++}: |t_i|\leq \delta,\ i=1,\ldots,m, k\geq k_0\}$ of the origin in the extended dual space.
From KKT's condition follows

$$\Phi(x^*,\lambda^*,0^m,k)=0^{n+m}, \quad \forall k>0.$$

Let us consider the Jacobian $\nabla_{x\hat{\lambda}}\Phi(x,\hat{\lambda},t,k)$ at $x=x^*,\hat{\lambda}=\lambda^*,t=0^m$. We obtain

$$\nabla_{x,\hat{\lambda}}\Phi(x^*,\lambda^*,0^m,k)=$$
$$=\begin{bmatrix}\nabla^2 L(x^*,\lambda^*) & -\nabla c(x^*)^T\\ \nabla c(x^*) & -k^{-1}I^m\end{bmatrix}=\nabla\Phi_k.$$

Along with $\nabla\Phi_k$, let us consider the following matrix

$$\nabla\Phi_\infty=\begin{bmatrix}\nabla^2_{xx}L(x^*,\lambda^*) & -\nabla c(x^*)^T\\ \nabla c(x^*) & 0\end{bmatrix}.$$

From (4.138) we have $\|\Phi_\infty^{-1}\|\leq\rho_0^{-1}$.
From Lemma 4.2, we have $\|(\nabla\Phi_k)^{-1}\|\leq\rho^{-1}$, and $\rho>0$ is independent on $k\geq k_0$.
From the second implicit function theorem follows the existence of a vector–function $\hat{y}(t,k)=(\hat{x}(t,k),\hat{\lambda}(t,k))$ that

$$\Phi(\hat{x}(t,k),\hat{\lambda}(t,k),t,k)\equiv 0, \forall(t,k)\in\Lambda(0,\delta,k_0), \tag{4.168}$$

or

$$\nabla f(\hat{x}(t,k))-\sum_{i=1}^{m}\hat{\lambda}_i(t,k)\nabla c_i(\hat{x}(t,k))\equiv 0 \tag{4.169}$$

$$c(\hat{x}(t,k))+t-\frac{\hat{\lambda}(t,k)-\lambda^*}{k}\equiv 0. \tag{4.170}$$

(2) By differentiating the identities (4.169)–(4.170) in t under the fixed $k\geq k_0$, we obtain

$$\nabla_t \hat{y}(t,k) = \begin{bmatrix} \nabla_t \hat{x}(\cdot) \\ \nabla_t \hat{\lambda}_r(\cdot) \end{bmatrix} = \begin{pmatrix} \nabla^2_{xx} L(\cdot) & -\nabla c^T(\cdot) \\ \nabla c(\cdot) & -k^{-1} I^m \end{pmatrix}^{-1} \begin{pmatrix} O^{n,m} \\ -I^m \end{pmatrix} =$$

$$\nabla^2_{x,\hat{\lambda}} \Phi(\cdot)^{-1} \begin{pmatrix} O^{n,m} \\ -I^m \end{pmatrix},$$

where $O^{n,m}$ is a zero matrix in $\mathbb{R}^{n\times m}$ and I^m is the identical matrix in $\mathbb{R}^m$. Then, using Newton–Leibniz formula, we obtain

$$\begin{bmatrix} \hat{x}(t,k) - x^* \\ \hat{\lambda}(t,k) - \lambda^* \end{bmatrix} = \begin{bmatrix} x(t,k) - x(0,k) \\ \lambda(t,k) - \lambda(0,k) \end{bmatrix} =$$

$$= \int_0^1 \nabla_{x,\hat{\lambda}} \Phi^{-1}(x(\tau t,k); \lambda(\tau t,k)) \begin{bmatrix} O^{n,m} \\ -I^m \end{bmatrix} [t] d\tau$$

$$= \int_0^1 (\nabla \Phi_k(\cdot))^{-1} \begin{bmatrix} O^{n,m} \\ -I^m \end{bmatrix} [t] d\tau. \tag{4.171}$$

Due to the continuity of $\nabla^2 f$, $\nabla^2 c_i$, $i = 1, \ldots, m$, for small enough $\delta > 0$, we have

$$||(\nabla \Phi_k(\cdot))^{-1}|| \le 2\rho^{-1}, \forall (\lambda, k) \in \Lambda(0^m, \delta, k_0) =$$

$$\{(t,k) \in \mathbb{R}^m \times \mathbb{R}_{++} : |t_i| \le \delta,\ i = 1, \ldots, m,\ k \ge k_0\} \tag{4.172}$$

Therefore, there is $c = 2\rho^{-1} > 0$ independent on $(t,k) \in \Lambda(0, \delta, k_0)$ that

$$||\nabla_t \hat{y}(t,k)|| \le c, \quad \forall (t,k) \in \Lambda(0, \delta, k_0). \tag{4.173}$$

The bound (4.167) follows from (4.171)–(4.173).

(3) From the bound (4.167) follows

$$\nabla^2_{xx} \mathscr{L}(\hat{x}, \hat{\lambda}, k) \approx \nabla^2 L(x^*, \lambda^*) + k \nabla c^T(x^*) \nabla c(x^*).$$

From the second-order sufficient condition (4.73)–(4.74) and Debreu's Lemma follows existence of $\mu > 0$ that

$$\langle \nabla^2_{xx} \mathscr{L}(\hat{x}, \hat{\lambda}, k) y, y \rangle \ge \mu \langle y, y \rangle,\ \forall y \in \mathbb{R}^n$$

for any $k \ge k_0$.
So $\hat{x} = \hat{x}(\lambda, k)$ is a local minimizer of $\mathscr{L}(x, \lambda, k)$.
Using arguments similar to those used to prove that

$$x^* = \arg\min\{\mathscr{L}(x, \lambda^*, k) \mid x \in \mathbb{R}^n\},$$

one can show that for $k_0 > 0$ large enough and any $k \ge k_0$

$$\hat{x} \equiv \hat{x}(\hat{\lambda}, k) = \operatorname{argmin}\{\mathscr{L}(x, \lambda, k) \mid x \in \mathbb{R}^n\}.$$

We would like to point out that a vector $\bar{y} = (\bar{x}, \bar{\lambda})$, which satisfies the Lagrange system of Equations (4.10)–(4.11), is not what necessarily defines a local solution (4.8). In particular, $\bar{x}$ can be a local or even a global maximizer of f on Ω. The following remark is a consequence of the Theorem 4.16

Remark 4.1. Under conditions of Theorem 4.16 for large enough $k_0 > 0$ and small enough $\delta > 0$, any vector

$$\hat{y} \in \hat{Y}_{\delta,k_0} = \{\hat{y} = (\hat{x}, \hat{\lambda}) = (\hat{x}(\lambda,k), \hat{\lambda}(\lambda,k)) : (\lambda,k) \in \Lambda(\lambda^*,\delta,k_0)\}, \quad (4.174)$$

which satisfies the Lagrange system (4.10)–(4.11) is the primal–dual solution.

4.11 Primal–Dual AL Method for ECO

The LMM (4.165)–(4.166) require at each step solving an unconstrained minimization problem (4.165), which is, generally speaking, an infinite procedure.

In this section, we introduce and analyze primal–dual augmented Lagrangian (PDAL) method, which requires at each step solving one linear system of equations. It generates a primal–dual sequence, which locally converges to the primal–dual solution with quadratic rate.

First, let us introduce the merit function $\nu : \mathbb{R}^{n+m} \to \mathbb{R}_+$, defined by formula

$$\nu(y) = \max\{||\nabla_x L(x,\lambda)||, ||c(x)||\}. \quad (4.175)$$

It follows from (4.175) that $\nu(y) \geq 0$. Keeping in mind Remark 4.1 for any $y \in \hat{Y}_{\delta,k_0}$ and large enough $k_0 > 0$, we have

$$\nu(y) = 0 \Leftrightarrow y = y^*. \quad (4.176)$$

We will use $\nu(y)$ for measuring the distance from $y \in B(y^*,\delta) = \{y \in \mathbb{R}^{n+m} : \|y - y^*\| \leq \delta\}$ to the primal–dual solution $y^* = (x^*, \lambda^*)$.

The following Lemma shows that the merit function $\nu(y)$ locally has properties similar to the norm of a gradient of a strongly convex function with Lipschitz continuous gradient.

Lemma 4.4. *If the second-order sufficient optimality conditions (4.73)–*(4.74) *hold and Lipschitz condition* (4.130) *is satisfied, then there exists* $0 < l < L < \infty$ *and small enough* $\delta > 0$ *that for any* $y \in B(y^*,\delta)$ *the following bounds hold*

$$l\|y - y^*\| \leq \nu(y) \leq L\|y - y^*\|. \quad (4.177)$$

Proof. The right inequality (4.177) follows from $\nu(y^*) = 0$ and Lipschitz condition (4.130).

On the other hand,

$$\|\nabla_x L(x,\lambda)\| \leq \nu(y), \; \|c(x)\| \leq \nu(y). \quad (4.178)$$

By linearizing Lagrange system (4.10)–(4.11) at $y^* = (x^*, \lambda^*)$, we obtain for $y = (x, \lambda) \in B(y^*, \delta)$ the following system

$$\begin{pmatrix} \nabla^2_{xx} L(x^*, \lambda^*) & -\nabla^T c(x^*) \\ \nabla c(x^*) & O^{m,m} \end{pmatrix} \begin{pmatrix} x - x^* \\ \lambda - \lambda^* \end{pmatrix} = \nabla \Phi_\infty (y - y^*)$$
$$= \begin{pmatrix} \nabla_x L(x, \lambda) + o(\|x - x^*\|) e_n \\ c(x) + o(\|x - x^*\|) e_m \end{pmatrix}, \tag{4.179}$$

where $e_r = (1, \ldots, 1)^T \in \mathbb{R}^r$.

We knew already that under the second-order sufficient conditions (4.73)–(4.74), the matrix $\nabla \Phi_\infty$ is nonsingular and there is $\rho_0 > 0$ that

$$\|\nabla \Phi_\infty^{-1}\| \leq \rho_0^{-1}. \tag{4.180}$$

From (4.179) follows

$$\begin{pmatrix} x - x^* \\ \lambda - \lambda^* \end{pmatrix} = \nabla \Phi_\infty^{-1} \begin{pmatrix} \nabla_x L(x, \lambda) + o(\|x - x^*\|) e_n \\ c(x) + o(\|x - x^*\|) e_m \end{pmatrix}$$

Keeping in mind (4.180), we obtain

$$\|y - y^*\| \leq \rho_0^{-1} \nu(y) + o(\|x - x^*\|).$$

Therefore, for any $y \in B(y^*, \delta)$, we have $\nu(y) \geq l \|y - y^*\|$, where $l = 0.5 \rho_0$. □

We will use bounds (4.177) to control the penalty parameter $k > 0$.

Let us fix $(\lambda, k) \in \Lambda(\lambda^*, \delta, k_0)$; then one step of the AL method (4.158)–(4.159) is equivalent to solving the following nonlinear PD system

$$\nabla_x L(\hat{x}, \hat{\lambda}) = \nabla f(\hat{x}) - \sum_{i=1}^m \hat{\lambda}_i \nabla c_i(\hat{x}) = 0 \tag{4.181}$$

$$\hat{\lambda} = \lambda - k c(\hat{x}) \tag{4.182}$$

for $(\hat{x}, \hat{\lambda})$.

Application of Newton's method to the PD system (4.181)–(4.182) with $y = (x, \lambda) \in B(y^*, \delta)$ as a starting point leads to the PDAL method.

By linearizing (4.181)–(4.182) at $y = (x, \lambda)$, we obtain the following linear PD system for $\Delta y = (\Delta x, \Delta \lambda)$

$$\nabla f(x) + \nabla^2 f(x) \nabla x - \sum_{i=1}^m (\lambda_i + \triangle \lambda_i) \left(\nabla c_i(x) + \nabla^2 c_i(x) \triangle x \right) = 0 \tag{4.183}$$

$$\lambda + \triangle \lambda = \lambda - k (c(x) + \nabla c(x) \triangle x) \tag{4.184}$$

Ignoring terms of the second and higher orders, we can rewrite the system (4.183)–(4.184) as follows

$$\nabla \Phi_k(x,\lambda)\triangle y = \begin{pmatrix} \nabla^2_{xx}L(x,\lambda) & -\nabla^T c(x) \\ \nabla c(x) & k^{-1}I^m \end{pmatrix} \begin{pmatrix} \triangle x \\ \triangle \lambda \end{pmatrix} =$$
$$= \begin{pmatrix} -\nabla_x L(x,\lambda) \\ -c(x) \end{pmatrix} = -\nabla N(y). \tag{4.185}$$

The critical issue is how to update the penalty parameter $k > 0$ step by step to guarantee convergence of the PDAL method with quadratic rate. It turned out that by taking the penalty parameter as inverse of the merit function, one can guarantee convergence of the primal–dual sequence to the primal–dual solution $y^* = (x^*;\lambda^*)$ with quadratic rate from any $y \in \hat{Y}_{\delta,k_0}$ as a starting point.

We are ready to describe the PDAL method.

Let $y \in \hat{Y}_{\delta,k_0}$ be the starting point and $\varepsilon > 0$ be the required accuracy. The PDAL method consists of the following operations:

1. If $\nu(y) \leq \varepsilon$, then $y^* := y$, else
2. Find the primal–dual direction $\overline{\triangle y} = (\overline{\triangle x}, \overline{\triangle \lambda})$ from the linear PD system (4.185)
3. Find the new primal–dual vector $\bar{y} = (\bar{x}, \bar{\lambda})$:

$$\bar{x} = x + \overline{\triangle x}, \quad \bar{\lambda} = \lambda + \overline{\triangle \lambda} \tag{4.186}$$

4. Update the scaling parameter

$$\bar{k} := (\nu(\bar{y}))^{-1} \tag{4.187}$$

5. Set

$$y := \bar{y}, \; k := \bar{k} \text{ and go to 1}. \tag{4.188}$$

To prove quadratic convergence rate of the PDAL method, we need the following lemma.

Lemma 4.5. *If the second-order sufficient optimality condition (4.73)–(4.74) holds and Lipschitz condition* (4.130) *is satisfied, then there is small enough $\delta > 0$ and large enough $k_0 > 0$ that the matrices*

$$\nabla \Phi_\infty(x,\lambda) = \begin{pmatrix} \nabla^2 L(x,\lambda) & -\nabla^T c(x) \\ \nabla c(x) & O^{m,m} \end{pmatrix} \text{ and } \nabla \Phi_k(x,\lambda) = \begin{pmatrix} \nabla^2 L(x,\lambda) & -\nabla^T c(x) \\ \nabla c(x) & k^{-1}I^m \end{pmatrix}$$

are nonsingular and there is $\rho > 0$ independent on $y \in B(y^*,\delta)$ such that the following bound

$$\max\{||\nabla \Phi_\infty^{-1}(y)||, ||\nabla \Phi_k^{-1}(y)||\} \leq \rho^{-1} \tag{4.189}$$

holds for any $y \in B(y^*,\delta)$ and $k \geq k_0$.

Proof. We have seen already that $||\Phi_\infty^{-1}(y^*)|| \leq \rho_0^{-1}$.

From Lemma 4.2 follows existence $0 < \rho < \rho_0$ independent on $k \geq k_0$ that the following bound

$$\max\{\left\|\nabla \Phi_\infty^{-1}(y)\right\|, \left\|\nabla \Phi_k^{-1}(y)\right\|\} \leq \rho^{-1}, \; \forall y \in B(y^*,\delta)$$

holds. □

Now we are ready to prove the quadratic convergence of the PDAL method 1.-5.

Theorem 4.17. *If the second-order sufficient optimality condition (4.73)–(4.74) holds and Lipschitz condition (4.130) is satisfied, then there exist a small enough $\delta > 0$ and a large enough $k_0 > 0$ such that for any starting point $y_0 \in \hat{Y}_{\delta,k_0}$ the PDAL method 1.-5. generates primal–dual sequence, which converges to the primal–dual solution $y^* = (x^*; \lambda^*)$ and the following bound*

$$||\bar{y} - y^*|| \leq c||y - y^*||^2$$

holds, where $c > 0$ is independent on $y \in B(y^*, \delta)$.

Proof. We find the primal–dual Newton direction $\overline{\triangle y} = (\overline{\triangle x}, \overline{\triangle \lambda})$ from the PD linear system (4.185) Now let us consider Newton's step applied to Lagrange system (4.125) at the same point $y = (x, \lambda)$. We find the Newton direction $\widehat{\triangle y} = (\widehat{\triangle x}, \widehat{\triangle \lambda})$ from the following system of linear equations

$$\nabla N(y)\widehat{\triangle y} = -N(y).$$

In the new approximation $\hat{y} = (\hat{x}, \hat{\lambda})$, we obtain the formula

$$\hat{y} = y + \widehat{\triangle y}.$$

From Theorem 4.14 follows existence $c_1 > 0$ independent of $y \in B(y^*, \delta)$ that the following bound holds

$$||\hat{y} - y^*|| \leq c_1||y - y^*||^2. \tag{4.190}$$

Let us prove that similar bound holds for $\bar{y} = (\bar{x}, \bar{\lambda})$. We have

$$\begin{aligned} ||\bar{y} - y^*|| &= ||y + \overline{\triangle y} - y^*|| = ||y + \widehat{\triangle y} + \overline{\triangle y} - \widehat{\triangle y} - y^*|| \\ &\leq ||\hat{y} - y^*|| + ||\overline{\triangle y} - \widehat{\triangle y}||. \end{aligned}$$

For $||\overline{\triangle y} - \widehat{\triangle y}||$, we obtain

$$\begin{aligned} ||\overline{\triangle y} - \widehat{\triangle y}|| &= ||(\nabla \Phi_k^{-1}(y) - \nabla \Phi_\infty^{-1}(y))N(y)|| \\ &\leq ||(\nabla \Phi_k^{-1}(y) - \nabla \Phi_\infty^{-1}(y))||||N(y)||. \end{aligned} \tag{4.191}$$

Using Lemma 4.2 and keeping in mind

$$||(\nabla \Phi_k(y) - \nabla \Phi_\infty(y))|| = k^{-1}$$

from (4.191), we obtain

$$||\overline{\triangle y} - \widehat{\triangle y}|| \leq 2k^{-1}\rho^2||N(y)||. \tag{4.192}$$

In view of $\nabla_x L(x^*, \lambda^*) = 0$, $c(x^*) = 0$ and Lipschitz condition for ∇f and ∇c_i, there exists $L_0 > 0$ such that

$$||N(y)|| \leq L_0||y - y^*||, \forall y \in B(y^*, \delta) . \tag{4.193}$$

Using (4.177), (4.187), (4.192), and (4.193), we obtain

$$\begin{aligned} ||\overline{\triangle y} - \widehat{\triangle y}|| &\leq \rho^2 L_0 \nu(y)||y - y^*|| \\ &\leq \rho^2 L L_0 ||y - y^*||^2 \\ &= c_2||y - y^*||^2 . \end{aligned} \tag{4.194}$$

From (4.190) and (4.194), we obtain

$$||\bar{y} - y^*|| \leq ||\hat{y} - y^*|| + ||\triangle y - \triangle \bar{y}|| \leq c||y - y^*||^2,$$

where $c = 2\max\{c_1, c_2\}$ is independent on $y \in B(y^*, \delta)$. □

Notes

4.1 For the first-order optimality condition in ECO, see Bertsekas (1999); Boyd and Vanderberghe (2004); Fiacco and McCormick (1990); B. Polyak (1987).

4.2 For the second-order necessary and sufficient optimality condition in ECO, see Bertsekas (1999); Boyd and Vanderberghe (2004); B. Polyak (1987); for the eigenvalues theorem, see Polyak (1973).

4.3 For the optimality condition with both inequality constrains and equations, see Bertsekas (1999); Fiacco and McCormick (1990); B. Polyak (1987).

4.4 For the duality in constrained optimization, see Bertsekas (1982, 1999); Fiacco and McCormick (1990); Grossman and Kaplan (1981); Ioffe and Tichomirov (1968); Ioffe and Tihomirov (2009); Rockafellar (1973); Rockafellar and Wets (2009); Rockafellar (1976a).

4.5 For the Courant's penalty method, see Courant (1943); for Tikhonov's regularization, see Tikhonov (1963); Tikhonov and Arsenin (1977); for their equivalence, see Polyak (2016).

4.6 For the gradient method for ECO, see B. Polyak (1970, 1987) and references therein.

4.7 For the Newton's method for nonlinear systems, see Boyd and Vanderberghe (2004); Dennis and Schnabel (1996); Kantorovich and Akilow (1964); B. Polyak (1987) and references therein.

4.8 For the Newton's method for ECO, see Bertsekas (1999); Boyd and Vanderberghe (2004); B. Polyak (1970); B. Polyak and Tret'yakov (1973); B. Polyak (1987).

4.9 The augmented Lagrangians were introduced in Hestenes (1969) and independent in Powell (1969); see also Antipin (1977, 1979); Bertsekas (1982, 1976);

DiPillo and Grippo (1979); Goldshtein and Tretiakov (1989); Grossman and Kaplan (1981); Haarhoff and Buys (1970); B. Polyak (1970, 1987); Rockafellar (1973, 1976,a).

4.10 For the multipliers method and dual quadratic prox, see Rockafellar (1973, 1976,a) and references therein. Also see Goldfarb et al. (1999); Goldshtein and Tretiakov (1989); Griva and Polyak (2013); Kort and Bertsekas (1973); B. Polyak and Tret'yakov (1973); B. Polyak (1987).

4.11 For the primal–dual AL method, see Polyak (2009).

Chapter 5
Basics in Linear and Convex Optimization

5.0 Introduction

In the first part of this chapter we cover basic LP facts, including LP duality, special LP structure, as well as two main tools for solving LP: simplex and interior point methods (IPMs).

In the second part we cover basic convex optimization (CO) facts, including KKT's Theorem and CO duality. We emphasize the role of the LF transform and CO duality in the convergence analysis of SUMT methods.

In particular, SUMT is viewed as interior regularization for the dual problem. It allows to prove convergence of SUMT methods in a unified manner and establish error bounds for these methods.

We conclude the chapter by considering three first-order classical CO methods: gradient projection (GP), conditional gradient (CG), and primal–dual feasible direction (PDFD) methods.

Often computing second-order information for large-scaled convex optimization problems is not realistic; therefore speeding up first-order methods is critical for important applications.

Along with classical GP method, we consider fast GP, which has a much better convergence rate.

The GP method is efficient if projection on the feasible set is an easy operation.

Often, however, projection on the feasible set is an operation similar to solving the original problem; then one needs other first-order methods.

In particular, the conditional gradient (CG) method, introduced for QP by Frank and Wolfe (1956), got a great deal of attention lately. The convergence and the rate of convergence were analyzed by E. Levitin and B. Polyak in their classical 1966 paper.

The CG method, however, has well-known limitation. Even for strongly convex and smooth enough objective function and simple constraints, it does not converge, generally speaking, faster than $O(k^{-1})$, where k is the number of steps.

R. A. Polyak, *Introduction to Continuous Optimization*, Springer Optimization and Its Applications 172, https://doi.org/10.1007/978-3-030-68713-7_5

For special feasible sets, however, the CG method can be very efficient. For a general convex optimization problem, it might require at each step solving a problem similar to the initial one.

In such case, the primal–dual feasible direction (PDFD) method (see Polyak (1966)) can be more efficient, because at each step it requires solving, usually, small QP problems.

5.1 Linear Programming

Before World War II L. V. Kantorovich (1912–1986) recognized that a number of real-world problems, associated with optimal distribution limited resources, from the mathematical stand point, are problems of finding min or max of a linear function under linear equality and inequality constraints. This was the beginning of linear programming (LP).

Kantorovich knew that the optimal solution one finds among the vertices of the polyhedron, which defines the feasible set. Moreover, Kantorovich found the optimality criteria for the solution–vertex. The criteria were based on the so-called objectively conditioned estimates (OCE), which is now known as shadow prices.

Unfortunately, Kantorovich's work was practically unknown for about 20 years. His book *The best use of Economic Resources* was published in 1959.

It was clear that Kantorovich not only knew the optimality criteria; he also knew how to use OCE to improve current feasible solution, if it is not the optimal one. The book, however, have not contained any well-grounded method for solving LP.

The simplex method for solving general LP problem has been introduced by George B. Dantzig (1914–2005) in 1947. The impact of his work on the LP theory and methods is hard to overestimate.

"To this day the simplex algorithm remains the primary computational tool in linear and mixed—integer optimization"; see Bixby (2012).

The LP was extremely active area of research from the late 1940s till the early 1960s. A number of fundamental results were obtained by G. Dantzig and his colleagues during these very productive for LP years; see, for example, Dantzig et al. (1954); G. Dantzig et al. (1956); Dantzig and Wolfe (1960); T. Koopmans (1949).

Not only G. Dantzig introduced the simplex algorithm; he together with his colleagues used it for solving real-world problems. One of the first such problem was Stigler Diet Problem with 21 constraints and 77 variable. It took 120 man-days to solve it by computational means, which those days were available. The results of the "golden years" of LP was summarized by G. Dantzig in his book *Linear programming and extensions*, Prinston University Press, Princeton (1963).

During the 1970s and 1980s, substantial progress was made in understanding the numerical aspects of simplex method. It was clear, however, from the early 1970s that the number of simplex method steps can grow exponentially together with the size of the problem. The existence of a polynomial method for LP was an important issue for quite some time.

In 1979 Leonid Khachiyan, using ellipsoid method, shows that LP can be solved in polynomial time. The result got a number of important theoretical implications, in particular, in combinatorial optimization.

Numerically, however, polynomial ellipsoid method was not able to compete with exponential simplex method.

Up to now much improved simplex codes are used in commercial applications, as well as, to crossover to a basis, if a code, based on IPM, is used in the first phase.

The fact that a number of mathematical problems can be formulated as LP was known for a long time.

In 1826 Fourier (1768–1830) pointed out that the bottom of a piece-wise linear surface can be reached fast by moving along the descent adges from one vertex to another.

This is exactly the basic idea of Dantzig's simplex method.

Without knowing Western developments in LP (the "iron curtain" has been already in place), Simon I. Zuchovitsky (1908–1994), using the Fourier's idea, in 1949 developed a method for finding the Chebyshev approximation for overdetermined systems of linear equations. Geometrically, it is exactly the problem of finding a bottom of a piece-wise linear surface. Such a problem, on the other hand, is equivalent to a particular LP; see Cheney and Goldstein (1958), Stiefel (1960).

It took another 10 years before S. Zuchovitsky learned from Stiefel's (1960) paper that his algorithm is, in fact, the simplex method for LP equivalent to finding the Chebyshev approximation for an overdetermined linear system.

This was the turning point in the life of S. I. Zuchovitsky and his students, including the author.

For a long time, simplex method was the only working tool for solving LP. The modern numerical realization of simplex method still remains an efficient instrument for solving large-scale LP.

The dominance of simplex method, however, ended at the beginning of the 1990s, when new efficient LP tools based on interior point methods were developed.

5.1.1 Primal and Dual LP Problems

The LP problem is defined by a matrix $A \in \mathbb{R}^{m \times n}$ and two column—vectors $c \in \mathbb{R}^n$, $b \in \mathbb{R}^m$. It consists of finding

$$\langle c, x^* \rangle = \max\{\langle c, x \rangle \mid x \in \Omega\}, \tag{5.1}$$

where

$$\Omega = \{x \in \mathbb{R}^n_+ : Ax \leq b\}. \tag{5.2}$$

The centerpiece of LP is the duality theory. Along with primal LP problem (5.1) always exists the co-called dual LP

$$\langle b, \lambda^* \rangle = \min\{\langle b, \lambda \rangle | \; \lambda \in Q\}, \tag{5.3}$$

where

$$Q = \{\lambda \in \mathbb{R}^m_+ : A^T \lambda \geq c\}. \tag{5.4}$$

There are three aspects associated with primal and dual LP: algebraic, geometric, and economic. Let us start with the economic aspect.

We consider an economy, which produces n products by consuming m resources.

The component c_j of the price–vector c defines the market price of one unit of product $1 \leq j \leq n$.

The component b_i of the resource vector b defines the availability of the resource $1 \leq i \leq m$.

The technological matrix $A = (A_1, \ldots, A_n)$ is made up of column–vectors A_j, elements of which indicate the amount of each resource required to produce one unit of product $1 \leq j \leq n$.

In fact, matrix A transforms the resource vector $b \in \mathbb{R}^m_+$ into production vector $x \in \mathbb{R}^n_+$. The component x_j of the production vector x defines how many units of product $1 \leq j \leq n$ one is going to produce.

The system (5.2) defines the primal feasible set. The primal LP (5.1) find such feasible production vector $x^* \in \mathbb{R}^n_+$, which maximizes the value (c, x) of the production output $x \in \Omega$.

Along with $x^* \in \mathbb{R}^n_+$ one finds the resource allocation vectors $A_1 x_1^*, \ldots, A_n x_n^*$, which show how much of a given resource vector b was allocated to each product (technology or sector of an economy) in the optimal solution.

If results of the "optimal" allocation are not satisfactory, the owner of the resource vector $b \in \mathbb{R}^m_+$ might consider selling the resources.

The price offer comes from the buyer. The only owner's requirement is to make a "fair deal": under any price vector $\lambda \in \mathbb{R}^m_+$ each column A_j cannot cost less than the given market price c_j of one unit of product j. In other words $(A_j, \lambda) \geq c_j$, $1 \leq j \leq n$, that is, the price vector λ has to satisfy (5.4).

Obviously, under constraints (5.4) the buyer will minimize its expenses. It leads to the dual LP (5.3).

Exercise 5.1. Formulate rules one has to follow to write down the dual LP problem using the primal LP information.

Both primal and dual LP are based on the same information, therefore the existence of their solutions, their optimal values, and the primal and dual optimal vectors are closely connected. The LP duality theory establishes such connections.

The LP duality is due to John von Neumann (1903–1957), who formulated the Duality Theorems in a conversation with George Dantzig in 1947.

First, let us prove a simple, but a critical Lemma.

Lemma 5.1 (Weak Duality). *For any primal feasible vector x and dual feasible vector λ the following inequality holds*

$$\langle c, x \rangle \leq \langle b, \lambda \rangle.$$

Proof. For a primal feasible solution x we have

$$b - Ax \geq 0, \ x \geq 0. \tag{5.5}$$

For a dual feasible solution λ we have

$$A^T\lambda - c \geq 0, \ \lambda \geq 0. \tag{5.6}$$

Let us multiply the first inequality of (5.5) by $\lambda \in \mathbb{R}^n_+$ and the first inequality of (5.6) by $x \in \mathbb{R}^n_+$. We obtain

$$\langle \lambda, b - Ax \rangle \geq 0; \ \langle A^T\lambda - c, x \rangle \geq 0$$

or

$$\langle b, \lambda \rangle \geq \langle \lambda, Ax \rangle = \langle A^T\lambda, x \rangle \geq \langle c, x \rangle.$$

□

Corollary 5.1. *If $\bar{x}$ and $\bar{\lambda}$ are primal and dual feasible solution, for which*

$$\langle b, \bar{\lambda} \rangle = \langle c, \bar{x} \rangle \tag{5.7}$$

then $\bar{x} = x^*$ and $\bar{\lambda} = \lambda^*$.

From Lemma 5.1 follows $\langle c, x \rangle \leq \langle b, \bar{\lambda} \rangle$ for any $x \in \Omega$, but for $\bar{x}$ we have (5.7), so $\bar{x}$ is the maximizer in (5.1). On the other hand, for any $\lambda \in Q$ we have $\langle b, \lambda \rangle \geq \langle c, \bar{x} \rangle$, but for $\lambda = \bar{\lambda}$ we have (5.7); therefore $\bar{\lambda}$ is the minimizer in (5.3).

The standard primal LP (5.1) can be rewritten in the canonical form by introducing the nonnegative slack vector $x_{(s)} = (x_{n+1}, \ldots, x_{n+m})$. We have

$$\langle c, x \rangle = \max\{\langle c, x \rangle | Ax = b, x \in \mathbb{R}^{n+m}_+\} \tag{5.8}$$

where $c := (c; 0, \ldots, 0)$, $x := (x; x_{(s)}) \in \mathbb{R}^{n+m}_+$ $A := (A; I^m) : \mathbb{R}^{m \times (n+m)}$ and $I : \mathbb{R}^m \to \mathbb{R}^m$—identical matrix.

Exercise 5.2. Show the equivalence of the standard LP (5.1) and canonical LP (5.8).

5.1.2 Optimality Condition for LP Problem

We assume that the solution set $X^* = \{x \in \Omega : \langle c, x \rangle = \langle c, x^* \rangle\}$ is bounded. It follows from Corollary 2.4 that by adding one constraint $\langle c, x \rangle \leq N$ with N large enough we obtain LP equivalent to (5.1) with a bounded feasible set. For N large enough the extra constraint does not effect the solution set X^*. Therefore, from this point on, we assume that Ω is a polytop.

We recall that $x \in \Omega$ is a vertex if $x \neq (1-\lambda)x_1 + \lambda x_2$ for any two vectors x_1 and x_2 from Ω and any $0 < \lambda < 1$.

Lemma 5.2. *If Ω is a polytop, then there is a vertex $x^* \in X^*$.*

Proof. Let $x_1, \ldots, x_M$ be the set of vertices of Ω. From Proposition 2.2 follows that for any given $x \in \Omega$, there exist such $\mu_i \geq 0$, $i = 1, \ldots, M$ that $x = \sum_{i=1}^{M} \mu_i x_i$ and $\sum_{i=1}^{M} \mu_i = 1$.

Let $\langle c, x_1 \rangle \geq \langle c, x_i \rangle$, $i = 2, .., M$, then

$$\langle c, x \rangle = \sum_{i=1}^{M} \mu_i \langle c, x_i \rangle \leq \sum_{i=1}^{M} \mu_i \langle c, x_1 \rangle = \langle c, x_1 \rangle$$

for any $x \in \Omega$.

So among the solutions of LP always exists a vertex $x^* \in X^*$. □

Let us consider the following LP in the canonical form

$$\langle c, x^* \rangle = \max\{\langle c, x \rangle \mid Ax = b, x \in \mathbb{R}_+^n\}, \tag{5.9}$$

where A is $m \times n$ matrix $c \in \mathbb{R}^n$, $b \in \mathbb{R}^m$ and $\operatorname{rank} A = m < n$.

Feasible vector $x = (x_1, \ldots, x_n)$ is a basic solution if columns A_j of the matrix A in (5.9), which correspond to positive components x_j, are linear independent. It is clear that a basic solution $x = (x_1, \ldots, x_n)$ cannot have more than m positive components, but it can have less than m. In such case the solution is degenerate, because x belongs to more than n hyperplanes.

The geometric equivalent of a basic solution is a vertex.

Lemma 5.3. *Feasible vector $x \in \Omega$ is a vertex if and only if x is a basic solution in LP (5.8).*

Proof. Let $x = (x_1, \ldots, x_m, 0, \ldots, 0)$ be a vertex. If x is not a basic solution, then vectors $A_1, \ldots, A_m$ are linear dependent, i.e., there are $\alpha_1, \ldots, \alpha_m$ such that

$$\alpha_1 A_1 + \ldots \alpha_m A_m = 0$$

and not all α_i are zeros. Let $\alpha_1 \neq 0$, then for small $t > 0$, we obtain two feasible vectors $\bar{x} = (x_1 + t\alpha_1, \ldots, x_m + t\alpha_m, 0, \ldots, 0) \neq 0$ and $\bar{\bar{x}} = (x_1 - t\alpha_1, \ldots, x_m - t\alpha_m, 0, \ldots, 0)$ that $x = \frac{1}{2}(\bar{x} + \bar{\bar{x}}) \in \Omega$, which means that x is not a vertex.

Conversely, assuming that $x = (x_1, \ldots, x_m, 0, \ldots, 0)$ is a basic solution but not a vertex, then there exist $\bar{x} = (\bar{x}_1, \ldots, \bar{x}_m, 0 \ldots, 0)$ and $\bar{\bar{x}} = (\bar{\bar{x}}_1, \ldots, \bar{\bar{x}}_m, 0..0)$, $\bar{x} \neq \bar{\bar{x}}$ and $x = \frac{1}{2}(\bar{x} + \bar{\bar{x}})$. From the feasibility of $\bar{x}$ and $\bar{\bar{x}}$ follows

$$\bar{x}_1 A_1 + \ldots \bar{x}_m A_m = b$$

and

$$\bar{\bar{x}}_1 A_1 + \ldots + \bar{\bar{x}}_m A_m = b.$$

Therefore,

$$(\bar{x}_1 - \bar{\bar{x}}_1) A_1 + \ldots + (\bar{x}_m - \bar{\bar{x}}_m) A_m = 0,$$

but for linear independent vectors $A_1, \ldots, A_m$, it is possible only if $\bar{x} = \bar{\bar{x}}$.

It follows from Lemma 5.3 that moving from one vertex the adjacent one is equivalent to moving from a basic solution to another one which is obtained from the original one by replacing one column.

Exercise 5.3. Show that the following LP

$$\langle b,\lambda^*\rangle = \min\{\langle b,\lambda\rangle \mid A^T\lambda \geq c, \lambda \in \mathbb{R}^m\} \tag{5.10}$$

is the dual to the LP (5.9).

Our goal now is to establish the optimality criteria for basic solution $\bar{x} \in \mathbb{R}^n_+$ in LP (5.8).

Without loss of generality we can assume $\bar{x} = (\bar{x}_1,\ldots,\bar{x}_m;0,\ldots,0)$. It means that columns $A_1,\ldots,A_m$ are linear independent or $\operatorname{rank}[A_1,\ldots,A_m] = \operatorname{rank} B = m$. Let us split the row price–vector $c = (c_B,c_N) = (c_1,\ldots,c_m;c_{m+1},\ldots,c_n)$ and the matrix $A = (B;N)$ accordingly.

Theorem 5.1. *For row vector $\bar{x} = (\bar{x}_B,\bar{x}_N)$ to be the optimal basic solution in LP (5.8), it is necessary and sufficient existence of column vector $\bar{\lambda} \in \mathbb{R}^m$ that*

$$a)\Delta_B = c_B - \bar{\lambda}^T B = 0 \text{ and } b)\Delta_N = c_N - \bar{\lambda}^T N \leq 0. \tag{5.11}$$

Proof. Let us start with sufficiency. If (5.11) holds, then along with primal basic solution $\bar{x} = (\bar{x}_B = B^{-1}b;0)$, there exists dual feasible solution $\bar{\lambda}^T = c_B B^{-1}$ that

$$\bar{z} = (c,\bar{x}) = (c_B,\bar{x}_B) = c_B B^{-1} b = \bar{\lambda}^T b = \bar{w}.$$

Therefore, from Corollary 5.1 follows $\bar{x} = x^*$, $\bar{\lambda} = \lambda^*$.

Conversely, let $\bar{x} = (\bar{x}_B;0)$ be the primal basic optimal solution of (5.8). Then the system (5.11a) has a unique solution $\bar{\lambda}^T = c_B B^{-1}$. We have to show that $\bar{\lambda} \in \mathbb{R}^m$ is such that (5.11b) holds. Let assume that (5.11b) is not true, then there exists $m+1 \leq j_0 \leq n$ that

$$\Delta_{j_0} = c_{j_0} - \bar{\lambda}^T A_{j_0} > 0. \tag{5.12}$$

There is a unique representation of column A_{j_0}:

$$Bx_{B,j_0} = x_{1j_0}A_1 + \ldots + x_{mj_0}A_m = A_{j_0}$$

through the basis B.

In other words

$$x_{B,j_0} = B^{-1}A_{j_0}$$

or

$$-Bx_{B,j_0} + A_{j_0} = 0. \tag{5.13}$$

From (5.13) follows: to produce a unit of product j_0 one has to "reduce" the production of basic products by $x_{1j_0},\ldots,x_{mj_0}$ units each. Such reduction leads to a "loss" of $(c_B,x_{B,j_0}) = c_B B^{-1} A_{j_0} = \bar{\lambda}^T A_{j_0}$ versus "gain" of c_{j_0} as compare with basic solution $(\bar{x}_B,0)$.

Therefore, from (5.12) follows that $\bar{x}$ cannot be an optimal solution.

The contradiction shows that (5.11b) holds. □

Now we apply the optimality criteria given by Theorem 5.1 to prove the "truncated" Duality Theorem.

Let $c \in \mathbb{R}^n$, $b \in \mathbb{R}^m$, $A \in \mathbb{R}^{m\times n}$ we consider the following primal LP:

$$\langle c,x^*\rangle = \max\{\langle c,x\rangle | Ax \leq b\}. \tag{5.14}$$

First, we present $\Omega = \{x \in \mathbb{R}^n : Ax \leq b\}$ in a standard form (5.8).

Let $x_i = x_j^{'} - x_j^{''}$ and $x_j^{'} \geq 0$, $x_j^{''} \geq 0$, $1 \leq j \leq n$, then the primal LP can be rewritten as follows:

$$\langle c,x^*\rangle = \max\{\langle c,x^{'}\rangle - \langle c,x^{''}\rangle | Ax^{'} - Ax^{''} \leq b, x^{'} \in \mathbb{R}^n_+, x^{''} \in \mathbb{R}^n_+\}. \tag{5.15}$$

For the dual LP we have

$$\langle b,\lambda^*\rangle = \min\{\langle b,\lambda\rangle | A^T\lambda \geq c, -A^T\lambda \geq -c, \lambda \in \mathbb{R}^m_+\}.$$

Combining $A^T\lambda \geq c$ and $A^T\lambda \leq c$, we obtain the following dual problem:

$$w^* = \langle b,\lambda^*\rangle = \min\{\langle b,\lambda\rangle | A^T\lambda = c, \lambda \in \mathbb{R}^m_+\}. \tag{5.16}$$

Now we rewrite LP (5.15) in the canonical form (5.8) by introducing the slack vector $x_{(s)} = (x_{2n+1},\ldots,x_{2n+m}) \in \mathbb{R}^m_+$.

Let $A := [A; -A, I]$, where $I : \mathbb{R}^m \to \mathbb{R}^m$ is identical matrix in $\mathbb{R}^m$, $c := (c, -c, 0)$, $0 \in \mathbb{R}^m$, and $x = (x^{'}, -x^{''}, x_{(s)}) \in \mathbb{R}^{2n+m}_+$. Then

$$z^* = \langle c,x^*\rangle = \max\{\langle c,x\rangle | Ax = b, x \geq 0\} \tag{5.17}$$

is the canonical form of the primal LP (5.14).

The following Theorem we call "truncated" version of the standard first Duality Theorem because it is a one-way statement: primal solution exists ⇒ dual solution exists, and the optimal values are the same.

Theorem 5.2. *If the primal LP (5.14) has a solution, then the dual LP (5.16) also has a solution and the optimal objective functions value are the same, that is*

$$z^* = w^*.$$

Proof. The proof is based on the optimality criteria (5.11).

Let $\bar{x} = (\bar{x}^{'}_B, \bar{x}^{''}_B, \bar{x}^s_+, 0) \in \mathbb{R}^{2n+m}_+$ be the optimal basic solution for primal LP (5.14) and $B = (A^{'}_B; A^{''}_B, I_0) \in \mathbb{R}^{m\times m}$ be the correspondent basic matrix. Sub-matrix $A^{'}_B$ is made up of columns A_j of A, which correspond to $\bar{x}^{'}_j > 0$; sub-matrix $A^{''}_B$ is made up of the columns $-A_j$ of $-A$, which correspond $\bar{x}^{''}_j > 0$; and I_0 is made up of columns $e_j = (0,\ldots,1,\ldots,0)^T$, which correspond to $\bar{x}_{2n+j} > 0$.

From existence of the primal optimal solution $\bar{x}$ and Theorem 5.1 follows existence uniquely defined by the following system column vector $\bar{\lambda}$:

$$c_B = \bar{\lambda}^T B. \tag{5.18}$$

Vector $\bar{\lambda}$ is the price—vector for the resources.

From (5.18) follows $\lambda_j = 0$ for $j \in I_0$. From the optimality of the primal solution $\bar{x} = (\bar{x}', \bar{x}'', \bar{x}^s_+; 0) \in \mathbb{R}^{2n+m}_+$ follows that for any nonbasic column A_j, we have

$$\bar{\lambda}^T A_j \geq c_j, A_j \notin B. \tag{5.19}$$

From (5.19) follows $\bar{\lambda}_j \geq 0, j \notin I_0, 2n < j \leq 2n+m$, that is, $\bar{\lambda} \in \mathbb{R}^m_+$. For $A_j \in B$, we have $\langle \bar{\lambda}, A_j \rangle = c_j$ therefore $-\langle \bar{\lambda}, A_j \rangle = -c_j$. If $-A_j \in B$ then $-\langle \bar{\lambda}, A_j \rangle = -c_j''$, so $\langle \bar{\lambda}, A_j'' \rangle = c_j''$. For any $j : A_j \notin B$ and $-A_j \notin B$ we obtain $\langle A_j, \bar{\lambda} \rangle \geq c_j$ and $\langle -A_j, \bar{\lambda} \rangle \geq -c_j$ or $\langle A_j, \bar{\lambda} \rangle \leq c_j$, that is, $c_j = \bar{\lambda}^T A_j,\ j \notin B$. In other word $c = A^T \bar{\lambda}$ and $\bar{\lambda} \in \mathbb{R}^m_+$. Therefore vector $\bar{\lambda}$ is feasible for the dual LP (5.16).

Moreover, from (5.18) and $\operatorname{rank} B = m$ follows

$$\bar{\lambda}^T = c_B B^{-1}.$$

Therefore

$$\bar{\lambda}^T b = c_B B^{-1} b = c_B \bar{x}_B, \tag{5.20}$$

where $\bar{x}_B = B^{-1} b \in \mathbb{R}^m_+$.

So $\bar{x} = (\bar{x}_B, 0)$ is primal feasible, $\bar{\lambda}$ is dual feasible, and $\langle c, \bar{x} \rangle = \langle b, \bar{\lambda} \rangle$. It follows from the Corollary 5.1 that $\bar{x} = x^*, \bar{y} = y^*$. □

The Farkas lemma follows directly from Theorem 5.2.

5.1.3 Farkas Lemma

We consider m vectors $a_1, .., a_m$ from $\mathbb{R}^n$, which define the polyhedron cone

$$C = \{x : \langle a_i, x \rangle \leq 0, i = 1, \ldots, m\}. \tag{5.21}$$

Lemma 5.4 (Farkas). *For a given vector $a \in \mathbb{R}^n$ the following inequality holds:*

$$\langle a, x \rangle \leq 0, \forall x \in C \tag{5.22}$$

if and only if there is $\lambda \in \mathbb{R}^m_+$ that

$$a = \sum_{i=1}^{m} \lambda_i a_i. \tag{5.23}$$

Proof. If such $\lambda \in \mathbb{R}^m_+$ exists then from (5.21) follows

$$\langle a,x\rangle = \sum_{i=1}^{m} \lambda \langle a_i,x\rangle \leq 0, \ \forall x \in C,$$

that is, (5.22) holds.

To show existence of $\lambda \in \mathbb{R}^m_+$ that (5.23) holds, we consider the following LP:

$$z^* = \langle a,x^*\rangle = \max\{\langle a,x\rangle | x \in C\}. \tag{5.24}$$

It follows from (5.22) that LP (5.24) has a solution; then from Theorem 5.2 follows that the dual to LP (5.24) problem

$$w^* = \langle b,\lambda^*\rangle = \langle 0,\lambda^*\rangle = \min\{\langle 0,\lambda\rangle | A^T\lambda = a, \lambda \in \mathbb{R}^m_+\}$$

has a solution, that is, (5.23) holds. □

For any $1 \leq i \leq m$ the half space $C_i = \{x : \langle a_i,x\rangle \leq 0\}$ is a cone.

The conjugate cone $C_i^* = \{u_i \in \mathbb{R}^n : u_i = \lambda_i a_i, \lambda_i \geq 0\}$. The conjugate to the polyhedron cone $C = C_1 \cap \ldots \cap C_m$ is

$$C^* = \{y \in \mathbb{R}^n : \langle y,x\rangle \leq 0, \forall x \in C\}.$$

Exercise 5.4. Show that the

$$C^* = C_1^* + \ldots + C_m^*.$$

5.2 The Karush–Kuhn–Tucker's Theorem

The KKT's Theorem is one of the most important results in convex optimization. It establishes the first-order optimality condition for convex optimization problems with inequality constrains. For many years it was known as Kuhn–Tucker's Theorem proven by Harold Kuhn (1925–2014) and Albert Tucker (1905–1995) in the early 1950s.

Since the mid-1970s, the Theorem was called Karush-Kuhn-Tucker's (KKT's) Theorem in recognition of William Karush's (1917–1997) contribution. He proved the Theorem in his master's thesis (MS) in 1939. Harold Kuhn on several occasions gave credit to Karush for his contribution, which Kuhn became aware of in 1974 from Takayama's monograph *Mathematical Economics*. The Karush's MS also contains what is known as Fritz John's Theorem, a result that appeared in F. John's paper in 1949.

We proved KKT's Theorem in Chapter 1 using some basic facts from convex analysis. Now we present a proof, which is based on Farkas Lemma. In particular, it allows eliminate Slater condition in case of linear constraints.

Let $f : \mathbb{R}^n \to \mathbb{R}$ be convex and all $c_i : \mathbb{R}^n \to \mathbb{R}$, $i = 1,\ldots,m$ are concave. We consider the following convex optimization (CO) problem of finding:

$$f(x^*) = \min\{f(x) | \, x \in \Omega\}, \tag{5.25}$$

where

$$\Omega = \{x \in \mathbb{R}^n : c_i(x) \geq 0, i = 1, \ldots, m\} \tag{5.26}$$

is the feasible set.

We assume that Slater condition

$$\exists x_0 \in \Omega : c_i(x_0) > 0, i = 1, \ldots, m \tag{5.27}$$

holds, that is, Ω has an interior point.

Exercise 5.5. Show that any local minimum of problem (5.25) is the solution of the problem.

Theorem 5.3 (Karush–Kuhn–Tucker). *If f and all $-c_i, i = 1, \ldots, m$ are convex, differentiable and Slater condition (5.27) holds; then for $x^* \in \Omega$ to be a solution of (5.25), it is necessary and sufficient existence of Lagrange multipliers vector $\lambda^* = (\lambda_1^*, \ldots, \lambda_m^*) \in \mathbb{R}_+^n$ that*

$$\nabla_x L(x^*, \lambda^*) = \nabla f(x^*) - \sum_{i=1}^{m} \lambda_i^* \nabla c_i(x^*) = 0 \tag{5.28}$$

holds and complementarity condition

$$\lambda_i^* c_i(x^*) = 0, i = 1, \ldots, m \tag{5.29}$$

is satisfied.

Proof. We start with necessary condition. For $x^* \in \Omega$ we consider the set of active constraints $I(x^*) = \{i : c_i(x^*) = 0\}$ and the feasible direction cone $D_0(x^*) = \{d : \langle \nabla c_i(x^*), d \rangle > 0, i \in I(x^*)\}$ at $x^* \in \Omega$. The fact that x^* is the solution of (5.25) means that for any $d \in D_0(x^*)$ we have $\langle \nabla f(x^*), d \rangle \geq 0$.

Now we will show that it remains true for $D(x^*) = \{d : \langle \nabla c_i(x^*), d \rangle \geq 0, i \in I(x^*)\}$ – the closure of $D_0(x^*)$.

Assuming that it is not true we can find direction

$$\bar{d} : \langle \nabla f(x^*), \bar{d} \rangle < 0, \langle \nabla c_i(x^*), \bar{d} \rangle \geq 0, i \in I(x^*).$$

From the Slater condition follows existence of

$$\bar{\bar{d}} : \langle \nabla c_i(x^*), \bar{\bar{d}} \rangle > 0, i \in I(x^*).$$

Therefore, for a small enough $\tau > 0$ and $d^* = \bar{d} + \tau \bar{\bar{d}}$ we have

$$\langle \nabla f(x^*), d^* \rangle = \langle \nabla f(x^*), \bar{d} \rangle + \tau \langle \nabla f(x^*), \bar{\bar{d}} \rangle < 0 \tag{5.30}$$

and at the same time for all $i \in I(x^*)$ we have

$$\langle \nabla c_i(x^*), d^* \rangle = \langle \nabla c_i(x^*), \bar{d} \rangle + \tau \langle \nabla c_i(x^*), \bar{\bar{d}} \rangle > 0. \tag{5.31}$$

It follows from (5.30) and (5.31) that for $t > 0$ small enough $f(x^* + td^*) < f(x^*)$ and $x^* + td^* \in \Omega$, which contradicts our assumption that x^* is a solution of (5.25). Therefore we have

$$\langle \nabla f(x^*), d \rangle \geq 0 \tag{5.32}$$

for all $d \in D(x^*)$.

Invoking Farkas lemma we obtain

$$\nabla f(x^*) = \sum_{i \in I(x^*)} \lambda_i^* \nabla c_i(x^*), \quad \lambda_i^* \geq 0, i \in I(x^*).$$

Let $\lambda_i^* = 0, i \notin I(x^*)$, then

$$\nabla_x L(x^*, \lambda^*) = \nabla f(x^*) - \sum_{i=1}^{m} \lambda_i^* \nabla c_i(x^*) = 0$$

and (5.29) holds.

In other words if x^* is a solution in (5.26), then there exists $\lambda^* \in \mathbb{R}_+^m$ that (5.28) holds, and complementarity condition (5.29) is satisfied.

Now let us show that (5.28)–(5.29) is sufficient for x^* to be a solution of (5.25) From (5.29) follows

$$\nabla f(x^*) = \sum_{i \in I(x^*)} \lambda_i^* \nabla c_i(x^*), \tag{5.33}$$

from convexity f for any $x \in \Omega$ we have

$$f(x) - f(x^*) \geq \langle \nabla f(x^*), x - x^* \rangle = \langle \sum_{i \in I(x^*)} \lambda_i^* \nabla c_i(x^*), x - x^* \rangle. \tag{5.34}$$

From concavity $c_i(x)$ follows $c_i(x) - c_i(x^*) \leq \langle \nabla c_i(x^*), x - x^* \rangle$. Therefore for any $x \in \Omega$ we have

$$\langle \nabla c_i(x^*), x - x^* \rangle \geq c_i(x) \geq 0, i \in I(x^*).$$

Keeping in mind $\lambda_i^* \geq 0$ we obtain

$$\left\langle \sum_{i \in I(x^*)} \lambda_i^* \nabla c_i(x^*), x - x^* \right\rangle \geq 0, \forall x \in \Omega.$$

Therefore from (5.34) we have $f(x) - f(x^*) \geq 0, \forall x \in \Omega$. So x^* is a local minimum in (5.25). From Exercise 5.5 follows that x^* solves of (5.25). □

Remark 5.1. The proof of KKT's Theorem is based on Farkas Lemma. The Farkas Lemma follows from truncated LP duality, which, in turn, is based on the LP optimality criteria. Therefore KKT's Theorem can be used for proving duality results in both linear and convex optimization.

The Lagrangian $L : \mathbb{R}^n \times \mathbb{R}^m_+$ for inequality constrained optimization problems (5.25) naturally appeared in the course of proving KKT's Theorem. It is the key instrument in constrained optimization.

Exercise 5.6. Consider convex optimization problem

$$f(x^*) = \{f(x)|\ c_i(x) \geq 0, i = 1,\dots,m, x \in \mathbb{R}^n_+\}. \tag{5.35}$$

Show that if there is $x_0 \in \mathbb{R}^n_+$ that $c_i(x_0) > 0$, $i = 1,\dots,m$ then under conditions of Theorem 5.3 for x^* to be a solution of (5.35), it is necessary and sufficient existence $\lambda^* \in \mathbb{R}^m_+$ that the following condition holds:

$$\nabla_{x_j} L(x^*,\lambda^*) \geq 0,\ \ x_j^* \geq 0,\ \ x_j^* \cdot \nabla_{x_j} L(x^*,\lambda^*) = 0$$

$$\nabla_{\lambda_i} L(x^*,\lambda^*) \geq 0, \lambda_i^* \geq 0,\ \ \lambda_i^* \cdot \nabla_{\lambda_i} L(x^*,\lambda^*) = 0.$$

Another important optimality criteria for convex optimization is given by the following theorem.

Theorem 5.4. *Under conditions of Theorem 5.3 vector $x^* \in \Omega$ is a solution of the problem (5.25) if and only if there exists a Lagrange multipliers vector $\lambda^* \in \mathbb{R}^m_+$, that the pair $(x^*;\lambda^*)$ is a saddle point of Lagrangian L, that is, for $\forall x \in \mathbb{R}^n$ and $\forall \lambda \in \mathbb{R}^m_+$ the following inequalities hold:*

$$L(x^*,\lambda) \leq L(x^*,\lambda^*) \leq L(x,\lambda^*). \tag{5.36}$$

Proof. If x^* is the solution of (5.25), then from Theorem 5.3 follows existence $\lambda^* \in \mathbb{R}^m_+$ that (5.28) and (5.29) holds.

From convexity f, concavity of all c_i, $i = 1,\dots m$ and (5.28) follows

$$f(x^*) - \sum_{i=1}^m \lambda_i^* c_i(x^*) \leq f(x) - \sum_{i=1}^m \lambda_i^* c_i(x), \forall x \in \mathbb{R}^n \tag{5.37}$$

From (5.29) for all $\lambda \in \mathbb{R}^m_+$ follows

$$f(x^*) - \sum_{i=1}^m \lambda_i c_i(x^*) \leq f(x^*) - \sum_{i=1}^m \lambda_i^* c_i(x^*). \tag{5.38}$$

From (5.37) and (5.38) follows (5.36).

Conversely, let assume that (x^*,λ^*) is a saddle point of the Lagrangian $L(x,\lambda)$, that is, for $\forall x \in \mathbb{R}^n$ and $\lambda \in \mathbb{R}^m_+$ the following inequalities hold true:

$$f(x^*) - \sum_{i=1}^m \lambda_i c_i(x^*) \leq f(x^*) - \sum_{i=1}^m \lambda_i^* c_i(x^*) \leq f(x) - \sum_{i=1}^m \lambda_i^* c_i(x). \tag{5.39}$$

From the left inequality follows

$$\sum_{i=1}^m \lambda_i c_i(x^*) \geq \sum_{i=1}^m \lambda_i^* c_i(x^*), \forall \lambda \in \mathbb{R}^m_+. \tag{5.40}$$

Therefore $c_i(x^*) \geq 0, i = 1, \ldots, m$, because if there is $i_0 : c_{i_0}(x^*) < 0$, then by taking λ_{i_0} large enough and $\lambda_i = 0, i \neq i_0$ we obtain inequality opposite to (5.40). So $\sum_{i=1}^m \lambda_i^* c_i(x^*) \geq 0$.

By taking $\lambda = 0$ from the left inequality of (5.39) we obtain $\sum_{i=1}^m \lambda_i^* c_i(x^*) \leq 0$, therefore $\sum_{i=1}^m \lambda_i^* c_i(c^*) = 0$.

From the right inequality (5.39) and complementarity condition, we get

$$f(x^*) \leq f(x) - \sum \lambda_i^* c_i(x). \tag{5.41}$$

For any $x \in \Omega$ we have $c_i(x) \geq 0$, also $\lambda_i^* \geq 0$; therefore from (5.41) we have

$$f(x^*) \leq f(x), \forall x \in \Omega,$$

that is, x^* solves (5.25). □

Exercise 5.7. Show that under condition of Exercise 5.6 for x^* to be a solution of (5.35), it is necessary and sufficient existence of $\lambda^* \in \mathbb{R}_+^m$ that the pair (x^*, λ^*) is a saddle point of L, that is, for any $x \in \mathbb{R}_+^n$ and any $\lambda \in \mathbb{R}_+^m$ the following holds:

$$L(x^*, \lambda) \leq L(x^*, \lambda^*) \leq L(x, \lambda^*).$$

Slater condition is weaker and much more convenient than the regularity condition (4.30), which requires linear independence gradients of the active at x^* constraints.

Under Slater condition, however, Lagrange multipliers vector $\lambda^* \in \mathbb{R}_+^m$ is not unique, but the set of optimal Lagrange multipliers

$$\Lambda^* = \{\lambda \in \mathbb{R}_+^m : \nabla f(x^*) - \sum_{i=1}^m \lambda_i \nabla c_i(x^*) = 0\} \tag{5.42}$$

is bounded. The set Λ^* we call KKT's polytop.

Lemma 5.5. *Under conditions of Theorem 5.3, the Λ^* is not empty and bounded.*

Proof. Due to KKT's Theorem, we have to prove only the boundedness of Λ^*. Assuming the opposite we will find a sequence $\{\lambda_s\}_{s \in N} \subset \Lambda^*$ that

$$\max_{1 \leq i \leq m} \lambda_{s,i} = \lambda_{s,i_s} \to \infty.$$

Also for any $\lambda \in \Lambda^*$ we have $\lambda_i = 0$, $i \bar{\in} I^* = \{i : c_i(x^*) = 0\}$. Therefore for any $\lambda \in \Lambda^*$ we have

$$\nabla f(x^*) - \sum_{i \in I^*} \lambda_i \nabla c_i(x^*) = 0.$$

Dividing

$$\nabla f(x^*) - \sum_{i \in I^*} \lambda_{s,i} \nabla c_i(x^*) = 0 \tag{5.43}$$

by λ_{s,i_s} we obtain

$$(\lambda_{s,i_s})^{-1}\nabla f(x^*) - \sum_{i \in I^*} \frac{\lambda_{s,i}}{\lambda_{s,i_s}} \nabla c_i(x^*) = 0. \tag{5.44}$$

Let

$$\lambda_{s,i} := \lambda_{s,i}(\lambda_{s,i_s})^{-1}, \ i = 1,\dots,m,$$

then the nonnegative sequence $\{\lambda_s = ((\lambda_{s,i}\lambda_{s,i_s}^{-1}), \ i = 1,\dots,m)\}$ is bounded; therefore there is a limit point. Let $\lambda_s \to \bar{\lambda}$, then by taking the limit in (5.44), we obtain

$$\sum_{i \in I^*} \bar{\lambda}_i \nabla c_i(x^*) = 0 \tag{5.45}$$

and not all $\bar{\lambda}_i = 0$. It contradicts the Slater condition (5.27). The contradiction proves boundedness of Λ^*. □

If x^* is a regular minimizer, then gradients $\nabla c_i(x^*), i \in I(x^*)$ are linear independent, and the KKT polytop Λ^* shrinks to a singleton $\lambda^* \in \mathbb{R}^m_+$.

We will call a subset $I \subset I(x^*)$ minimal if

$$\nabla f(x^*) = \sum_{i \in I} \lambda_i \nabla c_i(x^*) \tag{5.46}$$

and for any $i \in I$ we have

$$\nabla f(x^*) \neq \sum_{I \setminus i} \lambda_i \nabla c_i(x^*). \tag{5.47}$$

There is a one to one correspondence between minimal sets and vertices of the KKT polytop Λ^*.

Exercise 5.8. Show that vector $\lambda \in \Lambda^*$ is a vertex of Λ^* if and only if I is minimal set.

5.3 The KKT's Theorem for Convex Optimization with Linear Constraints

Let $f : \mathbb{R}^n \to \mathbb{R}$ be convex, matrix $A : \mathbb{R}^n \to \mathbb{R}^m$ and $b \in \mathbb{R}^m$.

We consider the following convex optimization problem

$$f(x^*) = \{f(x) | x \in \Omega\}, \tag{5.48}$$

where $\Omega = \{x \in \mathbb{R}^n_+ : c(x) = Ax - b \geq 0\}$ and $X^* = \{x \in \Omega : f(x) = f(x^*)\}$.

Theorem 5.5. *Let $f : \mathbb{R}^n \to \mathbb{R}$ be convex and differentiable, then for $x^* \in \Omega$ to be a solution of (5.48) it is necessary and sufficient the existence of a vector $\lambda^* \in \mathbb{R}^m_+$ that for the Lagrangian $L : \mathbb{R}^n_+ \times \mathbb{R}^m_+ \to \mathbb{R}$*

$$L(x,\lambda) = f(x) - \langle \lambda, Ax - b\rangle = f(x) - \sum_{i=1}^{m} \lambda_i[(a_i,x) - b_i]$$

the following inequalities

$$\nabla_x L(x^*,\lambda^*) = \nabla f(x^*) - \sum_{i=1}^{m} \lambda_i^* a_i \geq 0, \tag{5.49}$$

$$\nabla_\lambda L(x^*,\lambda^*) = c(x^*) = Ax^* - b \geq 0 \tag{5.50}$$

hold and complementarity conditions

$$\langle x^*, \nabla_x L(x^*,\lambda^*)\rangle = 0 \tag{5.51}$$

$$\langle \lambda^*, \nabla_\lambda L(x^*,\lambda^*)\rangle = \langle \lambda^*, c(x^*)\rangle = 0 \tag{5.52}$$

are satisfied.

Proof. Let x^* be the solution of (5.48), $I(x^*) = \{i : c_i(x^*) = 0\}$, $J(x^*) = \{x_j^* = 0\}$ be the active constraints set and $e_j = (0,\ldots,1,0,..,0) \in \mathbb{R}^n_+$.

Then,

$$\forall d : \langle a_i, d\rangle \geq 0, i \in I(x^*);\ \ \langle e_j, d\rangle \geq 0, j \in J(x^*) \tag{5.53}$$

we have

$$\langle \nabla f(x^*), d\rangle \geq 0,$$

otherwise x^* can't be a solution of (5.48). It follows from Farkas Lemma that there are $\lambda_i^* \geq 0, i \in I(x^*)$ and $\mu_j^* \geq 0,\ j \in J(x^*)$ that

$$\nabla f(x^*) = \sum_{i \in I(x^*)} \lambda_i^* a_i + \sum_{j \in J(x^*)} \mu_j^* e_j.$$

Let $\lambda_i^* = 0, i \notin I(x^*)$ and $\mu^* = 0,\ j \notin J(x^*)$, then

$$\nabla f(x^*) - \sum_{i=1}^{m} \lambda_i^* a_i - \sum_{j=1}^{n} \mu_j^* e_j = 0$$

or

$$\nabla_x L(x^*,\lambda^*) = \nabla f(x^*) - A^T\lambda^* = \sum_{j=1}^{n} \mu_j^* e_j \geq 0$$

and the complementarity conditions

$$\lambda_i^* c_i(x^*) = 0, i = 1,\ldots,m, \quad x_j^*\mu_j^* = x_j^*(\nabla f(x) - A^T\lambda^*)_j = 0, j = 1,\ldots,n$$

are satisfied. □

Exercise 5.9. Show that under condition of Theorem 5.5 for $x^* \in \mathbb{R}^m_+$ to be a solution of (5.48), it is necessary and sufficient existence $\lambda^* \in \mathbb{R}^m_+$ that the pair (x^*,λ^*) is a saddle point of L, that is, for any $x \in \mathbb{R}^n_+$ and any $\lambda \in \mathbb{R}^m_+$ the following inequalities hold:

$$L(x^*,\lambda) \leq L(x^*,\lambda^*) \leq L(x,\lambda^*).$$

5.4 Duality in Convex Optimization

It follows from KKT's Theorem that for $x^* \in \Omega$ to be the primal solution, it is necessary and sufficient the existence of Lagrange multipliers vector $\lambda^* \in \mathbb{R}^m_+$ that the pair $(x^*;\lambda^*)$ is a saddle point of the Lagrangian, that is

$$L(x^*,\lambda) \leq L(x^*,\lambda^*) \leq L(x,\lambda^*), \forall x \in \mathbb{R}^n; \forall \lambda \in \mathbb{R}^m_+. \tag{5.54}$$

In other words, we have

$$\min_{x\in\mathbb{R}^n}\max_{\lambda\in\mathbb{R}^m_+} L(x,\lambda) = \max_{\lambda\in\mathbb{R}^m_+}\min_{x\in\mathbb{R}^n} L(x,\lambda) = L(x^*,\lambda^*).$$

Let

$$p(x) = \sup_{\lambda\in\mathbb{R}^m_+} L(x,\lambda),$$

then

$$p(x) = \begin{cases} f(x), & if\, x \in \Omega \\ \infty, & if\, x \notin \Omega \end{cases} \tag{5.55}$$

The primal problem (5.25) is equivalent to the following unconstrained minimization problem:

$$p(x^*) = \min\{p(x)|x \in \mathbb{R}^n\}. \tag{5.56}$$

Let us consider the dual function

$$d(\lambda) = \inf_{x\in\mathbb{R}^n} L(x,\lambda)$$

(for some $\lambda \in \mathbb{R}^q_+$ it is possible $d(\lambda) = -\infty$). The dual problem consists of finding

$$d(\lambda^*) = \max\{d(\lambda)|\lambda \in \mathbb{R}^m_+\}. \tag{5.57}$$

Lemma 5.6 (Weak Duality). *For any primal feasible $x \in \Omega$ and any dual feasible $\lambda \in \mathbb{R}^q_+$ the following inequality holds:*

$$p(x) \geq d(\lambda). \tag{5.58}$$

Proof. For any given $\lambda \in \mathbb{R}^m_+$ we have

$$d(\lambda) = \inf_{x \in \mathbb{R}^n} L(x,\lambda) \leq L(x,\lambda) = f(x) - \sum_{i=1}^{m} \lambda_i c_i(x).$$

If $x \in \mathbb{R}^n$ is primal feasible, then $c_i(x) \geq 0$; therefore for any $\lambda \in \mathbb{R}^m_+$ we have

$$d(\lambda) \leq f(x) - \sum_{i=1}^{m} \lambda_i c_i(x) \leq f(x).$$

□

Corollary 5.2. *If $\bar{x}$—primal feasible, $\bar{\lambda}$—dual feasible and*

$$p(\bar{x}) = f(\bar{x}) = d(\bar{\lambda}),$$

then $\bar{x} = x^*$, $\bar{\lambda} = \lambda^*$.

In fact, for any primal feasible x, we have $p(x) \geq d(\bar{\lambda})$, and for $\bar{x}$ we have $p(\bar{x}) = d(\bar{\lambda})$; therefore $\bar{x} = x^*$. On the other hand, for any $\lambda \in \mathbb{R}^m_+$ we have $p(\bar{x}) \geq d(\lambda)$, and for $\bar{\lambda} \in \mathbb{R}^m_+$ we have $p(\bar{x}) = d(\bar{\lambda})$; therefore $\bar{\lambda} = \lambda^*$.

Theorem 5.6 (Duality). *Under condition of Theorem 5.3 the following holds:*

(1) if x^* is a primal solution, then there exists a dual solution $\lambda^* \in \mathbb{R}^m_+$ and

$$f(x^*) = p(x^*) = d(\lambda^*); \tag{5.59}$$

(2) the primal–dual feasible pair $(\bar{x},\bar{\lambda}) : L(\bar{x},\bar{\lambda}) = d(\bar{\lambda})$ is an optimal one if and only if the complementarity condition holds, that is

$$\bar{\lambda}_i c_i(\bar{x}) = 0, i = 1,\ldots,m \tag{5.60}$$

Proof.

(1) If x^* is a primal solution, then from KKT's Theorem follows existence of $\lambda^* \in \mathbb{R}^m_+$ that for $\forall x \in \mathbb{R}^n$ and $\forall \lambda \in \mathbb{R}^m_+$ we have

$$f(x^*) - \sum_{i=1}^{m} \lambda_i c_i(x^*) \leq f(x^*) - \sum_{i=1}^{m} \lambda_i^* c_i(x^*) \leq f(x) - \sum_{i=1}^{m} \lambda_i^* c_i(x). \tag{5.61}$$

Therefore

$$d(\lambda^*) = \inf_{x \in \mathbb{R}^n} L(x,\lambda^*) = L(x^*,\lambda^*) \geq L(x^*,\lambda) \geq \inf_{x \in \mathbb{R}^n} L(x,\lambda) = d(\lambda),\ \forall \lambda \in \mathbb{R}^m_+,$$

so $\lambda^* \in \mathbb{R}^m_+$ is the dual solution.

Also from the left inequality (5.61) for $\lambda = 0$, we have $\sum_{i=1}^{m} \lambda_i^* c_i(x^*) \leq 0$, therefore

$$\lambda_i^* c_i(x^*) = 0, i = 1,\ldots,m,$$

hence

$$p(x^*) = f(x^*) = L(x^*,\lambda^*) = \inf_{x \in \mathbb{R}^n} L(x,\lambda^*) = d(\lambda^*).$$

(2) If the primal–dual pair $(\bar{x},\bar{\lambda})$ is the optimal one, then from KKT's Theorem follows $f(\bar{x})-\sum_{i=1}^{m}\bar{\lambda}c_i(\bar{x})=d(\bar{\lambda})$; therefore the complementarity condition (5.60) holds.
Conversely, assuming that the primal–dual feasible pair $(\bar{x},\bar{\lambda})$ is such that : $L(\bar{x},\bar{\lambda})=d(\bar{\lambda})$ and the complementarity condition (5.60) satisfied, then

$$d(\bar{\lambda})=L(\bar{x},\bar{\lambda})=f(\bar{x})-\sum_{i=1}^{m}\bar{\lambda}_i\bar{c}_i(x)=f(\bar{x}).$$

The optimality of the pair $(\bar{x},\bar{\lambda})$ follows from Corollary 5.2. □

In the absence of Slater's condition Theorem 5.6, generally speaking, is not true.

Example 5.1. $f(x_1^*,x_2^*)=\min\{f(x_1,x_2)=x_1+x_2^2|3-x_1\geq 0\}$, the Lagrangian $L(x,\lambda)=x_1+x_2^2-\lambda(3-x_1)=(1+\lambda)x_1+x_2^2-3\lambda$. For any $\lambda\geq 0$ we have $\inf_x L(x,\lambda)=-\infty$; therefore neither the primal nor the dual solution exists.

Example 5.2. $f(x^*)=\min\{f(x)=e^{-x_1}|x_1\geq 0\}$, the Lagrangian $L(x,\lambda)=e^{-x_1}-\lambda x_1$. Then for $\lambda>0$ $\inf_x L(x,\lambda)=-\infty$, for $\lambda=0$ we have $\inf_x L(x,0)=0$. So the primal has no solution, but the dual solution $\lambda^*=0$ exists and $d(\lambda^*)=0$.

Example 5.3. $f(x^*)=\min\{f(x)=x_1+x_2|-x_1^2-x_2^2\geq 0$, the Lagrangian $L(x,\lambda)=x_1+x_2+\lambda(x_1^2+x_2^2)$, $\nabla_{x_1}L(x,\lambda)=1+2\lambda x_1=0, x_1=-\frac{1}{2\lambda}$, $\nabla_{x_2}L(x,\lambda)=1+2\lambda x_2=0$, $x_2=-\frac{1}{2\lambda}$, $d(\lambda)=\min_x L(x,\lambda)=-\frac{1}{2\lambda}$.
The primal problem has a solution $x_1^*=x_2^*=0$; the dual $\sup_{\lambda>0}d(\lambda)$ does not.

5.5 Wolfe's Duality

If f and all $-c_i$ $i=1,\ldots,m$- are convex and differentiable, then the dual problem can be written as follows (P. Wolfe duality):

$$L(x^*,\lambda^*)=\max\{L(x,\lambda)|\nabla_x L(x,\lambda)=0,\lambda\in\mathbb{R}_+^m\}. \tag{5.62}$$

It follows from (5.62) that we are maximizing Lagrangian in both primal and dual variables. Due to constraints $\nabla_x L(x,\lambda)=0$, however, the problem (5.62) is, in fact, the standard dual

$$\max_{\lambda\in\mathbb{R}_+^m}\min_{x\in\mathbb{R}^n}L(x,\lambda)=\max_{\lambda\in\mathbb{R}_+^m}d(\lambda),$$

because for any given $\lambda\in\mathbb{R}_+^m$ the system $\nabla_x L(x,\lambda)=0$ is sufficient condition for the primal vector to be the Lagrangian minimizer in $x\in\mathbb{R}^n$.

Obviously, the duality results from Theorem 5.6 is true for P. Wolfe's duality.

Sometimes the primal problem is written as follows:

$$f(x^*)=\min\{f(x)|c_i(x)\geq 0,i=1,\ldots,m,x\in Q\},$$

where Q is convex set, in particular, $Q = \mathbb{R}^n_+$. Let $p(x) = \sup\{L(x,\lambda)|\lambda \in \mathbb{R}^m_+\}$ then the primal problem is

$$p(x^*) = \min\{p(x)|x \in \mathbb{R}^n_+\}. \tag{5.63}$$

Let $d(\lambda) = \inf\{L(x,\lambda)|x \in \mathbb{R}^n_+\}$, then the dual is

$$d(\lambda^*) = \max\{d(\lambda)|\lambda \in \mathbb{R}^m_+\}. \tag{5.64}$$

Let us consider Wolfe's duality for the following convex optimization problem:

$$f(x^*) = \min\{f(x)|c_i(x) \geq 0, i = 1,\dots,m,\ x \in \mathbb{R}^n_+\}. \tag{5.65}$$

Exercise 5.10. Show that under conditions of Theorem 5.3 for $x^* \in \mathbb{R}^n_+$ to be a solution of (5.65), it is necessary and sufficient the existence of $\lambda^* \in \mathbb{R}^m_+$ that the pair $(x^*,\lambda^*) \in \mathbb{R}^n_+ \times \mathbb{R}^m_+$ solves the following problem:

$$L(x^*,\lambda^*) = \max\{L(x,\lambda)|\nabla_x L(x,\lambda) \geq 0, \langle x, \nabla_x L(x,\lambda)\rangle = 0, x \in \mathbb{R}^n_+, \lambda \in \mathbb{R}^m_+\}.$$

5.6 LP Duality

In this section we consider LP duality based on the KKT's Theorem. It follows from Remark 5.1 that such approach is legitimate.

Let us recall the Primal LP problem

$$\langle c, x^*\rangle = \max\{\langle c, x\rangle | Ax \leq b, x \in \mathbb{R}^n_+\}, \tag{5.66}$$

where $A \in \mathbb{R}^{m\times n}$, $c \in \mathbb{R}^n$, $b \in \mathbb{R}^m$.

From Theorem 5.5 follows that $x^* \in \mathbb{R}^n_+$ is a solution of LP (5.66) if and only if there exists $\lambda^* \in \mathbb{R}^q_+$ such that

$$L(x,\lambda^*) \leq L(x^*,\lambda^*) \leq L(x^*,\lambda), x \in \mathbb{R}^n_+, \lambda \in \mathbb{R}^m_+, \tag{5.67}$$

where

$$L(x,\lambda) = \langle c, x\rangle + \langle \lambda, b - Ax\rangle$$

is the Lagrangian for LP problem (5.66).

Let consider

$$p(x) = \inf_{\lambda\in\mathbb{R}^m_+} L(x,\lambda) = \inf_{\lambda\in\mathbb{R}^m_+} \{\langle c,x\rangle + \langle \lambda, b - Ax\rangle\}.$$

Then

$$p(x) = \begin{cases} \langle c, x\rangle, & \text{if } Ax \leq b \\ -\infty, & \text{if } \exists i_0 : (Ax-b)_{i_0} > 0. \end{cases} \tag{5.68}$$

The primal LP consists of finding

$$\max\{p(x)|x \in \mathbb{R}^n_+\}. \tag{5.69}$$

Let consider the dual function

$$d(\lambda) = \sup_{x\in\mathbb{R}^n_+} L(x,\lambda) = \sup_{x\in\mathbb{R}^n_+} \{\langle b,\lambda\rangle + \langle c - A^T\lambda, x\rangle\}.$$

Then

$$d(\lambda) = \begin{cases} \langle b,\lambda\rangle, & \text{if } A^T\lambda \geq c \\ \infty, & \text{if } \exists j_0 : (c - A^T\lambda)j_0 > 0 \end{cases} \tag{5.70}$$

and the dual LP consists of finding

$$d(\lambda^*) = \min\{d(\lambda)|\lambda \in \mathbb{R}^m_+\} = \min\{\langle b,\lambda\rangle | A^T\lambda \geq c, \lambda \in \mathbb{R}^m_+\}. \tag{5.71}$$

Exercise 5.11. Show that the dual to the dual LP (5.71) is identical to the primal LP (5.66).

Theorem 5.7 (First Duality Theorem). *If one of the dual pair of LP has a solution, then the other has a solution and the optimal values are the same.*

Proof. If the primal LP has a solution $x^* \in \mathbb{R}^n_+$, then from Theorem 5.5 follows existence $\lambda^* \in \mathbb{R}^m_+$ that (5.67) holds, that is, for any $x \in \mathbb{R}^n_+$ and any $\lambda \in \mathbb{R}^m_+$, we have

$$\langle c,x\rangle + \langle\lambda^*, b - Ax\rangle \leq \langle c,x^*\rangle + \langle\lambda^*, b - Ax^*\rangle \leq \langle c,x^*\rangle + \langle\lambda, b - Ax^*\rangle. \tag{5.72}$$

From the right inequality (5.72) follows $Ax^* \leq b$, because of (5.68). Therefore x^* is primal feasible and $\langle\lambda^*, b - Ax^*\rangle \geq 0$.

From the right inequality (5.72) for $\lambda = 0$ follows

$$\langle\lambda^*, b - Ax^*\rangle \leq 0.$$

Therefore,

$$\langle\lambda^*, Ax^* - b\rangle = 0.$$

From the left inequality (5.72) follows $c - A^T\lambda^* \leq 0$, because of (5.70).

Therefore, λ^* is dual feasible. By taking $x = 0$ in the left inequality (5.72) we obtain

$$\langle b,\lambda^*\rangle \leq \langle c,x^*\rangle.$$

For the primal–dual feasible pair $(x^*;\lambda^*)$ from the weak LP duality follows $\langle c,x^*\rangle \leq \langle b,\lambda^*\rangle$. Therefore,

$$\langle c,x^*\rangle = \langle b,\lambda^*\rangle. \tag{5.73}$$

Invoking Corollary 5.1 we conclude that λ^* is the dual solution.

Conversely, let λ^* be the dual solution then the existence of primal solution x^* and (5.73) follows from Exercise 5.11. □

The LP duality statement is symmetric:

$$\text{primal solution exists} \Leftrightarrow \text{dual solution exists}$$

and their optimal value are equal. It is possible due to the result of Exercise 5.11, which is not the case in convex optimization.

Exercise 5.12. Show that if primal and dual LP have feasible solutions, then they have optimal solutions as well.

Exercise 5.13. Show that if one of the dual pair of LP has unbounded solution, then the feasible set of the dual LP is empty.

The following example shows that both the primal and the dual feasible sets can be empty.

Primal

$$\begin{aligned} 2x_1 + 2x_2 &\to \max \\ x_1 - x_2 &\leq -2 \\ x_1 + x_2 &\leq -2 \\ x_1 \geq 0, x_2 &\geq 0. \end{aligned}$$

Dual

$$\begin{aligned} -2\lambda_1 - 2\lambda_2 &\to \min \\ \lambda_1 - \lambda_2 &\geq 2 \\ -\lambda_1 + \lambda_2 &\geq 2 \\ \lambda_1 \geq 0, \lambda_2 &\geq 0. \end{aligned}$$

Theorem 5.8 (Second Duality Theorem). *The following complementarity condition*

$$(a)\langle \bar{x}, A^T\bar{\lambda} - c\rangle = 0; \qquad (b)\langle \bar{\lambda}, b - A\bar{x}\rangle = 0 \tag{5.74}$$

is necessary and sufficient for the primal–dual feasible pair $(\bar{x}, \bar{\lambda})$ *to be the primal and the dual optimal solution.*

Proof. From (5.74) follows rl

$$\langle c, \bar{x}\rangle = \langle b, \bar{\lambda}\rangle.$$

From Corollary 5.1 follows $\bar{x} = x^*$ and $\lambda = \lambda^*$.

Conversely, if x^* is the primal solution and λ^* is the dual solution, then from Theorem 5.7 follows (5.73).

From the primal and dual the feasibility follows

$$A^T\lambda^* - c \geq 0, x^* \geq 0, b - Ax^* \geq 0, \lambda^* \geq 0$$

or

$$\langle c, x^* \rangle \leq \langle A^T\lambda^*, x^* \rangle = \langle \lambda^*, Ax^* \rangle \leq \langle b, \lambda^* \rangle. \tag{5.75}$$

In view (5.73) there is no inequalities in (5.75), therefore

$$\langle c - A^T\lambda^*, x^* \rangle = 0 \text{ and } \langle Ax^* - b, \lambda^* \rangle = 0$$

□

Exercise 5.14. Show that solving the primal and dual LP is equivalent to solving the following system of linear inequalities:

$$Ax \leq b, x \geq 0, A^T\lambda \geq c, \lambda \geq 0$$

$$\langle b, \lambda \rangle \leq \langle c, x \rangle.$$

5.7 Some Structural LP Properties

We consider the following LP

$$\langle c, x^* \rangle = \min\{\langle c, x \rangle | x \in \Omega\}, \tag{5.76}$$

where $\Omega = \{x \in \mathbb{R}^n_+ : Ax \leq b\} \neq \emptyset$.

Recession cone of the level set $\mathscr{L}_f(\alpha) = \{x \in \mathbb{R}^n : \langle c, x \rangle \leq \alpha\}$ is

$$RC(\mathscr{L}_f(\alpha)) = \{d : \langle c, d \rangle \leq 0\}.$$

Recession cone of the feasible set Ω is

$$PC(\Omega) = \{d \in \mathbb{R}^n_+ : Ad \leq 0\}.$$

The LP (5.76) has a solution if

$$RC(\Omega) \cap RC(\mathscr{L}_f(\alpha)) = \emptyset.$$

In short the LP (5.76) has a solution if there is L that $(c, x) \geq L, \forall x \in \Omega$.

We assume that the primal optimal set

$$X^* = \{x \in \Omega : \langle c, x \rangle = \langle c, x^* \rangle\}$$

is bounded.

Then from Corollary 2.4 follows boundedness of

$$\Omega := \Omega \cap \{x \in \mathbb{R}^n : \langle c,x \rangle \geq L\}.$$

Also, for small enough L the extra constraint does not effect X^*. Therefore, to simplify our considerations we assume that Ω is a polytop.

Let $I_{\mathrm{v}}^* = \{i : \langle \mathrm{v}_i, c \rangle = f^*\}$ be the set of optimal vertices, then

$$X^* = \{x : x = \sum_{i \in I_{\mathrm{v}}^*} \lambda_i \mathrm{v}_i,\ \lambda_i \geq 0,\ \sum_{i \in I_{\mathrm{v}}^*} \lambda_i = 1, i \in I_{\mathrm{v}}^*\}$$

and

$$\langle c,x \rangle = \sum_{i \in I_{\mathrm{v}}^*} \lambda_i \langle c, \mathrm{v}_i \rangle = f^*, \forall x \in X^*.$$

The LP (5.76) has a unique solution if among vectors a_i of the active set $I^* = \{i : \langle a_i, x^* \rangle = b_i\}$, there are n linear independent and the correspondent $\lambda_i^* > 0$.

Let $d(x,X^*) = \min\{\|x-u\| \ |u \in X^*\}$. We consider the following LP:

$$\langle c,x^* \rangle = \min\{\langle c,x \rangle |\, x \in \Omega\} \tag{5.77}$$

with a bounded feasible set $\Omega = \{x \in \mathbb{R}^n : Ax \leq b\}$.

Lemma 5.7. *There exists $\rho > 0$ that for any $x \in \Omega$ we have*

$$\rho d(x,X^*) \leq \langle c,x \rangle - \langle c,x^* \rangle. \tag{5.78}$$

Proof. For a bounded Ω with M vertices v_i, $i = 1,\ldots,M$ we have

$$\Omega = \{x : x = \sum_{i=1}^{M} \lambda_i \mathrm{v}_i, \lambda_i \geq 0, \sum_{i=1}^{M} \lambda_i = 1\}.$$

For the optimal set X^* we have

$$X^* = \{x^* : x^* = \sum_{i \in I_{\mathrm{v}}^*} \mu_i^* \mathrm{v}_i, \mu_i^* \geq 0, \sum_{i \in I_{\mathrm{v}}^*} \mu_i^* = 1\}.$$

Let consider

$$\rho = \min_{i \notin I_{\mathrm{v}}^*} \frac{\langle c, \mathrm{v}_i \rangle - f^*}{d(\mathrm{v}_i, X^*)}. \tag{5.79}$$

Obviously for $i \notin I_{\mathrm{v}}^*$ we have $d(\mathrm{v}_i,X^*) > 0$ and $\langle c, \mathrm{v}_i \rangle - f^* > 0$; therefore there is $\rho > 0$ such that from (5.79) follows

$$d(\mathrm{v}_i,X^*) \leq \rho^{-1}(\langle c, \mathrm{v}_i \rangle - f^*),\ i \notin I_{\mathrm{v}}^*. \tag{5.80}$$

From item 2 of Exercise 2.27 follows convexity of $d(x,X^*)$; therefore from Jensen's inequality and (5.80) follows

$$
\begin{aligned}
d(x,X^*) &= d(\sum_{i=1}^{M}\mu_i \mathrm{v}_i, X^*) \le \sum_{i=1}^{M}\mu_i d(\mathrm{v}_i, X^*) = \sum_{i\notin I_\mathrm{v}^*}\mu_i d(\mathrm{v}_i, X^*)\\
&\le \rho^{-1}\sum_{i\notin I_\mathrm{v}^*}\mu_i(\langle c,\mathrm{v}_i\rangle - f^*) = \rho^{-1}\sum_{i=1}^{M}\mu_i(\langle c,\mathrm{v}_i\rangle - f^*)\\
&= \rho^{-1}(\langle c,\sum_{i=1}^{M}\mu_i \mathrm{v}_i\rangle - f^*) = \rho^{-1}(\langle c,x\rangle - f^*),
\end{aligned}
$$

therefore (5.78) holds. □

Corollary 5.3. *It follows from Lemma 5.7 that in case of unique LP solution; we have*

$$\langle c,x\rangle - f^* = \rho\|x - x^*\|, \forall x \in \Omega. \tag{5.81}$$

Corollary 5.4. *From* (5.78) *follows, that for any sequence* $\{x_s\}_{s\in N}$*:* $\lim_{s\to\infty}\langle c,x_s\rangle = \langle c,x^*\rangle$ *we have* $\lim_{s\to\infty} d(x_s,X^*) = 0$.

Let

$$[a]_+ = \begin{cases} a, \text{ if } & a \ge 0\\ 0, \text{ if } & a < 0\end{cases} \tag{5.82}$$

We consider LP (5.77).

Lemma 5.8 (A. Hoffman). *Let* $\Omega = \{x \in \mathbb{R}^n : \langle a_i,x\rangle \le b_i, i = 1,\dots,m\} \ne \emptyset$*, then there exists* $\rho > 0$ *such that*

$$\rho d(x,\Omega) \le \sum_{i=1}^{m}[\langle a_i,x_i\rangle - b]_+, \forall x \in \mathbb{R}^n. \tag{5.83}$$

Proof. The bound (5.83) means that the sum of constraints violation is an upper bound for a distance from any $x \in \mathbb{R}^n$ to a polyhedron Ω. The bound (5.83) is trivially satisfied for any $x \in \Omega$; therefore we have to show that (5.83) is true for any $x \notin \Omega$.

Let us introduce $2m$ extra variables $y_i \ge 0$ and $z_i \ge 0$, $i = 1,\dots,m$ and consider the following LP:

$$\min\left\{\sum_{i=1}^{m} y_i | \langle a_i,x\rangle - b_i = y_i - z_i, y_i \ge 0, z_i \ge 0, i = 1,\dots,m\right\}. \tag{5.84}$$

For any $x \in \mathbb{R}^n$ vector $\{x,y,z\}$, where $y_i = [\langle a_i,x\rangle - b_i]_+$, $z_i = [b_i - \langle a_i,x\rangle]_+$, is feasible for LP (5.84).

For any $x \in \Omega$ the solution for LP (5.84) has the form $(x,0,[b - Ax]_+)$. Therefore, application of the bound (5.78) to problem (5.84) leads (5.83).v

Corollary 5.5. *If* X^* *is the solution set of LP (5.76) then there is* $\rho > 0$

$$\rho d(x,X^*) \leq \sum_{i=1}^{m} [\langle a_i,x\rangle - b_i]_+ + [\langle c,x\rangle - f^*]_+. \tag{5.85}$$

To prove the bound (5.78) it is sufficient to present the solution set X^* as follows

$$X^* = \{x : \langle a_i,x\rangle \leq b, i=1,\dots,m, \langle c,x\rangle \leq f^*\}.$$

Application of Hoffman's lemma leads to (5.85).

It follows from (5.85) that any sequence $\{x_s\}_{s\in N} : \langle c,x_s\rangle \to f^*$ and $(\langle a_i,x_s\rangle - b_i)_+ \to 0$ converges to X^*.

Exercise 5.15. Let $A \in \mathbb{R}^{m\times n}$, $m > n$, we consider the following overdetermined system of linear equations $Ax = b$. Show that finding Chebyshev approximation

$$f(x^*) = \min\{\max_{1\leq i\leq m} |\langle a_i,x\rangle - b_i| : x \in \mathbb{R}^n\}$$

is equivalent to solving an LP. Why such LP has a solution.

Exercise 5.16. Show that finding

$$f(x^*) = \min\left\{\sum_{i=1}^{m} |\langle a_i,x\rangle - b_i| : x \in \mathbb{R}^n\right\}$$

equivalent to solving an LP. Why such LP has a solution. Write the LP and consider its dual.

5.8 Simplex Method

There are few numerical realization of simplex method, but the basic idea is the same. It consists of starting from a vertex (basic solution) and moving from one vertex to an adjacent one with better objective function value until the optimal solution is found.

The current format of the simplex method, which we describe below, is a version of George B. Dantzig's algorithm, introduced in 1947.

Over the last 70 years, finding the best numerical realization of the simplex method was a subject of substantial effort by both mathematicians and software developers; see Robert Bixby (2012). The current commercial software for solving LP and mixed integer programming reflect the effort. It allowed to speed up the solution process by six orders of magnitude.

We describe simplex method using canonical LP format (5.8).

We recall, the matrix $A : \mathbb{R}^n \to \mathbb{R}^m$ $(n > m)$ transforms the resource vector $b \in \mathbb{R}^m$ into a production vector $x \in \mathbb{R}^n_+$. The components c_j of the price vector $c \in \mathbb{R}^n$, define the market price of one unit if product $1 \leq j \leq n$.

In other words we are dealing with the following LP:

$$\langle c, x^* \rangle = \max\{\langle c, x\rangle | Ax = b, x \in \mathbb{R}^n_+\} \tag{5.86}$$
$$= \max\{\langle c, x\rangle | \sum_{j=1}^{m} x_j A_j = b, x \in \mathbb{R}^n_+\}.$$

We assume that rank$A = m$, so the row vectors $a_i = (a_{11}, \ldots, a_{in})$, $i = 1, \ldots, m$ are linear independent. Therefore the matrix A has m linear independent columns $A_{j_1}, \ldots, A_{j_m}$. Without losing the generality, we assume that it is $A_1, \ldots, A_m$.

Let us split the matrix A into two matrices $B \in \mathbb{R}^{m\times m}$ (rank$B = m$) and $N \in \mathbb{R}^{m\times n-m}$. The price vector $c \in \mathbb{R}^n$ and production vector $x \in \mathbb{R}^n$ we split accordingly, so the LP (5.86) can be written as follows:

$$z = \langle c_B, x_B\rangle + \langle c_N, x_N\rangle \to \max \tag{5.87}$$

subject to

$$Bx_B + Nx_N = b \tag{5.88}$$

$$x_B \geq 0, x_N \geq 0. \tag{5.89}$$

After multiplying system (5.88) by B^{-1} we obtain

$$Ix_B + B^{-1}Nx_N = Ix_B + \bar{N}x_N = B^{-1}b = \bar{b} \tag{5.90}$$

or

$$e_1x_1 + \ldots e_mx_m + \bar{A}_{n-m+1}x_{n-m+1} + \ldots + \bar{A}_nx_n = \bar{b},$$

where $e_i = (0, \ldots, 1, \ldots, 0)^T$, $i = 1, \ldots, m$, $\bar{A}_j = B^{-1}A_j$, $n - m + 1 \leq j \leq n$.

Let $\bar{b} \in \mathbb{R}^m_+$, then $\bar{x} = (\bar{x}_B = \bar{b}; \bar{x}_N = 0)^T$ is a basic solution.

The correspondent objective function value

$$\bar{z} = \langle c_B, \bar{x}_B\rangle + \langle c_N, 0\rangle = c_B^T B^{-1} b = \bar{\lambda}^T b, \tag{5.91}$$

where $\bar{\lambda}^T = c_B^T B^{-1}$ is the shadow prices vector.

It follows from

$$\bar{\lambda}^T B = c_B^T,$$

that for each basic products the cost of resources required to produce one unit of product $1 \leq j \leq m$ is equal to the market price of this unit, that is

$$\bar{\lambda}^T A_j = c_j, 1 \leq j \leq m \tag{5.92}$$

(1) If

$$\Delta_j = \bar{\lambda}^T A_j - c_j = c_B^T \bar{A}_j - c_j \geq 0, n - m + 1 \leq j \leq n, \tag{5.93}$$

then among the nonbasic products no one is "profitable." It means that $\bar{x} = (\bar{x}_B, 0)^T$ is the optimal solution and $\bar{z} = \langle c_B, \bar{x}_B\rangle$ is the optimal function value. In fact, it follows from (5.92) and (5.93) that $\bar{\lambda}$ is dual feasible. Vector $\bar{x}$ is primal feasible and $\langle c, \bar{x}\rangle = \langle b, \bar{\lambda}\rangle$; therefore from Corollary 5.2 follows $\bar{x} = x^*$, $\bar{\lambda} = \lambda^*$.

(2) If there is $m+1 \leq j_0 \leq n$ that $\Delta_{j_0} = \bar{\lambda}^T A_{j_0} - c_{j_0} < 0$, then under the current prices $\bar{\lambda} \in \mathbb{R}^n_+$, the cost to produce one unit of product j_0 is less that the market price for this unit.
Therefore, by introducing column A_{j_0} into the basis, we can find a better basic solution. Each extra unit of product j_0 increases the primal objective function by $|\Delta_{j_0}|$.
If there are several products with $\Delta_j < 0$, $m+1 \leq j \leq n$, then we take j_0 : $c_{j_0} - \bar{\lambda}^T A_{j_0} = \max\{c_j - \lambda^T A_j > 0\}$, $m+1 \leq j \leq n$, which delivers max profit per unit. If matrix N has several such columns, we will take the one with the smallest index $m+1 \leq j_0 \leq n$.
(3) For A_{j_0} we have

$$A_{j_0} = \bar{a}_{1j_0}e_1 + \ldots + \bar{a}_{ij_0}e_i + \ldots + \bar{a}_{mj}e_m,$$

therefore for any $t > 0$ we have

$$tA_{j_0} - t\bar{a}_{1j_0}e_1 - \ldots - t\bar{a}_{ij_0}e_i - \ldots - t\bar{a}_{mj_0}e_m = 0. \tag{5.94}$$

By adding (5.90) and (5.94) we obtain

$$(\bar{x}_1 - t\bar{a}_{1j_0})e_1 + \ldots + tA_{j_0} + \ldots + (\bar{x}_m - t\bar{a}_{mj_0})e_m = \bar{b}. \tag{5.95}$$

The system (5.95) holds for any $t \geq 0$, so if $\forall \bar{a}_{ij_0} \leq 0$, $1 \leq i \leq m$, then by increasing $t > 0$ vector $x(t) = (\bar{x}_1 - t\bar{a}_{ij_0}, \ldots, t, \ldots, \bar{x}_m - t\bar{a}_{mj_0})^T$ remains feasible and $\max z = +\infty$. It follows from Exercise 5.13 that the dual feasible set is empty. The process ends.
Let us assume that there is $\bar{a}_{ij_0} > 0$.
(4) Find

$$t_0 = \min\left\{\frac{x_i}{a_{ij_0}} : a_{ij_0} > 0\right\} = \min\left\{\frac{\bar{b}_i}{a_{ij_0}} > 0\right\} = \frac{\bar{b}_{i_0}}{a_{i_0j_0}}.$$

Element $a_{i_0j_0}$ is called pivot; it indicates that the basic column i_0 will be replaced by nonbasic column j_0 because $\bar{x}_{i_0} - t_0 a_{i_0j_0} = 0$.
In other words we have a new basic solution $\bar{x} := (\bar{x}_B = \bar{b}; \bar{x}_N = 0)^T$ where

$$\bar{x}_B = \left(\bar{x}_1 - \frac{b_{i_0}}{a_{i_0j_0}}\bar{a}_{ij_0}, \ldots, \frac{b_{i_0}}{a_{i_0j_0}}, \ldots, \bar{x}_m - \frac{b_{i_0}}{a_{i_0j_0}}\bar{a}_{mj_0}\right)^T$$

and a new basic $\bar{B} = [A_1, \ldots, A_{j_0}, \ldots, A_m]$ so

$$\bar{B}\bar{x} = \bar{x}_1 A_1 + \ldots + \bar{x}_{j_0}A_{j_0} + \ldots. + \bar{x}_m A_m = \bar{b}.$$

The main task is to show how to get $\bar{B}^{-1}$ from B^{-1}.

We have

$$A_{j_0} = a_{1j_0}A_1 + \ldots + a_{i_0j_0}A_{i_0} \cdots + a_{mj_0}A_m,$$

therefore

$$A_{i_0} = -\frac{a_{1j_0}}{a_{i_0j_0}}A_1 + \ldots + \frac{1}{a_{i_0j_0}}A_{j_0} \ldots - \frac{a_{mj_0}}{a_{i_0j_0}}A_m = \bar{B}a,$$

where $a = (-a_{1j_0}/a_{i_0j_0}, \ldots, 1/a_{i_0j_0} \cdots - a_{mj_0}/a_{i_0j_0})^T = (a_1, \ldots, a_{i_0}, \ldots, a_m)^T$.

For all A_i, $i = 1, \ldots, m, i \neq i_0$ we have

$$A_i = (A_1 \cdot 0, \ldots A_i \cdot 1, \ldots, A_m \cdot 0) = \bar{B}e_i.$$

Therefore, for the old basis B and the new basis $\bar{B}$ we have

$$B = (A_1, \ldots, A_i, \ldots, A_m) \Rightarrow \bar{B} = [e_1, \ldots, a, \ldots, e_m],$$

$$B = \bar{B}E_{i_0}, \tag{5.96}$$

where

$$E_{i_0} = \begin{bmatrix} 1 & a_1 & \\ & 1 \;\; \vdots & \\ & a_m & 1 \end{bmatrix}$$

is an elementary matrix, which coincides with identical matrix I in $\mathbb{R}^m$ except for one column $e_{i_0} = (0, \ldots, 1, \ldots, 0)^T$ replaced by column a.

By multiplying (5.96), from the left on $\bar{B}^{-1}$ and from the right on B^{-1} we obtain

$$\bar{B}^{-1} = E_{i_0}B^{-1}. \tag{5.97}$$

For the elementary, matrix E_{i_0} we have

$$E_{i_0} = I + [0, \ldots, a - e_{i_0}, \ldots, 0] = I + [0, \ldots, \bar{a}, \ldots, 0],$$

where $\bar{a} = (a_1, \ldots, a_{i_0} - 1, \ldots, a_m)^T$. Therefore from (5.97) follows

$$\bar{B}^{-1} = B^{-1} + [0, \ldots, \bar{a}, \ldots, 0]B^{-1}.$$

It completes the simplex method step.

It takes $o(m^2)$ operations to find an inverse to the new basis.

The presented version of simplex method shows that one has to keep only the inverse of the basis. The nonbasic columns can be called one by one to find a "profitable" column. It reduces substantially the amount of information one has to keep at once. Keeping in mind that in practice it takes $O(m)$ steps to get the optimal basis, it became clear why simplex method is so efficient. It justifies Fourier's prediction.

5.9 Interior Point Methods

For about 40 years, simplex method was the tool of choice for solving LP; it was one of the ten best algorithms in the twentieth century. In 1973 V. Klee (1925–2007) and J. Minty show, however, that for a particular objective function on a slightly perturbed unit cube, simplex method may visit all vertices before reaching the optimal one. It means that the number of steps required for simplex method to solve an LP may grow exponentially with the size of the problem.

For many years the existence of polynomial method for solving LP was an open question.

In the mid-1970s N. Shor and independently A. Nemirovski and D. Yudin (1919–2006) introduced the ellipsoid method.

In 1979 L. Khachiyan (1952–2005) used ellipsoid method for solving the primal–dual system of linear inequalities, which is equivalent to LP. He introduced a new parameter L – the length of the binary encoding of the input data. Then he proved that finding $\varepsilon = 2^{-L}$ approximation to the solution of the primal–dual system is enough for finding LP solution in polynomial time.

After an incredible enthusiasm for its indisputable theoretical value, the practical performance of the ellipsoid method was a disappointment. Turned out, simplex method practically always outperformed ellipsoid method.

Simplex method lost its dominance only in the early 1990s after N. Karmarkar introduced in 1984 his projective transformation method for LP calculation and Gill et al. (1986) found that it is closely connected to Newton log–barrier method. It reignited interest to R. Frisch's (1895–1973) log–barrier function introduced in 1955 and P. Huard's interior distance function introduced in 1964.

Jim Renegar (1988) proved that path, following method, based on interior distance function, finds LP solution with accuracy $\varepsilon = 2^{-L}$ in $O(\sqrt{n}L)$ steps versus $O(nL)$ steps of N. Karmarkar's method. It was a big step forward.

Almost at the same time Clovis Gonzaga (1989) proved that path – following IPM based on log–barrier function finds solution with accuracy $\varepsilon = 2^{-L}$ in $O(n^3L)$ operations. Similar result was obtained by Vaidya (1990).

On the other hand, independently E. Barnes from IBM (1986) and R. Vanderbei et al. from AT&T (1986) found a simplified version of N. Karmarkar's method – the affine scaling (AS) algorithm.

As it turned out, the AS algorithm has been introduced 20 year earlier by Ilya Dikin (1936–2008), who was a student of L.V. Kantorovich.

Unfortunately, Dikin's result (1967) for many years was unknown not only to the Western, but to the Eastern optimization community as well.

The AS method played an important role in understanding the practicality of interior pointy methods.

Adler et al. (1989) implemented N. Karmarkar's algorithm and reported very encouraging numerical results.

After decades of research, the primal–dual IPMs emerged as the most numerically efficient approach for LP calculation. A detailed description of the primal–dual IPM results one can find in a very informative book by Wright (1997).

Since the early 1990s, most of the working IPMs codes were based on Mehrotra's (1990), (1992) predictor–corrector algorithm, which uses higher-order approximation of central path, foreshadowed by Megiddo (1989) and further developed by Monteiro et al. (1990).

In the 1990s I. Lustig, R. Marsten, and D. Shanno developed computational aspects of the primal–dual IPMs, implemented and numerically tested the algorithms. Their software for some time was the tool of choice for solving large-scale LP. Currently advanced software, based on IPM, along with advance software, based on simplex, are two basic tools for solving LP.

In the late 1980s, Robert Bixby founded a company CPLEX Optimization. The company made available commercial software for solving LP, QP, MIP, and other optimization problems. Currently GUROBI produces advance software for solving optimization problems with wide areas of commercial applications; see Bixby (2012).

In the late 1980s Yu. Nesterov and A. Nemirovski developed their remarkable self-concordance (SC) theory.

The SC theory allowed understanding the IPMs complexity for a wide classes of convex optimization problems from a unique and general view point. We will cover the basic SC results in Chapter 6.

In the following section we briefly discuss IPMs mainly to show their similarities, differences, and limitations.

5.9.1 *Newton Log– Barrier Method for LP*

We start with Newton's log–barrier method for the following LP:

$$\langle c, x^* \rangle = \min\{\langle c, x \rangle \mid Ax = b, x \in \mathbb{R}^m_+\},$$

where $A : \mathbb{R}^n \to \mathbb{R}^m$, $c \in \mathbb{R}^n$, $b \in \mathbb{R}^m$, $n > m$.

The classical log–barrier method for the given LP at each step finds

$$F(x(t),t) = \min\{F(x,t) | Ax = b\}, \tag{5.98}$$

where $F(x,t) = \langle c, x \rangle - t \sum_{j=1}^n \ln x_j$ and $t > 0$.

It is well-known that $\lim_{t \to 0} \langle c, x(t) \rangle = \langle c, x^* \rangle$. It took almost 20 years to recognize that instead of solving (5.98), it is enough to perform one Newton step toward minimizer $x(t_0)$.

If the starting point x_0 is in the Newton area for the minimizer $x(t_0)$, then approximation x_0 is called "warm" start.

It turns out, after one Newton step toward $x(t_0)$ from the "warm" start x_0, new approximation x_1 is a "warm" start for the minimizer $x(t_1)$, if $t_1 = t_0(1 - \frac{\alpha}{\sqrt{n}})$, where $\alpha > 0$ is a constant independent on n.

Moreover, the bound

$$\Delta(x_1) = \langle c, x_1 - x^* \rangle \leq (1 - \frac{\alpha}{\sqrt{n}}) \langle c, x_0 - x^* \rangle = (1 - \frac{\alpha}{\sqrt{n}}) \Delta(x_0) \tag{5.99}$$

holds.

Polynomial LP complexity follows immediately from (5.99), because it takes $O(\sqrt{n})$ steps to reduce Δ_0 at least in half. Newton steps requires $O(n^{2.5})$ operations, which leads to the overall complexity $O(n^3 \ln \varepsilon^{-1})$, where $\varepsilon > 0$ is the desire accuracy.

In case of $\varepsilon = 2^{-L}$ the overall complexity bound is $O(n^3 L)$, which is the best known complexity result for LP so far.

Let us describe Newton log–barrier method for solving (5.98). Newton direction $d = x - x_0$ one obtains by solving the following problem:

$$\tilde{F}(x_0,t_0,d) = F(x_0,t_0) + \langle \nabla_x F(x_0,t_0), d\rangle + \frac{1}{2}\langle \nabla^2_{xx} F(x_0,t_0)d, d\rangle \to \min \quad (5.100)$$

subject to

$$Ad = 0. \quad (5.101)$$

The first term of the objective function does not effect the solution. Therefore, keeping in mind $\nabla_x F(x_0,t_0) = c - t_0 X_0^{-1} e$ and $\nabla_{xx} F(x_0,t_0) = X_0^{-2} t_0$, where $X_0 = \operatorname{diag}[x_{0j}]_{j=1}^n$ and $e = (1,\ldots,1)^T \in \mathbb{R}^n$, for Lagrangian, which corresponds to the problem (5.100)–(5.101), we have

$$L(d,\lambda) = \langle c - t_0 X_0^{-1} e, d\rangle + \frac{1}{2}\langle X_0^{-2} d, d\rangle t_0 - \langle \lambda, Ad\rangle.$$

Let us consider the correspondent Lagrange system

$$\nabla_d L(d,\lambda) = (c - t_0 X_0^{-1} e) + X_0^{-2} d - A^T \lambda = 0 \quad (5.102)$$

$$\nabla_\lambda L(d,\lambda) = Ad = 0. \quad (5.103)$$

From (5.102) we find

$$d = -X_0^2 (c - t_0 X_0^{-1} e) + X_0^2 A^T \lambda. \quad (5.104)$$

From (5.103) we obtain

$$0 = Ad = -AX_0^2 (c - t_0 X_0^{-1} e) + AX_0^2 A^T \lambda.$$

So, for the dual vector λ we have the following system:

$$AX_0^2 A^T \lambda = AX_0^2 (c - t_0 X_0^{-1} e) \quad (5.105)$$

Let $B = AX_0$, then system (5.105) can be rewritten as the following least square (LS) problem:

$$BB^T \lambda = B(X_0 c - t_0 e). \quad (5.106)$$

Solving (5.106) requires $O(n^{2.5})$ operations, therefore $O(n^3 \ln \varepsilon^{-1})$ is the complexity of the Newton log–barrier method for LP.

In case of degenerate LP the system (5.106) is rank deficient, which requires close attention, in particular, is the final phase.

The Newton log–barrier method follows the primal trajectory $\{x(t)\}$, and it is considered as primal IPM.

In the following section, we consider primal–dual IPM. It is also Newton type method, which follows the primal–dual trajectory. In contrast to the primal IPM, the primal and dual vectors are "equal partners." The primal–dual type methods turned out to be the most efficient IPM for LP; see Wright (1997).

5.9.2 Primal–Dual Interior Point Method

Let us consider the Lagrangian for problem (5.98)

$$L(x,\lambda,t) = L(\cdot) = F(x,t) - \langle \lambda, Ax - b \rangle \tag{5.107}$$

and correspondent Lagrangian system of equations

$$\begin{cases} \nabla_x L(\cdot) = c - t\sum_{j=1}^n x_j^{-1} e_j - A^T\lambda = 0 \\ Ax = b, \end{cases} \tag{5.108}$$

where $e_j = (0,\ldots,1,\ldots,0)^T$.

By introducing diagonal matrix $X = \text{diag}\{x_j\}_{j=1}^n$, slack vector $s = c - A^T\lambda = tX^{-1}e$, diagonal matrix $S = \text{diag}\{s_j\}_{j=1}^n$, and vector $e = (1,\ldots,1)^T \in \mathbb{R}^n$, we can rewrite (5.108) as follows:

$$\begin{cases} c - A^T\lambda = s \\ Ax - b = 0 \\ SXe = te. \end{cases} \tag{5.109}$$

The third system is the co-called centering equations:

$$s_j x_j = t, \; j = 1,\ldots,n. \tag{5.110}$$

They are the only nonlinear part of system (5.109).

Also for $x \in \mathbb{R}^n_{++}$ from (5.110) follows $s \in \mathbb{R}^n_{++}$. In other words, the system (5.109) defines the interior primal–dual trajectory $\{x(t),\lambda(t)\}$ and a slack vector–function $s(t) \in \mathbb{R}^n_{++}$ that

$$\langle x(t), s(t) \rangle = tn, \forall t > 0. \tag{5.111}$$

Therefore $\lim_{t\to 0} x_j(t)s_j(t) = 0$, hence

$$\lim_{t\to 0}\langle c, x(t)\rangle = \langle c, x^*\rangle, \; \lim_{t\to 0}\langle b, \lambda(t)\rangle = \langle b, \lambda^*\rangle.$$

The main IPM idea is tracking the central path $(x(t),\lambda(t),s(t))$ by alternating one Newton step for solving (5.109) with special update of the parameter $t > 0$.

Let us describe Newton step for solving (5.109), assuming that vectors $x_0 \in \mathbb{R}^n_{++}$, $\lambda_0 \in \mathbb{R}^m$, $s_0 \in \mathbb{R}^n_{++}$ are from an appropriate neighborhood of $(x(t_0),\lambda(t_0),s(t_0))$. In other words having (x_0, λ_0, s_0)

$$\begin{cases} A^T\lambda_0 + s_0 = c \\ Ax_0 - b = 0 \\ X_0S_0e \cong te \end{cases} \tag{5.112}$$

we would like to find $x = x_0 + \Delta x$; $\lambda = \lambda_0 + \Delta\lambda$; and $s = s_0 + \Delta s$.

For the triple $(\Delta x, \Delta\lambda, \Delta s)$, we obtain the following system:

$$\begin{cases} A^T(\lambda_0 + \Delta\lambda) + s_0 + \Delta s = c \\ A(x_0 + \Delta x) = b \\ (x_{j,0} + \Delta x_j)(s_{j,0} + \Delta s_j) = t, j = 1, \ldots, n. \end{cases} \tag{5.113}$$

The third system in (5.113) we can rewrite as follows:

$$x_{j,0} \cdot \Delta s_j + s_{j,0}\Delta x_j = t - \Delta x_j \cdot \Delta s_j - x_{j,0} \cdot s_{j,0} \quad j = 1, \ldots, n.$$

By ignoring the second order terms, we can rewrite (5.113) as follows:

$$\begin{cases} A^T\Delta\lambda + \Delta s = c - A^T\lambda_0 - s_0 = 0 \\ A\Delta x = b - Ax_0 = 0 \\ s_{j,0}\Delta x_j + x_{j,0}\Delta s_j = t - x_{j,0} \cdot s_{j,0}, \; j = 1, \ldots, n \end{cases}$$

or

$$\begin{cases} A^T\Delta\lambda + \Delta s = 0 \\ A\Delta x = 0 \\ S_0\Delta x + X_0\Delta s = te - X_0S_0e, \end{cases} \tag{5.114}$$

where $X_0 = \mathrm{diag}\{x_{j,0}\}_{j=1}^n$, $S_0 = \mathrm{diag}\{s_{j,0}\}_{j=1}^n$.

From the first system in (5.114) we have

$$\Delta s = -A^T\Delta\lambda.$$

By substituting Δs into the third system (5.114), we obtain

$$S_0\Delta x - X_0A^T\Delta\lambda = te - X_0S_0e.$$

Multiplying the latter from the left by AS_0^{-1}, we obtain

$$A\Delta x - AS_0^{-1}X_0A\Delta\lambda = AS_0^{-1}(te - X_0S_0e).$$

Keeping in mind $A\Delta x = 0$, we obtain the following LS system for finding $\Delta\lambda$:

$$AD_0A^T\Delta\lambda = AS_0^{-1}\mathrm{v}(t), \tag{5.115}$$

where $D_0 = S_0^{-1}X_0$, $\mathrm{v}(t) = (X_0S_0e - te)$.

Then, from the first system in (5.113), we obtain $\Delta s = -A^T\Delta\lambda$, and from the last system in (5.114), we have $\Delta x = -S_0^{-1}\mathrm{v}(t) - D_0\Delta s$.

From the complementarity condition $x_j^* s_j^* = 0,\ j = 1,\dots,n$ for $x_j^* > 0$ we have $s_j^* = 0$.

Therefore, finding elements $d_{j,0} = x_{j,0} \cdot s_{j,0}^{-1}$ of D_0 for $j \notin J(x^*) = \{j : x_j^* = 0\}$ gets problematic in the neighborhood of (x^*, λ^*, s^*).

For degenerate LP solving the LS system (5.115) is also problematic due to the rank deficiency of the matrix AD_0A^T. Therefore, special technique is used for solving system (5.113).

Solving the LS systems (5.105) or (5.115) is the main numerical task per step in both Newton log–barrier method and primal–dual IPM. The solutions of the correspondent systems define directions, in which the primal and the primal–dual interior approximations are updated.

These methods require different mathematical analysis. It will be discussed in Chapter 6.

Exercise 5.17. Describe Newton log–barrier method for the dual LP

$$\langle b, \lambda^* \rangle = \max\{\langle b, \lambda \rangle \,|\, A^T\lambda \leq c\}.$$

5.9.3 *Affine Scaling Method*

The affine scaling method historically was the first IPM. It was introduced in 1967 by Ilya Dikin. We consider the canonical LP problem

$$\langle c, x^* \rangle = \max\{\langle c, x \rangle \,|\, Ax = b, x \in \mathbb{R}^n_+\}, \tag{5.116}$$

where $A : \mathbb{R}^n \to \mathbb{R}^m$, $c \in \mathbb{R}^n$, $b \in \mathbb{R}^m$, $n > m$ and the correspondent dual problem

$$\langle b, \lambda^* \rangle = \min\{\langle b, \lambda \rangle \,|\, A^T\lambda \geq c\}. \tag{5.117}$$

We start with $x_0 \in \mathbb{R}^n_{++}$, $0 < t < 1$, and $X_0 = \text{diag}(x_{j,0})_{j=1}^n$. The ellipsoid

$$E(x_0, t) = \{x : \langle X_0^{-2}(x - x_0), x - x_0 \rangle - t^2 = 0\} \subset \mathbb{R}^n_{++}$$

is called Dikin's ellipsoid (see Figure 5.1).

The main operation at each step of the AS method is solving the following problem:

$$\begin{aligned} &\langle c, d \rangle \Rightarrow \max \\ &\text{s. t. } Ad = 0, \langle X_0^{-2} d, d \rangle = t^2, \end{aligned} \tag{5.118}$$

where $d = x - x_0$.

By introducing Lagrange multipliers vector $\lambda = (\lambda_1, \dots, \lambda_m)$ and $\mu \in \mathbb{R}^1$, we obtain Lagrangian

$$L(d,\lambda,\mu) = \langle c,d\rangle - \langle\lambda,Ad\rangle - \frac{1}{2}\mu[\langle X_0^{-2}d,d\rangle - t^2)],$$

which corresponds to problem (5.118).

The correspondent Lagrangian system

$$\nabla_d L(\cdot) = c - A^T\lambda - \mu X_0^{-2}d = 0 \tag{5.119}$$

$$\nabla_\lambda L(\cdot) = Ad = 0 \tag{5.120}$$

$$\nabla_\mu L(\cdot) = \langle X_0^{-2}d,d\rangle - t^2 = 0 \tag{5.121}$$

has to be solved for d, λ, and μ.

From (5.119) follows

$$d = \mu^{-1}X_0^2(c - A^T\lambda) = \mu^{-1}X_0^2 r(\lambda), \tag{5.122}$$

where $r(\lambda) = c - A^T\lambda$. From (5.120) and (5.122), we obtain

$$0 = Ad = \mu^{-1}[AX_0^2A^T\lambda - AX_0^2c].$$

Let $B = AX_0$, then the Lagrange multipliers vector $\lambda_0 \in \mathbb{R}^m$ one finds by solving the following least square system:

$$BB^T\lambda = BX_0c. \tag{5.123}$$

From (5.121) and (5.122) we obtain

$$\mu^{-2}\langle X_0^{-2}X_0^2r(\lambda_0), X_0^2r(\lambda_0)\rangle = t^2$$

or

$$\mu^{-1} = t\|X_0r(\lambda_0)\|^{-1}.$$

Then

$$x_1 = x_0 + d = x_0 + t\frac{X_0^2r(\lambda_0)}{\|X_0r(\lambda_0)\|}. \tag{5.124}$$

By reiterating (5.124), we obtain the AS method

$$x_{s+1} = x_s + t\frac{X_s^2r(\lambda_s)}{\|X_sr(\lambda_s)\|}. \tag{5.125}$$

The main issue in regard of AS method (5.125) is the choice of $0 < t < 1$.

The practical performance of the AS method is better for the “long step” $t \approx 0.9$. The convergence for such step, however, was not established.

T. Tsuchiya in 1991 proved convergence of AS method for $t = 1/8$ without any nondegeneracy assumption.

I. Dikin in 1992 proved convergence of AS method for $t = 0.5$ assuming uniqueness of the primal optimal solution. Then in 1995 T. Tsuchiya and M. Muramatsu established AS convergence for $t = 2/3$ without any nondegeneracy assumptions.

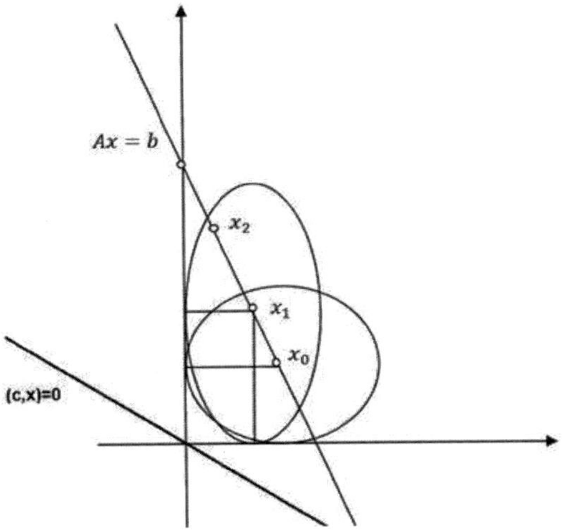

Fig. 5.1 Linear programming

From convergence of the AS method follows $\lim_{s\to\infty}(x_s, r(\lambda_s)) = 0$, which due to (5.125) leads to numerical instability in the final phase.

5.10 SUMT as Dual Interior Regularization

The log–barrier function, introduced by R. Frisch in 1955, hyperbolic barrier by C. Carroll in 1961, and the interior distance by P.Huard in the mid-1960s were all attempts to replace a constrained optimization problem by a sequence of unconstrained optimization problems.

Later this approach has been extensively studied by Anthony Fiacco (1928–2013) and Garth Mc Cormick (1935–2008) and incorporated in their classical SUMT.

In this section we show that SUMT is equivalent to interior regularization for the dual problems.

The equivalence is used for proving convergence and establishing the error bounds. We also discuss the shortcomings of SUMT and their dual equivalent.

We assume:

A. Convex optimization problem (5.25) has a non-empty and bounded optimal set X^*.
B. The Slater condition for the problem (5.25) holds.

It follows from Corollary 2.4 and assumption A that, by adding one new constraint $c_0(x) = N - f(x) \geq 0$, we obtain bounded primal feasible set and for large enough N the extra constraint does not effect the optimal set X^*. Therefore we will assume that Ω is bounded.

In view of B, the optimal dual set $\Lambda^* = \{\lambda \in \mathbb{R}^n : d(\lambda) = d(\lambda^*)\}$ is bounded too.

The log–barrier function for the problem (5.25) is defined as follows:

$$F(x,k) = f(x) - k^{-1}\sum_{i=1}^{m} \pi(kc_i(x)),$$

where $\pi(t) = \ln t$, $\pi(t) = -\infty$ for $t \leq 0$.

5.10.1 Log–Barrier Method and Its Dual Equivalent

Due to convexity f and concavity c_i $i = 1,\dots,m$ the function F is convex in x for any $k > 0$.

From Slater condition, convexity f, concavity c_i, and boundedness Ω follows that the recession cone of Ω is empty, that is, for any $x \in \Omega$, $k > 0$ and $0 \neq d \in \mathbb{R}^n$, we have

$$\lim_{t\to\infty} F(x+td,k) = \infty. \tag{5.126}$$

Therefore, for any $k > 0$ there exists

$$x(k) : \nabla_x F(x(k),k) = 0. \tag{5.127}$$

Theorem 5.9. *If A and B hold and f, $c_i \in C^1$, $i = 1,\dots,m$, then interior log–barrier method* (5.127) *is equivalent to the interior regularization method*

$$\lambda(k) = \operatorname{argmax}\{d(u) + k^{-1}\sum_{i=1}^{m} \ln u_i : u \in \mathbb{R}^m_+\} \tag{5.128}$$

for the dual problem and the error bound

$$\max\{f(x(k)) - f(x^*), d(\lambda^*) - d(\lambda(k))\} = mk^{-1} \tag{5.129}$$

holds.

Proof. From (5.126) follows existence $x(k) : F(x(k),k) = \min\{F(x,k) : x \in \mathbb{R}^n\}$ for any $k > 0$.

Therefore

$$\nabla_x F(x(k),k) = \nabla f(x(k)) - \sum_{i=1}^{m} \pi'(kc_i(x(k))\nabla c_i(x(k)) = 0. \tag{5.130}$$

Let

$$\lambda_i(k) = \pi'(kc_i(x(k)) = (kc_i(x(k)))^{-1},\ i = 1,..,m. \tag{5.131}$$

Then, from (5.130) and (5.131) follows

$$\nabla_x F(x(k),k) = \nabla_x L(x(k),\lambda(k)) = 0,$$

therefore $d(\lambda(k)) = L(x(k),\lambda(k))$. From $\pi''(t) = -t^2 < 0$ follows existence of the inverse function π'^{-1} and from (5.131) we have $kc(x(k)) = \pi'^{-1}(\lambda_i(k))$.

Using LF identity, we obtain

$$c_i(x(k)) = k^{-1}\pi^{*'}(\lambda_i(k)), \tag{5.132}$$

where $\pi^*(s) = \inf_{t>0}\{st - \ln t\} = 1 + \ln s$. The subdifferential $\partial d(\lambda(k))$ contains $-c(x(k)) = -(c_1(x(k)),\ldots,c_m(x(k)))^T$, that is

$$0 \in \partial d(\lambda(k)) + c(x(k)). \tag{5.133}$$

From (5.132) and (5.133) follows

$$0 \in \partial d(\lambda(k)) + k^{-1}\sum_{i=1}^{m} \pi^{*'}(\lambda_i(k))e_i. \tag{5.134}$$

The latter inclusion is the optimality criteria for $\lambda(k)$ to be the maximizer in (5.128).

The maximizer $\lambda(k)$ is unique due to the strict concavity of the objective function in (5.128).

Thus, SUMT with log–barrier function $F(x,k)$ is equivalent to the interior regularization method (5.128) for the dual problem.

For the primal interior trajectory $\{x(k)\}_{k=k_0>0}^{\infty}$ and the dual interior trajectory $\{\lambda(k)\}_{k=k_0>0}^{\infty}$ we have

$$f(x(k)) \geq f(x^*) = d(\lambda^*) \geq d(\lambda(k)) = L(x(k),\lambda(k)) = f(x(k)) - \langle c(x(k)),\lambda(k)\rangle.$$

From (5.131) follows $\lambda_i(k)c_i(x(k)) = k^{-1}$, $i = 1,\ldots,m$; hence for the primal–dual gap, we obtain

$$f(x(k)) - d(\lambda(k)) = \langle c(x(k)),\lambda(k)\rangle = mk^{-1},$$

which leads to the error bound (5.129). □

To solve problem (5.25) with accuracy $\varepsilon > 0$, it is enough to take $k = m\varepsilon^{-1}$ and find an approximation for $x(k)$, starting with $x_0 \in \operatorname{int}\Omega$.

First of all finding $x_0 \in \operatorname{int}\Omega$ can be a problem similar to the initial one, then finding

$$x(k): \; F(x(k),k) = \min\{F(x,k)|x \in \mathbb{R}^m\} \tag{5.135}$$

for $k = m\varepsilon^{-1}$, when $\varepsilon > 0$ is small enough is a rather difficult task.

Let us discuss it with more details.

To simplify considerations we assume that second-order sufficient optimality condition is satisfied; then x^* and λ^* are unique. We have

$$\nabla_x F(x(k),k) = \nabla F(x(\cdot),\cdot) = \nabla f(x(\cdot)) - \sum_{i=1}^{m} \pi'(kc_i(x(k)))\nabla c_i(x(\cdot)) = 0,$$

or

$$\nabla_x F(x(\cdot),\cdot) = \nabla f(x(\cdot)) - \sum_{i=1}^{m} \lambda_i(\cdot) c_i(x(\cdot)).$$

Let us consider the Hessian

$$\begin{aligned}\nabla^2_{xx} F(x(\cdot),\cdot) &= \nabla^2 f(x(\cdot)) - \sum_{i=1}^{m} \lambda_i(\cdot)\nabla^2_{xx} c_i(x(\cdot)) \\ &\quad + \sum_{i=1}^{m} \nabla c_i^T(x(\cdot))(kc_i(x(\cdot)))^{-1}(\lambda_i(\cdot))\nabla c_i(x(\cdot)) \\ &= \nabla_{xx} L(x(\cdot),\lambda(\cdot)) + \nabla c^T(x(\cdot))\Lambda(\cdot)C^{-1}(x(\cdot))\nabla c(x(\cdot))\end{aligned}$$

where $\Lambda(\cdot) = \text{diag}(\lambda_i(\cdot))_{i=1}^m$, $C(x(\cdot)) = \text{diag}(c_i(x(\cdot))_{i=1}^m$.

From (5.129) for $\varepsilon > 0$ small enough follows

$$\nabla^2_{xx} F(x(\cdot),\cdot) \approx \nabla^2_{xx} L(x^*,\lambda^*) + \nabla c^T(x^*)\Lambda^* C^{-1}(x(\cdot))\nabla c(x^*). \tag{5.136}$$

Let $I^* = \{i : c_i(x^*) = 0\} = \{1,\ldots,r\}$, $r < m$ be the active set, then $\lambda_i^* > 0$, $i = 1,\ldots,r$ and $\lambda_i^* = 0$ for $i \geq r+1,\ldots,m$. So, we can rewrite (5.136) as follows

$$\nabla^2_{xx} F(x(\cdot),\cdot) \approx \nabla^2_{xx} L(x^*,\lambda^*) + \nabla c^T_{(r)}(x^*)\Lambda^*_{(r)} C^{-1}_{(r)}(x(\cdot))\nabla c_{(r)}(x^*), \tag{5.137}$$

where $\nabla c_{(r)}(x) = J(c_{(r)}(x))|_{x=x^*}$ is the Jacobian of $c_{(r)}(x) = (c_1(x),\ldots,c_r(x))^T$, $\Lambda^*_{(r)} = \text{diag}(\lambda_i^*)_{i=1}^r$, $C(x(\cdot)) = \text{diag}(c_i(x(\cdot)))_{i=1}^r$.

From (5.137) and Debreu's Lemma with $A = \nabla^2_{xx} L(x^*,\lambda^*)$ and $C = \Lambda^{*\frac{1}{2}}_{(r)} \nabla c_{(r)}(x^*)$ follows

$$\text{mineigval}\, \nabla^2_{xx} F(x(\cdot),\cdot) = \mu > 0.$$

From $\lim_{k\to\infty} \lambda_i^*/c_i(x(k)) = \infty$, $i = 1,\ldots,r$ follows

$$\lim_{k\to\infty} \text{maxeigval}\, \nabla^2_{xx} F(x(\cdot),\cdot) = \infty,$$

therefore

$$\lim_{k\to\infty} \text{cond}\, \nabla^2_{xx} F(x(k),k) = 0.$$

So, the condition number of log–barrier Hassian $\nabla^2_{xx} F$ vanishes, when the primal–dual sequence approaches the solution.

It means finding an accurate approximation for $x(k)$, when $k > 0$ large enough, is practically an impossible task.

This is an intrinsic feature of the log–barrier method.

There are two basic ways to cure the problem.

First, the interior point methods, which, for some classes of CO, can mitigate the ill-conditioning phenomenon. The correspondent results will be covered in Chapter 6.

Second, the exterior point methods, based on nonlinear rescaling theory, which is considered in Chapters 7 and 8, or Lagrangian transformation theory, which is considered in Chapter 9.

In case of log–barrier function, the situation is symmetric, that is, both the primal log–barrier method (5.127) and the dual interior regularization method (5.128) are using the same log–barrier function.

It is not the case for other constraints transformations used in SUMT.

5.10.2 Hyperbolic Barrier as Dual Parabolic Regularization

The hyperbolic barrier

$$\pi(t) = \begin{cases} -t^{-1}, & t > 0 \\ -\infty, & t \leq 0, \end{cases}$$

has been introduced by C. Carroll in 1961. It leads to the following hyperbolic penalty function:

$$C(x,k) = f(x) - k^{-1}\sum_{i=1}^{m}\pi(kc_i(x)) = f(x) + k^{-1}\sum_{i=1}^{m}(kc_i(x))^{-1},$$

which is convex in $x \in \mathbb{R}^n$ for any $k > 0$. For the primal minimizer, we obtain

$$x(k) : \nabla_x C(x(k),k) = \nabla f(x(k)) - \sum_{i=1}^{m}\pi'(kc_i(x(k)))\nabla c_i(x(k)) = 0. \tag{5.138}$$

For the Lagrange multipliers vector, we have

$$\lambda(k) = (\lambda_i(k) = \pi'(kc_i(x(k)) = (kc_i(x(k)))^{-2},\ i = 1,\dots,m). \tag{5.139}$$

We will show later that vectors $\lambda(k)$, $k \geq 1$ are bounded. Let $L = \max_{i,k}\lambda_i(k)$.

Theorem 5.10. *If A and B hold and $f,\ c_i \in C^1$, $i = 1,..,m$, then hyperbolic barrier method* (5.138) *is equivalent to the parabolic regularization method*

$$d(\lambda(k)) + 2k^{-1}\sum_{i=1}^{m}\sqrt{\lambda_i(k)} = \max\{d(u) + 2k^{-1}\sum_{i=1}^{m}\sqrt{u_i} : u \in \mathbb{R}^m_+\} \tag{5.140}$$

and the error bound

$$\max\{f(x(k)) - f(x^*), d(\lambda^*) - d(\lambda(k))\} \leq m\sqrt{L}k^{-1}. \tag{5.141}$$

holds.

Proof. From (5.138) and (5.139) follows

$$\nabla_x C(x(k),k) = \nabla_x L(x(k),\lambda(k)) = 0,$$

therefore $d(\lambda(k)) = L(x(k),\lambda(k))$.

Then $\pi''(t) = -2t^{-3} < 0$, $\forall t > 0$; therefore π'^{-1} exist.

Using LF identity, from (5.139) we obtain

$$c_i(x(k)) = k^{-1}\pi'^{-1}(\lambda_i(k)) = k^{-1}\pi^{*'}(\lambda_i(k)),\ i = 1,\dots,m,$$

where $\pi^*(s) = \inf_t\{st - \pi(t)\} = 2\sqrt{s}$.

The subgradient $-c(x(k)) \in \partial d(\lambda(k))$ that is

$$0 \in \partial d(\lambda(k)) + c(x(k)) = \partial d(\lambda(k)) + k^{-1}\sum_{i=1}^{m} \pi^{*'}(\lambda_i(k))e_i. \tag{5.142}$$

The latter inclusion is the optimality condition for the interior regularization method (5.140) for the dual problem.

Thus, the hyperbolic barrier method (5.138) is equivalent to the parabolic regularization method (5.140), and $D(u,k) = d(u) + 2k^{-1}\sum_{i=1}^{m}\sqrt{u_i}$ is strictly concave.

Due to the strict concavity of $D(u,k)$ in u from (5.140) follows

$$d(\lambda(k)) + 2k^{-1}\sum_{i=1}^{m}\sqrt{\lambda_i(k)} > d(\lambda(k+1)) + 2k^{-1}\sum_{i=1}^{m}\sqrt{\lambda_i(k+1)}$$

and

$$d(\lambda(k+1)) + 2(k+1)^{-1}\sum_{i=1}^{m}\sqrt{\lambda_i(k+1)} > d(\lambda(k)) + 2(k+1)^{-1}\sum_{i=1}^{m}\sqrt{\lambda_i(k)}.$$

By adding the inequalities, we obtain

$$\sum_{i=1}^{m}\sqrt{\lambda_i(k)} > \sum_{k=1}^{m}\sqrt{\lambda_i(k+1)},\ \ k \geq 1.$$

Therefore, the sequence $\{\lambda(k)\}_{k=1}^{\infty}$ is bounded, so there exists $L = \max_{i,k}\lambda_i(k) > 0$. From (5.139) for any $k \geq 1$ and $i = 1,\dots,m$, we have

$$\lambda_i(k)c_i^2(x(k)) = k^{-2}$$

or

$$(\lambda_i(k)c_i(x(k)))^2 = k^{-2}\lambda_i(k) \leq k^{-2}L.$$

Therefore,

$$\langle\lambda(k), c(x(k))\rangle \leq m\sqrt{L}k^{-1}.$$

For the primal interior sequence $\{x(k)\}_{k\in N}$ and dual interior sequence $\{\lambda(k)\}_{k\in N}$, we have

$$f(x(k)) \geq f(x^*) = d(\lambda^*) \geq L(x(k), \lambda(k)) = d(\lambda(k)),$$

therefore

$$f(x(k)) - d(\lambda(k)) = (c(x(k)), \lambda(k))) \leq m\sqrt{L}k^{-1},$$

which leads to (5.141). □

The bounds (5.129) and (5.140) are fundamentally different because of $L > 0$, which can be very large for problems where Slater condition is "barely" satisfied, that is, when the primal feasible set is not "well defined."

This is one of the reasons why log–barrier function is so important.

5.10.3 Exponential Penalty as Dual Regularization with Shannon's Entropy Function

Exponential penalty $\pi(t) = -e^{-t}$ was used by Motzkin in 1952 to transform a systems of linear inequalities into an unconstrained convex optimization problem in order to use unconstrained minimization technique for solving linear inequalities.

The exponential transformation $\pi(t) = -e^{-t}$ leads to the exponential penalty function

$$M(x,k) = f(x) - k^{-1}\sum_{i=1}^{m} \pi(kc_i(x)) = f(x) + k^{-1}\sum_{i=1}^{m} e^{-kc_i(x)},$$

which is for any $k > 0$ convex in $x \in \mathbb{R}^n$, the recession cone of Ω is empty; therefore the primal minimizer $x(k)$ exists and

$$x(k) : \nabla_x M(x(k),k) = \nabla f(x(k)) - \sum_{i=1}^{m} e^{-kc_i(x(k))}\nabla c_i(x(k)) = 0. \tag{5.143}$$

Let us introduce the Lagrange multipliers vector

$$\lambda(k) = (\lambda_i(k) = \pi^{'}(c_i(x(k)) = e^{-kc_i(x(k))},\ i = 1, \ldots, m). \tag{5.144}$$

From (5.143) and (5.144) we have

$$\nabla_x M(x(k),k) = \nabla_x L(x(k), \lambda(k)) = 0.$$

Therefore, from convexity $L(x,\lambda(k))$ in $x \in \mathbb{R}^n$ follows $d(\lambda(k)) = \min\{L(x,\lambda(k)) |x \in \mathbb{R}^n\} = L(x(k),\lambda(k))$ and $-c(x(k)) \in \partial d(\lambda(k))$, hence

$$0 \in c(x(k)) + \partial d(\lambda(k)). \tag{5.145}$$

From $\pi^{''}(t) = -e^{-t} \neq 0$ follows the existence $\pi^{'-1}$, therefore, using LF identity, from (5.144) we obtain

$$c_i(x(k)) = k^{-1}\pi'^{-1}(\lambda_i(k)) = k^{-1}\pi^{*'}(\lambda_i(k)),\ i = 1,\dots,m.$$

Inclusion (5.145) we can rewrite as follows:

$$\partial d(\lambda(k)) + k^{-1}\sum_{i=1}^{m}\pi^{*'}(\lambda_i(k))e_i = 0.$$

Keeping in mind $\pi^*(s) = \inf_t\{st - \pi(t)\} = \inf\{st + e^{-t}\} = -s\ln s + s$ from the latter inclusion, we obtain

$$d(\lambda(k)) - k^{-1}\sum_{i=1}^{m}\lambda_i(k)(\ln(\lambda_i(k) - 1)) = \max\{d(u) - k^{-1}r(u) : u \in \mathbb{R}^m_+\}. \quad (5.146)$$

It means that the exponential penalty method (5.143) is equivalent to the interior regularization method (5.146) with Shannon's type entropy function $r(u) = \sum_{i=1}^{m} u_i(\ln u_i - 1)$, used for the regularization.

The convergence of the dual sequence $\{\lambda(k)\}_{k\in N}$ can be proven using arguments similar to those used in Theorem 5.10.

We conclude the section by considering smoothing technique for convex optimization.

Exercise 5.18. Find error bounds for $x(k)$ and $\lambda(k)$.

5.10.4 Log–Sigmoid Method as Dual Regularization with Fermi–Dirac's Entropy Function

It follows from Karush–Kuhn–Tucker's Theorem that under Slater condition for x^* to be a solution of (5.25), it is necessary and sufficient existence $\lambda^* \in \mathbb{R}^m$ that the pair $(x^*;\lambda^*)$ is the saddle point of the Lagrangian, that is, (5.36) hold.

From the right inequality of (5.36) and complementarity condition, we obtain

$$f(x^*) \le f(x) - \sum_{i=1}^{m}\lambda_i^*\min\{c_i(x),0\} \le$$

$$f(x) - (\max_{1\le i\le m}\lambda_i^*)\sum_{i=1}^{m}\min\{c_i(x),0\}$$

for any $x \in \mathbb{R}^n$. Therefore, for any $r > \max_{1\le i\le m}\lambda_i^*$, we have

$$f(x^*) \le f(x) - r\sum_{i=1}^{m}\min\{c_i(x),0\}, \forall x \in \mathbb{R}^n. \quad (5.147)$$

The function

$$Q(x,r) = f(x) - r\sum_{i=1}^{m}\min\{c_i(x),0\}$$

is called exact penalty function.

From concavity c_i, $i = 1,\dots,m$ follows concavity $q_i(x) = \min\{c_i(x),0\}$. From convexity f and concavity q_i, $i = 1,\dots,m$ follows convexity $Q(x,r)$ in $x \in \mathbb{R}^n$.

From (5.147) follows that solving (5.25) is equivalent to solving the following unconstrained minimization problem:

$$f(x^*) = Q(x^*,r) = \min\{Q(x,r) : x \in \mathbb{R}^n\}. \tag{5.148}$$

The function $Q(x,r)$ is non-smooth at x^*. The smoothing technique (see Chen and Mangasarian (1995)) replaces Q by a sequence of smooth functions, which approximate $Q(x,r)$.

The log–sigmoid (LS) function $\pi : \mathbb{R} \to \mathbb{R}$, which is defined by

$$\pi(t) = \ln S(t,1) = \ln(1+e^{-t})^{-1},$$

is one of such functions. We collect the log–sigmoid properties in the following assertion:

Assertion 5.1 *The following statements hold:*

1 $\pi(t) = t - \ln(1+e^t) < 0$, $\pi(0) = -\ln 2$
2. $\pi'(t) = (1+e^t)^{-1} > 0$, $\pi'(0) = 2^{-1}$
3. $\pi''(t) = -e^t(1+e^t)^{-2} < 0$, $\pi''(0) = -2^{-2}$.

Smooth penalty method employs scaled LS function

$$k^{-1}\pi(kt) = t - k^{-1}\ln(1+e^{kt}), \tag{5.149}$$

which is a smooth approximation of $q(t) = \min\{t,0\}$.

In particular, from (5.149) follows

$$0 < q(t) - k^{-1}\pi(kt) < k^{-1}\ln 2. \tag{5.150}$$

It means that by increasing $k > 0$ the approximation can be made as accurate as one wants.

Smooth penalty function $P : \mathbb{R}^n \times \mathbb{R}_{++} \to \mathbb{R}$, defined by

$$P(x,k) = f(x) - k^{-1}\sum_{i=1}^{m}\pi(kc_i(x)), \tag{5.151}$$

is the main instrument in the smoothing technique.

From Assertion 5.1 follows that P is as smooth as f and c_i, $i = 1,..,m$.

The LS method at each step finds

$$x(k) : P(x(k),k) = \min\{P(x,k) : x \in \mathbb{R}^n\} \tag{5.152}$$

and increases $k > 0$ if the obtained accuracy is not satisfactory.

Without loss of generality, we assume that f is bounded from below. Such assumption does not restrict the generality, because the original objective function f can be replaced by an equivalent $f(x) := \ln(1+e^{f(x)}) \geq 0$.

Boundedness of Ω together with Slater condition, convexity f, and concavity c_i, $i = 1,\dots,m$ make the recession cone of Ω empty, that is

$$\lim_{t\to\infty} P(x+td,k) = \infty$$

for any $k > 0$, $d \neq 0$ and $d \in \mathbb{R}^n$ and any $x \in \Omega$.

Therefore, minimizer $x(k)$ in (5.152) exists for any $k > 0$, that is

$$\nabla_x P(x(k),k) = \nabla f(x(k)) - \sum_{i=1}^{m} \pi^{'}(kc_i(x(k)))\nabla c_i(x(k)) =$$

$$= \nabla f(x(k)) - \sum_{i=1}^{m} (1+e^{kc_i(x(k))})^{-1}\nabla c_i(x(k)) = 0.$$

Let

$$\lambda_i(k) = (1+e^{kc_i(x(k))})^{-1},\ i = 1,\dots,m, \tag{5.153}$$

then

$$\nabla_x P(x(k);k) = \nabla f(x(k)) - \sum_{i=1}^{m} \lambda_i(k)\nabla c_i(x(k)) = 0.$$

From (5.153) follows $\lambda_i(k) \leq 1$ for any $k > 0$. Therefore, generally speaking, one can't expect finding a good approximation for optimal Lagrange multipliers, no matter how large the penalty parameter $k > 0$ is.

If the dual sequence $\{\lambda(k)\}_{k\in N}$ does not converges to $\lambda^* \in L^*$, then in view of the last equation one can't expect convergence of the primal sequence $\{x(k)\}_{k\in N}$ to $x^* \in X^*$ either.

To guarantee convergence of the LS method, we have to modify $P(x,k)$. Let $0 < \alpha < 0.5$ and

$$P(x,k) := P_\alpha(x,k) = f(x) - k^{-1+\alpha} \sum_{i=1}^{m} \pi(kc_i(x)). \tag{5.154}$$

It is easy to see that the modification does not effect the existence of $x(k)$. Therefore for any $k > 0$ there exists

$$x(k) : \nabla_x P(x(k),k) = \nabla f(x(k)) - k^\alpha \sum \pi^{'}(kc(x(k)))\nabla c_i(x(k)) = 0. \tag{5.155}$$

Theorem 5.11. *If A and B hold and f, $c_i \in C^1$, $i = 1,\dots,m$, then the LS method (5.155) is equivalent to an interior regularization method*

$$d(\lambda(k))+k^{-1}\sum_{i=1}^{m}\pi^*(k^{-\alpha}\lambda_i(k))=$$

$$\max\{d(u)+k^{-1}\sum_{i=1}^{m}\pi^*(k^{-\alpha}u_i):0\le u_i\le k^{\alpha},\ i=1,\dots,m\}.$$

Proof. Let

$$\lambda_i(k)=k^{\alpha}\pi^{'}(kc_i(x(k)))=k^{\alpha}(1+e^{kc_i(x(k))})^{-1},\ i=1,\dots,m. \tag{5.156}$$

From (5.155) and (5.156) follows

$$\begin{aligned}\nabla_x P(x(k),k)=&\nabla f(x(k))-\sum_{i=1}^{m}\lambda_i(k)\nabla c_i(x(k))=\\ &\nabla_x L(x(k),\lambda(k))=0.\end{aligned} \tag{5.157}$$

From (5.156) we have

$$\pi^{'}(kc_i(x(k))=k^{-\alpha}\lambda_i(k). \tag{5.158}$$

Due to $\pi^{''}(t)<0$ there exists $\pi^{'-1}$, therefore

$$c_i(x(k))=k^{-1}\pi^{'-1}(k^{-\alpha}\lambda_i(k)).$$

Using LF identity, we obtain

$$c_i(x(k))=k^{-1}\pi^{*'}(k^{-\alpha}\lambda_i(k)), \tag{5.159}$$

where

$$\pi^*(s)=\inf_t\{st-\pi(t)\}=-[(1-s)\ln(1-s)+s\ln s]$$

is Fermi–Dirac (FD) entropy function; see Ray and Majumder (2014).

From (5.157) follows $d(\lambda(k))=L(x(k),\lambda(k))$, also the subdifferential $\partial d(\lambda(k))$ contains $-c(x(k))$, that is

$$0\in c(x(k))+\partial d(\lambda(k)). \tag{5.160}$$

Combining (5.159) and (5.160), we obtain

$$0\in\partial d(\lambda(k))+k^{-1}\sum_{i=1}^{m}\pi^{*'}(k^{-\alpha}\lambda_i(k))e_i. \tag{5.161}$$

The inclusion (5.161) is the optimality criteria for the following problem:

$$d(\lambda(k))+k^{-1}\sum_{i=1}^{m}\pi^*(k^{-\alpha}\lambda_i(k))=$$

$$\max\{d(u)+k^{-1}r(u):0\le u_i\le k^{\alpha},\ i=1,..,m\}, \tag{5.162}$$

where $r(u) = \sum_{i=1}^{m} \pi^*(k^{-\alpha} u_i)$.

In other words the LS method (5.155)–(5.156) is equivalent to the interior regularization method (5.162) with FD entropy function used for dual regularization. The FD function is strongly concave inside the cube $\{u \in \mathbb{R}^m : 0 \leq u_i \leq k^\alpha, i = 1, \dots, m\}$.

It follows from (5.162) that for any regularization sequence $\{k_s\}_{s \in N}$, the Lagrange multipliers $0 < \lambda_i(k_s) < k_s^\alpha$, $i = 1, \dots, m$ can be any positive number, which underlines the importance of modification (5.154).

Theorem 5.12. *Under conditions of Theorem 5.11 for any regularization sequence $\{k_s\}_{s \in N}$, the primal sequence*

$$\{x_s\}_{s \in N} : \nabla_x P(x_s, k_s) = \nabla f(x_s) - \sum_{i=1}^{m} \lambda_{i,s} \nabla c_i(x_s) = 0 \tag{5.163}$$

and the dual sequence

$$\{\lambda_s\}_{s \in N} : d(\lambda_s) + k_s^{-1} r(\lambda_s) =$$

$$\max\{d(u) + k_s^{-1} r(u) : 0 \leq u_i \leq k^\alpha,\ i = 1, \dots, m\} \tag{5.164}$$

the following statements hold:

(1) (a) $d(\lambda_{s+1}) > d(\lambda_s)$; b) $r(\lambda_{s+1}) < r(\lambda_s)$;
(2) $\lim_{s \to \infty} d(\lambda_s) = d(\lambda^*)$ and $\lambda^* = \operatorname{argmin}\{r(\lambda) : \lambda \in L^*\}$;
(3) the primal–dual sequence $\{x_s, \lambda_s\}_{s \in N}$ is bounded, and any limit point is the primal–dual solution.

Proof.

(1) From (5.164) and strong concavity $r(u)$ follows

$$d(\lambda_{s+1}) + k_{s+1}^{-1} r(\lambda_{s+1}) > d(\lambda_s) + k_{s+1}^{-1} r(\lambda_s) \tag{5.165}$$

and

$$d(\lambda_s) + k_s^{-1} r(\lambda_s) > d(\lambda_{s+1}) + k_s^{-1} r(\lambda_{s+1}). \tag{5.166}$$

Therefore,

$$(k_{s+1}^{-1} - k_s^{-1})(r(\lambda_{s+1}) - r(\lambda_s)) > 0.$$

From $k_{s+1} > k_s$ and last inequality follows $r(\lambda_{s+1}) < r(\lambda_s)$, therefore from (5.165) follows

$$d(\lambda_{s+1}) > d(\lambda_s) + k_{s+1}^{-1}(r(\lambda_s) - r(\lambda_{s+1})) > d(\lambda_s). \tag{5.167}$$

(2) The monotone increasing sequence $\{d(\lambda_s)\}_{s \in N}$ is bounded by $f(x^*)$. Therefore, there is $\lim_{s \to \infty} d(\lambda_s) = \bar{d} \leq f(x^*) = d(\lambda^*)$.
From (5.164) follows

$$d(\lambda_s) + k_s^{-1} r(\lambda_s) \geq d(\lambda^*) + k_s^{-1} r(\lambda^*). \tag{5.168}$$

From (5.167) follows $\{\lambda_s\}_{s\in N} \subset \Lambda(\lambda_0) = \{\lambda \in \mathbb{R}^m_+ : d(\lambda) \geq d(\lambda_0)\}$.
The set $\Lambda(\lambda_0)$ is bounded due to the boundedness of L^* and concavity d. Therefore, the dual sequence $\{\lambda_s\}_{s\in N}$ is bounded, and there exists subsequence $\{\lambda_{s_i}\}_{s_i\in N} \subset \{\lambda_s\}_{s\in N}$ that $\lim_{s_i\to\infty} \lambda_{s_i} = \bar{\lambda}$. By taking the limit in (5.168), we obtain $d(\bar{\lambda}) \geq d(\lambda^*)$, that is $d(\bar{\lambda}) = d(\lambda^*)$.
From $\lim_{s\to\infty} d(\lambda_{s_i}) = d(\lambda^*)$ and from (1a) follows $\lim_{s\to\infty} d(\lambda_s) = d(\lambda^*)$.
From (5.168) follows

$$d(\lambda^*) - d(\lambda_s) \leq k_s^{-1}(r(\lambda^*) - r(\lambda_s)),\ \forall \lambda^* \in L^*, \tag{5.169}$$

therefore (5.169) is true for $\lambda^* = \operatorname{argmin}\{r(\lambda)|\lambda \in L^*\}$.

(3) First, we show that the primal sequence $\{x_s\}_{s\in N}$ is bounded. For a given approximation x_s let consider two sets of indices $I_+(x_s) = \{i : c_i(x_s) \geq 0\}$ and $I_-(x_s) = \{i : c_i(x_s) < 0\}$.
Then, keeping in mind $f(x_s) \geq 0$, we obtain

$$\begin{aligned} P(x_s,k_s) &= f(x_s) + k_s^{-1+\alpha} \sum_{i\in I_-(x_s)} \ln(1+e^{-k_sc_i(x_s)}) \\ &\quad + k_s^{-1+\alpha} \sum_{i\in I_+(x_s)} \ln(1+e^{-k_sc_i(x_s)}) \\ &\geq f(x_s) - k_s^{\alpha} \sum_{i\in I_-(x_s)} c_i(x_s) + k_s^{-1+\alpha} \sum_{i\in I_-(x_s)} \ln(1+e^{k_sc_i(x_s)}) \\ &\geq f(x_s) - k_s^{\alpha} \sum_{i\in I_-(x_s)} c_i(x_s) \geq -k_s^{\alpha} \sum_{i\in I_-(x_s)} c_i(x_s). \end{aligned} \tag{5.170}$$

On the other hand,

$$P(x_s,k_s) \leq P(x^*,k_s) = f(x^*) - k_s^{-1+\alpha} \sum_{i=1}^{m} \pi(k_sc_i(x^*))$$

$$= f(x^*) + k_s^{-1+\alpha} \sum_{i=1}^{m} \ln(1+e^{-k_sc_i(x^*)}) \leq f(x^*) + k_s^{-1+\alpha} m \ln 2. \tag{5.171}$$

From (5.170) and (5.171) follows

$$k_s^{\alpha} \sum_{i\in I_-(x_s)} |c_i(x_s)| \leq f(x^*) + k_s^{-1+\alpha} m \ln 2. \tag{5.172}$$

Therefore, for any $s \geq 1$ we have

$$\max_{i\in I_-(x_s)} |c_i(x_s)| \leq k_s^{-\alpha} f(x^*) + k_s^{-1} m \ln 2. \tag{5.173}$$

From boundedness of Ω, concavity c_i, $i = 1,\ldots,m$ and (5.173) follows boundedness of $\{x_s\}_{s\in N}$. Thus, the primal–dual sequence $\{x_s, \lambda_s\}_{s\in N}$ is bounded.

Therefore, there exists primal–dual converging subsequence $\{x_{s_i}, \lambda_{s_i}\}_{s_i \in N}$:

$$\bar{x} = \lim_{i\to\infty} x_{s_i};\ \bar{\lambda} = \lim_{i\to\infty} \lambda_{s_i}.$$

From (5.156) for $i : c_i(\bar{x}) > 0$ follows $\bar{\lambda}_i = 0$ and $\bar{\lambda}_i \geq 0$ for $i : c_i(\bar{x}) = 0$.

From (5.157) follows $\nabla_x L(\bar{x}, \bar{\lambda}) = 0$, therefore $(\bar{x}, \bar{\lambda})$ is KKT pair, that is $\bar{x} = x^*$, $\bar{\lambda} = \lambda^*$. □

5.10.5 Interior Distance Functions

In the mid-1960s, P. Huard introduced interior distance functions (IDFs) and developed interior center methods (ICMs) for solving constrained optimization problems. Later IDFs and correspondent interior center methods were incorporated into SUMT and studied by A. Fiacco and G. McCormick (1968) as well as (Grossman and Kaplan, 1981), Mifflin (1976), (Polak, 1971), just to mention a few.

The ICMs consist of finding at each step a central (in a sense) point of the relaxation feasible set (RFS) and updating the set by using new value of the objective function. The RFS is the intersection of the feasible set with the relaxation (level) set of the objective function at the attained level. The "center" is sought as a minimum of the IDF.

We consider convex optimization problem (5.25)–(5.26) and assume that conditions A. and B. are satisfied. Without losing generality, we can assume also that $\inf f(x) \geq 0$.

Let $x_0 \in \text{int}\, \Omega$ and $\tau_0 = f(x_0)$, then from the boundedness X^* follows boundedness of the relaxation feasible set (RFS) $\Omega(\tau_0) = \Omega \cap \{x : f(x) \leq \tau_0\}$ and $\text{int}\,\Omega(\tau_0) \neq \emptyset$.

Let $T = \{\tau : \tau_0 > \tau > \tau^* = f(x^*)\}$ be the interval between the initial and the optimal objective function value. For any $\tau_0 < \tau < \tau^*$ the set $\Omega(\tau)$ is bounded.

The IDFs F and $H : \Omega(\tau) \times T \to \mathbb{R}^1$ are defined by formulas

$$F(x,\tau) = \begin{cases} -m\ln(\tau - f(x)) - \sum_{i=1}^m \ln c_i(x) & ,x \in \Omega(\tau_0) \\ +\infty & ,x \notin \Omega(\tau_0) \end{cases}$$

$$H(x,\tau) = \begin{cases} m(\tau - f(x))^{-1} + \sum_{i=1}^m c_i^{-1}(x) & ,x \in \Omega(\tau_0) \\ +\infty & ,x \notin \Omega(\tau_0). \end{cases}$$

Due to convexity f and concavity c_i, $i = 1, \ldots, m$ both F and H are closed convex functions.

The ICM consists of finding the "center" of the RFS $\Omega(\tau_0)$ by solving the unconstrained optimization problem

$$\hat{x} = \hat{x}(\tau) = \text{argmin}\{F(x,\tau)/x \in \mathbb{R}^n\} \tag{5.174}$$

following by the parameter τ update by formula $\hat{\tau} = f(\hat{x})$.

From (5.174) follows

$$\nabla_x F(\hat{x}, \tau) = \nabla f(\hat{x}) - \sum_{i=1}^{m} \frac{\tau - f(\hat{x})}{m c_i(\hat{x})} \nabla c_i(\hat{x}) =$$

$$\nabla f(\hat{x}) - \sum_{i-1}^{m} \hat{\lambda}_i \nabla c_i(\hat{x}) = \nabla_x L(\hat{x}, \hat{\lambda}) = 0, \tag{5.175}$$

where

$$\hat{\lambda}_i = (\tau - f(\hat{x}))(m c_i(\hat{x}))^{-1},\ i = 1, \ldots, m.$$

Therefore,

$$d(\hat{\lambda}) = L(\hat{x}, \hat{\lambda}) = \min\{L(x, \hat{\lambda}) |\ x \in \mathbb{R}^n\}.$$

Thus, the ICM generates the following sequence $\{x_{s+1}, \lambda_{s+1}, \mu_{s+1}\}_{s \in N}$:

$$\begin{aligned} &(a) \quad x_{s+1} : \nabla_x F(x_{s+1}, \tau_s) = 0 \\ &(b) \quad \lambda_{s+1} := (\lambda_{i,s+1} = (\tau_s - f(x_{s+1}))(m c_i(x_{s+1}))^{-1}, i = 1, \ldots, m) \\ &(c) \quad \mu_{s+1} = f(x_{s+1}) \end{aligned} \tag{5.176}$$

Theorem 5.13. *If f, c_i, $i = 1, \ldots, m$ continuously differentiable and assumptions A and B hold, then*

1. *the primal sequence $\{x_s\}_{s \in N} \subset \operatorname{int} \Omega$ is monotone decreasing in value, that is, $f(x_{s+1}) < f(x_s), \forall s \geq 1$ and*

$$\lim_{s \to \infty} (\tau_s - f(x_{s+1})) = 0;$$

2. *the dual sequence $\{\lambda_s\}_{s \in N} \subset \mathbb{R}^m_{++}$ is generated by the dual interior regularization method*

$$\begin{aligned} &d(\lambda_{s+1}) + \frac{\tau_s - f(x_{s+1})}{m} \sum_{i=1}^{m} \ln \lambda_{i,s+1} \\ &= \max\left\{ d(u) + \frac{\tau_s - f(x_{s+1})}{m} \sum_{i=1}^{m} \ln u_i | u \in \mathbb{R}^m_+ \right\}; \end{aligned} \tag{5.177}$$

3. *the primal and dual sequences converge in value to the primal and dual solution, that is*

$$\lim_{s \to \infty} f(x_s) = f(x^*) = \lim_{s \to \infty} d(\lambda_s) = d(\lambda^*);$$

4. *the following error bound*

$$\max\{f(x_s) - f(x^*), d(\lambda^*) - d(\lambda_s)\} = \tau_s - f(x_{s+1})$$

holds.

Proof.

(1) For a given τ_s minimizer x_{s+1} exists due to boundedness of $\Omega(\tau_s)$ and $F(x, \tau_s) \to \infty$ $x \to \partial\Omega(\tau_s)$.
(2) It follows from (5.176) that $f(x_{s+1}) < f(x_s) = \tau_s$ and $\lim_{s\to\infty}(\tau_s - f(x_{s+1})) = 0$, because otherwise $\lim_{s\to\infty}(\tau_s - f(x_{s+1})) = \delta > 0$, then $f(x_s) \to -\infty$, which is impossible due to $\{x_s\}_{s\in N} \subset \Omega(\tau_0)$. It follows from (5.176) that

$$\nabla f(x_{s+1}) - \sum_{i=1}^{m} \frac{\tau_s - f(x_{s+1})}{mc_i(x_{s+1})} \nabla c_i(x_{s+1}) = 0. \tag{5.178}$$

Let us introduce Lagrange multipliers

$$\lambda_{i,s+1} = \frac{\tau_s - f(x_{s+1})}{mc_i(x_{s+1})} > 0, i = 1, \dots, m. \tag{5.179}$$

Then, from (5.178), we have

$$\nabla_x F(x_{s+1}, \tau_s) = \nabla_x L(x_{s+1}, \lambda_{s+1}) = 0. \tag{5.180}$$

From (5.180) follows

$$d(\lambda_{s+1}) = L(x_{s+1}, \lambda_{s+1}) = \min_{x\in\mathbb{R}^n} L(x, \lambda_{s+1}),$$

where the dual function d is closed and concave.
From (5.179) follows

$$c_i(x_{s+1}) = \frac{\tau_s - f(x_{s+1})}{\lambda_{i,s+1}}, i = 1, \dots, m. \tag{5.181}$$

Keeping in mind

$$-c(x_{s+1}) \in \partial d(\lambda_{s+1}),$$

where $\partial d(\lambda_{s+1})$ is subdifferential of the dual function d at $\lambda = \lambda_{s+1}$, from (5.181), we obtain

$$0 \in \partial d(\lambda_{s+1}) + \frac{\tau_s - f(x_{s+1})}{m} \sum_{i=1}^{m} \lambda_{i,s+1}^{-1} e_i,$$

where $e_i = (0, \dots, 1, \dots, 0) \in \mathbb{R}^m_+$.
The latter inclusion is the optimality condition for λ_{s+1} to be the maximizer in (5.177).
In other words a step of the ICM (5.176) is equivalent to a step of interior regularization method (5.177) for the dual problem with regularization function

$$r(u, \cdot) = m^{-1}(\tau_s - f(x_{s+1})) \sum_{i=1}^{m} \ln u_i,$$

which is closed and concave.

(3) From (5.179) follows

$$c_i(x_{s+1})\lambda_{i,s+1} = m^{-1}(\tau_s - f(x_{s+1})), i = 1,\ldots,m. \tag{5.182}$$

by summing up (5.182) we obtain

$$\langle c(x_s),\lambda_s\rangle = \tau_s - f(x_{s+1}).$$

Thus, the interior center method (5.176) generates primal interior sequence $\{x_s\}_{s\in N}$. The equivalent interior regularization method (5.177) generates dual interior sequence $\{\lambda_s\}_{s\in N}$ and the asymptotic complementarity condition

$$\lim_{s\to\infty}\langle c(x_s),\lambda_s\rangle = \lim_{s\to\infty}(\tau_{s-1} - f(x_s)) = 0 \tag{5.183}$$

holds.

(4) Hence,

$$\lim_{s\to\infty} f(x_s) = f(x^*),\ \lim_{s\to\infty} d(\lambda_s) = d(\lambda^*)$$

and the following error bound holds

$$f(x_s) - d(\lambda_s) = \langle c(x_s),\lambda_s\rangle = (\tau_{s-1} - f(x_s)) \to 0.$$ □

Exercise 5.19.

(1) Describe ICM method using interior distance function $H(x,\tau)$;
(2) show that ICM method for the primal problem is equivalent to interior regularization method for the dual;
(3) prove convergence of the primal and dual sequence to the primal and dual solution in value and establish error bounds.

Along with the above convergence properties, the IDFs have their well-known drawbacks.

First, IDFs are singular at the solution.

Second, the condition number of their Hessians is vanishing in the neighborhood of the solution, which reduces the efficiency of unconstrained optimization methods.

Third, although approximations for the Lagrange multipliers are available as a by-product of the ICM, they cannot be effectively used to speed up the computational process.

Let $\hat{x} = \hat{x}(\tau) : \nabla_x F(\hat{x},\tau) = \nabla f(\hat{x}) - \sum \frac{\tau - f(\hat{x})}{mc_i(\hat{x})}\nabla c_i(\hat{x}) = 0$, $\hat{\lambda}_i = \hat{\lambda}_i(\tau) = (\tau - f(\hat{x}))$ $(mc_i(\hat{x}))^{-1}$, $i = 1,\ldots,m$.

Then, for the Hessian of $F(x,\tau)$ we have

$$\begin{aligned}\nabla^2_{xx}F(x,\tau)_{/x=\hat{x}} &= (\tau - f(\hat{x})^{-1}\Bigg[\frac{1}{\tau - f(\hat{x})}\nabla f(\hat{x})\nabla f^T(\hat{x}) + \nabla^2 f(\hat{x})\\ &\quad -\sum_{i=1}^{m}\frac{(\tau - f(\hat{x}))}{m}\frac{\nabla^2 c_i(\hat{x})}{c_i(\hat{x})} + \sum_{i=1}^{m}\frac{(\tau - f(\hat{x}))}{mc_i^2(\hat{x})}\nabla c_i(\hat{x})\nabla c_i(\hat{x})^T\Bigg]\\ &= (\tau - f(\hat{x}))^{-1}\Big[\nabla^2_{xx}L(\hat{x},\hat{\lambda}) + \nabla c(\hat{x})^T C^{-1}(\hat{x})\hat{\Lambda}(\tau)\nabla c(\hat{x})\\ &\quad + (\tau - f(\hat{x}))^{-1}\nabla f(\hat{x})\nabla f^T(\hat{x})\Big],\end{aligned}$$

where $\nabla c(x)$ is Jacobian of $c(x) = (c_1(x),..,c_m(x))^T$, $C(x) = (\text{diag } c_i(x))_{i=1}^m$ and $\hat{\Lambda}(\tau) = (\text{diag } \hat{\lambda}_i(\tau) = (\tau - f(\hat{x}))(m(c_i(\hat{x}))^{-1})_1^m$ are diagonal matrices.

In view of $\hat{x} = \hat{x}(\tau) \to x^*$, $\hat{\lambda} = \hat{\lambda}(\tau) \to \lambda^*$, for τ close to τ^*, we obtain

$$\begin{aligned}\nabla^2_{xx}F(\hat{x},\tau) &\approx (\tau - f(x^*))^{-1}\left[\nabla^2_{xx}L(x^*,\lambda^*) + \nabla c(x^*)^T\Lambda^* C^{-1}(\hat{x})\nabla c(x^*)\right.\\ &\quad \left. + (\tau - f(x^*))^{-1}\nabla f^T(x^*)\nabla f(x^*)\right].\end{aligned}$$

Let $I(x^*) = \{i : c_i(x^*) = 0\} = \{1,\ldots,r\}$ the set of active constraints, $c_{(r)}(x) = (c_1(x),\ldots,c_r(x))^T$ and $\nabla c_{(r)}(x)$ is Jacobian of $c_{(r)}(x)$.

From K-K-T's condition we have

$$\nabla f(x^*) = \sum_{i=1}^{r}\lambda_i^*\nabla c_i(x^*),$$

hence

$$\forall y : \nabla c_{(r)}(x^*)y = 0 \Rightarrow \langle\nabla f(x^*), y\rangle = 0.$$

Therefore, for $\forall y : \nabla c_{(r)}(x^*)y = 0$ and

$$\langle\nabla^2_{xx}F(\hat{x},\tau)y,y\rangle \approx (\tau - f(x^*))^{-1}.$$

$$\left\langle\left(\nabla^2_{xx}L(x^*,\lambda^*) + \nabla c^T_{(r)}(x^*)\Lambda^*_{(r)}C^{-1}_{(r)}(\hat{x})\nabla c_{(r)}(x^*)\right)y,y\right\rangle, \tag{5.184}$$

where

$$\Lambda^*_{(r)} = (\text{diag } \lambda_i^*)_{i=1}^r,\ C_{(r)}(x) = (\text{diag } c_i(x))_{i=1}^r.$$

From $\hat{x} = \hat{x}(\tau) \to x^*$ follows $c_i(\hat{x}(\tau)) \to 0,\ i = 1,\ldots,r$, therefore

$$\lim_{\tau\to f(x^*)} M_i(\tau) = \lim_{\tau\to f(x^*)}\hat{\lambda}_i(\tau)c_i^{-1}(\hat{x}(\tau)) = \infty, \quad i = 1,\ldots,r. \tag{5.185}$$

For a given $\tau_0 > \tau > \tau^* = f(x^*)$ close enough to τ^* due to the Debreu's lemma with $A = \nabla^2_{xx}L(x^*,\lambda^*)$ and $C = \Lambda^{*\frac{1}{2}}_{(r)}\nabla c_{(c)}(x^*)$, there exists $\mu(\tau) > 0$:

$$\text{mineigenval } \nabla^2_{xx}F(\hat{x},\tau) = \mu(\tau)(\tau - f(x^*))^{-1}.$$

From (5.185) follows $M(\tau) = \min\{M_i(\tau) :\ 1 \le i \le r\} \to \infty$, therefore

$$\text{maxeigenval}\, \nabla^2_{xx} F(\hat{x}, \tau) \geq M(\tau)\,(\tau - f(x^*))^{-1}.$$

Hence

$$\lim_{x(\tau)\to x^*} \text{cond}\, \nabla^2_{xx} F(\hat{x}(\tau), \tau) = \mu(\tau) M^{-1}(\tau) = 0.$$

In other words, the condition number of the IDF Hessian vanishes, when $x(\tau) \to x^*$. It means, from some point on, finding an accurate approximation for $\hat{x}$ is practically impossible and so is finding an approximation for $f(x^*)$ with high accuracy.

The ill-conditioning of the Hessian $\nabla^2_{xx} F(\hat{x}, \tau)$ is much more critical in nonlinear optimization than in LP. In case of LP, the term $\nabla^2_{xx} L(x, \lambda)$ in the expression for the Hessian $\nabla^2_{xx} F(\hat{x}, \tau)$ disappears, and by rescaling the data properly, one can, to some extent, eliminate the ill-conditioning effect. It is not the case in NLP.

Partially, the difficulties can be overcome by using primal–dual IPM, which we consider in the following section.

The NR theory and methods, which we consider in Chapters 7–9, allow, to a large extend, eliminate the ill-conditioning.

The equivalence of the primal SUMT and dual interior regularization methods not only allows to prove convergence in a unified and simple manner and establish the error bounds but also provides important information about dual feasible solution, which can be used for a stopping criteria.

5.11 Primal–Dual IPM for Convex Optimization

We consider the convex optimization problem (5.25)–(5.26). From (5.131)–(5.132) follows that each SUMT step is equivalent to solving the following primal–dual system for $\hat{x}$ and $\hat{\lambda}$ under fixed $\mu = k^{-1}$.

$$\begin{aligned} \nabla_x L(\hat{x}, \hat{\lambda}) = \nabla f(\hat{x}) - \textstyle\sum_{i=1}^m \hat{\lambda}_i \nabla c_i(\hat{x}) &= 0. \\ \hat{\lambda}_i c_i(\hat{x}) &= \mu, \quad i = 1, \ldots, m, \end{aligned} \tag{5.186}$$

Solving the nonlinear PD system (5.186) is, generally speaking, an infinite procedure. Moreover, the solution $\hat{x}(\mu)$, $\hat{\lambda}(\mu)$ of (5.186) converges to the primal–dual solution (x^*, λ^*) only when $\mu \to 0$.

The main idea of the primal–dual interior–point methods is instead of solving system (5.186), one performs Newton step toward the solution of system (5.186) followed by the barrier parameter update. For a given approximation $y = (x, \lambda)$ and the barrier parameter $\mu > 0$, the application of Newton's method to the nonlinear PD system (5.186) leads to the following linear PD system:

$$\begin{bmatrix} \nabla^2_{xx} L(x, \lambda) & -\nabla c(x)^T \\ \Lambda \nabla c(x) & C(x) \end{bmatrix} \begin{bmatrix} \Delta x \\ \Delta \lambda \end{bmatrix} = \begin{bmatrix} -\nabla L(x, \lambda) \\ -\Lambda c(x) + \mu e \end{bmatrix}, \tag{5.187}$$

where $C(x) = \text{diag}\,(c_i(x))_{i=1}^m$, $\Lambda = \text{diag}\,(\lambda_i)_{i=1}^m$, and $e = (1, \ldots, 1)^T \in \mathbb{R}^m$.

The system (5.187) finds Newton direction $\Delta y = (\Delta x, \Delta\lambda)$, which is used to update the current approximation $y = (x, \lambda)$:

$$\bar{x} = x + \alpha\Delta x; \quad \bar{\lambda} = \lambda + \alpha\Delta\lambda. \tag{5.188}$$

The step length $\alpha > 0$ one has to determine in such a way that a new approximation $\bar{y} = (\bar{x}, \bar{\lambda})$ not only remains primal and dual interior, but also stays in the area where Newton method for PD system (5.186) is well defined for the updated barrier parameter.

For some classes of convex optimization including LP and QP, it is possible to take $\alpha = 1$ in (5.188) and update the barrier parameter μ by the formula $\bar{\mu} = \mu(1 - \rho/\sqrt{n})$, where $0 < \rho < 1$ is independent on n. The new approximation belongs to the solution neighborhood of the system (5.186) with μ replaced by $\bar{\mu}$. Moreover, each step reduces the primal–dual gap by the same factor $(1 - \rho/\sqrt{n})$. This leads to polynomial complexity of the primal–dual interior–point methods for LP and QP. The results will be covered in Chapter 6. The results are taking place for well-structured convex optimization, which means that the feasible set and the epigraph of the objective function can be equipped with SC barrier.

If a convex optimization is not well structured, establishing polynomial complexity of the path following methods becomes problematic, if not impossible. Nevertheless, the primal–dual interior–point approach remains productive for some problems.

The main challenge associated with the interior–point method (5.187)–(5.188) is keeping the trajectory inside of the feasible set. It requires selecting an appropriate $\alpha > 0$. For convex optimization with a large number of constraints finding such α is a significant computational burden. The following modification (see Griva et al. (2008)) of the IPM sometimes helps to overcome the difficulties. The convex optimization problem (5.25)–(5.26) one can rewrite as follows:

$$f(x) = \min\{f(x)|c(x) - w = 0,\ w \in \mathbb{R}^m_+\}. \tag{5.189}$$

The problem (5.189) along with equations has very simple inequality constraints $w_i \geq 0$, $i = 1, \ldots, m$.

The log–barrier term of $F(x, w, \mu) = f(x) - \mu\sum_{i=1}^m \ln w_i$ is used to handle the nonnegativity of the slack vector $w \in \mathbb{R}^m_+$.

One SUMT's step consists of solving the following ECO problem:

$$\begin{aligned} &\min F(x, w, \mu), \\ &\text{s.t. } c(x) - w = 0. \end{aligned} \tag{5.190}$$

Then, one has to reduce the barrier parameter $\mu > 0$.

The Lagrangian for (5.190) is defined by formula

$$L(x, w, \lambda, \mu) = f(x) - \mu\sum_{i=1}^m \log w_i - \sum_{i=1}^m \lambda_i(c_i(x) - w_i).$$

Let $W = \text{diag}\,(w_i)_{i=1}^m$ and $\Lambda = \text{diag}\,(\lambda_i)_{i=1}^m$, then the following Lagrange system

$$\begin{aligned}\nabla_x L(\cdot) &= \nabla f(x) - \nabla c(x)^T \lambda = 0\\ \nabla_w L(\cdot) &= -\mu e + W\Lambda e = 0\\ \nabla_\lambda L(\cdot) &= c(x) - w = 0\end{aligned} \tag{5.191}$$

corresponds to ECO (5.190).

Instead of solving system (5.191), we perform one Newton step following by the barrier parameter $\mu > 0$ update.

Application of Newton's method to nonlinear PD system (5.191) leads to the following linear PD system for finding the Newton directions $\Delta y = (\Delta x, \Delta w, \Delta\lambda)^T$

$$\begin{bmatrix}\nabla^2_{xx}L(x,\lambda) & 0 & -\nabla c(x)^T\\ 0 & \Lambda & W\\ \nabla c(x) & -I_m & 0\end{bmatrix}\begin{bmatrix}\Delta x\\ \Delta w\\ \Delta\lambda\end{bmatrix} = \begin{bmatrix}-\nabla f(x) + \nabla c(x)^T\lambda\\ \mu e - W\Lambda e\\ -c(x) + w\end{bmatrix}, \tag{5.192}$$

where $\nabla^2_{xx}L(x,\lambda) = \nabla^2 f(x) - \sum_{i=1}^m \lambda_i \nabla^2 c_i(x)$ is the Hessian in x of the Lagrangian $L(x,\lambda)$.

Our first step is show that under second-order sufficient optimality optimality condition (4.73)–(4.74), the system (5.192) is well defined in the neighborhood of the primal–dual solution.

Lemma 5.9. *If the second-order sufficient conditions* (4.73)–(4.74) *hold, and* $y^* = (x^*, w^*, \lambda^*)$ *is a solution to the system (5.191), then the matrix*

$$D(y^*) = \begin{bmatrix}\nabla^2_{xx}L(x^*,\lambda^*) & 0 & -\nabla c(x^*)^T\\ 0 & \Lambda^* & W^*\\ \nabla c(x^*) & -I_m & 0\end{bmatrix}$$

is nonsingular; hence there exists $M > 0$ *such that*

$$\|D^{-1}(y^*)\| \le M. \tag{5.193}$$

Proof. The nonsingularity of $D(y^*)$ follows from the implication

$$D(y^*)y = 0 \Rightarrow y = 0, \tag{5.194}$$

where

$$y = \begin{pmatrix}u\\ w\\ \lambda\end{pmatrix} \in \mathbb{R}^{n+2m}.$$

The system $D(y^*)y = 0$ can be rewritten as follows:

$$\nabla^2_{xx}L(x^*,\lambda^*)u - \nabla c(x^*)^T\lambda = 0, \tag{5.195}$$

$$\Lambda^* w + W^* \lambda = 0, \tag{5.196}$$

$$\nabla c(x^*) u - w = 0. \tag{5.197}$$

After splitting the system (5.196) on active and passive constraints, we obtain

$$\lambda_i^* w_i + w_i^* \lambda_i = 0, \quad i = 1, \ldots, r, \quad \lambda_i^* w_i + w_i^* \lambda_i = 0, \quad i = r+1, \ldots m. \tag{5.198}$$

Keeping in mind

$$\lambda_i^* > 0, \quad w_i^* = 0, \quad i = 1, \ldots, r \text{ and } \lambda_i^* = 0, \quad w_i^* > 0 \quad i = r+1, \ldots, m$$

from (5.198), we obtain

$$w_i = 0, \quad i = 1, \ldots, r \text{ and } \lambda_i = 0, \quad i = r+1, \ldots, m. \tag{5.199}$$

In other words, $w_A = (w_1, \ldots, w_r) = 0^r$ and $\lambda_p = (\lambda_{r+1}, \ldots, \lambda_m) = 0^{n-r}$, then

$$\Lambda_A^* w_A + W_A^* \lambda_A = \Lambda_A^* w_A = 0, \quad \Lambda_P^* w_P + W_P^* \lambda_P = W_P^* \lambda_P = 0. \tag{5.200}$$

From $W_A^* = 0$ and $\Lambda_P^* = 0$, strict complementarity $\Lambda_A^* > 0$ and $W_P^* > 0$ and (5.200) follows

$$w_A = 0^r, \quad \lambda_P = 0^{m-r}. \tag{5.201}$$

Let us revisit (5.195) and (5.197). From $w_A = 0^r$, $\lambda_p = 0^{m-r}$ follows

$$\nabla^2_{xx} L(x^*, \lambda^*) u - \nabla c_{(r)}(x^*)^T \lambda_A = 0, \tag{5.202}$$

$$\nabla c_{(r)}(x^*) u - w_A = \nabla c_{(r)}(x^*) u = 0. \tag{5.203}$$

After multiplying (5.202) by u, we have

$$\langle \nabla^2_{xx} L(x^*, \lambda^*) u, u \rangle = \langle u, \nabla c_{(r)}(x^*)^T \lambda_A) \rangle = \langle \nabla c_{(r)} u, \lambda_A \rangle = 0. \tag{5.204}$$

Due to the second-order sufficient conditions for any u, that satisfies (5.203), we have

$$0 = \langle \nabla^2_{xx} L(x^*, \lambda^*) u, u \rangle \geq \mu \langle u, u \rangle$$

and $\mu > 0$, which implies $u = 0$. From (5.202) follows

$$\nabla c_{(r)}(x^*)^T \lambda_A = 0$$

and from $\operatorname{rank} \nabla c_{(r)}(x^*) = r$ follows $\lambda_A = 0^r$, therefore $\lambda = (\lambda_A, \lambda_p) = 0^m$. Then, from (5.197) follows $w = 0^m$.

In other words, the implication (5.194) holds; therefore $D(y^*)$ is not singular. □

Lemma 5.10. *If the second-order sufficient optimality conditions* (4.73)–(4.74) *hold and Lipschitz conditions (4.130) is satisfied, then there exists* $\varepsilon_0 > 0$ *and* $M > 0$*, that for any* $y = (x, w, \lambda) \in B(y^*, \varepsilon_0)$ *the matrix*

$$D(y) = \begin{bmatrix} \nabla^2_{xx} L(x,\lambda) & 0 & -\nabla c(x)^T \\ 0 & \Lambda & W \\ \nabla c(x) & -I_m & 0 \end{bmatrix}$$

has an inverse and the following bound:

$$\|D^{-1}(y)\| \leq 2M \tag{5.205}$$

holds.

The proof is similar to the proof of Lemma 4.4

Let us consider the following merit function:

$$\nu(y) \equiv \nu(x,w,\lambda) = \max\{\|\nabla_x L(x,\lambda)\|, \|c(x) - w\|, \|W\Lambda e\|\}.$$

From KKT's Theorem follows

$$\nu(y) = 0 \Leftrightarrow y = y^*. \tag{5.206}$$

□

Lemma below is similar to Lemma 4.4 and can be proven using the same arguments

Lemma 5.11. *If the second-order sufficient optimality conditions* (4.73)–(4.74) *hold and Lipschitz conditions (4.130) is satisfied, then there exists* $\varepsilon_0 > 0$ *and* $0 < l < L$ *that for any* $y \in B(y^*, \varepsilon_0)$ *the following bounds:*

$$l\|y - y^*\| \leq \nu(y) \leq L\|y - y^*\| \tag{5.207}$$

hold.

Let $\delta > 0$ be small enough and $0 < \varepsilon \ll \delta$ is the required accuracy. We are ready to describe the primal–dual interior point method; see Griva et al. (2008)

PDIPM.

Step 0. Let $y \in S(y^*, \delta) = \{y : \|y - y^*\| \leq \delta\}$ *be the initial approximation.*
Step 1. If $\nu(x,w,\lambda) \leq \varepsilon$*, Output* (x,λ) *as a solution.*
Step 2. Calculate the barrier parameter $\mu = \min\{\theta\mu, \nu^2(x,w,\lambda)\}$, $0 < \theta < 1$.
Step 3. Find Δx, Δw *and* $\Delta\lambda$ *from (5.192).*
Step 4. Calculate the step lengths α_P *and* α_D *by formulas*

$$\kappa = \max\{\bar{\kappa}, 1 - \nu(x,w,\lambda)\}, \quad 0 < \bar{\kappa} < 1$$

$$\alpha_P = \min_{1 \leq i \leq m}\left\{1; \, -\kappa \frac{(w^s)_i}{(\Delta w^s)_i} : (\Delta w^s)_i < 0\right\},$$

$$\alpha_D = \min_{1 \leq i \leq m}\left\{1; \, -\kappa \frac{(\lambda^s)_i}{(\Delta \lambda^s)_i} : (\Delta \lambda^s)_i < 0\right\},$$

Step 5. Update the primal–dual pair by the formulas

$$\hat{x} := x + \alpha_P \Delta x, \quad \hat{w} := w + \alpha_P \Delta w. \quad \hat{\lambda} := \lambda + \alpha_D \Delta \lambda.$$

Step 6. Goto Step 1.

The following Theorem establishes the local convergence of the PDIPM.

Theorem 5.14. *If the second-order sufficient optimality conditions* (4.73)–(4.74) *hold and Lipschitz conditions (4.130) is satisfied, then there exists* $\varepsilon_0 > 0$ *that from any starting point* $y = (x, w, \lambda) \in B(y^*, \varepsilon_0)$, *the following bound holds*

$$\|\hat{y} - y^*\| \leq c\|y - y^*\|^2,$$

c and is independent of $y \in B(y^*, \varepsilon_0)$.

Theorem 5.14 one can prove using arguments similar to those used in the proof of Theorem 4.17.

5.12 Gradient Projection Method

5.12.1 Convergence of the GP Method

Let $f : \mathbb{R}^n \to \mathbb{R}$ be a convex function and $\Omega \in \mathbb{R}^n$ be a closed convex set.

We are concerned with the following convex optimization problem

$$f(x^*) = \min\{f(x)|x \in \Omega\}. \tag{5.208}$$

In this section we consider gradient projection (GP) method introduced and analyzed independently in the mid-1960s by A. Goldstein and E. Levitin and B. Polyak; see also J. Rosen (1960), (1961).

Along with classical GP method, we consider fast GP and establish convergence rate for both methods under various assumption on the input data.

The critical point of the method is projection on a close convex set (see Section 2.3.2).

For each $y \in \mathbb{R}^n$, there exists a unique point $x \in \Omega$ that

$$\|y - x\| \leq \|y - z\|, \ \forall z \in \Omega,$$

that is

$$x = P_\Omega(y) \tag{5.209}$$

and $P_\Omega : \mathbb{R}^n \to \Omega$ defined by (5.209) is called projection operator.

Let $y_0 \notin \Omega$, then projection $x_0 = P_\Omega(y_0)$ of y_0 on Ω is the solution of the following convex optimization problem

$$\pi(x_0) \equiv \pi(P_\Omega(y_0)) = \min\{\pi(x) = \frac{1}{2}\|x - y_0\|^2|x \in \Omega\}. \tag{5.210}$$

Minimizer $x_0 = P_\Omega(y_0)$ always exists and it is unique because $\pi(x)$ is strongly convex in $x \in \mathbb{R}^n$.

Due to $\nabla\pi(x_0) = x_0 - y_0$, vector x_0 is the minimizer in (5.210) if and only if

$$\langle x_0 - y_0, x - x_0\rangle \geq 0 \tag{5.211}$$

holds for all $x \in \Omega$.

For a given $x_0 = P_\Omega(y_0) \in \Omega$ the following set

$$N_\Omega(x_0) = \{g \in \mathbb{R}^n : \langle g, x - x_0\rangle \leq 0,\ \forall x \in \Omega\}$$

is the normal cone to Ω at $x_0 \in \Omega$.

It follows from (5.211) that $y_0 - x_0 \in N_\Omega(y_0)$.

For $\bar{x} \in \Omega$ to be the solution for the problem (5.208), it is necessary and sufficient that

$$\langle \nabla f(\bar{x}), x - \bar{x}\rangle \geq 0,\ \forall x \in \Omega \tag{5.212}$$

or

$$-\nabla f(\bar{x}) \in N_\Omega(\bar{x}).$$

Thus, for $\bar{x}$ to be the solution in (5.208), it is necessary and sufficient

$$\bar{x} = P_\Omega(\bar{x} - t\nabla f(\bar{x})), \forall t \geq 0. \tag{5.213}$$

The optimality condition (5.213) shows that for any $t > 0$ the solution $\bar{x}$ is a fixed point of the map $P_\Omega(I - t\nabla f) : \Omega \to \Omega$.

In other words (5.213) is the optimality criteria for GP method

$$x_{s+1} = P_\Omega(x_s - t\nabla f(x_s)). \tag{5.214}$$

Our next step is to study the convergence properties of GP method (5.214).

We assume that gradient ∇f satisfies Lipschitz condition

$$\|\nabla f(x_1) - \nabla f(x_2)\| \leq L_f\|x_1 - x_2\|. \tag{5.215}$$

From (5.215) follows

$$f(x) \leq f(x_0) + \langle \nabla f(x_0), x - x_0\rangle + \frac{L_f}{2}\|x - x_0\|^2.$$

The convergence results of GP method (5.214) are similar to those of the gradient method for unconstrained optimization (see Chapter 3). The proof is based on the following Lemma 5.12 similar to Lemma 3.1; see Beck and Teboulle (2009).

Let us consider quadratic approximation $\psi : \mathbb{R}^n \to \mathbb{R}$ of f at $x \in \mathbb{R}^n$ given by

$$\psi(X, x) = f(x) + \langle X - x, \nabla f(x)\rangle + \frac{L_f}{2}\|X - x\|^2.$$

For $L \geq L_f$ and a given $x \in \Omega$ we have

$$f(X) \leq \psi(X,x) = f(x) + \langle X - x, \nabla f(x) \rangle + \frac{L}{2}\|X - x\|^2$$

for any $X \in \Omega$.

Let

$$x_\Omega^L = \text{argmin}\{\langle X - x, \nabla f(x) \rangle + \frac{L}{2}\|X - x\|^2 | X \in \Omega\}, \tag{5.216}$$

then, using the optimality criteria (5.212), we obtain

$$\langle \nabla_X \psi(x_\Omega^L), X - x_\Omega^L \rangle = \langle \nabla f(x) + L(x_\Omega^L - x), X - x_\Omega^L \rangle \geq 0, \ \forall X \in \Omega.$$

The latter inequality we can rewrite as follows:

$$L\langle x_\Omega^L - (x - L^{-1}\nabla f(x)), X - x_\Omega^L \rangle \geq 0, \ \forall X \in \Omega.$$

From (5.212) with $\bar{x} = x_\Omega^L$ and $\bar{y} = x - L^{-1}\nabla f(x)$ follows

$$x_\Omega^L = P_\Omega(x - L^{-1}\nabla f(x)) = P_\Omega(\bar{y}). \tag{5.217}$$

By reiterating (5.217) we obtain the GP method (5.214) with $t = L^{-1}$.

Lemma 5.12. *For a given $x \in \Omega$ and $L > 0$ such that*

$$f(x_\Omega^L) \leq \psi(x_\Omega^L, x) \tag{5.218}$$

the following inequality holds for any $X \in \Omega$:

$$f(X) - f(x_\Omega^L) \geq \frac{L}{2}\|x_\Omega^L - x\|^2 + L\langle x - X, x_\Omega^L - x \rangle. \tag{5.219}$$

Proof. From (5.218) and convexity f follows

$$f(X) - f(x_\Omega^L) \geq f(X) - \psi(x_\Omega^L, x) = f(X) - f(x) - \langle x_\Omega^L - x, \nabla f(x) \rangle - \frac{L}{2}\|x_\Omega^L - x\|^2$$

$$\geq f(x) + \langle \nabla f(x), X - x \rangle - f(x) - \langle x_\Omega^L - x, \nabla f(x) \rangle - \frac{L}{2}\|x_\Omega^L - x\|^2 + L\|x_\Omega^L - x\|^2$$

$$- L\|x_\Omega^L - x\|^2 = \frac{L}{2}\|x_\Omega^L - x\|^2 + \langle \nabla f(x), X - x_\Omega^L \rangle - L\|x_\Omega^L - x\|^2. \tag{5.220}$$

Using the optimality criteria for x_Ω^L in (5.216), we obtain

$$\langle \nabla f(x) + L(x_\Omega^L - x), X - x_\Omega^L \rangle \geq 0, \ \forall X \in \Omega.$$

Combining the latter inequality with (5.220), we obtain

$$f(X) - f(x_\Omega^L) \geq \frac{L}{2}\|x_\Omega^L - x\|^2 - L\langle x_\Omega^L - x, X - x_\Omega^L \rangle - L\langle x_\Omega^L - x, x_\Omega^L - x \rangle =$$

$$\frac{L}{2}\|x_\Omega^L - x\|^2 + L\langle x_\Omega^L - x, x_\Omega^L - X \rangle - L\langle x_\Omega^L - x, x_\Omega^L - x \rangle = \frac{L}{2}\|x_\Omega^L - x\|^2 + L\langle x_\Omega^L - x, x - X \rangle.$$

□

From (5.216) follows that the GP method is, in fact, a linearization method, where the quadratic regularization is used to normalize the gradient direction; see Pshenichnyj (1994).

On the other hand, GP method has important features of quadratic prox method (see Section 3.5).

The following Theorem establishes convergence properties of the GP method.

Theorem 5.15. *Let Ω be a closed convex set and $f : \mathbb{R}^n \to \mathbb{R}$ be a convex function with Lipschitz continuous gradient, then for the sequence $\{x_s\}_{s\in N}$ generated by GP method (5.214) with $t = L^{-1}$ the following bound holds:*

$$\Delta_k = f(x_k) - f(x^*) \leq \frac{L}{2k} \|x_0 - x^*\|^2$$

Proof. Using (5.219) with $X = \lambda^*, x = x_s$and $x^L_\Omega = x_{s+1}$, we obtain

$$\begin{aligned}\frac{2}{L}(f(x^*) - f(x_{s+1})) &\geq \|x_{s+1} - x_s\|^2 + 2\langle x_s - x^*, x_{s+1} - x_s\rangle \\ &= \langle x_{s+1}, x_{s+1}\rangle - 2\langle x_{s+1}, x_s\rangle + \langle x_s, x_s\rangle + 2\langle x_s, x_{s+1}\rangle \\ &\quad - 2\langle x^*, x_{s+1}\rangle - 2\langle x_s, x_s\rangle + 2\langle x^*, x_s\rangle + \langle x^*, x^*\rangle - \langle x^*, x^*\rangle \\ &= \|x_{s+1} - x^*\|^2 - \|x_s - x^*\|^2\end{aligned}$$

Summing up the last inequality from $s = 0$ to $s = k-1$, we obtain

$$\frac{2}{L}(kf(x^*) - \sum_{s=0}^{k-1} f(x_{s+1})) \geq \|x^* - x_k\|^2 - \|x^* - x_0\|^2 \tag{5.221}$$

Using (5.219) with $X = x = x_s$, $x^L_+ = x_{s+1}$, we obtain

$$\frac{2}{L}(f(x_s) - f(x_{s+1})) \geq \|x_{s+1} - x_s\|^2$$

or

$$sf(x_s) - sf(x_{s+1}) \geq \frac{L}{2} s \|x_{s+1} - x_s\|^2 .$$

Therefore

$$sf(x_s) - (s+1)f(x_{s+1}) + f(x_{s+1}) \geq \frac{L}{2} s \|x_{s+1} - x_s\|^2$$

summing up the latter inequality from $s = 0$ to $s = k-1$, we obtain

$$-kf(x_k) + \sum_{s=0}^{k-1} f(x_{s+1}) \geq \frac{L}{2} \sum_{s=0}^{k-1} s \|x_{s+1} - x_s\|^2$$

From (5.221) we have

$$kf(x^*) - \sum_{s=0}^{k-1} f(x_{s+1}) \geq \frac{L}{2}\left[\|x^* - x_k\|^2 - \|x^* - x_0\|^2\right]$$

By adding last two inequalities, we obtain

$$k(f(x^*) - f(x_k)) \geq \frac{L}{2}\left[\sum_{s=0}^{k-1} s\,\|x_{s+1} - x_s\|^2 + \|x^* - x_k\|^2 - \|x^* - x_0\|^2\right]$$

that is

$$\Delta_k = f(x_k) - f(x^*) \leq \frac{L}{2k}\|x^* - x_0\|^2. \qquad \square \tag{5.222}$$

If follows from (5.222) that for a given accuracy $\varepsilon > 0$, it takes $k = \frac{L\|\lambda^* - \lambda_0\|^2}{2\varepsilon}$ steps of the GP method to get $\Delta_k \leq \varepsilon$.

In the following section, we consider the fast GP (FGP) method, which improves substantially the convergence rate of the GP method practically without extra computational work.

5.12.2 Fast GP Method

The FGP method is based on Yu. Nesterov gradient mapping approach, see Nesterov (1983) and Nesterov (2004), and closely related to the FISTA algorithm, see Beck and Teboulle (2009a) and Beck and Teboulle (2009).

The FGP generates an auxiliary sequence $\{x_k\}_{k\in N}$ and the main sequence $\{X_k\}_{k\in N}$. Vector x_k one can view as a predictor, while vector X_k is the corrector, i.e., the approximation at the step k.

FGP method

1. Input:
   - L - the upper bound for the Lipschitz constant of the gradient $\nabla f(x)$
   - $X_1 = X_0 \in \mathbb{R}^n$
   - $t_1 = 1$
2. Step k
   (a) update the step length $t_{k+1} = \frac{1+\sqrt{1+4t_k^2}}{2}$;
   (b) find predictor $x_{k+1} = X_k + (\frac{t_k - 1}{t_{k+1}})(X_k - X_{k-1})$;
   (c) find new corrector $X_{k+1} = x_\Omega^L(x_k) = \arg\min\{\psi(X, x_{k+1}) \mid X \in \Omega\}$

The correction phase c. produces the new approximation

$$X_{k+1} = \text{argmin}\left\{\langle X - x_{k+1}, \nabla f(x_{k+1})\rangle + \frac{L}{2}\|X - x_{k+1}\|^2 \mid X \in \Omega\right\} \tag{5.223}$$

From the optimality criteria for X_{k+1} follows the closed form solution for the problem (5.223)

$$X_{k+1} = P_\Omega \left(x_{k+1} - \frac{1}{L} \nabla f(x_{k+1}) \right) \tag{5.224}$$

It means that the correction phase of the FGP method is just a projected gradient step with step – length L^{-1} from the predictor point x_{k+1}.

First off all, it follows from b. that $t_k \geq \frac{1}{2}(k+1), \quad \forall k \geq 1$. In fact, it is true for $k = 1$. Assuming that $t_k \geq \frac{1}{2}(k+1)$ from b. we have

$$t_{k+1} = \frac{1}{2}\left(1 + \sqrt{1 + 4t_k^2}\right) \geq \frac{1}{2}\left(1 + \sqrt{1 + (k+1)^2}\right) \geq \frac{1}{2}(k+2)$$

Let $\Delta_k = f(X_k) - f(x^*)$, $y_k = tX_k + (t_k - 1)X_{k-1} - x^*$.

Exercise 5.20. Show that the following inequality:

$$2L^{-1}(t_k^2 \Delta_k - t_{k+1}^2 \Delta_{k+1}) \geq \|y_{k+1}\|^2 - \|y_k\|^2 \tag{5.225}$$

holds for any $k \geq 1$.
Hint. See proof of Theorem 3.8.

Theorem 5.16. *For the sequence* $\{X_k\}_{k \in N}$ *generated by FGP method a.– c., the following bound holds:*

$$\Delta_{k+1} \leq \frac{2L\|x_0 - x^*\|^2}{(k+2)^2} \tag{5.226}$$

Proof. From (5.225) follows

$$\begin{aligned} t_{k+1}^2 \Delta_{k+1} + \frac{L}{2}\|y_{k+1}\|^2 &\leq t_k^2 \Delta_k + \frac{L}{2}\|y_k\|^2 \\ &\leq t_{k-1}^2 \Delta_{k-1} + \frac{L}{2}\|y_{k-1}\|^2 \\ &\vdots \\ &\leq t_1^2 \Delta_1 + \frac{L}{2}\|y_1\|^2 \end{aligned} \tag{5.227}$$

Hence,

$$\begin{aligned} t_{k+1}^2 \Delta_{k+1} &\leq t_1^2 \Delta_1 + \frac{L}{2}\|y_1\|^2 - \frac{L}{2}\|y_{k+1}\|^2 \\ &\leq (f(X_1) - f(x^*)) + \frac{L}{2}\|X_1 - x^*\|^2 \end{aligned} \tag{5.228}$$

For $X = x^*$, $x_\Omega^L = X_1$ and $x = x_0$ it follows from (5.219) that

$$f(x^*) - f(X_1) \geq \frac{L}{2} \|X_1 - x_0\|^2 + L\langle x_0 - x^*, X_1 - x_0\rangle$$
$$= \frac{L}{2}\left[\|X_1 - x^*\|^2 - \|x_0 - x^*\|^2\right]$$

Therefore,

$$f(X_1) - f(x^*) \leq \frac{L}{2}\left[\|x_0 - x^*\|^2 - \|X_1 - x^*\|^2\right] \tag{5.229}$$

From (5.228) and (5.229) follows

$$t_{k+1}^2 \Delta_{k+1} \leq \frac{L}{2} \|x_0 - x^*\|^2$$

Keeping in mind $t_{k+1} \geq \frac{1}{2}(k+2)$, we obtain (5.226). □

We complete the proof of Theorem 5.16.

So, FGP method practically requires numerical effort similar to GP method per step, but the convergence rate is much better.

It follows from (5.226) that for a given accuracy $\varepsilon > 0$, it takes

$$k = \frac{\sqrt{2L}\|x_0 - x^*\|}{\sqrt{\varepsilon}}$$

steps of FGP method to get $\Delta_k \leq \varepsilon$.

Both the result and the proof are similar to correspondent results for unconstrained optimization.

The FGP method is particularly useful for CO, where projection on Ω is a relatively easy operation. We consider such applications in Sections 5.13 and 5.14.

5.12.3 GP Method for Strongly Convex Function

If on the top of Lipschitz condition (5.215) the gradient ∇f is a strongly monotone operator, that is, for any x any $y \in \mathbb{R}^n$ the following bound:

$$\langle \nabla f(x) - \nabla f(y), x - y\rangle \geq m\|x - y\|^2 \tag{5.230}$$

holds, then GP method

$$x_{s+1} = P_{\Omega}[x_s - t\nabla f(x_s)] \tag{5.231}$$

converges with Q-linear rate.

Theorem 5.17. *If conditions (5.215) and (5.230) are satisfied, then*

1. for $0 < t < 2/(m+L)$ the following bound holds

$$\|x_{s+1} - x^*\|^2 \le \left(1 - t\frac{2mL}{m+L}\right) \|x_s - x^*\|^2; \tag{5.232}$$

2. *for* $t = 2/(m+L)$ *we have*

$$\|x_{s+1} - x^*\| \le \left(\frac{1-\varkappa}{1+\varkappa}\right) \|x_s - \hat{x}^*\|, \tag{5.233}$$

where $\varkappa = mL^{-1}$ *is the condition number of the gradient operator* ∇f.

3. *for a given accuracy* $\varepsilon > 0$ *to obtain* $\Delta_s = \|x_s - x^*\| \le \varepsilon$ *it requires* $s = O(\varkappa^{-1} \ln \varepsilon^{-1})$ *steps.*

Proof.

1. From the optimality criteria (5.213) for any $t > 0$ follows

$$P_\Omega(x^* - t\nabla f(x^*)) = x^*.$$

From the nonexpansive property of the projection operator follows

$$\|x - x^*\|^2 = \|P_\Omega(x_s - t\nabla f(x_s)) - P_\Omega(x^* - t\nabla f(x^*))\|^2 \le$$

$$\|x_s - t\nabla f(x_s) - x^* + t\nabla f(x^*)\|^2$$

$$= \|x_s - x^*\|^2 - 2t\langle \nabla f(x_s) - \nabla f(x^*), x_s - x^*\rangle + t^2\|\nabla f(x_s) - \nabla f(x^*)\|. \tag{5.234}$$

From Theorem 2.5 we have

$$\langle \nabla f(x_s) - \nabla f(x^*), x_s - x^*\rangle \ge$$

$$\frac{mL}{m+L}\|x_s - x^*\|^2 + \frac{1}{m+L}\|\nabla f(x_s) - \nabla f(x^*)\|. \tag{5.235}$$

Therefore, from (5.234) and (5.235) we obtain

$$\|x_{s+1} - x^*\|^2 \le \|x_s - x^*\|^2 - 2t\frac{mL}{m+L}\|x_s - x^*\|^2$$

$$+t\left(t - \frac{2}{m+L}\right)\|\nabla f(x_s) - \nabla f(x^*)\|^2 \tag{5.236}$$

hence for $0 < t < \frac{2}{m+L}$ from (5.236) follows (5.232).

2. For $t = 2/(m+L)$ from (5.236) we have

$$\|x_{s+1} - x^*\| \le \sqrt{1 - \frac{4mL}{(m+L)^2}}\|x_s - x^*\|$$

or

$$\|x_{s+1} - x^*\| \le \left(\frac{L-m}{L+m}\right)\|x_s - x^*\|.$$

Keeping in mind $\varkappa = m/L$, we obtain

$$\Delta_k = \|X_k - x^*\| = \left(\frac{1-\varkappa}{1+\varkappa}\right) \|x_{k-1} - x^*\| \leq \left(\frac{1-\varkappa}{1+\varkappa}\right)^k \|x_0 - x^*\|.$$

3. For a given accuracy $\varepsilon > 0$ to obtain $\Delta_k \leq \varepsilon$ we need

$$k \ln \frac{1-\varkappa}{1+\varkappa} \leq \ln \frac{\varepsilon}{\|x_0 - x^*\|}$$

or

$$k \geq \frac{\ln \frac{\varepsilon}{\|x_0 - x_k\|}}{\ln \frac{1-\varkappa}{1+\varkappa}} = \frac{\ln \frac{\|x_0 - x^*\|}{\varepsilon}}{\ln \frac{1+\varkappa}{1-\varkappa}} = \frac{\ln \varepsilon^{-1} \|x_0 - x^*\|}{\ln\left(1 + \frac{2\varkappa}{1-\varkappa}\right)}.$$

Keeping in mind $\ln(1+x) \leq x$, $-1 < x < \infty$.
We obtain the following bound for the number at steps:

$$k = O(\varkappa^{-1} \ln \varepsilon^{-1}).$$

□

5.13 Quadratic Programming

A number of real-world applications as well as mathematical problems lead to the following quadratic programming (QP) problem:

$$f(x^*) = \max\{f(x) | x \in \Omega\}, \tag{5.237}$$

where $f(x) = \frac{1}{2}\langle Qx, x\rangle + \langle q, x\rangle$, $\Omega = \{x \in \mathbb{R}^n : b - Ax \geq 0\} \neq \emptyset$, $Q = Q^T : \mathbb{R}^n \to \mathbb{R}^n$; $A : \mathbb{R}^n \to \mathbb{R}^r$, q, $x \in \mathbb{R}^n$, $b \in \mathbb{R}^r$.

If $Q \prec 0$ then x^* exists and unique. If $Q \preceq 0$ and Ω closed and bounded, then x^* exists due to the Weierstrass theorem.

Finally, if Ω is unbounded as well as level set $\mathscr{L}_f(\alpha) = \{x \in \mathbb{R}^n : f(x) \geq \alpha\}$ for a given $\alpha \in \mathbb{R}^1$, then x^* exists if

$$RC(\Omega) \cap RC(\mathscr{L}_f(\alpha)) = \emptyset. \tag{5.238}$$

We also assume that $x^* = \text{argmax}\{f(x) | x \in \mathbb{R}^n\} \notin \Omega$.

Let us consider Lagrangian

$$L(x, \lambda) = \frac{1}{2}\langle Qx, x\rangle + \langle q, x\rangle + \langle \lambda, b - Ax\rangle$$

for the problem (5.237).

Due to KKT's Theorem, there is $\lambda^* \in \mathbb{R}^m_+$ that the pair $(x^*; \lambda^*)$ satisfies the following system

$$\nabla_x L(x,\lambda) = Qx + q - A^T\lambda = 0 \tag{5.239}$$

$$\nabla_\lambda L(x,\lambda) = b - Ax \geq 0 \tag{5.240}$$

and the complementarity condition

$$\langle \lambda, b - Ax \rangle = 0 \tag{5.241}$$

holds.

From (5.239) we have

$$Qx = A^T\lambda - q = s(\lambda). \tag{5.242}$$

Therefore,

$$\begin{aligned} -\langle Qx, x\rangle &= \langle x, q - A^T\lambda\rangle = \langle q, x\rangle - \langle x, A^T\lambda\rangle \\ &= \langle q, x\rangle - \langle Ax - b, \lambda\rangle - \langle b, \lambda\rangle = \langle q, x\rangle + \langle \lambda, b - Ax\rangle - \langle b, \lambda\rangle, \end{aligned}$$

so

$$\langle \lambda, b - Ax\rangle = -\langle Qx, x\rangle + \langle b, \lambda\rangle - \langle q, x\rangle.$$

Hence,

$$L(x,\lambda) = -\frac{1}{2}\langle Qx, x\rangle + \langle b, \lambda\rangle.$$

If $Q \prec 0$, then $-Q \succ 0$ and from (5.242) we have

$$x(\lambda) = Q^{-1}(A^T\lambda - q) = Q^{-1}s(\lambda).$$

The dual function is given by the following formula:

$$\begin{aligned} d(\lambda) = -\frac{1}{2}\langle Qx(\lambda), x(\lambda)\rangle + \langle b, \lambda\rangle = -\frac{1}{2}\langle Q^{-1}s(\lambda), s(\lambda)\rangle + \langle b, \lambda\rangle = \\ -\frac{1}{2}\langle Q^{-1}(A^T\lambda - q), A^T\lambda - q\rangle + \langle b, \lambda\rangle \end{aligned} \tag{5.243}$$

and the dual problem consists of finding

$$d(\lambda^*) = \min\{d(\lambda) \mid \lambda \in \mathbb{R}^m_+\}. \tag{5.244}$$

Exercise 5.21. Find the gradient ∇d; show that Lipschitz condition

$$\|\nabla d(u) - \nabla d(\mathrm{v})\| \leq L\|u - \mathrm{v}\|,\ u, \mathrm{v} \in \mathbb{R}^m_+ \tag{5.245}$$

holds and find $L > 0$.

5.13.1 Dual GP Method for QP

In this section we consider gradient projection method for the dual QP problem (5.244). It leads to the dual gradient projection (DGP) method.

Due to the Slater condition, the dual optimal set $\Lambda^* = \{\lambda \in \mathbb{R}^m_+ : d(\lambda) = d(\lambda^*)\}$ is bounded.

The optimality condition for $\lambda^* \in \Lambda^*$is given by the following inequality:

$$\langle \nabla d(\lambda^*), \Lambda - \lambda^* \rangle \geq 0 \quad \forall \Lambda \in \mathbb{R}^m_+$$

To formulate the DGP method, we consider the following quadratic approximation:

$$\psi_L(\Lambda, \lambda) = d(\lambda) + \langle \Lambda - \lambda, \nabla d(\lambda) \rangle + \frac{L}{2} \|\Lambda - \lambda\|^2$$

of $d(\lambda)$ at the point $\lambda \in \mathbb{R}^m_+$.

For a given $\lambda \in \mathbb{R}^m_+$ there exists a unique minimizer

$$\lambda^L_+ \equiv \lambda^L_+(\lambda) = \operatorname{argmin}\left\{\psi_L(\Lambda, \lambda) \mid \Lambda \in \mathbb{R}^m_+\right\}. \tag{5.246}$$

Let us fix $\lambda \in \mathbb{R}^m_+$, then the optimality criteria for $\lambda^L_+ \in \mathbb{R}^m_+$ is given by the following inequality:

$$\nabla_\Lambda \psi_L(\lambda^L_+, \lambda) = \nabla d(\lambda) + L(\lambda^L_+ - \lambda) \geq 0 \tag{5.247}$$

and complementarity condition

$$\left\langle \lambda^L_+, \nabla_\Lambda \psi_L(\lambda^L_+, \lambda) \right\rangle = 0. \tag{5.248}$$

The optimality conditions (5.247)–(5.248) yield the following closed form solution

$$\lambda^L_+ = \left[\lambda - L^{-1} \nabla d(\lambda)\right]_+ \tag{5.249}$$

for the problem (5.246), where $[a]_+ = ([a_i]_+, i = 1, \ldots, m)$ and

$$[a_i]_+ = \begin{cases} a_i & a_i \geq 0 \\ 0 & a_i < 0 \end{cases}.$$

In other words, λ^L_+ is projection of $(\lambda - L^{-1} \nabla d(\lambda))$ on $\mathbb{R}^m_+$.

The solution (5.249) for the problem (5.246) leads to the following DGP method:

$$\lambda_{s+1} = \left[\lambda_s - L^{-1} \nabla d(\lambda_s)\right]_+, \tag{5.250}$$

which is, in fact, a GP method for the dual QP.

On the other hand,

$$\lambda_+^L = \operatorname{argmin}\left\{ \langle \Lambda - \lambda, \nabla d(\lambda)\rangle + \frac{L}{2}\|\Lambda - \lambda\|^2 \mid \Lambda \in \mathbb{R}_+^m \right\} \tag{5.251}$$

therefore (5.250) has the flavor of a quadratic prox method.

Note that application of the GP method to the primal QP leads to finding

$$P_\Omega(x_s - t\nabla p(x_s)) = \operatorname{argmin}\{\|y - (x_s + t\nabla p(x_s))\| \mid y \in \Omega\}$$

at each step, which is a problem similar to the original QP.

The main operation per step in (5.250) is computing ∇d, which leads to matrix by vector multiplication.

Let us consider the convergence properties of the DGP method. Due to the Lipschitz condition (5.245) for a convex function $d : \mathbb{R}_+^m \longrightarrow \mathbb{R}$, the following bound holds:

$$d(\Lambda) - d(\lambda) - \langle \Lambda - \lambda, \nabla d(\lambda)\rangle \le \frac{L}{2}\|\Lambda - \lambda\|^2.$$

Therefore, for any pair $(\Lambda, \lambda) \in \mathbb{R}_+^m \times \mathbb{R}_+^m$ we have

$$d(\Lambda) \le \psi_L(\Lambda, \lambda) = d(\lambda) + \langle \Lambda - \lambda, \nabla d(\lambda)\rangle + \frac{L}{2}\|\Lambda - \lambda\|^2.$$

The following lemma, which is similar to Lemma 5.12, is taking place.

Lemma 5.13. *For any given $\lambda \in \mathbb{R}_+^m$ and $L > 0$ such that*

$$d(\lambda_+^L) \le \psi_L(\lambda_+^L, \lambda) \tag{5.252}$$

the following inequality holds for any $\Lambda \in \mathbb{R}_+^m$:

$$d(\Lambda) - d(\lambda_+^L) \ge \frac{L}{2}\left\|\lambda_+^L - \lambda\right\|^2 + L\langle \lambda - \Lambda, \lambda_+^L - \lambda\rangle \tag{5.253}$$

The following theorem establishes the convergence properties of the DPG method (5.250).

Theorem 5.18. *For the dual sequence $\{\lambda_s\}_{s\in N}$ generated by the DGP method (5.250), the following bound:*

$$\Delta_k = d(\lambda_k) - d(\lambda^*) \le \frac{L}{2k}\|\lambda_0 - \lambda^*\|^2 \tag{5.254}$$

holds.

The proof is practically identical to the proof of Theorem 5.15.

The main effort per step of the DGP method (5.250) is matrix by vector multiplication. It requires $O(nm)$ operations; therefore the complexity of the DGP method is given by following formula:

$$CompDGP = O(\varepsilon^{-1} mnL\|\lambda^* - \lambda_0\|^2).$$

In the following section, we consider the dual fast gradient projection (DFGP) method, which improves substantially both the convergence rate and complexity of the DGP.

5.13.2 Dual Fast Gradient Projection Method

Application of FGP method to the dual QP (5.244) leads to the dual fast gradient projection (DFGP) method for QP.

The DFGP generates an auxiliary sequence $\{\lambda_k\}_{k\in N}$ and the main sequence $\{\Lambda_k\}_{k\in N}$. Vector λ_k one can view as a predictor, while vector Λ_k is the corrector or approximation at step k.

DFGP Method

1. Input:
   - L – the upper bound for the Lipschitz constant of the gradient $\nabla d(\lambda)$
   - $\Lambda_1 = \Lambda_0 \in \mathbb{R}^m_+$
   - $t_1 = 1$
2. Step k
   (a) find step size $t_{k+1} = \frac{1+\sqrt{1+4t_k^2}}{2}$;
   (b) find the predictor $\lambda_{k+1} = \Lambda_k + (\frac{t_k-1}{t_{k+1}})(\Lambda_k - \Lambda_{k-1})$;
   (c) find new approximation

$$\Lambda_{k+1} = \arg\min\left\{\psi(\Lambda,\lambda_{k+1}) \mid \Lambda \in \mathbb{R}^m_+\right\}$$

$$= \arg\min\{\langle \Lambda - \lambda_{k+1}, \nabla f(\lambda_{k+1})\rangle + \frac{L}{2}\|\Lambda - \lambda_{k+1}\|^2 | \Lambda \in \mathbb{R}^m_+\}.$$

The correction phase c. produces new approximation

$$\Lambda_{k+1} = \operatorname{argmin}\left\{\langle \Lambda - \lambda_{k+1}, \nabla f(\lambda_{k+1})\rangle + \frac{L}{2}\|\Lambda - \lambda_{k+1}\|^2 \mid \Lambda \in \mathbb{R}^m_+\right\} \quad (5.255)$$

From the optimality criteria for Λ_{k+1} follows

$$\Lambda_{k+1} = [\lambda_{k+1} - \frac{1}{L}\nabla f(\lambda_{k+1})]_+ \quad (5.256)$$

It means that the correction phase of DFGP is just a projected gradient step with step–length L^{-1}.

The following Theorem is similar to Theorem 5.16.

Theorem 5.19. *For the sequence $\{\Lambda_k\}_{k\in N}$ generated by DFGP method (a) – (c), the following bound holds:*

$$\Delta_{k+1} \leq \frac{2L\|x_0 - x^*\|^2}{(k+2)^2} \tag{5.257}$$

The DFGP requires numerical effort per step similar to DPG method, but the convergence rate is much better.

It follows from (5.257) that for a given accuracy $\varepsilon > 0$ it takes

$$k = \frac{\sqrt{2L}\|x_0 - x^*\|}{\sqrt{\varepsilon}}$$

steps of DFGP to get $\Delta_k \leq \varepsilon$; therefore the overall complexity of the DFGP is

$$CompDFPG = O\left(mn\frac{\sqrt{L}\|\lambda_0 - \lambda^*\|}{\sqrt{\varepsilon}}\right).$$

The IPMs have a better complexity bound, but they require solving linear systems of equations at each step, which for large-scale QP can be a very difficult task.

The main operation at each step of DFGP is matrix by vector multiplication, which can be done in parallel. In this regard, the DFGP is suitable for solving large-scale QP.

If $d(\lambda)$ is strongly convex, then the convergence rate and complexity bounds can be improved (see Theorem 5.17).

5.14 Quadratic Programming Problems with Quadratic Constraints

A number of important applications as well as mathematical problems lead to quadratic optimization with quadratic constraints.

Let

$$Q_i = Q_i^T \in \mathbb{R}^{n\times n}, i = 0, 1, \ldots, m$$

be $m+1$ symmetric matrices. We assume

$$Q_0 \succeq 0 \text{ and } Q_i \preceq 0, i = 1, \ldots, m \tag{5.258}$$

$a_i \in \mathbb{R}^n$, $i = 1, \ldots, m$ given vectors and α_i, $i = 1, \ldots, m$ given m numbers.

We consider the following QP problem with quadratic constraints:

$$f(x^*) = \min\{f(x) \mid x \in \Omega\}, \tag{5.259}$$

where $\Omega = \{x \in \mathbb{R}^n : c_i(x) \geq 0,\ \ i = 1 \ldots, m\}$.

Let

$$f(x) = \frac{1}{2}\langle Q_0 x, x\rangle + \langle q_0, x\rangle,$$

$$c_i(x) = \frac{1}{2}\langle Q_i x, x\rangle + \langle q_i, x\rangle + \alpha_i,$$

and $c(x) = (c_1(x), \ldots, c_m(x))^T$.

We assume that Slater condition

$$\exists x_0 \in \mathbb{R}^n : c_i(x_0) > 0, i = 1, \ldots, m \tag{5.260}$$

holds and there is α_0 such that

$$RC(\mathscr{L}_f(\alpha_0)) \cap RC(\Omega) = \emptyset. \tag{5.261}$$

Then, there exists $\lambda^* \in \mathbb{R}^m_+$ that the pair (x^*, λ^*) is a saddle point of the Lagrangian

$$L(x,\lambda) = f(x) - \sum_{i=1}^{M} \lambda_i c_i(x) = \frac{1}{2}\langle Q_0 x, x\rangle + \langle q_0, x\rangle - \sum_{i=1}^{m} \lambda_i (\frac{1}{2}\langle Q_i x, x\rangle + \langle q_i, x\rangle + \alpha_i),$$

which corresponds to (5.259).

It follows from (5.258) that for any $\lambda \in \mathbb{R}^m_+$ Lagrangian $L(x,\lambda)$ is convex in $x \in \mathbb{R}^n$.

If, for example, $Q_0 \succ 0$, then the system

$$\nabla_x L(x,\lambda) = A(\lambda)x + q(\lambda) = 0,$$

where $A(\lambda) = Q_0 - \sum_{i=1}^m \lambda_i Q_i \succ 0$ and $q(\lambda) = q - \sum_{i=1}^m \lambda_i q_i$, has a unique solution

$$x(\lambda) = -(A(\lambda))^{-1} q(\lambda) = \text{argmin}\{L(x,\lambda) | x \in \mathbb{R}^n\}$$

and

$$\nabla_x L(x(\lambda), \lambda)) \equiv 0. \tag{5.262}$$

The dual function

$$d(\lambda) = \min\{L(x,\lambda) | x \in \Omega\} = L(x(\lambda),\lambda) = L(x(\cdot),\cdot)$$

is concave and smooth.

Moreover, the dual function has a Lipschitz continuous gradient

$$\nabla d(\lambda) = \nabla_x L(x(\cdot),\lambda) \cdot \nabla_\lambda x(\lambda) + \nabla_\lambda L(x(\cdot),\lambda).$$

From (5.262) follows

$$\nabla d(\lambda) = \nabla_\lambda L(x(\cdot),\cdot) = -c(x(\cdot)). \tag{5.263}$$

The dual problem

$$d(\lambda^*) = \max\{d(\lambda) | \lambda \in \mathbb{R}^m_+\} \tag{5.264}$$

has a very simple feasible set. Therefore, the gradient projection method for the dual problem could be efficient.

Exercise 5.22.

1. Find the upper bound for the Lipschitz constant $L > 0$ of the gradient $\nabla d(\lambda)$.
2. Consider fast gradient projected for the dual problem.
3. Estimate the complexity of the fast gradient projected method.

5.15 Conditional Gradient Method

Let $f : \mathbb{R}^n \to \mathbb{R}$ be a closed convex function and Ω be a convex, closed and bounded set in $\mathbb{R}^n$.

In this section we consider conditional gradient (CG) method. The CG method was introduced by M. Frank and P. Wolfe (1927–2016) in 1956 for solving QP. We consider CG method for the following CO problem:

$$f(x^*) = \min\{f(x)|x \in \Omega\}.$$

The CG method got a great deal of attention lately. It turned out that CG method is efficient for a number of important applications. A number of results on the CG method were obtained by Levitin and B. Polyak (1966); see also Demyanov and Rubinov (1970).

Let $x_0 \in \Omega$ be the starting point, and approximation $x_s \in \Omega$ has been found already.

The CG step consists of the following operations:

1. find $\hat{x}_s \in \Omega$:
$$\langle \nabla f(x_s), \hat{x}_s \rangle \leq \langle \nabla f(x_s), x \rangle, \quad \forall x \in \Omega.$$
2. find $x_{s+1} \in \Omega$:
$$(a)\ f(x_{s+1}) \leq f(t_s\hat{x}_s + (1-t_s)x_s) = f(x_s + t_s(\hat{x}_s - x_s)), \tag{5.265}$$
where
$$t_s = \min\{1, \gamma_s\langle \nabla f(x_s), x_s - \hat{x}_s\rangle \|x_s - \hat{x}_s\|^{-2}\} \tag{5.266}$$
$$0 < \varepsilon_1 \leq \gamma_s \leq \frac{2}{L}(1-\varepsilon_2),\ \ 0 < \varepsilon_2 < 1$$
or
$$(b)\ x_{s+1} = \operatorname{argmin}\{f(t\hat{x}_s + (1-t)x_s) |\ 0 \leq t \leq 1\};$$
or by solving the following problem:
$$(c)\ x_{s+1} = \operatorname{argmin}\{f(x)|x \in conv\{x_0, x_1, \ldots, x_s\}\}.$$

Due to 1. the CG method can be efficient if minimization of a linear function on Ω is an easy operation.

Let $\Delta_s = f(x_s) - f(x^*)$ and $D = \text{diam}\Omega = \max\{\|x - y\| | x \in \Omega, y \in \Omega\}$.
In what is following we assume that gradient ∇f satisfies Lipschitz condition

$$\|\nabla f(x) - \nabla f(y)\| \leq L\|x - y\|. \tag{5.267}$$

Let us update x_s to x_{s+1} using step – size $0 < t < 1$ and the following inequality:

$$f(x_{s+1}) \leq f(x_s + t(\hat{x}_s - x_s)) = f(\bar{x}_{s+1}). \tag{5.268}$$

Lemma 5.14. *For convex f with Lipschitz gradient the CG sequence $\{x_s\}_{s\in N}$ defined by* (5.268) *with $t = t_s = 2(2+s)^{-1}$ converges to x^* in value and*

$$\Delta_s = O(s^{-1}). \tag{5.269}$$

Proof. From (5.268) follows

$$f(x_{s+1}) - f(x^*) \leq f(\bar{x}_{s+1}) - f(x^*). \tag{5.270}$$

For $\bar{x}_{s+1} = x_s + t(\hat{x}_s - x_s)$ from (2.5) and (5.267) follows

$$f(x_{s+1}) \leq f(\bar{x}_{s+1}) \leq f(x_s) + \langle \nabla f(x_s), \bar{x}_{s+1} - x_s\rangle + \frac{L}{2}\|\bar{x}_{s+1} - x_s\|^2.$$

Therefore,

$$f(x_{s+1}) - f(x^*) \leq f(x_s) - f(x^*) + t\langle \nabla f(x_s), \hat{x}_s - x_s\rangle + \frac{L}{2}t^2\|\hat{x}_s - x_s\|.$$

or

$$\Delta_{s+1} \leq \Delta_s + t\langle \nabla f(x_s), \hat{x}_s - x_s\rangle + \frac{Lt^2}{2}\|\hat{x}_s - x_s\|^2. \tag{5.271}$$

From $\langle \nabla f(x_s), \hat{x}_s - x_s\rangle \leq \langle \nabla f(x_s), x^* - x_s\rangle$ and convexity f follows

$$\langle \nabla f(x_s), \hat{x}_s - x_s\rangle \leq f(x^*) - f(x_s) = -\Delta_s. \tag{5.272}$$

Therefore, from (5.271) and (5.272) we have

$$\Delta_{s+1} \leq (1-t)\Delta_s + \frac{Lt^2}{2}D^2. \tag{5.273}$$

Let us show that $\Delta_s \leq \frac{2LD^2}{s+2}$.
For $k = 1$ and $t_0 = 1$ from (5.273) we have

$$\Delta_1 \leq \frac{LD^2}{2} \leq \frac{2}{3}LD^2.$$

Let assume that for $k = s$ we have $\Delta_s \leq \frac{2LD^2}{s+2}$, then from (5.273) follows

$$\Delta_{s+1} \leq (1-t_s)\Delta_s + \frac{Lt_s^2}{2}D^2. \tag{5.274}$$

By taking $t_s = 2(s+2)^{-1}$ from (5.274), we obtain

$$\Delta_{s+1} \leq \left(1-\frac{2}{s+2}\right)\frac{2LD^2}{s+2} + \frac{LD^2}{2}\left(\frac{2}{s+2}\right)^2 = \frac{2(s+1)LD^2}{(s+2)^2} = \frac{2(s+1)(s+3)LD^2}{(s+2)^2(s+3)}.$$

Keeping in mind $(s+1)(s+3) \leq (s+2)^2$, we have

$$\Delta_{s+1} \leq \frac{2LD^2}{s+3}.$$

Therefore, (5.269) holds true. □

The CG method does not allow acceleration typical for GP method; moreover even strong convexity f along with Lipschitz condition for the gradient, generally speaking, does not lead to better then (5.269) convergence rate.

Let us consider the following example:

$$f(x) = \frac{L}{2}\langle x,x\rangle \Rightarrow \min \tag{5.275}$$

that is

$$x \in \Omega = \{x \in \mathbb{R}^n_+ : \sum_{i=1}^n x_i = C_0\},\ C_0 = \frac{C}{\sqrt{2}} \tag{5.276}$$

f is strongly convex and Ω is a simplex type set, with $diam\Omega = C$, so x^* exists and unique. Also $\nabla f(x) = Lx$ satisfies Lipschitz condition.

Let us consider Lagrangian for problem (5.275)–(5.276)

$$\mathcal{L}(x,\lambda) = \frac{L}{2}\langle x,x\rangle - \lambda(\sum_{i=1}^n x_i - C_0).$$

From KKT's Theorem follows existence λ^* that for the pair (x^*,λ^*) the following conditions hold:

$$(a)\ Lx_i^* - \lambda^* \geq 0, \quad (b)\ x_i^*(Lx_i^* - \lambda^*) = 0,\ i = 1,..,n. \tag{5.277}$$

Therefore, for any positive x_i^* we have $x_i^* = \lambda^* L^{-1} > 0$, hence $\lambda^* > 0$. From (5.277a) follows $x_i^* > 0$, $i = 1,\ldots,n$, hence

$$x_1^* = \ldots x_n^* = \lambda^* L^{-1}$$

and

$$x_i^* = C_0 n^{-1}, \quad i = 1,\ldots,n,$$

so

$$f(x^*) = \frac{LC_0^2}{2n}.$$

Let us apply CG method for solving (5.275)–(5.276) using $x_0 = C_0 e_1 = (C_0, 0, \ldots, 0)$ as a starting point.

By induction we can see that

$$x_s = \operatorname{conv}\{C_0 e_1, C_0 e_{i_1}, \ldots, C_0 e_{i_s}\}$$

for some index $(i_1, \ldots, i_s) \in \{1, \ldots, n\}$. Hence,

$$\begin{aligned} f(x_s) &\geq \min\{f(x) | x \in C_0 \operatorname{conv}\{e_1, e_{i_1}, .., e_{i_s}\}\} \\ &\geq \frac{LC_0^2}{2(s+1)} \end{aligned}$$

and

$$\Delta_s = f(x_s) - f(x^*) \geq \frac{LC_0^2}{2}\left(\frac{1}{s+1} - \frac{1}{n}\right).$$

Therefore, for any $n \geq 2(s+1)$ we obtain

$$\Delta_s \geq \frac{LC_0^2}{2}\left(\frac{1}{s+1} - \frac{1}{2(n+1)}\right) = \frac{LC_0^2}{4(s+1)} = \frac{LC^2}{8(s+1)}.$$

So, even strong convexity f cannot improve the CG bound $O(s^{-1})$.

The bound, however, can be improved by assuming the strong convexity of the set Ω.

We recall the convex set Ω is strongly convex if along with any x and y from Ω vector $\frac{x+y}{2} + z \in \Omega$, where $\|z\| = \varkappa \|x - y\|^2$ and $\varkappa > 0$ is small enough.

Theorem 5.20. *If f is convex and gradient ∇f satisfies Lipschitz condition, set Ω is strongly convex and $\|\nabla f(x)\| \geq \varepsilon > 0$, $\forall x \in \Omega$, then there is $0 < q < 1$ that*

$$\Delta_{s+1} \leq q \Delta_s, \quad \forall s \geq 1.$$

Proof. From gradient Lipschitz condition and (5.267) follows

$$\begin{aligned} f(x_{s+1}) - f(x_s) &\leq \langle \nabla f(x_s), x_{s+1} - x_s \rangle + \frac{L}{2}\|x_{s+1} - x_s\|^2 \\ &= -t_s \langle \nabla f(x_s), x_s - \hat{x}_s \rangle + \frac{Lt_s^2}{2}\|\hat{x}_s - x_s\|^2. \end{aligned} \tag{5.278}$$

If $t_s = 1$ then from (5.278) we obtain

$$\Delta_{s+1} - \Delta_s \leq \langle \nabla f(x_s), x_s - \hat{x}_s \rangle \left[\frac{L}{2} \frac{\|x_s - \hat{x}_s\|^2}{\langle \nabla f(x_s), x_s - \hat{x}_s \rangle} - 1\right]. \tag{5.279}$$

Also

$$\gamma_s \langle \nabla f(x_s), x_s - \hat{x}_s \rangle \|x_s - \hat{x}_s\|^{-2} \geq 1,$$

therefore from (5.279) follows

$$\Delta_{s+1} - \Delta_s \le \langle \nabla f(x_s), x_s - \hat{x}_s \rangle \left[\frac{L\gamma_s}{2} - 1 \right] \le -\varepsilon_2 \langle \nabla f(x_s), x_s - \hat{x}_s \rangle =$$

$$\varepsilon_2 \langle \nabla f(x_s), \hat{x}_s - x_s \rangle \le \varepsilon_2 \langle \nabla f(x_s), x^* - x_s \rangle \le \varepsilon_2 (f(x^*) - f(x_s)) = -\varepsilon_2 \Delta_s$$

or

$$\Delta_{s+1} \le (1 - \varepsilon_2)\Delta_s. \tag{5.280}$$

If $t_s < 1$ then

$$\frac{\gamma_s \langle \nabla f(x_s), x_s - \hat{x}_s \rangle}{\|x_s - \hat{x}_s\|^2} < 1,$$

or

$$\gamma_s < \frac{\|x_s - \hat{x}_s\|^2}{\langle \nabla f(x_s), x_s - \hat{x}_s \rangle}.$$

Using strong convexity Ω, we obtain

$$-\langle \nabla f(x_s), x_s - \hat{x}_s \rangle = 2 \left\langle \nabla f(x_s), \frac{x_s + \hat{x}_s}{2} - \varkappa \frac{\nabla f(x_s)}{\|\nabla f(x_s)\|} \|\hat{x}_s - x_s\|^2 \right\rangle$$

$$+2 \left\langle \nabla f(x_s), \varkappa \frac{\nabla f(x_s}{\|\nabla f(x_s)\|} \|\hat{x}_s - x_s\|^2 - x_s \right\rangle \ge 2\langle \nabla f(x_s), \hat{x}_s \rangle +$$

$$2\varkappa \|\nabla f(x_s)\| \|\hat{x}_s - x_s\|^2 - 2\langle \nabla f(x_s), x_s \rangle \ge 2\langle \nabla f(x_s), \hat{x}_s - x_s \rangle + 2\varkappa\varepsilon \|\hat{x}_s - x_s\|^2.$$

Therefore,

$$\langle \nabla f(x_s), x_s - \hat{x}_s \rangle \ge 2\varkappa\varepsilon \|\hat{x}_s - x_s\|^2. \tag{5.281}$$

From (5.280) follows

$$\Delta_{s+1} - \Delta_s \le -t_s \langle \nabla f(x_s), x_s - \hat{x}_s \rangle + \frac{Lt_s^2}{2} \|x_s - \hat{x}_s\|^2.$$

For $t_s < 1$ we have

$$\gamma_s < \|x_s - \hat{x}_s\|^2 \langle \nabla f(x_s), x_s - \hat{x}_s \rangle^{-1}$$

and

$$t_s = \gamma_s \langle \nabla f(x_s), x_s - \hat{x}_s \rangle \|x_s - \hat{x}_s\|^{-2}.$$

Therefore,

$$\Delta_{s+1} - \Delta_s \le -\gamma_s \langle \nabla f(x_s), x_s - \hat{x}_s \rangle^2 \|x_s - \hat{x}_s\|^{-2} + \frac{L}{2}\gamma_s^2 \langle \nabla f(x_s), x_s - \hat{x}_s \rangle^2 \|x_s - \hat{x}_s\|^{-2}$$

$$= \gamma_s \left[\frac{L}{2}\gamma_s - 1 \right] \langle \nabla f(x_s), x_s - \hat{x}_s \rangle^2 \|x_s - \hat{x}_s\|^{-2} = -\gamma_s \frac{\varepsilon_2}{2} \langle \nabla f(x_s), x_s - \hat{x}_s \rangle^2 \|x_s - \hat{x}_s\|^{-2}$$

or

$$\Delta_s - \Delta_{s+1} \ge \gamma_s \frac{\varepsilon_2}{2} \langle \nabla f(x_s), x_s - \hat{x}_s \rangle^2 \|x_s - \hat{x}_s\|^{-2}.$$

Combining the latter bound with (5.281), we obtain

$$\Delta_s - \Delta_{s+1} \geq \gamma_s \varepsilon_2 \varepsilon \varkappa \langle \nabla f(x_s), x_s - \hat{x}_s \rangle. \tag{5.282}$$

Then,

$$\langle \nabla f(x_s), x_s - \hat{x}_s \rangle = -\langle \nabla f(x_s), \hat{x}_s - x_s \rangle$$

$$\langle \nabla f(x_s), \hat{x}_s - x_s \rangle \leq \langle \nabla f(x_s), x^* - x_s \rangle \leq f(x^*) - f(x_s),$$

or

$$-\langle \nabla f(x_s), \hat{x}_s - x_s \rangle \geq f(x_s) - f(x^*) = \Delta_s.$$

From (5.282) and the last inequality follows

$$\Delta_s - \Delta_{s+1} \geq \gamma_s \varepsilon_2 \varepsilon \varkappa \Delta_s$$

or

$$\Delta_{s+1} \leq (1 - \gamma_s \varepsilon_2 \varepsilon \varkappa) \Delta_s.$$

□

If Ω is defined by a system of convex inequalities, then both GP and CG methods can't be useful, because they require at each step solving a convex optimization problem similar to the original one.

In the following section, we consider a method suitable for CO with Ω defined by a finite number of convex inequalities.

5.16 Primal–Dual Feasible Direction Method

The primal–dual feasible direction (PDFD) method was introduced by Polyak (1966). The PDFD method is a generalization for convex optimization of the classical Dantzig et al. (1956) method for simultaneous solution primal and dual LP. On the other hand, the PDFD is a version of feasible direction method.

Methods of feasible directions were independently developed in the early 1960s by Zoutendijk (1960) and Zukhovitsky et al. (1963, 1963a, 1965).

We are concerned with convex optimization problem (5.25)–(5.26).

As always, we assume that X^* is not empty and bounded and Slater condition (5.27) holds. Then the dual optimal set

$$L^* = \{\lambda \in \mathbb{R}^m_+ : d(\lambda) = d(\lambda^*)\}$$

is also bounded, where $d(\lambda) = \inf_{x \in \mathbb{R}} L(x, \lambda)$ and $L(x, \lambda) = f(x) - \sum_{i=1}^m \lambda_i c_i(x)$ is the Lagrangian for problem (5.25)–(5.26).

We will use P. Wolfe's duality

$$L(x^*, \lambda^*) = \max\{L(x, \lambda) | \nabla_x L(x, \lambda) = 0, \ \lambda \in \mathbb{R}^m_+\}. \tag{5.283}$$

Let us first describe the basic ideas of the PDFD method

(1) start with primal feasible vector $\bar{x} \in \Omega$;
(2) consider the active set $I(\bar{x}) = \{i : c_i(\bar{x}) = 0\}$ at $\bar{x}$;
(3) consider the reduced dual problem by setting $\bar{\lambda}_i = 0,\ i \notin I(\bar{x})$;
(4) the rest of the dual variables $\bar{\lambda}_i,\ i \notin I(\bar{x})$ we find by solving the following QP:

$$\min\left\{ \|\lambda_0 \nabla f(\bar{x}) - \sum_{i \notin I(\bar{x})} \lambda_i \nabla c_i(\bar{x})\|^2 | \lambda_0 + \sum_{i \in I(\bar{x})} \lambda_i = 1,\ \lambda_0 \geq 0,\ \lambda_i \geq 0,\ i \in I(\bar{x}) \right\}$$

$$= \|\hat{\lambda}_0 \nabla f(\bar{x}) - \sum_{i \in I(\bar{x})} \hat{\lambda}_i \nabla c_i(\bar{x})\| = D(\bar{x}; \hat{\lambda}); \tag{5.284}$$

(5) if $D(\bar{x}; \hat{\lambda}) = 0$, then $\hat{\lambda}_0 > 0$, because otherwise we have

$$\sum_{i \in I(\bar{x})} \hat{\lambda}_i \nabla c_i(\bar{x}) = 0,$$

which contradicts Slater condition;
(6) in view of $\hat{\lambda}_0 > 0$ and $\bar{\lambda}_i = 0,\ i \notin I(\bar{x})$ from $D(\bar{x}, \hat{\lambda}) = 0$ follows

$$\nabla f(\bar{x}) - \sum_{i \in I(\bar{x})} \bar{\lambda}_i \nabla f(\bar{x}) = 0,$$

where $\bar{\lambda}_i = \hat{\lambda}_i \cdot \hat{\lambda}_0^{-1},\ i \in I(\bar{x})$.

It means vector $\bar{x}$ is primal feasible, vector

$$\bar{\lambda} = \{\bar{\lambda}_i = \hat{\lambda}_i \cdot \hat{\lambda}_0^{-1},\ i \in I(\bar{x}),\ \bar{\lambda}_i = 0,\ i \notin I(\bar{x})\}$$

is dual feasible (see (5.283)) and complementarity condition

$$\bar{\lambda}_i c_i(\bar{x}) = 0, \quad i = 1, \ldots, m$$

holds, therefore $x^* = \bar{x},\ \lambda^* = \bar{\lambda}$.

In other words, by solving QP (5.284) is case of $D(\bar{x}, \hat{\lambda}) = 0$, we obtain feasible dual solution, which together with primal feasible $\bar{x}$ satisfies complementarity condition. Therefore $x^* = \bar{x},\ \lambda^* = \bar{\lambda}$

In case $D(\bar{x}, \hat{\lambda}) > 0$ the QP (5.284) solution $\hat{\lambda} \in \mathbb{R}^m_+$ defines primal feasible direction $\hat{\zeta} = \hat{\lambda}_0 \nabla f(\bar{x}) - \sum_{i \in I(\bar{x})} \hat{\lambda}_i \nabla c_i(\bar{x})$, moving along which from $\bar{x}$ one reduces $f(\bar{x})$ without leaving feasible set Ω, if the step–length $t > 0$ is not too large.

This is the main idea of the PDFD method. To show that $\hat{\zeta}$ is a feasible direction we need the following Lemma.

Let $a_i,\ i = 1, \ldots, m$ be a set of vectors in $\mathbb{R}^n$, $S_m = \{\lambda \in \mathbb{R}^m_+ : \sum_{i=1}^m \lambda_i = 1\}$-simplex in $\mathbb{R}^n$ and $\|x\| = (x^T x)^{\frac{1}{2}}$ is the Euclidean norm.

Lemma 5.15. *If*

$$\max_{1\le i\le m}\langle a_i,\hat{\zeta}\rangle = \min\{\max_{1\le i\le m}\langle a_i,\zeta\rangle \,|\, \|\zeta\|\le 1\} < 0, \tag{5.285}$$

then there is $\hat{\lambda}\in S_m$*:*

$$\|\sum_{i=1}^{m}\hat{\lambda}_i a_i\| = \min\{\|\sum_{i=1}^{m}\lambda_i a_i\| \,|\, \lambda\in S_m\}, \tag{5.286}$$

that

$$\hat{\zeta} = -(\sum_{i=1}^{m}\hat{\lambda}_i a_i)\|\sum_{i=1}^{m}\hat{\lambda} a_i\|^{-1} \tag{5.287}$$

solves (5.285).

Proof. Problem (5.285) is equivalent to

$$u \to \min \tag{5.288}$$

$$\langle a_i,\zeta\rangle \le u, \quad i=1,\dots,m \tag{5.289}$$

$$\|\zeta\|^2 \le 1. \tag{5.290}$$

For any solution $(\hat{\zeta};\hat{u})$ of the problem (5.288)–(5.290) we have $\|\hat{\zeta}\|=1$. In fact, assuming $\|\hat{\zeta}\|<1$ from (5.285) for a vector $\bar{\zeta}=\|\hat{\zeta}\|\|\hat{\zeta}\|^{-1}$ we obtain

$$\langle a_i,\bar{\zeta}\rangle < \langle a_i,\hat{\zeta}\rangle,\ i=1,\dots,m,$$

therefore $\hat{\zeta}$ can't be a solution in (5.285).

Hence, the problem (5.285) is equivalent

$$u \to \min \tag{5.291}$$

$$\langle a_i,\zeta\rangle \le u \tag{5.292}$$

$$\|\zeta\|^2 = 1. \tag{5.293}$$

The correspondent Lagrangian

$$L(\cdot) = L(\zeta,u,\lambda,\mu) = u + \sum_{i=1}^{m}\lambda_i[\langle a_i,\zeta\rangle - u] + \frac{1}{2}\mu(\|\zeta\|^2-1).$$

Let us consider the KKT's conditions for the optimal vector $(\hat{\zeta},\hat{u})$ and optimal Lagrange multipliers $\hat{\lambda}\in\mathbb{R}^m_+$, $\hat{\mu}\in\mathbb{R}$:

$$L'_u(\cdot) = 1 - \sum_{i=1}^{m}\hat{\lambda} = 0 \tag{5.294}$$

$$\nabla_\zeta L(\cdot) = \sum_{i=1}^{m} \hat{\lambda}_i a_i + \hat{\mu}\hat{\zeta} = 0 \tag{5.295}$$

$$\nabla_\lambda L(\cdot) = \langle a_i, \hat{\zeta}\rangle - \hat{u} \leq 0,\ i = 1, \ldots, m \tag{5.296}$$

and

$$\hat{\lambda}_i[\langle a_i, \hat{\zeta}\rangle - \hat{u}] = 0,\ i = 1, \ldots, m \tag{5.297}$$

$$L'_\mu(\cdot) = \|\hat{\zeta}\|^2 - 1 = 0.$$

First of all $\hat{\mu} \neq 0$, otherwise from (5.295) follows

$$\sum_{i=1}^{m} \hat{\lambda}_i a_i = 0,$$

which contradicts (5.285). Therefore from (5.295) follows

$$\hat{\zeta} = -\left(\sum_{i=1}^{m} \hat{\lambda}_i a_i\right)\hat{\mu}^{-1}.$$

From (5.293) follows $\hat{\mu} = \|\sum \hat{\lambda}_i a_i\|$, therefore for the dual function we have

$$d(\hat{\lambda}, \hat{\mu}) = \sum_{i=1}^{m} \langle \hat{\lambda}_i a_i, \hat{\zeta}\rangle = -\|\sum_{i=1}^{m} \hat{\lambda}_i a_i\|.$$

Therefore,

$$d(\hat{\lambda}, \hat{\mu}) = \max\{-\|\sum \lambda_i a_i\| | \lambda \in S_m\}$$

or

$$\min\{\|\sum_{i=1}^{m} \lambda_i a_i\| | x \in S_m\} = \|\sum_{i=1}^{m} \hat{\lambda}_i a_i\|$$

and

$$\hat{\zeta} = -\sum_{i=1}^{m} \hat{\lambda}_i a_i \|\sum_{i=1}^{m} \hat{\lambda}_i a_i\|^{-1}$$

solves problem (5.285) □

Now we are ready to describe the PDFD method.

Initialization. We start with $x_0 \in \Omega$, $\delta > 0$, $0 < \rho < 0.5$ and small enough $\varepsilon > 0$ - the required accuracy.

Let assume that x_s, λ_s and $\delta_s > 0$ have been found already.

A step of the PDFD method consists of the following operations:

1. set $x := x_s$, $\lambda := \lambda_s$ and $\delta := \delta_s$;
2. find $I(x, \delta) = \{1 \leq i \leq m :\ 0 \leq c_i(x) \leq \delta\}$; set $\lambda = 0$, $i \notin I(x, \delta)$ and solve the following reduced QP

$$\min\{\|\lambda_0 \nabla f(x) + \sum_{i \in I(x,\delta)} \lambda_i(-\nabla c_i(x)\| | \lambda_0 +$$

$$\sum_{i\in I(x,\delta)} \lambda_i = 1,\ \lambda_0 \geq 0,\ \lambda_i \geq 0,\ i \in I(x,\delta)\} =$$

$$= \|\bar{\lambda}_0 \nabla f(x) + \sum_{i\in I(x,\delta)} \bar{\lambda}_i(-\nabla c_i(x))\| = D_\delta(x,\bar{\lambda}) \tag{5.298}$$

3. if $D_\delta(x,\bar{\lambda}) \leq \varepsilon$ go 8; if $\varepsilon < D_\delta(x,\bar{\lambda}) \leq \delta$, then set $\delta_{s+1} = \delta_s/2$; if $D_\delta(x,\bar{\lambda}) > \delta$, then set $\delta_{s+1} = \delta\rho^2$.
4. find the dual vector

$$\lambda_{s+1} = (\lambda_{i,s+1} = 0,\ i \notin \bar{I}(x,\delta),\ \lambda_{i,s+1} = \bar{\lambda},\ i \in \bar{I}(x,\delta))$$

5. find the primal decent direction

$$\zeta_{s+1} = -\left(\lambda_{0,s+1}\nabla f(x) - \sum_{i\in I(x,\delta)} \lambda_{i,s+1}\nabla c_i(x)\right) D_\delta^{-1}(x,\bar{\lambda})$$

6. set $x_{s+1}(t) := x + t\zeta_{s+1}$ and find

$$t_{s+1} : f(x_{s+1}(t_{s+1})) = \min\{f(x_{s+1}(t))|\ x_{s+1}(t) \in \Omega\}$$

7. find primal vector

$$x_{s+1} = x + t_{s+1}\zeta_{s+1};$$

   set $s+1 := s$ go to 1
8. find

$$\langle \nabla f(x), \bar{\zeta}\rangle = \min\{\langle \nabla f(x), \zeta\rangle |\ \langle \nabla c(x), \zeta\rangle \geq 0,\ i \in I(x), \|\zeta\| \leq 1\}, \tag{5.299}$$

   and

$$D_0(x) = -\langle \nabla f(x), \bar{\zeta}\rangle$$

   if $D_0(x) \leq \varepsilon$, then find the dual vector $\bar{\lambda} = (\bar{\lambda}_i = 0,\ i \notin I(x),\ \lambda_i = \bar{\lambda}_i,\ i \in I(x)$, where $\bar{\lambda}_i,\ i \in I(x)$ is the optimal dual solution of (5.299) output (x,λ) as the primal–dual solution.
   If $D_0(x) > \varepsilon$, then $x_{s+1} := x$, $\lambda_{s+1} := \bar{\lambda}$, $\delta_{s+1} := \delta\rho^2$, set $s+1 := s$ go to 1.

We assume that the following gradient Lipschitz condition

$$\|\nabla f(x) - \nabla f(y)\| \leq L_0\|x-y\|$$

$$\|\nabla c_i(x) - \nabla c_i(y)\| \leq L_i\|x-y\|,\ \ i = 1,\ldots,m \tag{5.300}$$

hold.

It follows from Lemma 5.15 that ζ_{s+1} from step 5. is a feasible direction. Therefore, (1)–(8) is a feasible direction method. Parameter $\delta > 0$ is critical, because it allows to avoid "zigzagging" between active constraint, which can compromise convergence. Therefore, for finding feasible direction the set $I(x)$ active at $x \in \Omega$ is replaced by $I(x, \delta)$.

Theorem 5.21. *If f and all $-c_i$, $i = 1, .., m$ are convex and Lipschitz condition (5.300) holds, then $\lim_{s\to\infty} f(x_s) = f(x^*)$, $\lim_{s\to\infty} d(\lambda_s) = d(\lambda^*)$, and any primal–dual converging subsequence $\{(x_{s_i}, \lambda_{s_i})\}_{i\in N}$ has the primal–dual solution $(x^*, \lambda^*) \in X^* \otimes L^*$ as its limit point.*

Proof. First, let us show that $\lim_{s\to\infty} \delta_s = 0$. Assuming the opposite, we obtain $\delta_s \geq \delta > 0$, $\forall s \geq 1$.

From Lemma 5.15 follows

$$\langle \nabla f(x_s), \zeta_s\rangle \leq -\delta, \ \langle \nabla c_i(x_s), \zeta_s\rangle \geq \delta, i \in I(x_s, \zeta_s). \tag{5.301}$$

Let us show that there is $\tau > 0$, that $t_s > \tau$, $s \geq 1$. In fact, from convexity of f follows monotonicity of directional derivative in t, that is, $(\nabla f(x_s + t\zeta_s), \zeta_s)$ is monotone increasing in t.

Let

$$t_s' > 0 : \langle \nabla f(x_s + t_s'\zeta_s), \zeta_s\rangle \geq -\delta/2, \tag{5.302}$$

then from (5.300) follows

$$\langle \nabla f(x_s + t_s'\zeta_s), \zeta_s\rangle - \langle \nabla f(x_s), \zeta_s\rangle = \langle \nabla f(x_s + t_s'\zeta_s) - \nabla f(x_s), \zeta_s\rangle \geq \delta/2.$$

From Lipschitz condition for ∇f follows

$$\delta/2 \leq \|\nabla f(x_s + t_s'\zeta_s) - \nabla f(x_s)\| \|\zeta_s\| \leq L_0 t_s' \|\zeta_s\|^2 \leq L_0 t_s'$$

or

$$t_s' \geq \delta/2L_0. \tag{5.303}$$

For concave functions c_i the directional derivative $\langle \nabla c_i(x_s + t\zeta_s), \zeta_s\rangle$ is a monotone decreasing function in $t > 0$.

Let $t_{i,s}' > 0 : \langle \nabla c_i(x_s + t_{i,s}'\zeta_s), \ \zeta_s\rangle \leq \delta/2$, then from (5.301) follows

$$\langle \nabla c_i(x_s) - \nabla c_i(x_s + t_{i,s}'\zeta_s), \zeta_s\rangle \geq \delta/2.$$

From Lipschitz condition for ∇c_i follows

$$L_i \|\zeta_s\|^2 t_{i,s}' \geq \delta/2$$

or

$$t_{i,s}' \geq \delta/2L_i. \tag{5.304}$$

Let $L = \max_{1\leq i\leq m} L_i$, then $t_{i,s}' \geq \delta/2L$.

For $i \in\notin I(x_s, \delta)$ we have $c_i(x_s) > \delta$. From (5.300) follows existence $\bar{L}_i$ that for any x and y from Ω we have

$$|c_i(x) - c_i(y)| \leq \bar{L}_i \|x - y\|. \tag{5.305}$$

We consider $t''_{i,s} : c_i(x_s + t''_{i,s}\zeta_s) = 0$, then $c_i(x_s) - c_i(x_s + t''_{i,s}\zeta_s) \geq \delta$. From (5.305) follows

$$\bar{L}_i t''_{i,s}\|\zeta_s\| \geq c_i(x_s) - c_i(x_s + t''_{i,s}\zeta_s) \geq \delta > 0$$

or $t''_{i,s} \geq \delta/L_i \geq \delta/\bar{L}$, $i \notin I(x_s,\delta)$, where $\bar{L} = \max\{\bar{L}_i,\ 1 \leq i \leq m\}$, therefore

$$t''_s = \min\{t''_{i,s} | 1 \leq i \leq m\} \geq \delta/2\bar{L}. \tag{5.306}$$

From (5.304) and (5.306) follows

$$f(x_s) - f(x_s + t_s\zeta_s) \geq -t_s\langle \nabla f(x_s + t_s\zeta_s), \zeta_s\rangle = t_s\delta/2.$$

Let $L_0 = \max\{L, \bar{L}\}$, then from (5.302) and (5.304) follows $t_s = \min\{t'_s, t''_s\} \geq 0.5\delta L_0^{-1}$.

Therefore,

$$f(x_s) - f(x_s + t_s\zeta_s) \geq \frac{\delta^2}{4L_0}.$$

It means that $\lim_{s\to\infty} f(x_s) = -\infty$, which is impossible.

Therefore, there is subsequence $\{\delta_{s_l}\}_{l\in N}$: $\lim_{l\to\infty}\delta_{s_l} = 0$ and

$$\|\bar{\lambda}_{0s_l}\nabla f(x_{s_l}) - \sum_{i\in J(x_{s_l},\delta_{s_l})} \lambda_{is_l}\nabla c_i(x_{s_l})\| \leq \delta_{s_l}. \tag{5.307}$$

Taking the limit in (5.307), we obtain

$$\|\hat{\lambda}_0\nabla f(\bar{x}) - \sum_{i\in I(\bar{x})} \hat{\lambda}_i\nabla c_i(\bar{x})\| = 0.$$

From Slater's condition follows $\hat{\lambda}_0 > 0$, then $(\bar{x},\bar{\lambda})$, where $\bar{\lambda}_i := \bar{\lambda}_i \cdot \bar{\lambda}_0^{-1}$, $i \in I(\bar{x})$ and $\lambda = 0$, $i \notin I(\bar{x})$ is primal–dual feasible solution.

Moreover, the pair $(\bar{x},\bar{\lambda})$ satisfies complementarity condition, so $x^* = \bar{x}$, $\lambda^* = \bar{\lambda}$. □

Let us validate the stopping criteria $D_0(x) \leq \varepsilon$. It follows from (5.299) that for any vector $y \in \Omega : \|y - x\| = 1$ we have

$$\|y - x\|^{-1}\langle \nabla f(x), y - x\rangle \geq -\varepsilon$$

or

$$\langle \nabla f(x), y - x\rangle \geq -\varepsilon\|y - x\|.$$

For $y = x^*$ we have

$$f(x^*) - f(x) \geq \langle \nabla f(x), x^* - x\rangle \geq -\varepsilon\|x^* - x\|.$$

Therefore,

$$f(x) \leq f(x^*) + \varepsilon\|x - x^*\|.$$

Keeping in mind convergence $x_s \to x^* \in X^*$, from $D_0(x) \leq \varepsilon$ follows

$$f(x) \leq f(x^*) + o(\varepsilon).$$

Notes

The first steps in LP were made by L.V. Kantorovich before World War II (see Kantorovich (1939, 1959)). Simplex method is due to G. Dantzig (see Dantzig (1963)); independently simplex method for LP equivalent to finding Chebyshev approximation for overdetermined linear system of equations was discovered by S. I. Zuchovitsky (see Zuchovitsky (1951)); see also Cheney and Goldstein (1958) and Stiefel (1960).

The LP duality theory is due to John von Neumann, who, to the best of our knowledge, have not published it in a separate paper.

First, we consider the optimality condition for an LP basic solution. From the optimality condition directly follows the "truncated" LP duality Theorem, which is used to prove Farkas Lemma (Farkas, 1902). The KKT's Theorem (see Kuhn and Tucker (1951)), which was proven in Ch1, using basic facts of convex analysis, also follows from Farkas's Lemma. Then, LP and convex optimization duality follows directly from KKT's Theorem.

For different aspects of duality theory in convex optimization, see Bertsekas (1982, 1999); Goldshtein and Tretiakov (1989); Grossman and Kaplan (1981); Ioffe and Tichomirov (1968); Ioffe and Tihomirov (2009); B. Polyak (1987); Rockafellar (1973, 1970); Rockafellar and Wets (2009), and references therein.

For structural LP properties, in particular, A. Hoffman's Lemma, see Hoffman (1952).

Ellipsoid method was discovered in mid-70's by Shor and independently by Nemirovski and Yudin, see Shor (1998) and Nemirovski and Yudin (1983). It was used to prove polynomial complexity of LP in Khachiyan (1979).

The IPM era started with N. Karmarkar paper (Karmarkar, 1984). Soon after Gill et al. (1986) found connection between N. Karmarkar's method and Newton log–barrier method for LP.

A simplification of N. Karmarkar's method, the AS algorithm, has been introduced independently by Barnes (1986); see also Barnes (1990) and Vanderbei et al. (1986); see also Vanderbei (1996). As it turned out, AS algorithm was introduced 20 years earlier by Dikin (1967); see also Dikin (1974). For AS convergence properties, see Jensen et al. (1996); Monteiro et al. (1990); Tsuchiya (1991); Tsuchiya and Muramatsu (1995); Tsuchiya (1996). N. Karmarkar's and AS methods were implemented, and encouraging numerical results were reported in Adler et al. (1989) and Adler and Monteiro (1991).

N. Karmarkar complexity bound was improved by Renegar (1988); see also Renegar (2001), Gonzaga (1989), Gonzaga (1992), and Vaidya (1990).

Thousands paper on IPMs were published in the late 80's and 90's. After decades of research, the primal–dual predictor–corrector method emerged as the most numerically efficient approach for LP calculation. For detailed description of the primal–dual LP methods, see very informative book by Wright (1997).

Most of the working IPMs codes are based on Mehrotra's predictor–corrector algorithm Mehrotra (1990); see also Mehrotra (1992)), which uses higher-order approximation of the central path foreshadowed by Megiddo (1989); see also Megiddo and Shub (1989) and further developed in Monteiro et al. (1990).

In the early 1990s, I. Lustig, R.Marsten, and D.Shanno developed computational aspects of the primal–dual algorithms and implemented and tested them Lustig et al. (1991, 1992, 1994). Eventually, they produced the tool of choice for LP calculation. For progress in LP, see Bixby (2012).

For different aspects of IPMs, see Anstreicher (1996); Freund (1991); Goldfarb and Todd (1988); Guler et al. (1993); Den Hertog et al. (1992); Jensen et al. (1994); Kojima et al. (1989, 1989a, 1991); Mizuno and Todd (1991); Pardalos and Resende (1996); Pardalos and Wolkowicz (1998); Pardalos et al. (1990); Potra (1994); Potra and Wright (2000); Roos and Vial (1988); Roos et al. (1997); Roos and Terlaky (1997); Todd (1989, 1993); Todd and Ye (1996); Vial (1992); Ye (1987); Ye and Todd (1990); Ye and Pardalos (1991); Ye (1991); Ye et al. (1993, 1994); Ye (1997); Zhang and Tapia (1992).

SUMT methods is a product of the 1950s and 1960s; the research on SUMT was summarized in Fiacco and McCormick (1990); see also Grossman and Kaplan (1981); Polak (1971); B. Polyak (1987). For the equivalence of SUMT and dual interior regularization, see Polyak (2016). For the primal–dual method for convex optimization, based on log–barrier function, see Griva et al. (2008).

Gradient projection method was introduced and studied by Rozen (1960, 1961), Goldstein (1964), and Levitin and B. Polyak (1966); see also references therein. For the fast gradient projection, see Nesterov (2004, 1983) and FISTA algorithm by Beck and Teboulle (2009, 2009a).

Conditional gradient for QP was introduced by Frank and Wolfe (1956) in 1956. It has been studied in Demyanov and Rubinov (1970) for general convex optimization. In the classical paper Levitin and B. Polyak (1966) convergence and convergence rate was established under various assumption on the input data. Recently CG method got a great deal of attention due to its efficiency for important applications; see Freund and Grigas (2016); Garber and Hazan (2014,2015) and references therein.

The method for simultaneous solution primal and the dual convex optimization problems was introduced Polyak (1966). It is, in fact, a version of the feasible direction method.

Other feasible directions methods were introduced in the early 1960s by Zoutendijk (1960) and independently by S. Zuchovitsky et al. in both finite and Gilbert spaces Zukhovitsky et al. (1963,a, 1965); see also Polak (1971).

Chapter 6
Self-Concordant Functions and IPM Complexity

6.0 Introduction

The introduction by N. Karmarkar of his projective transformation method for LP in 1984; finding by P. Gill et al. (1986) of a connection between Karmarkar's and Newton log-barrier methods; rediscovery of the affine scaling (AS) method by Barnes (1986) and independently by Vanderbei et al. (1986); and best complexity results by Renegar (1988) and Gonzaga (1989), as well as strong numerical results produced by Lustig et al. (1991, 1992, 1994), Adler et al. (1989) and Adler and Monteiro (1991), lead to a new field in modern optimization – the IPMs.

Thousands of papers were published in a relatively short time. The IPMs became the mainstream in modern optimization. The same complexity results were proven time and again, but what was lacking is a unified and general viewpoint on all these developments as well as understanding of the roots of the IPMs.

In the late 1980s, Yu. Nesterov and A. Nemirovski developed their remarkable SC theory, which allowed understanding the IPMs from a unique and general point of view. They also generalized the IPMs for a wide class of convex and semidefinite optimization problems; see Nesterov and Nemirovski (1994); Nesterov (2004).

In the heart of their theory is the notion of self-concordant functions.

A strictly convex and three times differentiable function is self-concordant if its LF invariant is bounded.

Boundedness of the LF invariant leads to the basic differential inequality, four sequential integration of which produces the basic properties of SC functions.

These properties along with properties of Newton decrement are in the heart of globally convergent damped Newton method for minimization of SC function.

All these results were covered in the first part of the chapter.

In the second part, the properties of SC barrier were used for establishing IPM complexity for both the LP and the QP calculation, as well as for QP with quadratic constraints, semidefinite optimization, and conic programming with Lorentz cone.

R. A. Polyak, *Introduction to Continuous Optimization*, Springer Optimization and Its Applications 172, https://doi.org/10.1007/978-3-030-68713-7_6

In the third part of the chapter, we consider the primal–dual aspects of the IPMs; see Wright (1997). Particular attention is given to the primal–dual predictor–corrector (PDPC) algorithm, which happened to be most efficient for LP calculation.

We conclude the chapter by considering the convergence rate and the complexity bound of the PDPC algorithm.

6.1 LF Invariant and SC Functions

Let us consider a closed convex function $F \in C^3$ defined on an open convex set $\operatorname{dom} F$. We fix $x \in \operatorname{dom} F$, a direction $u \in \mathbb{R}^n \setminus \{0\}$, and consider the restriction

$$f(t) = F(x + tu)$$

of F on the direction u, which is defined on $\operatorname{dom} f = \{t : x + tu \in \operatorname{dom} F\}$. Along with f, we consider its derivatives

$$\begin{aligned} f'(t) &= \langle \nabla F(x+tu), u \rangle , \\ f''(t) &= \langle \nabla^2 F(x+tu)u, u \rangle , \\ f'''(t) &= \langle \nabla^3 F(x+tu)[u]u, u \rangle , \end{aligned}$$

where ∇F is the gradient, $\nabla^2 F$ is Hessian of F, and

$$\nabla^3 F(x)[u] = \lim_{\tau \to +0} \tau^{-1} \left[\nabla^2 F(x + \tau u) - \nabla^2 F(x) \right] .$$

Then,

$$\begin{aligned} DF(x)[u] &:= \langle \nabla F(x), u \rangle = f'(0) , \\ D^2 F(x)[u,u] &:= \langle \nabla^2 F(x)u, u \rangle = f''(0) , \\ D^3 F(x)[u,u,u] &:= \langle \nabla^3 F(x)[u]u, u \rangle = f'''(0) . \end{aligned}$$

Definition 6.1. Function F is self-concordant (SC) if, for any $x \in \operatorname{dom} F$ and any $u \in \mathbb{R}^n \setminus \{0\}$, there exists $M > 0$ such that the following inequality

$$|D^3 F(x)[u,u,u]| \leq M \left(D^2 F(x)[u,u] \right)^{3/2} \tag{6.1}$$

holds.

In other words function F is self-concordant if for any $x \in \operatorname{dom} F$ there is $M > 0$ that for the restriction f the following inequality

$$\operatorname{LFINV}(f) = |f'''(t)| f''(t)^{-3/2} \leq M \tag{6.2}$$

holds for any $t \in \operatorname{dom} f$ and any $u \in \mathbb{R}^n / \{0\}$.

Let us consider the log-barrier function $F(x) = -\ln x$. Then $\operatorname{dom} F = \{x : x > 0\}$, $F'(x) = -x^{-1}$, $F''(x) = x^{-2}$, $F'''(x) = -2x^{-3}$ and

$$|F'''(x)| \leq 2\left(F''(x)\right)^{3/2}.$$

Therefore, the log-barrier function F is self-concordant with the constant $M = 2$, that is,

$$\mathrm{LFINV}(F) \leq 2. \tag{6.3}$$

We will use this constant in the definition of self-concordant functions.

The following function,

$$g(t) = \langle \nabla^2 F(x+tu)u, u\rangle^{-1/2} = \left(f''(t)\right)^{-1/2}$$

defined on $\operatorname{dom} f$, is critical in the SC theory.

Note

$$|g'(t)| = |-\frac{1}{2}f'''(t)f''(t)^{-3/2}| = \frac{1}{2}\mathrm{LFINV}(f). \tag{6.4}$$

The other important component of the SC theory is two local scaled norms of a vector $u \in \mathbb{R}^n$. The first is defined at each point $x \in \operatorname{dom} F$ by the formula

$$\|u\|_x = \langle \nabla^2 F(x)u, u\rangle^{1/2}.$$

In what follows, we assume that Hessian $\nabla^2 F(x)$ is positive definite on $\operatorname{dom} F$. Therefore, for any $x \in \operatorname{dom} F$, we can define the second scaled norm:

$$\|\mathrm{v}\|_x^* = \left\langle \left(\nabla^2 F(x)\right)^{-1} \mathrm{v}, \mathrm{v}\right\rangle^{1/2}.$$

The following Cauchy–Schwartz (CS) inequality for scaled norms will be useful later.

Let us consider a positive definite symmetric $n \times n$ matrix A, then $A^{1/2}$ exists and

$$\begin{aligned}
|\langle u, \mathrm{v}\rangle| &= \left|\left\langle A^{1/2}u, A^{-1/2}\mathrm{v}\right\rangle\right| \leq \left\|A^{1/2}u\right\| \left\|A^{-1/2}\mathrm{v}\right\| \\
&= \left\langle A^{1/2}u, A^{1/2}u\right\rangle^{1/2} \left\langle A^{-1/2}\mathrm{v}, A^{-1/2}\mathrm{v}\right\rangle^{1/2} \\
&= \langle Au, u\rangle^{1/2} \left\langle A^{-1}\mathrm{v}, \mathrm{v}\right\rangle^{1/2} = \|u\|_A \, \|\mathrm{v}\|_{A^{-1}} = \|u\|_A \|\mathrm{v}\|_A^*.
\end{aligned}$$

In particular, taking $A = \nabla^2 F(x)$, we obtain for any $u, \mathrm{v} \in \mathbb{R}^n$ the following CS inequality:

$$|\langle u, \mathrm{v}\rangle| \leq \|u\|_x \, \|\mathrm{v}\|_x^*.$$

For $F = -\ln t$ from 6.3 and 6.4 follows

$$|g'(t)| \leq 1, \quad \forall t \in \operatorname{dom} f. \tag{6.5}$$

We will see later that all basic properties of SC function can be established by four sequential integrations of differential inequality (6.5).

The exception is the following lemma, which we present without proof.

Lemma 6.1. *A function F is self-concordant if and only if for any $x \in \text{dom}F$ and any $u_1, u_2, u_3 \in \mathbb{R}^n \setminus \{0\}$ we have*

$$\left|D^3F(x)\left[u_1,u_2,u_3\right]\right| \le 2\prod_{i=1}^{3}\|u_i\|_x ,$$

where $D^3F(x)[u_1,u_2,u_3] = \langle \nabla^3 F(x)[u_1]u_2, u_3\rangle$.

The following theorem establishes one of the most important facts about SC functions, namely, any SC function is a barrier function on $\text{dom}\,F$. The opposite statement is, generally speaking, not true, i.e., not every barrier function is self-concordant. For example, the hyperbolic barrier $F(x) = x^{-1}$ defined on $\text{dom}\,F = \{x : x > 0\}$ is not a SC function.

Exercise 6.1.

1. Why is $F(x) = x^{-1}$ not a SC function?
2. Are $f_1(x) = e^x$ and $f_2(x) = |x|^p$, $\rho > 2$ SC functions?

Theorem 6.1. *Let F be a closed convex function on an open $\text{dom}\,F$, then, for any $\bar{x} \in \partial(\text{dom}\,F)$ and any sequence $\{x_s\}_{s\in N} \subset \text{dom}\,F$ such that $x_s \to \bar{x}$, we have*

$$\lim_{s\to\infty} F(x_s) = \infty\,. \tag{6.6}$$

Proof. For any given $x_0 \in \text{dom}\,F$, it follows from the convexity of F that

$$F(x_s) \ge F(x_0) + \langle \nabla F(x_0), x_s - x_0\rangle\,.$$

So, the sequence $\{F(x_s)\}_{s\in N}$ is bounded from below. If (6.6) is not true, then the sequence $\{F(x_s)\}_{s\in N}$ is bounded from above. Therefore, it has a limit point $\bar{F}$. Without loss of generality, we can assume that $w_s = (x_s, F(x_s)) \to \bar{w} = (\bar{x}, \bar{F})$.

Since F is a closed function, we have $\bar{w} \in \text{epi}F$, but it is impossible because $\bar{x} \notin \text{dom}\,F$. Therefore, for any sequence $\{x_s\}_{s\in N} \subset \text{dom}\,F : \lim_{s\to\infty} x_s = \bar{x} \in \partial(\text{dom}\,F)$, the barrier property (6.6) holds. It means that F is a barrier function on $\text{dom}\,F$. □

Exercise 6.2. Let $a > 0$ and let $\gamma(t)$ be a differentiable function defined on $[0,a)$ such that $\frac{d\gamma}{dt} \le 2(\gamma(t))^{3/2}$, $t \in [0,a)$, and $\gamma(0) = 0$. Then $\gamma(t) = 0$, $\forall t \in [0,a)$.

Another important property of the SC function F is given by the following theorem.

Theorem 6.2. *Let F be a SC function, and let $\text{dom}\,F$ contains no straight line, then Hessian $\nabla^2 F(x)$ is positive definite at any $x \in \text{dom}\,F$.*

Proof. Let us assume the opposite: there exist $x \in \operatorname{dom} F$ and $u \in \mathbb{R}^n$, $u \neq 0$, such that

$$\langle \nabla^2 F(x)u, u\rangle = f''(0) = 0\,.$$

Then, for a nonnegative function

$$\varphi(t) = f''(t)$$

we have $\varphi(0) = 0$. It follows from (6.2) with $M = 2$ that $\varphi'(t) \leq 2\varphi(t)^{3/2}$; therefore, due to Exercise 6.2 for any given $0 < a < \infty$, we have $\varphi(t) = 0$, $t \in [0, a)$, and hence $f(t)$ is linear. We assume now that there exists $\bar{t}$ such that $\bar{x} = x + \bar{t}u \in \partial(\operatorname{dom} F)$ and consider a sequence $\{t_s\}_{s\in N}$ such that $t_s \to \bar{t}$. Then,

$$(x + t_s u,\, F(x + t_s u)) \subset \operatorname{epi} F\,.$$

Without restricting the generality, we can assume that $F(x + t_s u) \to \bar{F}$. Then, keeping in mind that F is a closed function, we obtain $(\bar{x}, \bar{F}) \in \operatorname{epi} F$. On the other hand, $(\bar{x}, \bar{F}) \notin \operatorname{epi} F$ because $\operatorname{dom} F$ is an open set. Therefore, $x + tu \in \operatorname{dom} F$ for any $t \geq 0$. Repeating the considerations for direction $-u$, we obtain

$$x - tu \in \operatorname{dom} F \quad \text{for any } \; t \geq 0\,.$$

Therefore, the straight line

$$h = \{y = x + tu,\ -\infty < t < \infty\}$$

is contained in $\operatorname{dom} F$, which contradicts the assumption of the theorem. □

From this point on, we assume that $\operatorname{dom} F$ does not contain a straight line. Thus, $\nabla^2 F(x) \succ 0$ for any $x \in \operatorname{dom} F$. Therefore, for any $x \in \operatorname{dom} F$, and any $u \in \mathbb{R}^n \setminus \{0\}$, we have

$$\langle \nabla^2 F(x)u, u\rangle = \|u\|_x^2 > 0\,.$$

Also,

$$g(t) = \langle \nabla^2 F(x + tu)u, u\rangle^{-1/2} = \|u\|_{x+tu}^{-1} > 0\,, \quad t \in \operatorname{dom} f\,. \tag{6.7}$$

6.2 Basic Properties of SC Functions

In this section we show that practically all basic properties of a SC function can be obtained by four sequential integrations of (6.5).

First Integration

From (6.5), keeping in mind that $f''(t) > 0$ for any $s > 0$, we obtain

$$-\int_0^s dt \leq \int_0^s d\left(f''(t)^{-1/2}\right) \leq \int_0^s dt\,.$$

By taking the integrals, we have

$$f''(0)^{-1/2} - s \le f''(s)^{-1/2} \le f''(0)^{-1/2} + s \tag{6.8}$$

or

$$\left(f''(0)^{-1/2} + s\right)^{-2} \le f''(s) \le \left(f''(0)^{-1/2} - s\right)^{-2}. \tag{6.9}$$

The left inequality in (6.9) holds for all $s \ge 0$, while the right inequality holds only for $0 \le s < f''(0)^{-1/2}$.

Let us consider $x, y \in \operatorname{dom} F$, $y \ne x$, $u = y - x$, and $y(s) = x + s(y - x)$, $0 \le s \le 1$, so $y(0) = x$ and $y(1) = y$. Therefore,

$$f''(0) = \langle \nabla^2 F(x)(y-x), y-x \rangle = \|y-x\|_x^2$$

and

$$f''(0)^{1/2} = \|y-x\|_x .$$

Also,

$$f''(1) = \langle \nabla^2 F(y)(y-x), y-x \rangle = \|y-x\|_y^2$$

and

$$f''(1)^{1/2} = \|y-x\|_y .$$

From (6.8), for $s = 1$, we obtain

$$f''(0)^{-1/2} - 1 \le f''(1)^{-1/2} \le f''(0)^{-1/2} + 1$$

or

$$\frac{1}{\|y-x\|_x} - 1 \le \frac{1}{\|y-x\|_y} \le \frac{1}{\|y-x\|_x} + 1 .$$

From the right inequality, we have

$$\|y-x\|_y \ge \frac{\|y-x\|_x}{1 + \|y-x\|_x}. \tag{6.10}$$

If $\|y-x\|_x < 1$, then from the left inequality, we have

$$\|y-x\|_y \le \frac{\|y-x\|_x}{1 - \|y-x\|_x}. \tag{6.11}$$

It follows from (6.5) that

$$g(t) \geqslant g(0) - |t|, \quad t \in \operatorname{dom} f . \tag{6.12}$$

For $x + tu \in \operatorname{dom} F$ from (6.7), we have $g(t) > 0$. It follows from Theorem 6.1 that $F(x+tu) \to \infty$ when $x + tu \to \partial(\operatorname{dom} F)$. Therefore, $\langle \nabla^2 F(x+tu)u, u \rangle$ cannot be bounded when $x + tu \to \partial(\operatorname{dom} F)$. From (6.7) follows that $g(t) \to 0$ when $x + tu \to \partial(\operatorname{dom} F)$. From (6.12) for any $t : g(0) - |t| > 0$ follows

$$(-g(0), g(0)) = \left(-\|u\|_x^{-1}, \|u\|_x^{-1}\right) \subset \operatorname{dom} f.$$

Therefore, set

$$E^0(x,1) = \left\{y = x + tu : t^2 \|u\|_x^2 < 1\right\}$$

belongs to $\operatorname{dom} F$. In other words, ellipsoid

$$E(x,r) = \left\{y \in \mathbb{R}^n : \|y-x\|_x^2 \le r\right\},$$

which is called *Dikin's ellipsoid*, belongs to $\operatorname{dom} F$ for any $x \in \operatorname{dom} F$ and any $r < 1$.

One can expect that for any $y \in E(x,r)$ Hessians $\nabla^2 F(x)$ and $\nabla^2 F(y)$ are "close" enough if $0 < r < 1$ is small enough. The second integration allows to establish the correspondent bounds.

Second Integration

Let us fix $x \in \operatorname{dom} F$. We consider $y \in \operatorname{dom} F$ such that $y \neq x$ and direction $u \in \mathbb{R}^n \setminus \{0\}$. Let $y(t) = x + t(y-x)$, then for $t \ge 0$ and $y(t) \in \operatorname{dom} F$ we define

$$\psi(t) = \|u\|_{y(t)}^2.$$

From Lemma 6.1 follows

$$|\psi'(t)| = D^3F(y(t))[y-x,u,u] \le 2\|y-x\|_{y(t)} \|u\|_{y(t)}^2 = 2\|y-x\|_{y(t)} \psi(t).$$

First of all, $\|y(t)-x\|_x \le \|y-x\|_x$ for any $t \in [0,1]$. Keeping in mind that $y - x = t^{-1}(y(t)-x)$ and assuming $\|y-x\|_x < 1$ from (6.11) follows

$$\begin{aligned} |\psi'(t)| &\le \frac{2}{t}\|y(t)-x\|_{y(t)} \psi(t) \le \frac{2}{t}\frac{\|y(t)-x\|_x}{1-\|y(t)-x\|_x}\psi(t) \\ &\le 2\frac{\|y-x\|_x}{1-t\|y-x\|_x}\psi(t). \end{aligned}$$

Therefore, for $0 < t < \|y-x\|_x^{-1}$ we have

$$\frac{|\psi'(t)|}{\psi(t)} \le \frac{2\|y-x\|_x}{1-t\|y-x\|_x}.$$

By integration of the above inequality, for any $0 < s < \|y-x\|_x^{-1}$ we obtain

$$-2\int_0^s \frac{\|y-x\|_x}{1-t\|y-x\|_x}dt \le \int_0^s \frac{\psi'(t)}{\psi(t)}dt \le 2\int_0^s \frac{\|y-x\|_x}{1-t\|y-x\|_x}dt;$$

hence,

$$2\ln(1-s\|y-x\|_x) \le \ln\psi(s) - \ln\psi(0) \le -2\ln(1-s\|y-x\|_x).$$

For $s=1$ we have

$$\psi(0)\,(1-\|y-x\|_x)^2 \le \psi(1) \le \psi(0)\,(1-\|y-x\|_x)^{-2}\,.$$

Keeping in mind $\psi(0)=\langle\nabla^2F(x)u,u\rangle$, $\psi(1)=\langle\nabla^2F(y)u,u\rangle$, we obtain

$$(1-\|y-x\|_x)^2\,\langle\nabla^2F(x)u,u\rangle \le \langle\nabla^2F(y)u,u\rangle \le (1-\|y-x\|_x)^{-2}\,\langle\nabla^2F(x)u,u\rangle$$

for any $u\in\mathbb{R}^n\setminus\{0\}$. Therefore, the following matrix inequality holds

$$(1-\|y-x\|_x)^2\,\nabla^2F(x) \preccurlyeq \nabla^2F(y) \preccurlyeq \nabla^2F(x)\,(1-\|y-x\|_x)^{-2}\,, \tag{6.13}$$

where $A\succcurlyeq B$ means that $A-B$ is nonnegative definite. Note that (6.13) takes place for any fixed $x\in\operatorname{dom}F$ and any given $y\in\operatorname{dom}F$.

It is often important to know the upper and lower bounds for the matrix

$$G=\int_0^1 \nabla^2F(x+\tau(y-x))d\tau\,. \tag{6.14}$$

We consider (6.13) for $y_\tau := x+\tau(y-x)$. From the left inequality in (6.13), we have

$$G=\int_0^1 \nabla^2F(x+\tau(y-x))d\tau \succcurlyeq \nabla^2F(x)\int_0^1 (1-\tau\,\|y-x\|_x)^2\,d\tau\,.$$

Therefore, for $r=\|y-x\|_x<1$, we have

$$G\succcurlyeq \nabla^2F(x)\int_0^1(1-\tau r)^2d\tau=\nabla^2F(x)\left(1-r+\frac{r^2}{3}\right). \tag{6.15}$$

On the other hand, from the right inequality in (6.13), we obtain

$$G\preccurlyeq \nabla^2F(x)\int_0^1(1-\tau r)^{-2}d\tau=\nabla^2F(x)\frac{1}{1-r}\,, \tag{6.16}$$

i.e., for any $x\in\operatorname{dom}F$, the following inequalities hold:

$$\left(1-r+\frac{r^2}{3}\right)\nabla^2F(x)\preccurlyeq G\preccurlyeq \frac{1}{1-r}\nabla^2F(x)\,. \tag{6.17}$$

The first two integrations produced two very important facts:

1. At any point $x\in\operatorname{dom}F$ Dikin's ellipsoid

$$E(x,r)=\{y\in\mathbb{R}^n:\|y-x\|_x^2\le r\}$$

belongs to $\operatorname{dom}F$, for any $0\le r<1$.
2. From (6.13) for any $x\in\operatorname{dom}F$ and any $y\in E(x,r)$ follows

$$(1-r)^2\nabla^2 F(x) \preccurlyeq \nabla^2 F(y) \preccurlyeq \frac{1}{(1-r)^2}\nabla^2 F(x), \tag{6.18}$$

for any $0 \le r < 1$.

The bounds for the gradient $\nabla F(x)$, which is a monotone operator in $\mathbb{R}^n$, we establish by integration (6.9).

Third Integration

From (6.9), for $0 \le t \le f(0)^{-1/2} = \|y-x\|_x^{-1}$, we obtain

$$\int_0^s \left(f''(0)^{-1/2}+t\right)^{-2} dt \le \int_0^s f''(t)dt \le \int_0^s \left(f''(0)^{-1/2}-t\right)^{-2} dt,$$

or

$$f'(0) + f''(0)^{1/2}\left(1-\left(1+sf''(0)^{1/2}\right)^{-1}\right)$$
$$\le f'(s) \le f'(0) - f''(0)^{1/2}\left(1-\left(1-sf''(0)^{1/2}\right)^{-1}\right). \tag{6.19}$$

From the right inequality (6.19), for $s=1$, we have

$$f'(1)-f'(0) \le -f''(0)^{1/2}\left(1-\frac{1}{1-f''(0)^{1/2}}\right) = \frac{f''(0)}{1-f''(0)^{1/2}}.$$

Recalling the formulas for $f'(0)$, $f'(1)$, $f''(0)$, and $f''(1)$ for any x and $y \in \operatorname{dom} F$, we have

$$\langle \nabla F(y) - \nabla F(x), y-x\rangle \le \frac{\|y-x\|_x^2}{1-\|y-x\|_x}. \tag{6.20}$$

From the left inequality in (6.19), for $s=1$, we obtain

$$f'(1)-f'(0) \ge f''(0)^{1/2}\left(1-\frac{1}{1+f''(0)^{1/2}}\right) = \frac{f''(0)}{1+f''(0)^{1/2}}$$

or

$$\langle \nabla F(y) - \nabla F(x), y-x\rangle \ge \frac{\|y-x\|_x^2}{1+\|y-x\|_x}. \tag{6.21}$$

Now we establish the bounds for $F(y)-F(x)$ by integration (6.19).

Fourth Integration

The strictly convex function $\omega : (-1,\infty) \to [0,\infty]$ defined by the formula $\omega(t) = t - \ln(1+t)$ will play an important role later.

Along with $\omega(t)$, let us consider its LF conjugate

$$\omega^*(s) = \sup_{t>-1}\{st - \omega(t)\} = -s - \ln(1-s) = \omega(-s)\,.$$

Taking the integral of the right inequality (6.19), we obtain

$$\begin{aligned} f(s) &\le f(0) + f'(0)s - f''(0)^{1/2}s - \ln\left(1 - f''(0)^{1/2}s\right) \\ &= f(0) + f'(0)s + \omega^*\left(f''(0)^{1/2}s\right) \\ &= U(s)\,. \end{aligned} \tag{6.22}$$

In other words, $U(s)$ is an upper bound for $f(s)$ on the interval $[0, f''(0)^{-1/2})$. Recall that $f''(0)^{-1/2} = \|y-x\|_x^{-1} > 1$. For $s = 1$, we have

$$f(1) - f(0) \le f'(0) + \omega^*\left(f''(0)^{1/2}\right) = f'(0) + \omega^*\left(\|y-x\|_x\right)\,. \tag{6.23}$$

Keeping in mind $f(0) = F(x)$, $f(1) = F(y)$ from (6.23), we obtain

$$F(y) - F(x) \le \langle \nabla F(x), y-x\rangle + \omega^*\left(\|y-x\|_x\right)\,. \tag{6.24}$$

Integration of the left inequality in (6.19) leads to the lower bound $L(s)$ for $f(s)$, namely,

$$\begin{aligned} f(s) &\ge f(0) + f'(0)s + f''(0)^{1/2}s - \ln\left(1 + f''(0)^{1/2}s\right) \\ &= f(0) + f'(0)s + \omega\left(f''(0)^{1/2}s\right) \\ &= L(s)\,,\ \forall s \ge 0\,. \end{aligned} \tag{6.25}$$

For $s = 1$ we have

$$f(1) - f(0) \geqslant f'(0) + \omega\left(f''(0)^{1/2}\right)$$

or

$$F(y) - F(x) \ge \langle \nabla F(x), y-x\rangle + \omega\left(\|y-x\|_x\right)\,. \tag{6.26}$$

We conclude this section by considering the existence of the minimizer

$$x^* = \arg\min\{F(x) : x \in \operatorname{dom} F\} \tag{6.27}$$

for a self-concordant function F.

As always we assume that $\operatorname{dom} F$ does not contain a straight line. By Theorem 6.2 this assumption guarantees that Hessian $\nabla^2 F$ is positive definite, that is, F is strictly convex on the open set $\operatorname{dom} F$. Therefore, the existence of x^* is not a direct consequence of our assumption and will be proven later.

On the other hand, the assumption guarantees existence of the local norm $\|\mathrm{v}\|_x^* = \left\langle \left(\nabla^2 F(x)\right)^{-1}\mathrm{v}, \mathrm{v}\right\rangle^{1/2}$ for any $\mathrm{v} \in \mathbb{R}^n$ and any $x \in \operatorname{dom} F$. For $\mathrm{v} = \nabla F(x)$, we obtain the following scaled norm of the gradient $\nabla F(x)$:

$$\lambda(x) = \langle \nabla^2 F(x)^{-1} \nabla F(x), \nabla F(x) \rangle^{1/2} = \|\nabla F(x)\|_x^*,$$

which is Newton decrement (see Chapter 3). It plays an important role in Newton's method in general and in the SC theory in particular.

Theorem 6.3. *If $\lambda(x) < 1$ for some $x \in \operatorname{dom} F$ and $\operatorname{dom} F$ does not contain a straight line, then the minimizer x^* in (6.27) exists.*

Proof. Let $\mathrm{v} = \nabla F(x)$, x and $y \in \operatorname{dom} F$ and $u = y - x$, then it follows from CS inequality $|\langle u, \mathrm{v} \rangle| \leq \|\mathrm{v}\|_x^* \|u\|_x$ that

$$|\langle \nabla F(x), y - x \rangle| \leq \|\nabla F(x)\|_x^* \|y - x\|_x . \tag{6.28}$$

From (6.26), (6.28), and the formula for $\lambda(x)$, we have

$$F(y) - F(x) \geq -\lambda(x) \|y - x\|_x + \omega(\|y - x\|_x) .$$

Therefore, for any $y \in \mathscr{L}(x) = \{y \in \mathbb{R}^n : F(y) \leq F(x)\}$, we obtain

$$\omega(\|y - x\|_x) \leq \lambda(x) \|y - x\|_x ,$$

that is,

$$\|y - x\|_x^{-1} \omega(\|y - x\|_x) \leq \lambda(x) < 1 .$$

By the definition of ω,

$$1 - \frac{1}{\|y - x\|_x} \ln(1 + \|y - x\|_x) \leq \lambda(x) < 1 .$$

The function $\tau^{-1}\omega(\tau) = 1 - \tau^{-1}\ln(1+\tau)$ is monotone increasing in $\tau > 0$. Therefore, for a given $0 < \lambda(x) < 1$, the equation

$$1 - \lambda(x) = \tau^{-1} \ln(1 + \tau)$$

has a unique root $\bar{\tau} > 0$. Thus, for any $y \in \mathscr{L}(x)$, we have

$$\|y - x\|_x \leq \bar{\tau},$$

that is, the sublevel set $\mathscr{L}(x)$ at $x \in \operatorname{dom} F$ is bounded and closed due to the continuity of F. The existence of the minimizer x^* follows from Weierstrass theorem. The minimizer x^* is unique due to the strict convexity of F. □

The theorem presents a remarkable result: a local condition $\lambda(x) < 1$ guarantees the existence of x^*, which is a global property of F on the entire $\operatorname{dom} F$.

Let us briefly summarize the basic properties of the SC functions we established so far.

1. The SC function F is a barrier function on $\operatorname{dom} F$ and, if $\operatorname{dom} F$ does not contain a straight line, then Hessian $\nabla^2 F(x)$ is positive definite at any $x \in \operatorname{dom} F$.

2. For any $x \in \operatorname{dom} F$ and $0 < r < 1$, there is a Dikin's ellipsoid

$$E(x,r) = \left\{ y : \|y - x\|_x^2 \leq r \right\} \subset \operatorname{dom} F .$$

3. For any $x \in \operatorname{dom} F$ and small enough $0 < r < 1$, the function F is almost quadratic inside of the Dikin's ellipsoid $E(x,r)$ due to the bounds (6.18).
4. The gradient ∇F is a strictly monotone operator on $\operatorname{dom} F$ with upper and lower monotonicity bounds given by (6.20) and (6.21).
5. For any $x \in \operatorname{dom} F$ and given direction $u = y - x$, the restriction $f(s) = F(x + s(y - x))$ is bounded by $U(s)$ and $L(s)$ (see (6.22) and (6.25)).
6. Condition $0 < \lambda(x) < 1$ at any $x \in \operatorname{dom} F$ guarantees existence of unique minimizer x^* on $\operatorname{dom} F$.

It is quite remarkable that practically all nice and important properties of SC functions follow from a single differential inequality (6.5) – the boundedness of LFINV(f).

6.3 Newton's Method for Minimization of SC Functions

In this section, we consider Newton's method for finding x^* and estimating its global complexity. The driving force for this analysis is the upper and lower bounds of the restriction $f(s) = F(x + su)$ at any given $x \in \operatorname{dom} F$ in the direction $u = y - x$, that is,

$$\underset{s \geq 0}{\mathscr{L}(s)} \leq f(s) \leq \underset{0 \leq s \leq f''(0)^{-(1/2)}}{U(s)} . \tag{6.29}$$

Recall that for $x \in \operatorname{dom} F$ and given direction $u = y - x$, we have

$$\begin{aligned} &(a) \quad f(0) = F(x), \\ &(b) \quad f'(0) = \langle \nabla F(x), u \rangle, \\ &(c) \quad f''(0) = \langle \nabla^2 F(x) u, u \rangle = \|u\|_x^2 . \end{aligned} \tag{6.30}$$

If $x \neq x^*$, then there exists a direction, in which $F(x)$ is decreasing from x, that is, there exists $y \in \operatorname{dom} F$ such that $f'(0) = \langle \nabla F(x), u \rangle < 0$. We retain the assumption that $\operatorname{dom} F$ does not contain a straight line; therefore,

$$f''(0) = \|y - x\|_x^2 = d^2 > 0 \text{ for any } x, y \in \operatorname{dom} F \text{ and } y \neq x .$$

We would like to estimate the reduction of restriction $f(s) = F(x + s(y - x))$, as a result of one Newton step with $s = 0$ as a starting point.

The upper bound

$$U(s) = f(0) + f'(0)s - ds - \ln(1 - ds)$$

for $f(s)$ is a strongly convex function on $[0, d^{-1})$. We have $U'(0) = f'(0) < 0$ and $U'(s) \to \infty$ for $s \to d^{-1}$. Therefore, the equation

$$U'(s) = f'(0) - d + d(1 - ds)^{-1} = 0 \tag{6.31}$$

has a solution $\bar{s} \in [0, d^{-1})$, which is the unconstrained minimizer for $U(s)$.

From (6.31) follows

$$\bar{s} = -f'(0)d^{-2}\left(1 - f'(0)d^{-1}\right)^{-1} = \Delta(1+\lambda)^{-1},$$

where $\Delta = -f'(0)d^{-2}$ and $\lambda = -f'(0)d^{-1} < 1$. On the other hand, the unconstrained minimizer $\bar{s}$ is a result of one damped Newton step for finding $\min_{s \geq 0} U(s)$ with step length $t = (1+\lambda)^{-1}$ from $s = 0$ as a starting point. It is easy to see that

$$U\left((1+\lambda)^{-1}\Delta\right) = f(0) - \omega(\lambda).$$

From the right inequality in (6.29) follows

$$f(\bar{s}) = f\left((1+\lambda)^{-1}\Delta\right) \leq f(0) - \omega(\lambda). \tag{6.32}$$

Keeping in mind (6.30b) and (6.30c)) and taking the Newton direction $u = y - x = -(\nabla^2 F(x))^{-1}\nabla F(x)$, we obtain

$$\Delta = -\frac{f'(0)}{f''(0)} = -\frac{\langle \nabla F(x), u\rangle}{\langle \nabla^2 F(x)u, u\rangle} = 1.$$

In view of (6.30a)), we can rewrite (6.32) as follows:

$$F\left(x - (1+\lambda)^{-1}\nabla^2 F(x)^{-1}\nabla F(x)\right) \leq F(x) - \omega(\lambda). \tag{6.33}$$

In other words, finding an unconstrained minimizer of the upper bound $U(s)$ is equivalent to one damped Newton step

$$x_{s+1} = x_s - (1+\lambda(x_s))^{-1}\left(\nabla F(x_s)\right)^{-1}\nabla F(x_s) \tag{6.34}$$

for minimization of $F(x)$ on $\operatorname{dom} F$. Moreover, our considerations are independent from the starting point $x_0 \in \operatorname{dom} F$. Therefore, for a starting point $x_0 \in \operatorname{dom} F$ and any $s \geq 1$, we have

$$F(x_{s+1}) \leq F(x_s) - \omega(\lambda). \tag{6.35}$$

The last bound is universal, it is true for any $x \in \operatorname{dom} F$ as a starting point.

We consider Newton direction $u = -\nabla^2 F(x)^{-1}\nabla F(x)$ and restriction $f(s) = F(x + su)$, then

$$\begin{aligned}\lambda \equiv \lambda(x) &= -f'(0)f''(0)^{-1/2} \\ &= -\frac{\langle \nabla F(x), u\rangle}{\langle \nabla^2 F(x)u, u\rangle^{1/2}} \\ &= \langle \nabla F(x)^{-1}\nabla F(x), \nabla F(x)\rangle^{1/2} \\ &= \|\nabla F(x)\|_x^* .\end{aligned}$$

For the sequence $\{x_s\}_{s\in N}$ generated by (6.34), we have $0 < \lambda(x_s) < 1$. Function $\omega(t) = t - \ln(1+t)$ is monotone increasing in $t \geq 0$. Therefore, for a small $\beta > 0$ and $1 > \lambda(x) \geq \beta$, each damped Newton step (6.34) reduces $F(x)$ by a constant $\omega(\beta)$ (see (6.35)).

Therefore, for a given starting point $x_0 \in \operatorname{dom} F$, the number of damped Newton steps required for finding x^* is bounded by

$$N \leq (\omega(\beta))^{-1}(F(x_0) - F(x^*)) .$$

The bound (6.35), however, can be substantially improved for

$$x \in S(x^*, r) = \{x \in \operatorname{dom} F : F(x) - F(x^*) \leq r\},$$

when $0 < r < 1$ is small enough.

Let us consider the lower bound

$$L(s) = f(0) + f'(0)s + ds - \ln(1+ds) \leq f(s), \quad s \geq 0 .$$

Function $L(s)$ is strictly convex in $s \geq 0$; then, for $0 < \lambda = -f'(0)(f''(0))^{-1} = -f'(0)d^{-1} < 1$, we have

$$L'\left(\Delta(1-\lambda)^{-1}\right) = 0 .$$

Therefore,

$$\bar{\bar{s}} = \Delta(1-\lambda)^{-1} = \arg\min\{L(s) \mid s \geq 0\}$$

and

$$L(\bar{\bar{s}}) = f(0) - \omega(-\lambda) .$$

Along with $\bar{s}$ and $\bar{\bar{s}}$, we consider (see Fig. 6.1)

$$s^* = \operatorname{argmin}\{f(s) \mid s \geq 0\} .$$

For a small $0 < r < 1$ and $x \in S(x^*, r)$, we have $f(0) - f(s^*) < 1$; hence, $f(0) - f(\bar{s}) < 1$. The relative progress per step is more convenient to measure on the logarithmic scale

$$\kappa = \frac{\ln(f(\bar{s}) - f(s^*))}{\ln(f(0) - f(s^*))} .$$

From $\omega(\lambda) < f(0) - f(s^*) < 1$, we have $-\ln \omega(\lambda) > -\ln(f(0) - f(s^*))$ or $\ln(f(0) - f(s^*)) > \ln \omega(\lambda)$.

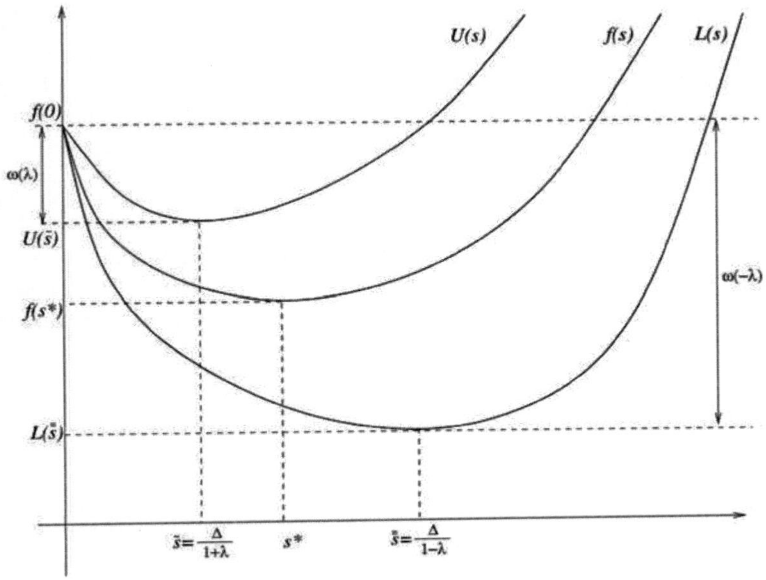

Fig. 6.1 The upper and lower bounds for $f(s)$

Also, $f(\bar{s}) - f(s^*) < f(0) - f(s^*) < 1$, $f(\bar{s}) \leq f(0) - \omega(\lambda)$ and $f(s^*) \geq f(0) - \omega(-\lambda)$. Therefore,

$$f(\bar{s}) - f(s^*) \leq f(0) - \omega(\lambda) - (f(0) - \omega(-\lambda)) = \omega(-\lambda) - \omega(\lambda).$$

Hence,

$$-\ln(f(\bar{s}) - f(s^*)) > -\ln(\omega(-\lambda) - \omega(\lambda))$$

or

$$\ln(f(\bar{s}) - f(s^*)) < \ln(\omega(-\lambda) - \omega(\lambda))$$

and

$$\begin{aligned}\kappa(\lambda) &\leq \frac{\ln(\omega(-\lambda) - \omega(\lambda))}{\ln \omega(\lambda)} \\ &= \frac{\ln\left(-2\lambda + \ln(1+\lambda)(1-\lambda)^{-1}\right)}{\ln(\lambda - \ln(1+\lambda))}.\end{aligned}$$

For $0 < \lambda \leq 0.5$, we have

$$\kappa(\lambda) \leq \frac{\ln\left(\frac{\lambda^3}{3} + \frac{\lambda^5}{5}\right)}{\ln\left(\frac{\lambda^2}{2} - \frac{\lambda^3}{3} + \frac{\lambda^4}{4}\right)}.$$

Therefore, $\lim_{\lambda\to 0}\kappa(\lambda)\le 3/2$. In particular, $\kappa(0.5)\approx 1.25$. Thus, the sequence $\{x_s\}_{s\in N}$ generated by damped Newton method (6.34) with $\lambda(x_s)=0.5$ converges in value with 1.25 Q-superlinear rate, that is, for $\Delta(x_s)=F(x_s)-F(x^*)<1$, we have $\Delta(x_{s+1})\le\Delta(x_s)^{1.25}$. Also, due to $\lim_{k\to\infty}\lambda(x_s)=0$ from some point on, method (6.34) practically turns into classical Newton method

$$x_{s+1}=x_s-\nabla^2 F(x_s)^{-1}\nabla F(x_s)\,, \tag{6.36}$$

which converges with quadratic rate. Instead of waiting for this to happen, there is a way of switching from (6.34) to (6.36) and guarantee that from this point on, only Newton's method (6.36) will be used. Using such a strategy, we can achieve quadratic convergence earlier.

Theorem 6.4. *Let $x\in \operatorname{dom} F$ and Newton decrement*

$$\lambda(x)=\langle\nabla^2 F(x)^{-1}\nabla F(x),\nabla F(x)\rangle^{1/2}<1\,,$$

then,

1. *the point*

$$\hat{x}=x-\nabla^2 F(x)^{-1}\nabla F(x) \tag{6.37}$$

   belongs to $\operatorname{dom} F$*;*
2. *the following bound holds*

$$\lambda(\hat{x})\le\left(\frac{\lambda(x)}{1-\lambda(x)}\right)^2. \tag{6.38}$$

Proof.

1. Let $p=\hat{x}-x=-\nabla^2 F(x)^{-1}\nabla F(x)$, then

$$\begin{aligned}\|p\|_x &= \langle\nabla^2 F(x)p,p\rangle^{1/2}=\langle\nabla F(x),\,\nabla^2 F(x)^{-1}\nabla F(x)\rangle^{1/2}\\ &= \|\nabla F(x)\|_x^*=\lambda(x)=\lambda<1\,;\end{aligned}$$

   therefore, $\hat{x}\in\operatorname{dom} F$.
2. First of all, note that if $A=A^T\succ 0$, $B=B^T\succ 0$ and $A\succcurlyeq B$, then

$$A^{-1}-B^{-1}=-A^{-1}(A-B)B^{-1}\preccurlyeq 0\,.$$

   For $y=\hat{x}$ from the left inequality in (6.13), we obtain

$$\begin{aligned}\lambda(\hat{x}) &= \|\nabla F(\hat{x})\|_{\hat{x}}^*\le(1-\|p\|_x)^{-1}\langle\nabla^2 F(x)^{-1}\nabla F(\hat{x}),\,\nabla F(\hat{x})\rangle^{1/2}\\ &= (1-\|p\|_x)^{-1}\,\|\nabla F(\hat{x})\|_x^*\,.\end{aligned}$$

   Then, from (6.37) follows

$$\nabla^2 F(x)\,(\hat{x}-x)+\nabla F(x)=0;$$

therefore,

$$\nabla F(\hat{x}) = \nabla F(\hat{x}) - \nabla F(x) - \nabla^2 F(x)(\hat{x} - x).$$

From (6.14) follows

$$\nabla F(\hat{x}) - \nabla F(x) = \int_0^1 \left(\nabla^2 F(x + \tau(\hat{x} - x))\right)(\hat{x} - x)d\tau = G(\hat{x} - x).$$

Hence,

$$\nabla F(\hat{x}) = \left(G - \nabla^2 F(x)\right)(\hat{x} - x) = \hat{G}(\hat{x} - x) = \hat{G}p$$

and $\hat{G}^T = \hat{G}$.

Using the CS inequality, we obtain

$$\begin{aligned} \|\nabla F(\hat{x})\|_x^{*2} &= \langle \nabla^2 F(x)^{-1}\hat{G}p, \hat{G}p\rangle = \langle \hat{G}\nabla^2 F(x)^{-1}\hat{G}p, p\rangle \\ &\leq \left\|\hat{G}\nabla^2 F(x)^{-1}\hat{G}p\right\|_x^* \|p\|_x . \end{aligned} \tag{6.39}$$

Let us consider

$$\begin{aligned} \left\|\hat{G}\nabla^2 F(x)^{-1}\hat{G}p\right\|_x^* &= \langle \hat{G}\nabla^2 F(x)^{-1}\hat{G}p, \nabla^2 F(x)^{-1}\hat{G}\nabla^2 F(x)^{-1}\hat{G}p\rangle^{1/2} \\ &= \left\langle H^2\nabla^2 F(x)^{-1/2}\hat{G}p, \nabla^2 F(x)^{-1/2}\hat{G}p\right\rangle^{1/2} \\ &\leq \|H\| \left\langle \nabla^2 F(x)^{-1/2}\hat{G}p, \nabla^2 F(x)^{-1/2}\hat{G}p\right\rangle^{1/2} \\ &= \|H\| \langle \nabla^2 F(x)^{-1}\hat{G}p, \hat{G}p\rangle \\ &= \|H\| \langle \nabla^2 F(x)^{-1}\nabla F(\hat{x}), \nabla F(\hat{x})\rangle^{1/2} \\ &= \|H\| \, \|\nabla F(\hat{x})\|_x^* , \end{aligned}$$

where $H = \nabla^2 F(x)^{-1/2}\hat{G}\nabla F(x)^{-1/2}$.

From (6.39) and the last inequality, we obtain

$$\|\nabla F(\hat{x})\|_x^* \leq \|H\| \, \|p\|_x = \lambda\|H\| . \tag{6.40}$$

It follows from (6.17) that

$$\left(-\lambda + \frac{\lambda^2}{3}\right)\nabla^2 F(x) \preccurlyeq \hat{G} = G - \nabla^2 F(x) \preccurlyeq \frac{\lambda}{1-\lambda}\nabla^2 F(x).$$

Then,

$$\|H\| \leq \max\left\{\frac{\lambda}{1-\lambda}, -\lambda + \frac{\lambda^2}{3}\right\} = \frac{\lambda}{1-\lambda}.$$

From (6.37)–(6.40) follows

$$\lambda^2(\hat{x}) \leq \frac{1}{(1-\lambda)^2}\|\nabla F(\hat{x})\|_x^{*2} \leq \frac{1}{(1-\lambda)^2}\lambda^2\|H\|^2 \leq \frac{\lambda^4}{(1-\lambda)^4}$$

or

$$\lambda(\hat{x}) \leq \frac{\lambda^2}{(1-\lambda)^2}.\square$$

We saw already that $\lambda = \lambda(x) < 1$ is the main ingredient for convergence of damped Newton method (6.34). To retain the same condition for $\lambda(\hat{x})$, it is sufficient to require $\lambda(\hat{x}) \leq \lambda \leq \lambda^2/(1-\lambda)^2$. Function $\lambda/(1-\lambda)^2$ is positive and monotone increasing on $(0,1)$. Therefore, to find an upper bound for λ, it is enough to solve the equation $\lambda/(1-\lambda)^2 = 1$. In other words, for any $\lambda = \lambda(x) < \bar{\lambda} = \frac{3-\sqrt{5}}{2}$, we have $\lambda(\hat{x}) < \lambda < 1$.

D amped Newton method (6.34) follows three major stages in terms of the rate of convergence.

First, for $0 < \lambda < 1$, it reduces the function value by a constant at each step.

Second, for $\lambda > 0$ small enough, it converges with the superlinear rate.

Third, for $\lambda \approx 0$, it converges with quadratic rate.

The properties of SC functions make possible explicit characterization of the Newton area $N(x^*, \beta)$, that is, the neighborhood of x^*, where Newton's method converges with the quadratic rate. The Newton area is defined as follows:

$$N(x^*, \beta) = \left\{ x : \lambda(x) = \|\nabla F(x)\|_x^* \leq \beta < \bar{\lambda} = \frac{3-\sqrt{5}}{2} \right\}. \tag{6.41}$$

It is possible to speed up damped Newton method (6.34) by using a switching strategy. For a given $0 < \beta < \bar{\lambda} = (3-\sqrt{5})/2$, one uses damped Newton method (6.34) if $\lambda(x_s) > \beta$ and the "pure" Newton method (6.36) when $\lambda(x_s) \leq \beta$.

We can measure the distance from the current approximation $x \in \operatorname{dom} F$ to the solution by $\Delta(x) = F(x) - F(x^*)$, $d(x) = \|x - x^*\|_x$ or $d_*(x) = \|x - x^*\|_{x^*}$.

The following theorem establishes useful bounds for $\Delta(x)$ and $d(x)$:

Theorem 6.5. *Let $\lambda(x) < 1$, then*

$$\omega(d(x)) \leq \Delta(x) \leq \omega^*(d(x)) \tag{6.42}$$

$$\omega'(\lambda(x)) \leq d(x) \leq \omega^{*'}(\lambda(x)). \tag{6.43}$$

Proof. By setting $y = x$ and $x = x^*$ from (6.24) and (6.26), we obtain (6.42). By setting $y = x$ and $x = x^*$ in (6.21) and, keeping in mind BCS inequality and $\nabla F(x^*) = 0$, we obtain

$$\frac{\|x - x^*\|_x^2}{1 + \|x - x^*\|_x} \leq \langle \nabla F(x) - \nabla F(x^*), x - x^* \rangle \leq \|\nabla F(x)\|_x^* \|x - x^*\|_x.$$

Let $d := d(x) = \|x - x^*\|_x$ and $\lambda := \lambda(x) = \|\nabla F(x)\|_x^*$, then we can rewrite the last inequality as follows:

$$\frac{d^2}{1+d} \leq \langle \nabla F(x), x - x^* \rangle \leq \lambda d.$$

Therefore, $d(1+d)^{-1} \leq \lambda$, that is,

$$d \leq \lambda(1-\lambda)^{-1} = \omega^{*'}(\lambda),$$

which is the right-hand side of (6.43). If $d > 1$, then the left-hand side of (6.43)

$$d \geq \omega'(\lambda) = \frac{\lambda}{1+\lambda}$$

is trivial.

Let us consider $0 < d < 1$, then from

$$\nabla F(x) = \nabla F(x) - \nabla F(x^*) = \int_0^1 \nabla^2 F(x^* + \tau(x - x^*))(x - x^*) d\tau = G(x - x^*),$$

where

$$G = \int_0^1 \nabla^2 F(x^* + \tau(x - x^*)) d\tau$$

and $G^T = G$, using BCS inequality, we obtain

$$\begin{aligned} \lambda^2(x) &= \langle \nabla^2 F(x)^{-1} \nabla F(x), \nabla F(x) \rangle \\ &= \langle \nabla^2 F(x)^{-1} G(x - x^*), G(x - x^*) \rangle \\ &\leq \left\| G \nabla^2 F(x)^{-1} G(x - x^*) \right\|_x^* \|x - x^*\|_x . \end{aligned}$$

Applying arguments similar to those we used in the proof of Theorem 6.4, we obtain

$$\left\| G \nabla^2 F(x)^{-1} G(x - x^*) \right\|_x^* \leq \|H\| \lambda(x),$$

where $H = \nabla^2 F(x)^{-1/2} G \nabla^2 F(x)^{-1/2}$. Therefore, $\lambda = \lambda(x) \leq \|H\| \, \|x - x^*\|_x$.

For $x \in \text{dom} F : d(x) = \|x - x^*\|_x = d = r < 1$, from (6.17)

$$G \preccurlyeq \frac{1}{1-d} \nabla^2 F(x).$$

Hence, $\|H\| \leq 1/(1-d)$ and

$$\lambda^2 \leq \frac{d^2}{(1-d)^2} = \left(\omega^{*'}(d) \right)^2$$

or $\lambda \leq \omega^{*'}(d)$. The function $\omega'(t) = t(1+t)^{-1}$ is monotone increasing. Thus,

$$\omega'(\lambda) \leq \omega' \left(\omega^{*'}(d) \right) = d$$

which is the left-hand side of (6.43). □

6.4 SC Barrier

In the previous section, we have seen that the self-concordant function F is a barrier function on $\operatorname{dom} F$. The properties of SC functions make Newton's method efficient for finding

$$x^* = \arg\min\{F(x) : x \in \operatorname{dom} F\}, \tag{6.44}$$

which is, in fact, an unconstrained optimization problem.

In this section, we consider Newton's method for constrained optimization.

We are concerned with the following constrained optimization problem

$$\min\{\langle c, x\rangle : x \in \Omega\}, \tag{6.45}$$

where Ω is a closed convex set. Let F be a self-concordant function with $\operatorname{dom} F = \operatorname{int} \Omega$. With problem (6.45), we associate the following function:

$$b(x,t) = \langle c, x\rangle t + F(x), \ t > 0.$$

From Definition 6.1 follows that $b(x,t)$ is self-concordant; therefore, $b(x,t)$ is a barrier function on Ω, that is, for any given $t > 0$, we have

$$\lim_{x \to \partial\Omega} b(x,t) = \infty.$$

Moreover, if for a given $t > 0$ there exists $x \in \operatorname{dom} F$ such that

$$\lambda(x,t) = \langle \nabla^2_{xx} b(x,t)^{-1} \nabla_x b(x,t), \nabla_x b(x,t)\rangle < 1,$$

then from Theorem 6.3 follows existence of the minimizer

$$x^*(t) = \arg\min\{b(x,t) : x \in \operatorname{dom} F\} \tag{6.46}$$

and (6.46) is, in fact, an unconstrained optimization problem.

The main idea of SUMT (see [Chapter 5]) consists of using the central trajectory $\{x^*(t)\}$, $t \in [t_0, \infty)$ for finding

$$x^* = \lim_{t \to \infty} x^*(t).$$

For any given $t > 0$, finding $x^*(t)$ is equivalent to solving for x the central path equation

$$\nabla b(x,t) = tc + \nabla F(x) = 0. \tag{6.47}$$

Generally speaking, solving system (6.47) requires an infinite number of operations. It makes the "pure" SUMT-type methods impractical.

The main idea of IPM consists of tracing the central path $\{x^*(t)\}$, $t \in [t_0, \infty)$. For a given $t > 0$, instead of solving (6.47), one performs Newton step toward $x^*(t)$ and then updates the penalty parameter $t > 0$.

Function $b(x,t)$ is a self-concordant on $\mathrm{dom}F$ for any given $t > 0$; therefore, Newton's method

$$\hat{x} = x - \left(\nabla^2_{xx} b(x,t)\right)^{-1} \nabla_x b(x,t) \tag{6.48}$$

converges to $x^*(t)$ with quadratic rate from any starting point $x \in N(x^*(t),\beta)$, where

$$N(x^*(t),\beta) = \left\{ x : \lambda(x,t) = \|\nabla_x b(x,t)\|^*_x \le \beta < \frac{3-\sqrt{5}}{2} \right\}$$

is the Newton area for the minimizer $x^*(t)$ (see (6.41)).

Let $x \in N(x^*(t),\beta)$, then IPM step consists of finding $\hat{x}$ by (6.48) and replacing t by $\hat{t} = t + \Delta = t(1 + (\Delta/t))$, where the value $\Delta > 0$ is our main concern in this section. Two things have to happen for IPM to be efficient.

First, $\Delta > 0$ has to be not too small to guarantee that the parameter $t > 0$ increases with a linear rate.

Second, the new approximation $\hat{x}$ has to be close enough to $x^*(\hat{t})$. To be more precise, $\hat{x}$ has to be in the Newton area $N(x^*(\hat{t}),\beta)$.

Let us first analyze how large the penalty increment Δ can be to guarantee that $\hat{x} \in N(x^*(\hat{t}),\beta)$.

We start with few preliminary considerations, which lead to the notion of ν - self-concordant barrier.

From (6.47) we obtain

$$c = -t^{-1} \nabla F(x^*(t)). \tag{6.49}$$

Note that for any $x \in \mathrm{dom}F$, we have $\nabla^2_{xx} b(x,t) = \nabla^2_{xx} b(x,\hat{t})$. To guarantee $x = x^*(t) \in N(x^*(\hat{t}),\beta)$, we need

$$\lambda(x,\beta) = \|\hat{t}c + \nabla F(x)\|^*_x = \Delta \|c\|^*_x = \frac{\Delta}{t} \|\nabla F(x)\|^*_x \le \beta$$

or

$$\frac{\Delta}{t} \le \beta \left(\|\nabla F(x)\|^*_x\right)^{-1}.$$

Thus, to guarantee the penalty parameter growth with linear rate, the norm $\|\nabla F(x)\|^*_x$ has to be uniformly bounded from above along the central trajectory $x^*(t)$, $t \in [0,\infty)$.

To make our assumption more practical, we assume that $\|\nabla F(x)\|^*_x$ is uniformly bounded on $\mathrm{dom}F$, so there is $\nu > 0$, such that

$$\lambda^2(x) = \|\nabla F(x)\|^{*2}_x \le \nu, \text{ for any } x \in \mathrm{dom}F. \tag{6.50}$$

In other words, the self-concordant function F needs an extra property, which would guarantee (6.50).

Definition 6.2. Function F is called a ν-self-concordant barrier on $\mathrm{dom}F$ if for any $x \in \mathrm{dom}F$ the following bound

$$\sup_{u\in\mathbb{R}^n}\left[2\langle\nabla F(x),u\rangle-\langle\nabla^2F(x)u,u\rangle\right]\leq\nu \tag{6.51}$$

holds.

The value $\nu>0$ is called the *parameter of the barrier*.

For positive definite $\nabla^2F(x)$, Definition 6.2 is equivalent to (6.50). Inequality (6.50) is also equivalent to

$$\langle\nabla F(x),u\rangle^2\leq\nu\langle\nabla^2F(x)u,u\rangle,\ \forall u\in\mathbb{R}^n,x\in\operatorname{dom}F\,. \tag{6.52}$$

Exercise 6.3. Let $\operatorname{dom}F$ contains no straight line. Show that (6.51) is equivalent to (6.52).

For the restriction $f(t)=F(x+tu)$, inequality (6.51) can be rewritten as follows:

$$\left(f'(0)\right)^2\leq\nu f''(0)\,. \tag{6.53}$$

Examples

1. Let us consider $F(x)=-\ln x$, $\operatorname{dom}F=\left\{x\in\mathbb{R}^1,\ x>0\right\}$. We have $F'(x)=-1/x$, $F''(x)=1/x^2$ and $F'(x)^2F''(x)^{-1}=1$. Therefore, $F(x)$ is a self-concordant barrier with $\nu=1$.
2. Let us consider the concave quadratic function $\varphi(x)=b+\langle a,x\rangle-\frac{1}{2}\langle Ax,x\rangle$ and define $F(x)=-\ln\varphi(x)$ on $\operatorname{dom}F=\{x\in\mathbb{R}^n:\varphi(x)>0\}$. Then,

$$\begin{aligned}\langle\nabla F(x),u\rangle&=-\frac{1}{\varphi(x)}[\langle a,u\rangle-\langle Ax,u\rangle],\\ \langle\nabla^2F(x)u,u\rangle&=\frac{1}{\varphi^2(x)}\left[\langle a,u\rangle-\langle Ax,u\rangle\right]^2+\frac{1}{\varphi(x)}\langle Au,u\rangle\geq\langle\nabla F(x),u\rangle^2.\end{aligned}$$

   Therefore,

$$2\langle\nabla F(x),u\rangle-\langle\nabla^2F(x)u,u\rangle\leq2\langle\nabla F(x),u\rangle-\langle\nabla F(x),u\rangle^2\leq1$$

   and $F(x)$ is a ν-self-concordant barrier with $\nu=1$.
3. The hyperbolic barrier $c(x)=x^{-1}$ is not a self-concordant function; therefore, it cannot be a ν-self-concordant barrier. Also, there is no $\nu>0$ such that $c'(x)^2c''(x)^{-1}=(2x)^{-1}<\nu$ for all $x\in\operatorname{dom}c=\{x:x>0\}$.

Exercise 6.4.

1. Consider $\psi(t)=\ln(t+1)$; is $-\psi(t)$ a ν—self-concordant barrier?
2. Show that $\varphi(s)=-\psi^*(s)=-\inf_t\{st-\psi(t)\}$ is a self-concordant function. Is $\varphi(s)$ a ν—self-concordant barrier?

Now we are going to show that (6.51) guarantees the existence of a small enough $\beta > 0$ such that the penalty parameter can grow with linear rate and after each Newton step (6.48) approximation $\hat{x} \in N(x^*(\hat{t}), \beta)$, if $x \in N(x^*(t), \beta)$.

As always we assume that $\operatorname{dom} F$ contains no straight line, then $\nabla^2 F(x)$ is positive definite on $\operatorname{dom} F$. We start with $t > 0$ and $x \in N(x^*(t), \beta)$, that is,

$$\lambda = \lambda(x,t) = \|tc + \nabla F(x)\|_x^* \leq \beta \tag{6.54}$$

and $\beta < (3 - \sqrt{5})/2$.

First, we find the upper bound for

$$\lambda(x,\hat{t}) = \|(t+\Delta)c + \nabla F(x)\|_x^* \leq |\Delta| \, \|c\|_x^* + \|tc + \nabla F(x)\|_x^* . \tag{6.55}$$

Keeping in mind (6.50), we obtain

$$\begin{aligned} t\|c\|_x^* &= \|tc + \nabla F(x) - \nabla F(x)\|_x^* \leq \|tc + \nabla F(x)\|_x^* + \|\nabla F(x)\|_x^* \\ &\leq \lambda(x,t) + \|\nabla F(x)\|_x^* \leq \lambda(x,t) + \sqrt{\nu} = \lambda + \sqrt{\nu} \leq \beta + \sqrt{\nu} . \end{aligned}$$

Therefore,

$$\|c\|_x^* \leq \frac{1}{t}\,(\beta + \sqrt{\nu})$$

and from (6.54) and (6.55) follows

$$\lambda(x,\hat{t}) \leq \frac{|\Delta|}{t}\,(\beta + \sqrt{\nu}) + \beta . \tag{6.56}$$

Let $x \in N(x^*(\hat{t}), \beta)$, then for

$$\hat{x} = x - \left(\nabla_{xx}^2 b(x,\hat{t})\right)^{-1} \nabla_x b(x,\hat{t}) \tag{6.57}$$

from (6.38), we obtain

$$\lambda(\hat{x},\hat{t}) \leq \left(\frac{\lambda(x,\hat{t})}{1-\lambda(x,\hat{t})}\right)^2 .$$

Inequality

$$\frac{\lambda(x,\hat{t})}{1-\lambda(x,\hat{t})} \leq \sqrt{\beta} \tag{6.58}$$

is a sufficient condition for $\hat{x} \in N(x^*(\hat{t}), \beta)$.

Function $\lambda(1-\lambda)^{-1}$ is a monotone increasing on $[0,1)$. Therefore, to find the upper bound for $0 \leq \lambda < 1$, as a function of $\beta > 0$, we have to solve equation $\lambda(1-\lambda)^{-1} = \sqrt{\beta}$ for λ.

We obtain $\lambda(x,\hat{t}) = \sqrt{\beta}\left(1+\sqrt{\beta}\right)^{-1}$; then it follows from (6.56) that $x \in N(x^*(\hat{t}), \beta)$, if the penalty increment Δ satisfies the following inequality

$$\frac{\Delta}{t}\,(\beta + \sqrt{\nu}) + \beta \leq \sqrt{\beta}(1+\sqrt{\beta})^{-1} .$$

Therefore,

$$\frac{\Delta}{t} \le \frac{\sqrt{\beta}\left(1+\sqrt{\beta}\right)^{-1} - \beta}{\beta + \sqrt{\nu}}.$$

It means that if $x \in N(x^*(t), \beta)$ and $\hat{x}$ is found by (6.57), then for any penalty increment $\Delta > 0$, which satisfies the last inequality, we have $\hat{x} \in N(x^*(\hat{t}), \beta)$.

For $\beta = 0.125$, we can take

$$\frac{\Delta}{t} = \frac{1}{1+8\sqrt{\nu}} = \chi .$$

Let $t > 0$ is given, and $\hat{t} = \left(1 + 1/(1+8\sqrt{\nu})\right) t$, then for any x from $N(x^*(t), 0.125)$, the vector $\hat{x}$ defined by (6.57) belongs to $N(x^*(\hat{t}), 0.125)$.

Therefore, each Newton step allows an increase of the penalty parameter $t > 0$ by a factor of $\left(1 + 1/(1+8\sqrt{\nu})\right)$, and the choice of $\beta > 0$ guarantees that the new pair $(\hat{x}, \hat{t})$ satisfies the centering condition

$$\lambda(\hat{x}, \hat{t}) = \|\nabla_x b(\hat{x}, \hat{t})\|_{\hat{x}}^* = \|\hat{t}c + \nabla F(\hat{x})\|_{\hat{x}}^* \le \beta ,$$

provided such condition is satisfied for the old pair (x, t), that is, (6.54) holds.

Let us now consider the path-following method.

6.5 Path-Following Method

Initialization

We pick accuracy $\varepsilon > 0$ and choose the barrier parameter $\nu > 0$, parameters $0 < \beta \le (3-\sqrt{5})/2$ and $\chi = (1+8\sqrt{\nu})^{-1}$. Set $t > 0$, and find $\hat{x} \in N(x^*(t), \beta)$, that is,

$$x : \lambda(x, t) = \|tc + \nabla F(x)\|_x^* \le \beta$$

1. Perform Newton's step

$$\hat{x} = x - \nabla_{xx}^2 b(x,t)^{-1} \nabla_x b(x,t).$$

2. If $\varepsilon t \ge \nu + \frac{(\beta+\sqrt{\nu})\beta}{1-\beta}$, then stop, output $x^* = x$.
3. $\hat{t} := t(1+\chi)$.
4. Set $x := \hat{x}$, $t := \hat{t}$
   Go to step 1.

For any given $\varepsilon > 0$, the path-following method allows finding an approximation $x \in \operatorname{dom} F$ such that $\langle c, x\rangle - \langle c, x^*\rangle \le \varepsilon$ in a finite number of steps. To estimate the number of steps, we need some information about the global structure of $\operatorname{dom} F$. The properties of ν-self-concordant functions provide such information.

Theorem 6.6. *Let $F(x)$ be a ν-self-concordant barrier, then, for any x and y from* $\operatorname{dom} F$*, we have*

$$\langle \nabla F(x), y-x\rangle < \nu. \tag{6.59}$$

If $\langle \nabla F(x), y-x\rangle \geq 0$, then

$$\langle \nabla F(y) - \nabla F(x), y-x\rangle \geq \frac{\langle \nabla F(x), y-x\rangle^2}{\nu - \langle \nabla F(x), y-x\rangle}. \tag{6.60}$$

Proof. Let x and y belong to $\operatorname{dom} F$. We consider

$$\varphi(t) = f'(t) = \langle \nabla F(x+t(y-x)), y-x\rangle,\ 0 \leq t \leq 1.$$

If $\varphi(0) < 0$, then (6.59) is trivial, so let $\varphi(0) \geq 0$. If $\varphi(0) = 0$, then (6.60) is just the definition of a convex function.

From the definition of the ν-self-concordant barrier (6.52) follows

$$\begin{aligned}\varphi'(t) &= \langle \nabla^2 F(x+t(y-x))(y-x), y-x\rangle \\ &\geq \frac{1}{\nu}\langle \nabla F(x+t(y-x)), y-x\rangle^2 = \frac{1}{\nu}\varphi^2(t).\end{aligned} \tag{6.61}$$

Therefore, function $\varphi(t)$ is increasing on $[0,1]$ and positive. From (6.61) follows

$$\int_0^t \frac{\varphi'(\tau)}{\varphi^2(\tau)}\, d\tau \geq \frac{1}{\nu}\int_0^t d\tau$$

or

$$-\frac{1}{\varphi(t)} + \frac{1}{\varphi(0)} \geq \frac{1}{\nu} t. \tag{6.62}$$

Thus, $\varphi(0) = \langle \nabla F(x), y-x\rangle < \nu/t$ for all $t \in (0,1]$. For $t=1$ we obtain (6.59). Also, from (6.62) follows

$$\varphi(t) \geq \frac{\nu\varphi(0)}{\nu - t\varphi(0)}.$$

Therefore,

$$\varphi(t) - \varphi(0) \geq \frac{t}{\nu}\varphi(t)\varphi(0) \geq \frac{t\varphi^2(0)}{\nu - t\varphi(0)}, \quad \forall t \in [0,1].$$

For $t = 1$, we have

$$\varphi(1) - \varphi(0) \geq \frac{\varphi(0)^2}{\nu - \varphi(0)},$$

which is, in fact, (6.60).

Now we are ready to prove the main result.

Theorem 6.7. *For any $t > 0$, we have*

$$\langle c, x^*(t)\rangle - \langle c, x^*\rangle \leq \frac{\nu}{t}.$$

If a point $x \in \operatorname{dom} F$ *satisfies the centering condition, that is,* $x \in N(x^*(t), \beta)$, *then*

$$\langle c, x\rangle - \langle c, x^*\rangle \leq \frac{1}{t}\left(\nu + \frac{(\beta+\sqrt{\nu})\,\beta}{1-\beta}\right).$$

Proof. From (6.49) and (6.59), we have

$$\langle c, x^*(t) - x^*\rangle = \frac{1}{t}\langle \nabla F(x^*(t)), x^* - x^*(t)\rangle \leq \frac{\nu}{t}.$$

We assume that $x \in N(x^*(t), \beta)$, i.e.,

$$\lambda(x,t) = \|\nabla_x b(x,t)\|_x^* = \|tc + \nabla F(x)\|_x^* \leq \beta. \tag{6.63}$$

Then, using (6.47) and BSC inequality, we obtain

$$\begin{aligned} t\langle c, x - x^*(t)\rangle &= \langle \nabla_x b(x,t) - \nabla F(x), x - x^*(t)\rangle \\ &\leq \|\nabla_x b(x,t) - \nabla F(x)\|_x^* \|x - x^*(t)\|_x. \end{aligned}$$

Let $\lambda = \lambda(x,t)$, then using (6.50), (6.63) we obtain

$$\|\nabla_x b(x,t) - \nabla F(x)\|_x^* \leq \|\nabla_x b(x,t)\|_x^* + \|\nabla F(x)\|_x^* \leq (\lambda + \sqrt{\nu}).$$

Using the right-hand side of (6.43), we obtain

$$\|x - x^*(t)\|_x \leq \omega^{*\prime}(\lambda(x,t)) = \frac{\lambda(x,t)}{1-\lambda(x,t)} = \frac{\lambda}{1-\lambda}.$$

Keeping in mind $x \in N(x^*(t), \beta)$, we have $\lambda \leq \beta < 1$. Therefore,

$$t|\langle c, x - x^*(t)\rangle| \leq t\|c\|_x^* \|x - x^*(t)\|_x \leq (\lambda + \sqrt{\nu})\,\frac{\lambda}{1-\lambda} \leq \frac{(\beta+\sqrt{\nu})\,\beta}{1-\beta}.$$

Finally,

$$\langle c, x - x^*\rangle = \langle c, x - x^*(t)\rangle + \langle c, x^*(t) - x^*\rangle \leq \frac{1}{t}\left(\nu + \frac{(\beta+\sqrt{\nu})\,\beta}{1-\beta}\right).$$

□

Therefore, for $t \geq \left(\nu + \frac{(\beta+\sqrt{\nu})\beta}{1-\beta}\right)\varepsilon^{-1}$, we obtain $\langle c, x - x^*\rangle \leq \varepsilon$, which justifies the stopping criteria 2 of the path-following method.

Now, we can estimate the number of steps required to obtain $\Delta(x) = F(x) - F(x^*) \leq \varepsilon$.

For $\beta = 0.125$, we have

$$\Delta(x) \leq \frac{1}{t}\left(\nu + \frac{(0.125+\sqrt{\nu})\,0.125}{0.875}\right) = \frac{1}{t}\left(\nu + \frac{1}{56}\left(1 + 8\sqrt{\nu}\right)\right).$$

Each step of the path-following method allows increase of the penalty parameter by a factor of $(1+1/(1+8\sqrt{\nu}))$. Hence, it requires $\approx 8\sqrt{\nu}$ steps to increase t at least twice. For $\nu \geq 1$, we have

$$\Delta(x) \leq t^{-1}\left(\nu + \frac{1}{56}(1+8\sqrt{\nu})\right) \leq (\nu+\sqrt{\nu})\,t^{-1}.$$

Therefore, to find $x : \Delta(x) \leq \varepsilon$ we need $t \geq (\nu+\sqrt{\nu})\varepsilon^{-1}$. It requires $O(\sqrt{\nu})s$ steps to get $t = O(2^s)$; therefore, for $t \geq (\nu+\sqrt{\nu})\varepsilon^{-1}$, we need $s = O\left(\ln \frac{\nu+\sqrt{\nu}}{\varepsilon}\right)$ steps. Thus, the total number of steps of the path-following method is

$$N = O\left(\sqrt{\nu}\ln\frac{\nu+\sqrt{\nu}}{\varepsilon}\right).$$

In order to initiate the path-following method, one has to find $x \in N(x^*(0),\beta)$. It can be done by using damped Newton method applied for minimization F.

Definition 6.3. Let $F(x)$ be a ν-self-concordant barrier for the set dom$F$. The point

$$x^* = \arg\min\{F(x) \mid x \in \operatorname{dom} F\}$$

is called analytic center of the convex set dom$F$ generated by the barrier function F.

Due to Theorem 6.3 the existence of $x \in \operatorname{dom} F$: $\lambda(x) < 1$ is sufficient for existence of x^*.

Theorem 6.8. *If the analytic center of the* ν*-self-concordant barrier exists, then for any* $x \in$ dom$F$ *we have*

$$\|x - x^*\|_{x^*} \leq \nu + 2\sqrt{\nu}. \tag{6.64}$$

Proof. For any $y \in$ dom$F$, we have

$$\langle \nabla F(x^*), y - x^* \rangle \geq 0. \tag{6.65}$$

Let $r = \|y - x^*\|_{x^*}$ and $r > \sqrt{\nu}$, then $\alpha = \sqrt{\nu}/r < 1$. Consider $y(\alpha) = x^* + \alpha(y - x^*)$. It follows from (6.65) and (6.21) that

$$\begin{aligned}
&\langle \nabla F(y(\alpha)), y - x^* \rangle \geq \langle \nabla F(y(\alpha)) - \nabla F(x^*), y - x^* \rangle \\
&= \frac{1}{\alpha}\langle \nabla F(y(\alpha)) - \nabla F(x^*), y(\alpha) - x^* \rangle \\
&\geq \frac{1}{\alpha}\frac{\|y(\alpha) - x^*\|_{x^*}^2}{1 + \|y(\alpha) - x^*\|_{x^*}} = \frac{\alpha\|y - x^*\|_{x^*}^2}{1 + \alpha\|y - x^*\|_{x^*}} > \frac{r\sqrt{\nu}}{1+\sqrt{\nu}}.
\end{aligned}$$

On the other hand, from (6.59), we have

$$(1-\alpha)\langle \nabla F(y(\alpha)), y - x^* \rangle = \langle \nabla F(y(\alpha)), y - y(\alpha) \rangle \leq \nu.$$

Therefore,

$$\nu \geq (1-\alpha)\langle \nabla F(y(\alpha)), y - x^* \rangle \geq \left(1 - \frac{\sqrt{\nu}}{r}\right) \frac{r\sqrt{\nu}}{1+\sqrt{\nu}}$$

or

$$r \leq \nu + 2\sqrt{\nu}. \tag{6.66}$$

□

Also, for any $x \in \mathbb{R}^n$ such that $\|x - x^*\|_{x^*} \leq 1$, we have $x \in \mathrm{dom} F$.

Inequality (6.66) presents another remarkable result about ν-self-concordant functions F: the asphericity of the set $\mathrm{dom} F$ with respect to x^* computed in the norm $\| \ \|_{x^*}$ does not exceed $\nu + 2\sqrt{\nu}$.

For a given $\beta > 0$ small enough, the approximation $x \in N(x^*, \beta)$ can be found by damped Newton method.

1. Choose $x \in \mathrm{dom} F$
2. Set $\hat{x} := x - (1 + \|\nabla F(x)\|_x^*)^{-1} \nabla^2 F(x)^{-1} \nabla F(x)$
3. Set $x := \hat{x}$ and check the inequality

$$\|\nabla F(x)\|_x^* \leq \beta.$$

If it is satisfied, then $x \in N(x^*, \beta)$, otherwise go to 2.

6.6 Applications of ν-SC Barrier. IPM Complexity for LP and QP

We consider applications of ν-self-concordant functions to linear programming and quadratic programming with quadratic constraints, as well as to semidefinite optimization.

We start by proving the following four technical lemmas.

First, we show that self-concordance is an affine invariant property.

We consider a linear operator $y = R(x) = Ax + b : \mathbb{R}^n \to \mathbb{R}^m$ and assume that $F(y)$ is a self-concordant function on $\mathrm{dom} F$ with a constant $M > 0$, i.e.,

$$|D^3 F(y)[u,u,u]| \leq M D^2 F(y)[u,u] = M\langle \nabla^2 F(y) u, u \rangle^{3/2}. \tag{6.67}$$

Lemma 6.2. *Function* $f(x) = F(R(x)) = F(Ax + b)$ *is self-concordant on* $\mathrm{dom} f = \{x : Ax + b \in \mathrm{dom} F\}$ *with the same constant* $M > 0$, *i.e.,*

$$|D^3 f(x)[u,u,u]| \leq M D^2 f(x)[u,u], \ x \in \mathrm{dom} f$$

Proof. Let v $= Au$, then

$$\begin{aligned} Df(x)[u] &= \langle \nabla F(R(x+tu)), u\rangle_{|t=0} = \langle A^T \nabla F(Rx), u\rangle \\ &= \langle \nabla F(R(x)), Au\rangle = \langle \nabla F(y), \mathrm{v}\rangle\,, \\ D^2 f(x)[u,u] &= \langle A^T \nabla^2 F(R(x+tu))Au, u\rangle_{|t=0} \\ &= \langle \nabla^2 F(R(x))Au, Au\rangle = \langle \nabla^2 F(y)\mathrm{v}, \mathrm{v}\rangle\,, \\ D^3 f(x)[u,u,u] &= D^3 F(R(x+tu))[Au,Au,Au]]_{|t=0} = D^3 F(y)[\mathrm{v},\mathrm{v},\mathrm{v}]\,. \end{aligned}$$

From (6.67) we get

$$|D^3 f(x)[u,u,u]| = |D^3 F(y)[\mathrm{v},\mathrm{v},\mathrm{v}]| \le M\langle \nabla^2 F(y)\mathrm{v}, \mathrm{v}\rangle 3/2 = M(D^2 f(x)[u,u])^{3/2}.$$

□

Lemma 6.3. *If $F(y)$ is a ν–self-concordant barrier on* dom F*, then* $f(x) = F(R(x))$ *is a ν–self-concordant barrier on* $\operatorname{dom} f = \{x \in \mathbb{R}^n \mid y = Ax + b \in \operatorname{dom} F\}$.

Proof. It follows from Lemma 6.2 that $f(x)$ is a self-concordant function on dom f.

Keeping in mind that v $= Au$, $\langle \nabla f(x), u\rangle = \langle \nabla F(y), Au\rangle$, $\langle \nabla^2 f(x)u, u\rangle = \langle \nabla^2 F(y)Au, Au\rangle$ and using the definition of a ν-self-concordant barrier (6.51), we obtain

$$\begin{aligned} &\max_{u\in\mathbb{R}^n}[2\langle \nabla f(x), u\rangle - \langle \nabla^2 f(x)u, u\rangle] \\ &= \max_{u\in\mathbb{R}^n}[2\langle \nabla F(y), Au\rangle - \langle \nabla^2 F(y)Au, Au\rangle] \\ &\le \max_{\mathrm{v}\in\mathbb{R}^m}[2\langle \nabla F(y), \mathrm{v}\rangle - \langle \nabla^2 F(y)\mathrm{v}, \mathrm{v}\rangle] \le \nu\,. \end{aligned}$$

□

Lemma 6.4.

(1) *Let F_i, $i = 1, \ldots, r$ be self-concordant functions defined on* dom F_i*, with constants* $M_i > 0$*. Then,*

$$F(x) = F_i(x) + \cdots + F_r(x)$$

is a self-concordant function defined on $\operatorname{dom} F = \bigcap_{i=1}^r \operatorname{dom} F_i$ *with a constant* $M = \max_{1\le i\le r} M_i$.

(2) *If F_i are ν_i-self-concordant barriers, then F is a ν-self-concordant barrier with*

$$\nu \le \nu_1 + \cdots + \nu_r\,.$$

Proof.

(1) The function F is closed and convex as the sum of closed convex functions. Any $x \in \operatorname{dom} F$ belongs to dom F_i; therefore, $\operatorname{dom} F = \bigcap_{i=1}^r \operatorname{dom} F_i$. Recall that

$$f(t) = F(x+tu) = \sum_{i=1}^r F_i(x+tu) = \sum_{i=1}^r f_i(t)\,.$$

It follows from Definition 6.1 that

$$|f'''(0)| \le \sum_{i=1}^{r} |f_i'''(0)| \le \sum_{i=1}^{r} M_i (f_i''(0))^{3/2}.$$

Let $\omega_i = f_i''(0) > 0$, then $\omega_i > 0$ and

$$\frac{|f'''(0)|}{(f''(0))^{3/2}} \le \frac{\sum_{i=1}^r M_i f_i''(0)^{3/2}}{(\sum_{i=1}^r f_i''(0))^{3/2}} = \frac{\sum_{i=1}^r M_i \omega_i^{3/2}}{(\sum_{i=1}^r \omega_i)^{3/2}}. \tag{6.68}$$

The right-hand side remains the same if one replaces ω_i by $\omega_i t$ for any $t > 0$. Therefore, we can assume that $\sum_{i=1}^r \omega_i = 1$ and for

$$M = \max\left\{\sum_{i=1}^{r} M_i \omega_i^{3/2} : \sum_{i=1}^{r} \omega_i = 1,\ \omega_i \ge 0\right\} = \max\{M_i \mid i = 1, \dots, r\}$$

from (6.68), we obtain

$$|f'''(0)| \le M (f''(0))^{3/2}.$$

(2) For $F(x) = F_1(x) + \cdots + F_r(x)$, we have

$$\begin{aligned} &\max_{u \in \mathbb{R}^n} \{2\langle \nabla F(x), u\rangle - \langle \nabla^2 F(x) u, u\rangle\} \\ &= \max_{u \in \mathbb{R}^n} \left\{\sum_{i=1}^{r} (2\langle \nabla F_i(x), u\rangle - \langle \nabla^2 F_i(x) u, u\rangle)\right\} \\ &\le \sum_{i=1}^{r} \max_{u \in \mathbb{R}^n} \{2\langle \nabla F_i(x), u\rangle - \langle \nabla^2 F_i(x) u, u\rangle\} = \nu_1 + \cdots + \nu_r. \end{aligned}$$ □

The complexity bound

$$N = O\left(\sqrt{\nu} \ln \frac{\nu + \sqrt{\nu}}{\varepsilon}\right)$$

of the path-following method is a function of the positive parameter ν. Therefore, it is important to find the upper and lower bounds for this parameter. We consider a few important constrained optimization problems and estimate the bounds for $\nu > 0$.

Let us start with a ν–self-concordant barrier $f(t)$ on an interval $(a, b) \in \mathbb{R}^1$, $a < b < \infty$.

Lemma 6.5. *If $f(t)$ is a ν-self-concordant barrier, then*

$$\nu \ge \kappa = \sup_{t \in (a,b)} \frac{f'(t)^2}{f''(t)} \ge 1. \tag{6.69}$$

Proof. It follows from (6.52) that $\nu \geq \kappa$, for a barrier function $f(t)$. From (6.5) we have $\lim_{t\to-\beta} f(t) = \infty$. Therefore, there exists $\bar{t} \in (a,b)$ such that $f'(t) > 0$, $t \in [\bar{t},\beta)$.

Let us consider $\varphi(t) = (f'(t))^2/f''(t)$, $t \in [\bar{t},\beta)$. We have

$$\varphi'(t) = 2f'(t) - \left(\frac{f'(t)}{f''(t)}\right)^2 f'''(t) = f'(t)\left(2 - \frac{f'(t)}{\sqrt{f''(t)}}\,\frac{f'''(t)}{[f''(t)]^{3/2}}\right).$$

Suppose that $\kappa < 1$, and keeping in mind that

$$\varphi(t) \leq \kappa,\ \frac{f'(t)}{\sqrt{f''(t)}} \leq \sqrt{\kappa}$$

from Definition 6.1, we obtain

$$\varphi'(t) \geq 2\left(1 - \sqrt{k}\right) f'(t).$$

Thus, for $t \in [\bar{t},\beta)$, we have

$$\varphi(t) \geq \varphi(\bar{t}) + 2\left(1 - \sqrt{k}\right)(f(t) - f(\bar{t})).$$

This is impossible because $f(t)$ is a barrier and $\varphi(t)$ is bounded from above due to (6.53); therefore, $\nu \geq 1$. □

Let Ω be a closed convex set with nonempty interior and $\bar{x} \in \operatorname{int}\Omega$. We recall that d is a recession direction for F at $\bar{x} \in \Omega$ if $\bar{x}+td \in \Omega$ and $f(t) = F(\bar{x}+td)$ is a monotone decreasing function of $t > 0$.

Let us assume that there is a nontrivial set of recession directions $(d_1,\ldots,d_k)$ of Ω, that is,

$$\bar{x} + t\,d_i \in \Omega, \forall t > 0$$

and $F(\bar{x}+td_i)$, $i = 1,\ldots,k$ are monotone decreasing of $t > 0$.

Lemma 6.6. *If positive numbers $\beta_1,\ldots,\beta_k$ satisfy $\bar{x} - \beta_i d_i \notin \operatorname{int} Q$, $i = 1,\ldots,k$ and positive numbers $\alpha_1,\ldots,\alpha_k$ satisfy $\bar{y} = \bar{x} - \sum_{i=1}^{k} \alpha_i d_i \in \Omega$, then the parameter ν of any self-concordant barrier for Ω satisfies inequality*

$$\nu \geq \sum_{i=1}^{k} \frac{\alpha_i}{\beta_i}.$$

Proof. Let $F(x)$ be a ν-self-concordant barrier for the set Ω. We consider a recession direction d and the restriction $f(t) = F(\bar{x}+td)$; then we have

$$\lambda = -f'(0)f''(0)^{1/2} = -\langle\nabla F(\bar{x}),d\rangle\langle\nabla^2 F(\bar{x})d,d\rangle^{1/2} \geq 1,$$

otherwise the restriction $f(t)$ attains its minimum (see Theorem 6.3).

Therefore, for each recession direction d_i, we have

$$\frac{-\langle \nabla F(x), d_i\rangle}{\langle \nabla^2 F(x) d_i, d_i\rangle^{1/2}} \geq 1,$$

that is,

$$-\langle \nabla F(\bar{x}), d_i\rangle \geq \langle \nabla^2 F(\bar{x}) d_i, d_i\rangle^{1/2} = \|d_i\|_{\bar{x}}.$$

It follows from $\bar{x} \in \Omega$ that $E(\bar{x},1) \subset \Omega$ but $\bar{x} - \beta_i d_i \notin \operatorname{int} \Omega$; therefore, $\beta_i \|d_i\|_{\bar{x}} \geq 1$ and $\|d_i\|_{\bar{x}} \geq 1/(\beta_i)$. We recall that $\bar{y} - \bar{x} = -\sum \alpha_i d_i \in \Omega$. From Theorem 6.6, we obtain

$$\begin{aligned} \nu > \langle \nabla F(\bar{x}), \bar{y} - \bar{x}\rangle &= \langle \nabla F(\bar{x}), -\sum \alpha_i d_i\rangle = -\sum \alpha_i \langle \nabla F(\bar{x}), d_i\rangle \\ &\geq \sum_{i=1}^{k} \alpha_i \|d_i\|_{\bar{x}} \geq \sum_{i=1}^{k} \frac{\alpha_i}{\beta_i}. \end{aligned} \tag{6.70}$$

□

6.6.1 Linear and Quadratic Optimization

The linear programming problem consists of finding

$$\langle c, x^*\rangle = \min\left\{ \langle c, x\rangle : Ax = b,\, x \in \mathbb{R}^n_+ \right\}, \tag{6.71}$$

where A is an $m \times n$ matrix.

The correspondent barrier is defined as follows:

$$F(x) = -\sum_{i=1}^{n} \ln x_i .$$

It follows from Lemma 6.4 that $F(x)$ is a ν-self-concordant barrier with $\nu = n$. From Lemmas 6.2 and 6.3 follows that the restriction of this barrier to the affine subspace $\{x : Ax = b\}$ is n-self-concordant barrier. It means that

$O(\sqrt{n}\ln n/\varepsilon)$ iteration is enough to find approximation for $\langle c, x^*\rangle$ with accuracy $\varepsilon > 0$.

This is the main theoretical result of the interior point methods for LP.

It follows from Lemma 6.6 that for $\bar{x} = e \equiv (1,\dots,1)^T \in \operatorname{int}\mathbb{R}^n_+$ and $d_i = e_i = (0,\dots,1,\dots,0)$ we have $\beta_i = 1$ and $\alpha_i = 1$. Therefore,

$$\nu \geq \sum_{i=1}^{n} \frac{\alpha_i}{\beta_i} = n .$$

It means that it is impossible to improve the lower bound for ν on the entire set $\mathbb{R}^n_+$.

Now we consider an important class of quadratic optimization problems with quadratic constraints

$$\min\{f(x) : x \in \Omega\}, \tag{6.72}$$

where $f(x) = \alpha + \langle a,x\rangle + 1/2\langle Ax,x\rangle$

$$\Omega = \left\{c_i(x) = q_i + \langle b_i,x\rangle + \frac{1}{2}\langle B_i x,x\rangle \geq 0, \quad i = 1,\ldots,m\right\},$$

and A and all $-B_i$ are nonnegative definite $n \times n$ matrices. We can rewrite (6.72) as follows:

$$\min \tau \tag{6.73}$$

s.t.

$$\tau - f(x) \geq 0 \tag{6.74}$$

$$c_i(x) \geq 0, \qquad i = 1,\ldots,m. \tag{6.75}$$

It follows from Example 2 (see Section 6.4) and Lemma 6.4 that the function

$$F(x,\tau) = -\ln(\tau - f(x)) - \sum_{i=1}^{m} \ln c_i(x)$$

is a ν-self-concordant barrier with $\nu = m+1$. Therefore, the complexity bound for problem (6.73)–(6.75) is $O(\sqrt{m+1}\ln(m+1)/\varepsilon)$. This is the number of Newton steps required to obtain an approximation for $f(x^*)$ with accuracy $\varepsilon > 0$. The bound does not depend on the dimension of $\mathbb{R}^n$.

The dimension n will show up when we estimate the number of arithmetic operations. For finding such a bound, the number of steps has to be multiplied by the number of arithmetic operations required to solve a linear system of equations for finding Newton direction Δx.

6.6.2 The Lorentz Cone

The cone $K = \{(x,x_{n+1}) \in \mathbb{R}^{n+1} : x_{n+1} \geq \|x\|\}$ is called a Lorentz, or ice-cream, cone. The corresponding barrier function

$$F(x,x_{n+1}) = -\ln\left(x_{n+1}^2 - \|x\|^2\right)$$

is a 2-self-concordant barrier for the cone K. The proof follows from the definition of ν-self-concordant barriers. Let us fix $(x,x_{n+1}) = (x_1,\ldots,x_n,x_{n+1})$, a nonzero direction $u = (u,\ldots,u_n,u_{n+1})$, and consider the function

$$\varphi(t) = (x_{n+1} - tu_{n+1})^2 - \|x - tu\|^2 .$$

We have to compute derivatives of $f(t) = F(x+tu) = -\ln \varphi(t)$ at $t = 0$. We thus have

$$\varphi'(t)_{|t=0} = \varphi' = 2\left(x_{n+1}u_{n+1} - \langle x,u\rangle\right), \quad \varphi''(t)_{|t=0} = \varphi'' = 2\left(u_{n+1}^2 - \|u\|^2\right).$$

Then for $f' = f'(0)$, $f'' = f''(0)$, and $f''' = f'''(0)$, we have

$$f' = -\frac{\varphi'}{\varphi}, \quad f'' = \left(\frac{\varphi'}{\varphi}\right)^2 - \frac{\varphi''}{\varphi}, \quad f''' = 3\frac{\varphi'\varphi''}{\varphi^2} - 2\left(\frac{\varphi'}{\varphi}\right)^3.$$

First, let us show that $(f')^2 \leq 2f''$, which is equivalent to $\varphi'^2 \geq 2\varphi\varphi''$ or

$$\left(x_{n+1}u_{n+1} - \langle x,u\rangle\right)^2 \geq \left(x_{n+1}^2 - \|x\|^2\right)\left(u_{n+1}^2 - \|u\|^2\right).$$

We assume that $u_{n+1} > \|u\|$, otherwise the last inequality is trivial. We have assume that sign $u_{n+1} =$ sign (x,u) and $(x,u) = \|x\|\ \|u\|$, which makes the expression in the brackets of the left-hand side as small as possible. Then we have

$$\begin{aligned}
&\left(x_{n+1}u_{n+1} - \|x\|\ \|u\|\right)^2 \\
&= x_{n+1}^2 u_{n+1}^2 - 2x_{n+1}u_{n+1}\|x\|\ \|u\| + \|x\|^2\|u\|^2 \\
&\geq \left(x_{n+1}^2 - \|x\|^2\right)\left(u_{n+1}^2 - \|u\|^2\right).
\end{aligned}$$

In other words, if F is a self-concordant function, then due to $f'^2 \leq 2f''$, it is a 2-self-concordant barrier.

Let us show that F is a self-concordant function. Note that $0 \leq \gamma = \varphi\varphi''/\varphi'^2 \leq 1/2$. Then,

$$\frac{|f'''|}{(f'')^{3/2}} = 2\frac{\left[1 - \frac{3}{2}\varphi\varphi''/\varphi'^2\right]}{\left[1 - \varphi\varphi''/\varphi'^2\right]^{3/2}} = 2\frac{(1 - \frac{3}{2}\gamma)}{(1-\gamma)^{3/2}} \leq 2$$

for any $\gamma \in [0, \frac{1}{2}]$.

Thus, $F(x,t) = -\ln(t^2 - \|x\|^2)$ is a 2-self-concordant barrier for the Lorentz cone

$$K = \left\{(x, x_{n+1}) \in \mathbb{R}^{n+1} \mid x_{n+1} \geq \|x\|\right\}.$$

It is easy to show, using Lemma 6.6, that for the Lorentz cone $\nu \geq 2$. Therefore, the self-concordant barrier parameter $\nu = 2$ is optimal for the Lorentz cone.

6.6.3 Semidefinite Optimization

Semidefinite programming (SDP) is an extension of linear programming in which symmetric positive semidefinite matrices are the decision variables. Semidefinite programming has a multitude of applications to physical problems, optimal control, combinatorial optimization, etc.

Let $X = \{X_{ij}\}_{i,j=1}^{n}$ be a symmetric $n \times n$ matrix, that is, $X = X^T$. We consider a linear space $S^{n\times n}$ of symmetric matrices with the inner product

$$\langle X,Y\rangle_{\mathrm{F}} = \sum_{i=1}^{n}\sum_{j=1}^{n} X_{ij}Y_{ij}\,.$$

The norm $\|X\|_{\mathrm{F}} = \langle X,X\rangle^{1/2}$ is called Frobemius norm of matrix X. The following identity will be helpful later:

$$\begin{aligned}
\langle X,Y^2\rangle_{\mathrm{F}} &= \sum_{i=1}^{n}\sum_{j=1}^{n} X_{ij}\left(\sum_{k=1}^{n} Y_{ik}Y_{kj}\right)\\
&= \sum_{k=1}^{n}\sum_{j=1}^{n} Y_{kj}\sum_{i=1}^{n} X_{ji}Y_{ik}\\
&= \sum_{k=1}^{n}\sum_{j=1}^{n} Y_{kj}(XY)_{jk} = \sum_{k=1}^{n}(YXY)_{kk} = \mathrm{Tr}(YXY)\\
&= \langle YXY, I_n\rangle_{\mathrm{F}}\,.
\end{aligned} \tag{6.76}$$

We consider a cone K_n of positive semidefinite matrices $X \succ 0$, that is, $u^T Xu > 0$ for any $u \in \mathbb{R}^n \setminus \{0\}$. The semidefinite optimization problem consists of finding

$$\langle C,X^*\rangle_{\mathrm{F}} = \min\left\{\langle C,X\rangle_{\mathrm{F}} : \langle A_i,X\rangle_{\mathrm{F}} = b_i, \quad i = 1,\ldots,n,\ X \succeq 0\right\}. \tag{6.77}$$

Let us consider

$$F(X) = -\ln\det X = -\ln\prod_{i=1}^{n}\lambda_i(X)$$

where $\{\lambda_i(X)\}_{i=1}^{n}$ is the set of eigenvalues of X.

For any matrix $X \in K_n$, the eigenvalues are positive; therefore, $\mathrm{dom}\,F = \mathrm{int}\,K_n$.

To show that F is a ν-self-concordant barrier on K_n, we have to check the standard inequalities (6.2) and (6.51). Let us consider $\Delta \in S^{n,n}$ and $X \in \mathrm{int}\,K_n$ such that $X + \Delta \in \mathrm{int}\,K_n$, then

$$\begin{aligned}
F(X+\Delta) - F(X) &= -\ln\det(X+\Delta) + \ln\det X\\
&= -\ln\det X^{1/2}\left(I_n + X^{-1/2}\Delta X^{-1/2}\right)X^{1/2} + \ln\det X\\
&= -\ln\det\left(I_n + X^{-1/2}\Delta X^{-1/2}\right)\\
&= -\ln\prod_{i=1}^{n}\lambda_i\left(I_n + X^{-1/2}\Delta X^{-1/2}\right).
\end{aligned}$$

Then, using the inequality between geometric and arithmetic means

$$\sqrt[n]{\prod_{i=1}^{n}\lambda_i} \le \frac{1}{n}\sum_{i=1}^{n}\lambda_i\,,$$

we obtain

$$\prod_{i=1}^{n} \lambda_i(B) \leq \left(\frac{1}{n} \operatorname{Tr} B\right)^n ,$$

where $B = I_n + X^{-1/2} \Delta X^{-1/2}$. Therefore, applying identity (6.76) for matrices $X^{-1/2}$ and Δ, we obtain

$$\begin{aligned} F(X+\Delta) - F(X) &\geq -\ln\left[\frac{1}{n} \operatorname{Tr}\left(I_n + X^{-1/2} \Delta X^{-1/2}\right)\right]^n \\ &= -n \ln\left[1 + \frac{1}{n}\left\langle I_n, X^{-1/2} \Delta X^{-1/2}\right\rangle_{\mathrm{F}}\right]. \end{aligned}$$

Keeping in mind $-\ln(1+x) \geq -x$ and using again the identity (6.76), we obtain

$$F(X+\Delta) - F(X) \geq -\left\langle I_n, X^{-1/2} \Delta X^{-1/2}\right\rangle_{\mathrm{F}} = -\left\langle X^{-1}, \Delta\right\rangle_{\mathrm{F}}.$$

Thus, $-X^{-1} \in \partial F(X)$; therefore, for strictly convex F, we have $\nabla F(x) = -X^{-1}$.

Let us now consider matrix $A = B - I_n = X^{-1/2} \Delta X^{-1/2}$ and eigenvalues $\lambda_i = \lambda_i(A)$ of A. Then

$$\langle \nabla F(X), \Delta \rangle_F = -\left\langle X^{-1}, \Delta\right\rangle_F = -\langle I_n, A \rangle_F = -\operatorname{Tr} A = -\sum_{i=1}^{n} \lambda_i(A).$$

On the other hand, we consider

$$\begin{aligned} f(t) &= F(X + t\Delta) = -\ln\det(X + t\Delta) \\ &= -\ln\det\left(\sqrt{X}(I_n + tA)\sqrt{X}\right) \\ &= -\ln\det X - \ln\det(I_n + tA) \\ &= -\ln\det X - \ln\prod_{i=1}^{n}(1 + t\lambda_i(A)) \\ &= -\ln\det X - \sum_{i=1}^{n}\ln(1 + t\lambda_i(A)). \end{aligned}$$

Therefore,

$$\begin{aligned} \langle \nabla F(X), \Delta \rangle_{\mathrm{F}} &= \langle \nabla F(X + t\Delta), \Delta \rangle_{\mathrm{F}|t=0} \\ &= f'(0) = \sum_{i=1}^{n} \lambda_i(A)(1 + t\lambda_i(A))^{-1}_{|t=0} = -\sum_{i=1}^{n} \lambda_i(A). \end{aligned}$$

To find $\langle \nabla^2 F(X)\Delta, \Delta \rangle_{\mathrm{F}}$, we first consider

$$\varphi(t) = \langle \nabla F(X + t\Delta, \Delta \rangle_{\mathrm{F}}, \quad 0 \leq t \leq 1.$$

Then,

$$\begin{aligned}\varphi(t)-\varphi(0) &= \left\langle X^{-1}-(X+t\Delta)^{-1},\Delta\right\rangle_{\mathrm{F}}\\ &= \left\langle (X+t\Delta)^{-1}[(X+t\Delta)-X]X^{-1},\Delta\right\rangle_{\mathrm{F}}\\ &= t\left\langle (X+t\Delta)^{-1}\Delta X^{-1},\Delta\right\rangle_{\mathrm{F}},\end{aligned}$$

or

$$\varphi'(0)=\langle \nabla^2 F(X)\Delta,\Delta\rangle_{\mathrm{F}}=\left\langle X^{-1}\Delta X^{-1},\Delta\right\rangle_{\mathrm{F}}.$$

On the other hand,

$$\begin{aligned}&\langle \nabla^2 F(X)\Delta,\Delta\rangle_{\mathrm{F}} = \langle \nabla^2 F(X+t\Delta)\Delta,\Delta\rangle_{\mathrm{F}|t=0}\\ &= f''(0)=(f'(t))_{|t=0}=\sum_{i=1}^n \lambda_i^2(A)(1+t\lambda_i(A))^{-2}_{|t=0}=\sum_{i=1}^n \lambda_i^2(A).\end{aligned}$$

Finally,

$$\begin{aligned}&D^3F(X)[\Delta,\Delta,\Delta]=D^3F(X+t\Delta)[\Delta,\Delta,\Delta]_{|t=0}\\ &= f'''(0)=(f''(t))'_{|t=0}=-2\sum_{i=1}^n \lambda_i^3(A)(1+t\lambda_i(A))^{-3}_{|t=0}=-2\sum_{i=1}^n \lambda_i^3(A).\end{aligned}$$

We recall that the function $F(X)$ is self-concordant if for any $\Delta\in S^{n,n}$ we have

$$|f'''(0)|\le 2f''(0)^{3/2},$$

i.e.,

$$\left|\sum_{i=1}^n \lambda_i^3(A)\right|\le \left(\sum \lambda_i^2(A)\right)^{3/2},$$

which is a standard inequality.

To show that the self-concordant function F is a n-self-concordant barrier, we have to check the inequality

$$f'(0)^2\le nf''(0)$$

or

$$\left(\sum_{i=1}^n \lambda_i\right)^2\le n\sum_{i=1}^n \lambda_i^2,$$

which is also a standard inequality. Therefore, F is a n-self-concordant barrier and can be used in the path-following method, which we consider later.

Lemma 6.7. *For any self-concordant barrier F for the cone K_n, we have $\nu\ge n$.*

Proof. We consider $\bar{X}=I_n\in \operatorname{int} K_n$ and the direction $P_i=e_ie_i^T$, $i=1,\dots,n$, where $e_i=(0,\dots,1,\dots,0)\in\mathbb{R}^n$. For any $t\ge 0$, we have $I_n+tP_i\in \operatorname{int} K_n$. Also, $I_n-P_i\notin \operatorname{int} K_n$ and $I_n-\sum_{i=1}^n P_i=0\in K_n$. Therefore, the conditions of Lemma 6.6 are satisfied with $\alpha_i=\beta_i=1$, $i=1,\dots,n$ and $\nu\ge\sum_{i=1}^n \alpha_i/\beta_i=n$. Thus, the parameter $\nu=n$ is optimal for the ν-self-concordant function F. □

Let us now consider the path-following method for problem (6.77). It follows from Lemma 6.3 that the restriction of F onto the set

$$\mathscr{L} = \{X : \langle A_i, X\rangle_F = b_i, \quad i = 1,\dots,m\}$$

is a n-self-concordant barrier. Therefore, the complexity bound of the path-following method is $O(\sqrt{n}\ln(n/\varepsilon))$ steps.

To estimate the arithmetical cost of each iteration, we describe one Newton step for problem (6.77). We have to minimize the barrier function

$$b(X,t) = t\langle C, X\rangle_F + F(X) = t\langle C, X\rangle_F - \ln\det X$$

s.t.

$$\langle A_i, X\rangle_F = b_i, \quad i = 1,\dots,m.$$

We start with $X \succ 0 : \langle A_i, X\rangle_F = b_i,\ i = 1,\dots,m$. To find the Newton direction Δ, we have to solve the following problem

$$\min\Big\{\langle \nabla_x b(x,t), \Delta\rangle_F + \frac{1}{2}\langle \nabla^2_{xx} b(x,t)\Delta, \Delta\rangle_F \tag{6.78}$$
$$= \left\langle tC - X^{-1}, \Delta\right\rangle_F + \frac{1}{2}\left\langle X^{-1}\Delta X^{-1}, \Delta\right\rangle_F : \langle A_i, \Delta\rangle_F = 0,\ i = 1,\dots,m\Big\}$$

where $X \succ 0$.

Problem (6.78) is equivalent to the system of equations

$$U + X^{-1}\Delta X^{-1} = \sum_{i=1}^m \lambda_i A_i \tag{6.79}$$

$$\langle A_i, \Delta\rangle_F = 0, \tag{6.80}$$

where $U = tC - X^{-1}$. From (6.79), we get

$$\Delta = X\left[-U + \sum_{i=1}^m \lambda_i A_i\right] X. \tag{6.81}$$

After substituting Δ from (6.81) into (6.80), we obtain

$$\sum_{j=1}^m \lambda_i \langle A_i, XA_jX\rangle_F = \langle A_i, XUX\rangle_F, \quad i = 1,\dots,m. \tag{6.82}$$

System (6.82) can be rewritten in the following form:

$$D\lambda = d, \tag{6.83}$$

where $D = (d_{ij})^n_{i=1,j}$ and $d_{ij} = \langle A_i, XA_jX\rangle_F$ and $d_j = \langle U, XA_jX\rangle_F$. The cost of computing the Newton direction Δ has four components:

(1) Computing XA_jX, $j = 1, \dots, m$ – $O(mn^3)$ operations
(2) Computing D and d – $O(m^2n^2)$ operations
(3) Solving system (6.83) – $O(m^3)$ operations
(4) Finding Δ from (6.81) – $O(mn^2)$ operations

Exercise 6.5. Find complexity bound for one Newton step.

6.7 Primal–Dual Predictor–Corrector for LP

We saw in Section 6.6 that $O(\sqrt{n}\ln(n\varepsilon^{-1}))$ is the complexity bound for the primal path-following method for LP. It is the best bound so far in terms of the number of steps required for finding an approximation for the solution in value with accuracy $\varepsilon > 0$.

In this section, we consider the primal–dual predictor–corrector (PC) method for LP, which has the same complexity bound. There are three main reasons, however, for our choice.

First, primal–dual PC has proven to be very efficient in practice.

Second, the primal–dual approach originally introduced for LP calculations later became very productive in nonlinear and semidefinite programming.

Third, we would like to set the stage for the primal–dual approach, which will be discussed later for nonlinear optimization.

Let us consider the standard LP

$$\langle c, x^* \rangle = \min\{\langle c, x \rangle : Ax = b,\ x \geq 0\}, \tag{6.84}$$

where $c,\ x \in \mathbb{R}^n$, $b \in \mathbb{R}^m$ and $A \in \mathbb{R}^m \times \mathbb{R}^n$ is a full rank matrix, that is, rank $A = m < n$. The dual LP finds

$$\langle b, \lambda^* \rangle = \max\left\{\langle b, \lambda \rangle : A^T\lambda \leq c\right\} \tag{6.85}$$

and can be rewritten as follows:

$$\langle b, \lambda^* \rangle = \max\left\{\langle b, \lambda \rangle : A^T\lambda + s = c,\ s \geq 0\right\}. \tag{6.86}$$

From the primal–dual system

$$Ax = b, \quad x \geq 0, \quad s = c - A^T\lambda \geq 0 \tag{6.87}$$

follows weak duality

$$\langle c, x \rangle \geq \langle b, \lambda \rangle.$$

Therefore, by adding one inequality

$$\langle c, x \rangle \leq \langle b, \lambda \rangle, \tag{6.88}$$

we replace the primal (6.84) and dual (6.86) LPs by the following primal–dual system:

$$Ax = b, \tag{6.89}$$

$$s = c - A^T\lambda \geq 0, \tag{6.90}$$

$$x \geq 0, \tag{6.91}$$

$$\langle c,x\rangle \leq \langle b,\lambda\rangle . \tag{6.92}$$

In other words, one can solve the primal and dual LP problem by finding nonnegative vectors x and s from the following system of equations

$$\begin{pmatrix} A^T\lambda + s - c \\ Ax - b \\ XSe \end{pmatrix} = \begin{pmatrix} 0 \\ 0 \\ 0 \end{pmatrix} \tag{6.93}$$

where $X = \mathrm{diag}(x_i)_{i=1}^n$, $S = \mathrm{diag}(s_i)_{i=1}^n$, $e = (1,\ldots,1) \in \mathbb{R}^n$.

System (6.93) is, in fact, the KKT condition for $w = (x,\lambda,s) : x \in \mathbb{R}^n_+$, $s \in \mathbb{R}^n_+$ to be the solution of the primal and dual LP. System (6.93) is nonlinear due to the complementarity condition $XSe = 0$.

Let

$$\Omega = \left\{(x,\lambda,s) : Ax = b, A^T\lambda + s = c,\ x \in \mathbb{R}^n_+,\ s \in \mathbb{R}^n_+\right\}$$

and

$$\Omega_0 = \left\{w = (x,\lambda,s) : Ax = b, A^T\lambda + s = c,\ x \in \mathbb{R}^n_{++},\ s \in \mathbb{R}^n_{++}\right\}.$$

Application of Newton's method to system (6.93) with $(x,\lambda,s) \in \Omega_0$ as a starting point leads to the following primal–dual linear system for finding Newton direction $\Delta w = (\Delta x, \Delta\lambda, \Delta s)$:

$$\begin{pmatrix} 0 & A^T & I \\ A & 0 & 0 \\ S & 0 & X \end{pmatrix} \begin{pmatrix} \Delta x \\ \Delta\lambda \\ \Delta s \end{pmatrix} = \begin{pmatrix} 0 \\ 0 \\ XSe \end{pmatrix}. \tag{6.94}$$

Direction Δw leads toward the *KKT* point. It will play an important role later.

The Newton step consists of finding $\hat{w} = w + \alpha\Delta w$, that is,

$$\hat{x} = x + \alpha\Delta x,\ \hat{\lambda} = \lambda + \alpha\Delta\lambda,\ \hat{s} = s + \alpha\Delta s. \tag{6.95}$$

The step length $\alpha > 0$ has to be such that $\hat{w} \in \Omega_0$. On the other hand, in order to make reasonable progress in Newton direction Δw, we need $\alpha > 0$ to be not too small. If w is too close to the boundary of Ω, our desire to stay in Ω_0 might lead to very small $\alpha > 0$. Thus, we have to modify system (6.93) to guarantee that Newton sequence given by (6.95) stays away from the boundary Ω until we get very close to the solution. Application of a log-barrier function for problem (6.84) leads to such modification.

We saw in Chapter 5 that primal problem (6.84) can be replaced by a sequence of problems

$$F(x,t) = \langle c,x\rangle - t\sum_{i=1}^{n} \ln x_i \to \min \tag{6.96}$$

$$Ax = b\,. \tag{6.97}$$

Solution $x(t)$ of problem (6.96)–(6.97) converges to x^* when $t \to 0$. Let us consider optimality condition for problem (6.96)–(6.97). We have

$$\nabla_x F(x,t) = c - tX^{-1}e - A^T\lambda = 0\,,$$
$$Ax = b\,.$$

By introducing $S = tX^{-1}$, we obtain the following system:

$$A^T\lambda + Se - c = 0\,, \tag{6.98}$$

$$Ax - b = 0\,, \tag{6.99}$$

$$x \geq 0,\ \ s \geq 0\,. \tag{6.100}$$

$$XSe = te \tag{6.101}$$

Let us consider $t_0 > 0$, fix $0 < t < t_0$, and solve system (6.98)–(6.101) for (x,λ,s).

The solution vector $w(t) = (x(t),\lambda(t),s(t))$ belongs to the central trajectory $\{w(t),\, t \in [t_0,0)\}$, which is called the central path. It follows from results of Chapter 5 that

$$\lim_{t\to 0} w(t) = w^* = (x^*,\lambda^*,s^*)\,. \tag{6.102}$$

Every path-following method is tracking the central path $\{w(t)\}$, $t \in (0,t_0])$.

Two things are critical for the quality of any path-following method:

1. First, one has to guarantee a reasonable reduction of the barrier parameter at each step.
2. Second, the primal–dual approximation has to be rather close to the central path to avoid any "contact" with the boundary of Ω.

For a given $w = (x,\lambda,s)$, x and $s \in \mathbb{R}^n_{++}$, we compute $t = n^{-1}(x,s)$. If $t = 0$, then $w = w^*$.

If $t > 0$, then we consider the primal–dual direction $\Delta w = (\Delta x,\, \Delta\lambda,\, \Delta s)$ defined by the following primal–dual system

$$\begin{pmatrix} 0 & A^T & I \\ A & 0 & 0 \\ S & 0 & X \end{pmatrix} \begin{pmatrix} \Delta x \\ \Delta\lambda \\ \Delta s \end{pmatrix} = \begin{pmatrix} 0 \\ 0 \\ -XSe + \sigma te \end{pmatrix} \tag{6.103}$$

where $\sigma \in [0,1]$ is called the *centering parameter*.

The predictor–corrector (PC) method uses only $\sigma = 0$ or $\sigma = 1$. For $\sigma = 0$, system (6.103) turns into (6.94). It finds direction $\Delta w_0 = (\Delta x_0, \Delta \lambda_0, \Delta s_0)$ towards the KKT point defined by (6.93).

For $\sigma = 1$, it finds the centering direction $\Delta w_1 = (\Delta x_1, \Delta \lambda_1, \Delta s_1)$ towards the solution of (6.98)–(6.101).

The PC method alternates the *predictor* step ($\sigma = 0$) to reduce the barrier parameter $t > 0$, with *corrector* step ($\sigma = 1$) to improve the centrality.

For a given $0 < \theta < 1$, let us consider the neighborhood

$$N(\theta) = \{w = (x, \lambda, s) \in \Omega_0 : \|XSe - te\| \leq \theta t\},$$

of the central path $\{w(t)\}, t \in (0, t_0]$.

Note that for $0 < \theta_1 < \theta_2 < 1$, we have $N(\theta_1) \subset N(\theta_2)$.

To describe the PC method, we need some preliminary results. Let $w = (x, \lambda, s) \in \Omega_0$, then $t = n^{-1} \sum_{i=1}^{n} x_i s_i$.

The predictor direction $\Delta w = (\Delta x, \Delta \lambda, \Delta s)$ we find from (6.103) with $\sigma = 0$, that is, by solving system (6.94). Let $0 < \alpha < 1$ be the step length in the direction Δw. We consider

$$w(\alpha) = (x(\alpha), \lambda(\alpha), s(\alpha)) = (x + \alpha \Delta x, \lambda + \alpha \Delta \lambda, s + \alpha \Delta s) \tag{6.104}$$

and

$$t(\alpha) = n^{-1} \sum_{i=1}^{n} x_i(\alpha) s_i(\alpha). \tag{6.105}$$

Lemma 6.8. *If Δw is the solution of system (6.94), then*

$$\langle \Delta x \ , \Delta s \rangle = 0, \tag{6.106}$$

$$t(\alpha) = (1 - \alpha) t. \tag{6.107}$$

Proof. From the first two equations of system (6.94), we have

$$A^T \Delta \lambda + \Delta s = 0,$$
$$A \Delta x = 0.$$

Multiplying the first equation by Δx and keeping in mind the second one, we obtain

$$\langle \Delta x, A^T \Delta \lambda \rangle + \langle \Delta x, \Delta s \rangle = \langle A \Delta x, \Delta \lambda \rangle + \langle \Delta x, \Delta s \rangle = \langle \Delta x, \Delta s \rangle = 0,$$

that is, we have (6.106).

From up the third equation (6.103) with $\sigma = 0$, we obtain

$$\langle s, \Delta x \rangle + \langle x, \Delta s \rangle = -\langle x, s \rangle. \tag{6.108}$$

From (6.105) we have

$$t(\alpha) = n^{-1} \langle x + \alpha \Delta x, s + \alpha \Delta s \rangle = n^{-1} [\langle x, s \rangle + \alpha (\langle s, \Delta x \rangle + \langle x, \Delta s \rangle) + \alpha^2 \langle \Delta x, \Delta s \rangle].$$

Then, from (6.106) and (6.108) follows (6.107). □

We would like to have $0 < \alpha < 1$ as large as possible. On the other hand, we have to keep the approximation $w(\alpha)$ far enough from the boundary Ω

We consider two neighborhoods of the central path: $N(0.25)$ and $N(0.5)$. The choice of the neighborhood sizes will be clear later. Without losing generality, we can assume $x_1 s_1 = \min\{x_i s_i \mid 1 \leq i \leq n\}$. Then, from the definition $N(\theta)$, we have

$$|x_1 s_1 - t| \leq \|XSe - te\| \leq \theta t,$$

that is,

$$\min_{1 \leq i \leq n} x_i s_i = x_1 s_1 \geq (1 - \theta)t. \tag{6.109}$$

Now we need the following technical lemma.

Lemma 6.9. *Let u and v be two vectors from $\mathbb{R}^n$ with $\langle u, \mathrm{v} \rangle \geq 0$. Then,*

$$\|UVe\| \leq 2^{-(3/2)} \|u + \mathrm{v}\|^2, \tag{6.110}$$

where $U = \mathrm{diag}(u_i)_{i=1}^n$, $V = \mathrm{diag}(\mathrm{v}_i)_{i=1}^n$.

Proof. Let $I_+ = \{i : u_i \mathrm{v}_i \geq 0\}$, $I_- = \{i : u_i \mathrm{v}_i < 0\}$. It follows from $\langle u, \mathrm{v} \rangle \geq 0$ that

$$\sum_{i \in I_+} u_i \mathrm{v}_i \geq \sum_{i \in I_-} |u_i \mathrm{v}_i|. \tag{6.111}$$

Also, for any $i \in I_+$, we have

$$\sqrt{u_i \mathrm{v}_i} = \sqrt{|u_i|\,|\mathrm{v}_i|} \leq \frac{1}{2}(|u_i| + |\mathrm{v}_i|) = \frac{1}{2}|u_i + \mathrm{v}_i|. \tag{6.112}$$

Now from (6.111), we have

$$\|UVe\| = \left(\sum_{i \in I_+} (u_i \mathrm{v}_i)^2 + \sum_{i \in I_-} (u_i \mathrm{v}_i)^2 \right)^{1/2} \leq \left(2 \sum_{i \in I_+} (u_i \mathrm{v}_i)^2 \right)^{1/2} \leq \sqrt{2} \left(\sum_{i \in I_+} u_i \mathrm{v}_i \right).$$

It follows from (6.112) that

$$\sqrt{2} \left(\sum_{i \in I_+} u_i \mathrm{v}_i \right) \leq \frac{\sqrt{2}}{4} \sum_{i \in I_+} (u_i + \mathrm{v}_i)^2 \leq 2^{-3/2} \sum_{i \in 1}^{n} (u_i + \mathrm{v}_i)^2 = 2^{-3/2} \|u + \mathrm{v}\|^2.$$

□

We are ready to describe the PC method.

Initialization

Let $w = (x, \lambda, s) \in N(0.25)$ and $\varepsilon > 0$ is the given accuracy.

1. Find $t = n^{-1} \sum_{i=1}^n x_i s_i$. If $0 < t \leq \varepsilon$, then $w^* := w$.

2. Predictor step: solve system (6.103) with $\sigma = 0$ for finding direction $\Delta w_0 = (\Delta x_0, \Delta\lambda_0, \Delta s_0)$ toward the KKT point.
3. Find $\alpha = \max\{0 < \tau \le 1 : (x + \tau\Delta x_0, \lambda + \tau\Delta\lambda_0, s + \tau\Delta s_0) \in N(0.5)\}$.
4. Set $w := w + \alpha\Delta w$.
5. Corrector step: solve system (6.103) with $\sigma = 1$ for finding the centering direction $\Delta w_1 = (\Delta x_1, \Delta\lambda_1, \Delta s_1)$.
6. Set $w := w + \Delta w_1$. Go to 1.

It follows from (6.107) that each predictor step reduces $t > 0$ by a factor $(1-\alpha)$. Our purpose now is to find the lower bound for $\alpha > 0$, keeping in mind that we start the predictor step from $w \in N(0.25)$ and

$$\alpha = \max\{\tau : w + \tau\Delta w \in N(0.5)\}. \tag{6.113}$$

We start with the following lemma.

Lemma 6.10. *If $w \in N(\theta)$ and $\Delta w = (\Delta x, \Delta\lambda, \Delta s)$ is found from system (6.103), then*

$$\|\Delta X \Delta S e\| \le \frac{\theta^2 + n(1-\sigma)^2}{2^{3/2}(1-\theta)}\, t. \tag{6.114}$$

Proof. We consider a diagonal matrix $D = S^{-1/2}X^{1/2}$. Let us multiply the third equation in (6.103) by $(XS)^{-1/2}$. We then obtain

$$D^{-1}\Delta x + D\Delta s = (XS)^{-1/2}(-XSe + \sigma te).$$

Now, we apply (6.110) with $u = D^{-1}\Delta x$, $v = D\Delta s$ to obtain

$$\begin{aligned} \|\Delta X \Delta S e\| &= \left\|\left(D^{-1}\Delta X\right)(D\Delta S)e\right\| \le 2^{-3/2}\left\|D^{-1}\Delta x + D\Delta s\right\|^2 \\ &= 2^{-3/2}\left\|(XS)^{-1/2}(-XSe + \sigma te)\right\|^2 \le 2^{-3/2}\frac{\|XSe - \sigma te\|^2}{\min_{1\le i\le n} x_i s_i} \end{aligned} \tag{6.115}$$

Since $w = (x, \lambda, s) \in N(\theta)$ and keeping in mind (6.109), we have

$$\min_{1\le i\le n} x_i s_i \ge (1-\theta)t \ \text{ and } \ \|XSe - te\| \le \theta t.$$

Then,

$$\begin{aligned} \|XSe - \sigma te\|^2 &= \|(XSe - te) + (1-\sigma)te\|^2 \\ &= \|XSe - te\|^2 + 2(1-\sigma)t\langle e, XSe - te\rangle + (1-\sigma)^2 t^2 \langle e, e\rangle. \end{aligned}$$

Note that $\langle XSe - te, e\rangle = 0$. From $w \in N(\theta)$ follows

$$\|XS - \sigma te\|^2 = \|XSe - te\|^2 + (1-\sigma)^2 t^2 n \le \theta^2 t^2 + (1-\sigma)^2 t^2 n.$$

Therefore, from (6.115) we obtain

$$\|\Delta X \Delta S e\| \le 2^{-3/2} \frac{\|XSe - \sigma t e\|^2}{\min_{1\le i\le n} x_i s_i} \le \frac{\theta^2 + n(1-\sigma)^2}{2^{3/2}(1-\theta)} t .$$

□

For $\theta = 0.25$ and $\sigma = 0$, we have

$$\|\Delta X \Delta S e\| \le \frac{1+16n}{3\sqrt{2}} t .$$

The following lemma establishes the lower bound for the step length $\alpha > 0$.

Lemma 6.11. *Let $w = (x, \lambda, s) \in N(0.25)$ and $\Delta w = (\Delta x, \Delta\lambda, \Delta s)$ be solution of (6.94), then $w(\alpha) \in N(0.5)$ for all $\alpha \in [0, \bar{\alpha}]$, where*

$$\bar{\alpha} = \min\left\{0.5, \left(\frac{t}{8\|\Delta X \Delta S e\|}\right)^{1/2}\right\} . \tag{6.116}$$

Thus, the predictor step is at least $\bar{\alpha}$, and the value of the parameter t in one step can be reduced at least by a factor of $(1-\bar{\alpha})$.

Proof. From the third equation in (6.94) and (6.107), we get

$$\begin{aligned} x_i(\alpha)s_i(\alpha) - t(\alpha) &= x_i s_i + \alpha(s_i \Delta x_i + x_i \Delta s_i) + \alpha^2 \Delta x_i \Delta s_i - (1-\alpha)t \\ &= x_i s_i (1-\alpha) + \alpha^2 \Delta x_i \Delta s_i - (1-\alpha)t . \end{aligned}$$

Considering diagonal matrices $X(\alpha)$ and $S(\alpha)$ with the corresponding components, we obtain

$$\begin{aligned} \|X(\alpha)S(\alpha)e - t(\alpha)e\| &= \|XSe(1-\alpha) - (1-\alpha)te + \alpha^2 \Delta X \Delta S e\| \\ &\le (1-\alpha)\|XSe - te\| + \alpha^2 \|\Delta X \Delta S e\| . \end{aligned} \tag{6.117}$$

For the predictor step, we have $w \in N(0.25)$ as a starting point. Therefore,

$$\|XSe - te\| \le 0.25t .$$

Assuming $\alpha = (t/(8\|\Delta X \Delta S e\|))^{1/2} \le 0.5$, we obtain from (6.117)

$$\begin{aligned} &\|X(\alpha)S(\alpha)e - t(\alpha)e\| \le (1-\alpha)\|XSe - te\| \\ &\quad + \frac{t}{8\|\Delta X \Delta S e\|}\|\Delta X \Delta S e\| \le 0.25(1-\alpha)t + 0.125t \\ &\le 0.25(1-\alpha)t + 0.25(1-\alpha)t = 0.5(1-\alpha)t \\ &= 0.5t(\alpha) . \end{aligned} \tag{6.118}$$

Therefore, $w(\alpha) = (x(\alpha), \lambda(\alpha), s(\alpha)) \in N(0.5)$. Also, it follows from (6.118) that

$$x_i(\alpha)s_i(\alpha) - t(\alpha) \ge -0.5t(\alpha)$$

or

$$x_i(\alpha)s_i(\alpha) \geq 0.5t(\alpha) > 0, \quad i = 1, \ldots, n.$$

Thus, $w(\alpha)$ satisfies both the proximity condition for $N_2(0.5)$ and the strict feasibility condition. It means that if $\alpha = (t/(8\|\Delta X \Delta Se\|))^{1/2} \leq 0.5$, then the approximation $w(\alpha)$ belongs to $N_2(0.5)$ and is strictly feasible.

Thus, $w(\alpha)$ can be a starting point for the corrector step. □

Let us now show that a corrector step with a starting point $w = (x, \lambda, s) \in N(0.5)$ will put the approximation $\bar{w} = w + \Delta w$ back on the central path.

Solving system (6.103) with $\sigma = 1$, we obtain $\Delta w = (\Delta x, \Delta \lambda, \Delta s)$. The result of the corrector step is

$$(x(1), \lambda(1), s(1)) = (x, \lambda, s) + (\Delta x, \Delta \lambda, \Delta s).$$

Keeping in mind $\langle \Delta x, \Delta s \rangle = 0$, we obtain

$$\begin{aligned} \langle x(1), s(1) \rangle &= \langle x + \Delta x, s + \Delta s \rangle = \langle x, s \rangle + \langle s, \Delta x \rangle + \langle x, \Delta s \rangle + \langle \Delta x, \Delta s \rangle \\ &= \langle x, s \rangle + \langle s, \Delta x \rangle + \langle x, \Delta s \rangle. \end{aligned}$$

Using the third equation in (6.103) with $\sigma = 1$, we have

$$\langle s, \Delta x \rangle + \langle x, \Delta s \rangle = -\langle x, s \rangle + tn.$$

Therefore, $\langle x(1), s(1) \rangle = t(1)n$, i.e. $t = t(1)$. To show that $(x(1), s(1), \lambda(1)) \in N(0.25)$, let us consider

$$\|X(\alpha)S(\alpha)e - t(\alpha)e\| = |1 - \alpha| \, \|XSe - t(\alpha)e\| + \alpha^2 \|\Delta X \Delta Se\|.$$

Using Lemma 6.10, we obtain

$$\|X(\alpha)S(\alpha)e - t(\alpha)e\| \leq |1 - \alpha| \, \|XSe - t(\alpha)e\| + \alpha^2 \left[\frac{\theta^2 + n(1-\sigma)^2}{2^{3/2}(1-\theta)} \right] t.$$

Therefore, for $\alpha = 1$, $\sigma = 1$ and $\theta = 0.5$, we have

$$\|X(1)S(1)e - te\| \leq \frac{1}{4\sqrt{2}} t < \frac{t}{4}.$$

Hence, as a result of the corrector step, we obtain an approximation $w = (x, \lambda, s)$, which belongs to $N(0.25)$.

The only thing that remains is to estimate $\bar{\alpha}$ from below, i.e., to find the lower bound for $(t(8\|\Delta X \Delta Se\|)^{-1})^{1/2}$. Keeping in mind that $\sigma = 0$ and using the bound (6.114) for $\|\Delta X \Delta Se\|$ from Lemma 6.10 for $\theta = 0.25$, we have

$$\bar{\alpha} \geq \left(\frac{2^{3/2}(1-\theta)}{8(\theta^2 + n)} \right)^{1/2} \geq \left(\frac{0.24}{n} \right)^{1/2} > \frac{0.4}{\sqrt{n}}.$$

Therefore, from (6.107) we have

$$t_{k+1} = t_k \left(1 - \frac{0.4}{\sqrt{n}}\right)$$

for every predictor step.

Notes

The SC theory was developed by Yurii Nesterov and Arkadi Nemirovski in the late 1980s; see Nesterov and Nemirovski (1994) and Nesterov (2004). Not only SC theory provide a unified approach for analyzing IPMs, but it also extends the IPM approach for a wide class of CO problems and semidefinite optimization; see also Wolkowicz et al. (2000) and references therein.

The primal–dual aspects of IPMs were covered in an informative book Wright (1997) by S. Wright; see also Jensen et al. (1994), Kojima et al. (1989a), Mehrotra (1990, 1992) and Megiddo (1989), Megiddo and Shub (1989), Monteiro et al. (1990), Potra (1994), Potra and Wright (2000).

For the computational aspects of IPMs, see Gill et al. (1986), Lustig et al. (1991, 1992, 1994), Pardalos et al. (1990), Potra (1994), Potra and Wright (2000).

Other aspects of IPMs were considered in the following books: Renegar (2001), Roos et al. (1997), and Ye (1997).

For the key role of the LF invariant in the SC theory, see Polyak (2016) .

Chapter 7
Nonlinear Rescaling: Theory and Methods

7.0 Introduction

The first result on Nonlinear Rescaling (NR) theory and methods was obtained in the early 1980s. The purpose was finding an alternative for SUMT free from its well-known shortcomings (see Chapter 5). Roughly speaking, for inequality-constrained optimization, the NR is to SUMT as AL is to Courant's penalty method for ECO. Unfortunately, the NR results were published almost 10 years later.

The nonlinear rescaling principle consists of transforming the objective function and/or the constraints of a given constrained optimization problem into an equivalent one and using classical Lagrangian for equivalent problem (LEP) for both theoretical analysis and numerical methods.

One of the NR schemes, which leads to a wide class of NR methods, employs smooth, strictly concave, and monotone increasing functions $\psi \in \Psi$ with particular properties to transform the original set of constraints into an equivalent set. The transformation is scaled by a positive scaling (penalty) parameter. The LEP is the main NR instrument.

Each NR step finds an approximation for LEP's primal minimizer, which is used for LM update. The scaling parameter can be fixed or updated from step to step.

In the early 1990s, Teboulle (1992) systematically studied proximal point methods with entropy-like distance for maximization concave functions on $\mathbb{R}^n_+$.

The NR and prox methods are coming from different parts of the world of optimization and, as it seemed, have nothing in common. In fact, they turned out to be equivalent; see Polyak and Teboulle (1997).

The interplay between the Lagrangian for the original problem and the LEP is the key element in proving the equivalence. Thus, any exterior primal NR method is equivalent to interior dual proximal point method with first-order φ – divergence entropy distance. The kernel of the distance $\varphi = -\psi^*$, where ψ^* is LF transform of ψ.

R. A. Polyak, *Introduction to Continuous Optimization*, Springer Optimization and Its Applications 172, https://doi.org/10.1007/978-3-030-68713-7_7

The fundamental difference between SUMT and NR is the active role of the Lagrange multipliers in the latter case. The NR methods converge under any fixed positive scaling parameter, just due to the LM update. Moreover, the convergence is taking place under very mild assumptions on the input data.

Important advantage of NR methods is their convergence in the absence of convexity assumptions. If the input data is smooth enough and the second-order sufficient optimality condition is satisfied, then NR method generates primal–dual sequence, which converges to the primal–dual solution with Q-linear rate under a fixed, but large enough scaling parameter, no matter whether the problem is convex or not.

By increasing the scaling parameter from step to step, one can achieve Q-superlinear convergence rate, while SUMT methods under the same conditions converge with sublinear rate (see Chapter 5).

Another important advantage of LEPs is their strong convexity and smoothness in the neighborhood of the primal solution. Therefore, regularized Newton method is particularly suitable for finding approximations for primal minimizer.

The NR framework increases substantially the efficiency of Newton's method toward the end of calculations, because from some point on, the approximation for the primal minimizer remains in the Newton area after each LM update.

It leads for the nondegenerate constrained optimization problems to the "hot" start phenomenon. Let $\varepsilon > 0$ be the required accuracy; then from some point on, only $O(\ln\ln\varepsilon^{-1})$ Newton's steps is enough for LM update, which reduces the distance from current approximation to the primal–dual solution by a factor $0 < \gamma < 1$ that is inversely proportional to the scaling parameter $k > 0$.

In the second part of the chapter, one scaling parameter for all constraints is replaced by a vector of scaling parameters, one for each constraint.

The correspondent NR method, at each step, finds an approximation for the primal LEP's minimizer, which is used for the LM update. Then the new LM vector is used for scaling parameters' update.

The LM vector is updated traditionally, while scaling parameters are updated inversely proportional to the correspondent Lagrange multipliers.

Such "dynamic" scaling parameters' update for exponential transformation was first suggested by Tseng and Bertsekas (1993).

According to the authors, the convergence analysis of such NR methods even for a particular exponential transformation is "proven to be surprisingly difficult."

For such scheme the convergence is proven for a wide class of constraints transformations. The key arguments are based on the equivalence of such NR methods and interior quadratic prox for the dual problem in rescaled from step to step dual space.

The equivalence allows to prove convergence of the NR methods in value under very mild assumptions on the input data and establish convergence rate under some extra assumption.

In the third part of the chapter, we concentrate on the primal–dual aspect of the NR.

The standard NR step is equivalent to solving a primal–dual (PD) nonlinear system of equations made up by the optimality criteria for the primal minimizer and formulas for the LM update.

Application of Newton's method to the PD system leads to the primal–dual NR (PDNR) methods.

Under the second-order sufficient optimality condition, such PDNR method converges globally with asymptotic quadratic rate.

We conclude the chapter by considering Nonlinear Rescaling–Augmented Lagrangian (NRAL) method for convex optimization with both inequality constraints and equations. The NR technique is used for inequality constraints and AL for equations.

7.1 Nonlinear Rescaling

7.1.1 Preliminaries

Throughout the chapter we are mainly concerned with the following convex optimization (CO) problem.

Let $f:\mathbb{R}^n\rightarrow\mathbb{R}$ be convex and all c_i: $\mathbb{R}^n\rightarrow\mathbb{R}$ $i=1,\dots,m$ be concave. The primal CO problem finds

$$f(x^*)=\inf\{f(x)|x\in\Omega\}, \tag{P}$$

where

$$\Omega=\{x\in\mathbb{R}^n : c_i(x)\geq 0,\ i=1,\dots,m\}. \tag{7.1}$$

We assume

A. The primal optimal set

$$X^*=\{x\in\Omega : f(x)=f(x^*)\}$$

is not empty and bounded.

B. Slater condition

$$\exists x_0\in\Omega : c_i(x_0)>0, i=1,\dots,m$$

holds.

Then, the dual optimal set

$$L^*=\{\lambda\in\mathbb{R}^m_+ :\ d(\lambda)=d(\lambda^*)\}$$

is not empty and bounded too, where

$$d(\lambda)=\inf_{x\in\mathbb{R}^n}L(x,\lambda)=\inf_{x\in\mathbb{R}^n}(f(x)-\sum_{i=1}^m\lambda_i c_i(x))$$

is dual function and

$$d(\lambda^*) = \sup\{d(\lambda) | \lambda \in \mathbb{R}^m_+\}, \tag{D}$$

is the dual problem.

If Ω is unbounded, then due to the assumption A, by adding one extra constraint

$$c_0(x) = N - f(x) \geq 0$$

to the set of constraints (7.1), we obtain (see Theorem 2.7) closed, convex, and bounded set

$$\Omega := \{x \in \mathbb{R}^n : c_i(x) \geq 0,\ i = 0, 1, \ldots, m\}.$$

For large enough $N > 0$, the extra constraint does not affect the optimal set X^*. Therefore, to simplify considerations we assume that Ω is a bounded and closed set.

7.1.2 Constraints Transformation and Lagrangian for the Equivalent Problem: Local Properties

Let Ψ be a class of smooth functions $\psi : (a,b) \to \mathbb{R}$, $-\infty < a < 0 < b < \infty$ with following properties:

(1) $\psi(0) = 0$;
(2) (a) $\psi'(0) = 1$, (b) $\psi'(t) > 0$, (c) $\psi'(t) \leq d/t$, $d > 0$, $t > 0$;
(3) $-m_0^{-1} \leq \psi''(t) < 0$; $t \in [a, b]$
(4) $\psi''(t) \leq -M_0^{-1}$, $\forall t \in [a, 0]$, and $0 < m_0 < M_0 < \infty$

From (1)–(2) for any $k > 0$ follows

$$\Omega = \{x \in \mathbb{R}^n : c_i(x) \geq 0,\ i = 1, \ldots, m\} = \{x \in \mathbb{R}^n : k^{-1}\psi(kc_i(x)) \geq 0,\ i = 1, \ldots, m\}.$$

Therefore, problem (7.1) is equivalent to

$$f(x^*) = \min\{f(x) | k^{-1}\psi(kc_i(x)) \geq 0,\ i = 1, \ldots, m\} \tag{7.2}$$

for any $k > 0$.

The LEP $\mathcal{L} : \mathbb{R}^n \times \mathbb{R}^m_+ \times \mathbb{R}_{++} \to \mathbb{R}$ for (7.2)

$$\mathcal{L}(x, \lambda, k) = f(x) - k^{-1} \sum_{i=1}^{m} \lambda_i \psi(kc_i(x))$$

is the main NR tool.

The properties of $\mathcal{L}$ at the KKT's pair (x^*, λ^*), which are critical for NR theory and methods, we collect in the following assertion.

Assertion 7.1 *For any $k > 0$ and any KKT's pair (x^*, λ^*), the following properties*

1° $\mathscr{L}(x^*,\lambda^*,k) = f(x^*)$
2° $\nabla_x \mathscr{L}(x^*,\lambda^*,k) = \nabla f(x^*) - \sum_{i=1}^m \lambda_i^* \nabla c_i(x^*) = \nabla_x L(x^*,\lambda^*) = 0$
3° $\nabla_{xx}^2 \mathscr{L}(x^*,\lambda^*,k) = \nabla_{xx}^2 L(x^*,\lambda^*) + k\nabla c^T(x^*)\Lambda^*\nabla c(x^*)$,

hold, where $\nabla c(x^*) = J(c(x^*))$ is the Jacobian of $c(x) = (c_1(x),\dots,c_m(x))^T$, $\Lambda^* = \mathrm{diag}(\lambda^*)_{i=1}^m$.

Assertion 7.1 follows directly from the complementarity condition (2.57) and formulas for $\mathscr{L}$, $\nabla_x \mathscr{L}$, and $\nabla_{xx}^2 \mathscr{L}$.

Remark 7.1. The properties 1° – 3° show the fundamental difference between NR and SUMT. To see it, let us consider the following problem, which is equivalent to (7.1):

$$N(x^*,x^*) = \min\{N(x,x^*) |\, x \in \mathbb{R}^n\},$$

where $N(x,x^*) = \max\{f(x) - f(x^*),\ -c_i(x),\ i = 1,\dots,m\}$ is convex, but not smooth at x^* (see Fig. 7.1).

For simplicity let us assume $f(x^*) = 0$; then $N(x) = \max\{f(x),\ -c_i(x),\ i = 1,\dots,m\}$ and (7.1) is equivalent to

$$N(x^*) = \min\{N(x) | x \in \mathbb{R}^n\} = 0.$$

The log-barrier function

$$F(x,k) = f(x) - k^{-1}\sum_{i=1}^m \ln c_i(x)$$

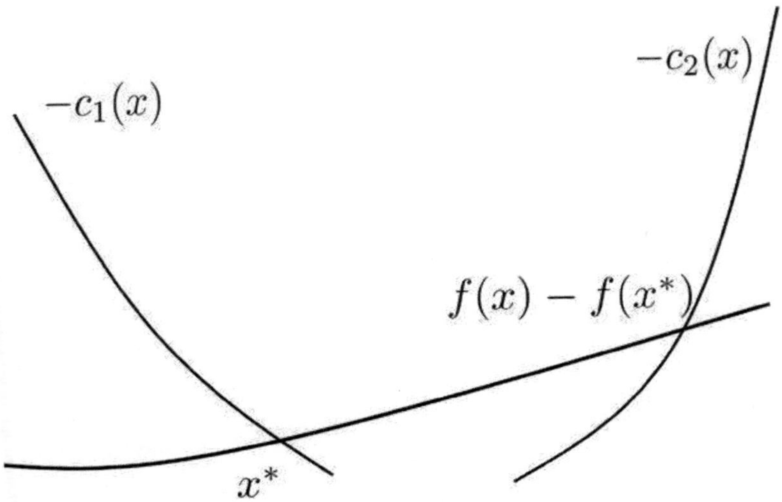

Fig. 7.1 Equivalent unconstrained optimization problem

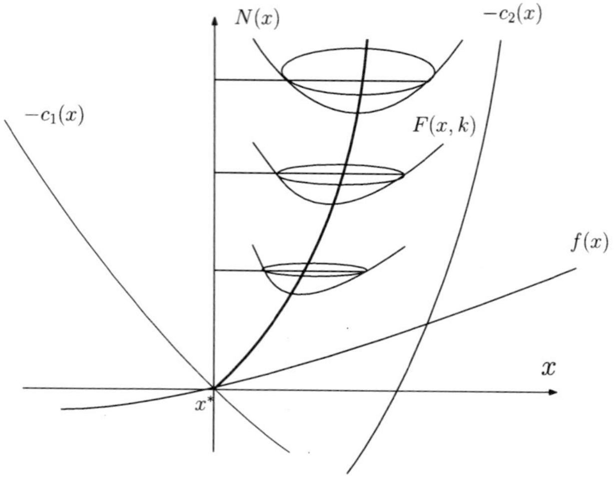

Fig. 7.2 Log-barrier function as smooth approximation of $N(x)$

one can consider as a smooth approximation of $N(x)$.

For any fixed $k > 0$, however, we have (see Fig. 7.2)

$$\lim_{x \to x^*} |F(x,k) - N(x)| = \infty.$$

On the other hand, for any fixed $k > 0$, the LEP $\mathcal{L}(x, \lambda^*, k)$ is an exact smooth approximation of $N(x)$ at $x = x^*$, that is (see Fig. 7.3),

$$\lim_{x \to x^*} [\mathcal{L}(x, \lambda^*, k) - N(x)] = 0.$$

Obviously, LM vector $\lambda^* \in \mathbb{R}^m_+$ is unknown a priori, but, as we will see later, a good approximation for λ^* can be relatively easily found by using unconstrained smooth optimization techniques.

7.1.3 Primal Transformations and Dual Kernels

We will use the following transformations:

- exponential $\hat{\psi}_1(t) = 1 - e^{-t}$;
- logarithmic MBF $\hat{\psi}_2(t) = \ln(t+1)$;
- hyperbolic MBF $\hat{\psi}_3(t) = t/(t+1)$;
- log-sigmoid $\hat{\psi}_4(t) = 2(\ln 2 + t - \ln(1+e^t))$;
- Chen–Harker–Kanzow–Smale (CHKS) $\hat{\psi}_5(t) = t - \sqrt{t^2 + 4\eta} + 2\sqrt{\eta}$, $\eta > 0$.

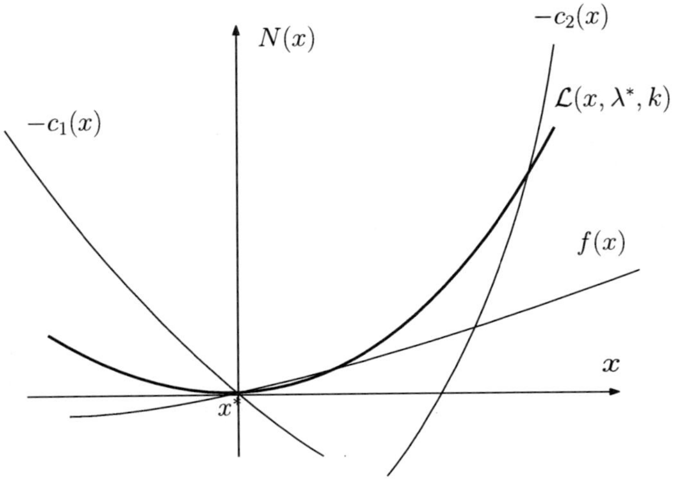

Fig. 7.3 Exact smooth approximation

Exponential transformation for solving systems of linear inequalities was used by Motzkin (1952).

For constrained optimization, exponential transformation was used by Kort and Bertsekas (1973), Tseng and Bertsekas (1993); see also Bertsekas (1982).

For logarithmic and hyperbolic MBF, see Polyak (1987, 1992) and references therein; for log-sigmoid and CHKS transformations see Polyak (2001, 2002).

Unfortunately, none of the transformations above belongs to Ψ.

Transformations $\hat{\psi}_1, \hat{\psi}_2, \hat{\psi}_3$ do not satisfy (3) $(m_0 = 0)$, while transformations $\hat{\psi}_4$ and $\hat{\psi}_5$ do not satisfy 4) $(M_0 = \infty)$.

A slight modification of $\hat{\psi}_i, i = 1,\ldots,5$, however, leads to $\psi_i \in \Psi$.

Let $-1 < \tau < 0$; we consider the following truncated transformations $\psi_i : \mathbb{R} \to \mathbb{R}$:

$$\psi_i(t) := \begin{cases} \hat{\psi}_i(t), \tau \le t < \infty \\ q_i(t), -\infty < t \le \tau, \end{cases} \tag{7.3}$$

where $q_i(t) = a_i t^2 + b_i t + c_i$ and $a_i = 0.5\hat{\psi}_i''(\tau)$, $b_i = \hat{\psi}_i'(\tau) - \tau\hat{\psi}''(\tau)$, $c_i = \hat{\psi}_i'(\tau) - \tau\hat{\psi}_i'(\tau) + 0.5\tau^2\hat{\psi}_i''(\tau), i = 1,\ldots,5$.

Such modification for ψ_2 was suggested by Ben-Tal et al. (1992), just to eliminate the singularity of ψ_2; see also Ben-Tal and Zibulevsky (1997).

It is easy to check properties (1)–(4) for truncated transformations $\psi_i, i = 1,\ldots,5$.

Along with transformations $\psi \in \Psi$, their LF transform plays an important role in NR theory and methods. Therefore, along with $\psi_i \in \Psi$, $i = 1,\ldots,5$, let us consider their LF transform:

$$\psi_i^*(s) := \begin{cases} \hat{\psi}_i^*(s), & s \leq \hat{\psi}_i'(\tau) \\ q_i^*(s) = (4a_i)^{-1}(s-b_i)^2 - c_i, & s \geq \hat{\psi}_i'(\tau), i = 1,\dots,5, \end{cases} \tag{7.4}$$

where $\hat{\psi}^*{}_i(s) = \inf_t\{st - \hat{\psi}_i(t)\}$.

With the class Ψ of transformations ψ, we associate the class Φ of kernels

$$\varphi \in \Phi = \{\varphi = -\psi^* : \psi \in \Psi\}.$$

Using properties 2b) and 4), one can find $0 < \theta_i < 1$ that

$$\hat{\psi}_i'(\tau) - \hat{\psi}_i'(0) = -\hat{\psi}_i''(\tau\theta_i)(-\tau) \geq -\tau M_0^{-1}, i = 1,\dots,5$$

or

$$\hat{\psi}_i'(\tau) \geq 1 - \tau M_0^{-1} = 1 + |\tau| M_0^{-1}.$$

Therefore, from (7.4) for any $0 < s \leq 1 + |\tau| M_0^{-1}$, we have

$$\varphi_i(s) = \hat{\varphi}_i(s) = -\hat{\psi}_i^*(s) = \inf_t\{st - \hat{\psi}_i(t)\}. \tag{7.5}$$

Moreover, the following kernels are infinitely differentiable on $]0, 1 + |\tau| M_0^{-1}[$:

- exponential $\hat{\varphi}_1(s) = s\ln s - s + 1, \hat{\varphi}_1(0) = 1$;
- logarithmic MBF $\hat{\varphi}_2(s) = -\ln s + s - 1$;
- hyperbolic MBF $\hat{\varphi}_3(s) = -2\sqrt{s} + s + 1, \hat{\varphi}_3(0) = 1$;
- Fermi–Dirac $\hat{\varphi}_4(s) = (2-s)\ln(2-s) + s\ln s, \hat{\varphi}_4(0) = 2\ln 2$;
- CMKS $\hat{\varphi}_5(s) = -2\sqrt{\eta}(\sqrt{(2-s)s} - 1), \hat{\varphi}_5(0) = 2\sqrt{\eta}$.

We say kernel $\varphi \in \Phi$ is well defined if $\varphi(0) < \infty$. So $\varphi_1, \varphi_3, \varphi_4, \varphi_5$ are all well defined, while the MBF kernel φ_2 is not well defined. Kernel φ_2, however, is a self-concordant function. Moreover, on (τ, b) the negate of MBF transform $-\psi_2(t) = -\ln(t+1)$ is a self-concordant barrier.

To simplify the notations, we omit indices of ψ and φ. The kernel properties follow from (1) to (4) of $\psi \in \Psi$ and (2.75). We collect them in the following assertion:

Assertion 7.2 *The kernels $\varphi \in \Phi$ are strictly convex on $\mathbb{R}^m_+$, are twice continuously differentiable, and possess the following properties:*

1. $\varphi(s) \geq 0, \forall s \in]0,\infty[$, $\min_{s\geq 0} \varphi(s) = \varphi(1) = 0$;
2. *(a)* $\lim_{s\to 0^+} \varphi'(s) = -\infty$, *(b) $\varphi'(s)$ is monotone increasing and (c)* $\varphi'(1) = 0$;
3. *(a)* $\lim_{s\to 0^+} \varphi''(s) = \infty$, *(b)*$\varphi''(s) \geq m_0 > 0, \forall s \in [\psi'(b), \psi'(a)]$, *(c)* $\varphi''(s) \leq M_0 < \infty, \forall s \in]0,\infty[$.

Exercise 7.1. Show that if Ω is bounded and Slater condition holds, then for any $\psi \in \Psi$, any fixed $k > 0$ and $\lambda \in \mathbb{R}^m_{++}$, the RC of Ω is empty, that is, for any $x \in \Omega$

$$\lim_{t\to\infty} \mathscr{L}(x + td, \lambda, k) = \infty \tag{7.6}$$

for any $d \neq 0 \in \mathbb{R}^n$.

7.1.4 NR Method and Dual Prox with φ-Divergence Distance

In this section we describe the general NR method and establish its equivalence to the interior prox method with first-order entropy-like φ-divergence distance for the dual problem. The equivalence is the key in convergence analysis of NR methods.

Let $\psi \in \Psi$, $\lambda \in \mathbb{R}^m_{++}$, and $k > 0$. The NR step consists of finding the primal minimizer

$$\hat{x} :\equiv \hat{x}(\lambda, k) : \nabla_x \mathscr{L}(\hat{x}, \lambda, k) = 0 \tag{7.7}$$

followed by Lagrange multipliers update:

$$\hat{\lambda} \equiv \hat{\lambda}(\lambda, k) = (\hat{\lambda}_1, \ldots, \hat{\lambda}_m) : \hat{\lambda}_i = \lambda_i \psi'(kc_i(\hat{x})),\ i = 1, \ldots, m. \tag{7.8}$$

Theorem 7.1. *If conditions A and B hold and f, $c_i \in C^1$, $i = 1, \ldots, m$, then the NR method (7.7)–(7.8) is*

(1) well defined;
(2) equivalent to the following prox method

$$d(\hat{\lambda}) - k^{-1} D(\hat{\lambda}, \lambda) = \max\{d(u) - k^{-1} D(u, \lambda) | u \in \mathbb{R}^m_{++}\}, \tag{7.9}$$

where $D(u, \lambda) = \sum_{i=1}^m \lambda_i \varphi(u_i / \lambda_i)$ is first-order φ-divergence entropy-like distance function based on kernel $\varphi = -\psi^*$.

Proof.

(1) Due to the properties (1)–(2) of ψ, convexity f and concavity c_i, $i = 1, \ldots, m$, the LEP $\mathscr{L}$ is convex in x. From Exercise 7.1 follows (7.6).
Therefore, for a given $(\lambda, k) \in \mathbb{R}^{m+1}_{++}$, there exist $\hat{x} \equiv \hat{x}(\lambda, k)$ defined by (7.7) and $\hat{\lambda} \equiv \hat{\lambda}(\lambda, k)$ defined by (7.8).
From (2) of ψ and (7.8) follows $\lambda \in \mathbb{R}^m_{++} \Rightarrow \hat{\lambda} \in \mathbb{R}^m_{++}$; therefore NR method (7.7)–(7.8) is well defined.

(2) From (7.7) and (7.8) follows

$$\nabla_x \mathscr{L}(\hat{x}, \hat{\lambda}, k) = \nabla f(\hat{x}) - \sum_{i=1}^m \lambda_i \psi'(kc_i(\hat{x})) \nabla c_i(\hat{x}) = \nabla_x L(\hat{x}, \hat{\lambda}) = 0;$$

therefore

$$d(\hat{\lambda}) = \min_{x \in \mathbb{R}} L(x, \hat{\lambda}) = L(\hat{x}, \hat{\lambda}).$$

The subdifferential $\partial d(\hat{\lambda})$ contains $-c(\hat{x})$, that is,

$$0 \in c(\hat{x}) + \partial d(\hat{\lambda}). \tag{7.10}$$

From (7.8) follows $\psi'(kc_i(\hat{x})) = \hat{\lambda}_i / \lambda_i$, $i = 1, \ldots, m$.
Due to the property (3) of ψ, there exists an inverse ψ'^{-1}. From LF identity follows

$$c_i(\hat{x}) = k^{-1}\psi'^{-1}(\hat{\lambda}_i/\lambda_i) = k^{-1}\psi^{*'}(\hat{\lambda}_i/\lambda_i). \tag{7.11}$$

Let $\varphi = -\psi^*$; then from (7.10) and (7.11) follows

$$0 \in \partial d(\hat{\lambda}) + k^{-1}\sum_{i=1}^{m} \psi^{*'}\left(\hat{\lambda}_i/\lambda_i\right) e_i = \partial d(\hat{\lambda}) - k^{-1}\sum_{i=1}^{m} \varphi'(\hat{\lambda}_i/\lambda_i) e_i, \tag{7.12}$$

where $e_i = (0,\ldots,1,\ldots 0)^T$.

The inclusion (7.12) is the optimality criterion for $\hat{\lambda}$ to be a solution of the problem (7.9). □

Remark 7.2. It follows from 1° and 2° of Assertion 7.1 that for any $k > 0$ we have $x^* = x(\lambda^*, k)$ and $\lambda^* = \lambda(\lambda^*, k)$, that is, $\lambda^* \in \mathbb{R}^m_+$ is a fixed point of the mapping $\lambda \to \hat{\lambda}(\lambda, k)$.

Remark 7.3. In view of (7.5), from some point on, only original transformation $\hat{\psi}$ is used in NR (7.7)–(7.8) and only original kernels $\hat{\varphi}$ are used in the correspondent prox method (7.9).

From (7.8) for $i : c_i(\hat{x}) > 0$ follows $\hat{\lambda}_i < \lambda_i$ and for $i : c_i(\hat{x}) < 0$ follows $\hat{\lambda}_i > \lambda_i$ (see Fig. 7.4).

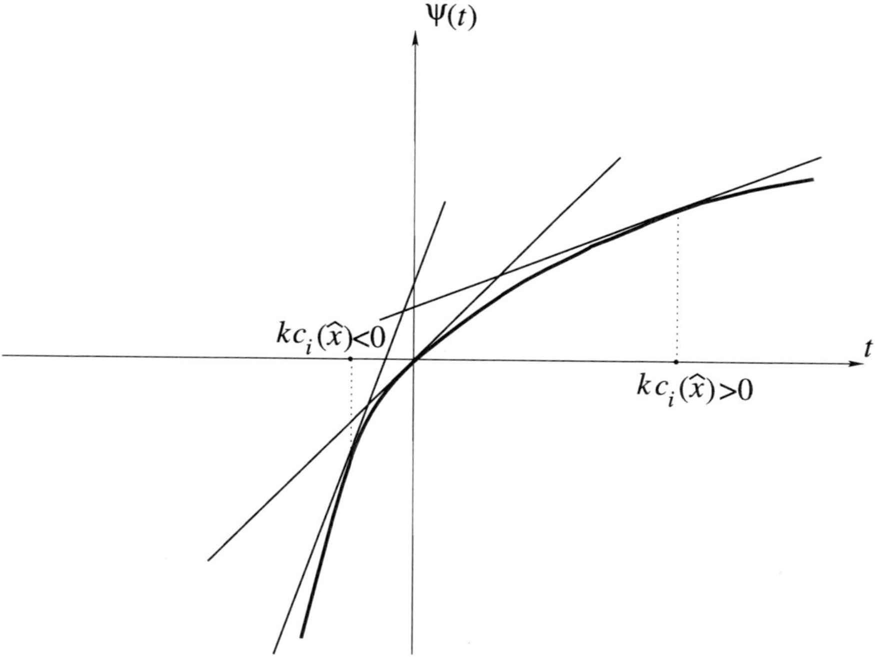

Fig. 7.4 NR as pricing mechanism

The NR method (7.7)–(7.8) is a pricing mechanism for establishing equilibrium between the objective function reduction and penalty for constraint violation. The equilibrium is given by KKT's Theorem.

Exercise 7.2. Consider few real-life applications, and explain NR pricing mechanisms.

7.1.5 Q-Linear Convergence Rate

In this section, we establish rate of convergence results for NR method in the absence of convexity assumption.

First, we show that if f and all c_i are C^2 functions and second-order sufficient optimality conditions (4.73)–(4.74) hold, then the NR method converges with Q-linear rate. Moreover, the ratio $0 < \gamma < 1$ is inversely proportional to the penalty parameter; therefore, it can be made as small as one wants by choosing a fixed, but large enough scaling parameter $k > 0$.

Second, LEP $\mathscr{L}$ has several important properties, which classical Lagrangian L for the original problem does not possess.

In particular, for any fixed and large enough scaling parameter $k > 0$, the LEP $\mathscr{L}(x,\lambda^*,k)$ is strongly convex in the neighborhood of x^*, whether f and all $-c_i$ are convex or not. It remains true for any Lagrange multiplier vector $\lambda \in B(\lambda^*,\varepsilon) = \{\lambda \in \mathbb{R}_+ : \|\lambda - \lambda^*\| \leq \varepsilon\}$, when $\varepsilon > 0$ is small enough.

Third, if LEP $\mathscr{L}$ is nonconvex in x, then for a given $\lambda \in \mathbb{R}^m_{++}$ and $k > 0$ large enough, one has to find an approximation for the primal minimizer **only once**. Further, each NR step requires finding an approximation for primal minimizer of a **strongly convex** and smooth enough LEP $\mathscr{L}$, followed by the LM update.

Fourth, the dual function, which is based on the LEP, is as smooth as f and c_i, $i = 1,\ldots,m$.

To describe the convergence result, we have to characterize the extended dual domain, where the main facts of the basic theorem are taking place.

Let $0 < \delta < \min_{1\leq i\leq r} \lambda_i^*$ be small enough, and let $k_0 > 0$ be large enough. We split the extended dual domain into active and passive parts, that is,

$$\Lambda(\cdot) = \Lambda(\lambda^*,\delta,k_0) = \Lambda_{(r)}(\cdot) \otimes \Lambda_{(m-r)}(\cdot),$$

where

$$\Lambda_{(r)}(\cdot) \equiv \Lambda(\lambda^*_{(r)},\delta,k_0) = \{(\lambda_{(r)},k_0) : \lambda_i \geq \delta,\ |\lambda_i - \lambda_i^*| \leq \delta k,\\ i = 1,\ldots,r, k \geq k_0\},$$

is the extended dual domain for the active constraints and

$$\Lambda_{(m-r)}(\cdot) \equiv \Lambda(\lambda^*_{(m-r)},\delta,k_0) = \{(\lambda_{(m-r)},k) : 0 < \lambda_i < \delta k,\\ i = r+1,\ldots,m, k \geq k_0\}.$$

is the extended dual domain for the passive constraints (see Fig. 7.5).

For a vector $a \in \mathbb{R}^n$, we use the following norm $\|a\| = \max_{1\le i\le n} |a_i|$ and for a matrix $A \in \mathbb{R}^{n\times n}$ the correspondent norm $\|A\| = \max_{1\le i\le n} \sum_{j=1}^{n} |a_{ij}|$.

The following theorem establishes the basic convergence results of the NR method for any $\psi \in \Psi$.

Theorem 7.2. *If f, $c_i \in C^2$, $i = 1,\dots,m$ and the second-order sufficient optimality conditions* (4.73)–(4.74) *are satisfied, then for any $(\lambda,k) \in \Lambda(\lambda^*,\delta,k_0)$:*

(1) there exist

$$\hat{x} = \hat{x}(\lambda,k) : \nabla_x \mathscr{L}(\hat{x},\lambda,k) = 0$$

and

$$\hat{\lambda}(\lambda,k) : (\hat{\lambda}_i = \psi'(kc_i(\hat{x}))\lambda_i, \quad i = 1,\dots,m);$$

(2) the bounds

$$\|\hat{x} - x^*\| \le (c/k)\|\lambda - \lambda^*\|, \quad \|\hat{\lambda} - \lambda^*\| \le (c/k)\|\lambda - \lambda^*\| \tag{7.13}$$

hold and $c > 0$ is independent on $k \ge k_0$; also

$$\hat{x}(\lambda^*,k) = x^* \text{ and } \hat{\lambda}(\lambda^*,k) = \lambda^*,$$

that is, λ^* is a fixed point of the map $\lambda \to \hat{\lambda}(\lambda,k)$;

(3) LEP $\mathscr{L}$ is strongly convex in the neighborhood of $\hat{x}$.

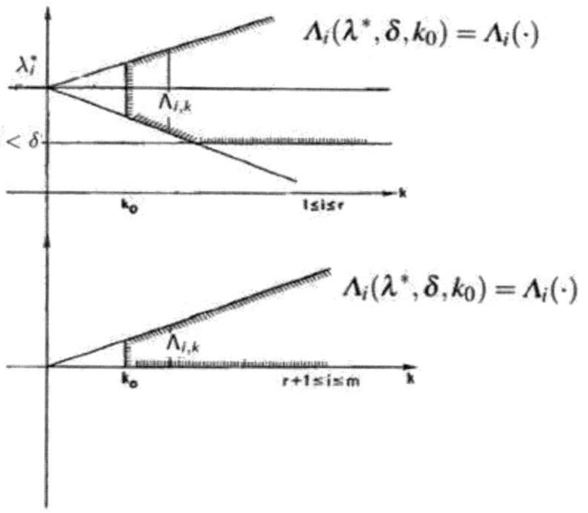

$$\Lambda(\lambda^*,\delta,k_0) = \Lambda(\cdot) = \Lambda_1(\cdot)\otimes\ldots\otimes\Lambda_r(\cdot)\otimes\Lambda_{r+1}\otimes\ldots\otimes\Lambda_m(\cdot)$$
$$\Lambda_k = \Lambda_{1,k}\otimes\ldots\otimes\Lambda_{r,k}\times\Lambda_{r+1,k}\ldots\otimes\Lambda_{m,k}$$

Fig. 7.5 Dual subsets

Proof.

(1) To simplify considerations we introduce vector $t = (t_1, \ldots, t_r, t_{r+1}, \ldots, t_m)^T = (t_{(r)}, t_{(m-r)})^T$, with components $t_i = (\lambda_i - \lambda_i^*)k^{-1}$, $i = 1, \ldots, m$. It transforms the extended dual set $\Lambda(\cdot)$ into the neighborhood

$$S(0^m, \delta, k_0) = S_{(r)}(0^r, \delta, k_0) \otimes S_{(m-r)}(0^{m-r}, \delta, k_0)$$

of the origin of the extended dual space, where

$$S_{(r)}(0^r, \delta, k) = \{(t_{(r)}, k) : |t_i| \leq \delta, t_i \geq (\delta - \lambda_i^*)k^{-1},\ i = 1, \ldots, r, k \geq k_0\}$$

$$S_{(m-r)}(0^{m-r}, \delta, k) = \{(t_{(m-r)}, k) : 0 \leq t_i \leq \delta,\ i = r+1, \ldots, m, k \geq k_0\}.$$

Let us consider vector-function h: $\mathbb{R}^{n+m-r+1} \to \mathbb{R}^n$ defined by formula

$$h(x, t_{(m-r)}, k) = k \sum_{i=r+1}^{m} t_i \psi'(kc_i(x)) \nabla c_i(x);$$

then

$$\nabla_t h(x, t_{(m-r)}, k) = [0^{n,r} \quad k(\nabla c_{(m-r)}(x))^T \psi'(kc_{(m-r)}(x))],$$

where

$$\psi'(kc_{(m-r)}(x)) = \operatorname{diag}(\psi'(kc_i(x)))_{i=r+1}^{m}$$

and $\nabla c_{(m-r)}(x)$ is the Jacobian of $c_{(m-r)}(x) = (c_{r+1}(x), \ldots . c_m(x))^T$.
Also,

$$\nabla_x h(x, t_{(m-r)}, k) = k^2 \sum_{i=r+1}^{m} t_i \psi''(kc_i(x)) \nabla c_i^T(x) \nabla c_i(x) + k \sum_{i=r+1}^{m} t_i \psi'(kc_i(x)) \nabla^2 c_i(x);$$

then, for any given $k > 0$, we have

$$h(x^*, 0^{m-r}, k) = 0^n, \quad \nabla_x h(x^*, 0^{m-r}, k) = 0^{n,n}.$$

Now we consider a map $\Phi : \mathbb{R}^{n+m+r+1} \to \mathbb{R}^{n+r}$, given by

$$\Phi(x, \hat{\lambda}_{(r)}, t, k) = \begin{pmatrix} \nabla f(x) - \sum_{i=1}^{r} \hat{\lambda}_i \nabla c_i(x) - h(x, t_{(m-r)}, k) \\ (t_i + k^{-1}\lambda_i^*) \psi'(kc_i(x)) - k^{-1}\hat{\lambda}_i,\ i = 1, \ldots, r \end{pmatrix}.$$

For any given $k > 0$ from KKT's Theorem follows

$$\Phi(x^*, \lambda_{(r)}^*, 0^m, k) = \begin{bmatrix} \nabla f(x^*) - \sum_{i=1}^{r} \lambda_i^* \nabla c_i(x^*) \\ k^{-1}(\lambda_i^* - \lambda_i^*),\ i = 1, \ldots, r \end{bmatrix} = \begin{bmatrix} 0^n \\ 0^r \end{bmatrix}.$$

Also, for Jacobian $\nabla_{x\hat{\lambda}_{(r)}} \Phi(\cdot)$, we have

$$\nabla_{x\hat{\lambda}_{(r)}} \Phi(x, \hat{\lambda}_{(r)}, t, k) = \nabla_{x\hat{\lambda}_{(r)}} \Phi(\cdot) = \begin{bmatrix} \nabla^2_{xx} L(\cdot) + \nabla_x h(\cdot) & -\nabla c^T_{(r)}(\cdot) \\ \Lambda_{(r)} \Psi''\left(kc_{(r)}(\cdot)\right) \nabla c_{(r)}(\cdot) & -k^{-1} I^r \end{bmatrix},$$

where $\Lambda_{(r)} = \mathrm{diag}(\lambda_i)_{i=1}^r$, $\Psi''\left(kc_{(r)}(\cdot)\right) = \mathrm{diag}\left(\psi''(kc_i(\cdot))\right)_{i=1}^r$, and $\nabla c_{(r)}(\cdot)$ is the Jacobian of $c_{(r)}(x) = (c_1(x), \ldots, c_r(x))^T$.

Then,

$$\nabla_{x\hat{\lambda}_{(r)}} \Phi(x^*, \lambda^*_{(r)}, 0^m, k) = \begin{bmatrix} \nabla^2_{xx} L(x^*, \lambda^*) & -\nabla c^T_{(r)}(x^*) \\ \Lambda^*_{(r)} \Psi''\left(kc_{(r)}(x^*)\right) \nabla c_{(r)}(x^*) & -k^{-1} I^r \end{bmatrix}$$

$$= \begin{bmatrix} \nabla^2_{xx} L & -\nabla c^T_{(r)} \\ \psi''(0) \Lambda^*_{(r)} \nabla c_{(r)} & -k^{-1} I^r \end{bmatrix} = \nabla \Phi_k,$$

where $\Lambda^*_{(r)} = \mathrm{diag}(\lambda^*_i)_{i=1}^r$, $\Psi''\left(kc_{(r)}(x^*)\right) = \psi''(0) I^r$, and I^r is identical matrix in $\mathbb{R}^r$.

From $\psi''(0) < 0$, $\lambda^*_i > 0$, $i = 1, \ldots, r$, second-order sufficient optimality conditions (4.73)–(4.74) from Lemma 5 of the Appendix it follows existence of $\rho > 0$, that for $k_0 > 0$ large enough and any $k \geq k_0$, we have

$$\left\| \nabla \Phi_k^{-1} \right\| \leq \rho \,. \tag{7.14}$$

Let $k_0 > 0$ be large enough, $\infty > k_1 > k_0$, and $K = \{0 \in \mathbb{R}^n\} \times [k_0, k_1]$.

We consider the following neighborhood

$$S(K, \delta) = \{(t, k) : |t_i| \leq \delta,\ t_i \geq (\delta - \lambda^*_i) k^{-1},\ i = 1, \ldots, r,$$
$$0 \leq t_i \leq \delta,\ i = r+1, \ldots, m\}$$

of K.

Due to the second implicit function theorem, see Bertsekas (1982), the system

$$\Phi(x, \hat{\lambda}_{(r)}, t, k) = 0$$

defines on $S(K, \delta)$ a unique pair of vector-functions $x(t,k) = (x_i(t,k),\ i = 1, \ldots, n)$ and $\hat{\lambda}_{(r)}(t,k) = (\hat{\lambda}_i(t,k),\ i = 1, \ldots, r)$, such that $x(0,k) = x^*$, $\hat{\lambda}_{(r)}(0,k) = \lambda^*_{(r)}$ and

$$\Phi(x(t,k), \hat{\lambda}_{(r)}(t,k), t, k) \equiv 0. \tag{7.15}$$

Identity (7.15) can be rewritten as follows:

$$\nabla f(x(t,k)) - \sum_{i=1}^{r} \hat{\lambda}_i(t,k) \nabla c_i(x(t,k)) - h(x(t,k), t, k) \equiv 0 \tag{7.16}$$

$$\hat{\lambda}_{(r)}(t,k) = (\hat{\lambda}_i(t,k) = (kt_i + \lambda^*_i) \psi'(kc_i(x(t,k))),\ i = 1, \ldots, r)\,. \tag{7.17}$$

Also,

$$\hat{\lambda}_{(m-r)}(t,k) = (\hat{\lambda}_i(t,k) = kt_i\psi'(kc_i(x(t,k))),\ i = r+1,\ldots,m)\,. \tag{7.18}$$

Let $\hat{\lambda}(t,k) = (\hat{\lambda}_{(r)}(t,k), \hat{\lambda}_{(m-r)}(t,k))$; then from the definition of $h(x,t_{(m-r)},k)$ and (7.16)–(7.18) follows

$$\nabla_x \mathscr{L}(x(t,k),\lambda,k) = \nabla_x L(x(t,k),\hat{\lambda}(t,k)) = 0\,.$$

So, we proved the first part of the theorem.

(2) To establish bound (7.13), we start with Lagrange multipliers, which correspond to the passive constraints.

From the second-order optimality condition follows existence $\sigma > 0$ that $c_i(x(0,k)) = c_i(x^*) \geq \sigma$, $i = r+1,\ldots,m$.

For any given small $\varepsilon > 0$, there is $\delta > 0$ small enough that

$$\|x(t,k) - x(0,k)\| = \|x(t,k) - x^*\| \leq \varepsilon,$$

holds for $\forall\,(t,k) \in S(K,\delta)$.

Therefore, for the passive constraints, we have

$$c_i(x(t,k)) \geq 0.5\sigma, \quad i = r+1,\ldots,m, \quad \forall\,(t,k) \in S(K,\delta)\,. \tag{7.19}$$

From property $(2c)$ of ψ and (7.19) follows

$$\hat{\lambda}_i = \lambda_i\psi'(kc_i(x(t,k))) \leq \lambda_i\psi'(0.5k\sigma) \leq \frac{2d}{\sigma k}\lambda_i\,, \quad i = r+1,\ldots,m, \tag{7.20}$$

where $2d/\sigma > 0$ is independent from $k \geq k_0$.

To prove bound (7.13), let us first estimate $\|\nabla_t x(t,k)\|$ and $\|\nabla_t \hat{\lambda}_{(r)}(t,k)\|$ for $t = 0^m$.

For any given $k_0 < k < k_1$ and $t = 0^m$, we have

$$x(t,k)|_{t=0^m} = x^*, \quad \hat{\lambda}_{(r)}(t,k)|_{t=0^m} = \lambda^*_{(r)}, \quad \nabla_t x(t,k)|_{t=0^m} = \nabla_t x(0^m,k),$$

$$\nabla_t \hat{\lambda}_{(r)}(t,k)|_{t=0^m} = \nabla_t \hat{\lambda}_{(r)}(0^m,k), \nabla^2_{xx} L(x(\cdot),\hat{\lambda}_{(r)}(\cdot))|_{t=0^m} = \nabla_{xx} L(x^*,\lambda^*) = \nabla^2_{xx} L,$$

$$\nabla c_{(r)}(x(\cdot))|_{t=0^m} = \nabla c_{(r)}(x^*) = \nabla c_{(r)}, \Psi'\left(kc_{(r)}(x(\cdot))\right)|_{t=0^m} = \Psi'\left(kc_{(r)}(x^*)\right) = I^r,$$

$$\Psi''\left(kc_{(r)}(x(\cdot))\right)|_{t=0^m} = \Psi''\left(kc_{(r)}(x^*)\right) = \psi''(0)I^r,$$

$$\nabla_x h(x(\cdot),\cdot)|_{t=0^m} = 0^{n,n}, \nabla_t h(x(\cdot),\cdot)|_{t=0^m} = [0^{n,r} \quad k\nabla c^T_{(m-r)}(x^*)\Psi'\left(kc_{(m-r)}(x^*)\right)],$$

where $\Psi'(kc_{(m-r)}(x^*)) = \mathrm{diag}(\psi'(kc_i(x^*)))^m_{i=r+1}$.

From $c_i(x^*) \geq \sigma > 0$, $i = r+1,\ldots,m$ and property 2c) of ψ, we obtain

$$\|k\nabla c^T_{(m-r)}(x^*)\Psi'\left(kc_{(m-r)}(x^*)\right)\| \leq d\sigma^{-1}\left(\|c^T_{(m-r)}(x^*)\|\right). \tag{7.21}$$

By differentiating the identities (7.16)–(7.17) in t for Jacobians $\nabla_t x(t,k)$ and $\nabla_t \hat{\lambda}_{(r)}(t,k)$ at $t = 0^m$, we have

$$\begin{bmatrix} \nabla^2_{xx} L & -\nabla c^T_{(r)} \\ \psi''(0)\Lambda^*_{(r)}\nabla c_{(r)} & k^{-1} I^r \end{bmatrix} \begin{bmatrix} \nabla_t x(0^m,k) \\ \nabla_t \lambda_{(r)}(0^m,k) \end{bmatrix}$$

$$= \begin{bmatrix} 0^{n,r} & k\nabla c^T_{(m-r)}(x^*)\Psi'\left(kc_{(m-r)}(x^*)\right) \\ I^r & 0^{r,m-r} \end{bmatrix}.$$

Therefore,

$$\begin{bmatrix} \nabla_t x(0^m,k) \\ \nabla_t \lambda_{(r)}(0^m,k) \end{bmatrix} = \nabla \Phi_k^{-1} \begin{bmatrix} 0^{n,r} & k\nabla c^T_{(m-r)}(x^*)\Psi'\left(kc_{(m-r)}(x^*)\right) \\ I^r & 0^{r,m-r} \end{bmatrix} \tag{7.22}$$

From (7.14), (7.21), and (7.22) follows

$$\max\left\{\|\nabla_t \hat{x}(0,k)\|, \|\nabla_t \hat{\lambda}(0,k)\|\right\} \le \rho \max\left\{1, d\sigma^{-1}\|\nabla c^T_{(m-r)}(x^*)\|\right\}.$$

Thus, for $\delta > 0$ small enough and any $(t,k) \in S(K,\delta)$, the following bound

$$\max\{\|\nabla_t x(\tau t,k)\|, \|\nabla_t \hat{\lambda}(\tau t,k)\|\} \le 2\rho \max\{1, d\sigma^{-1}\|\nabla c^T_{(m-r)}(x^*)\|\} = c_0 \tag{7.23}$$

holds for any $0 \le \tau \le 1$. Therefore, keeping in mind (2.8), we obtain

$$\begin{bmatrix} x(t,k) - x^* \\ \hat{\lambda}_{(r)}(t,k) - \lambda^* \end{bmatrix} = \begin{bmatrix} x(t,k) - x(0^m,k) \\ \hat{\lambda}_{(r)}(t,k) - \hat{\lambda}_{(r)}(0^m,k) \end{bmatrix}$$

$$= \int_0^1 \begin{bmatrix} \nabla_t x(\tau t,k) \\ \nabla_t \hat{\lambda}(\tau t,k) \end{bmatrix} t d\tau. \tag{7.24}$$

From (7.23) and (7.24) follows

$$\max\left\{\|x(t,k) - x^*\|,\ \|\hat{\lambda}(t,k) - \lambda^*\|\right\} \le c_0\|t\| = c_0 k^{-1}\|\lambda - \lambda^*\|.$$

Let

$$\hat{x}(\lambda,k) = x\left(\frac{\lambda - \lambda^*}{k}, k\right),\ \hat{\lambda}(\lambda,k) = \left(\hat{\lambda}_{(r)}\left(\frac{\lambda - \lambda^*}{k}, k\right),\ \hat{\lambda}_{(m-r)}\left(\frac{\lambda - \lambda^*}{k}, k\right)\right).$$

Then, for $c = \max\{2d\sigma^{-1}, c_0\}$, which is independent on $k \ge k_0$, we obtain

$$\max\left\{\|\hat{x}(\lambda,k) - x^*\|,\ \|\hat{\lambda}(\lambda,k) - \lambda^*\|\right\} \le ck^{-1}\|\lambda - \lambda^*\|.$$

(3) To prove the third part of the basic theorem, we consider

$$\nabla^2_{xx}\mathscr{L}(\hat{x},\lambda,k) = \nabla^2_{xx}L(\hat{x},\hat{\lambda})$$

$$- k\nabla c^T_{(r)}(\hat{x})\Psi''\left(kc_{(r)}(\hat{x})\right)\hat{\Lambda}_{(r)}\left(\Psi'\left(kc_{(r)}(\hat{x})\right)\right)^{-1}\nabla c_{(r)}(\hat{x})$$

$$- k\nabla c^T_{(m-r)}(\hat{x})\Psi''\left(kc_{(m-r)}(\hat{x})\right)\Lambda_{(m-r)}\nabla c_{(m-r)}(\hat{x}) \tag{7.25}$$

Again, from (7.25) and the bound (7.13) for $k_0 > 0$ large enough and any $k \geq k_0$, there exists such small $\delta > 0$, that for any $(\lambda,k) \in \Lambda(\lambda,\delta,k)$, we have

$$\nabla^2_{xx}\mathscr{L}(\hat{x},\lambda,k) \approx \nabla^2_{xx}L(x^*,\lambda^*) - k\psi''(0)\nabla c^T_{(r)}(x^*)\Lambda^*_{(r)}\nabla c_{(r)}(x^*).$$

From $\psi''(0) < 0$, the second-order sufficient optimality condition (4.74), and Debreu's lemma follows that Hessian $\nabla^2_{xx}\mathscr{L}(\hat{x},\lambda,k)$ is positive definite, that is, the LEP $\mathscr{L}(x,\lambda,k)$ is strongly convex in x in the neighborhood of $\hat{x} = \hat{x}(\lambda,k)$, $\forall(\lambda,k) \in \Lambda(\cdot)$. □

The NR method generates primal–dual sequence $\{x_s,\lambda_s\}_{s\in N}$ by the following formulas:

$$x_{s+1} : \nabla_x\mathscr{L}(x_{s+1},\lambda_s,k) = 0\,, \tag{7.26}$$

$$\lambda_{s+1} : \left(\lambda_{i,s+1} = \psi'\left(kc_i\left(x_{s+1}\right)\right)\lambda_{i,s},\quad i = 1,\ldots,m\right). \tag{7.27}$$

It follows directly from Theorem 7.2 that

$$\|x_s - x^*\| \leq \left(\frac{c}{k}\right)^s \|\lambda_0 - \lambda^*\|$$

$$\|\lambda_s - \lambda^*\| \leq \left(\frac{c}{k}\right)^s \|\lambda_0 - \lambda^*\|$$

By increasing $k \geq k_0$ from step to step, one can make the rate of convergence Q-superlinear.

Unfortunately, the NR method (7.26)–(7.27) requires solving system (7.26) at each step, which, generally speaking, is an infinite procedure.

7.1.6 Stopping Criteria

To make the NR method practical, one has to introduce a stopping criterion, which allows replacing x_{s+1} by its approximation while preserving bound (7.13).

For a given $\alpha > 0$, let us consider

$$\tilde{x} = \tilde{x}(\lambda,k) : \|\nabla_x\mathscr{L}(\tilde{x},\lambda,k)\| \leq \frac{\alpha}{k}\|\tilde{\lambda} - \lambda\| \tag{7.28}$$

and

$$\tilde{\lambda} = \tilde{\lambda}(\lambda,k) = \left(\tilde{\lambda}_i = \psi'(kc_i(\tilde{x}))\,\lambda_i,\quad i = 1,\ldots,m\right). \tag{7.29}$$

It turned out that the basic fact of Theorem 7.2 remains true for the primal–dual approximation $(\widetilde{x}, \widetilde{\lambda})$.

Theorem 7.3. *If* $f, c_i \in C^2$, $i = 1, \ldots, m$, $\alpha > 0$, *and the second-order sufficient optimality conditions* (4.73)–(4.74) *are satisfied, then for any* $(\lambda, k) \in \Lambda(\lambda, \delta, k_0)$, *we have*

(1) for the pair $(\tilde{x}, \tilde{\lambda})$, *which satisfies* (7.28)–(7.29), *the bounds*

$$\|\tilde{x} - x^*\| \le \frac{c}{k}(1+2\alpha)\|\lambda - \lambda^*\|, \ \|\tilde{\lambda} - \lambda^*\| \le \frac{c}{k}(1+2\alpha)\|\lambda - \lambda^*\| \qquad (7.30)$$

hold and $c > 0$ *is independent on* $\alpha > 0$ *and* $k \ge k_0$*;*
(2) the LEP $\mathscr{L}(x, \lambda, k)$ *is strongly convex in the neighborhood of* $\tilde{x} = \tilde{x}(\lambda, k)$.

Proof.

(1) For a small enough $\delta > 0$ and large enough $k_0 > 0$, we define the following sets:

$$\Lambda(\lambda, \theta, \delta, k_0) = \Lambda(\lambda, \delta, k_0) \otimes \{\theta \in \mathbb{R}^n : \|\theta\| \le \delta\} =$$

$$\Lambda_{(r)}(\lambda, \delta, k_0) \otimes \Lambda_{(m-r)}(\lambda, \delta, k_0) \otimes \{\theta \in \mathbb{R}^n : \|\theta\| \le \delta\}$$
$$= \{\lambda \in \mathbb{R}^m_+ : \lambda_i \ge \delta, \ |\lambda_i - \lambda_i^*| \le \delta k, \quad i = 1, \ldots, r, k \ge k_0 > 0\} \otimes$$
$$\{0 \le \lambda_i < \delta k, \ i = r+1, \ldots, m, k \ge k_0 > 0\} \otimes \{\theta \in \mathbb{R}^n : \|\theta\| \le \delta\}.$$

By introducing vector $t = (t_1, \ldots, t_r, t_{r+1}, \ldots, t_m)^T$ with $t_i = (\lambda_i - \lambda_i^*)k^{-1}$, $i = 1, \ldots, m$, we transform the set $\Lambda(\lambda, \theta, \delta, k)$ into the neighborhood

$$S(0^m, 0^n, \delta, k) = S_{(r)}(0, \delta, k_0) \otimes S_{(m-r)}(0, \delta, k_0) \otimes \{\theta \in \mathbb{R}^n : |\theta_i| \le \delta, i = 1, \ldots, n\}$$

of the origin.
Let us consider vector-function $h : \mathbb{R}^{n+m-r+1} \to \mathbb{R}^n$ defined as follows:

$$h(x, t_{(m-r)}, k) = k \sum_{i=r+1}^{m} t_i \psi'(kc_i(x)) \nabla c_i(x).$$

Then,

$$\nabla_t h(x, t_{(m-r)}, k) = [0^{n,r} \quad k(\nabla c_{(m-r)}(x))^T \Psi'(kc_{(m-r)}(x))],$$

where

$$\Psi'(kc_{(m-r)}(x)) = \mathrm{diag}(\psi'(kc_i(x)))_{i=r+1}^{m}.$$

Also,

$$\nabla_x h(x, t_{(m-r)}, k) = k^2 \sum_{i=r+1}^{m} t_i \psi''(kc_i(x) \nabla c_i^T(x) \nabla c_i(x) + k \sum_{i=r+1}^{m} t_i \psi(kc_i(x)) \nabla c_i^2(x);$$

therefore, $h(x, 0^{m-r}, k) = 0^n$ and $\nabla_x h(x, 0^{m-r}, k) = 0^{n,n}$.

We consider the map $\Phi : \mathbb{R}^{2n+r+m+1} \to \mathbb{R}^{n+r}$ defined by formula

$$\Phi(x, \tilde{\lambda}_{(r)}, t, \theta, k) = \begin{pmatrix} \nabla f(x) - \sum_{i=1}^{r} \tilde{\lambda}_i \nabla c_i(x) - h(x, t_{(m-r)}, k) - \theta \\ (t_i + k^{-1}\lambda_i^*)\, \psi'(kc_i(x)) - k^{-1}\tilde{\lambda}_i, \quad i = 1, \ldots, r \end{pmatrix}.$$

Then, from KKT's Theorem for any $k \in [k_0, k_1]$ follows

$$\Phi(x^*, \lambda^*_{(r)}, 0^m, 0^n, k) = \begin{bmatrix} \nabla f(x^*) - \sum_{i=1}^{r} \lambda_i^* \nabla c_i(x^*) - h(x^*, 0^{m-r}, k) \\ k^{-1}(\lambda_i^* - \lambda_i^*), \quad i = 1, \ldots, r \end{bmatrix} = \begin{bmatrix} 0^n \\ 0^r \end{bmatrix}.$$

Further,

$$\nabla_{x, \tilde{\lambda}_{(r)}} \Phi(x^*, \lambda^*_{(r)}, 0^m, 0^n, k) = \begin{bmatrix} \nabla^2_{xx} L(x^*, \lambda^*) & -\nabla c^T_{(r)}(x^*) \\ \psi''(0) \Lambda^*_{(r)} \nabla c_{(r)}(x^*) & -k^{-1} I^r \end{bmatrix} = \nabla \Phi_k.$$

Therefore, due to the second implicit function theorem (see Bertsekas (1982)), there exist $\varepsilon > 0$, $\delta > 0$ and uniquely defined on $S(0^m, 0^n, \delta, k)$ two vector-functions

$$x(\cdot) = x(t, \theta, k) = (x_1(t, \theta, k), \ldots, x_n(t, \theta, k))$$

and

$$\tilde{\lambda}_{(r)}(.) = \tilde{\lambda}_{(r)}(t, \theta, k) = (\tilde{\lambda}_1(t, \theta, k), \ldots, \tilde{\lambda}_r(t, \theta, k)),$$

such that

$$\|x(t, \theta, k) - x^*\| \le \varepsilon, \ \|\tilde{\lambda}_{(r)}(t, \theta, k) - \lambda_r^*\| \le \varepsilon, \ \forall (t, \theta, k) \in S(0^m, 0^n, k, \delta)$$

and the following identities hold:

$$\nabla f(x(\cdot)) - \sum_{i=1}^{r} \tilde{\lambda}_i \nabla c_i(x(\cdot)) - h(x(\cdot), t, k) - \theta \equiv 0^n \tag{7.31}$$

$$\tilde{\lambda}_i(\cdot) \equiv \tilde{\lambda}_i(t, \theta, k) = (kt_i + \lambda_i^*) \psi'(kc_i(x(\cdot))), \quad i = 1, \ldots, r. \tag{7.32}$$

For the passive constraints, we have

$$c_i(x^*) \ge \sigma > 0, \quad i = r+1 \ldots, m.$$

Therefore, for $\delta > 0$ small enough, we obtain

$$c_i(x(t, \theta, k)) \ge 0.5\sigma, \quad \forall (t, \theta, k) \in S(0^m, 0^n, \delta, k).$$

From $(2c)$ of ψ, we have

$$\tilde{\lambda}_i = \lambda_i \psi'(kc_i(x(t, \theta, k))) \le \frac{2d}{\sigma k} \lambda_i \qquad i = r+1, \ldots, m, \tag{7.33}$$

where d and $\sigma > 0$ is independent on $k \in [k_0, k_1]$.
The next step is to estimate the norm of the Jacobians

$$\nabla_{t,\theta} x(t,\theta,k) \in \mathbb{R}^{n,m+n} \text{ and } \nabla_{t,\theta} \tilde{\lambda}(t,\theta,k) \in \mathbb{R}^{r,m+n}$$

at the origin $t = 0^m$, $\theta = 0^n$.
Using arguments similar to those we used in Theorem 7.2, we obtain

$$\begin{bmatrix} \nabla_{t,\theta} x(0^m,0^n,k) \\ \nabla_{t,\theta} \tilde{\lambda}_{(r)}(0^m,0^n,k) \end{bmatrix} = \tag{7.34}$$

$$= \begin{bmatrix} \nabla^2 L(x^*,\lambda^*) & -\nabla c_{(r)}^T(x^*) \\ \Lambda_{(r)}^* \nabla c_{(r)}(x^*) & -k^{-1} I^r \end{bmatrix}^{-1} \begin{bmatrix} 0^{n,r} & k\nabla c_{(m-r)}^T(x^*)\psi'\left(kc_{(m-r)}(x^*)\right) & I^n \\ I^r & 0^{r,m-r} & 0^{r,n} \end{bmatrix}.$$

From (7.14), (7.21), and (7.34) follows

$$\begin{aligned} &\max\left\{\|\nabla_{t,\theta} x(0^m,0^n,k)\|,\ \|\nabla_{t,\theta}\tilde{\lambda}_{(r)}(0^m,0^n,k)\|\right\} \\ &\qquad \le \rho \max\left\{1, d\sigma^{-1}\|\nabla c_{(m-r)}^T(x^*)\|\right\}. \end{aligned} \tag{7.35}$$

Therefore, there exists $\delta > 0$ small enough that for any $0 \le \tau \le 1$ and any $(t,\theta,k) \in S(0^m,0^n,k,\delta)$, we have

$$\begin{aligned} &\max\left\{\|\nabla_{t,\theta} x(\tau t,\tau\theta,k)\|,\ \|\nabla_{t,\theta}\tilde{\lambda}_{(r)}(\tau t,\tau\theta,k)\|\right\} \\ &\le 2\rho \max\left\{1,\ d\sigma^{-1}\|\nabla_{(m-r)} c(x^*)\|\right\} = c_0. \end{aligned} \tag{7.36}$$

Thus,

$$\begin{bmatrix} x(t,\theta,k) - x^* \\ \tilde{\lambda}_{(r)}(t,\theta,k) - \lambda_{(r)}^* \end{bmatrix} = \begin{bmatrix} x(t,\theta,k) - x(0^m,0^n,k) \\ \tilde{\lambda}_{(r)}(t,\theta,k) - \lambda_{(r)}(0^m,0^n,k) \end{bmatrix}$$

$$= \int_0^1 \begin{bmatrix} \nabla_{t,\theta} x(\tau t,\tau\theta,k) \\ \nabla_{t,\theta}\hat{\lambda}(\tau t,\tau\theta,k) \end{bmatrix} \begin{bmatrix} t \\ \theta \end{bmatrix} d\tau. \tag{7.37}$$

From (7.34)–(7.37) follows

$$\max\left\{\|x(t,\theta,k) - x^*\|,\ \|\tilde{\lambda}_{(r)}(t,\theta,k) - \lambda^*\|\right\}$$

$$\le c_0[k^{-1}\|\lambda - \lambda^*\| + \|\theta\|].$$

Let

$$\bar{x} = \bar{x}\left(\frac{\lambda-\lambda^*}{k},\theta,k\right), \bar{\lambda} = \left(\frac{\lambda-\lambda^*}{k},\theta,k\right) =$$

$$\left(\tilde{\lambda}_{(r)}\left(\frac{\lambda-\lambda^*}{k},\theta,k\right), \tilde{\lambda}_{(m-r)}\left(\frac{\lambda-\lambda^*}{k},\theta,k\right)\right).$$

Then, for $c = \max\{2d\sigma^{-1}, c_0\}$, we obtain

$$\|\tilde{x}-x^*\| = \left\|\tilde{x}\left(\frac{\lambda-\lambda^*}{k}, \theta, k\right) - x^*\right\| \le \frac{c}{k}\|\lambda-\lambda^*\| + \|\theta\| \quad (7.38)$$

$$\|\tilde{\lambda}-\lambda^*\| = \left\|\tilde{\lambda}\left(\frac{\lambda-\lambda^*}{k}, \theta, k\right) - \lambda^*\right\| \le \frac{c}{k}\|\lambda-\lambda^*\| + \|\theta\|. \quad (7.39)$$

From (7.28) we have

$$\|\nabla_x \mathscr{L}(\tilde{x},\lambda,k)\| = \|\theta\| \le \frac{\alpha}{k}\|\tilde{\lambda}-\lambda\|. \quad (7.40)$$

From (7.38)–(7.40) follows

$$\|\tilde{x}-x^*\| \le \frac{c}{k}\|\lambda-\lambda^*\| + \frac{\alpha}{k}\|\tilde{\lambda}-\lambda\|$$
$$\|\tilde{\lambda}-\lambda^*\| \le \frac{c}{k}\|\lambda-\lambda^*\| + \frac{\alpha}{k}\|\tilde{\lambda}-\lambda\|.$$

Then,

$$\|\tilde{\lambda}-\lambda^*\| \le \frac{c}{k}\|\lambda-\lambda^*\| + \frac{\alpha}{k}\|\tilde{\lambda}-\lambda\| \le \frac{c}{k}\|\lambda-\lambda^*\| + \frac{\alpha}{k}[\|\tilde{\lambda}-\lambda^*\| + \|\lambda-\lambda^*\|]$$

or

$$\left(1-\frac{\alpha}{k}\right)\|\tilde{\lambda}-\lambda^*\| \le \frac{c+\alpha}{k}\|\lambda-\lambda^*\|.$$

For $\alpha > 0$ and $k_0 > c+2\alpha$ and any $k > k_0$, we have

$$\|\tilde{\lambda}-\lambda^*\| \le \frac{c+2\alpha}{k}\|\lambda-\lambda^*\|.$$

Using (7.38) and (7.40) and the last bound, we obtain

$$\|\tilde{x}-x^*\| \le \frac{c}{k}\|\lambda-\lambda^*\| + \|\theta\| \le \frac{c}{k}\|\lambda-\lambda^*\| + \frac{\alpha}{k}\|\tilde{\lambda}-\lambda\| \le$$
$$\frac{c}{k}\|\lambda-\lambda^*\| + \frac{\alpha}{k}[\|\tilde{\lambda}-\lambda^*\| + |\lambda-\lambda^*\|] =$$
$$= \frac{c}{k}\|\lambda-\lambda^*\| + \frac{\alpha}{k}\|\lambda-\lambda^*\| + \frac{\alpha}{k}\frac{c+2\alpha}{k}\|\lambda-\lambda^*\|.$$

Therefore, again for $k_0 > c+2\alpha$ and any $k \ge k_0$, we obtain

$$\|\tilde{x}-x^*\| \le \frac{c}{k}(1+2\alpha)\|\lambda-\lambda^*\|.$$

(2) The statement follows from the formula for the Hessian $\nabla_{xx}\mathscr{L}(\tilde{x},\tilde{\lambda})$, bound (7.30), and Debreu's Lemma, see Debreu (1952). □

7.1.7 Newton NR Method and "Hot" Start Phenomenon

The main operation at each NR step is finding an approximation for the primal minimizer. Any efficient and numerically stable method for unconstrained minimization can be used for it; see, for example, Dennis and Schnabel (1996).

We use regularized Newton method (RNM), due to its global convergence with asymptotic quadratic rate. Moreover, RNM is particularly efficient in the NR framework due to stability of the Newton area in the neighborhood of the primal solution.

Finding primal approximation $\tilde{x}$ by RNM from (7.28) and dual approximation $\tilde{\lambda}$ from (7.29) leads to Newton NR method.

Before describing this method, let us make few comments.

First, we will use the merit function $\nu : \mathbb{R}^n \times \mathbb{R}^m_+ \to \mathbb{R}$ defined via

$$\nu(y) \equiv \nu(x,\lambda) = \max\{\|\nabla_x L(x,\lambda)\|, \sum_{i=1}^{m} \lambda_i |c_i(x)| \; - c_i(x), i = 1,\ldots,m\} \qquad (7.41)$$

to measure primal infeasibility and complementarity; we also use it as a regularization parameter, and for the stopping criteria.

Obviously, from (7.41) follows $\nu(y) \geq 0$, $\forall y \in \mathbb{R}^n \times \mathbb{R}^m_+$, and from KKT's Theorem follows

$$\nu(y) = 0 \Leftrightarrow y = y^*.$$

Let $\varepsilon_0 > 0$ be small enough. The following lemma is similar to Lemma 4.4 and can be proven using similar arguments.

Lemma 7.1. *Under the second-order sufficient optimality conditions (4.73)–(4.74) and Lipschitz condition (4.130), there are* $0 < l < L < \infty$ *and* $\varepsilon_0 > 0$ *that for any* $y \in B(y^*, \varepsilon_0)$ *the following bounds*

$$l\|y - y^*\| \leq \nu(y) \leq L\|y - y^*\| \qquad (7.42)$$

hold.

Second, for finding $\tilde{x}$ from (7.28), we use damped regularized Newton method

$$x := x + tn(x), \qquad (7.43)$$

where $n(x)$ is the solution of the following regularized Newton system

$$H(x,\lambda,k)n(x) = (\nabla_{xx}\mathscr{L}(x,\lambda,k) + \nu(y)I^n)n(x) = -\nabla_x\mathscr{L}(x,\lambda,k). \qquad (7.44)$$

Third, $n(x)$ is unique, because matrix

$$H(x,\lambda,k) = \nabla^2_{xx}\mathscr{L}(x,\lambda,k) + \nu(y)I^n \qquad (7.45)$$

is positive definite for any $(\lambda,k) \in \mathbb{R}^m_+ \times \mathbb{R}_{++}$.

In fact, from (7.41) follows existence of $\nu_0 > 0$, that for any $y \notin B(y^*, \varepsilon_0)$

$$v(y) \geq v_0, . \tag{7.46}$$

From convexity f, concavity c_i, properties (1)–(4) of $\psi \in \Psi$, and the definition of $\mathscr{L}$ follows convexity $\mathscr{L}(x,\lambda,k)$ in x for any given $(\lambda,k) \in \mathbb{R}^m_+ \times \mathbb{R}_{++}$.

From (7.45)–(7.46) follows existence $m_1 \geq v_0$, that

$$\langle H(x,\lambda,k)u,u\rangle \geq m_1\langle u,u\rangle, \ \ \forall u \in \mathbb{R}^n$$

for any $y \notin B(y^*, \varepsilon_0)$.

For $y \in B(y^*, \varepsilon_0)$, the merit function v is vanishing, but for $k_0 > 0$ large enough, from Debreu's Lemma follows existence $m_2 > 0$, that

$$\langle H(x,\lambda,k)u,u\rangle \geq m_2\langle u,u\rangle, \ \forall u \in \mathbb{R}^n. \tag{7.47}$$

for any $(\lambda,k) \in \Lambda(\lambda^*,\delta,k_0)$.

Hence, there is $k_0 > 0$ large enough and $0 < m_0 = \min\{m_1,m_2\}$, that for any $k \geq k_0$ and $(\lambda,k) \in \mathbb{R}^m_+ \times \mathbb{R}_{++}$, we have

$$\langle H(x,\lambda,k)u,u\rangle \geq m_0\langle u,u\rangle, \ \forall u \in \mathbb{R}^n, \forall x \in \mathbb{R}^n. \tag{7.48}$$

From (7.44) follows that $n(x)$ is a decent direction for LEP $\mathscr{L}$ in $x \in \mathbb{R}^n$ for any fixed $(\lambda,k) \in \Lambda(\lambda^*,\delta,k_0)$.

In fact,

$$\langle \nabla_x \mathscr{L}(x,\lambda,k), n(x)\rangle = -\langle H(x,\lambda,k)n(x), n(x)\rangle \leq -m_0\langle n(x), n(x)\rangle.$$

Fourth, for adjusting step length $t > 0$, we use the following backtracking line search:

$$\mathscr{L}(x+tn(x),\lambda,k) \leq \mathscr{L}(x,\lambda,k) + 0.5t\langle \nabla_x \mathscr{L}(x,\lambda,k), n(x)\rangle \tag{7.49}$$

starting with $t = 1$.

The backtracking line search consists of the following steps:

1. For $t(x) = t = 1$ check (7.49). If (7.49) holds go to 2. If (7.49) does not hold, set $t := t/2$ and repeat it till (7.49) holds; then go to 2.
2. Set $t(x) = t$, $x := x + t(x)n(x)$.

Fifth, the critical part of Newton NR method is the bound

$$\max\{\|\tilde{x} - x^*\|, \|\tilde{\lambda} - \lambda^*\|\} \leq ck^{-1}(1+2\alpha)\|\lambda - \lambda^*\|, \tag{7.50}$$

which is item (1) of Theorem 7.3. It holds for any $\tilde{x}$ and $\tilde{\lambda}$ defined by (7.28)–(7.29).

Parameters $c > 0$ and $k_0 > 0$ are unknown a priori. Let $\gamma = ck^{-1}(1+2\alpha)$; then by taking $0 < \gamma \leq 0.5$ from (7.50), we obtain

$$\max\{\|\tilde{x} - x^*\|, \|\tilde{\lambda} - \lambda^*\|\} \leq \gamma\|\lambda - \lambda^*\| \leq 0.5\|\lambda - \lambda^*\| \tag{7.51}$$

for any $k \geq k_0 := \max\{k_0, c(1+2\alpha)\gamma^{-1}\}$ and $\lambda \in \Lambda(\lambda^*, k_0, \delta)$.

The key element of Newton NR method is finding $\tilde{x}$ and adjusting $k_0 > 0$ to guarantee (7.51). Obviously, one can't check (7.51) without y^*.

It follows, however, from (7.51) that

$$\|\tilde{y} - y^*\| = \max\{\|\tilde{x} - x^*\|, \|\tilde{\lambda} - \lambda^*\|\} \leq \gamma\|\lambda - \lambda^*\| \leq \gamma\|y - y^*\|.$$

For any $y \in B(y^*, \varepsilon)$ from the left inequality (7.42) follows

$$\|y - y^*\| \leq l^{-1}\nu(y).$$

Therefore, if $\tilde{y} \in B(y^*, \varepsilon)$, then

$$\|\tilde{y} - y^*\| \leq \gamma l^{-1}\nu(\tilde{y}).$$

From the right inequality (7.42) follows

$$L^{-1}\nu(\tilde{y}) \leq \|\tilde{y} - y^*\|.$$

Hence,

$$\nu(\tilde{y}) \leq \gamma\frac{L}{l}\nu(y).$$

For $\tilde{\gamma} = \frac{L}{l}\gamma$, we have

$$\nu(\tilde{y}) \leq \tilde{\gamma}\nu(y). \tag{7.52}$$

For a given $0 < \tilde{\gamma} \leq 0.5$, the inequality (7.52) can be checked. If (7.52) does not hold, then the scaling parameter $k > 0$ should be increased.

Each step of the Newton NR method requires finding an approximation $\tilde{x}$ for the primal minimizer from (7.28) followed by the LM vector update by (7.29). For finding approximation $\tilde{x}$ we introduce procedure called **Minimizer** (x, λ, k).

The procedure takes (x, λ, k) as an input and produces $(\tilde{x}, \tilde{\lambda}, \tilde{k})$, which satisfies (7.28)–(7.29) as an output.

Let us describe **Minimizer** (x, λ, k).

begin

0. Initialization $x \in \mathbb{R}^n$, $\lambda \in \mathbb{R}^m_{++}$, $k > 0$ large enough, $0 < \varkappa < 0.5$ and $0 < \alpha < 0.25$;
1. find $n(x)$ from (7.44);
2. find $t(x)$ using backtracking technique;
3. **set** $\tilde{x} := x + t(x)n(x)$ and $\tilde{\lambda} = (\tilde{\lambda}_i = \lambda_i\psi'(kc_i(\tilde{x})), i = 1, \ldots, m)$
4. **while** $\|\nabla_x\mathscr{L}(\tilde{x}, \lambda, k)\| > \alpha k^{-1}\|\tilde{\lambda} - \lambda\|$ **set** $x := \tilde{x}$, go to 1;
5. **while** $\nu(\tilde{y}) > 0.5\nu(y)$ **set** $x := \tilde{x}$, $\tilde{k} := k(1 + \varkappa)$ go to 1;
6. output $y := \tilde{y}$.

end

Now we are ready to describe the **Newton NR** method.

0. Initialization $x \in \mathbb{R}^n$, $\lambda = (1,\dots,1)^T \in \mathbb{R}^m$, $k > 0$ large enough, $0 < \varkappa < 0.5$, $0 < \gamma \leq 0.5$, $\varepsilon > 0$ – required accuracy;
1. if $\nu(y) \leq \varepsilon$ stop and output the solution $y^* = y$;
2. apply **Minimizer** (x, λ, k);
3. if $\nu(\tilde{y}) \leq \gamma \nu(y)$, then $x := \tilde{x}$, $\lambda := \tilde{\lambda}$, go to 1, else $x := \tilde{x}$, $\lambda := \tilde{\lambda}$, $k := \tilde{k}(1+\varkappa)$ go to 2.

There exists $s_0 > 0$, that for any $s \geq s_0$ approximation $\tilde{x}_s$ for the primal minimizer $\hat{x}_s$ belongs to Newton area (see Fig. 8.5) of the following minimizer $\hat{x}_{s+1}$. Then, for finding

$$\tilde{x}_{s+1} : \|\nabla_x \mathscr{L}(\tilde{x}_{s+1}, \tilde{\lambda}_{s+1}, k)\| \leq \frac{\alpha}{k} \|\tilde{\lambda}_{s+1} - \tilde{\lambda}_s\|$$

, Newton's method is used from $\tilde{x}_s$ as a starting point.

It means that for any $s \geq s_0$ it requires at most $O(\ln \ln \varepsilon^{-1})$ Newton steps for finding primal approximation with accuracy $\varepsilon > 0$.

Application of the Newton NR method for LP calculations in case of nondegenerate primal and dual LP requires for any $s \geq s_0$ at most $O(\ln \ln \varepsilon^{-1}) = O(\ln \ln 2^L) = O(\ln L) = O(\ln n)$ Newton steps for each LM vector update. Each update shrinks the distance from $\tilde{y}_s$ to y^* by $0 < \gamma \leq 0.5$.

We call step s_0 the "hot" start. Obviously one has to characterize the neighborhood of primal and dual solutions, when the "hot" start occurs. It has been done for QP by Melman and Polyak (1996) It is a rather technical issue, which we are not going to discuss here.

On the practical level, the "hot" start has been systematically observed on a number of both test and real-life nonlinear optimization problems, as well as on QP and LP (see Chapter 11).

Due to the "hot" start phenomenon, the Newton NR method produced solutions with high accuracy, often requiring much less Newton steps than the NR theory predicts, see Fig. 7.6.

Exercise 7.3. Consider the NR method for solving the following LP:

$$\langle a, x^* \rangle = \min\{\langle a, x \rangle | c(x) = Ax - b \geq 0\},$$

where $a \in \mathbb{R}^n, b \in \mathbb{R}^m, A : \mathbb{R}^n \to \mathbb{R}^m$, $m > n$.

Exercise 7.4. Under the assumption that x^* is nondegenerate solution, estimate the convergence rate of the NR method.

Exercise 7.5. Consider the Newton NR method for LP. Develop a code based on Newton NR method.

Exercise 7.6. What is the main difference between IPM and Newton NR methods. Consider standard LP test problems, and show the "hot" start phenomenon.

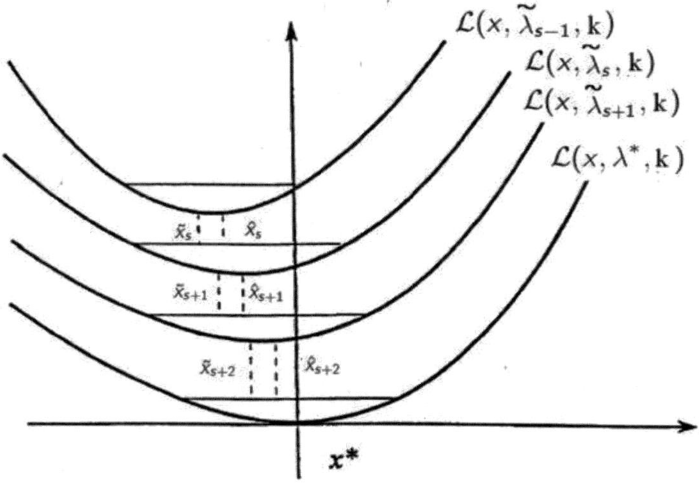

Fig. 7.6 "Hot" start

7.2 NR with "Dynamic" Scaling Parameters

7.2.0 Introduction

The equivalence of the AL multipliers method and quadratic prox for equality constraints was established by Rockafellar (1973, 1976), and references therein. It is one of the centerpieces of AL theory (see Chapter 4).

In this section we show that similar equivalence exists between NR with "dynamic" scaling parameter update and interior quadratic prox (IQP) for the dual problem.

We used $\psi \in \Psi$ to transform constraints of a given constrained optimization problem into an equivalent set of constraints. The transformation is scaled by a vector of positive scaling parameters, one for each constraint. Unconstrained minimization of LEP $\mathscr{L}$ in primal space followed by both LM vector and scaling parameters vector update forms a general NR multiplier method with "dynamic" scaling parameter update.

The scaling parameters are updated inversely proportional to Lagrange multipliers – formula suggested by P. Tseng and D. Bertsekas for the exponential transformation in 1993.

The NR multiplier method with "dynamic" scaling parameter update leads to prox method with second-order φ-divergence distance for the dual problem. The equivalence is the basic tool for convergence analysis.

The analysis for NR method with "dynamic" scaling parameter updates turned out to be rather difficult, even for a particular exponential transformation.

In the mid-1990s Ben-Tal and Zibulevsky (1997) proved for a particular class of constraint transformations that primal and dual sequences generated by NR methods are bounded and any convergent subsequence has the primal–dual solution as its limit point.

At the end of the 1990s, Auslender et al. (1999) obtained similar results for inexact version of the multiplier method with "dynamic" scaling parameter updates. For the regularized logarithmic MBF kernel $\varphi(t) = 0.5\nu(t-1)^2 + \mu(-\ln t + t - 1)$, the authors proved global convergence of the dual sequence with convergence rate $O\left((ks)^{-1}\right)$, where $k > 0$ is the scaling parameter and s is the number of steps.

The regularization, which provides strong convexity of the dual kernel, was an important element of their analysis. Unfortunately, such modification of the dual kernel, in some instances, leads to substantial difficulties, when it comes to finding the primal transformation, which is LF conjugate of the dual kernel. For example, in case of exponential transformation, it leads to solving a transcendent equation.

In this section we consider an alternative approach. The strong convexity of the dual kernels on $\mathbb{R}_+$ is guaranteed by replacing the original transformations $\psi \in \Psi$, with their truncated versions (see Section 7.1.2).

Properties (1)–(4) of truncated transformations $\psi \in \Psi$ lead to equivalence of the primal exterior NR method and dual interior quadratic prox (IQP), which is the key element of the convergence analysis.

Then we show that under strict complementarity conditions the IQP converges with $o\left((ks)^{-1}\right)$ rate.

Under the second-order sufficient optimality condition, the NR method converges with Q-linear rate without unbounded increase of the scaling parameters, which correspond to the active constraints. It means that Q-linear rate can be achieved without compromising the condition number of the LEP's Hessian. Moreover, the results are taking place, whether f and - c_i, $i = 1, \ldots, m$ are convex or not.

When applied for LP calculations, the NR method with "dynamic" scaling parameter updates converges in value with quadratic rate for any $\psi \in \Psi$, which corresponds to the well-defined dual kernel, if one of the dual LPs has a unique solution.

In the final part of the chapter, we introduce the primal–dual NR (PDNR) method based on NR with "dynamic" scaling parameter update.

Along with the global convergence, we show that under the second-order sufficient optimality condition, PDNR converges with asymptotic quadratic rate.

We conclude the chapter by considering Nonlinear Rescaling–Augmented Lagrangian (NRAL) method for problems with both inequality constraints and equations.

7.2.1 Nonlinear Rescaling as Interior Quadratic Prox

For a given vector $\mathbf{k} = (k_1, \ldots, k_m) \in \mathbb{R}^m_{++}$ from properties (1) and (2) of $\psi \in \Psi$ follows

$$c_i(x) \geq 0 \Leftrightarrow k_i^{-1}\psi(k_i c_i(x)) \geq 0, \quad i \equiv 1 \ldots, m. \tag{7.53}$$

Therefore, problem

$$f(x^*) = \min\{f(x)/k_i^{-1}\psi(k_ic_i(x)) \geq 0,\ i = 1,\ldots,m\} \tag{7.54}$$

is equivalent to the original primal problem (P).

LEP $\mathscr{L} : \mathbb{R}^n \times \mathbb{R}^m_{++} \times \mathbb{R}^m_{++} \to R$, which is given by

$$\mathscr{L}(x,\lambda,\mathbf{k}) = f(x) - \sum_{i=1}^{q} \lambda_i k_i^{-1}\psi(k_ic_i(x)),$$

is our basic tool.

The following proposition establishes the basic LEP $\mathscr{L}$ properties at any KKT pair (x^*,λ^*).

Assertion 7.3 *For any $\mathbf{k} \in \mathbb{R}^m_{++}$ and any KKT pair (x^*,λ^*), the following properties*

1. $\mathscr{L}(x^*,\lambda^*,\mathbf{k}) = L(x^*,\lambda^*) = f(x^*)$
2. $\nabla_x\mathscr{L}(x^*,\lambda^*,\mathbf{k}) = \nabla_x L(x^*,\lambda^*) = \nabla f(x^*) - \sum_{i=1}^m \lambda_i^*\nabla c_i(x^*) = 0$
3. $\nabla^2_{xx}\mathscr{L}(x^*,\lambda^*,\mathbf{k}) = \nabla^2_{xx}L(x^*,\lambda^*) + 0.5\nabla c(x^*)^T K\Lambda^*\nabla c(x^*)$, *where* $K = \mathrm{diag}(k_i)_{i=1}^m$, $\Lambda^* = \mathrm{diag}(\lambda_i^*)_{i=1}^m$

hold.

Exercise 7.7. Verify properties (1)–(3).

Remark 7.4. In view of $\lambda_i^* = 0,\ i = r+1\ldots,m$ from 3. follows

$$\nabla^2_{xx}\mathscr{L}(x^*,\lambda^*,\mathbf{k}) = \nabla^2_{xx}L(x^*,\lambda^*) + 0.5\nabla c_{(r)}(x^*)^T K_r\Lambda_r^*\nabla c_{(r)}(x^*),$$

where $K_r = \mathrm{diag}(k_i)_{i=1}^r$, $\Lambda_r^* = \mathrm{diag}(\lambda_i^*)_{i=1}^r$. Let $k > 0$ and $K_r = k\Lambda_r^{*-1}$; then

$$\nabla^2_{xx}\mathscr{L}(x^*,\lambda^*,\mathbf{k}) = \nabla^2_{xx}L(x^*,\lambda^*) + 0.5k\nabla c_{(r)}^T(x^*)\nabla c_{(r)}(x^*),$$

which is identical to Hessian of Quadratic Augmented Lagrangian. It clarifies why the scaling parameters are updated inversely proportional to the Lagrange multipliers.

We are ready to describe the NR method with "dynamic" scaling vector update.

Let $\psi \in \Psi$, $\lambda \in \mathbb{R}^m_{++}$, $\mathbf{k} \in \mathbb{R}^m_{++}$, and $k > 0$. The NR step consists of the following operations:

1. find primal minimizer

$$\hat{x} \equiv \hat{x}(\lambda,\mathbf{k}) : \nabla_x\mathscr{L}(\hat{x},\lambda,\mathbf{k}) = 0; \tag{7.55}$$

2. update Lagrange multipliers

$$\hat{\lambda} \equiv \hat{\lambda}(\lambda,\mathbf{k}) = (\hat{\lambda}_1,\ldots,\hat{\lambda}_m) : \hat{\lambda}_i = \lambda_i\psi'(k_ic_i(\hat{x})),\quad i = 1,\ldots,m \tag{7.56}$$

3. update the scaling vector

$$\hat{\mathbf{k}} \equiv \hat{\mathbf{k}}(\hat{\lambda}, k) = (\hat{k}_1, \ldots, \hat{k}_m) : \hat{k}_i = k\hat{\lambda}_i^{-1}, \quad i = 1, \ldots, m. \tag{7.57}$$

Theorem 7.4. *If conditions A and B hold and $f, c_i \in C^1$, $i = 1, \ldots, m$, then NR method (7.55)–(7.57) is:*

(1) well defined;
(2) equivalent to the following prox-method

$$d(\hat{\lambda}) - k^{-1}D_2(\hat{\lambda}, \lambda) = \max\{d(u) - k^{-1}D_2(u, \lambda) | \; u \in \mathbb{R}^m_{++}\}, \tag{7.58}$$

where $D_2(u, \lambda) = \sum_{i=1}^m \lambda_i^2 \varphi(u_i/\lambda_i)$ is the second-order φ-divergence entropy-like distance function based on kernel $\varphi = -\psi^*$, where ψ^* is LF transform of $\psi \in \Psi$.

Proof.

(1) From properties (1) and (2) of $\psi \in \Psi$, convexity f, concavity c_i, $i = 1, \ldots, m$ follows convexity $\mathscr{L}$ in x. From Exercise 7.1 follows that for any $x \in \Omega$, any $(\lambda, \mathbf{k}) \in \mathbb{R}^m_{++} \times \mathbb{R}^m_{++}$, and any $d \neq 0$, we have

$$\lim_{t \to \infty} \mathscr{L}(x + td, \lambda, \mathbf{k}) = \infty.$$

Therefore, the solution $\hat{x}$ of the system (7.55) exists for any given $(\lambda, \mathbf{k}) \in \mathbb{R}^m_{++} \times \mathbb{R}^m_{++}$.
From property (2) of $\psi \in \Psi$ and (7.56) follows $\lambda \in \mathbb{R}^m_{++} \Rightarrow \hat{\lambda} \in \mathbb{R}^m_{++}$ and $\hat{\mathbf{k}} = k(\hat{\Lambda})^{-1}e \in \mathbb{R}^m_{++}$, where $\hat{\Lambda} = \mathrm{diag}(\hat{\lambda}_i)_{i=1}^m$ and $e = (1, \ldots, 1)^T \in \mathbb{R}^m$.
So NR method (7.55)–(7.56) is well defined.

(2) From (7.55) and (7.56) follows

$$\nabla_x \mathscr{L}(\hat{x}, \hat{\lambda}, \kappa) = \nabla f(\hat{x}) - \sum_{i=1}^m \lambda_i \psi'(k_i c_i(\hat{x})) \nabla c_i(\hat{x}) = \nabla_x L(\hat{x}, \hat{\lambda}) = 0.$$

Therefore,

$$d(\hat{\lambda}) = \min_{x \in \mathbb{R}^n} L(x, \hat{\lambda}) = L(\hat{x}, \hat{\lambda}).$$

The subdifferential $\partial d(\hat{\lambda})$ contains $-c(\hat{x})$; hence,

$$0 \in c(\hat{x}) + \partial d(\hat{\lambda}). \tag{7.59}$$

From (7.56) follows $\psi'(k_i c_i(\hat{x})) = \hat{\lambda}_i/\lambda_i$, $i = 1, \ldots, m$.

From property (3) of $\psi \in \Psi$ follows existence of the inverse ψ'^{-1}. Using LF identity, we can rewrite (7.56) as follows:

$$c_i(\hat{x}) = k_i^{-1}\psi_i'^{-1}(\hat{\lambda}_i/\lambda_i) = k_i^{-1}\psi^{*'}(\hat{\lambda}_i/\lambda_i), \; i = 1, \ldots, m. \tag{7.60}$$

By introducing kernel $\varphi = -\psi^*$ and keeping in mind (7.60), the inclusion (7.59) can be rewritten as follows:

$$0 \in \partial d(\hat{\lambda}) - k^{-1} \sum_{i=1}^{m} \lambda_i \varphi'(\hat{\lambda}_i/\lambda_i) e_i, \tag{7.61}$$

where $e_i = (0,\dots,1,\dots,0)^T \in \mathbb{R}^m$. The inclusion (7.61) is the optimality criteria for $\hat{\lambda}$ to be solution of (7.58). □

Our next step is to show that NR method (7.55)–(7.57) is equivalent to the IQP for the dual problem.

Theorem 7.5. *Under conditions A and B and f, $c_i \in C^1$, $i = 1,\dots,m$ NR method (7.55)–(7.57) is equivalent to IQP for the dual problem.*

Proof. From Theorem 7.4 follows existence $(\hat{x}, \hat{\lambda})$ for any $k > 0$ and any $(\lambda, \Bbbk) \in \mathbb{R}^m_{++} \times \mathbb{R}^m_{++}$.

For any $1 \le i \le m$ from the property 2.(a) of $\psi \in \Psi$, the update formula (7.56) and the mean value formula it follows that

$$\hat{\lambda}_i - \lambda_i = \lambda_i(\psi'(k_i c_i(\hat{x})) - \psi'(0)) = \lambda_i k_i \psi''(\theta_{i,\lambda} k_i c_i(\hat{x})) c_i(\hat{x}), \tag{7.62}$$

where $0 < \theta_{i,\lambda} < 1$. From the update formula (7.57) and (7.62), we have

$$\hat{\lambda} = \lambda + k\psi''_{[\lambda]}(\cdot) c(\hat{x}), \tag{7.63}$$

where

$$\psi''_{[\lambda]}(\cdot) = \mathrm{diag}(\psi''(\theta_{i,\lambda} k_i c_i(\hat{x})))_{i=1}^m.$$

Let $R_\lambda = -\psi''_{[\lambda]}(\cdot)$; then system (7.63) can be rewritten as follows:

$$\hat{\lambda} = \lambda + kR_\lambda(-c(\hat{x})). \tag{7.64}$$

From (3) and (4) of $\psi \in \Psi$ follows

$$M_0^{-1} I^m \le R_\lambda \le m_0^{-1} I^m,$$

$0 < m_0 < M_0 < \infty$.

Our next step is to show that (7.63) is an IQP method for the dual problem in the rescaled from step to step dual space.

System (7.63) can be rewritten as follows:

$$-c(\hat{x}) - k^{-1} R_\lambda^{-1}(\hat{\lambda} - \lambda) = 0, \tag{7.65}$$

which is the optimality criteria for vector $\hat{\lambda} \in \mathbb{R}^m_{++}$ to be solution of the following quadratic prox

$$d(\hat{\lambda}) - \frac{1}{2} k^{-1} \|\hat{\lambda} - \lambda\|^2_{R_\lambda^{-1}} = \max\{d(u) - \frac{1}{2} k^{-1} \|u - \lambda\|^2_{R_\lambda^{-1}} \mid u \in \mathbb{R}^m_{++}\}. \tag{7.66}$$

Thus, NR (7.55)–(7.57) is equivalent to the IQP (7.66) for the dual problem. □

Remark 7.5. From properties (3) and (4) of $\psi \in \Psi$ follows

$$m_0 I^m \prec R_\lambda^{-1} \prec M_0 I^m.$$

From $-c(\hat{x}) \in \partial d(\hat{\lambda})$ follows that (7.63) is an implicit rescaled subgradient method for the dual problem.

Remark 7.6. If $\hat{x}$ is unique, then $-c(\hat{x}) = \nabla d(\hat{\lambda})$ and from (7.64) follows

$$\hat{\lambda} = \lambda + kR_\lambda \nabla d(\hat{\lambda}).$$

In other words NR method (7.55)–(7.57) is equivalent to implicit Euler method for the following system of ordinary equations:

$$\frac{d\lambda}{dt} = kR_\lambda \nabla d(\lambda), \;\; \lambda(0) = \lambda_0.$$

7.2.2 Convergence of the NR Method

We consider convergence of the NR method (7.55)–(7.57) for a wide class of constraint transformation $\psi \in \Psi$ under mild assumption on the input data. Our analysis is based on the equivalence of the NR method (7.55)–(7.57) and IQP (7.66).

Let $\{x_s, \lambda_s\}_{s\in N}$ be the primal–dual sequence and $\{\Bbbk_s\}_{s\in N}$ be the sequence of scaling vectors generated by NR method (7.55)–(7.57) and $\Delta(\lambda_0) = d(\lambda^*) - d(\lambda_0)$.

The dual level set $\Lambda_0 = \{\lambda \in \mathbb{R}^m_+ : d(\lambda) \geq d(\lambda^0)\}$ is bounded due to concavity of d and boundedness of the dual optimal set L^*, which follows from Slater condition.

For x_s we consider two sets of indices $I_s^- = \{i : c_i(x_s) < 0\}$ and $I_s^+ = \{i : c_i(x_s) \geq 0\}$ and the maximum constraint violation

$$\mathrm{v}_s = \max\{-c_i(x_s)/i \in I_s^-\}$$

at step s.

Along with the dual sequence $\{\lambda_s\}_{s\in N}$, we consider the corresponding convex and bounded dual level sets $\Lambda_s = \{\lambda \in \mathbb{R}^m_+ : d(\lambda) \geq d(\lambda_s)\}$ and their boundaries $\partial\Lambda_s = \{\lambda \in \Lambda_s : d(\lambda) = d(\lambda_s)\}$.

Theorem 7.6. *Let assumptions A and B be satisfied and $f,\ c_i \in C^1,\ i = 1,\ldots,m$; then*

(1) the dual sequence $\{\lambda_s\}_{s\in N}$ is bounded and monotone increasing in value and the following bound

$$d(\lambda_{s+1}) - d(\lambda_s) \geq m_0 k^{-1} \|\lambda_{s+1} - \lambda_s\|^2, \; s \geq 1$$

holds;

(2) for the primal–dual sequence $\{x_s, \lambda_s\}_{s\in N}$*, the following bound*

$$d(\lambda_{s+1}) - d(\lambda_s) \geq kmM^{-2} \sum_{i\in I^-_{s+1}} c_i^2(x_{s+1})$$

holds and $\lim_{s\to\infty} \nu_s = 0$*;*

(3) every converging primal–dual subsequence has the primal–dual solution as its limit point;

(4) the entire dual sequence $\{\lambda_s\}_{s\in N}$ *converges to the dual solution in value, that is,*

$$\lim_{s\to\infty} d(\lambda_s) = d(\lambda^*);$$

(5)

$$\lim_{s\to\infty} d_H(\partial\Lambda_s, L^*) = 0, \tag{7.67}$$

where $d_H(\partial\Lambda_s, L^*)$ *is Hausdorff distance between the dual solution set* L^* *and the boundaries at the dual level set* $\Lambda_s = \{\lambda \in \mathbb{R}^m : d(\lambda) \geq d(\lambda_s)\}$ *.*

Proof.

(1) From (7.66) with $\hat{\lambda} = \lambda_{s+1}$, $\lambda = \lambda_s$ follows

$$d(\lambda_{s+1}) \geq d(\lambda_s) + \frac{1}{2}k^{-1}||\lambda_{s+1} - \lambda_s||^2_{R_s^{-1}}, \tag{7.68}$$

where $R_s = \text{diag}(-\psi''(\theta_{i,s}k_{i,s}c_i(x_{s+1})))$.
So the dual sequence $\{\lambda_s\}_{s\in N}$ is monotone increasing in value.
Therefore, the dual sequence $\{\lambda_s\}_{s\in N} \subset \Lambda_0$ is bounded. We start by finding the lower bound for $d(\lambda_{s+1}) - d(\lambda_s)$.
From concavity of d and $-c(x_{s+1}) \in \partial d(\lambda_{s+1})$ follows $d(\lambda) - d(\lambda_{s+1}) \leq (-c(x_{s+1}), \lambda - \lambda_{s+1})$ or

$$d(\lambda_{s+1}) - d(\lambda) \geq (c(x_{s+1}), \lambda - \lambda_{s+1}),\ \lambda \in \mathbb{R}^m_+. \tag{7.69}$$

From $\psi''(\cdot) \neq 0$ follows existence of ψ'^{-1}; then from the update formula (7.56), we have

$$c_i(x_{s+1}) = (k_{i,s})^{-1}\psi'^{-1}(\lambda_{i,s+1}/\lambda_{i,s}).$$

Using LF identity $\psi'^{-1} = \psi^{*\prime}$, we obtain

$$c_i(x_{s+1}) = (k_{i,s})^{-1}\psi^{*\prime}(\lambda_{i,s+1}/\lambda_{i,s}), \quad i = 1, \ldots, m. \tag{7.70}$$

Using $\psi^{*\prime}(1) = \psi^{*\prime}(\lambda_{i,s}/\lambda_{i,s}) = 0$ from (7.69) and (7.70) for $\lambda = \lambda_s$, we obtain

$$d(\lambda_{s+1}) - d(\lambda_s) \geq$$

$$\sum_{i=1}^m (k_{i,s})^{-1}\left(\psi^{*\prime}\left(\frac{\lambda_{i,s+1}}{\lambda_{i,s}}\right) - \psi^{*\prime}\left(\frac{\lambda_{i,s}}{\lambda_{i,s}}\right)\right)(\lambda_{i,s} - \lambda_{i,s+1}). \tag{7.71}$$

Using $\psi'(1) = 0$, mean value formula, and $\varphi''(\cdot) = -\psi^{*''}(\cdot)$, we have

$$\psi^{*\prime}\left(\frac{\lambda_{i,s+1}}{\lambda_{i,s}}\right) - \psi^{*\prime}\left(\frac{\lambda_{i,s}}{\lambda_{i,s}}\right) = -\psi^{*\prime\prime}(\cdot)(\lambda_{i,s})^{-1}(\lambda_{i,s} - \lambda_{i,s+1})$$

$$= \varphi''(\cdot)(\lambda_{i,s})^{-1}(\lambda_{i,s} - \lambda_{i,s+1}), i = 1, \ldots, m.$$

From (7.71) and the latter system follows

$$d(\lambda_{s+1}) - d(\lambda_s) \geq \sum_{i=1}^{m} (k_{i,s}\lambda_{i,s})^{-1}\varphi''(\cdot)(\lambda_{i,s} - \lambda_{i,s+1})^2.$$

Keeping in mind the update formula (7.57) and 3.b) from Assertion 7.2, we obtain the following bound

$$d(\lambda_{s+1}) - d(\lambda_s) \geq m_0 k^{-1} \|\lambda_s - \lambda_{s+1}\|^2, \forall s \geq 1, \tag{7.72}$$

which is typical for classical quadratic prox.

(2) Let us consider set $I_{s+1}^- = \{i : c_i(x_{s+1}) < 0\}$ of constraints violated at the point x_{s+1}.
From $\psi^{*\prime}(1) = \psi^{*\prime}(\lambda_{i,s}/\lambda_{i,s}) = 0$, mean value formula, property 3.c), and Assertion 7.2 follows

$$-c_i(x_{s+1}) = (k_{i,s})^{-1}\left[\psi^{*\prime}\left(\frac{\lambda_{i,s}}{\lambda_{i,s}}\right) - \psi^{*\prime}\left(\frac{\lambda_{i,s+1}}{\lambda_{i,s}}\right)\right] =$$

$$(\lambda_{i,s}k_{i,s})^{-1}\left(-\psi^{*\prime\prime}(\cdot)\right)(\lambda_{i,s+1}-\lambda_{i,s}) \leq k^{-1}\varphi''(\cdot)|\lambda_{i,s+1} - \lambda_{i,s}| \leq k^{-1}M_0|\lambda_{i,s+1}-\lambda_{i,s}|$$

or

$$|\lambda_{i,s+1} - \lambda_{i,s}| \geq kM_0^{-1}(-c_i(x_{s+1})), \quad i \in I_{s+1}^-.$$

Combining the last inequality with (7.72), we obtain

$$d(\lambda_{s+1}) - d(\lambda_s) \geq km_0M_0^{-2} \sum_{i \in I_{s+1}^-} c_i^2(x_{s+1}). \tag{7.73}$$

For the maximum constraint violation at the step $l+1$ from (7.73) follows

$$d(\lambda_{l+1}) - d(\lambda_l) \geq km_0M_0^{-2}\mathrm{v}_{l+1}^2. \tag{7.74}$$

Summing up (7.73) from $l = 1$ to $l = s$, we obtain

$$\Delta(\lambda_0) = d(\lambda^*) - d(\lambda_0) \geq d(\lambda_{s+1}) - d(\lambda_0) \geq km_0M_0^{-2}\sum_{l=0}^{s}\mathrm{v}_{s+1}^2.$$

Therefore, $\lim_{s\to\infty} \mathrm{v}_s \to 0$, which leads to primal asymptotic feasibility. For the best in s steps maximum constraint violation $\bar{\mathrm{v}}_s = \min\{\mathrm{v}_l | 1 \leq l \leq s\}$, we obtain

$$\bar{\nu}_s \leq M_0 \sqrt{\Delta(\lambda_0) m_0^{-1}} (ks)^{-0.5} = o\left((ks)^{-0.5}\right). \quad (7.75)$$

Due to (7.72) sequence $\{d(\lambda_s)\}_{s\in N}$ is monotone increasing and $d(\lambda_s) \leq f(x^*)$, $s \geq 1$; therefore, there is $\lim_{s\to\infty} d(\lambda_s) = \bar{d} \leq f(x^*)$. Again from (7.72) we have

$$\lim_{s\to\infty} \|\lambda_{s+1} - \lambda_s\| = 0. \quad (7.76)$$

(3) From boundedness $\{\lambda_s\}_{s\in N}$ follows existence of a converging subsequence $\{\lambda_{s_l}\}_{l\in N}$.
From (7.76) we have

$$\lim_{s_l\to\infty} \lambda_{s_l} = \lim_{s_l\to\infty} \lambda_{s_l+1} = \bar{\lambda}. \quad (7.77)$$

From assumption **A** and $\lim_{s\to\infty} \nu_s = 0$ follows boundedness of the primal sequence $\{x_s\}_{s\in N}$. Without losing generality we can assume that $\lim x_{s_l} = \lim_{s_l\to\infty} x_{s_l+1} = \bar{x}$.
Let us consider two sets of indices: $I_+ = \{i : \bar{\lambda}_i > 0\}$ and $I_0 = \{i : \bar{\lambda}_i = 0\}$.
From (7.77) and update formula (7.57), we have

$$\lim_{s_l\to\infty} k_{i,s_l} = k \lim_{s_l\to\infty} (\lambda_{i,s_l})^{-1} = k(\bar{\lambda}_i)^{-1}, \quad i \in I_+.$$

From (7.56), we obtain

$$c_i(x_{s_l+1}) = k^{-1}\lambda_{i,s_l}\psi'^{-1}(\lambda_{i,s_l+1}/\lambda_{i,s_l}) = -k^{-1}\lambda_{i,s_l}\varphi'^{-1}(\lambda_{i,s_l+1}/\lambda_{i,s_l}). \quad (7.78)$$

By passing (7.78) to the limit, we obtain

$$c_i(\bar{x}) = -k^{-1}\bar{\lambda}_i\varphi'(1) = 0, \quad i \in I_+ \quad (7.79)$$

(4) From (7.55)–(7.56) follows $\nabla_x \mathscr{L}(x_{s_l+1}, \lambda_{s_l}, \mathbf{k}_{s_l}) = \nabla_x L(x_{s_l+1}, \lambda_{s_l+1}) = 0$.

$$\lim_{s_l\to\infty} \nabla_x L(x_{s_l+1}, \lambda_{s_l+1}) = \nabla_x L(\bar{x}, \bar{\lambda}) = 0$$

and

$$\lim_{s\to\infty} \lambda_{i,s_l} c_i(x_{s_l}) = \bar{\lambda}_i c_i(\bar{x}) = 0, \quad i = 1, \ldots, m. \quad (7.80)$$

From $\lim_{s\to\infty} \nu_s = 0$ follows $\lim c_i(x_{s_l+1}) = c_i(\bar{x}) \geq 0$, $i \in I_0$.
Thus, the KKT's conditions are satisfied for $(\bar{x}, \bar{\lambda})$; therefore, $\bar{x} = x^*$, $\bar{\lambda} = \lambda^*$.
From the dual monotonicity follows convergence of the entire dual sequence in value, that is, $\lim_{s\to\infty} d(\lambda_s) = d(\lambda^*)$.
From (7.80) we obtain

$$d(\lambda^*) = \lim_{s\to\infty} d(\lambda_{s_l}) = \lim_{s\to\infty} L(x_{s_l}, \lambda_{s_l}) = \lim_{s\to\infty} f(x_{s_l}) = f(x^*). \quad (7.81)$$

So we proved the item (4).

(5) item (5) follows from (30) of Appendix and $\lim_{s\to\infty} d(\lambda_s) = d(x^*)$. □

Remark 7.7. The bound (7.73) is critical for the convergence analysis. It indicates how the NR method translates the primal constraint violation into the increase of the dual function value.

From (7.73) follows

$$d(\lambda^*) - d(\lambda_0) \geq d(x_s) - d(x_0) \geq km_0 M_0^{-2} \sum_{i=1}^{s} \sum_{i \in I_{s+1}^-} c_i^2(x_{s+1}).$$

It shows that all constraints, which are violated at each step, contribute into the increase in the dual objective function value. Therefore, the constraint violation has to vanish. It is the way the pricing mechanism manifests itself in the NR method.

We would like to emphasize that for original transformations $\hat{\psi}_1 - \hat{\psi}_5$ either $m = 0$ or $M = \infty$, which makes the estimation (7.73) trivial and useless.

Remark 7.8. It follows from (7.73) that for any given $\tau < 0$ and any $i = 1, \ldots, m$ the inequality $c_i(x_{s+1}) \leq \tau < 0$ is possible only for a finite number of steps. Therefore, from some point on only original transformations $\hat{\psi}_1 - \hat{\psi}_5$ are used in the NR method and only original kernels $\hat{\varphi}_1 - \hat{\varphi}_5$ are used in the dual IQP.

Transformations $\hat{\psi}_1$–$\hat{\psi}_5$ for $t \geq \tau$ are infinitely differentiable and so is LEP $\mathscr{L}$ if the input data possesses the correspondent property. This is an important advantage, because it allows using Newton's or RNM methods for primal minimization.

7.2.3 Rate of Convergence

The complementarity condition is satisfied in the strict form if

$$\max\{\lambda_i^*, c_i(x^*)\} > 0, \quad i = 1, \ldots, m. \tag{7.82}$$

Theorem 7.7. *Under assumption of Theorem 7.6 and strict complementarity condition (7.82) for the primal–dual sequence $\{x_s, \lambda_s\}_{s \in N}$, generated by NR method (7.55)–(7.57), the following bound*

$$d(\lambda^*) - d(\lambda_s) = o(ks)^{-1}$$

holds for any given $k > 0$.

Proof. Let $I^* = \{i : c_i(x^*) = 0\} = \{1, \ldots, r\}$ be the active constraint set; then from (7.82) for the passive constraints follows $\min\{c_i(x^*) \mid i \bar{\in} I^*\} = \sigma > 0$. Therefore, there is s_0 that $c_i(x_s) \geq \frac{\sigma}{2}$, $s \geq s_0$, $i \bar{\in} I^*$. From (2c) and (7.56) and any $i \bar{\in} I^* = \{1, \ldots, r\}$ we have

$$\text{a)} \lambda_{i,s+1} = \lambda_{i,s} \psi'(k_{i,s} c_i(x_{s+1})) \leq d\lambda_{i,s}(k_{i,s} c_i(x_{s+1}))^{-1} = 2d(\sigma k)^{-1}(\lambda_{i,s})^2. \tag{7.83}$$

We consider the Lagrangian for the equivalent problem

$$\begin{aligned}\mathscr{L}(x_{s+1},\lambda_s,\kappa_s) =& f(x_{s+1}) - k^{-1}\sum_{i=1}^{r}(\lambda_{i,s})^2\psi(k_{i,s}c_i(x_{s+1})) \\ & -k^{-1}\sum_{i=r+1}^{m}(\lambda_{i,s})^2\psi(k_{i,s}c_i(x_{s+1}))\,.\end{aligned} \tag{7.84}$$

Let us estimate the last term. Using $\psi(0)=0$, mean value formula, and (7.56) , we obtain

$$\begin{aligned}k^{-1}\sum_{i=r+1}^{m}(\lambda_{i,s})^2\psi(k_{i,s}c_i(x_{s+1})) &= k^{-1}\sum_{i=r+1}^{m}(\lambda_{i,s})^2(\psi(k_{i,s}c_i(x_{s+1}))-\psi(0)) \\ &= k^{-1}\sum_{i=r+1}^{m}(\lambda_{i,s})^2 k_{i,s}c_i(x_{s+1})\psi'(\theta_{i,s}k_{i,s}c_i(x_{s+1}))\end{aligned}$$

and $0<\theta_{i,s}<1$. For $k_{i,s}\to\infty$ and $c_i(x_{s+1})\geq 0.5\sigma$, we obtain $\theta_{i,s}\to 1$. Therefore, for s_0 large enough and any $s\geq s_0$, we have $\theta_{i,s}\geq 0.5$. Thus, from 2c) and (7.57) follows

$$\begin{aligned}k^{-1}\sum_{i=r+1}^{m}(\lambda_{i,s})^2\psi(k_{i,s}c_i(x_{s+1})) &= \sum_{i=r+1}^{m}\lambda_{i,s}c(x_{s+1})\psi'(\theta_{i,s}k_{i,s}c(x_{s+1})) \\ &\leq d\sum_{i=r+1}^{m}\lambda_{i,s}c(x_{s+1})(\theta_{i,s}k_{i,s}c_i(x_{s+1}))^{-1} \\ &\leq 2d\sum_{i=r+1}^{m}\lambda_{i,s}(k_{i,s})^{-1} = 2dk^{-1}\sum_{i=r+1}^{m}(\lambda_{i,s})^2.\end{aligned} \tag{7.85}$$

It follows from (7.83a) and (7.85) that terms of LEP, corresponding to the passive constraints, converge to zero with quadratic rate. Therefore, there are $s_0>0$ that for any $s\geq s_0$ such terms became negligibly small. Thus, instead of $\mathscr{L}(x,\lambda,\mathbf{k})$, we can consider the truncated LEP associated only with active constraints

$$\mathscr{L}(x,\lambda,\mathbf{k}) := f(x) - \sum_{i=1}^{r}(k_{i,s})^{-1}(\lambda_{i,s})\psi(k_{i,s}c_i(x)).$$

Then $L(x,\lambda):=f(x)-\sum_{i=1}^{r}\lambda_i c_i(x)$ is the correspondent truncated Lagrangian for the original problem (P). Accordingly, instead of the original dual function and the original second-order φ-divergence distance, we consider $d(\lambda):=\inf_{x\in\mathbb{R}^n}L(x,\lambda)$ and $D_2(u,\mathrm{v}):=\sum_{i=1}^{r}\mathrm{v}_i^2\varphi(u_i/\mathrm{v}_i)$.

For simplicity, we retain original notations for the truncated Lagrangian, equivalent problem, correspondent dual function, and the second-order φ-divergence distance.

We also assume that $\{\lambda_s\}_{s\in N}$ is the truncated dual sequence $\{\lambda_s=(\lambda_{1,s},\ldots,\lambda_{r,s})\}_{s\in N}$.

Then, the NR step is equivalent to

$$d(\hat{\lambda})-k^{-1}D_2(\hat{\lambda},\lambda)=\max\{d(u)-k^{-1}D_2(u,\lambda)|\ u\in\mathbb{R}^r_{++}\}. \tag{7.86}$$

From (7.72) follows

$$d(\lambda^*) - d(\lambda_s) - (d(\lambda^*) - d(\lambda_{s+1})) \geq m_0 k^{-1} \|\lambda_s - \lambda_{s+1}\|^2$$

or

$$\Delta_s - \Delta_{s+1} \geq m_0 k^{-1} \|\lambda_s - \lambda_{s+1}\|^2, \tag{7.87}$$

where $\Delta_s = d(\lambda^*) - d(\lambda_s) > 0$. Using the concavity of d, we obtain

$$d(\lambda) - d(\lambda_{s+1}) \leq \langle -c(x_{s+1}), \lambda - \lambda_{s+1} \rangle.$$

From (7.65) follows $-c(x_{s+1}) = (kR_s)^{-1}(\lambda_{s+1} - \lambda_s)$; therefore,

$$d(\lambda_{s+1}) - d(\lambda) \geq -(kR_s)^{-1} \langle \lambda_{s+1} - \lambda_s, \lambda - \lambda_{s+1} \rangle.$$

So, for $\lambda = \lambda^*$ we have

$$(kR_s)^{-1} \langle \lambda_{s+1} - \lambda_s, \lambda^* - \lambda_{s+1} \rangle \geq d(\lambda^*) - d(\lambda_{s+1}) = \Delta_{s+1}$$

or

$$(kR_s)^{-1} \langle \lambda_{s+1} - \lambda_s, \lambda^* - \lambda_s \rangle - (kR_s)^{-1} \|\lambda_{s+1} - \lambda_s\|^2 \geq \Delta_{s+1}.$$

Hence,

$$\|R_s^{-1}\| \cdot \|\lambda_{s+1} - \lambda_s\| \cdot \|\lambda^* - \lambda_s\| \geq k\Delta_{s+1}.$$

From Remark 7.5 follows $m_0 < \|R_s^{-1}\| \leq M_0$; therefore,

$$\|\lambda_{s+1} - \lambda_s\| \geq \frac{1}{M_0} k\Delta_{s+1} \|\lambda_s - \lambda^*\|^{-1}. \tag{7.88}$$

From (7.87) and (7.88) follows

$$\Delta_s - \Delta_{s+1} \geq \frac{m_0}{M_0^2} k\Delta_{s+1}^2 \|\lambda_s - \lambda^*\|^{-2}$$

or

$$\Delta_s \geq \Delta_{s+1} \left(1 + \frac{m_0}{M_0^2} k\Delta_{s+1} \|\lambda_s - \lambda^*\|^{-2} \right). \tag{7.89}$$

If $m_0 > 1$, then

$$\Delta_s \geq \Delta_{s+1} \left(1 + \frac{m_0}{M_0^2} k\Delta_{s+1} \|\lambda_s - \lambda^*\|^2 \right) \geq \Delta_{s+1} \left(1 + \frac{k\Delta_{s+1}}{M_0^2} \|\lambda_s - \lambda^*\|^2 \right).$$

By inverting the last inequality, we obtain

$$\Delta_s^{-1} \leq \Delta_{s+1}^{-1} \left(1 + \frac{1}{M_0^2} k\Delta_{s+1} \|\lambda_s - \lambda^*\|^{-2} \right)^{-1}. \tag{7.90}$$

Further, from (7.66) with $\hat{\lambda} = \lambda_{s+1}$ and $\lambda = \lambda_s$, we obtain

$$d(\lambda_{s+1}) - 0.5k^{-1}\|\lambda_{s+1} - \lambda_s\|^2_{R_s^{-1}} \geq d(\lambda^*) - 0.5k^{-1}\|\lambda^* - \lambda_s\|^2_{R_s^{-1}}$$

or

$$\Delta_{s+1} \leq 0.5k^{-1}\|\lambda_s - \lambda^*\|^2_{R_s^{-1}} \leq 0.5k^{-1}M_0^2\|\lambda_s - \lambda^*\|^2.$$

Therefore,

$$M_0^{-2}k\Delta_{s+1}\|\lambda^* - \lambda_s\|^{-2} \leq 0.5. \tag{7.91}$$

It is easy to see that $(1+t)^{-1} \leq 1 - 0.5t$, for $0 \leq t \leq 1$. Therefore, for $t = \frac{1}{M_0^2}k\Delta_{s+1}\|\lambda_{s+1} - \lambda^*\|^{-2} < 0.5$ from (7.90), we obtain

$$\Delta_s^{-1} \leq \Delta_{s+1}^{-1}\left(1 - 0.5\frac{1}{M_0^2}k\Delta_{s+1}\|\lambda_{s+1} - \lambda^*\|^{-2}\right)$$

or

$$\Delta_i^{-1} \leq \Delta_{i+1}^{-1} - 0.5\frac{1}{M_0^2}k\|\lambda_{i+1} - \lambda^*\|^{-2}, \quad i = 0, \ldots, s-1. \tag{7.92}$$

Summing up (7.92) for $i = 1, \ldots, s-1$, we obtain

$$\Delta_s^{-1} \geq \Delta_s^{-1} - \Delta_0^{-1} \geq 0.5\frac{1}{M_0^2}k\sum_{i=1}^{s-1}\|\lambda_i - \lambda^*\|^{-2}.$$

By inverting the last inequality, we obtain

$$\Delta_s = d(\lambda^*) - d(\lambda_s) \leq \frac{2M_0^2}{k\sum_{i=0}^{s-1}\|\lambda_i - \lambda^*\|^{-2}}$$

or

$$ks\Delta_s = \frac{2M_0^2}{s^{-1}\sum_{i=0}^{s-1}\|\lambda_i - \lambda^*\|^{-2}}.$$

From $\|\lambda_s - \lambda^*\| \to 0$ follows $\|\lambda_s - \lambda^*\|^{-2} \to \infty$. Using the Silverman–Toeplitz theorem, we have $\lim_{s\to\infty} s^{-1}\sum_{i=1}^{s}\|\lambda_i - \lambda^*\|^{-2} = \infty$. Therefore, there exists $\alpha_s \to 0$ such that

$$\Delta_s = M_0^2\alpha_s\frac{1}{ks} = o\left((ks)^{-1}\right). \tag{7.93}$$

If $m_0 \leq 1$, then from (7.89) follows

$$\Delta_s^{-1} \leq \Delta_{s+1}^{-1}\left(1 + \frac{m_0}{M_0^2}k\Delta_{s+1}\|\lambda_s - \lambda^*\|^2\right)^{-1}.$$

From (7.91) we have

$$\frac{m_0}{M_0^2}k\Delta_{s+1}\|\lambda_s - \lambda^*\|^2 \leq 0.5m_0 \leq 0.5.$$

Therefore

$$\Delta_s \leq \frac{2m_0^{-1}M_0^2}{k\sum_{i=0}^{s-1}\|\lambda_i - \lambda^*\|^2},$$

which leads to $\Delta_s = o((ks)^{-1})$. □

The estimation (7.93) can be strengthened. Under the standard second- order optimality conditions, NR method (7.55)–(7.57) converges with Q-linear rate if $k > 0$ is fixed, but large enough.

First of all, due to the standard second-order optimality conditions, the primal–dual solution is unique. Therefore, the primal–dual sequence $\{x_s, \lambda_s\}_{s\in N}$ converges to the primal–dual solution (x^*, λ^*), for which the complementarity conditions are satisfied in a strict form (7.82). Therefore, the Lagrange multipliers for the passive constraints converge to zero quadratically.

For the active constraints from (7.57), we have $\lim_{s\to\infty} k_{i,s} = k(\lambda_i^*)^{-1}, i = 1,\ldots,r$, that is, the scaling parameters, which correspond to the active constraints, grow linearly with $k > 0$. Therefore, the technique used for NR with fixed scaling parameter can be applied for NR method (7.55)–(7.57).

For a given small enough $\delta > 0$, we define the extended neighborhood of λ^* as follows:

$$\begin{gathered} D(\lambda^*, \mathbb{k}, \delta) = \\ \{(\lambda, \mathbb{k}) \in \mathbb{R}^m_+ \times \mathbb{R}^m_+ : \lambda_i \geq \delta,\, |\lambda_i - \lambda_i^*| \leq \delta k_i,\ i = 1,\ldots,r;\ k_i \geq k_0\}. \end{gathered} \tag{7.94}$$

The following proposition can be proven using arguments from Theorem 7.2

Proposition 7.1. *If f and all $c_i \in C^2$ and second-order sufficient optimality conditions hold, then there exist small $\delta > 0$ and large $k_0 > 0$, that for any $(\lambda, \mathbb{k}) \in D(\cdot)$, the following statements hold:*

1. *There exists $\hat{x} = \hat{x}(\lambda, \mathbb{k}) : \nabla_x \mathscr{L}(\hat{x}, \lambda, \mathbb{k}) = 0$*

$$\nabla_x \mathscr{L}(\hat{x}, \lambda, \mathbb{k}) = 0$$

and

$$\hat{\lambda}_i = \lambda_i \psi'(k_i c_i(\hat{x})),\ \hat{k}_i = k\hat{\lambda}_i^{-1},\ i = 1,\ldots,m.$$

2. *For the pair $(\hat{x}, \hat{\lambda})$ the following bound*

$$\max\{\|\hat{x} - x^*\|,\ \|\hat{\lambda} - \lambda^*\|\} \leq ck^{-1}\|\lambda - \lambda^*\| \tag{7.95}$$

holds and $c > 0$ is independent on $k \geq k_0$.

3. *The Lagrangian $\mathscr{L}(x, \lambda, \mathbb{k})$ is strongly convex in the neighborhood of $\hat{x}$.*

To make NR method (7.55)–(7.57) practical one, can use the following stopping criterion.

For a given $\sigma > 0$, let us consider the sequence $\{\bar{x}_s, \bar{\lambda}_s, \bar{\mathbb{k}}_s\}_{s\in N}$ generated by the following formulas.

$$\bar{x}_{s+1} : \|\nabla_x \mathscr{L}(\bar{x}_{s+1}, \bar{\lambda}_s, \bar{\mathbb{k}}_s)\| \leq \sigma k^{-1} \|\Psi'\left(k(\bar{\lambda}_s)^{-1} c(\bar{x}_{s+1})\right)\bar{\lambda}_s - \bar{\lambda}_s\| \tag{7.96}$$

$$\bar{\lambda}_{s+1} = \Psi'\left(k(\bar{\lambda}_s)^{-1}c(\bar{x}_{s+1})\right)\bar{\lambda}_s, \tag{7.97}$$

where

$$\Psi'\left(k(\bar{\lambda}_s)^{-1}c(\bar{x}_{s+1})\right) = \operatorname{diag}\left(\psi(k(\bar{\lambda}_{i,s})^{-1}c_i(\bar{x}_{s+1}))\right)_{i=1}^{m}$$

and

$$\bar{\mathbb{k}}_{s+1} = (\bar{k}_{i,s+1} = k(\lambda_{i,s+1})^{-1}, i = 1,\ldots,m). \tag{7.98}$$

The following proposition can be proven using arguments from Theorem 7.3.

Proposition 7.2. *If the second-order sufficient optimality conditions* (4.73)–(4.74) *hold, f and all $c_i \in C^2$, $i = 1,\ldots,m$, then there is $k_0 > 0$ large enough, that for the primal–dual sequence $\{\bar{x}_s, \bar{\lambda}_s\}_{s\in N}$, generated by formulas (7.96)–(7.98), the following bounds hold for any $s \geq 0$ and $c > 0$ is independent on $k \geq k_0$*

$$\begin{aligned} \|\bar{x}_{s+1} - x^*\| &\leq c(1+\sigma)k^{-1}\|\bar{\lambda}_s - \lambda^*\|, \\ \|\bar{\lambda}_{s+1} - \lambda^*\| &\leq c(1+\sigma)k^{-1}\|\bar{\lambda}_s - \lambda^*\|. \end{aligned} \tag{7.99}$$

In the next section, we apply the NR method (7.55)–(7.57) for LP. The convergence under very mild assumption follows from Theorem 7.6. Under the dual uniqueness, we prove the global quadratic convergence rate. The key ingredients of the proof are the A. Hoffman-type lemma and the properties of the well-defined kernels $\varphi \in \boldsymbol{\varphi}$.

7.2.4 Nonlinear Rescaling for LP

Let $A : \mathbb{R}^n \to \mathbb{R}^m$, $a \in \mathbb{R}^n$, $b \in \mathbb{R}^m$. We consider the following primal LP

$$\langle a, x^* \rangle = \min\{\langle a, x\rangle \mid c_i(x) = (Ax - b)_i = \langle a_i, x\rangle - b_i \geq 0,\ i = 1,\ldots,m\}, \tag{7.100}$$

; then for the dual LP, we have

$$\langle b, \lambda^* \rangle = \max\{\langle b, \lambda\rangle \mid A^T\lambda - a = 0,\ \lambda_i \geq 0,\ i = 1,\ldots,m\}. \tag{7.101}$$

We assume that primal and dual feasible sets are bounded.

We consider $\psi \in \Psi$, which corresponds to the well-defined kernel $\varphi \in \boldsymbol{\varphi}$, that is, $0 < \varphi(0) < \infty$.

The NR method (7.55)–(7.57), when applied to (7.100), produces three sequences $\{x_s\}_{s\in N} \subset \mathbb{R}^n$, $\{\lambda_s\}_{s\in N} \subset \mathbb{R}^m_{++}$, and $\{\mathbb{k}_s\}_{s\in N} \in \mathbb{R}^m_{++}$:

$$x_{s+1} : \nabla_x \mathscr{L}(x_{s+1}, \lambda_s, \mathbb{k}_s) = a - \sum_{i=1}^{m} \lambda_{i,s}\psi'\left(k_{i,s}c_i(x_{s+1})\right)a_i = 0, \tag{7.102}$$

$$\lambda_{s+1} : \lambda_{i,s+1} = \lambda_{i,s}\psi'\left(k_{i,s}c_i(x_{s+1})\right),\ i = 1,\ldots,m \tag{7.103}$$

$$\mathbb{k}_{s+1} : k_{i,s+1} = k(\lambda_{i,s+1})^{-1},\ i = 1,\ldots,m. \tag{7.104}$$

From boundedness X^* and L^* and Theorem 7.6 follows

$$\lim_{s\to\infty}\langle b,\lambda_s\rangle = \langle b,\lambda^*\rangle .$$

From Lemma 5.7 follows existence $\alpha > 0$ that

$$\langle b,\lambda^*\rangle - \langle b,\lambda_s\rangle \geq \alpha\rho(\lambda_s,L^*) . \tag{7.105}$$

Therefore $\lim_{s\to\infty}\rho(\lambda_s,L^*) = 0$.

If λ^* is a unique dual solution, then from Corollary 5.3 follows existence of $\alpha > 0$ that

$$\langle b,\lambda^*\rangle - \langle b,\lambda\rangle = \alpha\|\lambda - \lambda^*\| \tag{7.106}$$

holds for $\forall\lambda \in L = \{\lambda : A^T\lambda = a,\ \lambda \in \mathbb{R}^m_+\}$.

Theorem 7.8. *If the dual problem (7.101) has a unique solution, then for any well-defined kernel $\varphi \in \boldsymbol{\varphi}$, the dual sequence $\{\lambda_s\}_{s\in N}$ converges in value quadratically*

$$\langle b,\lambda^*\rangle - \langle b,\lambda_{s+1}\rangle \leq ck^{-1}[\langle b,\lambda^*\rangle - \langle b,\lambda_s\rangle]^2 \tag{7.107}$$

and $c > 0$ is independent on $k > 0$.

Proof. It follows from (7.102)–(7.104) that

$$\nabla_x\mathscr{L}(x_{s+1},\lambda_s,\Bbbk_s) = a - \sum_{i=1}^m \lambda_{i,s+1}a_i = a - A^T\lambda_{s+1} = 0 \tag{7.108}$$

and $\lambda_{s+1} \in \mathbb{R}^m_{++}$. In other words, the NR method generates a dual interior point sequence $\{\lambda_s\}_{s\in N}$.

From (7.108) we obtain

$$0 = \langle a - A^T\lambda_{s+1}, x_{s+1}\rangle = \langle a, x_{s+1}\rangle - \sum_{i=1}^m \lambda_{i,s+1}c_i(x_{s+1}) - \langle b,\lambda_{s+1}\rangle$$

or

$$\langle b,\lambda_{s+1}\rangle = L(x_{s+1},\lambda_{s+1}) .$$

The multipliers method (7.102)–(7.104) is equivalent to the following interior prox for the dual problem

$$\lambda_{s+1} = \arg\max\left\{\langle b,\lambda\rangle - k^{-1}\sum_{i=1}^m(\lambda_{i,s})^2\varphi\left(\frac{\lambda_i}{\lambda_{i,s}}\right) \mid A^T\lambda - a = 0\right\} . \tag{7.109}$$

Keeping in mind Remark 7.8, we can assume, without restricting the generality, that only well-defined kernels φ_i, $i = 1,3,4,5$, which corresponds to the original transformations $\hat{\psi}_1$, $i = 1,3,4,5$, are used in the method (7.109).

From (7.109), taking into account $\lambda^* \in \mathbb{R}^m_+$ and $A^T\lambda^* = a$, we obtain

$$\langle b, \lambda_{s+1}\rangle - k^{-1}\sum_{i=1}^{m}(\lambda_{i,s})^2\varphi\left(\frac{\lambda_{i,s+1}}{\lambda_{i,s}}\right) \geq \langle b, \lambda^*\rangle - k^{-1}\sum_{i=1}^{m}(\lambda_{i,s})^2\varphi\left(\frac{\lambda_i^*}{\lambda_{i,s}}\right).$$

Keeping in mind $k^{-1}\sum_{i=1}^{m}(\lambda_{i,s})^2\varphi\left(\frac{\lambda_{i,s+1}}{\lambda_{i,s}}\right) \geq 0$, we have

$$k^{-1}\sum_{i=1}^{m}(\lambda_{i,s})^2\varphi\left(\frac{\lambda_i^*}{\lambda_{i,s}}\right) \geq \langle b, \lambda^*\rangle - \langle b, \lambda_{s+1}\rangle = \Delta_{s+1}. \tag{7.110}$$

Let us assume that $\lambda_i^* > 0$, $i = 1, \ldots, r$; $\lambda_i^* = 0$, $i = r+1, \ldots, m$; then $\varphi\left(\frac{\lambda_i^*}{\lambda_{i,s}}\right) = \varphi(0) < \infty$, $i = r+1, \ldots, m$. Keeping in mind $\varphi(1) = \varphi'(1) = 0$, $i = 1, \ldots, r$, we obtain

$$k^{-1}\sum_{i=1}^{m}(\lambda_{i,s})^2\varphi\left(\frac{\lambda_i^*}{\lambda_{i,s}}\right) =$$

$$k^{-1}\left[\sum_{i=1}^{r}(\lambda_{i,s})^2\left(\varphi\left(\frac{\lambda_i^*}{\lambda_{i,s}}\right) - \varphi\left(\frac{\lambda_{i,s}}{\lambda_{i,s}}\right)\right) + \varphi(0)\sum_{i=r+1}^{m}(\lambda_i^* - \lambda_{i,s})^2\right] =$$

$$k^{-1}\left[\sum_{I=1}^{r}\varphi''(\cdot)(\lambda_i^* - \lambda_{i,s})^2 + \varphi(0)\sum_{i=r+1}^{m}(\lambda_i^* - \lambda_{i,s})^2\right]. \tag{7.111}$$

Taking into account $m_0 \leq \varphi''(\cdot) \leq M_0$ for $\varphi_0 = \max\{M_0, \varphi(0)\}$ from (7.111) follows

$$k^{-1}\sum_{i=1}^{m}\lambda_{i,s}^2\varphi\left(\frac{\lambda_i^*}{\lambda_{i,s}}\right) \leq \varphi_0 k^{-1}\|\lambda^* - \lambda_s\|^2. \tag{7.112}$$

From (7.106) for $\lambda = \lambda_s$, we have

$$\Delta_s = \alpha\|\lambda_s - \lambda^*\|. \tag{7.113}$$

Combining (7.110), (7.112), and (7.113), we obtain

$$\Delta_{s+1} \leq \varphi_0 k^{-1}\|\lambda^* - \lambda_s\|^2 \leq \varphi_0 k^{-1}\alpha^{-2}\Delta_s^2 = ck^{-1}\Delta_s^2, \tag{7.114}$$

where $c = \varphi_0\alpha^{-2}$. □

Remark 7.9. Theorem 7.8 is valid for the NR method (7.55)–(7.57) with exponential, LS, CHKS, and hyperbolic MBF transformations because correspondent kernels $\varphi_1, \varphi_3 - \varphi_5$ are well defined.

7.3 Primal–Dual NR Method for Convex Optimization

7.3.0 Introduction

We introduce and analyze primal–dual NR (PDNR) method, which is based on the NR technique with "dynamic" scaling parameter update. The PD approach reduces the ill-conditioning effect and, at the same time, improves the convergence rate up to quadratic.

Each step of the NR method is equivalent to solving nonlinear primal–dual (PD) system of equations. The system is comprised of the optimality criteria for the primal minimizer and formulas for the Lagrange multipliers update.

Application of Newton's method for the nonlinear PD system leads to the PDNR method.

The PDNR requires solving at each step a linear PD system of equations for finding primal and dual directions. It generates such primal–dual sequence that, under the second-order sufficient optimality conditions, converges locally to the PD solution with quadratic rate.

The PDNR method does not require finding the primal minimizer at each step and allows the unbounded increase of the penalty parameter without compromising numerical stability.

The PDNR is also free from any stringent conditions on accepting the Newton direction, which is typical for inequality-constrained optimization.

There are three important features that make PDNR free from such restrictions.

First, LEP $\mathcal{L}$ is defined on the entire primal space.

Second, after a few Lagrange multipliers updates, the terms of LEP $\mathcal{L}$, which correspond to the passive constraints, become negligibly small due to the quadratic convergence to zero of the Lagrange multipliers for the passive constraints.

Therefore, those terms became irrelevant for finding Newton direction, and there is no need to enforce nonnegativity of the correspondent Lagrange multipliers.

Third, the NR method is an exterior point method in the primal space. Therefore, there is no need to enforce the nonnegativity of the slack variables for the active constraints, which is typical for the interior point methods.

Due to the quadratic convergence to zero Lagrange multipliers for the passive constraints after few Lagrange multipliers updates, the PD direction becomes close to the Newton directions for the Lagrange system of equations, which corresponds to active constraints.

From this point on, PDNR method practically turns into Newton's method for solving Lagrange system of equations for only active constraints. It leads to asymptotic convergence with quadratic rate.

7.3.1 Local Convergence of the PDNR

One step of the NR method (7.55)–(7.57) maps the given triple $(x,\lambda,\mathbf{k})\in\mathbb{R}^n\times\mathbb{R}^m_{++}\times\mathbb{R}^m_{++}$ into the triple $(\widehat{x},\widehat{\lambda},\widehat{\mathbf{k}})\in\mathbb{R}^n\times\mathbb{R}^m_{++}\times\mathbb{R}^m_{++}$ defined by formulas:

$$\begin{aligned}\widehat{x}:\ \nabla_x\mathscr{L}(\widehat{x},\lambda,\mathbf{k}) &= \nabla f(\widehat{x})-\textstyle\sum_{i=1}^m\psi'(k_ic_i(\widehat{x}))\lambda_i\nabla c_i(\widehat{x})\\ &=\nabla f(\widehat{x})-\textstyle\sum_{i=1}^m\widehat{\lambda}_i\nabla c_i(\widehat{x})=0,\end{aligned}\tag{7.115}$$

$$\widehat{\lambda}_L=\lambda_L\psi'(k_ic_i(\widehat{x})),\quad i=1,\ldots,m,\tag{7.116}$$

$$\widehat{\mathbf{k}}:\ \widehat{k}_i=k\widehat{\lambda}_i^{-1},\quad i=1,\ldots,m.\tag{7.117}$$

By removing the scaling vector $\mathbf{k}$ update formula (7.117) from the system (7.115)–(7.117), we obtain the PDNR system:

$$\nabla_x L(\widehat{x},\widehat{\lambda})=\nabla f(\widehat{x})-\sum_{i=1}^m\widehat{\lambda}\nabla c_i(\widehat{x})=0,\tag{7.118}$$

$$\widehat{\lambda}=\Psi'(\mathbf{k}c(\widehat{x}))\lambda,\tag{7.119}$$

where $\Psi'(\mathbf{k}c(\widehat{x}))=\operatorname{diag}(\psi'(k_ic_i(\widehat{x})))_{i=1}^m$.

The PD system (7.118)–(7.119) maps LM vector $\lambda\in\mathbb{R}^m_{++}$ and scaling vector $\mathbf{k}\in\mathbb{R}^m_{++}$ into a new PD pair $(\hat{x}=\hat{x}(\lambda,\mathbf{k});\hat{\lambda}=\hat{\lambda}(\lambda,\mathbf{k}))$, while the penalty parameter $k>0$ is fixed.

The contractibility of the corresponding map is critical for both convergence and rate of convergence. To understand conditions under which the corresponding map is contractive and to find the contractibility bounds, one has to analyze the PD map.

It should be emphasized that neither the primal NR sequence $\{x_s\}_{s\in N}$ nor the dual sequence $\{\lambda_s\}_{s\in N}$ provides sufficient information for this analysis.

Only the PD system (7.118)–(7.119) has all necessary ingredients for such analysis. It reflects the important observation that for any multipliers method, neither the primal nor the dual sequences control the computational process. The computational process is governed by the PD system. In fact, it follows from the proofs of Theorem 7.2 and 7.3.

In this section, we use the specific properties of the PD system (7.118)–(7.119) for developing the PDNR method and show its local convergence with quadratic rate. Then we describe the globally convergent PDNR.

Let us make few preliminary remarks.

From the second-order sufficient optimality condition follows uniqueness of x^* and λ^*; therefore there exists $\tau^*>0$ that

(a) $\min\{c_i(x^*)\mid r+1\le i\le m\}\ge\tau^*$ and (b) $\min\{\lambda_i^*\mid 1\le i\le r\}\ge\tau^*$.

From Theorem 7.6 follows existence of $k_0>0$ and $s_0>1$ that for any $k\ge k_0$ and $s\ge s_0$, NR method generates primal–dual sequence $\{x_s,\lambda_s\}_{s\in N}$ that:

$$\text{(a) }\min\{c_i(x_s)\mid r+1\le i\le m\}\ge 0.5\tau^*\quad\text{and}\quad\text{(b) }\min\{\lambda_{i,s}\mid 1\le i\le r\}\ge 0.5\tau^*.\tag{7.120}$$

Using (7.96)–(7.97) and the property 2.c) of $\psi \in \Psi$, we obtain

$$\bar{\lambda}_{i,s+1} = \psi'(k_{i,s}c_i(\bar{x}_{s+1}))\bar{\lambda}_{i,s} \leq 2d(k\tau^*)^{-1}(\bar{\lambda}_{i,s})^2, \quad r+1 \leq i \leq m.$$

Hence, for any fixed $k > \max\{k_0, 2d(\tau^*)^{-1}\}$, we have

$$\bar{\lambda}_{i,s+1} \leq (\bar{\lambda}_{i,s})^2 \tag{7.121}$$

for any $r+1 \leq i \leq m$ and $s \geq s_0$.

So for a given accuracy $0 < \varepsilon << 1$, in at most $s = O(\ln \ln \varepsilon^{-1})$ Lagrange multipliers updates, the Lagrange multipliers for the passive constraints will be of the order of $o(\varepsilon^2)$. From this point on, terms related to the passive constraints will be ignored.

Therefore, the PD system (7.118)–(7.119) will be actually reduced to the following system for $\widehat{x}$ and $\widehat{\lambda} = (\widehat{\lambda}_1, \ldots, \widehat{\lambda}_r)$:

$$\nabla_x L(\widehat{x}, \widehat{\lambda}) = \nabla f(\widehat{x}) - \sum_{i=1}^{r} \widehat{\lambda}_i \nabla c_i(\widehat{x}) = 0, \tag{7.122}$$

$$\widehat{\lambda}_i = \psi'(k_i c_i(\widehat{x}))\lambda_i, \quad i = 1, \ldots, r. \tag{7.123}$$

To simplify notation, we keep $L(x,\lambda)$ for the truncated Lagrangian, that is, $L(x,\lambda) = f(x) - \sum_{i=1}^{r} \lambda_i c_i(x)$ and $c(x) = (c_1(x), \ldots, c_r(x))^T$.

We use the following norm $\|x\| = \max_{1<i\leq n} |x_i|$ and correspondent matrix norm $\|A\| = \max_{1\leq i\leq n}(\sum_{j=1}^{n} |a_{ij}|)$. For a given $\varepsilon_0 > 0$, we define the ε_0-neighborhood $B(y^*, \varepsilon_0) = \{y = (x,\lambda) \in \mathbb{R}^n \times \mathbb{R}^m_+ : \|y - y^*\| \leq \varepsilon_0\}$ of $y^* = (x^*, \lambda^*)$.

We will measure the distance between the current approximation $y = (x,\lambda) \in \mathbb{R}^n \times \mathbb{R}^m_+$ and solution $y^* = (x^*, \lambda^*)$ using the merit function $\nu(y)$ defined by (7.41).

Let us consider Newton's method for solving PD system (7.122)–(7.123). To find the Newton direction $(\Delta x, \Delta\lambda)$, we have to linearize the system (7.122)–(7.123) at $y = (x,\lambda)$.

We start with system (7.123). Due to property (3) of ψ, the inverse ψ'^{-1} exists. Therefore, using the LF identity $\psi'^{-1} = \psi^{*\prime}$, keeping in mind $\varphi = -\psi^*$ and (7.116), we can rewrite (7.123) as follows:

$$c_i(\widehat{x}) = k_i^{-1}\psi'^{-1}(\widehat{\lambda}_i/\lambda_i) = k_i^{-1}\psi^{*\prime}(\widehat{\lambda}_i/\lambda_i) = -k_i^{-1}\varphi'(\widehat{\lambda}_i/\lambda_i).$$

Assuming $\widehat{x} = x + \Delta x$ and $\widehat{\lambda} = \lambda + \Delta\lambda$, keeping in mind $\varphi'(1) = 0$, and ignoring terms of second and higher order, we obtain

$$c_i(\widehat{x}) = c_i(x) + \nabla c_i(x)\Delta x =$$

$$-k_i^{-1}\varphi'((\lambda_i + \Delta\lambda_i)/\lambda_i) = -k_i^{-1}(\varphi'(1) + \lambda_i^{-1}\varphi''(1)\Delta\lambda_i) =$$

$$-(k_i\lambda_i)^{-1}\varphi''(1)\Delta\lambda_i = -k^{-1}\varphi''(1)\Delta\lambda_i, \quad i = 1, \ldots, r,$$

or

$$c_i(x)+\nabla c_i(x)\Delta x+k^{-1}\varphi''(1)\Delta\lambda_i=0,\quad i=1,\ldots,r.$$

Now we linearize the system (7.122) at $y=(x,\lambda)$. We have

$$\nabla f(x)+\nabla^2 f(x)\Delta x-\sum_{i=1}^{r}(\lambda_i+\Delta\lambda_i)(\nabla c_i(x)+\nabla^2 c_i(x)\Delta x)=0.$$

Again, ignoring terms of second and higher orders, we obtain the following linear PD system:

$$\nabla^2_{xx}L(x,\lambda)\Delta x-\nabla c^T(x)\Delta\lambda=-\nabla_x L(x,\lambda), \tag{7.124}$$

$$\nabla c(x)\Delta x+k^{-1}\varphi''(1)I^r\Delta\lambda=-c(x), \tag{7.125}$$

where I^r is the identical matrix in R^r and $\nabla c(x)=J(c(x))$ is Jacobian of the vector-function $c(x)=(c_1(x)\ldots.c_r(x))^T$. Let

$$\nabla\Phi_k(x,\lambda)\equiv\nabla\Phi_k(y)\equiv\nabla\Phi_k(\cdot)\equiv\begin{bmatrix}\nabla^2_{xx}L(\cdot) & -\nabla c^T(\cdot)\\ \nabla c(\cdot) & k^{-1}\varphi''(1)I^r\end{bmatrix}.$$

Then, we can rewrite the system (7.124)–(7.125) as follows:

$$\nabla\Phi_k(\cdot)\begin{bmatrix}\Delta x\\ \Delta\lambda\end{bmatrix}=\begin{bmatrix}-\nabla_x L(\cdot)\\ \nabla_\lambda L(\cdot)\end{bmatrix}=\begin{bmatrix}-\nabla_x L(\cdot)\\ -c(\cdot)\end{bmatrix}=-N(y).$$

The local PDNR method consists of the following operations:

(1) Find the Newton PD direction $\Delta y=(\Delta x,\Delta\lambda)$ from the following system

$$\nabla\Phi_k(\cdot)\begin{bmatrix}\Delta x\\ \Delta\lambda\end{bmatrix}=-N(y). \tag{7.126}$$

(2) Find the new PD vector $\widehat{y}=(\widehat{x},\widehat{\lambda})$:

$$\widehat{x}:=x+\Delta x,\quad \widehat{\lambda}:=\lambda+\Delta\lambda. \tag{7.127}$$

(3) Update the penalty parameter

$$\widehat{k}=(\nu(\widehat{y}))^{-1}. \tag{7.128}$$

Along with $\nabla\Phi_k(\cdot)$, we consider

$$\nabla\Phi_\infty(y)=\nabla\Phi_\infty(\cdot)=\begin{bmatrix}\nabla^2 L(\cdot) & -\nabla c^T(\cdot)\\ \nabla c(\cdot) & 0\end{bmatrix}.$$

The following theorem, which is similar to Theorem 4.17, establishes local quadratic convergence of the PDNR method.

Theorem 7.9. *Under the second-order sufficient optimality conditions (4.73)–(4.74) and Lipschitz condition (4.130), there exists small enough $\varepsilon_0>0$, that for any $y=$*

$(x,\lambda)\in B(y^*,\varepsilon_0)$ as a starting point the PDNR method (7.126)–(7.128) generates such $\hat{y}=(\hat{x},\hat{\lambda})\in B(y^*,\varepsilon_0)$, that the following bound

$$\|\widehat{y}-y^*\|\leq c\|y-y^*\|^2 \tag{7.129}$$

holds and $c>0$ is independent on $\forall y\in B(y^*,\varepsilon_0)$.

Proof. We find the PD direction $\Delta y=(\Delta x,\Delta\lambda)$ from the system (7.126), which can be rewritten as follows:

$$\nabla\Phi_k(y)\Delta y=-N(y).$$

Then, for the new approximation $\hat{y}=(\hat{x},\hat{\lambda})$, we have

$$\hat{x}=x+\Delta x,\quad \hat{\lambda}=\lambda+\Delta\lambda,$$

where $\Delta y=(\Delta x,\Delta\lambda)$ -solution of (7.126).

We recall that $L(x,\lambda)=f(x)-\sum_{i=1}^r\lambda_i c_i(x)$ and $c(x)=(c_i(x),\ldots,c_r(x))^T$.

Let us consider Newton's method for Lagrange system

$$\nabla_x L(x,\lambda)=\nabla f(x)-\nabla c(x)^T\lambda=0, \tag{7.130}$$

$$c(x)=0, \tag{7.131}$$

which corresponds to the active constraints, starting from the same point $y=(x,\lambda)$.

The Newton direction $\Delta\bar{y}=(\Delta\bar{x},\Delta\bar{\lambda})$ one obtains from the following linear PD system:

$$\nabla\Phi_\infty(y)\Delta\bar{y}=-N(y).$$

The new approximation for the system (7.130)–(7.131) is

$$\bar{y}=y+\Delta\bar{y}.$$

From Theorem 4.14 follows

$$\|\bar{y}-y^*\|\leq c_1\|y-y^*\|^2. \tag{7.132}$$

Now let us show that a similar bound holds for $\|\widehat{y}-y^*\|$. We have

$$\|\widehat{y}-y^*\|=\|y+\Delta y-y^*\|=\|y+\Delta\bar{y}+\Delta y-\Delta\bar{y}-y^*\|$$
$$\leq\|\bar{y}-y^*\|+\|\Delta y-\Delta\bar{y}\|.$$

For $\|\Delta y-\Delta\bar{y}\|$, we obtain

$$\|\Delta y-\Delta\bar{y}\|=\|(\nabla\Phi_k^{-1}(y)-\nabla\Phi_\infty^{-1}(y))N(y)\|\leq\|\nabla\Phi_k^{-1}(y)-\nabla\Phi_\infty^{-1}(y)\|\|N(y)\|.$$

From Lemma 4.5, we have $\max\{\|\nabla\Phi_k^{-1}(y)\|,\|\nabla\Phi_\infty^{-1}(y)\|\}\leq 2c_0$.

Also, $\|\nabla\Phi_k(y)-\nabla\Phi_\infty(y)\|=k^{-1}\varphi''(1)$; therefore from Lemma 4.2 with $A=\nabla\Phi_k(y), B=\nabla\Phi_\infty(y)$ follows

$$\|\Delta y - \Delta \bar{y}\| \leq 2k^{-1}\varphi''(1)c_0^2\|N(y)\|. \tag{7.133}$$

In view of $\nabla_x L(x^*, \lambda^*) = 0, c(x^*) = 0$ and the Lipschitz condition (4.130), we have

$$\|N(y)\| \leq L_0\|y - y^*\| \quad \forall y \in B(y^*, \varepsilon_0).$$

Using right inequality (7.42), (7.128), and (7.133), we obtain

$$\begin{aligned}\|\Delta y - \Delta \bar{y}\| &\leq 2\varphi''(1)c_0^2\nu(y)L_0\|y - y^*\| \\ &\leq 2\varphi''(1)c_0^2 L_0 L\|y - y^*\|^2.\end{aligned}$$

Therefore, for $c_2 = 2\varphi''(1)c_0^2 L_0 L$, which is independent on $y \in B(y^*, \varepsilon_0)$, we have

$$\|\Delta y - \Delta \bar{y}\| \leq c_2\|y - y^*\|^2. \tag{7.134}$$

Using (7.129) and (7.134), for $c = 2\max\{c_1, c_2\}$, we obtain

$$\|\widehat{y} - y^*\| \leq \|\bar{y} - y^*\| + \|\Delta y - \Delta \bar{y}\| \leq c\|y - y^*\|^2, \quad \forall y \in B(y^*, \varepsilon_0)$$

and $c > 0$ is independent on $y \in B(y^*, \varepsilon_0)$. □

7.3.2 Global Convergence of the PDNR

In this section we describe globally convergent PDNR method and prove its asymptotic quadratic convergence rate under conditions of Theorem 7.9.

The globally convergent PDNR method, roughly speaking, works as Newton's NR method in the initial phase and as the PDNR (7.126)–(7.128) in the final phase.

The merit function $\nu(y)$ is critical for PDNR. It is used:

(1) for penalty parameter $k > 0$ update;
(2) for controlling accuracy at each step as well as for the overall stopping criteria;
(3) for identifying "small" and "large" Lagrange multipliers at each PDNR step;
(4) for deciding whether the primal or the primal–dual direction has to be used at the current step;
(5) for regularization of the Hessian $\nabla^2_{xx}L(x, \lambda)$.

We would like to emphasize that PDNR is not a mechanical combination of two different methods; it is a unified procedure.

Each step of the PDNR method consists of finding the PD direction $\Delta y = (\Delta x, \Delta \lambda)$. We obtain Δy by applying Newton's method to the nonlinear PD system (7.118)–(7.119) solving for $\hat{x}$ and $\hat{\lambda}$. Then, we use either the obtained PD Newton direction Δy to find a new PD approximation $\bar{y} := y + \Delta y$ for $\hat{y}$ or the primal Newton direction Δx to minimize LEP $\mathscr{L}$ in x. The choice at each step depends on the reduction of the merit function $\nu(y)$ per step. If $y \in B(y^*, \varepsilon_0)$, then according to Theorem 4.14, for the PD pair $\bar{y} = (\bar{x}, \bar{\lambda})$, we have

$$\|\bar{y}-y^*\| \le c\|y-y^*\|^2.$$

From the left inequality (7.42), we obtain $\|\bar{y}-y^*\| \le l^{-1}\nu(y)$; therefore,

$$\|\bar{y}-y^*\| \le c\|y-y^*\|^2 \le cl^{-2}(\nu(y))^2.$$

Using the right inequality (7.42), we have

$$\nu(\bar{y}) \le L\|\bar{y}-y^*\| \le cL\|y-y^*\|^2 \le cLl^{-2}(\nu(y))^2.$$

Also, $\nu(y) \le L\|y-y^*\| \le L\varepsilon_0$, $\forall y \in B(y^*,\varepsilon_0)$. Thus, for small enough $\varepsilon_0 > 0$ and $y \in B(y^*,\varepsilon_0)$, we obtain

$$\nu(\bar{y}) \le \nu(y)^{1.5}.$$

Therefore, if the PD step produces at least a 1.5-superlinear reduction of the merit function, then the PD direction is accepted, and we obtain new PD approximation $y := y+\Delta y$; otherwise, we use the primal direction Δx to minimize LEP $\mathscr{L}$ in x.

The important part of the PDNR method is the way the PD system (7.118)–(7.119) is linearized. Let us start with $y=(x,\lambda) \in \mathbb{R}^n \times \mathbb{R}^m_{++}$ and compute $\nu(y)$.

By linearizing the system (7.118), we obtain

$$\nabla^2_{xx}L(x,\lambda)\Delta x - \nabla c^T(x)\Delta\lambda = -\nabla_x L(x,\lambda). \tag{7.135}$$

Let us split system (7.119) into two sub-systems. The first is associated with the set $I_l(y)=\{i:\ \lambda_i > \nu(y)\}$ of so-called "large" Lagrange multipliers, the second is associated with the set $I_s(y)=\{i:\ \lambda_i \le \nu(y)\}$ of "small" Lagrange multipliers. Therefore, $I_l(y) \cap I_s(y) = \varnothing$ and $I_l(y) \cup I_s(y) = \{1,\dots,m\}$.

Let us consider the following two sub-systems:

$$\widehat{\lambda}_i = \psi'(k_i c_i(\widehat{x}))\lambda_i, \quad i \in I_l(y), \tag{7.136}$$

and

$$\widehat{\lambda}_i = \psi'(k_i c_i(\widehat{x}))\lambda_i, \quad i \in I_s(y). \tag{7.137}$$

The system (7.136) can be rewritten as follows:

$$k_i c_i(\widehat{x}) = \psi'^{-1}(\widehat{\lambda}_i/\lambda_i) = -\varphi'(\widehat{\lambda}_i/\lambda_i),\ i \in I_l(y).$$

Let $\widehat{x} = x+\Delta x$ and $\widehat{\lambda} = \lambda + \Delta\lambda$; then

$$c_i(x) + \nabla c_i(x)\Delta x = -k_i^{-1}\varphi'(1+\Delta\lambda_i/\lambda_i), \quad i \in I_l(y).$$

Taking into account $\varphi'(1)=0$ and ignoring terms of the second and higher order, we obtain

$$c_i(x) + \nabla c_i(x)\Delta x = -(k_i\lambda_i)^{-1}\varphi''(1)\Delta\lambda_i = -k^{-1}\varphi''(1)\Delta\lambda_i, \quad i \in I_l(y). \tag{7.138}$$

Let $c_{[l]}(x)$ be column vector-function correspondent to constraints associated with "large" Lagrange multipliers, that is, $c_{[l]}(x) = (c_i(x),\ i\in I_l(y))$, $\nabla c_{[l]}(x) = J(c_{[l]}(x))$ is the correspondent Jacobian and $\Delta\lambda_{[l]} = (\Delta\lambda_i,\ i\in I_l(y))$ is Newton direction associated with "large" Lagrange multipliers. Then the system (7.138) can be rewritten as follows:

$$\nabla c_{[l]}(x)\Delta x + k^{-1}\varphi''(1)\Delta\lambda_{[l]} = -c_{[l]}(x). \tag{7.139}$$

Now let us linearize system (7.137). Ignoring terms of the second and higher order, we obtain

$$\begin{aligned}\widehat{\lambda}_i &= \lambda_i + \Delta\lambda_i = \psi'(k_i(c_i(x) + \nabla c_i(x)\Delta x))\lambda_i \\ &= \psi'(k_i c_i(x))\lambda_i + k\psi''(k_i c_i(x))\lambda_i \Delta c_i(x)\Delta x \\ &= \bar{\lambda}_i + k\psi''(k_i c_i(x))\lambda_i \nabla c_i(x)\Delta x, \quad i\in I_s(y),\end{aligned} \tag{7.140}$$

where $\bar{\lambda}_i = \psi'(k_i c_i(x))\lambda_i$, $i \in I_s(y)$.

Let $c_{[s]}(x)$ be column vector-function associated with "small" Lagrange multiplier, $\nabla c_{[s]}(x) = J(c_{[s]}(x))$ the corresponding Jacobian, $\lambda_{[s]} = (\lambda_i,\ i\in I_s(y))$ vector of "small" Lagrange multipliers; and $\Delta\lambda_{[s]} = (\nabla\lambda_i,\ i\in I_s(y))$ the corresponding Newton direction. Then (7.140) can be rewritten as follows:

$$-k\Psi''(\mathbf{k}_{[\mathbf{s}]}c_{[s]}(x))\Lambda_{[s]}\Delta c_{[s]}(x)\Delta x + \Delta\lambda_{[s]} = \bar{\lambda}_{[s]} - \lambda_{[s]}, \tag{7.141}$$

where $\bar{\lambda}_{[s]} = \Psi'(\mathbf{k}_{[\mathbf{s}]}c_{[s]}(x))\lambda_{[s]}$, $\Psi'(\mathbf{k}_{[\mathbf{s}]}c_{[s]}(x)) = \operatorname{diag}\left(\psi'(k_i c_i(x))\right)_{i\in I_s(y)}$, $\Lambda_s = \operatorname{diag}(\lambda_i)_{i\in I_s(y)}\Psi''(\mathbf{k}_{[\mathbf{s}]}c_{[s]}(x)) = \operatorname{diag}\left(\psi''(k\lambda_i^{-1}c_i(x))\right)_{i\in I_s(y)}$.

Combining (7.135), (7.140), and (7.141), we obtain the following system for finding the PD direction $\Delta y = (\Delta x, \Delta\lambda)^T$, where $\Delta\lambda = (\Delta\lambda_l, \Delta\lambda_s)^T$ and I_l and I_S are identical matrices in spaces of "large" and "small" Lagrange multipliers

$$\begin{aligned} M(x,\lambda)\Delta y &= \begin{bmatrix} \nabla^2_{xx}L(x,\lambda) & -\nabla c_l^T(x) & -\nabla c_s^T(x) \\ \nabla c_l(x) & k^{-1}\varphi''(1)I_l & 0 \\ -k\Psi''(\mathbf{k}_{[\mathbf{s}]}c_{[s]}(x))\Lambda_{[s]}\nabla c_{[s]}(x) & 0 & I_S \end{bmatrix} \begin{bmatrix} \Delta x \\ \Delta\lambda_l \\ \Delta\lambda_s \end{bmatrix} \\ &= \begin{bmatrix} -\nabla_x L(x,\lambda) \\ -c_{[l]}(x) \\ \bar{\lambda}_s - \lambda_s \end{bmatrix}. \end{aligned} \tag{7.142}$$

To guarantee existence of the PD direction, Δy for any $(x,\lambda) \in \mathbb{R}^n \times \mathbb{R}^m_+$ we replace system (7.142) with the following regularized system, where I^n is identical matrix in $\mathbb{R}^n$

$$M_k(x,\lambda)\Delta y = \begin{bmatrix} \nabla^2_{xx}L(x,\lambda)+k^{-1}I^n & -\nabla c^T_{[l]}(x) & -\nabla c^T_{[s]}(x) \\ \nabla c_{[l]}(x) & k^{-1}\varphi''(1)I_l & 0 \\ -k\Psi''(\mathbf{k}_{[\mathbf{s}]}c_{[s]}(x))\Lambda_{[s]}\nabla c_{[s]}(x) & 0 & I_S \end{bmatrix} \begin{bmatrix} \Delta x \\ \Delta\lambda_l \\ \Delta\lambda_s \end{bmatrix}$$
$$= \begin{bmatrix} -\nabla_x L(x,\lambda) \\ -c_{[l]}(x) \\ \bar{\lambda}_s - \lambda_s \end{bmatrix}. \tag{7.143}$$

Finding PDNR direction $\nabla y = (\Delta x, \Delta\lambda)$ from system (7.143) we call **PDNRD** (x,λ) procedure.

Now we are ready to describe the PDNR method.

0. Initialization: We choose an initial primal approximation $x_0 \in \mathbb{R}^n$, Lagrange multipliers vector $\lambda_0 = (\lambda_{0,1},\ldots,\lambda_{0,m}) \in \mathbb{R}^m_{++}$, large enough penalty parameter $k > 0$, and scaling parameters vector $\mathbf{k}_0 = \mathrm{k}\lambda_0^{-1}e$, where $\lambda_0^{-1} = \mathrm{diag}(\lambda_{0,i}^{-1})_{i=1}^m$, $e = (1,\ldots,1)^T \in \mathbb{R}^m$.

Let $\varepsilon > 0$ be the overall accuracy. We choose parameters $0 < \eta < 0.5$, $0 < \sigma \leq 1$, and $0.25 < \theta < 0.5$. Set $x := x_0, \lambda = \lambda_0$, $\nu := \nu(x,\lambda)$, $\lambda_c := \lambda_0$, λ_c – current Lagrange multipliers vector and current scaling vector $\mathbf{k}_c := \mathbf{k}_0$.

1. If $\nu \leq \varepsilon$ then stop. **Output:** x, λ.

2. Find direction $(\Delta x, \Delta\lambda) :=$**PDNRD**$(x,\lambda)$, $\lambda := \lambda_c$. Set $\widehat{x} := x + \Delta x$, $\widehat{\lambda} := \lambda + \Delta\lambda$.

3. If $\nu(\widehat{x},\widehat{\lambda}) \leq \min\{\nu^{2-\theta}, (1-\theta)\nu\}$, set $x := \widehat{x}$, $\lambda := \widehat{\lambda}$, $\nu := \nu(x,\lambda)$, $k := \max\{\nu^{-1}, k\}$, Goto 1, else

4. Start with $t = 1$ decrease $0 < t \leq 1$ using backtracking line search until $\mathscr{L}(x + t\Delta x, \lambda_c, \mathbf{k}_c) - \mathscr{L}(x, \lambda_c, \mathbf{k}_c) \leq \eta t (\nabla\mathscr{L}(x,\lambda_c,\mathbf{k}_c), \Delta x)$.

5. Set $\widehat{x} := x + t\Delta x, \widehat{\lambda} := \Psi'(\mathbf{k}_c c(\widehat{x}))\lambda_c$.

6. If $\|\nabla_x \mathscr{L}(\widehat{x},\widehat{\lambda},\mathbf{k}_c)\| \leq \frac{\sigma}{k}\|\widehat{\lambda} - \lambda_c\|$, Goto 7, else Goto 2;

7. If $\nu(\hat{x},\widehat{\lambda}) \leq (1-\theta)\nu$, set $x := \widehat{x}$, $\lambda_c := \widehat{\lambda}$, $\lambda := \lambda_c$, $\nu := \nu(x,\lambda)$, $\mathbf{k}_c := (k_{i,c} = k\lambda_{i,c}^{-1}, i = 1,\ldots,m)$, Goto 1, else

8. Set $k := k(1+\theta)$, Goto 2.

The following theorem proves global convergence of the PDNR method and establishes its asymptotic quadratic rate.

Theorem 7.10. *If second-order sufficient optimality condition (4.73)–(4.74) holds and Lipschitz condition (4.130) is satisfied, then PDEP method generates a sequence $\{y_s = (x_s,\lambda_s)\}_{s\in N}$ globally convergent to the PD solution with asymptotic quadratic rate.*

Proof The existence of $\tau > 0$, that for any $y \notin B(y^*,\varepsilon_0)$ we have $\nu(y) \geq \tau$, follows from left inequality (7.42). Therefore, from (7.128) we have $k^{-1} = \nu(y) \geq \tau$.

After substitution of $\Delta\lambda_l$ and $\Delta\lambda_s$ into the first system of (7.143), we obtain

$$H_k(y)\Delta x = -\nabla_x \mathscr{L}(x,\lambda,k) = -\nabla_x L(x,\bar{\lambda}), \tag{7.144}$$

where

$$\begin{aligned} H_k(y) =& \nabla^2_{xx}L(x,\lambda)+k^{-1}I^n \\ &+k(\varphi''(1))^{-1}\nabla c^T_{[l]}(x)\nabla c_{[l]}(x)-k\nabla c^T_{[s]}(x)\Lambda_s\Psi''(\mathbf{k}_{[s]}c_{[s]}(x))\nabla c_{[s]}(x) \end{aligned} \tag{7.145}$$

and $\bar{\lambda}=(\bar{\lambda}_l,\bar{\lambda}_s)$, $\bar{\lambda}_l=\lambda_l-k(\psi''(1))^{-1}c_{[l]}(x)$, $\bar{\lambda}_s=(\lambda_i=\psi'(k_ic_i(x))\lambda_i,\ i\in I_s(y))$.

Due to the convexity f and concavity $c_i, i=1,\cdots,m$, the Hessian $\nabla^2_{xx}L(x,\lambda)$ is nonnegative definite. It follows from 3. of ψ that $\Psi''(\mathbf{k}_{[s]}c_{[s]}(x))$ is a diagonal matrix with negative entries. It follows from Assertion 7.2 that $\varphi''(1)>0$. Therefore, the third and the fourth terms in (7.145) are nonnegative definite matrices. Keeping in mind the second term in (7.145), we conclude that the symmetric matrix $H_k(y)$ is positive definite. Moreover, due to $k^{-1}\geq\tau>0$, the matrix $H_k(y)$ has, uniformly bounded from below, mineigenvalue $H_k(y)\geq\tau>0, \forall y\notin B(y^*,\varepsilon_0)$.

For any $y\in B(y^*,\varepsilon_0)$, there exists $\rho>0$ that mineigenvalue $H_k(y)\geq\rho>0$ due to Derbeu's Lemma, the second-order optimality condition, and Lipschitz condition (4.130). Therefore, for Δx we have

$$\mathscr{L}(x+t\Delta x,\lambda,\mathbf{k})-\mathscr{L}(x,\lambda,\mathbf{k})\leq\eta t(\nabla_x\mathscr{L}(x,\lambda,\mathbf{k}),\Delta x)\leq-\sigma t\eta\|\Delta x\|_2^2,$$

where $\sigma=\min\{\tau,\rho\}$. Therefore, Δx defined from (7.143) is a descent direction for minimization $\mathscr{L}(x,\lambda,\mathbf{k})$ in x.

Due to boundedness of $\mathscr{L}$ from below, the primal sequence converges to $\widehat{x}=\widehat{x}(\lambda,\mathbf{k}):\nabla_x\mathscr{L}(\widehat{x},\lambda,\mathbf{k})=\nabla_xL(\widehat{x},\widehat{\lambda})=0$.

Again, keeping in mind the second-order optimality condition (4.73)–(4.74) and Lipschitz condition (4.130), it follows from Theorem 7.3 that for PD approximation $(\bar{x}_{s+1},\bar{\lambda}_{s+1})$ bound (7.40) holds. Therefore, after $s_0=O(\ln\varepsilon_0^{-1})$ Lagrange multipliers and scaling vector updates, we find PD approximation $y\in B(y^*,\varepsilon_0)$.

Let $0<\varepsilon<<\varepsilon_0<1$ be the desired accuracy.

Keeping in mind properties 2.c) for $\psi\in\Psi$, as well as (7.120), after $s_1=O(\ln\ln\varepsilon^{-1})$ updates, we obtain

$$\max\{\|k\Psi''(k\lambda_{[s]}^{-1}c_{[s]}(x))\Lambda_{[s]}\|,\|\bar{\lambda}_s-\lambda_s\|\}=o(\varepsilon^2),\quad i\in I_s. \tag{7.146}$$

For any $y\in B(y^*,\varepsilon_0)$, the term $\|\nabla c_{[s]}(x)\|$ is bounded. The boundedness of $\|\Delta x\|$ follows from boundedness of $\|\nabla_xL(x,\bar{\lambda})\|$ and mineigenvalue $H_k(y)\geq\sigma>0\ \forall y\in B(y^*,\varepsilon_0)$.

Let us consider the third part of the system (7.143), which is associated with the "small" Lagrange multipliers

$$k\psi''(k\lambda_{[s]}{}^{-1}c_{[s]}(x))\Lambda_{[s]}\nabla c_{[s]}(x)\Delta x+\Delta\lambda_s=\bar{\lambda}_s-\lambda_s.$$

It follows from (7.146) that $\|\Delta\lambda_s\|=o(\varepsilon^2)$, which means that after $s=\max\{s_0,s_1\}\leq O(\ln\varepsilon^{-1})$ LM updates, the part of the system (7.143) associated with "small" Lagrange multipliers becomes irrelevant for the calculation of Newton direction. In fact, the system (7.143) is reduced to the following system:

$$\bar{M}_k(x,\lambda)\Delta\bar{y} = \bar{N}(x,\lambda), \tag{7.147}$$

where $\Delta\bar{y} = (\Delta x, \Delta\lambda_l)^T, \bar{N}(x,\lambda) = (-\nabla_x L(x,\lambda) - c_l(x))^T$, and

$$\bar{M}_k(x,\lambda) = \begin{bmatrix} \nabla^2_{xx}L(x,\lambda) + k^{-1}I^n & -\nabla c^T_{[l]}(x) \\ \nabla c_{[l]}(x) & k^{-1}\varphi''(1)I_{[l]} \end{bmatrix}.$$

At this point we have $y\in B(y^*,\varepsilon_0)$; therefore, it follows from right-hand side (7.42) that $\nu(y)\leq L\varepsilon_0$. Hence, for small $\varepsilon_0 > 0$ and $|\lambda_i - \lambda_i^*|\leq\varepsilon_0$ follows $\lambda_i\geq\nu(y), i \in I_{[l]}(y) = I^*$. On the other hand, we have $\nu(y) > \lambda_i = O(\varepsilon^2)$, $i\in I_s(y)$; otherwise, we obtain $\nu(y)\leq O(\varepsilon^2)$ and from (7.42) follows $\|y - y^*\| = o(\varepsilon^2)$.

So, if after $s = O(\ln\varepsilon^{-1})$ Lagrange multipliers updates we have not solved the problem with a given accuracy $\varepsilon > 0$, then $I_{[l]}(y) = I^* = \{1,\cdots,r\}$ and $I_{[s]}(y) = I_0^* = \{r+1,\ldots,m\}$ and PDNR method (0)–(8) turns into local PDNR (7.126)–(7.128) with matrix

$$\bar{M}_k(y) = \bar{M}_k(x,\lambda) = \begin{bmatrix} \nabla^2_{xx}L(x,\lambda) + k^{-1}I^n & -\nabla c^T_{(r)}(x) \\ \nabla c_{(r)}(x) & k^{-1}\varphi''(1)I^r \end{bmatrix}$$

used instead of $\nabla\Phi_k(\cdot)$ and $L(x,\lambda) = f(x) - \sum_{i=1}^r \lambda_i c_i(x_i)$ is the truncated Lagrangian.

Therefore, we have

$$||\Delta y - \Delta\bar{y}|| = ||(\bar{M}_k^{-1}(y) - \nabla\Phi_\infty^{-1}(y))\bar{N}(y)|| \leq ||\bar{M}_k^{-1}(y) - \nabla\Phi_\infty^{-1}(y)||||\bar{N}(y)||$$

and $\|\bar{M}_k(y) - \nabla\Phi_\infty(y)\| \leq k^{-1}(1+\varphi''(1))$.

From Lemma 4.4 we have $\max\{\|\bar{M}_k^{-1}(y)\|, \|\nabla\Phi_\infty^{-1}(y)\|\}\leq 2c_0$. Then

$$\begin{aligned} \|\Delta y - \Delta\bar{y}\| &\leq 2{c_0}^2k^{-1}(1+\varphi''(1))\|\bar{N}(y)\| = 2c_0^2\nu(y)(1+\varphi''(1))\|\bar{N}(y)\| \\ &\leq 2(1+\varphi''(1))c_0^2M_0L\|y-y^*\|^2 = c_3\|y-y^*\|^2, \end{aligned} \tag{7.148}$$

where $c_3 > 0$ is independent on $y\in B(y^*,\varepsilon_0)$.

Using bound (7.148) instead of (7.134), we conclude that the PD sequence generated by PDNR method converges to the PD solution (x^*,λ^*) with asymptotic quadratic rate. □

In the initial phase, the PDNR is similar to regularized Newton's NR method. It allows to reach $B(y^*,\varepsilon_0)$ in $O(\ln\varepsilon_0^{-1})$ Lagrange multipliers updates without compromising the condition number of the Hessian, $\nabla^2_{xx}\mathscr{L}(x,\lambda,\mathbf{k})$.

Once $y \in B(y^*,\varepsilon_0)$, the penalty parameter, which is inversely proportional to the merit function, grows extremely fast. The unbounded increase of the scaling parameter, however, at this point does not compromise the numerical stability, because instead of unconstrained minimization, the PDNR solves the PD system by Newton's method.

Moreover, the PD direction Δy becomes close to the Newton direction (see (7.148)) for the Lagrange system of equations corresponding to the active constraints. It guarantees asymptotic quadratic convergence.

The situation in some sense recalls the damped regularized Newton's method for unconstrained smooth convex optimization.

7.4 Nonlinear Rescaling and Augmented Lagrangian

7.4.0 Introduction

The idea of combining barrier and penalty functions for solving constrained optimization problems with both inequality and equality constraints has been suggested by Fiacco and McCormick nearly 50 years ago.

Their interior–exterior point method uses log-barrier function to treat inequality constraints and quadratic penalty function to handle equations. In such construction the barrier–penalty parameter is the only tool, which controls the computations. To guarantee convergence the parameter has to be unboundedly increased, which leads to well-known difficulties.

We consider an alternative approach called Nonlinear Rescaling–Augmented Lagrangian (NRAL), which is a combination of NR for inequality-constrained and AL for equations. The NRAL approach is a generalization of the MBAL method; see Goldfarb et al. (1999).

The NRAL leads to a multipliers method, which alternates finding primal minimizer (or its approximation) of the LEP $\mathscr{L}$ with LM update. The penalty–barrier parameter $k > 0$ can be fixed or updated from step to step.

The NRAL method eliminates major drawbacks of the barrier–penalty approach. The LEP $\mathscr{L}$ exists at the solution and inherits the smoothness of the objective function and constraints in a neighborhood of solution. Under the second-order sufficient optimality conditions and a fixed, but large enough scaling parameter $k > 0$, the LEP $\mathscr{L}$ is strongly convex in the neighborhood of primal minimizer. Therefore, the primal minimizer is unique and, based on LEP $\mathscr{L}$ dual function, is smooth and the dual problem has several important properties.

Also, under the second-order sufficient optimality condition, the multipliers method converges with Q-linear rate when the penalty–barrier parameter is fixed, but large enough. The Q-linear convergence rate is taking place, whether the problem is convex or not.

The condition number of LEP's Hessian remains stable in the neighborhood of the primal solution. It keeps stable the area, where Newton's method is well defined, which is critical for efficiency of the NRAL method.

7.4.1 Problem Formulation and Basic Assumptions

We consider nonlinear programming problem (4.48), assuming that $f, c_i, \dots, c_q$, and $g_1, \dots, g_r$ are C^2 functions.

The classical Lagrangian $L: \mathbb{R}^n \times \mathbb{R}^q_+ \times \mathbb{R}^r \to \mathbb{R}$, which corresponds to (4.48), is given by the following formula:

$$L(x,\lambda,u) = f(x) - \langle \lambda, c(x) \rangle - \langle u, g(x) \rangle = f(x) - \sum_{i=1}^{q} \lambda_i c_i(x) - \sum_{j=1}^{r} u_j g_j(x),$$

where $\lambda \in \mathbb{R}^q_+$, $u \in \mathbb{R}^r$, $c(x) = (c_1(x), \dots, c_q(x))^T$, and $g(x) = (g_1(x), \dots, g_r(x))^T$, $\mathbb{R}^p_+$ is the nonnegative orthant in $\mathbb{R}^p$ and $\mathbb{R}^p_{++}$ is its interior.

Let x^* be the local minimizer of problem (4.48) and $I^* = \{i : c_i(x^*) = 0\} = \{1, \dots, m\}$ $m < q$ be the set of inequality constraints active at x^*.

We assume that the second-order sufficient optimality conditions (4.49) and (4.62) are satisfied.

7.4.2 Lagrangian for the Equivalent Problem

Let $\psi \in \Psi$ and $k > 0$; then

$$c_i(x) \geq 0 \Leftrightarrow k^{-1}\psi(kc_1(x)) \geq 0, \; i = 1, \dots, q.$$

Let $\pi(t) = \frac{1}{2}t^2$; then

$$g_j(x) = 0 \Leftrightarrow 0.5k^{-1}\pi(kg_j(x)) = 0, \; j = 1, \dots, r.$$

Therefore problem (4.48) is equivalent to

$$f(x^*) + 0.5k^{-1}\sum_{j=1}^{r} \pi(kg_j(x^*)) = \min\{f(x) + 0.5k^{-1}\sum_{j=1}^{r} \pi(kg_j(x))|$$

$$k^{-1}\psi(kc_i(x)) \geq 0, \; i = 1, \dots, q, \; g_j(x) = 0, \; j = 1, \dots, r\}. \tag{7.149}$$

The LEP $\mathscr{L}: \mathbb{R}^n \times \mathbb{R}^q_+ \times \mathbb{R}^r \times \mathbb{R}_{++} \to \mathbb{R}$

$$\mathscr{L}(x,\lambda,u,k) = f(x) + \frac{1}{2}k\sum_{j=1}^{r} g_j^2(x) - k^{-1}\sum_{i=1}^{q} \lambda_i \psi(kc_i(x)) - \sum_{j=1}^{r} u_j g_j(x), \tag{7.150}$$

is our main tool.

In the following proposition, we collected the LEP's $\mathscr{L}$ properties at the KKT's point $(x^*, w^*) = (x^*, \lambda^*, u^*)$. The properties show the fundamental difference $\mathscr{L}$ from the classical penalty–barrier function

$$F(x,k) = f(x) - k^{-1}\sum_{i=1}^{q}\ln c_i(x) + 0.5k\sum_{j=1}^{r} g_j^2(x)$$

used by Fiacco and McCormick in SUMT.

Proposition 7.3. *For any given $k > 0$, the LEP $\mathscr{L}$ has the following properties:*

1° $\mathscr{L}(x^*,\lambda^*,u^*,k) = f(x^*)$
2° $\nabla_x\mathscr{L}(x^*,\lambda^*,u^*,k) =$

$$\nabla f(x^*) - \sum_{i=1}^{q}\lambda_i^*\psi'(kc_i(x^*))\nabla c_i(x^*) - \sum_{j=1}^{r}(u_j^* - kg_j(x^*))\nabla g_j(x^*) =$$

$$\nabla f(x^*) - \sum_{i=1}^{q}\lambda_i^*\nabla c_i(x^*) - \sum_{j=1}^{r}u_j^*\nabla g_j(x^*) = \nabla_x L(x^*,\lambda^*,u^*)$$

3° $\nabla_{xx}^2\mathscr{L}(x^*,\lambda^*,u^*,k) =$

$$\nabla_{xx}^2 L(x^*,\lambda^*,u^*) + k\nabla c_{(m)}(x^*)^T\Lambda_m^*\nabla c_{(m)}(x^*) + k\nabla g^T(x^*)\nabla g(x^*),$$

where $\nabla c_{(m)}(x) = J(c_{(m)}(x))$*-Jacobian of the vector-function*

$$c_{(m)}(x) = (c_1(x),\ldots,c_m(x))^T,$$

$\nabla g(x^*) = J(g(x^*))$*-Jacobian of vector-function* $g(x) = (g_1(x),\ldots g_r(x))^T$*, and* $\Lambda_m^* = \mathrm{diag}[\lambda_i^*]_{i=1}^m$.

Let $B(x^*,\varepsilon) = \{x\in\mathbb{R}^n : \|x-x^*\|\le\varepsilon\}$ and $B(w^*,\varepsilon) = \{w=(\lambda,u) : \lambda\in\mathbb{R}_+^q,\ u\in\mathbb{R}^r,\ \|w-w^*\|\le\varepsilon\}$.

The following Lemma characterizes the convexity property of $\mathscr{L}$ in x at the neighborhood of the primal–dual solution (x^*,w^*).

Lemma 7.2. *If all f, c_i $i=1,\ldots,q$ and g_j, $j=1,\ldots,r$ are from C^2, then under second-order sufficient optimality conditions* (4.49) *and* (4.62)*, there exist $k_0 > 0$, $\varepsilon > 0$, and $\mu > 0$, that for $\forall x\in B(x^*,\varepsilon)$ and $\forall w\in B(w^*,\varepsilon)$, the following*

$$\langle\nabla_{xx}^2\mathscr{L}(x,w,k)y,y\rangle \ge \mu\langle y,y\rangle,\ \forall y\in\mathbb{R}^n \tag{7.151}$$

holds for any $k\ge k_0$.

Proof. From (4.49), (4.62) property 3° of $\mathscr{L}$, and Derbeu's Lemma with $A = \nabla_{xx}^2 L(x^*,w^*)$ and $C = [\Lambda^{*\frac{1}{2}}\nabla c(x^*);\nabla g(x^*)]$ follows existence $\mu_0 > 0$ such that

$$\langle\nabla_{xx}^2\mathscr{L}(x^*,w^*,k)y,y\rangle \ge \mu_0\langle y,y\rangle,\ \forall y\in\mathbb{R}^n \tag{7.152}$$

for any $k\ge k_0$. Then there exists $0<\mu<\mu_0$ that (7.151) follows from (7.152) and smoothness of f, c_i, g_j. □

Remark 7.10. It follows from 2° and Lemma 7.2 that for any $k \geq k_0$ map $w = (\lambda, u) \to \hat{w} = (\hat{\lambda} = \lambda \psi'(kc(\hat{x})),\ \hat{u} = u - kg(\hat{x}))$ has a fixed point $w^* = (\lambda^*, u^*)$, which is the dual solution.

7.4.3 Multipliers Method

Let $w = (\lambda, u)$, $\lambda \in \mathbb{R}^q_{++}$, $u \in \mathbb{R}^r$, and $k > 0$. The new primal–dual approximation $(\hat{x}, \hat{w}) = (\hat{x}, \hat{\lambda}, \hat{u})$ one finds by the following formulas:

$$\hat{x} : \nabla_x \mathscr{L}(\hat{x}, w, k) =$$

$$\nabla f(\hat{x}) - \sum_{i=1}^{q} \lambda_i \psi'(kc_i(\hat{x})) \nabla c_i(\hat{x}) - \sum_{j=1}^{r} (u_j - kg_j(\hat{x})) \nabla g_j(\hat{x}) =$$

$$\nabla f(\hat{x}) - \sum_{i=1}^{q} \hat{\lambda}_i \nabla c_i(\hat{x}) - \sum_{j=1}^{r} \hat{u}_j \nabla g_j(\hat{x}) = \nabla_x L(\hat{x}, \hat{\lambda}, \hat{u}) = 0 \tag{7.153}$$

$$\hat{\lambda} = (\hat{\lambda}_i = \lambda_i \psi'(kc_i(\hat{x})),\ i = 1, \ldots, q) \tag{7.154}$$

$$\hat{u} = (\hat{u}_j = u_j - kg_j(\hat{x}),\ j = 1, \ldots, r). \tag{7.155}$$

For the basic theorem, we recall the extended dual domain, where the basic results are taking place.

Let $0 < \delta < \min_{1 \leq i \leq r} \lambda_i^*$ be small enough and $k_0 > 0$ be large enough. We split the extended dual set $\Lambda(\cdot) = \Lambda(\lambda, \delta, k) = \Lambda_{(m)}(\cdot) \otimes \Lambda_{(q-m)}(\cdot)$, associated with inequality constraints, on two subsets. The first,

$$\Lambda_{(m)}(\cdot) \equiv \Lambda(\lambda^*_{(m)}, k, \delta) = \{(\lambda_{(m)}, k) : \lambda_i \geq \delta, |\lambda_i - \lambda_i^*| \leq \delta k,\ i = 1, \ldots, m;\ k \geq k_0\},$$

corresponds to the active constraints.

The second,

$$\Lambda_{(q-m)} \equiv \Lambda(\lambda^*_{(q-m)}, k_0, \delta) = \{(\lambda_{(q-m)}, k) : 0 \leq \lambda_i \leq \delta k,\ i = m+1, \ldots, q,\ ;\ k \geq k_0\},$$

corresponds to the passive constraints.

The dual domain, which corresponds to the equality constraints, is defined as follows:

$$U(u^*, k_0, \delta) = \{(u, k) : |u_i - u_i^*| \leq \delta k,\ i = 1, \ldots, r,\ k \geq k_0\}.$$

Then

$$W(\lambda^*, u^*, k_0, \delta) \equiv W(w^*, k_0, \delta) = \Lambda(\lambda^*, k_0, \delta) \otimes U(u^*, k_0, \delta)$$

is the dual domain of our interest.

Now we can formulate the basic theorem for the NRAL multipliers method.

Theorem 7.11. *Let all f, c_i, $i=1,\dots,q$, g_j, $j=1,\dots,r$ be from C^2 and the second-order sufficient optimality conditions (4.49) and (4.62) be satisfied; then there exist small $\delta>0$ and large $k_0>0$, that for any $(w,k)=(\lambda,u,k)\in W(w^*,k_0,\delta)$, the following statements hold*

(1) there exist

$$\hat{x}\equiv\hat{x}(w,k):\nabla_x\mathscr{L}(\hat{x},w,k)=\nabla_x L(\hat{x},\hat{w})=0 \tag{7.156}$$

and

$$\hat{\lambda}=(\hat{\lambda}_i=\lambda_i\psi'(kc_i(\hat{x})),\ i=1,\dots,q) \tag{7.157}$$

$$\hat{u}=(\hat{u}_j=u_j-kg_j(\hat{x}),\ j=1,\dots,r) \tag{7.158}$$

(2) for the triple $(\hat{x},\hat{\lambda},\hat{u})=(\hat{x},\hat{w})$ the following bound

$$\max\{\|\hat{x}-x^*\|,\|\hat{w}-w^*\|\}\le ck^{-1}\|w-w^*\| \tag{7.159}$$

holds and $c>0$ is independent on $k\ge k_0$.
(3) LEP $\mathscr{L}$ is strongly convex in x in the neighborhood of $\hat{x}$.

Theorem 7.11 can be proven by combining arguments used in the proofs of Theorems 4.16 and 7.2.

We are leaving it for the reader.

The LEP $\mathscr{L}$, generally speaking, is not convex in $x\in\mathbb{R}^n$; however, it is strongly convex at the minimizer $\hat{x}$ and in its neighborhood.

Therefore, once an approximation for $\hat{x}$ is found for a fixed, but large enough $k>0$, then after updating Lagrange multipliers by (7.157) and (7.158), the LEP $\mathscr{L}$ becomes strongly convex in x .

In other words, each new step requires finding an approximation for the primal minimizer $\hat{x}$ from a starting point, which is in the area where LEP $\mathscr{L}$ is strongly convex in x.

Finding $\hat{x}$ from (7.157) is, generally speaking, an infinite procedure, so to make the multipliers method practical, we introduce the following stopping criteria similar to (7.28). Instead of $(\hat{x},\hat{w})$ one finds $(\bar{x},\bar{w})$ by the following formulas:

$$\bar{x}:\|\nabla_x\mathscr{L}(\bar{x},w,k)\|\le\frac{\alpha}{k}\|\bar{w}-w\|$$

$$\bar{\lambda}=(\bar{\lambda}_i=\lambda_i\psi'(kc_j(\bar{x})),\ j=1,\dots,q)$$

$$\bar{u}=(\bar{u}_i=u_i-kg_j(\bar{x}),\ i=1,\dots,r).$$

Exercise 7.8. Using results of Theorem 7.3 shows that there exist $c>0$ and $\alpha>0$ independent on $k\ge k_0$ that for the vector $(\bar{x},\bar{w})\equiv(\bar{x},\bar{\lambda},\bar{u})$ the following bound

$$\max\{\|\bar{x}-x^*\|,\|\bar{w}-w^*\|\}\le\frac{c}{k}(1+2\alpha)\|w-w^*\|$$

holds.

7.4.4 NRAL and the Dual Prox

If f and all $-c_i$, $i = 1, \dots, q$ are convex and g_j, $j = 1, .., r$ are linear, then (4.48) is a convex optimization problem.

Let $H = \{x \in \mathbb{R}^n : g(x) = Ax - a = 0\}$ be a linear manifold defined by full rank $r < n$ matrix $A : \mathbb{R}^n \to \mathbb{R}^r$ and vector $a \in \mathbb{R}^r$; then $\Omega = \{x \in \mathbb{R}^n : c_i(x) \geq 0, \ i = 1, \dots, q\} \cap H$ is convex set.

We assume:

A. the primal optimal set

$$X^* = \{x \in \Omega : f(x) = f(x^*)\}$$

is not empty and bounded

B. $reint\Omega \neq \emptyset$, that is there exists

$$x_0 \in H : c_i(x_0) > 0, \ i = 1, \dots, q.$$

Let us assume that Ω is bounded. If it is not the case, then we can add one constraint $c_0(x) = N - f(x) \geq 0$, which for a large enough N does not affect X^*, but it keeps Ω bounded.

Exercise 7.9. Show that from convexity f and all $-c_i$, properties (1)–(4) of $\psi \in \Psi$ and assumptions A and B follow: for any $w \in \mathbb{R}^q_+ \times \mathbb{R}^r$ and any $k > 0$ the recession cone of Ω is empty, that is,

$$\lim_{t \to \infty} \mathcal{L}(x + td, w, k) = \infty \tag{7.160}$$

for any $x \in \Omega$, any $w \in \mathbb{R}^q_{++} \times \mathbb{R}^r$, and any nontrivial direction $d : Ad = 0$.

Let $\lambda \in \mathbb{R}^q_{++}$ and $u \in \mathbb{R}^r$ be fixed; then from (7.160) follows existence of

$$\hat{x} \equiv \hat{x}(w) : \nabla_x \mathcal{L}(\hat{x}, w, k) = 0. \tag{7.161}$$

From (7.153)–(7.155) follows

$$\min\{\mathcal{L}(x, w, k) | x \in \mathbb{R}^n\} = \mathcal{L}(\hat{x}(w, k), w, k) = L(\hat{x}, \hat{w}) =$$

$$\min\{L(x, \hat{w}) | x \in \mathbb{R}^n\} = d(\hat{w}),$$

where the dual function d is closed and concave, while the dual problem

$$d(w^*) = \max\{d(w) | \lambda \in \mathbb{R}^q_+, \ u \in \mathbb{R}^r\}, \tag{7.162}$$

is convex.

From Assumption B follows boundedness of the dual optimal set

$$W^* = \{w = (\lambda, u) \in \mathbb{R}^q_+ \times \mathbb{R}^r : d(w) = d(w^*)\}.$$

If $\hat{x}(w) = (\hat{x}_1(w), \ldots, \hat{x}_n(w))^T$ is unique and all f, $-c_i$, $i = 1, \ldots, q$ are C^1 functions, then the dual function d is smooth and

$$\nabla d(\hat{w}) = \nabla_x L(\hat{x}(w), w) \nabla_w \hat{x}(w) + \nabla_w L(\hat{x}(w), w). \tag{7.163}$$

From (7.161) and (7.163) follows

$$\nabla d(\hat{w}) = \nabla_w L(\hat{x}(w), w) = \begin{bmatrix} -c(\hat{x}(w)) \\ -g(\hat{x}(w)) \end{bmatrix}.$$

In case if $\hat{x}(w)$ is not unique, then

$$G(w) = \begin{bmatrix} -c(\hat{x}(w)) \\ -g(\hat{x}(w)) \end{bmatrix} \in \partial d(w), \tag{7.164}$$

where $\partial d(w)$ is the subdifferential of d at $w \in \mathbb{R}^q_+ \times \mathbb{R}^r$.

In fact, let $\hat{w} \in \mathbb{R}^q_+ \times \mathbb{R}^r$

$$\hat{w} : d(\hat{w}) = L(\hat{x}, \hat{w}) = \min_{x \in \mathbb{R}^n} L(x, \hat{w});$$

then we obtain

$$d(\hat{w}) = \min\{f(x) - \sum_{i=1}^{q} \hat{\lambda}_i c_i(x) - \sum_{j=1}^{r} \hat{u}_j g(x) | x \in \mathbb{R}^n\} \leq$$

$$f(x(w)) - \sum_{i=1}^{q} \hat{\lambda}_i c_i(x(w)) - \sum_{j=1}^{r} \hat{u}_j g(x(w))$$

$$d(\hat{w}) = \min\{L(x, \hat{w}) | x \in \mathbb{R}^n\} = L(\hat{x}(w), \hat{w}) = L(\hat{x}, \hat{w}) \leq L(x(w), \hat{w}) = L(x(w), w) -$$
$$\langle c(x(w)), \hat{\lambda} - \lambda \rangle - \langle g(x(w)), \hat{u} - u \rangle = d(w) - \langle G(x(w)), \hat{w} - w \rangle.$$

Thus,

$$d(\hat{w}) - d(w) \leq \langle -G(x(w)), \hat{w} - w \rangle, \quad \forall \hat{w} \in \mathbb{R}^q_+ \times \mathbb{R}^r;$$

therefore (7.164) holds. Note that $\partial d(w)$ is a convex and bounded set.

The following equivalence theorem is important for convergence analysis.

Theorem 7.12. *If conditions A and B are satisfied and all f, c_i are continuously differentiable functions, then the multipliers method (7.153)–(7.155) is:*

(1) well defined;
(2) equivalent to the following proximal point method

$$d(\hat{w}) - k^{-1}[d_1(\hat{\lambda}, \lambda) + d_2(\hat{u} - u)] \tag{7.165}$$

$$= \max\{d(W) - k^{-1}\left(\sum_{i=1}^{q} \lambda_i \varphi(\Lambda_i/\lambda_i) + 1/2\|U - u\|^2\right) | \Lambda \in \mathbb{R}^q_+,\ U \in \mathbb{R}^r\},$$

where $d(w) = \min_{x\in\mathbb{R}^n} L(x,w)$ *is the dual function;* $d_1(\Lambda,\lambda) = \sum_{i=1}^q \lambda_i \varphi(\Lambda_i/\lambda_i)$ *is the entropy-like distance, which corresponds to inequality constraints; and* $d_2(U,u) = \frac{1}{2}\|U-u\|^2$ *is the quadratic distance, which corresponds to equations.*

Proof.

(1) From Exercise 7.9 follows existence of $\hat{x} \equiv \hat{x}(w,k)$:

$$\hat{x} : \nabla_x \mathscr{L}(\hat{x},w,k) = 0,$$

which for convex optimization problem type (4.48) means

$$\mathscr{L}(\hat{x},w,k) = \min\{\mathscr{L}(x,w,k)|x \in \mathbb{R}^n\}.$$

Then,

$$\nabla_x \mathscr{L}(\hat{x},w,k) = \nabla f(\hat{x}) - \sum_{i=1}^q \lambda_i \psi'(kc_i(\hat{x}))\nabla c_i(\hat{x}) - \sum_{i=1}^r (u_j - kg_j(\hat{x}))\nabla g_j(\hat{x})$$

$$\nabla f(\hat{x}) - \sum_{i=1}^q \hat{\lambda}_i \nabla c_i(\hat{x}) - \sum_{j=1}^r \hat{u}_j \nabla g_j(\hat{x}) = \nabla_x L(\hat{x},\hat{w}) = 0.$$

From $\psi' > 0$ follows $\lambda \in \mathbb{R}^q_{++} \Rightarrow \hat{\lambda} \in \mathbb{R}^q_{++}$; therefore, method (7.157)–(7.159) is well defined.

(2) From convexity $\mathscr{L}$ in x and (7.161) follows

$$d(\hat{w}) = \min\{L(x,\hat{w})|x \in \mathbb{R}^n\},$$

and the dual function d is closed and concave.
Let us consider the dual problem

$$d(w^*) = \max\{d(w)|\lambda \in \mathbb{R}^q_+,\ u \in \mathbb{R}^r\} = \max\{d(w)|w \in \mathbb{R}^q_+ \times \mathbb{R}^r\}. \quad (7.166)$$

From (7.157) follows

$$c_i(\hat{x}) = k^{-1}\psi'^{-1}(\hat{\lambda}_i/\lambda_i),\ i = 1,\ldots,q. \quad (7.167)$$

The inverse ψ'^{-1} exists due to $\psi'' < 0$. From LF identity $\psi'^{-1} = \psi^{*'}$ and $\varphi = -\psi^*$ follows

$$c_i(\hat{x}) + k^{-1}\varphi'(\hat{\lambda}_i/\lambda_i) = 0,\ \ i = 1,\ldots,q. \quad (7.168)$$

From (7.158) we have

$$0 = -g_j(\hat{x}) = k^{-1}(\hat{u}_j - u_j)\ \ j = 1,\ldots,r. \quad (7.169)$$

Therefore, from (7.164), (7.168), and (7.169) follows

$$0 \in \begin{pmatrix} c(\hat{x}) \\ g(\hat{x}) \end{pmatrix} + \partial d(\hat{w}) \quad (7.170)$$

or

$$0 \in \partial d(\hat{w}) - k^{-1} \left[\sum_{i=1}^{q} \varphi'(\hat{\lambda}_i/\lambda_i)e_i - \sum_{j=1}^{r} (\hat{u}_j - u_j)e_j \right], \tag{7.171}$$

where $e_i = (0,\ldots,\overset{i}{1},\ldots,0) \in \mathbb{R}^q_+$, $e_j = (0,\ldots,\overset{j}{1},..,0) \in \mathbb{R}^r$.
The inclusion (7.171) is the optimality condition for vector $\hat{w}$ in convex optimization problem (7.165).
In other words, the multipliers (7.156)–(7.158) method is equivalent to the proximal point method (7.165) with distance function, which is a sum of first-order entropy-like distance $d_1(\Lambda,\lambda)$ for inequality constraint and quadratic distance $d_2(U,u)$ for equations. □

Exercise 7.10. Establish convergence property of NRAL method (7.156)–(7.158) under assumptions A and B.

Notes

The first NR results were obtained in 1980–1981 for discrete minimax in an effort to find an alternative to sub-gradient-type methods for non-smooth optimization. At this point it was clear that the sub-gradient-type ellipsoid method is not efficient for LP calculations. The NR framework provides an opportunity to use progress in smooth unconstrained optimization for discrete minimax. Unfortunately, the results obtained were published much later; see Polyak (1988).

It became clear that practically every SUMT method can be transformed into an NR alternative with much better properties. The classical barrier and distance functions were the first best candidates for such transformation; see Polyak (1987), Polyak (1992), Polyak (1992a), and Polyak (1997). For the equivalence of MBF and the dual prox with the entropy-like distance see Polyak and Teboulle (1997) and also Iusem et al. (1994)

The "modified" in the titles comes from Russian "Modified Lagrangian," which stands for Augmented Lagrangian; see Goldshtein and Tretiakov (1989). For non-smooth optimization in general and discrete minimax in particular, see, for example, Ermoliev and Shor (1967), Nemirovski and Yudin (1983), Nesterov and Nemirovski (1994), Nesterov (2004), Nesterov (1984), B. Polyak (1967), B. Polyak (1987), Shor (1998), and references therein.

The Truncated MBF transformation was introduced in Ben-Tal et al. (1992). The Newton MBF method and the "hot" start phenomenon were first discussed in Polyak (1992). The MBF complexity for QP was established in Melman and Polyak (1996). The MBF theory and methods for non-generate LP were developed in Polyak (1992a). Interesting MBF convergence results for LP were obtained by M. Powell; see Powell (1995) and also Jensen and Polyak (1994).

The NR with "dynamic" scaling parameter update for exponential transformation was introduced in Tseng and Bertsekas (1993). The correspondent proximal point methods with second-order entropy-like distance functions were studied in Auslender et al. (1999), Ben-Tal and Zibulevsky (1997), and Tseng and Bertsekas (1993); see also references therein.

The NR methods as an alternative to the smoothing techniques were introduced in Polyak (2001). For "dynamic" version of NR and correspondent proximal point method with the second-order Fermi–Dirac entropy distance function, see Polyak (2002). For convergence properties of the proximal point methods with second-order entropy-like distance, see Polyak (2002, 2006) and Griva and Polyak (2013) and references therein.

The primal–dual NR methods were developed in the late 1990s; see Griva and Polyak (2004, 2006, 2013). Local quadratic convergence for the primal–dual NR method was established in Polyak (2008) and Polyak (2009).

The MBAL method was introduced and its Q-linear convergence rate has been proven in Goldfarb et al. (1999). The primal–dual version of the MBAL method was introduced in Griva and Polyak (2008); it was also proven 1.5 Q-superlinear convergence rate under the second-order sufficient optimality condition.

Chapter 8
Realizations of the NR Principle

8.0 Introduction

We concentrate on three important realizations of NR principle: modified log-barrier function (MBF), exterior distance function (EDF), and log-sigmoid (LS) Lagrangian. The correspondent NR methods can be viewed as alternatives to both SUMT and IPMs (see Chapters 5 and 6).

There are few distinct features, which make NR approach fundamentally different from SUMT.

First, the Lagrange multipliers presented in NR explicitly. They are critical for the convergence and convergence rate. In SUMT they appear only as a byproduct of sequential unconstrained minimization and have no computational impact.

Second, any LEP $\mathscr{L}$ is not singular at the primal solution, which makes Newton's method for finding an approximation for the primal minimizer particularly efficient.

Third, the role of the scaling (penalty) parameter is fundamentally different in SUMT and NR. In contrast to SUMT, any NR realizations do not require unbounded increase of the scaling parameter to guarantee convergence.

Each NR Realization Is Primal Exterior and Dual Interior Point Method

The Lagrange multipliers can be viewed as prices for constraints violation. Each NR realization leads to a pricing mechanism for finding equilibrium between the objective function reduction and the penalty for constraints violation.

The pricing mechanism translates primal constraints violation into dual objective function increase by the value of entropy-like distance between two sequential iterates.

Therefore, for every NR realization, the distance vanishes together with constraints violation.

Finally, under the second-order sufficient optimality condition and a fixed, but large enough scaling parameter, the LEP's Hessian is positive definite at the primal solution, and its condition number is stable at the neighborhood of the primal

R. A. Polyak, *Introduction to Continuous Optimization*, Springer Optimization and Its Applications 172, https://doi.org/10.1007/978-3-030-68713-7_8

minimizer for any Lagrange multipliers vector from the neighborhood of the dual solution. It leads to the "hot" start phenomenon.

8.1 Modified Barrier Functions

8.1.0 Introduction

Classical barrier functions (CBFs) were introduced by Frisch (1955) and Carroll (1961). The purpose was to replace a constrained optimization problem by a sequence of unconstrained problems. Later CBFs were extensively studied by Fiacco and McCormick and incorporated them into SUMT (1968, 1990).

The interest in log-barrier function was revived in the mid-1980s after Gill et al. (1986) found that Karmarkar's method (1984) is closely connected to Newton log-barrier method for LP calculation (see Chapter 5).

Then, C. Gonzaga and J. Renerar, using path-following technique based on classical log-barrier and interior distance functions, improved substantially N. Karmarkar's complexity bound for LP calculations.

Practically at the same time, the self-concordance (SC) theory was developed by Nesterov and Nemirovski (1994).

Not only IPMs have polynomial complexity, they also produced good numerical results for both LP and some NLP problems.

Still, there are some important issues, which cannot be addressed in the framework of CBFs and SC theory.

First, not every convex optimization problem can be equipped with SC barrier.

Second, the barrier parameter is the only driving force in SUMT and later in the IPMs. The dual information is not used at all or, in case of primal–dual IPM, has impact neither on the convergence rate nor the complexity bounds.

Third, the IPMs do not take advantage of the local, near the solution structure of constrained optimization problem.

In particular, the second-order sufficient optimality conditions have little or no effect on both the rate of convergence and complexity of the IPMs.

On the other hand, the classical Lagrangian (CL), which is the main tool in constrained optimization, along with some important properties, has essential limitations as well.

First, the unconstrained primal minimizer of the CL under fixed optimal Lagrange multipliers might not exist, even for a convex optimization problems under the second-order sufficient optimality condition.

For LP the set of primal minimizers of the CL under fixed optimal dual solution is the entire primal space. In other words, we cannot find the primal solution by minimizing CL even if we know the optimal Lagrange multipliers.

Second, the objective function for the dual problem, which is based on the CL, generally speaking, is not smooth, no matter how smooth are the original functions.

Third, for non-convex optimization, the basic duality results are not true, even under standard second-order sufficient optimality conditions.

The MBF was introduced in the early 1980s as particular realizations of the NR principle. Being a classical Lagrangian for the equivalent problem, the MBF combines the best properties of both the CL and CBF and, at the same time, are free from their basic drawbacks. We will concentrate on the logarithmic MBF.

First, MBF exists at the primal solution and as smooth as the given data on the extended primal feasible set.

Second, the condition number of the MBF Hessian remains stable, when the primal–dual approximation approaches the primal–dual solution.

Third, under the second-order sufficient optimality conditions, that is, for a nondegenerate constrained optimization problems, the dual function, which is based on MBF, is as smooth as the given data and the basic facts of the duality theory hold, whether the constrained optimization problem is convex or not.

The fundamental difference between modified and classical barrier method is reflected in their convergence properties. The MBF methods converge not due to the unbounded increase of the barrier parameter but rather due to the Lagrange multipliers update, while the barrier parameter can be fixed or updated from step to step.

In particular, for convex optimization, the MBF generates a primal–dual sequence, which converges in value to the primal–dual solution for any fixed positive penalty parameter.

Moreover, Powell (1995) proved that for LP the MBF method generates such primal sequence, which convergence to the Chebyshev center of the primal optimal face under any fixed positive barrier parameter.

Under the second-order sufficient optimality condition, the MBF generates primal–dual sequence, which converges to the primal–dual solution with Q-linear rate under a fixed, but large enough penalty parameter, whether the constrained optimization problem is convex or not.

For convex optimization the key component of the convergence proof is the equivalence of the MBF and the dual interior prox method with Kullback–Leibler φ–divergence distance; see Polyak and Teboulle (1997).

The Lagrange multipliers update formula is, in fact, the multiplicative method for the dual problem, which has similarities (see Eggermont (1990)) with EM algorithm (see Shepp and Vardi (1982)).

The dual problem associated with MBF has some important properties, which can be used for developing second-order multiplier methods with up to quadratic rate.

The numerical realization of the MBF method leads to the Newton MBF method. Newton's method is used for finding an approximation for the primal minimizer which is used for the Lagrange multipliers update. The way MBF method updates the dual variables is fundamental for convergence, convergence rate, and complexity bound; in particular, it leads to the "hot" start phenomenon.

For any nondegenerate convex optimization problem, there exists a point on the primal MBF trajectory, the so-called "hot" start, from which it requires at most

$O(\ln\ln\varepsilon^{-1})$ Newton steps for finding an approximation for the primal minimizer with given accuracy $\varepsilon > 0$.

Moreover, the new approximation belongs to the Newton area of the primal MBF minimizer after each LM vector update (see Fig. 7.6). Each update shrinks the distance to the primal–dual solution by a factor $0 < \gamma < 1$, which is inversely proportional to the scaling parameter $k > 0$.

Due to the "hot" start, the Newton MBF requires substantially less Newton steps per extra digit of accuracy toward the end of calculation, which is a fundamental departure from SUMT.

Moreover, the universal SC properties of the shifted log-barrier function far from the solution can be combined with excellent MBF properties near the solution.

Over last 30 years, the NR method with truncated MBF transformation has been widely tested and produced strong numerical results, in particular, for large-scale constrained optimization problems (see Chapter 11). One of the best constrained optimization solvers, PENNON (see Kočvara and Stingl (2005, 2007, 2015)), is based on the NR technique with truncated MBF transformation.

8.1.1 Logarithmic MBF

We are back to convex optimization (CO) problem (P).

Also for any $k > 0$, we have

$$\Omega = \{x \in \mathbb{R}^n :\ c_i(x) \geq 0,\ i = 1,\dots,m\} =$$

$$\{x \in \mathbb{R}^n : k^{-1}\ln(kc_i(x)+1) \geq 0\,,\ i = 1,\dots,m\}.$$

The logarithmic MBF $F : \mathbb{R}^n \times \mathbb{R}^m_+ \times \mathbb{R}^1_+ \to \mathbb{R}^1$

$$F(x,\lambda,k) = \begin{cases} f(x) - k^{-1}\sum_{i=1}^m \lambda_i \ln(kc_j(x)+1), & \text{if} \quad x \in \operatorname{int}\,\Omega_k \\ \infty, & \text{if} \quad x \notin \operatorname{int}\,\Omega_k \end{cases}$$

is the Lagrangian for the equivalence problem.

For a given $k > 0$, we consider the extended feasible set

$$\Omega_k = \{x \in \mathbb{R}^n : kc_i(x) + 1 \geq 0\,,\ i = 1,\dots,m\} \supset \Omega\,.$$

If c_i, $i = 1,\dots,m$ concave, then boundedness of Ω implies boundedness of Ω_k for any fixed $k > 0$ (see Theorem 2.7).

If the primal problem is a non-convex, then boundedness of Ω does not imply boundedness of Ω_k. In the non-convex case, we will use the following growth assumption:

$$\exists k_0 > 0 \;\text{ and }\; \tau > 0 : \max\left\{\max_{1\leq i\leq m} c_i(x) | x \in \Omega_{k_0}\right\} = \theta(k_0) \leq \tau\,. \tag{8.1}$$

It is clear that $\theta(k)$ is a monotone decreasing function of $k > 0$. So if (8.1) is true for some $k_0 > 0$, it will be true for any $k \geq k_0$.

In case of CO, it is easy to see that MBF $F(x,\lambda,k)$ is convex in x for any given $\lambda \in \mathbb{R}^m_{++}$ and $k > 0$.

The MBF is not singular at the primal solution together with its derivatives of any order, and for any KKT's pair (x^*,λ^*), it possesses the following properties, which are critical for the MBF theory and methods:

1°
$$F(x^*,\lambda^*,k) = f(x^*) \quad \forall\, k > 0 .$$

From the complementarity condition for any KKT's pair (x^*,λ^*) and any $k > 0$ follows

2°
$$\nabla_x F(x^*,\lambda^*,k) = \nabla f(x^*) - \sum_{i=1}^{m} \lambda_i^* \nabla c_i(x^*) = \nabla_x L(x^*,\lambda^*) = 0.$$

For the MBF's Hessian, we have

3°
$$\nabla^2_{xx} F(x^*,\lambda^*,k) = \nabla^2_{xx} L(x^*,\lambda^*) + k \nabla c^{\mathrm{T}}_{(r)}(x^*) \Lambda_r^* \nabla c_{(r)}(x^*) ,$$

where $\Lambda_r^* = [\mathrm{diag}(\lambda_i)]_{i=1}^r$.

Under the second-order sufficient optimality condition from Debreu's lemma with $A = \nabla^2_{xx} L(x^*,\lambda^*)$ and $C = \Lambda_r^{*\frac{1}{2}} \nabla c_{(r)}(x^*)$ follows existence $\mu > 0$ that

4°
$$\langle \nabla^2_{xx} F(x^*,\lambda^*,k) y, y \rangle \geq \mu \langle y, y \rangle \quad \forall\, y \in \mathbb{R}^n$$

holds for $\forall k \geq k_0$ when $k_0 > 0$ large enough.

Moreover

5°
$$F(x^*,\lambda^*,k) = \min\{F(x,\lambda^*,k) | x \in \mathbb{R}^n\}.$$

We would like to point out that property 5° is not true even for a convex programming problem under the second-order sufficient optimality conditions if the classical Lagrangian $L(x,\lambda)$ is used instead of MBF, whereas 5° holds for MBF even if the primal problem (P) is non-convex, but sufficient second-order optimality condition is satisfied and $k > 0$ is large enough.

The Lagrange multipliers and the specific role of the penalty parameter together with the extension of the feasible set give rise to properties 1°–5° and allow to obtain for CO problems convergence results under minimum assumption on the input data and any positive scaling parameter $k > 0$.

Exercise 8.1. Construct MBF $C(x,\lambda,k)$, which corresponds to hyperbolic barrier $C(x,k) = f(x) + \sum_{i=1}^m c_i^{-1}(x)$ and show properties 1°–5°.

8.1.2 Convergence of the Logarithmic MBF Method

For a given $k > 0$ and $\lambda_0 = e = (1,\dots,1)^T \in \mathbb{R}^m_{++}$, MBF method generates the following primal–dual sequence $\{x_s, \lambda_s\}_{s\in N}$:

$$x_{s+1} : \nabla_x F(x_{s+1}, \lambda_s, k) =$$

$$\nabla f(x_{s+1}) - \sum_{i=1}^{m} \lambda_{i,s}(kc_i(x_{s+1})+1)^{-1}\nabla c_i(x_{s+1}) = \nabla_x L(x_{s+1}, \lambda_{s+1}) = 0 \qquad (8.2)$$

$$\lambda_{s+1} : \lambda_{i,s+1} = \lambda_{i,s}(kc(x_{s+1})+1)^{-1},\ i = 1,\dots,m. \qquad (8.3)$$

Exercise 8.2. Describe multiplicative method type (8.2)–(8.3) based on the hyperbolic MBF $C(x,\lambda,k)$.

From Slater condition follows boundedness of dual optimal set $L^* = \{\lambda \in \mathbb{R}^m_+ : d(\lambda) = d(\lambda^*)\}$. Therefore from concavity d follows boundedness of the dual level set

$$\Lambda_0 = \{\lambda \in \mathbb{R}^m_+ : d(\lambda) \geq d(\lambda_0)\}.$$

Along with the dual sequence $\{\lambda_s\}_{s\in N}$, we consider the corresponding convex and bounded dual level sets $\Lambda_s = \{\lambda \in \mathbb{R}^m_+ : d(\lambda) \geq d(\lambda_s)\}$ and their boundaries $\partial\Lambda_s = \{\lambda \in \Lambda_s : d(\lambda) = d(\lambda_s)\}$.

Theorem 8.1. *If f and all - c_i, $i = 1,\dots,m$ are convex and continuously differentiable functions, then under Assumptions A and B, any $k > 0$ and any $\lambda_0 \in \mathbb{R}^m_{++}$, the MBF method (8.2)–(8.3) generates such primal–dual sequence $\{x_s, \lambda_s\}_{s\in N}$ that*

(1) $d(\lambda_{s+1}) > d(\lambda_s)$, $s \geq 0$;
(2) $\lim_{s\to\infty} d(\lambda_s) = d(\lambda^*)$, $\lim_{s\to\infty} f(x_s) = f(x^*)$;
(3) for the Hausdorff distance $d_H(\partial\Lambda_s, L^*)$, we have

$$\lim_{s\to\infty} d_H(\partial\Lambda_s, L^*) = 0;$$

(4) there exists a subsequence $\{s_l\}_{l\in N}$ such that for $\bar{x}_l = \sum_{s=s_l}^{s_{l+1}} (s_{l+1}-s_l)^{-1} x_s$, we have $\lim_{l\to\infty} \bar{x}_l = \bar{x} \in X^*$, that is, the primal sequence converges to the primal solution in the ergodic sense.

Proof.

(1) From Theorem 7.1 follows that MBF method (8.2)–(8.3) is well defined and it is equivalent to the following proximal point method:

$$d(\lambda_{s+1}) - k^{-1}\sum_{i=1}^{m} \lambda_{i,s}\varphi(\lambda_{i,s+1}/\lambda_{i,s}) =$$

$$\max\{d(u) - k^{-1}\sum_{i=1}^{m}\lambda_{i,s}\varphi(u_i/\lambda_{i,s}) : u \in \mathbb{R}^m_{++}\}, \tag{8.4}$$

with the Kullback–Leibler (KL) φ-divergence entropy-like distance function

$$D(\lambda, u) = \sum_{i=1}^{m}\lambda_i\varphi(u_i/\lambda_i) = \sum_{i=1}^{m}[-\lambda_i \ln u_i/\lambda_i + u_i - \lambda_i],$$

which measures the divergence between two vectors λ and u from $\mathbb{R}^m_{++}$. Also MBF kernel $\varphi = -\psi^* = -\inf_{t>-1}\{st - \ln(t+1)\} = -\ln s + s - 1$ is a SC function, while ψ is a SC barrier.
The MBF kernel $\varphi(s) = -\ln s + s - 1$ is strictly convex on $\mathbb{R}_{++}$ and $\varphi'(1) = 0$; therefore $\min_{s>0}\varphi(s) = \varphi(1) = 0$. For the KL distance we have

(a) $D(\lambda, u) > 0$, $\forall \lambda \neq u \in \mathbb{R}^m_{++}$
(b) $D(\lambda, u) = 0 \Leftrightarrow \lambda = u$.

From (8.4) for $u = \lambda_s$ follows

$$d(\lambda_{s+1}) \geq d(\lambda_s) + k^{-1}\sum_{i=1}^{m}\lambda_{i,s}\varphi(\lambda_{i,s+1}/\lambda_{i,s}). \tag{8.5}$$

Therefore the dual sequence $\{d(\lambda_s)\}_{s\in N}$ is monotone increasing in value, unless $\varphi(\lambda_{i,s+1}/\lambda_{i,s}) = 0$ for all $i = 1,\ldots,m$, then $\lambda_{i,s+1} = \lambda_{i,s} > 0$ and from (8.3) follows $c_i(x_{s+1}) = 0$, $i = 1,\ldots,m$. It means for (x_{s+1}, λ_{s+1}) the KKT conditions are satisfied, so $x_{s+1} = x^*$, $\lambda_{s+1} = \lambda^*$.
Monotone increasing sequence $\{d(\lambda_s)\}_{s\in N}$ is bounded from above by $f(x^*)$; therefore there exists $\lim_{s\to\infty} d(\lambda_s) = \bar{d} \leq f(x^*)$.

(2) Our next step is to show that $\bar{d} = f(x^*)$.
From $-c(x_{s+1}) \in \partial d(\lambda_{s+1})$ and concavity of the dual function d follows

$$d(\lambda) - d(\lambda_{s+1}) \leq \langle -c(x_{s+1}), \lambda - \lambda_{s+1}\rangle, \ \forall \lambda \in \mathbb{R}^m_{++}.$$

For $\lambda = \lambda_s$ we have

$$d(\lambda_{s+1}) - d(\lambda_s) \geq \langle c(x_{s+1}), \lambda_s - \lambda_{s+1}\rangle. \tag{8.6}$$

From the update formula (8.3) follows

$$(\lambda_{i,s} - \lambda_{i,s+1}) = kc_i(x_{s+1})\lambda_{i,s+1}, \ i = 1,\ldots,m; \tag{8.7}$$

therefore from (8.6) and (8.7) we obtain

$$d(\lambda_{s+1}) - d(\lambda_s) \geq k\sum_{i=1}^{m}c_i^2(x_{s+1})\lambda_{i,s+1}. \tag{8.8}$$

It follows from the dual monotonicity (8.5) that the dual sequence $\{\lambda_s\}_{s\in N} \subset \Lambda_0$ is bounded. Therefore, there exists $L > 0 : \max_{i,s}\lambda_{i,s} = L$.

From (8.8) follows

$$d(\lambda_{s+1})-d(\lambda_s)\geq kL^{-1}\sum_{i=1}^{m}(c_i(x_{s+1})\lambda_{i,s+1})^2. \tag{8.9}$$

By summing up (8.9) from $s=1$ to $s=N$, we obtain

$$d(\lambda^*)-d(\lambda_0)\geq d(\lambda_{N+1})-d(\lambda_0)>kL^{-1}\sum_{s=1}^{N}\sum_{i=1}^{m}(c_i(x_{s+1})\lambda_{i,s+1})^2,$$

which leads to asymptotic complementarity

$$\lim_{s\to\infty}\langle\lambda_s,c(x_s)\rangle=0. \tag{8.10}$$

Summing up (8.5) from $s=1$ to $s=N$, we obtain

$$d(\lambda^*)-d(\lambda_0)\geq d(\lambda_N)-d(\lambda_0)\geq k^{-1}\sum_{s=1}^{N}D(\lambda_s,\lambda_{s+1}). \tag{8.11}$$

Therefore, $\lim_{s\to\infty}D(\lambda_s,\lambda_{s+1})=0$, which means that divergence (entropy) between two sequential LM vectors asymptotically disappears; therefore the map $\lambda\to\hat{\lambda}(\lambda,k)$ has a fixed point, which due to Remark 7.2 is λ^*.
We need, however, few more steps to prove it. Let us show that

$$D(\lambda^*,\lambda_s)>D(\lambda^*,\lambda_{s+1}),\ \forall s\geq 0 \tag{8.12}$$

unless $\lambda_s=\lambda_{s+1}$, then, as we saw already, $\lambda_s=\lambda^*$.
We assume $x\ln x=0$ for $x=0$, then

$$D(\lambda^*,\lambda_s)-D(\lambda^*,\lambda_{s+1})=\sum_{i=1}^{m}\left(\lambda_i^*\ln\frac{\lambda_{i,s+1}}{\lambda_{i,s}}+\lambda_{i,s}-\lambda_{i,s+1}\right).$$

Invoking the update formula (8.3), we obtain

$$D(\lambda^*,\lambda_s)-D(\lambda^*,\lambda_{s+1})=\sum_{i=1}^{m}\lambda_i^*\ln(kc_i(x_{s+1})+1)^{-1}+k\sum_{i=1}^{m}\lambda_{i,s+1}c_i(x_{s+1}).$$

Keeping in mind $\ln(1+t)^{-1}=-\ln(1+t)\geq -t$ and (8.7), we have

$$D(\lambda^*,\lambda_s)-D(\lambda^*,\lambda_{s+1})\geq k\sum_{i=1}^{m}(\lambda_{i,s+1}-\lambda_i^*)c_i(x_{s+1})=$$

$$k\langle -c(x_{s+1}),\lambda^*-\lambda_{s+1}\rangle. \tag{8.13}$$

From concavity d and $-c(x_{s+1})\in\partial d(\lambda_{s+1})$ follows

$$0\leq d(\lambda^*)-d(\lambda_{s+1})\leq\langle -c(x_{s+1}),\lambda^*-\lambda_{s+1}\rangle. \tag{8.14}$$

Combining (8.13) and (8.14), we obtain

$$D(\lambda^*,\lambda_s)-D(\lambda^*,\lambda_{s+1})\geq k(d(\lambda^*)-d(\lambda_{s+1}))>0. \tag{8.15}$$

Let $d(\lambda^*)-\bar{d}=\rho>0$, and then summing up the last inequality from $s=0$ to $s=N$, we obtain

$$D(\lambda^*,\lambda_0)-D(\lambda^*,\lambda_{N+1})=\sum_{i=1}^{m}(\lambda_i^*\ln\lambda_{N+1}+\lambda_{i,0}-\lambda_{i,N+1})\geq kN\rho,$$

which is impossible for $N>0$ large enough.
Therefore, $\lim_{s\to\infty}d(\lambda_s)=\bar{d}=d(\lambda^*)$, which together with asymptotic complementarity (8.10) leads to

$$d(\lambda^*)=\lim_{s\to\infty}d(\lambda_s)=\lim_{s\to\infty}\left[f(x_s)-\langle\lambda_s,c(x_s)\rangle\right]=$$

$$\lim_{s\to\infty}f(x_s)=f(x^*). \tag{8.16}$$

(3) The bounded sequence $\{\lambda_s\}_{s\in N}$ has a converging subsequence $\{\lambda_{s_l}\}_{l\in N}$: $\lim_{l\to\infty}\lambda_{s_l}=\bar{\lambda}$. From dual convergence in value follows $\bar{\lambda}=\lambda^*\in L^*$.
From (8.5) follows $L^*\subset\ldots\subset\Lambda_{s+1}\subset\Lambda_s\subset\ldots\subset\Lambda_0$; therefore from (8.9) we obtain a monotone decreasing sequence $\{d_H(\partial\Lambda_s,L^*)\}_{s\in N}$, which has a limit

$$\lim_{s\to\infty}d_H(\partial\Lambda_s,L^*)=\tau\geq 0,$$

but $\tau>0$ is impossible due to the continuity of the dual function d and the convergence of the dual sequence in value; therefore (3) holds.

(4) The primal sequence $\{x_s\}_{s\in N}\subset\Omega_k$, is bounded. Therefore, there exists $\{x_{s_l}\}_{l\in N}$ $\subset\{x_s\}_{s\in N}$ that $\lim_{l\to\infty}x_{s_l}=\bar{x}$. Let us consider subset $I_+=\{i:\bar{\lambda}_i>0\}$ of indices, then from (8.10) follows $\lim_{s_l\to\infty}c_i(x_{s_l})=c_i(\bar{x})=0,\ \ i\in I_+$. Now we consider subset $I_0=\{i:\bar{\lambda}_i=0\}$. Without loss of generality, we can assume that $\lim_{l\to\infty}\lambda_{s_l}=\bar{\lambda}_i=0$, $i\in I_0$ and $\lambda_{i,s_{l+1}}\leq 0.5\lambda_{i,s_l}$, $i\in I_0$.
Using the update formula (8.3), we obtain

$$\lambda_{s_{l+1}}\prod_{s=s_l}^{s_{l+1}}(kc_i(x_s)+1)=\lambda_{i,s_l}\geq 2\lambda_{s_{l+1}},\ i\in I_0.$$

Invoking arithmetic–geometric means inequality, we have

$$\frac{1}{s_{l+1}-s_l}\sum_{s=s_l}^{s_{l+1}}(kc_i(x_s)+1)\geq\left(\prod_{s=s_l+1}^{s_{l+1}}(kc_i(x_s)+1)\right)^{1/(s_{l+1}-s_l)}\geq 2^{(1/s_{l+1}-s_l)}>1.$$

Therefore

$$\frac{k}{(s_{l+1}-s_l)}\sum_{s=s_l}^{s_{l+1}}c_i(x_s)>0\ i\in I_0.$$

From concavity c_i we obtain

$$c_i(\bar{x}_{l+1}) = c_i\left(\sum_{s=s_l+1}^{s_{l+1}} \frac{1}{s_{l+1}-s_l} x_s\right) \geq \frac{1}{s_{l+1}-s_l} \sum_{s=s_l+1}^{s_{l+1}} c_i(x_s) > 0,\ i \in I_0. \quad (8.17)$$

On the other hand, from convexity of f, we have

$$f(\bar{x}_{l+1}) \leq \frac{1}{s_{l+1}-s_l} \sum_{s=s_l+1}^{s_{l+1}} f(x_s). \quad (8.18)$$

From (8.17) follows $\lim_{l\to\infty} \bar{x}_l = \bar{x} \in \Omega$. From (8.16) follows

$$f(\bar{x}) = \lim_{l\to\infty} f(\bar{x}_l) \leq \lim_{s\to\infty} f(x_s) = \lim_{s\to\infty} d(\lambda_s) = d(\lambda^*) = f(x^*).$$

Thus, $f(\bar{x}) = f(x^*) = d(\lambda^*) = d(\bar{\lambda})$ and $\bar{x} = x^*$, $\bar{\lambda} = \lambda^*$. □

We conclude the section with few remarks.

Remark 8.1. The MBF method leads to the multiplicative method (8.3) for the dual problem. If the dual function d is smooth, then $\nabla d(\lambda_{s+1}) = -c(x_{s+1})$. Formula (8.3) can be rewritten as follows:

$$\lambda_{i,s+1} - \lambda_{i,s} = k\lambda_{i,s+1}[\nabla d(\lambda_{s+1})]_i,\ i = 1,\ldots,m, \quad (8.19)$$

which is, in fact, implicit Euler method for the following system of ordinary differential equations:

$$\frac{d\lambda}{dt} = k\lambda\nabla d(\lambda),\ \lambda(0) = \lambda_0. \quad (8.20)$$

Formula (8.19) is called by Eggermont (1990) implicit multiplicative algorithm.

The explicit multiplicative algorithm is given by the following formula:

$$\lambda_{i,s+1} - \lambda_{i,s} = k\lambda_{i,s}[\nabla d(\lambda_s)]_i,\ i = 1,\ldots,m. \quad (8.21)$$

It was used by P. Eggermont for solving nonnegative least square, by Daube-Witherspoon and Muehllehner (1986) for image space reconstruction (ISRA), and by Shepp and Vardi (1982) in their EM method for finding maximum likelihood reconstruction in emission tomography.

Remark 8.2. We conclude the section by considering the truncated logarithmic MBF transformation

$$\psi(t) = \begin{cases} \ln(t+1), & \text{for } \infty > t \geq \tau \\ at^2 + bt + c, & \text{for } -\infty < t \leq \tau, \end{cases}$$

where $-1 < \tau < 0$. The coefficients a, b, and c can be found from the following system:

$$a\tau^2 + b\tau + c = \ln(\tau + 1)$$

$$2a\tau + b = (\tau+1)^{-1}$$
$$2a = -(\tau+1)^{-2}.$$

For $\tau = -0.5$ we obtain the following truncated MBF transformation:

$$\psi(t) = \begin{cases} \ln(t+1), & \text{for } \infty > t \geq -0.5 \\ -2t^2 + 0.5 - \ln 2, & \text{for } -\infty < t \leq -0.5. \end{cases} \tag{8.22}$$

The NR method with truncated MBF transformation (8.22) is proven to be very efficient in a number of real-life applications, which we describe in Chapter 11. It is used in well-known constrained optimization solver PENNON.

8.1.3 Convergence Rate

The MBF convergence rate under the second-order sufficient optimality condition (4.73)–(4.74) is a consequence of Theorem 7.2 except for item (4) of the following Theorem:

Theorem 8.2. *If f, $c_i \in C^2$, $i = 1, \ldots, m$ and the second-order sufficient optimality condition* (4.73)–(4.74) *holds, then for* $\forall(\lambda, k) \in \Lambda(\lambda^*, k_0, \delta)$

(1) there exists

$$\hat{x} \equiv \hat{x}(\lambda, k) : \nabla_x F(\hat{x}, \lambda, k) = 0 \tag{8.23}$$

and

$$\hat{\lambda} \equiv \hat{\lambda}(\lambda, k) : \hat{\lambda}_i = \lambda_i (k c_i(\hat{x}) + 1)^{-1},\ i = 1, \ldots, m; \tag{8.24}$$

(2) the following bound

$$\max\{\|\hat{x} - x^*\|, \|\hat{\lambda} - \lambda^*\|\} \leq c/k \|\lambda - \lambda^*\| \tag{8.25}$$

holds and constant $c > 0$ is independent on $k \geq k_0$;
(3) MBF F is strongly convex in $B(\hat{x}, \varepsilon) = \{x \in \mathbb{R}^n : \|x - \hat{x}\| \leq \varepsilon\}$ for $\varepsilon > 0$ small enough;
(4) under assumption (8.1) the following holds

$$d_k(\lambda) = F(\hat{x}, \lambda, k) = \min\{F(x, \lambda, k) | x \in \mathbb{R}^n\},$$

and

$$d_k(\lambda^*) = \max\{d_k(\lambda) | \lambda \in \mathbb{R}^m_+\} = F(x^*, \lambda^*, k) = f(x^*),\ \forall k \geq k_0.$$

Proof. The items (1)–(3) follow from the correspondent results of Theorem 7.2.

Let us consider the item (4). From (8.23) and strong convexity F in $B(\hat{x}, \varepsilon)$ follows

$$F(\hat{x}, \lambda, k) \leq F(x, \lambda, k),\ \forall x \in B(\hat{x}, \varepsilon).$$

From the complementarity condition

$$\lambda_i^* c_i(x^*) = 0, \quad i = 1, \dots, m$$

follows

$$F(\hat{x}, \lambda, k) \le F(x^*, \lambda, k) = f(x^*) - k^{-1} \sum_{i=1}^{m} \lambda_i \ln(kc_i(x^*) + 1) \le$$

$$f(x^*) - k^{-1} \sum_{i=1}^{r} \lambda_i^* \ln(kc_i(x^*) + 1) = f(x^*), \ \forall \lambda \in \mathbb{R}^m_+. \tag{8.26}$$

If there is $\tilde{x} \in \Omega_k = \{x \in \mathbb{R}^n : c_i(x) \ge -k^{-1}, \ i = 1, \dots, m\}$ and a number $\rho > 0$ such that

$$F(\tilde{x}, \lambda, k) \le F(\hat{x}, \lambda, k) - \rho,$$

then from (8.26) follows

$$F(\tilde{x}, \lambda, k) \le f(x^*) - \rho.$$

Let $I_+(\tilde{x}) = \{i : c_i(\tilde{x}) > 0\}$, then from the last inequality follows

$$f(\tilde{x}) \le f(x^*) + k^{-1} \sum_{i \in I_+(\tilde{x})} \lambda_i \ln(kc_i(\tilde{x}) + 1) - \rho. \tag{8.27}$$

For $k_0 > 0$ large enough from (8.27), assumption (8.1) and $(\lambda, k) \in \Lambda(\lambda^*, k_0, \delta)$ follow

$$f(\tilde{x}) \le f(x^*) - 0.5\rho. \tag{8.28}$$

On other hand, we have $f(\tilde{x}) \ge \min\{f(x) | x \in \Omega_k\}$.

Under the second-order optimality condition (4.73)–(4.74) from the optimum perturbation Theorem 6 by Fiacco and McCormick (1990) follows

$$f(\tilde{x}) \ge f(x^*) - k^{-1} \sum_{i=1}^{r} \lambda_i^*;$$

therefore for any $k \ge k_0$ we have

$$f(\tilde{x}) \ge f(x^*) - 0.25\rho.$$

The contradiction with (8.28) leads to

$$F(\hat{x}, \lambda, k) = \min\{F(x, \lambda, k) | x \in \Omega_k\} \tag{8.29}$$

for any $(\lambda, k) \in \Lambda(\lambda^*, k_0, \delta)$.

From the definition of F and (8.29) follows

$$d_k(\lambda) = F(\hat{x}, \lambda, k) = \min\{F(x, \lambda, k) | x \in \mathbb{R}^n\}$$

and

$$d_k(\lambda^*) = F(x^*, \lambda^*, k) = \min\{F(x, \lambda^*, k)|x \in \mathbb{R}^n\} = f(x^*), \forall k \geq k_0.$$

□

Theorem 8.2 is invalid if instead of $F(x, \lambda, k)$ one uses the classical Lagrangian $L(x, \lambda)$ for the original problem.

Example 8.1. Let us consider the following example:

$$x^* = \text{argmin}\{x_1^2 - x_2^2 | f_i(x) = 2 - x_2 \geq 0, f_2(x) = x_2 \geq 0\} = (0, 2).$$

The correspondent classical Lagrangian $L(x, \lambda) = x_1^2 - x_2^2 - \lambda_1(2 - x_2) - \lambda_2 x_2$. Then $\lambda_1^* = 4, \lambda_2^* = 0, f_{(r)}(x) = f_1(x), L''_{xx}(x^*, \lambda^*) = \begin{pmatrix} 2 & 0 \\ 0 & -2 \end{pmatrix}, f'_{(r)}(x^*) = f'_1(x^*) = \begin{pmatrix} 0 \\ -1 \end{pmatrix}$ and $f'_{(r)}(x^*)y = 0 \Rightarrow y = \begin{pmatrix} y_1 \\ 0 \end{pmatrix}$, so $(L''_{xx}(x^*, \lambda^*)y, y) = 2y_1^2 \quad \forall y\colon f'_{(r)}y = 0$, that is, the second-order optimality condition is satisfied, but $\inf\{L(x, \lambda^*)|x \in \mathbb{R}^2\} = \inf\{x_1^2 - x_2^2 + 4x_2 - 8|x \in \mathbb{R}^2\} = -\infty$ and, moreover, $\inf\{L(x, \lambda)|x \in \mathbb{R}^2\} = -\infty$ for any $\lambda = (\lambda_1, \lambda_2) > 0$. Now let us consider the problem with equivalent set of constraints:

$$x^* = \text{argmin}\{x_1^2 - x_2^2 | k^{-1}\ln(k(2 - x_2) + 1) \geq 0,\ k^{-1}\ln(kx_2 + 1) \geq 0\}$$

and Lagrangian for the equivalent problem

$$F(x, \lambda, k) = x_1^2 - x_2^2 - k^{-1}\lambda_1 \ln(k(2 - x_2) + 1) - k^{-1}\lambda_2 \ln(kx_2 + 1).$$

Then

$$\begin{aligned} F''_{xx}(x^*, \lambda^*, k) &= L''_{xx}(x^*, \lambda^*) - \frac{4}{k}(\ln(k(2 - x_2^*) + 1))''_{xx} \\ &= \begin{pmatrix} 2 & 0 \\ 0 & -2 \end{pmatrix} - \frac{4}{k}\begin{pmatrix} 0 & 0 \\ 0 & -k^2 \end{pmatrix} = \begin{pmatrix} 2 & 0 \\ 0 & 4k - 2 \end{pmatrix}. \end{aligned}$$

So $\nabla_{xx}^2 F(x^*, \lambda^*, k)$ is positive definite for any $k > 0.5$ and $x^* = (0, 2) = \text{argmin}\{F(x, \lambda^*, k)|x \in \mathbb{R}^2\}$.

8.1.4 MBF and Duality Issues

The MBF $F(x, \lambda, k)$ is a LEP; therefore it preserves all properties of classical Lagrangians for the original problem in the convex case.

On the top of it, the MBF-based duality has some extra important facts, which we discuss in this section.

Assertion 8.1 *If f and all $-c_i$, $i = 1, \ldots, m$ are convex and Slater's condition holds, then $x^* \in \Omega$ is a solution of problem (P) for any $k > 0$ if and only if:*

(1) There exists a vector $\lambda^* \geq 0$ such that

$$(a)\,\lambda_i^* c_i(x^*) = 0,\ i = 1,\ldots,m; \quad (b)\,F(x,\lambda^*,k) \geq F(x^*,\lambda^*,k)\,\forall x \in \mathbb{R}^n. \quad (8.30)$$

(2) *The pair* (x^*,λ^*) *is a saddle point of the Lagrangian for the equivalent problem, that is*

$$F(x,\lambda^*,k) \geq F(x^*,\lambda^*,k) \geq F(x^*,\lambda,k),\ \forall x \in \mathbb{R}^n,\ \forall \lambda \in \mathbb{R}^m_+. \tag{8.31}$$

Let $\psi_k(x) = \sup_{\lambda \geq 0} F(x,\lambda,k)$, then

$$\psi_k(x) = \begin{cases} f(x), & \text{if } c_i(x) \geq 0,\ i = 1,\ldots,m \\ \infty, & \text{otherwise,} \end{cases}$$

and the problem (P) is equivalent to finding

$$\psi_k(x^*) = \min\{\psi_k(x) \mid x \in \mathbb{R}^n\}. \tag{8.32}$$

Let $d_k(\lambda) = \inf_{x \in \mathbb{R}^n} F(x,\lambda,k)$, then the dual problem consists of finding

$$d_k(\lambda^*) = \max\{d_k(\lambda) \mid \lambda \geq 0\}. \tag{8.33}$$

From the definitions of $\psi_k(x)$ and $d_k(\lambda)$ follows weak duality

$$f(x) = \psi_k(x) \geq d_k(\lambda), \quad \forall x \in \Omega, \quad \forall \lambda \in \mathbb{R}^m_+.$$

Therefore, if $\bar{x}$ and $\bar{\lambda}$ are primal and dual feasible solutions and $\psi_k(\bar{x}) = d_k(\bar{\lambda})$, then $\bar{x} = x^*$ and $\bar{\lambda} = \lambda^*$. The smoothness of the dual function $d_k(\lambda)$ depends on the convexity and smoothness of f and c_i, $i = 1,\ldots,m$.

If (P) is a convex programming problem, then for $\lambda \in \mathbb{R}^m_+$ and $k > 0$, the dual function d_k is as smooth as f and c_i, $i = 1,\ldots,m$ if, for example, f is strongly convex.

If f and $-c_i$, $i = 1,\ldots,r$, are non-convex but smooth enough, then the following lemma takes place.

Lemma 8.1. *Let* $f, c_i \in C^2$, $i = 1,\ldots,m$, *and the second-order sufficient optimality condition* (4.73)–(4.74) *holds, then for any fixed* $k \geq k_0$, *the concave function* d_k *is twice continuous differentiable in* $\lambda \in \Lambda_k$.

Proof. First of all, for any $k > 0$ the dual d_k is a concave whether f and $-c_i$, $i = 1,\ldots,m$ are convex or not. By Theorem 8.2 item (3), MBF $F(x,\lambda,k)$ is strongly convex in a neighborhood of $\hat{x} = \hat{x}(\lambda,k)$, $\forall (\lambda,k) \in \Lambda(\lambda^*,k_0,\delta)$. Therefore $\hat{x}(\lambda,k) = \hat{x}(\cdot)$ is a unique minimizer of $F(x,\lambda,k)$ in x. From smoothness of f and c_i, $i = 1,\ldots,m$ and item (4) of Theorem 8.2, the dual $d_k(\lambda) = F(\hat{x}(\lambda,k);\lambda,k) = F(\hat{x}(\cdot),\cdot) = \min\{F(x,\lambda,k) | \lambda \in \mathbb{R}^n\}$ is smooth in $\lambda \in \Lambda_k$; therefore there exists

$$\nabla_\lambda d_k(\lambda) = \nabla_x F(\hat{x}(\cdot);\cdot)\nabla_\lambda \hat{x}(\cdot) + \nabla_\lambda F(\hat{x}(\cdot);\cdot). \tag{8.34}$$

Since $\nabla_x F(\hat{x}(\cdot);\cdot) = 0$, we obtain

$$\nabla_\lambda d_k(\cdot) = \nabla_\lambda F(\hat{x}(\cdot);\cdot) = -k^{-1}(\ln(kc_1(\cdot)+1),\dots,\ln(kc_m(\cdot)+1)).$$

The matrix $\nabla^2_{xx}F(\hat{x}(\cdot),\cdot)$ is positive definite for $\forall\lambda \in \Lambda_k$; therefore system

$$\nabla_x F(x,\lambda,k) = 0$$

yields a unique vector function $\hat{x}(\lambda,k) = \hat{x}(\cdot)$ such that

$$\hat{x}(\lambda^*,k) = x^*.$$

From identity

$$\nabla_x F(\hat{x}(\cdot);\cdot) \equiv \nabla_x F(\hat{x}(\lambda,k);\lambda,k) \equiv 0,\ \ \forall\lambda \in \Lambda_k$$

follows

$$\nabla^2_{xx}F(\hat{x}(\cdot);\cdot)\nabla_\lambda\hat{x}(\cdot) + \nabla^2_{x\lambda}F(\hat{x}(\cdot);\cdot) = 0$$

or

$$\nabla_\lambda\hat{x}(\lambda,k) = \nabla_\lambda\hat{x}(\cdot) = -(\nabla^2_{xx}F(\hat{x}(\cdot),\cdot))^{-1}\cdot\nabla^2_{x\lambda}F(\hat{x}(\cdot),\cdot), \quad \forall\lambda \in \Lambda_k.$$

Therefore

$$\nabla^2_{\lambda\lambda}d_k(\cdot) = \nabla^2_{\lambda x}F_x(\hat{x}(\cdot);\cdot)\nabla_\lambda(\hat{x}(\cdot)) =$$
$$-\nabla^2_{\lambda x}F(\hat{x}(\cdot);\cdot)(\nabla^2_{xx}F(\hat{x}(\cdot);\cdot))^{-1}\nabla^2_{x\lambda}F(\hat{x}(\cdot);\cdot).$$

We set $\tilde{c}_k(\cdot) = \mathrm{diag}[(kc_i(\hat{x}(\cdot))+1)^{-1}]_{i=1}^m : \mathbb{R}^m \to \mathbb{R}^m$ and $\tilde{c}_k = \tilde{c}_k(x^*) = \mathrm{diag}[(kc_i(x^*)+1)^{-1}]_{i=1}^m : \mathbb{R}^m \to \mathbb{R}^m$, then $\nabla^2_{\lambda x}F(\cdot) = -\tilde{c}_k(\cdot)^T\nabla c(\cdot) = -\tilde{c}_k(\cdot)\nabla c(\cdot)$, $\nabla^2_{x\lambda}F(\cdot) = -\nabla c^{\mathrm{T}}(\cdot)\tilde{c}_k(\cdot)$.

Therefore

$$\nabla^2_{\lambda\lambda}d_k(\cdot) = -\tilde{c}_k(\cdot)\nabla c(\cdot)(\nabla^2_{xx}F(\cdot))^{-1}\nabla c^{\mathrm{T}}(\cdot)\tilde{c}_k(\cdot)$$

and

$$\nabla^2_{\lambda\lambda}d_k(\lambda^*) = -\tilde{c}_k\nabla c(x^*)(\nabla^2_{xx}F(x^*,\lambda^*,k))^{-1}\nabla c^{\mathrm{T}}(x^*)\tilde{c}_k =$$
$$-\tilde{c}_k\nabla c(\nabla_{xx}F^{-1})\nabla c^{\mathrm{T}}\tilde{c}_k.$$

□

The dual problem, which is based on MBF, possesses all well-known properties typically for Lagrangian duality in the convex case.

Theorem 8.3 (Duality, Convex Case). *Let f be convex and all $-c_i$, $i = 1,\dots,m$ be concave and Slater condition holds, then*

(1) existence of the primal solution x^* implies existence of the dual solution λ^* and

$$f(x^*) = \psi_k(x^*) = d_k(\lambda^*) = d(\lambda^*), \quad \forall k > 0.$$

(2) if f is strongly convex and $f,\ c_i \in C^2$, $i=1,\ldots,m$, then existence of the dual solution implies existence of the primal solution and the optimal values of the objective functions are the same.

(3) if $f, c_i \in C^2$, and the second-order sufficient optimality condition (4.73)–(4.74) satisfied, then for every $k \geq k_0$, the dual solution exists and the second-order sufficient optimality conditions for the dual problem hold.

Proof.

(1) Let x^* be a primal solution. From Assertion 8.1 follows existence of a vector $\lambda^* \geq 0$, which satisfies (8.30b); therefore

$$\begin{aligned} d_k(\lambda^*) &= \min_{x\in\mathbb{R}^n} F(x,\lambda^*,k) = F(x^*,\lambda^*,k) = f(x^*) \geq F(x^*,\lambda,k) \\ &\geq \min_{x\in\mathbb{R}^n} F(x,\lambda,k) = d_k(\lambda), \quad \forall\lambda \geq 0, \end{aligned}$$

thus λ^* is the dual optimal solution, and due to complementarity condition (8.30a), we have $f(x^*) = d_k(\lambda^*)$ for any $k > 0$.

(2) The assumption (2) implies that $F(x,\lambda,k)$ is strongly convex in $x \in \Omega_k$ for every $\lambda \in \mathbb{R}^m_+$ and $k > 0$. Therefore, the minimizer $x(\lambda,k) = \operatorname{argmin}\{F(x,\lambda,k) \mid x \in \mathbb{R}^n\}$ is unique, and due to the smoothness of f and c_i, $i=1,\ldots,m$, gradient ∇d_k exists for any $k > 0$ and given by (8.34).
Let $\bar{\lambda} \in \mathbb{R}^m_+$ be a dual solution and $\bar{x} = x(\bar{\lambda},k)$, then the optimality conditions for the dual problem (8.33) are

$$\begin{aligned} \nabla_{\lambda_i} d_k(\bar{\lambda}) &= -k^{-1}\ln(kc_i(\bar{x})+1) \leq 0 \Leftarrow \bar{\lambda}_i = 0, \\ \nabla_{\lambda_i} d_k(\bar{\lambda}) &= -k^{-1}\ln(kc_i(\bar{x})+1) = 0 \Leftarrow \bar{\lambda}_i > 0. \end{aligned}$$

From $\bar{\lambda}_i > 0$ follows $c_i(\bar{x}) = 0$ and from $\bar{\lambda}_i = 0$ follows $c_i(\bar{x}) \geq 0, \quad i=1,\ldots,m$, that is, $\bar{x} \in \Omega$, and for the pair $(\bar{x},\bar{\lambda})$, the complementarity conditions

$$c_i(\bar{x}) \cdot \bar{\lambda}_i = 0,\ i=1,\ldots,m,$$

hold. Therefore

$$d_k(\bar{\lambda}) = f(\bar{x}) - k^{-1}\sum_{i=1}^m \bar{\lambda}_i \ln(kc_i(\bar{x})+1) = f(\bar{x}),$$

that is, for the primal–dual feasible pair $(\bar{x},\bar{\lambda})$ and any $k > 0$, we have

$$d_k(\bar{\lambda}) = f(\bar{x});$$

hence $\bar{x} = x^*$ and $\bar{\lambda} = \lambda^*$.

(3) Since the second-order sufficient optimality condition is satisfied, then from Theorem 8.2 follows that $F(x,\lambda,k)$ is strongly convex in x for any $\lambda \in \Lambda_k$ and $k \geq k_0$; therefore the first part of the statement can be proven as in item 2).

We now show that the sufficient second-order optimality conditions hold for the dual problem (8.33). It means the gradients of active constraints of the dual problem (8.33) are linearly independent, the correspondent Lagrange multipliers are positive, and condition type (4.73)–(4.74) for the dual problem (8.33) is satisfied.

We first note that gradients of the active dual constraints $\lambda_i \geq 0,\ i = r+1,\dots,m$, are linear independent

$$(\overbrace{0,\dots,0}^{r},0,\dots,\overset{i}{1},\dots,0) = e_i\,, \quad i = r+1,\dots,m\,,$$

that is, condition type (4.73) holds true for the dual problem (8.33).

Now we show that condition (4.74) is satisfied for the dual problem.

Let us consider the Lagrangian $L(\lambda,\rho,k) = d_k(\lambda) + \sum_{i=1}^m \rho_i\lambda_i$ for the dual problem (8.33), then $\nabla^2_{\lambda\lambda}L(\lambda,\rho,k) = \nabla^2_{\lambda\lambda}d_k(\lambda)$.

Let $\mathrm{v} \in \mathbb{R}^m$, then $\langle \mathrm{v}, e_i\rangle = 0 \Rightarrow \mathrm{v}_i = 0,\ i = r+1,\dots,m$. Therefore, any vector $\mathrm{v} \in \mathbb{R}^m : \langle \mathrm{v}, e_i\rangle = 0,\ i = r+1,\dots,m$ has the form $\mathrm{v} = (\mathrm{v}_1,\dots,\mathrm{v}_r,0,\dots,0)$. From Theorem 8.2 item (3) follows that for any fixed $k \geq k_0$ the matrix $\nabla^2_{xx}F = \nabla^2_{xx}F(x^*,\lambda^*,k)$ is positive definite. Let mineigenval $\nabla^2_{xx}F = m_k > 0$ and maxeigenval $\nabla^2_{xx}F = M_k > 0$, then

$$m_k^{-1}\langle y,y\rangle \geq \langle \nabla^2_{xx}F^{-1}y,y\rangle \geq M_k^{-1}\langle y,y\rangle, \forall y \in \mathbb{R}^n,$$

that is

$$-m_k^{-1}\langle y,y\rangle \leq -\langle \nabla^2_{xx}F^{-1}y,y\rangle \leq -M_k^{-1}\langle y,y\rangle.$$

So for $\nabla^2_{\lambda\lambda}L(\lambda^*,\rho^*,k)$ we obtain

$$\begin{aligned}\langle \nabla^2_{\lambda\lambda}L(\lambda^*,\rho^*,k)\mathrm{v},\mathrm{v}\rangle &= \langle \nabla^2_{\lambda\lambda}d_k(\lambda^*)\mathrm{v},\mathrm{v}\rangle = \langle \tilde{c}_k\nabla\tilde{c}(-(\nabla^2_{xx}F)^{-1})\nabla c^{\mathrm{T}}\tilde{c}_k\mathrm{v},\mathrm{v}\rangle \\ &= \langle -(\nabla^2_{xx}F)^{-1}\nabla c^{\mathrm{T}}\tilde{\mathrm{v}},\nabla c^{\mathrm{T}}\tilde{\mathrm{v}}\rangle \leq -M_k^{-1}\langle \nabla c^{\mathrm{T}}\tilde{\mathrm{v}},\nabla c^{\mathrm{T}}\tilde{\mathrm{v}}\rangle,\end{aligned}$$

where $\tilde{\mathrm{v}} = \tilde{c}_k\mathrm{v} = ((kc_1(x^*)+1)^{-1}\mathrm{v}_1,\dots,(kc_r(x^*)+1)^{-1}\mathrm{v}_r,0,\dots,0) = (\mathrm{v}_1,\dots,\mathrm{v}_r,0,\dots,0) = (\mathrm{v}_{(r)},0,\dots,0)$. Thus

$$\begin{aligned}\langle \nabla^2_{\lambda\lambda}L(\lambda^*,\rho^*,k)\mathrm{v},\mathrm{v}\rangle &\leq -M_k^{-1}\langle \nabla c^{\mathrm{T}}_{(r)}(x^*)\mathrm{v}_{(r)},\nabla c^{\mathrm{T}}_{(r)}(x^*)\mathrm{v}_{(r)})\rangle \\ &= -M_k^{-1}\langle \nabla c_{(r)}(x^*)\nabla c^{\mathrm{T}}_{(r)}(x^*)\mathrm{v}_{(r)},\mathrm{v}_{(r)}\rangle\,.\end{aligned}$$

It follows from (4.73) that the Gram matrix $\nabla c_{(r)}(x^*)\nabla c^{\mathrm{T}}_{(r)}(x^*)$ is nonsingular, so mineigenval $\nabla c_{(r)}(x^*)\nabla c^{\mathrm{T}}_{(r)}(x^*) = \mu_0 > 0$. Hence, for $\mu_k = M_k^{-1}\mu_0 > 0$ we obtain

$$\langle \nabla^2_{\lambda\lambda}L(\lambda^*,\rho^*,k)\mathrm{v},\mathrm{v}\rangle \leq -\mu_k\|\mathrm{v}\|^2\,, \quad \forall \mathrm{v} : \langle \mathrm{v},e_i\rangle = 0\,, \quad i = r+1,\dots,m\,,$$

which is condition type (4.74) for the dual problem.

Further, $\rho_i^* = -\nabla_{\lambda_i}d_k(\lambda^*) = k^{-1}(kc_i(x^*)+1)^{-1} > 0,\ i = r+1,\dots,m$ that together with the linear independence gradients $e_i,\ i = r+1,\dots,m$, for the active constraints $\lambda_i \geq 0,\ i = r+1,\dots,m$, comprises the second-order sufficient optimality condition for the dual problem. The proof is complete. □

Remark 8.3. All facts of items (1) and (2) fail to be true if the convexity of the functions f and $-c_i$, $i=1,\dots,m$ is abandoned; moreover item (3) is, generally speaking, invalid even for the convex programming problem if the dual problem is based on the classical Lagrangian $L(x,\lambda)$.

On the other hand, results of items (1)–(3) are valid for $k \geq k_0$ even for non-convex optimization problem if the duality is based on the logarithmic MBF F.

Theorem 8.4 (Duality, Non-convex Case). *Let $f, c_i \in C^2$, $i=1,\dots,m$, and second-order sufficient optimality condition (4.73)–(4.74) is satisfied, then under assumption (8.1) there exists $k_0 > 0$ such that for any $k \geq k_0$ the following hold:*

(1) existence of the primal solution guarantees existence of the dual solution and

$$f(x^*) = d_k(\lambda^*);$$

(2) the second-order sufficient optimality condition is satisfied for the dual problem;
(3) the pair (x^*, λ^*) is a solution of the primal and dual problems if and only if it is a saddle point of F, that is, (8.31) holds.

Proof.

(1) Let x^* be the primal solution. From item (3) of Theorem 8.2 follows that $F(x,\lambda^*,k)$ is strongly convex in a neighborhood of x^*, and from the complementarity condition follows

$$\begin{aligned}\nabla_x F(x,\lambda^*,k)\big|_{x=x^*} &= \nabla f(x^*) - \sum_{i=1}^m \lambda_i^*(kc_i(x^*)+1)^{-1}\nabla c_i(x^*)\\ &= \nabla f(x^*) - \sum_{i=1}^m \lambda_i^* \nabla c_i(x^*) = 0.\end{aligned}$$

From item (4) of Theorem 8.2 follows

$$d_k(\lambda^*) = \min\{F(x,\lambda^*,k) \mid x \in \mathbb{R}^n\} = F(x^*,\lambda^*,k)$$

and there exists

$$\begin{aligned}\nabla_\lambda d_k(\lambda^*) &= -k^{-1}(\ln(kc_1(x^*)+1),\dots,\ln(kc_r(x^*)+1),\\ &\qquad \ln(kc_{r+1}(x^*)+1),\dots,\ln(kc_m(x^*)+1))\\ &= (0,\dots,0;-k^{-1}\ln(kc_{r+1}(x^*)+1),\dots,-k^{-1}\ln(kc_m(x^*)+1)).\end{aligned}$$

The dual problem is always convex whether f and $-c_i$, $i=1,\dots,m$ are convex or not. For $i: \lambda_i^* > 0$, we have $\nabla_{\lambda_i} d_k(\lambda^*) = 0, i=1,\dots,r$ and for $i=r+1,\dots,m$, due to $c_i(x^*) \geq \sigma > 0$, we have $\nabla_{\lambda_i} d_k(\lambda^*) \leq -k^{-1}\ln(k\sigma+1) < 0$. Therefore, the optimality condition for λ^* is satisfied, i.e., λ^* is the dual solution and the complementarity condition

$$\lambda_i^* c_i(x^*) = 0 \quad i=1,\dots,m \tag{8.35}$$

holds; therefore $d_k(\lambda^*) = f(x^*)$.

(2) The dual problem is convex, and it was shown already that λ^* is the dual solution. Therefore, the second-order sufficient optimality conditions for the dual problem, which have been proven in Theorem 8.3, are valid here.

(3) We first show that if (x^*, λ^*) is the primal–dual solution, then the pair is a saddle point of the MBF F for $k \geq k_0$.
In fact, if λ^* is the dual solution, then from item (4) of Theorem 8.2 follows $F(x^*, \lambda^*, k) \leq F(x, \lambda^*, k)$, $\forall x \in \mathbb{R}^n$. Further, for any $\lambda \in \mathbb{R}^m_+$, keeping in mind, complementarity condition (8.35), we obtain

$$F(x^*, \lambda, k) = f(x^*) - \sum_{i=1}^{m} \lambda_i \ln(kc_i(x^*) + 1) \leq F(x^*, \lambda^*, k).$$

So the pair (x^*, λ^*) is the saddle point of F.
Finally, let $(\bar{x}, \bar{\lambda})$ be a saddle point of F

$$f(x) - k^{-1} \sum_{i=1}^{m} \bar{\lambda}_i \ln(kc_i(x) + 1) \geq f(\bar{x}) - k^{-1} \sum_{i=1}^{m} \bar{\lambda}_i \ln(kc_i(\bar{x}) + 1)$$

$$\geq f(\bar{x}) - k^{-1} \sum_{i=1}^{m} \lambda_i \ln(kc_i(\bar{x}) + 1) \tag{8.36}$$
$$\forall x \in \mathbb{R}^n,\ \lambda \in \mathbb{R}^m_+,$$

then we have to show that $\bar{x} = x^*$ and $\bar{\lambda} = \lambda^*$.
The right inequality in (8.36) yields

$$\sum_{i=1}^{m} \bar{\lambda}_i \ln(kc_i(\bar{x}) + 1) \leq \sum_{i=1}^{m} \lambda_i \ln(kc_i(\bar{x}) + 1) \quad \forall \lambda \in \mathbb{R}^m_+. \tag{8.37}$$

It follows from (8.37) that $c_i(\bar{x}) \geq 0$, $i = 1, \ldots, m$. In fact, if there is i_0 such that $c_{i_0}(\bar{x}) < 0$, then for $k > 0 : 0 < kc_{i_0}(\bar{x}) + 1 < 1$, we have $\ln(kc_{i_0}(\bar{x}) + 1) < 0$. Therefore, by setting $\lambda_i = \bar{\lambda}_i$, $i \neq i_0$, and taking $\lambda_{i_0} > 0$ large enough, we obtain inequality opposite to (8.37).
Therefore $\bar{x}$ is a primal feasible solution. Consequently, $\ln(kc_i(\bar{x}) + 1) \geq 0$, $i = 1, \ldots, m$, and

$$\sum_{i=1}^{m} \bar{\lambda}_i \ln(kc_i(\bar{x}) + 1) \geq 0.$$

Since (8.36) holds true for any $\lambda \geq 0$, we can set $\lambda = 0$. Then, from the right inequality (8.36), we obtain $\sum_{i=1}^{m} \bar{\lambda}_i \ln(kc_i(\bar{x}) + 1) \leq 0$. Therefore, $\sum_{i=1}^{m} \bar{\lambda}_i \ln(kc_i(\bar{x}) + 1) = 0$, which together with $c_i(\bar{x}) \geq 0, i = 1, \ldots, m$ implies complementarity condition

$$\bar{\lambda}_i f_i(\bar{x}) = 0, \qquad i = 1, \ldots, m.$$

The left inequality in (8.36) yields

$$f(x) \geq f(\bar{x}) + k^{-1} \sum_{i=1}^{m} \bar{\lambda}_i \ln(kc_i(x)+1).$$

For every $x \in \Omega$, we have $\ln(kc_i(x)+1) \geq 0,\ i = 1,\ldots,m$. Therefore, $f(x) \geq f(\bar{x})$, $\forall x \in \Omega$, i.e., $\bar{x}$ is the minimum of $f(x)$ on Ω, and $\bar{x} = x^*$.
From the left inequality (8.36), we have

$$d_k(\bar{\lambda}) = \min\{F(x,\bar{\lambda},k)|x \in \mathbb{R}^n\} = f(\bar{x}) - k^{-1} \sum \bar{\lambda}_i \ln(kc_i(\bar{x})+1).$$

From the complementarity condition follows

$$d_k(\bar{\lambda}) = f(\bar{x});$$

therefore $\bar{x} = x^*$ and $\bar{\lambda} = \lambda^*$. □

Corollary 8.1. *If the conditions of Theorem 8.4 fulfilled, then the restriction*

$$\bar{d}_k(\lambda) = d_k(\lambda)\big|_{\lambda_{r+1}=0,\ldots,\lambda_m=0}$$

of the dual function to the manifold of the active constraints of the dual problem is strongly concave. □

The convexity and smoothness properties of dual function can be used for finding nonzero components of λ^* by applying smooth optimization methods to $\bar{d}_k(\lambda)$.

Exercise 8.3. Consider duality theory for the modified hyperbolic barrier

$$C(x,\lambda,k) = \begin{cases} f(x) + k^{-1} \sum_{i=1}^{m} \lambda_i[(kc_i(x)+1)^{-1} - 1], & \text{if } x \in \operatorname{int}\Omega_k \\ \infty, & \text{if } x \bar{\in} \operatorname{int}\Omega_k. \end{cases}$$

In the following section, we consider exterior distance function (EDF), which is a NR realization, where both the objective function and the constraint are rescaled.

In fact, the EDF is an alternative to the classical interior distance function (IDF). The EDF is to IDF as MBF is to classical log-barrier function.

8.2 Exterior Distance Functions

8.2.0 Introduction

Interior distance functions (IDFs) and interior center methods (ICMs) were developed by P. Huard in mid-1960s .

Later IDFs and correspondent ICMs were incorporated into SUMT and studied by Fiacco and McCormick (1990), Grossman and Kaplan (1981), Mifflin (1976), Polak (1971), and Jarre et al. (1988), just to mention a few.

At each step ICM finds a central (in a sense) point of the relaxation feasible set (RFS) and updates the level set using the new objective function value. The RFS is the intersection of the feasible set and the relaxation (level) set of the objective function at the attained level.

The "center" is sought as a minimizer of the IDF. It is a point in the RFS "most distant" from both the boundary of the objective function level set and the active constraints.

Interest in IDFs and correspondent center methods has grown dramatically after N. Karmarkar published in 1984 his projective scaling method. In fact, his potential function is an IDF and his method is a center method.

Mainly for this reason, the concept of center became extremely popular in the 1980s. Centering and reducing the objective function are two basic ideas behind the IPMs.

Centering means to stay away from the boundary of the RFS. An answer to the basic question: how far from the boundary one should stay in case of LP was given by Sonnevend (1986) through the definition of analytic center of a polytope. The central path is a curve formed by analytic centers.

Following the central path, Renegar (1988) introduced the first path-following algorithm with $O(\sqrt{n}L)$ number of iterations, versus $O(nL)$ iterations for the N. Karmarkar's method.

Soon after Gonzaga (1989) and Vaidya (1990) developed algorithms for LP, based on the centering ideas, with overall complexity $O(n^3L)$ arithmetic operations, which is the best known result so far.

After Yu. Nesterov and A. Nemirovski developed their SC theory, it becomes evident that path-following methods with polynomial complexity for convex optimization problems are possible if the RFS can be equipped with SC barrier.

If it is not the case, then one can use the classical IDF and correspondent ICM.

The classical IDF, however, has few well-known drawbacks:

(1) the IDF, its gradient, and the Hessian do not exist at the primal solution;
(2) the IDF unboundedly grows, and at the same time the condition number of the IDF Hessian vanishes, when the primal approximation approaches the solution.

The singularity of the IDF at the solution leads to numerical instability, in particular in the final phase. It means that from some point on, finding an accurate approximation for the IDF's minimizer is practically an impossible task.

In spite of a long history of IDFs and correspondent ICMs, the fundamental question still is: **how the main idea of the center methods, to stay away from the boundary, consistent with the main purpose of constrained optimization, finding a solution on the boundary**.

The issue was partially addressed in the 1980s, when the modified interior distance functions was introduced and correspondent theory and methods were developed, see Polyak (1987) (see also Polyak (1997)). The results, however, were obtained under the second-order sufficient optimality condition. We will cover the results in the second part of this paragraph.

In the first part, we address the issue by introducing the exterior distance function (EDF) and correspondent exterior point method (EPM).

The EDF is a LEP, which one obtains by transforming both the objective function and the constraints.

Thus, EDF is a particular realization of the NR principle.

The basic convergence results are taking place under minimum assumptions on the input data.

In contrast to the classical IDF, the EDF, its gradient, and Hessian are defined on an extension of the feasible set.

The EDF has two extra tools, which control the computational process: the positive barrier (scaling) parameter and the LM vector.

At each step the EPM finds an approximation for the EDF's primal minimizer and uses it for the Lagrange multipliers update. Both the "center" and the barrier parameter can be fixed or updated from step to step.

Under a fixed "center," the EDF resembles the MBF, but for an equivalent problem. The "center" is an extra tool. By changing the "center" from step to step it is possible to strengthen MBF convergence results.

Convergence due to the Lagrange multipliers update keeps the condition number of the EDF's Hessian stable, which is critical for numerical stability. This is a fundamental departure from classical IDF theory.

Under the second-order sufficient optimality condition, EPM converges with Q-linear rate even when both the "center" and the scaling parameter are fixed, but the parameter is large enough. By changing the scaling parameter and/or the "center" from step to step, one gets Q-superlinear convergence rate.

Also under- second order sufficient optimality condition for a fixed, but large enough scaling parameter and any fixed interior point as a "center," the EDF is strongly convex in the neighborhood of the primal minimizer no matter the objective function and the constraints are convex or not.

It makes Newton's or regularized Newton method particularly efficient in the final phase of the computation.

In the second part of the paragraph, we consider the modified interior distance functions (MIDFs) and their local and global properties. The properties can be used for developing modified center method (MCM).

Let $y \in \text{int } \Omega$, then the relaxation feasible set (RFS)

$$\Omega(y) = \{x \in \Omega : f(x) < f(y)\}$$

is convex and bounded for any given $y \in \text{int } \Omega$. It follows from assumption A, convexity f, concavity c_i, $i = 1, \ldots, m$, and Theorem 2.7.

Also, without losing generality, we can assume that $f(x) \geq 0$, because, otherwise, we can replace $f(x)$ by an equivalent objective function $f(x) := \ln(e^{f(x)} + 1) \geq 0$.

8.2.1 Exterior Distance

For a given $y \in \text{int } \Omega$, let us consider the following problem:

$$F(y,x^*) = \min\{F(y,x)|c_i(x) \geq 0, i = 1,\ldots,m\}, \tag{8.38}$$

where

$$F(y,x) = -\ln \Delta(y,x) = -\ln(f(y) - f(x)).$$

For any fixed $y \in \text{int } \Omega$ function F is convex and monotone decreasing together with f on $\Omega(y) = \{x \in \Omega : f(x) \leq f(y)\}$; therefore solution $x^* \in \Omega(y)$ belongs to X^* and vice versa: any $x^* \in X^*$ solves (8.38). Therefore, the original problem (P) and (8.38) are equivalent. From now on we consider the problem (8.38) instead of (P).

The correspondent to (8.38) Lagrangian $L_y : \mathbb{R}^n \times \mathbb{R}^m_+ \to \mathbb{R}$ is

$$L_y(x,\lambda) = F(y,x) - \sum_{i=1}^m \lambda_i c_i(x). \tag{8.39}$$

The correspondent dual function $d_y : \mathbb{R}^m_+ \to \mathbb{R}$ is

$$d_y(\lambda) = \inf_{x \in \mathbb{R}} L_y(x,\lambda)$$

and

$$d_y(\lambda^*) = \max\{d_y(\lambda)|\lambda \in \mathbb{R}^m_+\} \tag{8.40}$$

is the dual problem.

Let $k > 0$ be fixed, then the feasible set

$$\Omega = \{x \in \mathbb{R}^n : \ c_i(x) \geq 0, \ i = 1,\ldots,m\}$$

$$= \{x \in \mathbb{R}^n : \ k^{-1}\ln(kc_i(x)+1) \geq 0, \ i = 1,\ldots,m\}.$$

Therefore, for any given $y \in \text{int } \Omega$ and $k > 0$, the problem

$$F(y,x^*) = \min\{F(y,x)|k^{-1}\ln(kc_i(x)+1) \geq 0, \ i = 1,..,m\} \tag{8.41}$$

is equivalent to (8.38).

For any fixed $y \in \text{int } \Omega$ and $k > 0$, the following extension

$$\Omega_k(y) = \{x \in \mathbb{R}^n : c_i(x) \geq -k^{-1}, i = 1,\ldots,m, \ f(y) > f(x)\} \tag{8.42}$$

of the feasible set is convex and bounded due to convexity f, concavity c_i, $i = 1,\ldots,m$, boundedness $\Omega(y)$, and Theorem 2.7. For $\gamma > 0$ small enough, let us consider the set

$$\Omega_\gamma(y) = \{x \in \mathbb{R} : c_i(x) \geq \gamma > 0, \ i = 1,..,m, \ \ f(y) > f(x)\}, \tag{8.43}$$

which is a contraction of $\Omega(y)$; therefore it is convex and bounded.

The set $\Omega_\gamma(y)$ for small $\gamma > 0$ is not empty due to the Slater condition.

Let us fix $y \in \text{int } \Omega$ for now, then LEP $\mathcal{L}_y : \mathbb{R}^n \times \mathbb{R}^m_+ \times \mathbb{R}_{++} \to \mathbb{R}$

$$\mathscr{L}_y(x,\lambda,k) = F(y,x) - k^{-1}\sum_{i=1}^{m}\lambda_i \ln(kc_i(x)+1); \tag{8.44}$$

we call the exterior distance function (EDF).

Thus, EDF is a particular realization of the NR principle, when both the objective function and the constraints are transformed.

First of all, LEP $\mathscr{L}_y$ is convex in $x \in \Omega_k(y)$ for any $k > 0$ and $\lambda \in \mathbb{R}^m_+$.

Let us consider the second-order sufficient optimality conditions for the problem (8.38).

There exists $\mu > 0$ such that

$$\langle \nabla^2_{xx} L_y(x^*,\lambda^*)u,u\rangle \geq \mu\langle u,u\rangle, \forall u : \nabla c_{(r)}(x^*)u = 0 \tag{8.45}$$

and

$$\operatorname{rank}\nabla c_{(r)}(x^*) = r. \tag{8.46}$$

We conclude the section by pointing out EDF properties at (x^*,λ^*).

Proposition 8.1. *For any $k > 0$ and any KKT's point (x^*,λ^*), we have*

1^0 $\mathscr{L}_y(x^*,\lambda^*,k) = F(y,x^*) = -\ln(f(y)-f(x^*))$

or

$$f(x^*) = f(y) - e^{-F(x^*,y)};$$

2^0 $\nabla_x\mathscr{L}_y(x^*,\lambda^*,k) = \Delta^{-1}(x^*,y)\nabla f(x^*) - \sum_{i=1}^m (kc_i(x^*)+1)^{-1}\lambda_i^*\nabla c_i(x^*) =$

$$\Delta^{-1}(x^*,y)\nabla f(x^*) - \sum_{i=1}^{m}\lambda_i^*\nabla c_i(x^*) = \nabla_x L_y(x^*,\lambda^*) = 0;$$

3^0 $\nabla^2_{xx}\mathscr{L}_y(x^*,\lambda^*,k) = \Delta^{-2}(x^*,y)\nabla f(x^*)\nabla f^T(x^*) + \Delta^{-1}(x^*,y)\nabla^2 f(x^*)$

$$-\sum_{i=1}^{m}\lambda_i^*\nabla^2 c_i(x^*) + k\nabla c(x^*)^T\Lambda^*\nabla c(x^*) = \nabla^2_{xx}L_y(x^*,\lambda^*) + k\nabla c(x^*)^T\Lambda^*\nabla c(x^*),$$

where $\Lambda^*_{(r)} = \operatorname{diag}(\lambda_i^*)_{i=1}^r$, *and* $\lambda_i^* = 0$, $i = r+1,\ldots,m$.

Properties 1^0–3^0 follow from formulas for $\mathscr{L}_y$, $\nabla_x\mathscr{L}_y$, $\nabla^2_{xx}\mathscr{L}_y$ and complementarity condition

$$\lambda_i^* c_i(x^*) = 0, i = 1,\ldots,m. \tag{8.47}$$

The EDF (8.44) is fundamentally different from Huard's IDF

$$H(y,x) = -m\ln(\tau - f(x)) - \sum_{i=1}^{m}\ln c_i(x).$$

First, neither EDF $\mathscr{L}_y$ nor $\nabla_x\mathscr{L}_y$ and $\nabla^2_{xx}\mathscr{L}_y$ are singular at the primal solution.

Second, from 2^0 follows that for any $k > 0$ the optimal solution of (8.38) can be found by solving one smooth unconstrained optimization problem

$$\min_{x \in \mathbb{R}^n} \mathscr{L}_y(x, \lambda^*, k) = \mathscr{L}_y(x^*, \lambda^*, k). \tag{8.48}$$

It means that $\mathscr{L}_y(x, \lambda^*, k)$ is an exact smooth approximation for the following non-smooth problem

$$\min_{x \in \mathbb{R}^n} \max\{F(y,x) - F(y,x^*), -c_i(x),\ i = 1, \ldots, m\}, \tag{8.49}$$

which is equivalent to (8.38).

Third, from 3^0 for any $u \in \mathbb{R}^n$ follows

$$\langle \nabla^2_{xx} \mathscr{L}_y(x^*, \lambda^*, k)u, u \rangle = \langle (\nabla^2_{xx} L_y(x^*, \lambda^*) + k \nabla c_{(r)}(x^*)^T \Lambda^*_{(r)} \nabla c_{(r)}(x^*))u, u \rangle.$$

Proposition 8.2. *Under the second-order sufficient optimality condition (8.45)–(8.46), for $k_0 > 0$ large enough, there exists $0 < \rho < \mu$ such that*

$$\langle \nabla^2_{xx} \mathscr{L}_y(x^*, \lambda^*, k)u, u \rangle \geq \rho \langle u, u \rangle,\ \forall u \in \mathbb{R}^n$$

holds for any $k \geq k_0$.

Proposition 8.2 follows from the second-order sufficient optimality condition (8.45)–(8.46) and Debreu's lemma with

$$A = \nabla^2_{xx} L_y(x^*, \lambda^*)\ ,\ C = \Lambda^{*\frac{1}{2}}_{(r)} \nabla c_{(r)}(x^*).$$

In other words, for any $k \geq k_0$, the EDF $\mathscr{L}_y(x, \lambda^*, k)$ is strongly convex in the neighborhood of x^* whether f and all $-c_i$, $i = 1, \ldots, m$ are convex or not.

The EDF relates to the classical Huard's IDF $H(x, \tau)$ as MBF to the classical R. Frisch's log-barrier function $F(x,k) = f(x) - k^{-1} \sum_{i=1}^m \ln c_i(x)$.

Moreover, on the top of MBF properties, the EDF has one extra tool – the "center" – which can be used for improving convergence and convergence rate.

The EDF properties lead to a new multipliers method, which under minimum assumptions on the input data converges to the optimal primal and dual solution in value for any fixed scaling parameter $k > 0$, just due to the Lagrange multipliers update. This is another fundamental departure from the classical IDF theory and methods.

8.2.2 Exterior Point Method: Convergence and Convergence Rate

The EPM at each step finds the primal minimizer (or its approximation) of $\mathscr{L}_y$ following by Lagrange multipliers updates.

We start with $y \in \text{int}\ \Omega$ as a given fixed "center," fixed scaling parameter $k > 0$, and initial Lagrange multipliers vector $\lambda_0 \in \mathbb{R}^m_{++}$. Let us assume that the primal–dual approximation (x_s, λ_s) has been found already.

We find the next approximation (x_{s+1}, λ_{s+1}) as follows:

$$x_{s+1} : \nabla_x \mathscr{L}_y(x_{s+1}, \lambda_s, k)$$

$$= \Delta^{-1}(x_{s+1}, y)\nabla f(x_{s+1}) - \sum_{i=1}^{m} \lambda_{i,s} \psi'(kc_i(x_{s+1}))\nabla c_i(x_{s+1}) = 0 \quad (8.50)$$

$$\lambda_{s+1} : \lambda_{i,s+1} = \lambda_{i,s}\psi'(kc_i(x_{s+1})) = \lambda_{i,s}(kc_i(x_{s+1}) + 1)^{-1},\ i = 1, \ldots, m. \quad (8.51)$$

Method (8.50)–(8.51) is, in fact, MBF method for the equivalent problem (8.38); therefore the results of the previous paragraph are taking place for the EPM (8.50)–(8.51). For completeness we provide the correspondent statements.

Theorem 8.5. *If conditions A and B hold, f, $c_i \in C^1$, $i = 1, \ldots, m$, f is convex, and c_i, $i = 1, \ldots, m$ are concave, then EPM (8.50)–(8.51) is:*

(1) well defined;
(2) equivalent to the following proximal point method

$$d_y(\hat{\lambda}) - k^{-1}D(\hat{\lambda}, \lambda) = \max\{d_y(u) - k^{-1}D(u, \lambda) | u \in \mathbb{R}^m_+\}, \quad (8.52)$$

where

$$D(u, \lambda) = \sum_{i=1}^{m} \lambda_i \varphi(u_i/\lambda_i)$$

is the KL φ-divergence entropy-like distance based on the MBF kernel $\varphi = -\psi^*$, where ψ^* is LF transform of $\psi(t) = \ln(t+1)$.

The following Theorem establishes convergence of the EPM under minimum assumptions on the input data.

Theorem 8.6. *Under assumptions of Theorem 8.5 for given $y \in \operatorname{int} \Omega$, as a "center," and for any $k > 0$, the EPM (8.50)–(8.51) generates primal–dual sequence $\{x_s, \lambda_s\}_{s \in N}$:*

(1) $d_y(\lambda_{s+1}) > d_y(\lambda_s)$, $s \geq 0$
(2) $\lim_{s\to\infty} d_y(\lambda_s) = d_y(\lambda^*)$, $\lim_{s\to\infty} F(y, x_s) = F(y, x^*) = d_y(\lambda^*)$
(3) $\lim_{s\to\infty} d_H(\partial\Lambda_s, L^*) = 0$
(4) there exists a subsequence $\{s_l\}_{l \in N}$ such that for $\{\bar{x}_l = \sum_{s=s_l}^{s_{l+1}} (s_{l+1} - s_l)^{-1} x_s\}_{l \in N}$ we have $\lim_{l\to\infty} \bar{x}_l = \bar{x} \in X^*$, that is, the primal sequence converges to the primal solution in the ergodic sense.

The following Theorem establishes the convergence rate of the EPM.

Theorem 8.7. *If f, $c_i \in C^2$, $i = 1, \ldots, m$ and the second-order sufficient optimality condition (8.45)–(8.46) is satisfied, then there exist a small enough $\delta > 0$ and large enough $k_0 > 0$, that for any $(\lambda, k) \in \Lambda(\cdot)$, the following statements hold:*

(1) there exists

$$\hat{x} = \hat{x}(\lambda, k) := \nabla_x \mathscr{L}_y(\hat{x}, \lambda, k) = 0$$

and

$$\hat{\lambda} = (\hat{\lambda}_i = \lambda_i (kc_i(\hat{x}) + 1)^{-1},\ i = 1, \ldots, m).$$

(2) for $(\hat{x}, \hat{\lambda})$ *the following bound holds*

$$\max\{\|\hat{x} - x^*\|, \|\hat{\lambda} - \lambda^*\|\} \leq ck^{-1}\|\lambda - \lambda^*\|, \tag{8.53}$$

where $c > 0$ *is independent on* $k \geq k_0$. *Also* $x(\lambda^*, k) = x^*$ *and* $\hat{\lambda}(\lambda^*, k) = \lambda^*$, *that is,* λ^* *is a fixed point of the map* $\lambda \to \hat{\lambda}(\lambda, k)$.

(3) The EDF $\mathscr{L}_y$ *is strongly convex in the neighborhood of* $\hat{x}$.

The results of Theorem 8.7 follow from Theorem 7.2.

8.2.3 Stopping Criteria

The EPM (8.50)–(8.51) is an infinite procedure, which require, at each step, an infinite procedure as well.

The following result allows replacing x_{s+1} from (8.50) by an approximation $\bar{x}_{s+1}$, finding which requires a finite procedure. Such replacement does not compromise Q-linear convergence rate.

For a small enough $\alpha > 0$, let us consider the primal–dual approximation $(\bar{x}, \bar{\lambda})$:

$$\bar{x} = \bar{x}(\lambda, k) : \|\nabla_x \mathscr{L}_y(\bar{x}, \lambda, k)\| \leq \frac{\alpha}{k}\|\bar{\lambda} - \lambda\| \tag{8.54}$$

$$\bar{\lambda} = \bar{\lambda}(\lambda, k) = (\bar{\lambda}_i = \psi'(kc_i(\bar{x}))\lambda_i,\ i = 1, \ldots, m.) \tag{8.55}$$

Obviously $\bar{x}$ depends not only on $\lambda \in \mathbb{R}^m_{++}$ and $k > 0$ but also on y and α as well.

At this point $y \in \text{int}\ \Omega$ and $\alpha > 0$ are fixed; therefore to simplify notation we omitted y and α from the definition of $\bar{x}$ and $\bar{\lambda}$.

Theorem 8.8. *If* $f,\ c_i \in C^2$, $i = 1, \ldots, m$ *and the second-order sufficient optimality condition (8.45)–(8.46) is satisfied, then there exists small enough* $\alpha > 0$, $\delta > 0$ *and large enough* k_0, *that for any* $k \geq k_0$ *and any* $(\lambda, k) \in \Lambda(\lambda, k, \delta)$ *the following holds:*

(1) there exists $(\bar{x}, \bar{\lambda})$ *defined by (8.54)–(8.55);*

(2) there is $c > 0$ *independent on* $k \geq k_0$ *that the following bound*

$$\max\{\|\bar{x} - x^*\|, \|\bar{\lambda} - \lambda^*\|\} \leq \frac{c}{k}(1 + 2\alpha)\|\lambda - \lambda^*\| \tag{8.56}$$

holds;

(3) the Lagrangian $\mathscr{L}_y$ *is strongly convex at the neighborhood of* $\bar{x}$.

The Theorem 8.8 can be proven using arguments from Theorem 7.3

We conclude the section by considering the numerical realization of the EPM.

The EPM consists of inner and outer iteration. On the inner iteration, we find an approximation $\bar{x}$ for the primal minimizer, using the stopping criteria (8.54).

On the outer iteration, we update the Lagrange multipliers by (8.55), using the approximation $\bar{x}$.

For finding $\bar{x}$ any unconstrained minimization technique can be used. Fast gradient method and regularized Newton method are two possible candidates.

Under usual convexity and smoothness assumptions, both methods converge to the minimizer from any starting point, and for both methods, there exist complexity bounds (see Chapter 3).

To describe the numerical realization of EPM, we introduce the relaxation operator $R : \Omega_k \times \mathbb{R}^m_{++} \to \Omega_k \times \mathbb{R}^m_{++}$, which is defined as follows:

$$Ru = \bar{u} = (\bar{x}, \bar{\lambda}), \tag{8.57}$$

where $\bar{x}$ and $\bar{\lambda}$ are given by (8.54) and (8.55).

We also use the following merit function $\nu_y : \Omega_k \times \mathbb{R}^m_+ \to R$ defined for a given $y \in \operatorname{int}\Omega$:

$$\nu_y(u) \equiv \nu_y(x,\lambda) = \max\{\|\nabla_x L_y(x,\lambda)\|,\ \sum_{i=1}^m \lambda_i |c_i(x)|, -c_i(x),\ i = 1,\dots,m\}. \tag{8.58}$$

From (8.58) follows $\nu_y(u) \geq 0$, $\forall u \in \Omega_k \times \mathbb{R}^m_+$ and from KKT's Theorem follows

$$\nu_y(u) = 0 \Leftrightarrow u = u^* = (x^*;\lambda^*). \tag{8.59}$$

Moreover, under the second-order sufficient optimality condition and $f, c_i \in C^2$, $i = 1,\dots,m$, the merit function ν_y in the neighborhood of u^* is similar to the norm of a gradient of a strongly convex function with Lipschitz continuous gradient.

Let $\gamma > 0$ be small enough, $y \in \operatorname{int}\ \Omega_\gamma$ be the initial "center," $u = (x;\lambda) \in \Omega_k \times \mathbb{R}^m_{++}$ be the initial primal–dual approximation, $\Delta > 0$ be the reduction parameter for the objective function, $k > 0$ be the scaling parameter, and $\varepsilon > 0$ be the required accuracy.

The EPM consists of the following operations:

1. find $\bar{u} = (\bar{x};\bar{\lambda}) = Ru$
2. if $\nu_y(\bar{u}) \leq \varepsilon$, then $u^* = (x^*;\lambda^*) := (\bar{x};\bar{\lambda})$ else;
3. find $\bar{\tau} = \max\{0 \leq \tau \leq 1 : x(\tau) = y + \tau(\bar{x} - y) \in \Omega_\gamma,\ \langle \nabla f(x(\tau)), \bar{x} - y\rangle \leq 0\}$ and $x(\bar{\tau})$;
4. if

$$f(y) - f(x(\bar{\tau})) \geq \Delta, \tag{8.60}$$

then update the center $\bar{y} := 0.5(y + x(\bar{\tau}))$, set $x := \bar{x}$, $y := \bar{y}$ and go to 1;
else set $x := \bar{x}$; $\lambda := \bar{\lambda}$ and go to 1.

It follows from 3. and 4. that the sequence of centers is monotone decreasing in value; therefore from some point on the inequality (8.60) can't be satisfied, so from such point on the "center" $\bar{y}$ is fixed.

Hence, from this point on, the primal–dual sequence is generated only by the relaxation operator (8.57) and, under conditions of Theorem 8.8, the sequence converges to the primal–dual solution with Q-linear rate.

The EPM efficiency heavily depends on the efficiency of unconstrained minimization methods, used in operator R.

The absence of EDF's singularity at the solution leads to stability of its Hessian's condition number. It keeps the area stable, where Newton's method is well defined, which leads to the "hot" start phenomenon.

The extra tool, which EDF possesses, on the top of MBF properties, contribute to the numerical efficiency mainly because updating the center does not require much computational effort but can substantial reduce the objective function value.

It means that by updating the "center," one can to reach the "hot" start faster.

The interior point $\bar{x} \in \operatorname{int} \Omega$ can be found by applying the EPM to the following problem:

$$\bar{v} = \max\{v : c_i(x) - v \geq 0, x \in \mathbb{R}^n, i = 1, \ldots, m\} > 0,$$

starting with $x_0 \in \mathbb{R}^n$ and $v_0 : \min_{1 \leq i \leq m} c_i(x_0) > v_0$.

8.2.4 Modified Interior Distance Functions

We consider modified interior distance function (MIDF), which is another realization of the NR principle.

Let $y \in \operatorname{int} \Omega$ as fixed; we consider $x \in \Omega$: $\Delta(y,x) = f(y) - f(x) > 0$, then

$$\Omega(y) = \{x : c_i(x) \geq 0, \quad i = 1, \ldots, m;\ \Delta(y,x) > 0\},$$

is the relaxation feasible set (RFS) at the level $f(y)$.

Due to the boundedness of X^*, convexity f, $-c_i$, $i = 1, \ldots, m$, and Slater condition, the RFS $\Omega(y)$ is not empty, convex, and bounded and contains x^* for any $y \in \operatorname{int} \Omega$.

Thus, the original problem is equivalent to

$$f(x^*) = \min\{f(x)/x \in \Omega(y)\}. \tag{8.61}$$

For any given $k > 0$ and $y \in \operatorname{int} \Omega$ the RFS

$$\Omega(y) \equiv \Omega_k(y) = \{x : k^{-1}[\ln(kc_i(x) + \Delta(y,x)) - \ln \Delta(y,x)] \geq 0,$$
$$i = 1, \ldots, m;\ \Delta(y,x) > 0\}.$$

Therefore, for any given $k > 0$, $y \in \operatorname{int} \Omega$ and $F(y,x) = -\ln \Delta(y,x)$ problem (8.61) is equivalent to the following problem:

$$F(y,x^*) = \min\{F(y,x)/x \in \Omega_k(y)\}. \tag{8.62}$$

Assuming $\ln t = -\infty$ for $t \leq 0$, we define the MIDF $F : \mathbb{R}^n \times \operatorname{int}\Omega(y) \times \mathbb{R}^m_+ \times \mathbb{R}^1_{++} \to \mathbb{R}^1$ as the LEP for (8.62):

$$F(x,y,\lambda,k) = \left(-1 + k^{-1}\sum_{i=1}^{m}\lambda_i\right)\ln\Delta(y,x) - k^{-1}\sum_{i=1}^{m}\lambda_i\ln(kc_i(x) + \Delta(y,x)). \tag{8.63}$$

It follows from (8.63) that to keep $F(x,y,\lambda,k)$ convex in x for a given $\lambda \in \mathbb{R}^m_+$, one has to take $k > \sum_{i=1}^m \lambda_i$.

The MIDF $F(x,y,\lambda,k)$ is the modification of the classical Huard's IDF $H(x,\tau) = -m\ln(f(y) - f(x)) - \sum_{i=1}^m \ln c_i(x)$.

8.2.5 Local MIDF Properties

The MIDF is defined at the solution, and we will see that under the optimal Lagrange multipliers, one can find the primal solution by solving one smooth unconstrained optimization problem, when both the "center" and the scaling parameter are fixed.

For any given $k > 0$ and any $y \in \operatorname{int}\Omega$ as a "center," the following property

1° $F(x^*,y,\lambda^*,k) = -\ln\Delta(y,x^*)$ that is $f(x^*) = f(y) - \exp(-F(x^*,y,\lambda^*,k))$

holds.

Property 1° follows immediately from the definition of MIDF and the complementarity conditions

$$\lambda_i^* c_i(x^*) = 0, \quad i = 1,\dots,m.$$

Exercise 8.4. Construct MIDF $C(x,y,\lambda,k)$ and $Q(x,y,\lambda,k)$, which corresponds to classical IDF

$$C(x,\tau) = m(\tau - f(x))^{-1} + \sum_{i=1}^{m} c_i^{-1}(x)$$

and

$$Q(x,\tau) = (\tau - f(x))^{-m}\Pi_{i=1}^m c_i^{-1}(x).$$

It follows from 1° that for any KKT's pair (x^*,λ^*) the value of MIDF coincided with the optimal objected function value of (8.62) for any $k > 0$ and any $y \in \operatorname{int}\Omega$. It indicates that one can approach the solution by means other than those traditionally used in IPMs.

For any given scaling parameter $k > 0$ and any given $y \in \operatorname{int}\Omega$, as a "center," the following

2° $$\nabla_x F(x^*,y,\lambda^*,k) = \Delta^{-1}(y,x^*)\nabla_x L(x^*,\lambda^*) = 0$$

holds.

Property 2° follows from the formula for $\nabla_x F$ and complementarity condition.

If f and all $-c_i$ are convex, then for any given $y \in \text{int}\ \Omega$ as a "center" and any $k > \sum \lambda_i^*$, from 2° follows

3° $$F(x^*, y, \lambda^*, k) = \min\{F(x, y, \lambda^*, k)/x \in \mathbb{R}^n\}$$

Exercise 8.5. Show properties 1° and 2° for MIDF $C(x,y,\lambda,k)$ and $Q(x,y,\lambda,k)$.

Exercise 8.6. Prove 3° for MIDFs $F(x,y,\lambda,k) = F(\cdot)$, $C(\cdot)$ and $Q(\cdot)$.

In other words, having the optimal Lagrange multipliers, one can solve the original problem by solving one unconstrained optimization problem 3°. Property 3° holds for any given $y \in \text{int}\ \Omega$ as a "center" and any $k > \sum \lambda_i^*$; therefore one can view MIDF as a smooth exact penalty function.

If $F(x,y,\lambda,k)$ is strongly convex in x and we know a "good" approximation λ for λ^*, then

$$\hat{x} = \hat{x}(y, \lambda, k) = \text{argmin}\{F(x, y, \lambda, k)/x \in \mathbb{R}^n\}$$

is unique, and it is a "good" approximation for x^*, while both the "center" $y \in \text{int}\ \Omega$ and $k > \sum \lambda_i$ are fixed.

We will see later that one can use F to develop a method, which guarantee convergence due to the LM update rather than the center or the barrier parameter update.

Our goal is developing such a method. We start by describing the conditions, under which the MIDF $F(x,y,\lambda,k)$ is convex or strongly convex in x when both y and $k > 0$ are fixed.

The following proposition is the first step in this direction.

Proposition 8.3. *If f and $c_i \in C^2$, $i = 1, \ldots, m$, then for any $y \in$ int Ω, $k > 0$ and any K-K-T pair (x^*, λ^*), the following is true:*

4° $$\begin{aligned}\nabla^2_{xx} F(x^*, y, \lambda^*, k) &= \Delta^{-1}(y, x^*) \left[\nabla^2_{xx} L(x^*, \lambda^*)\right. \\ &\left.+ \Delta^{-1}(y, x^*)(k(\nabla c_{(r)}(x^*))^T \Lambda_r^* \nabla c_{(r)}(x^*) - \nabla c^T_{(r)}(x^*) \lambda^{*T}_{(r)} \lambda^*_{(r)} \nabla c_{(r)}(x^*))\right].\end{aligned}$$

The proof follows directly from the formula for the Hessian $\nabla^2_{xx} F(x^*, y, \lambda^*, k)$ and the complementarity condition.

Exercise 8.7. Show that Proposition 8.3 holds for $H(\cdot)$ and $Q(\cdot)$.

Theorem 8.9. *Let f and $c_i \in C^2$, $i = 1, \ldots, m$ and $k > \sum_{i=1}^m \lambda_i^*$, then*

(1) if f and all $-c_i$ are convex and f or one of $-c_i$, $i = 1, \ldots, r$ is strongly convex at x^*, then for any given $y \in \text{int}\,\Omega$ as a "center" $F(x, y, \lambda^*, k)$ is strongly convex in the neighborhood of x^*;

(2) if the second-order sufficient optimality condition (4.73)–(4.74) holds, then there exist $k_0 > 0$ that for a given $y \in \text{int}\,\Omega$ as a "center" and fixed $k > \Delta(y, x^*) k_0 + \sum_{i=1}^m \lambda_i^*$, there are such $0 < m_0 < M_0 < +\infty$ that

$$(a)\ \textit{mineigenval}\ \nabla^2_{xx}F(x^*,y,\lambda^*,k) \geq \Delta^{-1}(y,x^*)m_0$$
$$(b)\ \textit{maxeigenval}\ \nabla^2_{xx}F(x^*,y,\lambda^*,k) \leq \Delta^{-1}(y,x^*)M_0\,,$$

and for the condition number of the Hessian $\nabla^2_{xx}F(x^*,y,\lambda^*,k)$, *the following bound*

5° $$\varkappa(x^*,y,\lambda^*,k) \geq m_0 M_0^{-1}$$

holds.

Proof.

(1) Using Proposition 8.3 for any $\mathrm{v} \in \mathbb{R}^n$, we obtain

$$\begin{aligned}
&\langle \nabla^2_{xx}F(x^*,y,\lambda^*,k)\mathrm{v},\mathrm{v}\rangle = \Delta^{-1}(y,x^*)[\langle \nabla^2_{xx}L(x^*,\lambda^*)\mathrm{v},\mathrm{v}\rangle)\\
&+ \Delta^{-1}(y,x^*)\langle k(\nabla c_{(r)}(x^*))^T \Lambda^*_r \nabla c_{(r)}(x^*)\mathrm{v},\mathrm{v}\rangle\\
&- \langle (\nabla c_{(r)}(x^*))^T \lambda^{*T}_{(r)} \lambda^*_{(r)} \nabla c_{(r)}(x^*)\mathrm{v},\mathrm{v}\rangle]\\
&= \Delta^{-1}(y,x^*)\left[\langle \nabla^2_{xx}L(x^*,\lambda^*)\mathrm{v},\mathrm{v}\rangle\right.\\
&+ \Delta^{-1}(y,x^*)\left(k - \sum_{i=1}^m \lambda_i^*\right)\langle (\nabla c_{(r)}(x^*))^T \Lambda^*_r \nabla c_{(r)}(x^*)\mathrm{v},\mathrm{v}\rangle\\
&+ \Delta^{-1}(y,x^*)\left.\left(\left(\sum_{i=1}^m \lambda_i^*\right)\left(\sum_{i=1}^m \lambda_i^* \langle \nabla c_i(x^*),\mathrm{v}\rangle^2\right) - \left(\sum_{i=1}^m \lambda_i^* \langle \nabla c_i(x^*),\mathrm{v}\rangle\right)^2\right)\right]
\end{aligned}$$

Taking into account identity

$$\begin{aligned}
&\left(\sum_{i=1}^m \lambda_i^*\right)\left(\sum_{i=1}^m \lambda_i^* \langle \nabla c_i(x^*),\mathrm{v}\rangle^2\right) - \left(\sum_{i=1}^m \lambda_i^* \langle \nabla c_i(x^*),\mathrm{v}\rangle\right)^2\\
&= \frac{1}{2}\sum_{i=1}^m \sum_{j=1}^m \lambda_i^* \lambda_j^* \langle \nabla c_i(x^*) - \nabla c_j(x^*),\mathrm{v}\rangle^2 \geq 0,
\end{aligned} \tag{8.64}$$

we obtain

$$\begin{aligned}
&\langle \nabla^2_{xx}F(x^*,y,\lambda^*,k)\mathrm{v},\mathrm{v}\rangle \geq \Delta^{-1}(y,x^*)[(\nabla^2_{xx}L(x^*,\lambda^*)\\
&+ \Delta^{-1}(y,x^*)\left(k - \sum_{i=1}^m \lambda_i^*\right)\langle (\nabla c_{(r)}(x^*))^T \Lambda^*_r \nabla c_{(r)}(x^*)\mathrm{v},\mathrm{v}\rangle].
\end{aligned} \tag{8.65}$$

So, for a convex optimization problem (P), the MIDF $F(x,y,\lambda^*,k)$ is convex in x for any $y \in \operatorname{int}\Omega$ and $k \geq \sum_{i=1}^m \lambda_i^*$.
If f or one of $-c_i$, $i = 1,\ldots,r$ is strongly convex at x^*, then due to $\lambda_i^* > 0$, $i = 1,\ldots,r$, the classical Lagrangian $L(x,\lambda^*)$ is strongly convex in the neighborhood of x^*, and the matrix

$$M(x^*,y,\lambda^*,k) = \left(k - \sum_{i=1}^{m} \lambda_i^*\right) \Delta^{-1}(y,x^*) \left(\nabla c_{(r)}(x^*)\right)^T \Lambda_r^* \nabla c_{(r)}(x^*)$$

is nonnegative defined for any $y \in \text{int}\,\Omega$ as a "center" and any $k > \sum_{i=1}^m \lambda_i^*$. Therefore, $F(x,y,\lambda^*,k)$ is strongly convex in the neighborhood of x^*.

(2) Now, we consider the case when none of f and $-c_i$, $i = 1,\ldots,r$ are convex and $r < n$. From (8.65) and $k > \Delta(y,x^*)k_0 + \sum_{i=1}^m \lambda_i^*$ follows

$$\langle \nabla_{xx}^2 F(x^*,y,\lambda^*,k)\mathrm{v},\mathrm{v}\rangle \geq \Delta^{-1}(y,x^*)\langle[\nabla_{xx}^2 L(x^*,\lambda^*) + k_0(\nabla c_{(r)}(x^*))^T \Lambda_r^* \nabla c_{(r)}(x^*)]\mathrm{v},\mathrm{v}\rangle,\ \forall \mathrm{v} \in \mathbb{R}^n. \tag{8.66}$$

If the second-order sufficient optimality condition (4.73)–(4.74) is satisfied, then for $k_0 > 0$ large enough from Debreu's lemma with $A = \nabla_{xx}^2 L(x^*,\lambda^*)$ and $B = \Lambda^{*\frac{1}{2}} \nabla c_{(r)}(x^*)$ follows existence $m_0 > 0$, such that

$$\langle \nabla_{xx}^2 F(x^*,y,\lambda^*,k)\mathrm{v},\mathrm{v}\rangle \geq \Delta^{-1}(y,x^*) m_0 \langle \mathrm{v},\mathrm{v}\rangle,\ \forall \mathrm{v} \in \mathbb{R}^n. \tag{8.67}$$

Also from (8.65) for a given $y \in \text{int}\,\Omega$ as a "center," and $k > \Delta(y,x^*)k_0 + \sum_{i=1}^m \lambda_i^*$ follows existence $M_0 < \infty$

$$\langle \nabla_{xx}^2 F(x^*,y,\lambda^*,k)\mathrm{v},\mathrm{v}\rangle \leq \Delta^{-1}(y,x^*) M_0 \langle \mathrm{v},\mathrm{v}\rangle,\ \forall \mathrm{v} \in \mathbb{R}^n. \tag{8.68}$$

Therefore, the condition number of the MIDF Hessian $\nabla_{xx}^2 F(x^*,y,\lambda^*,k)$ at the primal–dual solution (x^*,λ^*) is

$$\varkappa(x^*,y,\lambda^*,k) \geq m_0 M_0^{-1}.$$

For f and $c_i \in C^2$, $i = 1,\ldots,m$ the condition number $\varkappa(x,y,\lambda,k)$ remains stable in the neighborhood of (x^*,λ^*).

In other words, the penalty parameter k not only retains convexity in x of $F(x,y,\lambda,k)$ in a convex case; it also provides "convexification" of $F(x,y,\lambda,k)$ in the neighborhood of the primal solution in a non-convex case.

Exercise 8.8. Show that results of Theorem 8.9 are true for MIDF $C(\cdot)$ and $Q(\cdot)$.

Although there are similarities between MBF and MIDF, there are substantial differences between these two realizations of the NR principle. In fact, the differences between MBF and MIDF are more substantial than those between the classical barrier function

$$F(x,k) = f(x) - k^{-1} \sum_{i=1}^{m} \ln c_i(x)$$

and classical interior distance functions

$$H(x,\tau) = -m(\ln(\tau - f(x)) - \sum_{i=1}^{m} \ln c_i(x).$$

First, MBF remains convex in x for any $k>0$ and any $\lambda\in\mathbb{R}^m_+$. It is not true for MIDF. The MIDF $F(x,y,\lambda,k)$ remains convex in x if $k>\sum_{i=1}^m\lambda_i$.

Second, MBF method converges to the primal–dual solution for any $k>0$ under very mild assumptions on the input data. It is not true for MIDF.

Third, for non-convex optimization, MBF is strongly convex in $B(x^*,\varepsilon)=\{x\in\mathbb{R}^n:\|x-x^*\|\leq\varepsilon\}$ if the second-order sufficient optimality condition is satisfied and $k>0$ is large enough. It is not true for MIDF.

To guarantee convexity $F(x,y,\lambda,k)$ in x under fixed $y\in\inf\Omega$ and $\lambda\in\mathbb{R}^m_{++}$ in case of CO, we have to take the barrier parameter $k>\sum_{i=1}^m\lambda_i$.

To guarantee strong convexity $F(x,y,\lambda^*,k)$ in x in $B(x^*,\varepsilon)$, under second-order sufficient optimality condition, the barrier parameter has to be $k\geq\Delta(y,x^*)k_0+\sum_{i=1}^m\lambda_i^*$ and $k_0>0$ be large enough.

On the top of the barrier parameter's role in MBF theory, in case of MIDF, the parameter is also responsible for preserving the convexity of MIDF in x even for convex optimization.

On the other hand, by changing the center and Lagrange multiplier from step to step, it is possible to develop MIDF methods with up to Q-superlinear rate under fixed barrier parameter, which is impossible in case of MBF.

Example 8.2. Let us consider the following example

$$x^*=\operatorname{argmin}\{f(x)=e^x\mid c(x)=x\geq 0\}=0\,.$$

Then MBF $F(\cdot)=F(x,\lambda,k)=e^x-k^{-1}\lambda\ln(kx+1)$, $F'_x(\cdot)=e^x-\lambda(kx+1)^{-1}$, $F''(\cdot)=e^x+\frac{\lambda k}{(kx+1)^2}$. Therefore, for any $\lambda>0$ and $k>0$, the MBF $F(x,\lambda,k)$ is strongly convex in $x\geq 0$. Also, $F'_x(x^*,\lambda,k)=e^x-\lambda(kx+1)^{-1}_{|x=0}=1-\lambda=0$. Thus, $\lambda^*=1$, and

$$x^*=\operatorname{argmin}F(x,\lambda^*,k)=0\,,\quad\forall k>0\,.$$

Now, let us consider the correspondent MIDF. We have

$$\begin{aligned}F(\cdot)=F(x,y,\lambda,k)&=-\ln(e^y-e^x)-k^{-1}\lambda\left[\ln(kx+e^y-e^x)-\ln(e^y-e^x)\right]\\&=-(1-k^{-1}\lambda)\ln(e^y-e^x)-k^{-1}\lambda\ln(kx+e^y-e^x)\,.\end{aligned}$$

Then

$$F'_x(\cdot)=(1-k^{-1}\lambda)e^x(e^y-e^x)^{-1}-k^{-1}\lambda(k-e^x)(kx+e^y-e^x)^{-1}\,.$$

For $\lambda^*=1$, any $k>1$ and any $y>0$, we have

$$F'_x(x,y,1,k)=0\Rightarrow x=x^*=0$$

and

$$F''_{xx}(x^*,y,\lambda^*,k)>0\,.$$

In contrast to MBF, the MIDF $F(x,y,\lambda,k)$ is not convex in x for any $k>0$ even for a convex optimization problem. For example, MIDF $F(x,1.5,\lambda^*,0.01)$ is strongly concave at $x=1$ because $F''_{xx}(1,1.5,\lambda^*,0.01)=-1.34$.

On the other hand, if the barrier parameter is chosen as Theorem 8.9 prescribes, then by updating the center and the Lagrange multipliers at each step, one can substantially improve the convergence rate as compared with MBF.

The MCM method will be discussed in the following section.

8.2.6 Modified Center Method

The extension $\Delta(y,x)$ of Ω and the choice of the barrier parameter together with the special way of updating Lagrange multipliers give rise to the modified center method.

From Theorem 8.9 follows that for solving a constrained optimization problem, it is enough finding a minimizer for a strongly convex and twice continuously differentiable in x function $F(x,y,\lambda^*,k)$, when both the "center" $y\in \operatorname{int}\Omega$ and the barrier parameter $k\geq \Delta(y,x^*)k_0+\sum_{i=1}^m \lambda_i^*$ are fixed and $k_0>0$ is large enough.

If Lagrange multipliers vector $\lambda\in\mathbb{R}^m_{++}$ "close enough" to λ^*, is given, then a "good" approximation for x^* one can find as a MIDF's minimizer

$$\hat{x}=\hat{x}(y,\lambda,k)=\operatorname{argmin}\{F(x,y,\lambda,k)/x\in\mathbb{R}^n\}, \tag{8.69}$$

when both y and k are fixed.

Having minimizer $\hat{x}$, one can find a better approximation $\hat{\lambda}$ for λ^* by changing neither the "center" y nor barrier parameter k, just by updating λ.

For the minimizer $\hat{x}$, we have

$$\nabla_x F(\hat{x},y,\lambda,k)=\left(1-k^{-1}\sum_{i=1}^m\lambda_i+k^{-1}\sum_{i=1}^m\hat{\lambda}_i\right)\nabla f(\hat{x})-\sum_{i=1}^m\hat{\lambda}_i\nabla c_i(\hat{x})=0, \tag{8.70}$$

where

$$\hat{\lambda}\equiv\hat{\lambda}(y,\lambda,k):=(\hat{\lambda}_i\equiv\hat{\lambda}_i(y,\lambda,k)=\lambda_i\Delta(y,\hat{x})\Delta_i^{-1}(\hat{x},y,k),\quad i=1,\ldots,m) \tag{8.71}$$

is the new LM vector and $\Delta_i(x,y,k)=kc_i(x)+\Delta(y,x)$, $i=1,\ldots,m$.

Formula (8.71) for the Lagrange multipliers update is critical for our further considerations.

First of all, from Theorem 8.9 follows $\hat{\lambda}(y,\lambda^*,k)=\lambda^*$ for any fixed $y\in\operatorname{int}\Omega$ as a "center" and any $k>\Delta(y,x^*)k_0+\sum_{i=1}^m\lambda_i^*$, that is, λ^* is a fixed point of the map $\lambda\rightarrow\hat{\lambda}(y,\lambda,k)$.

Second, for the new LM vector $\hat{\lambda}$, the following bound

$$\|\hat{\lambda}-\lambda^*\|\leq ck^{-1}\Delta(y,x^*)\|\lambda-\lambda^*\| \tag{8.72}$$

holds and $c > 0$ is independent on $y \in \text{int}\,\Omega$ and $k \geq \Delta(y,x^*)k_0 + \sum_{i=1}^m \lambda_i^*$, where $\|x\| = \|x\|_\infty = \max_{1 \leq i \leq n} |x_i|$.

Third, the bound type (8.72) holds for minimizer $\hat{x}$ as well, that is

$$\|\hat{x} - x^*\| \leq ck^{-1}\Delta(y,x^*)\|\lambda - \lambda^*\|. \tag{8.73}$$

In other words, finding a minimizer $\hat{x}$ and updating vector $\lambda \in \mathbb{R}^m_{++}$ by (8.71) are equivalent to applying to $\lambda \in \mathbb{R}^m_{++}$ an operator

$$C_{y,k} : C_{y,k}\lambda = \hat{\lambda}(y,\lambda,k) = \hat{\lambda}.$$

From $C_{y,k}\lambda^* = \lambda^*$ for a contractive operator $C_{y,k}$ follows

$$\|C_{y,k}\lambda - \lambda^*\| = \|C_{y,k}(\lambda - \lambda^*)\| < \|\lambda - \lambda^*\|.$$

The constructability of $C_{y,k}$ is defined by

$$\text{Contr}\, C_{y,k} = \gamma_{y,k} = ck^{-1}\Delta(y,x^*).$$

The constant $c > 0$ depends on the input data and the size of a given problem, but it is independent on y and k; therefore for a fixed $y \in \text{int}\,\Omega$, there is large enough $k_0 > 0$ that for any $k \geq \Delta(y,x^*)k_0 + \sum_{i=1}^m \lambda_i^*$, we have $\gamma_{y,k} < 1$.

Now, we describe the basic version of the modified center method.

We start with $y \in \text{int}\,\Omega$, $\lambda_0 = e = (1,\ldots,1)^T \in \mathbb{R}^m$ and $k > m$. Let us assume that the pair (x_s, λ_s) has been found already. Then by taking $k > \sum_{i=1}^m \lambda_{i,s}$ we find the next approximation (x_{s+1}, λ_{s+1}) by formulas

$$x_{s+1} : \nabla_x F(x_{s+1}, y, \lambda_s, k) = 0 \tag{8.74}$$

$$\lambda_{s+1} : \lambda_{i,s+1} = \lambda_{i,s}\Delta(y,x_{s+1})\Delta_i^{-1}(x_{s+1},y,k), \quad i = 1,\ldots,m. \tag{8.75}$$

In view of boundedness X^*, convexity f, and concavity c_i, $i = 1,\ldots,m$, the set $\Omega_k(y) = \{x : kc_i(x) + \Delta(y,x) \geq 0,\ i = 1,\ldots,m;\ \Delta(y,x) > 0\}$ is closed, convex, and bounded for any $y \in \text{int}\,\Omega$ and $k > 0$.

Also, $x \to \partial\Omega_k(y) \Rightarrow F(x,y,\lambda_s,k) \to \infty$; therefore, for any $\lambda_s \in \mathbb{R}^m_{++}$, $y \in \text{int}\,\Omega$ and $k > \sum \lambda_{i,s}$, MIDF $F(x,y,\lambda_s,k)$ is convex in $x \in \Omega_k(y)$. The minimizer

$$x_{s+1} = \text{argmin}\{F(x,y,\lambda_s,k)/x \in \mathbb{R}^n\}$$

exists due to the Weierstrass theorem.

For the minimizer x_{s+1}, we have

$$\begin{aligned} \nabla_x F(x_{s+1}, y, \lambda_s, k) = &\left(1 - k^{-1}\sum_{i=1}^m \lambda_{i,s} + k^{-1}\sum_{i=1}^m \lambda_{i,s+1}\right)\nabla f(x_{s+1}) \\ &- \sum_{i=1}^m \lambda_{i,s+1}\nabla c_i(x_{s+1}) = 0 \end{aligned} \tag{8.76}$$

and $\lambda_s \in \mathbb{R}^m_{++} \Rightarrow \lambda_{s+1} \in \mathbb{R}^m_{++}$.

Therefore, the MIDF method (8.74)–(8.75) is well defined.

Thus, starting with a vector $\lambda_0 \in \mathbb{R}^m_{++}$, one can guarantee that the Lagrange multipliers will remain positive up to the end of the process without any particular care, just due to the update formula (8.75).

The convergence and the rate of convergence of the modified center method (8.74)–(8.75) establish the following basic theorem.

8.2.7 Basic Theorem

The basic theorem establishes the contractibility properties of the operator $C_{y,k}$. We start by characterizing the domain, where the operator $C_{y,k}$ is defined and contractive. Let $\tau > 0$ be small enough and $y_0 : c_i(y_0) \geq \tau$, $i = 1,\ldots,m$. The subset $\Omega_\tau = \{x : c(x) \geq \tau e\} \cap \{x : \Delta(y_0,x) > 0\}$ of the RFS $\Omega(y_0)$ is bounded. For any $y \in \Omega_\tau$, we have $\Delta(y,x^*) > 0$. Along with the fixed $\tau > 0$, we consider a small enough $\delta > 0$ and a large enough $k_0 > 0$.

Let us describe the domain, where the operator $C_{y,k}$ is defined and contractive.

The sets

$$\Lambda^i_{y,k} = \left\{ \lambda_i : \lambda_i \geq \delta,\, |\lambda_i - \lambda^*| \leq \delta\Delta^{-1}(y,x^*)k,\, k \geqslant k_0\Delta(y,x^*) + \sum_{i=1}^m \lambda_i^* \right\},\ i = 1,\ldots,r$$

are associated with active constraints.

The sets

$$\Lambda^i_{y,k} = \left\{ \lambda_i : 0 \leq \lambda_i \leq \delta\Delta^{-1}(y,x^*)k,\, k \geqslant k_0\Delta(y,x^*) + \sum_{i=1}^m \lambda_i^* \right\},\ i = r+1,\ldots,m$$

are associated with the passive constraints.

The basic theorem below establishes condition, under which operator $C_{y,k}$ is a contractive one on

$$\Lambda_{y,k} = \Lambda^1_{y,k} \times \cdots \times \Lambda^r_{y,k} \times \cdots \times \Lambda^m_{y,k}.$$

Let us briefly describe the main idea of the proof. Using considerations similar to those we used in Theorem 7.2, one can show that for any $\lambda \in \Lambda_{y,k}$, there exists a MIDF minimizer $\hat{x} = \hat{x}(y,\lambda,k)$; therefore

$$\begin{aligned} \nabla_x F(\hat{x},y,\lambda,k) &= \nabla f(\hat{x}) - \sum_{i=1}^r \hat{\lambda}_i \nabla c_i(\hat{x}) - h(\hat{x},y,\lambda,k) + g(\hat{x},y,\lambda,k) \\ &= \bar{L}_x(\hat{x},\hat{\lambda}) - h(\hat{x},y,\lambda,k) + g(\hat{x},y,\lambda,k) = 0 \end{aligned} \tag{8.77}$$

where

$$\hat{\lambda}_i = \lambda_i \Delta(y,\hat{x})\Delta_i^{-1}(\hat{x},y,k), \quad i = 1,\ldots,r. \tag{8.78}$$

Let

$$\bar{L}(x,\lambda) = f(x) - \sum_{i=1}^{r} \lambda_i \nabla c_i(x),$$

$$h(\hat{x},y,\lambda,k) = \sum_{i=r+1}^{m} \lambda_i \Delta(y,\hat{x})\Delta_i^{-1}(\hat{x},y,k)\nabla c_i(\hat{x})$$

and

$$g(\hat{x},y,\lambda,k) = k^{-1}\sum_{i=1}^{m} \lambda_i(-1+\Delta(y,\hat{x})\Delta_i^{-1}(\hat{x},y,k))\nabla f(\hat{x}).$$

We consider (8.77) and (8.78) as a primal–dual system of equations for $\hat{x}$ and $\hat{\lambda}_{(r)}$.

It is easy to verify that $\hat{x} = x^*$ and $\hat{\lambda}_{(r)} = \lambda^*_{(r)}$ satisfy the system for a given $y \in \Omega_\tau$, $k > k_0\Delta(y,x^*) + \sum_{i=1}^m \lambda_i^*$, and $\lambda = \lambda^*$.

Using consideration similar to those used in the proof of Theorem 7.2, one can show that

the system (8.77)–(8.78) can be solved for $\hat{x}$ and $\hat{\lambda}_{(r)}$ under fixed $y \in \Omega_k, k \geq k_0\Delta(y,x^*) + \sum_{i=1}^m \lambda_i^*$ and $\lambda \in \Lambda_{y,k}$.

Having the solution $\hat{x} = \hat{x}(y,\lambda,k)$ and $\hat{\lambda}_{(r)} = \hat{\lambda}_{(r)}(y,\lambda,k)$, we find the Jacobians $\nabla_\lambda \hat{x}(\cdot) = J_\lambda(\hat{x}(y,\lambda,k))$ and $\nabla_\lambda \hat{\lambda}_{(r)}(\cdot) = J_\lambda(\hat{\lambda}_{(r)}(y,\lambda,k))$ and estimate the norms $\|\nabla_\lambda \hat{x}(y,\lambda^*,k)\|$ and $\|\nabla_\lambda \hat{\lambda}_{(r)}(y,\lambda^*,k)\|$.

It turns out that under second-order sufficient optimality conditions, there is such $k_0 > 0$ that for a fixed $y \in \Omega_\tau$ and $k \geq k_0\Delta(y,x^*) + \sum_{i=1}^m \lambda_i^*$, the following bound

$$\max\left\{\|\nabla_\lambda \hat{x}(y,\lambda^*,k)\|\,,\,\left\|\nabla_\lambda \lambda_{(r)}(y,\lambda^*,k)\right\|\right\} \leq c \tag{8.79}$$

holds and $c > 0$ is independent on y and k.

The bound

$$\max\{\|\hat{x}-x^*\|,\,\|\hat{\lambda}-\lambda^*\|\} \leq ck^{-1}\Delta(y,x^*)\|\lambda-\lambda^*\| = \gamma_{y,k}\|\lambda-\lambda^*\| \tag{8.80}$$

which is the main result, follows immediately from (8.79).

The independence $c > 0$ on y and k means that for any fixed $y \in \Omega_\tau$, there exists $k_0 > 0$ such that for any $k \geq k_0\Delta(y,x^*) + \sum_{i=1}^m \lambda_i^*$ the operator $C_{y,k}$ is contractive, that is, $0 < \gamma_{y,k} < 1$. Therefore

$$\lambda \in \Lambda_{y,k} \to C_{y,k}\lambda = \hat{\lambda} \in \Lambda_{y,k}.$$

Moreover, for any given $0 < \gamma < 1$ and a given "center" $y \in \Omega_\tau$, there is $k_0 > 0$ such that Contr $C_{y,k} \leq \gamma$ if $k \geq k_0\Delta(y,x^*) + \sum_{i=1}^m \lambda_i^*$. □

The following Theorem establishes the basic results for the modified center method.

Theorem 8.10.

(1) If X^* is bounded, f, $-c_i$, $i = 1,\ldots,m$ are convex, and the Slater condition holds, then for any given $y \in \Omega_\tau$, $\lambda \in \mathbb{R}^m_{++}$ and $k > k_0\Delta(y,x^*) + \sum_{i=1}^m \lambda_i$, there exists

$$\hat{x} = \hat{x}(y, \lambda, k) : \nabla_x F(\hat{x}, y, \lambda, k) = 0$$

and

$$\hat{\lambda} \equiv \hat{\lambda}(y, \lambda, k) = \left(\hat{\lambda}_i(y, \lambda, k) = \lambda_i \Delta(y, \hat{x}) \Delta_i^{-1}(\hat{x}, y, k), \quad i = 1, \ldots, m\right).$$

(2) *If* f, $c_i \in C^2$, $i = 1, \ldots, m$ *and second-order sufficient optimality conditions take place, then*

(a) *for any triple* $(y, \lambda, k) : \lambda \in \Lambda_{y,k}$, *there exists a minimizer* $\hat{x} = \hat{x}(y, \lambda, k)$ *such that*

$$\nabla_x F(\hat{x}, y, \lambda, k) = 0$$

and for the primal–dual pair $(\hat{x}, \hat{\lambda})$, *bound (8.80) holds and* $c > 0$ *is independent on* y *and* k;

(b) *for a given* $y \in \Omega_\tau$ *the MIDF* $F(x, y, \lambda, k)$ *is strongly convex in the neighborhood of* $\hat{x}$, *and there exists* $\hat{m} > 0$ *and* $\hat{M} < \infty$ *independent on* $\lambda \in \Lambda_{y,k}$ *and* $k > k_0 \Delta(y, x^*) + \sum_{i=1}^m \lambda_i^*$, *such that*

$$\text{mineigenval}\, \nabla^2_{xx} F(\hat{x}, y, \lambda, k) \geq \Delta^{-1}(y, x^*) \hat{m} \tag{8.81}$$

$$\text{maxeigenval}\, \nabla^2_{xx} F(\hat{x}, y, \lambda, k) \leq \Delta^{-1}(y, x^*) \hat{M} \tag{8.82}$$

8.3 Nonlinear Rescaling vs. Smoothing Technique

8.3.0 Introduction

Smoothing technique for constrained optimization employs a smooth approximation of $x_- = \min\{x, 0\}$ to transform a constrained optimization problem into a sequence of unconstrained minimization problems (see Chen and Mangasarian (1995)). The convergence of the unconstrained minimizers to the primal solution in value is due to the unbounded increase of the scaling parameter. So the smoothing technique is, in fact, a SUMT-type method with a smooth penalty function (see Chapter 5).

There are few well-known difficulties associated with SUMT; therefore we consider an alternative approach, which is a realization of the NR principle with "dynamic" scaling parameters (see Chapter 7).

We define the log-sigmoid (LS) transformation as follows:

$$\psi(t) = 2 \ln 2S(t, 1),$$

where $S(t, 1) = (1 + exp(-t))^{-1}$ is the sigmoid function, which is widely used in neural network literature (see Mangasarian (1993)).

The LS transformation is a modification of the LS smoothing function, used by Chen and Mangasarian (1995) for solving convex inequalities and linear complementarity problems (see also Kanzow (1996), Auslender et al. (1997) and references therein).

We use the LS transformation ψ for rescaling constraints of a given convex optimization problem into equivalent set of constraints. For rescaling we use a positive vector of scaling parameters one for each constraint.

The correspondent LEP we call LS Lagrangian, which is our main tool in this section.

The LS multipliers method finds primal LS Lagrangian minimizer (or its approximation) under fixed both LM and scaling parameters. Then, the minimizer is used for both LM and scaling parameters update.

There are four basic reasons for using the LS transformation and correspondent LEP $\mathscr{L}$:

(1) $\psi \in C^\infty$ on $(-\infty, \infty)$;
(2) the LS Lagrangian is as smooth as the input data;
(3) ψ' and ψ'' are bounded on $(-\infty, \infty)$.
(4) due to the properties of ψ and its LF transform ψ^*, the LS Lagrangian enjoys the best properties of both quadratic and nonquadratic augmented Lagrangians.

In the first part of this section, we provide convergence proof and estimate the number of steps needed for finding an ε - approximation for the solution in value.

There are two basic ingredients of the convergence analysis.

The first is the equivalence of the LS multipliers method and the interior prox method with second-order entropy-like φ-divergence distance function. The distance function is based on the LS kernel $\varphi = -\psi^*$, which happened to be Fermi–Dirac (FD) entropy function.

The second is the equivalence of the prox method with second-order φ-divergence distance to the interior quadratic prox (IQP) in the rescaled from step to step dual space.

Under the second-order sufficient optimality condition, LS multipliers method converges with Q-linear rate without unbounded increase of scaling parameters, which correspond to the active constraints.

In the second part of this section, we extend our analysis on a broad class of smoothing functions. In particular, we introduce truncated Chen–Harker–Kanzow–Smale (CHKS) smoothing function and show that the basic LS results remain true for the truncated CHKS transformation.

Then, we consider truncated exponential transformation. Convergence of the correspondent NR with "dynamic" scaling parameters update follows from the results obtained.

Finally, application of LS method for LP calculation generates dual sequence, which converges in value to the dual solution for nondegenerate LP with quadratic rate.

8.3.1 Log-Sigmoid Transformation and Its Modification

The LS transformation $\psi : \mathbb{R} \to (-\infty, 2\ln 2)$ is defined by formula (Fig. 8.1)

$$\psi(t) = 2\ln 2S(t,1) = 2\ln 2(1+e^{-t})^{-1} = 2(\ln 2 + t - \ln(1+e^{t})). \qquad (8.83)$$

We collect the basic properties of the LS transformation in the following proposition.

Proposition 8.4. *The LS transformation* ψ *has the following properties:*

A1. $\psi(0) = 0$;
A2. $0 < \psi'(t) = 2(1+e^{t})^{-1} < 2, \ \forall t \in (-\infty, +\infty), \ \psi'(0) = 1; \ \lim_{t\to\infty} \psi'(t) = 0$
A3. $-0.5 \le \psi''(t) = -2e^{t}(1+e^{t})^{-2} < 0, \ -\infty < t < +\infty.$

The LS transformation ψ is defined on $(-\infty, +\infty)$ together with its derivatives of any order, which makes LS transformation substantially different from transformations considered so far.

On the top of A3, which distinguishes ψ from both exponential and barrier transformations, it leads to an important property of ψ^* – the LF conjugate of ψ.

Due to A2, for any $s \in (0,2)$, the equation $st - \psi(t)' = 0$ can be solved for t, so $\psi'^{-1}(s) = t(s) = \ln(2s^{-1} - 1)$.

The LS conjugate

$$\psi^*(s) = st(s) - \psi(t(s)) = (s-2)\ln(2-s) - s\ln s$$

is, in fact, a Fermi–Dirac entropy function.

Assuming $t\ln t = 0$ for $t = 0$, we obtain $-2\ln 2 \le \psi^*(s) \le 0$ for $0 \le s \le 2$, and from $\psi^*(1) = 0$ follows $\max\{\psi^*(s) \mid 0 \le s \le 2\} = \psi^*(1) = 0$.

We call the function $\varphi : [0,2] \to [0, 2\ln 2]$, defined by formula

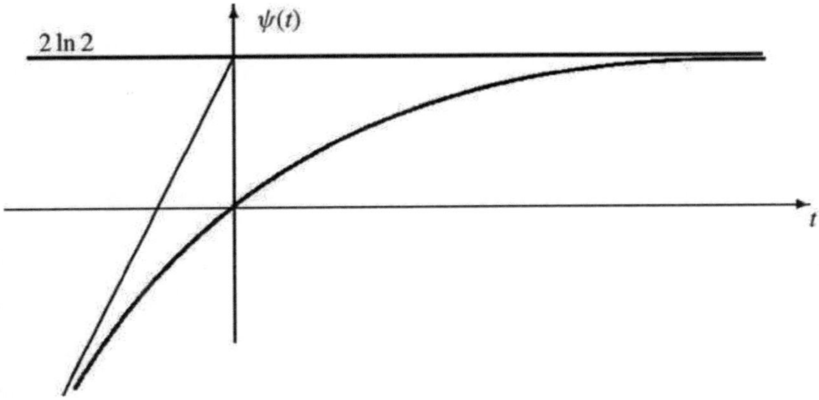

Fig. 8.1 LS transformation

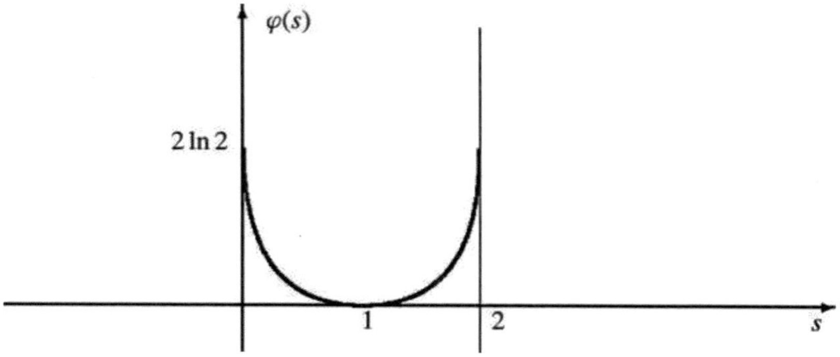

Fig. 8.2 Fermi–Dirac kernel

$$\varphi(s) = -\psi^*(s) = (2-s)\ln(2-s) + s\ln s, \tag{8.84}$$

a Fermi–Dirac (FD) kernel.

From (2.75) and A3 follows $\psi^{*''}(s) \le \psi^{*''}(1) = -2$, $\forall s \in]0,2[$; therefore

$$\varphi''(s) \ge 2, \ \ \forall s \in]0,2[, \tag{8.85}$$

which are important properties (see Fig. 8.2) that neither MBF nor exponential kernels possess.

The properties of FD kernel are given in the following proposition.

Proposition 8.5. *The FD kernel has the following properties:*

B1. $\varphi(s)$ *is nonnegative and strongly convex on* $]0,2[$ *and* $\varphi(1) = 0$*;*

B2. $\varphi'(s) = \ln s - \ln(2-s)$, $\varphi'(1) = 0$, $\lim_{s\to 0^+}\varphi'(s) = -\infty$, $\lim_{s\to 2^-}\varphi'(s) = +\infty$*;*

B3. $\varphi''(s) = s^{-1} + (2-s)^{-1} \ge 2$, $\forall s \in]0,2[$, $\lim_{s\to 0^+}\varphi''(s) = \lim_{s\to 2^-}\varphi''(s) = \infty$.

On the other hand, we have $\psi''(t) = -2e^t(1+e^t)^{-2}$; therefore $\lim_{t\to -\infty}\psi''(t) = 0$. Thus, LST ψ does not satisfy property (4) (see Section 7.1.2)

We conclude the section by introducing truncated LST ψ, which satisfy property (4).

There are two main reasons for ψ to be modified. First, from A2 follows

$$1 < \psi'(t) < 2, \ \forall t \in]-\infty, 0[. \tag{8.86}$$

The critical element in the NR multipliers method (7.55)–(7.57) is the formula (7.56) for the Lagrange multipliers update

$$\lambda_{i,s+1} = \lambda_{i,s}\psi'(k_{i,s}c_i(x_{s+1})), \ i = 1,\dots,m.$$

From (8.86) follows $\lambda_{i,s+1} < 2\lambda_{i,s}$, $i = 1,\ldots,m$. It means, that none of the Lagrange multiplier can be increased more then twice, independent of the constraints violation and the value of the scaling parameter $k_{i,s} > 0$.

On the other hand, due to (7.56), for any $i : c_i(x_{s+1}) > 0$, the Lagrange multiplier $\lambda_{i,s}$ can be practically reduced to zero if $c_i(x_{s+1})$ or $k_{i,s} > 0$ or both are large enough. Such asymmetry can compromise the convergence.

The second reason also related to (8.86). We have $s = \psi'(t) \to 2^- \Rightarrow \psi''(t) \to 0$ and $\psi''(t) < 0$, so $s \to 2^- \Rightarrow \psi^{*''}(s) = (\psi''(t))^{-1} \to -\infty$ or

$$s \to 2^- \Rightarrow \varphi''(s) \to \infty. \tag{8.87}$$

The property (8.87) can compromise the convergence of the dual interior prox, which is equivalent to the NR method.

Therefore, we will use the truncated version of LST (see Fig. 8.3), given by formula

$$\psi(t) := \begin{cases} \psi(t) = 2\ln 2S(t,1), \; t \geq -\ln 2 \\ q(t) = at^2 + bt + c, \; t \leq -\ln 2. \end{cases} \tag{8.88}$$

We find the parameters of the parabolic branch from the following system

$$\psi(-\ln 2) = q(-\ln 2), \; \psi'(-\ln 2) = q'(-\ln 2) \text{ and } \psi''(-\ln 2) = q''(-\ln 2),$$

to ensure that $\psi \in C^2$. By direct calculation, we obtain $a = -2/9$, $b = (4/3)(1 - (1/3)\ln 2)$, $c = (10/3)\ln 2 - (2/9)\ln^2 2 - 2\ln 3$.

So the properties A1. –A3. remain true for the truncated LS transformation $\bar{\psi}$.

For the first and second derivative, we have

$$\psi'(t) := \begin{cases} 2(1+e^t)^{-1}, & t \geq -\ln 2 \\ -4/9t + 4/3(1 - 1/3\ln 2), & t \leq -\ln 2. \end{cases}$$

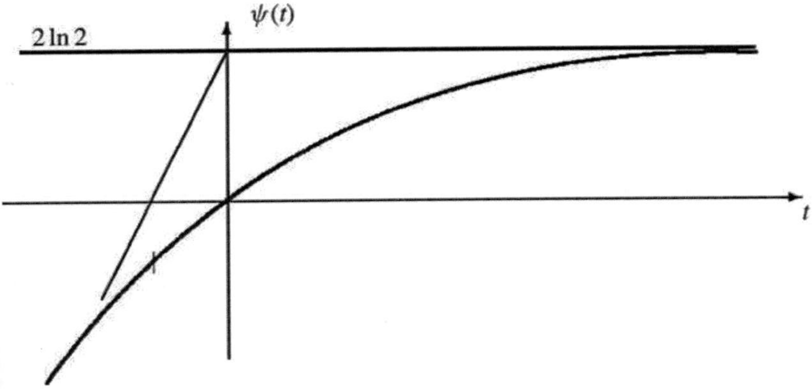

Fig. 8.3 Truncated LS transformation

$$\psi''(t) := \begin{cases} -2e^t(1+e^t)^{-2}, & t \geq -\ln 2 \\ -4/9, & t \leq -\ln 2. \end{cases}$$

and

A4. $\psi''(t) \leq -4/9$, $-\infty < t \leq 0$.

The correspondent Fenchel conjugate $\psi^* : (0,\infty) \to (-\infty, 0)$ is defined as follows:

$$\psi^*(s) := \begin{cases} \psi^*(s), & 0 < s \leq 4/3 = \psi'(-\ln 2) \\ q^*(s) = (4a)^{-1}(s-b)^2 - c, & 4/3 \leq s < \infty. \end{cases}$$

For the truncated Fermi–Dirac kernel $\varphi : (0,\infty) \to (0,\infty)$, we have

$$\varphi(s) := \begin{cases} -\psi^*(s), & 0 < s \leq 4/3 \\ -q^*(s), & 4/3 \leq s < \infty \end{cases}. \tag{8.89}$$

The truncated FD kernel φ has the following properties (see Fig. 8.4):

B1. $\varphi(s) \geq 0$, $s \in (0,\infty)$, $\varphi(1) = 0$;
B2. φ' is monotone increasing on $]0,+\infty[$ $\lim_{s\to 0^+} \varphi'(s) = -\infty$, $\lim_{s\to\infty} \varphi'(s) = +\infty$, $\varphi'(1) = 0$;
B3. $\varphi''(s) \geq 2$, $\forall s \in (0,\infty)$.

Also for the truncated FD kernel, we have

$$\varphi''(s) := \begin{cases} \varphi''(s), & 0 < s \leq 4/3 \\ -(2a)^{-1} = 2.25, & 4/3 \leq s < \infty. \end{cases}$$

So, along with B1–B3, the truncated FD kernel φ possesses the following property:

B4. $$\varphi''(s) = -\psi^{*''} \leq 2.25, \quad \forall s \in [1,\infty),$$

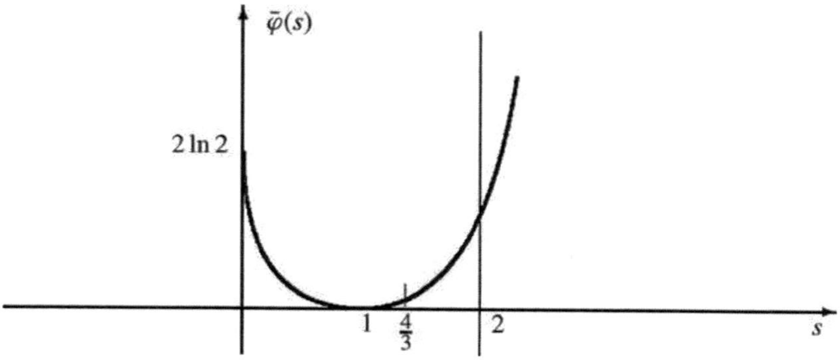

Fig. 8.4 Truncated Fermi–Dirac kernel

which plays an important role in the future analysis of LS multipliers method.

The truncated LS transformation relates to LS transformation as truncated MBF (8.22) relates to MBF. Motivations, however, for considering instead of MBF and LS and their truncate versions are fundamentally different.

We shall see later that the parabolic branch of ψ is used only a finite number of steps. In fact, for the scaling parameter $k > 0$ large enough, it can happen just once. Hence, eventually only the LS transformation governs the computational process in primal space, and only FD kernel $\varphi = -\psi^*$ does it in the dual space.

We retain ψ and φ as our notation, keeping in mind that we are dealing with truncated LS, which possesses properties A1–A4 and truncated FD kernel φ with properties B1–B4 .

8.3.2 Equivalent Problem and LS Lagrangian

For any given vector $\mathbf{k} = (k_1, \ldots, k_m) \in \mathbb{R}^m_{++}$, due to A1–A2, we have

$$c_i(x) \geq 0 \Leftrightarrow k_i^{-1}\psi(k_i c_i(x)) \geq 0, \quad i = 1, \ldots, m. \tag{8.90}$$

Therefore, problem

$$f(x^*) = \min\left\{f(x) \mid k_i^{-1}\psi(k_i c_i(x)) \geq 0, \quad i = 1, \ldots, m\right\} \tag{8.91}$$

is equivalent to the original problem (P).

The LEP $\mathscr{L} : \mathbb{R}^n \times \mathbb{R}^m_{++} \times \mathbb{R}^m_{++} \to \mathbb{R}$

$$\mathscr{L}(x, \lambda, \mathbf{k}) = f(x) - \sum_{i=1}^{m} \lambda_i k_i^{-1}\psi(k_i c_i(x)) \tag{8.92}$$

is called log-sigmoid Lagrangian (LSL).

If $k_1 = \cdots = k_m = k$, then

$$\mathscr{L}(x, \lambda, k) \equiv \mathscr{L}(x, \lambda, k) = f(x) - k^{-1}\sum_{i=1}^{m} \lambda_i \psi(k c_i(x)). \tag{8.93}$$

8.3.3 LS Multipliers Method as Interior Prox with Fermi–Dirac Entropy Distance

We use the truncated LS transformation, given by (8.88), in the framework of the NR method (7.55)–(7.57).

Let $\lambda_0 = e = (1, \ldots, 1)^T$, $\mathbf{k}_0 = (k_{1,0}, \ldots, k_{m,0})$, $k_{i,0} = k/\lambda_{i,0} = k$, $i = 1, \ldots, m$, and $k > 0$.

The LS multipliers method generates three sequences $\{x_s\}_{s\in N}$, $\{\lambda_s\}_{s\in N}$, and $\{\mathbf{k}_s\}_{s\in N}$ by formulas

$$x_{s+1} \ : \ \nabla_x \mathscr{L}(x_{s+1},\lambda_s,\mathbf{k}) = 0 \tag{8.94}$$

$$\lambda_{s+1} : \lambda_{i,s+1} = \lambda_{i,s}\psi'\left(k_{i,s}c_i\left(x_{s+1}\right)\right), \quad i=1,\ldots,m \tag{8.95}$$

$$\mathbf{k}_{s+1} : k_{i,s+1} = k\left(\lambda_{i,s+1}\right)^{-1}, \quad i=1,\ldots,m. \tag{8.96}$$

Function $D_2 : \mathbb{R}^m_+ \times \mathbb{R}^m_{++} \to \mathbb{R}_+$

$$D_2(u,\mathrm{v}) = \sum_{i=1}^{m} \mathrm{v}_i^2 \varphi(u_i/\mathrm{v}_i), \tag{8.97}$$

where φ is truncated FD kernel, given by (8.89), is called second-order Fermi–Dirac entropy distance function.

It follows from Theorem 7.4 that NR method (8.94)–(8.96) is equivalent to the following prox method:

$$\begin{aligned} d(\lambda_{s+1}) - k^{-1}D_2(\lambda_{s+1},\lambda_s) &= \max\left\{ d(\lambda) - k^{-1}\sum_{i=1}^{m}\left(\lambda_{i,s}\right)^2\varphi\left(\lambda_i/\lambda_{i,s}\right) \mid \lambda \in \mathbb{R}^m_+ \right\} \\ &= \max\left\{ d(\lambda) - k^{-1}D_2(\lambda,\lambda_s) \mid \lambda \in \mathbb{R}^m_+ \right\}. \end{aligned} \tag{8.98}$$

The LEP $\mathscr{L}(x,\lambda,k)$, given by (8.93), leads to the following multipliers method:

$$x_{s+1} : \nabla_x \mathscr{L}(x_{s+1},\lambda_s,k) = 0$$

$$\lambda_{s+1} : \lambda_{i,s+1} = \lambda_{i,s}\psi'\left(kc_i(x_{s+1})\right),\ i=1,\ldots,m,$$

which is equivalent (see Theorem 7.1) to the following interior prox:

$$\lambda_{s+1} = \max\{d(\lambda) - k^{-1}\sum_{i=1}^{m}\lambda_{i,s}\varphi(\lambda_i/\lambda_{i,s}) | \lambda \in \mathbb{R}_{++}\}$$

with first-order FD distance

$$D_1(u,\mathrm{v}) = \sum_{i=1}^{m} \mathrm{v}_i\,\varphi(u_i/\,\mathrm{v}_i).$$

Note that first- and second-order FD distances have the basic characteristics of distance functions:

(a) $D_i(u,\mathrm{v}) \geq 0$, $u \in \mathbb{R}^m_+$, $\mathrm{v} \in \mathbb{R}^m_{++}$, $i = 1,2$
(b) $D_i(u,\mathrm{v}) = 0 \Leftrightarrow u = \mathrm{v}$

The following theorem is a consequence of Theorems 7.4 and 7.5. We provide the proof to remind the basic facts, which we will need later.

Theorem 8.11. *If for a convex programming problem* (P) *assumptions A and B are satisfied, f and all c_i, $i = 1, \ldots, m$ are C^1 functions, then for any $k > 0$ and* $(\lambda_0, \mathbf{k}_0) \in \mathbb{R}^m_{++} \times \mathbb{R}^m_{++}$:

(1) the LS method (8.94)–(8.96) is well defined;
(2) the LS method (8.94)–(8.96) is equivalent to the proximal point method (8.98);
(3) the proximal point method (8.98) *is equivalent to the following interior quadratic prox (IQP):*

$$d(\lambda_{s+1}) - \frac{1}{2}k^{-1}\sum_{i=1}^{m}\varphi''(1+\theta_{i,s}(\lambda_{i,s+1}/\lambda_{i,s}-1))(\lambda_{i,s+1}-\lambda_s)^2$$
$$= \operatorname{argmax}\{d(\lambda) - 0.5k^{-1}\sum_{i=1}^{m}\varphi''(\cdot)_{i,s}(\lambda_i-\lambda_{i,s})^2 | \lambda \in \mathbb{R}^m_{++}\}, \tag{8.99}$$

where $0 < \theta_{i,s} < 1$.
(4) the IQP (8.99) *is equivalent to the following rescaled implicit subgradient method for the dual problem:*

$$\lambda_{i,s+1} = \lambda_{i,s} + k\psi''(\cdot)_{i,s}c_i(x_{s+1}),\ i = 1, .., m, \tag{8.100}$$

where $\psi''(\cdot)_{i,s} = \psi''(1+\theta_{i,s}(\lambda_{i,s+1}/\lambda_{i,s}-1))$ *and* $0 < \theta_{i,s} < 1$.

Proof.

(1) Keeping in mind Remark 7.1, we can assume boundedness of Ω, then from convexity f, concavity c_i, and Slater condition as well as A4 follows emptiness of $RC(\Omega)$. In other words, for any $x \in \Omega$ and $(\lambda, \mathbb{k}) \in \mathbb{R}^m_{++} \times \mathbb{R}^m_{++}$ follows

$$\lim_{t\to\infty} \mathscr{L}(x+td, \lambda, \mathbf{k}) = \infty$$

for any direction $d \neq 0$ from $\mathbb{R}^n$.

(2) Therefore, there exists x_{s+1}

$$\nabla_x \mathscr{L}(x_{s+1}, \lambda_s, \mathbf{k}) = \nabla f(x_{S+1}) - \sum_{i=1}^{m}\psi'(k_{i,s}c_i(x_{s+1}))\lambda_{i,s}\nabla c_i(x_{s+1}) =$$

$$\nabla f(x_{s+1}) - \sum_{i=1}^{m}\lambda_{i,s+1}\nabla c_i(x_{s+1}) = \nabla_x L(x_{s+1}, \lambda_{s+1}) = 0. \tag{8.101}$$

It means that $d(\lambda_{s+1}) = L(x_{s+1}, \lambda_{s+1}) = \min_{s\in\mathbb{R}^n} L(x, \lambda_{s+1})$.
Moreover, from A2 and (8.95) follows $\lambda_s \in \mathbb{R}^m_{++} \to \lambda_{s+1} \in \mathbb{R}^m_{++} \Rightarrow \mathbf{k}_{s+1} \in \mathbb{R}^m_{++}$; therefore LS multipliers method (8.94)-(8.96) is well defined.
Further

$$(-c_1(x_{s+1}), \ldots, -c_m(x_{s+1}))^T \in \partial d(\lambda_{s+1}), \tag{8.102}$$

where $\partial d(\lambda_{s+1})$ is the subdifferential of $d(\lambda)$ at λ_{s+1}. From (8.102) follows

$$0 \in \partial d(\lambda_{s+1}) + (c_1(x_{s+1}), \ldots, c_m(x_{s+1}))^T. \tag{8.103}$$

From the update formulas (8.95), we have

$$c_i\left(x_{s+1}\right)=\left(k_{i,s}\right)^{-1}\psi'^{-1}\left(\lambda_{i,s+1}/\lambda_{i,s}\right),\quad i=1,\ldots,m. \tag{8.104}$$

The existence of ψ'^{-1} is guaranteed by $\psi''(t)<0$, $-\infty<t<\infty$.
The inclusion (8.103) can be rewritten as follows:

$$0\in\partial d\left(\lambda_{s+1}\right)+$$

$$\left(\left(k_{1,s}\right)^{-1}\psi'^{-1}\left(\lambda_{1,s+1}/\lambda_{1,s}\right),\ldots,\left(k_{m,s}\right)^{-1}\psi'^{-1}\left(\lambda_{m,s+1}/\lambda_{m,s}\right)\right)^T. \tag{8.105}$$

Using LF identity $\psi'^{-1}=\psi^{*'}$, we obtain

$$c_i\left(x_{s+1}\right)=\left(k_{i,s}\right)^{-1}\psi^{*'}\left(\lambda_{i,s+1}/\lambda_{i,s}\right),\quad i=1,\ldots,m. \tag{8.106}$$

Using $\varphi=-\psi^*$, we can rewrite (8.105) as follows:

$$0\in\partial d\left(\lambda_{s+1}\right)-$$

$$\left(\left(k_{1,s}\right)^{-1}\varphi'\left(\lambda_{1,s+1}/\lambda_{1,s}\right),\ldots,\left(k_{m,s}\right)^{-1}\varphi'\left(\lambda_{m,s+1}/\lambda_{m,s}\right)\right)^T. \tag{8.107}$$

From B3, concavity d, and (8.96) follow that inclusion (8.107) is the optimality criteria for $\lambda_{s+1}\in\mathbb{R}^m_{++}$ to be an unconstrained maximizer in (8.98).

(3) From $\varphi(1)=\varphi'(1)=0$ follows

$$\varphi\left(\frac{u}{v}\right)=\varphi''\left(1+\theta_{u,v}\left(\frac{u}{v}-1\right)\right)\left(\frac{u}{v}-1\right)^2,$$

where $0<\theta_{u,v}<1$ due to strong convexity φ. By taking $u=\lambda$ and $v=\lambda_s$ from (8.107), we obtain (3).
The dual approximation λ_{s+1}, for any $s\geq 1$, is unique due to concavity d and strong concavity of the prox term in (8.99).

(4) from $\psi'(0)=1$ and the update formula (8.95) follows:

$$\lambda_{i,s+1}-\lambda_{i,s}=\lambda_{i,s}(\psi'(k_{i,s}c_i(x_{s+1}))-\psi'(0))=$$

$$\lambda_{i,s}k_{i,s}\psi''(\theta_{i,s}k_{i,s}c_i(x_{s+1}))c_i(x_{s+1}),\ \imath=1,\ldots,m. \tag{8.108}$$

Keeping in mind (8.96), we obtain

$$\lambda_{s+1}=\lambda_s+k\psi''(\cdot)_s c(x_{s+1}), \tag{8.109}$$

where $\psi''(\cdot)_s=\mathrm{diag}[\psi''(\theta_{i,s}k_{i,s}c_i(x_{x+1}))]_{i=1}^m$ and $0<\theta_{i,s}<1$.
If d is smooth, then $\nabla d(\lambda_{s+1})=-c(x_{s+1})$ and (8.109) is a rescaled implicit gradient method for the dual problem

$$\lambda_{s+1}=\lambda_s+k(-\psi''(\cdot)_s)\nabla d(\lambda_{s+1}).$$

Let us introduce the following diagonal matrix:

$$H_s = \text{diag}[\varphi''(1+\theta_{i,s}(\lambda_{i,s+1}/\lambda_{i,s}-1)]_{i=1}^m = \text{diag}[\varphi''(\cdot)_{i,s}]_{i=1}^m = \text{diag}[h_{i,s}]_{i=1}^m.$$

From $B3$ and $B4$ follows

$$(a) \quad 2 \le h_{i,s} \le 2.25,\ \forall s \in (0,\infty)$$

and from A3 and A4 follows (8.110)

$$(b) -0.5 \le \psi''(\cdot)_{i,s} \le -4/9, \quad -\infty < t < \infty.$$

Using H_s we can rewrite the prox method (8.99) as following IQP

$$d(\lambda_{s+1}) - \frac{1}{2}k^{-1}(\lambda_{s+1}-\lambda_s)^T H_s(\lambda_{s+1}-\lambda_s) =$$

$$= \max\{d(\lambda) - \frac{1}{2}k^{-1}(\lambda-\lambda_s)^T H_s(\lambda-\lambda_s)|\lambda \in \mathbb{R}^m_{++}\}. \tag{8.111}$$

In other words, the LS multipliers method (8.94)–(8.96) is equivalent to IQP (8.111) in the rescaling from step to step dual space.

Note the classical AL method is equivalent quadratic prox in the dual space.

8.3.4 Convergence of the LS Multipliers Method

The convergence of LS method (8.94)–(8.96) follows from Theorem 7.6. We provide a new proof with arguments, which will be used later.

Theorem 8.12. *Under conditions of Theorem 8.11 the LS multipliers method* (8.94)–(8.96) *generates primal–dual sequence* $\{x_s, \lambda_s\}_{s\in N}$*:*

(1) the dual sequence is monotone increasing in value, that is $d(\lambda_{s+1}) > d(\lambda_s)$, $s \ge 0$*;*

(2) the dual level sets $L_s = \{\lambda \in \mathbb{R}^m_+ : d(\lambda) \ge d(\lambda_s)\}$ *are bounded and the Hausdorff distances* $d_H(L^*, \partial L_s)$ *monotone decreasing, that is*

$$d_H(L^*, \partial L_{s+1}) < d_H(L^*, \partial L_s),\ \forall s \ge 0, \tag{8.112}$$

there exists $0 < L < \infty$ *that*

$$\max_{i,s}\{\lambda_{i,s}\} = L; \tag{8.113}$$

(3)

$$\lim_{s\to\infty} \|\lambda_{s+1}-\lambda_s\| = 0 \Rightarrow \lim_{s\to\infty} |\lambda_{i,s+1}-\lambda_{i,s}| = 0,\ i = 1,\ldots,m; \tag{8.114}$$

(4) asymptotic complementarity

$$\lim_{s\to\infty} \lambda_{i,s+1}c_i(x_{s+1}) = 0,\ i = 1,\ldots,m \tag{8.115}$$

holds;

(5) dual sequence converges to the optimal solution in value, that is

$$\lim_{s \to \infty} d(\lambda_s) = d(\lambda^*);$$

(6) $\lim d_H(L^*, \partial L_s) = 0$*;*

(7) primal sequence converges to the optimal solution in value, that is

$$\lim_{s \to \infty} f(x_s) = f(x^*).$$

Proof.

(1) From (8.111) with $\lambda = \lambda_s$ follows

$$\begin{aligned} d(\lambda_{s+1}) &\geq d(\lambda_s) + (2k)^{-1}(\lambda_{s+1} - \lambda_s)^T H_s(\lambda_{s+1} - \lambda_s) \\ &= d(\lambda_s) + (2k)^{-1}\|\lambda_{s+1} - \lambda_s\|^2_{H_s}, \end{aligned} \tag{8.116}$$

where $H_s = \text{diag}[h_{i,s}]_{i=1}^m$ and $2 \geq h_{i,s} \leq 2.25$.

(2) Therefore, $d(\lambda_{s+1}) > d(\lambda_s)$, unless $\lambda_{s+1} = \lambda_s = \lambda^*$.
From the boundedness of L^* follows boundedness dual level sets $L_s = \{\lambda \in \mathbb{R}^m_+ : d(\lambda) \geq d(\lambda_s)\}$, $s \geq 0$. From $d(\lambda_{s+1}) > d(\lambda_s)$ follows

$$L^* \subset \ldots L_{s+1} \subset L_s \ldots \subset L_0. \tag{8.117}$$

Therefore, (8.112) and (8.113) hold.

(3) Summing up (8.111) from $s = 0$ to $s = N$, we obtain

$$d(\lambda^*) - d(\lambda_0) > d(\lambda_{N+1}) - d(\lambda_0) \geq \frac{1}{2k} \sum_{s \in N} \|\lambda_{s+1} - \lambda_s\|^2_{H_s}. \tag{8.118}$$

Therefore, $\lim_{s \to \infty} \|\lambda_{s+1} - \lambda_s\|^2_{H_s} = 0$ and from (8.110a) and (8.118) follows item (3).

(4) From (8.108) follows

$$|\lambda_{i,s+1} - \lambda_{i,s}| = k\lambda^{-1}_{i,s+1}|\psi''(\cdot)||c_i(x_{s+1})\lambda_{i,s+1}|.$$

From (8.110b) and (8.113), we obtain

$$0 = \lim_{s \to \infty} (2k)^{-1} L|\lambda_{i,s+1} - \lambda_{i,s}| \geq \lim_{s \to \infty} |c_i(x_{s+1})\lambda_{i,s+1}|,\ i = 1, \ldots, m,$$

so item (4) holds.

(5) The sequence $\{d(\lambda_s)\}_{s \in N}$ is monotone increasing and bounded from above by $f(x^*)$; therefore there exists

$$\bar{d} = \lim_{s \to \infty} d(\lambda_s) \leq d(\lambda^*).$$

Our next step is to show that $\bar{d} = d(\lambda^*)$.
Let's assume the opposite; then there exists $\rho > 0$ that $f(\lambda^*) - d(\lambda_s) \geq \rho > 0$ for any $s \geq 1$.
Let us consider the optimality criteria for λ_{s+1} in (8.111). There exists subgradient $g_{s+1} \in \partial d(\lambda_{s+1})$, that

$$\langle \lambda - \lambda_{s+1}, kg_{s+1} - H_s(\lambda_{s+1} - \lambda_s) \rangle \leq 0, \ \forall \lambda \in \mathbb{R}^m_+. \tag{8.119}$$

Then

$$k\langle \lambda - \lambda_{s+1}, g_{s+1} \rangle \leq (\lambda - \lambda_{s+1})^T H_s(\lambda_{s+1} - \lambda_s). \tag{8.120}$$

From concavity d and (8.120) for any $\lambda \in \mathbb{R}^m_+$ follows

$$k(d(\lambda_{s+1}) - d(\lambda)) \geq k\langle g_{s+1}, \lambda_{s+1} - \lambda \rangle \geq (\lambda - \lambda_{s+1})^T H_s(\lambda_s - \lambda_{s+1}).$$

Using the three-point identity

$$\langle a - b, c - b \rangle = \frac{1}{2}[\|a - b\|^2 + \|b - c\|^2 - \|a - c\|^2]$$

with $a = H_s^{\frac{1}{2}}\lambda, b = H_s^{\frac{1}{2}}\lambda_{s+1}, c = H_s^{\frac{1}{2}}\lambda_s$, we obtain

$$k(d(\lambda_{s+1}) - d(\lambda)) \geq (\lambda - \lambda_{s+1})^T H_s(\lambda_s - \lambda_{s+1}) = \frac{1}{2}(\lambda - \lambda_{s+1})^T H_s(\lambda - \lambda_{s+1}) +$$

$$\frac{1}{2}(\lambda_{s+1} - \lambda_s)^T H_s(\lambda_{s+1} - \lambda_s) - \frac{1}{2}(\lambda - \lambda_s)^T H_s(\lambda - \lambda_s). \tag{8.121}$$

For any $\lambda^* \in L^*$ from (8.121) follows

$$2k(d(\lambda_{s+1}) - d(\lambda^*)) \geq \|\lambda^* - \lambda_{s+1}\|^2_{H_s} - \|\lambda^* - \lambda_s\|^2_{H_s} \tag{8.122}$$

or

$$2\rho k \leq k(d(\lambda^*) - d(\lambda_{s+1})) \leq \|\lambda^* - \lambda_s\|^2_{H_s} - \|\lambda^* - \lambda_{s+1}\|^2_{H_s}. \tag{8.123}$$

Then, keeping in mind (8.110a) and (8.113), we obtain

$$2\rho k \leq k(d(\lambda^*) - d(\lambda_{s+1}) \leq$$

$$\|\lambda^* - \lambda_s\|^2_{H_s} - \|\lambda^* - \lambda_{s+1}\|^2_{H_s} = \left[\sum_{i=1}^m h_{i,s}(\lambda_i^* - \lambda_{i,s})^2 - \sum_{i=1}^m h_{i,s}(\lambda_i^* - \lambda_{i,s+1})^2 \right] =$$

$$\sum_{i=1}^m 2h_{i,s}\lambda_i^*(\lambda_{i,s+1} - \lambda_{i,s}) + \sum_{i=1}^m h_{i,s}(\lambda_{i,s} + \lambda_{i,s+1})(\lambda_{i,s} - \lambda_{i,s+1}) \leq$$

$$4.5 \left[\sum_{i=1}^m \lambda_i^* |\lambda_{i,s+1} - \lambda_{i,s}| + \sum_{i=1}^m L|\lambda_{i,s} - \lambda_{i,s+1}| \right]. \tag{8.124}$$

From (8.114) follows that (8.124) impossible for $s > 0$ large enough. Therefore

$$\lim_{s\to\infty} d(\lambda_s) = d(\lambda^*). \tag{8.125}$$

(6) Item (6) follows directly from (8.125)
(7) From asymptotic complementarity (8.115) and (8.125) follows

$$d(\lambda^*) = \lim_{s\to\infty} d(\lambda_s) = \lim_{s\to\infty} L(x_s, \lambda_s) =$$

$$\lim_{s\to\infty}\left(f(x_s) - \sum_{i=1}^{m} \lambda_{i,s} c_i(x_s)\right) = \lim_{s\to\infty} f(x_s) = f(x^*)$$

□

8.3.5 The Upper Bound for the Number of Steps

Let's consider again the optimality criteria for $\lambda_{s+1} \in \mathbb{R}^m_{++}$ in (8.111). There exists a subgradient $g_{s+1} \in \partial d(\lambda_{s+1})$ that

$$g_{s+1} - k^{-1} H_s(\lambda_{s+1} - \lambda_s) = 0. \tag{8.126}$$

Let's consider interior ellipsoid

$$E(\lambda_s, r_s) = \{\lambda \in \mathbb{R}^m_{++} : (\lambda - \lambda_s)^T H_s(\lambda - \lambda_s) = r_s^2\} \subset \mathbb{R}^m_{++},$$

where $r_s^2 = (\lambda_{s+1} - \lambda_s)^T H_s(\lambda_{s+1} - \lambda_s)$.

Then, from (8.126) follows

$$d(\lambda_{s+1}) = \max\{d(\lambda) | \lambda \in E(\lambda_s, r_s)\}. \tag{8.127}$$

It means that LS multipliers method (8.94)–(8.96) is equivalent to the interior ellipsoid method (IEM) (8.127).

We will need the following Lemma.

Lemma 8.2. *Under condition of Theorem 8.12, the following bound holds*

$$d(\lambda^*) - d(\lambda_{s+1}) \le \left(1 - \frac{\|\lambda_{s+1} - \lambda_s\|_{H_s}}{O(\sqrt{m})}\right)(d(\lambda^*) - d(\lambda_s))$$

for any $s \ge 0$.

Proof. Let

$$\|\lambda^* - \lambda_s\| = \min\{\|\lambda - \lambda_s\| | \lambda \in L^*\}.$$

We consider segment $[\lambda^*, \lambda_s] = \{\lambda \in \mathbb{R}^m_{++} : \lambda = (1-t)\lambda_s + t\lambda^* | 0 \le t \le 1\}$ and let $\bar{\lambda}_s = [\lambda^*, \lambda_s] \bigcap \partial E(\lambda_s, r_s)$ where $r_s = \|\lambda_{s+1} - \lambda_s\|_{H_s}$ and $\partial E(\lambda_s, r_s) = \{\lambda : (\lambda - \lambda_s)^T H_s(\lambda - \lambda_s) = r_s^2\}$ is the boundary of the interior ellipsoid $E(\lambda_s, r_s)$ (see Fig 8.5).

From (8.126) follows $d(\lambda_{s+1}) \geq d(\bar{\lambda}_s)$ and

$$\|\bar{\lambda}_s - \lambda_s\|_{H_s} = \|\lambda_{s+1} - \lambda_s\|_{H_s} = r_s.$$

From concavity of $d(\lambda)$ follows $d(\bar{\lambda}_s) \geq \hat{d}(\bar{\lambda}_s)$. Therefore (see Fig 8.5) we have

$$\frac{\hat{d}(\bar{\lambda}_s)}{d(\lambda^*)} = \frac{\|\bar{\lambda}_s - \lambda_s\|}{\|\lambda^* - \lambda_s\|}. \tag{8.128}$$

From (8.110a) follows

$$2\|\bar{\lambda}_s - \lambda_s\|^2 \leq \|\bar{\lambda}_s - \lambda_s\|^2_{H_s} \leq \frac{9}{4}\|\bar{\lambda}_s - \lambda_s\|^2,$$

therefore

$$\|\bar{\lambda}_s - \lambda_s\| \geq \frac{2}{3}\|\bar{\lambda}_s - \lambda_s\|_{H_s}.$$

From boundedness of $L_0 = \{\lambda \in \mathbb{R}^m_+ : d(\lambda) \geq d(\lambda_0)\}$ and (8.117) follows existence $0 < L < \infty$ that $\|\lambda^* - \lambda_s\| \leq L\sqrt{m}$ for $\forall s \geq 0$.

Hence

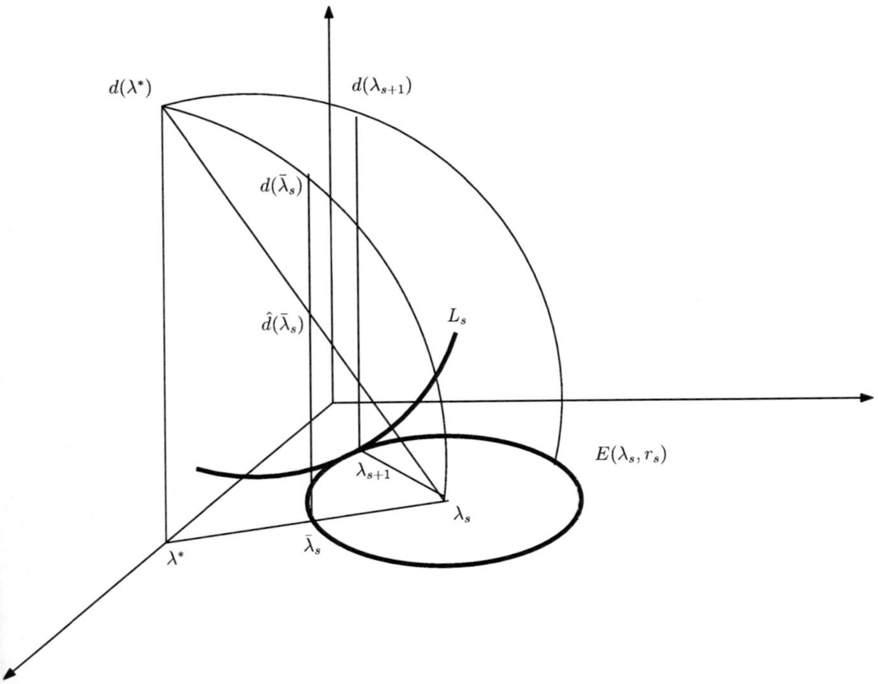

Fig. 8.5 Interior ellipsoid method

$$\frac{\hat{d}(\bar{\lambda}_s)}{d(\lambda^*)} \geq \frac{2}{3}\frac{\|\bar{\lambda}_s - \lambda_s\|_{H_s}}{L\sqrt{m}} = \frac{2}{3}\frac{\|\lambda_{s+1} - \lambda_s\|_{H_s}}{L\sqrt{m}} = \frac{\|\bar{\lambda}_s - \lambda_s\|_{H_s}}{O(\sqrt{m})} \tag{8.129}$$

or

$$\hat{d}(\bar{\lambda}_s) \geq d(\lambda^*)\frac{\|\bar{\lambda}_s - \lambda_s\|_{H_s}}{O(\sqrt{m})}. \tag{8.130}$$

From concavity d follows $d(\bar{\lambda}_s) \geq \hat{d}(\bar{\lambda}_s)$ and from (8.111) follows $d(\lambda_{s+1}) \geq d(\bar{\lambda}_s)$.

Therefore, from (8.129) and $d(\lambda_s) = 0$, which we assume without loss of generality, follows

$$d(\lambda^*) - d(\lambda_{s+1}) \leq d(\lambda^*) - d(\bar{\lambda}_s) \leq \tag{8.131}$$

$$(d(\lambda^*) - d(\lambda_s))\left(1 - \frac{\|\bar{\lambda}_{s+1} - \lambda_s\|_{H_s}}{O(\sqrt{m})}\right). \square$$

So far the scaling parameter $k > 0$ in the LS method (8.94)–(8.96) was fixed.

By increasing $k > 0$ from step to step, one can improve the convergence, but the procedure of finding primal minimizer will cost more.

In the rest of the section, we consider such method and estimate the number of steps required for finding $\varepsilon > 0$ approximation for $d(\lambda^*)$.

By replacing k for k_s in (8.96) and assuming that $k_{s+1} > k_s$, $s \geq 1$ from (8.111), we obtain

$$d(\lambda_s) - d(\lambda_{s-1}) \geq \frac{1}{2k_s}\|\lambda_s - \lambda_{s-1}\|^2_{H_{s-1}}. \tag{8.132}$$

For inequality (8.121) we can rewrite as follows:

$$-k_s d(\lambda) + k_s d(\lambda_s) \geq \frac{1}{2}\|\lambda - \lambda_s\|^2_{H_{s-1}} +$$

$$\frac{1}{2}\|\lambda_s - \lambda_{s-1}\|_{H_{s-1}} - \frac{1}{2}\|\lambda - \lambda_{s-1}\|^2_{H_{s-1}}. \tag{8.133}$$

We are ready to estimate the number of steps N the IQP method (8.111) required for finding $x_N : \Delta_N = d(\lambda^*) - d(\lambda_N) \leq \varepsilon$, where $1 >> \varepsilon > 0$ is the required accuracy. Let $\sigma_s = \sum_{l=0}^{s} k_l$. The following Theorem establishes the lower bound for σ_N that guarantees $\Delta_N \leq \varepsilon$.

Theorem 8.13. *Let $\{k_s\}_{s\in N}$ be a nondecreasing sequence; then under condition of Theorem 8.12 for a given accuracy $\varepsilon > 0$, we obtain $\Delta_N \leq \varepsilon$ for any N*

$$\sigma_N \geq O\left(\frac{m}{\varepsilon}\ln\frac{\Delta_0}{\varepsilon}\right). \tag{8.134}$$

Proof. Multiplying (8.132) by σ_{s-1}, we obtain

$$(\sigma_s - k_s)d(\lambda_s) - \sigma_{s-1}d(\lambda_{s-1}) \geq \sigma_{s-1}\frac{1}{2k_s}\|\lambda_s - \lambda_{s-1}\|^2_{H_{s-1}}, s \geq 1.$$

Summing up the last inequality over $s = 1, \ldots, N$, we have

$$\sigma_N d(\lambda_N) - \sum_{s=1}^{N} k_s d(\lambda_s) \geq \frac{1}{2}\sum_{s=1}^{N} \frac{\sigma_{s-1}}{k_s} \|\lambda_s - \lambda_{s-1}\|^2_{H_{s-1}}. \tag{8.135}$$

Summing up (8.133) over $s = 1,\ldots,N$, we obtain

$$-\sigma_N d(\lambda) + \sum_{s=1}^{N} k_s d(\lambda_s) \geq$$

$$\frac{1}{2}\left(\sum_{s=1}^{N}(\|\lambda-\lambda_s\|^2_{H_{s-1}} - \|\lambda-\lambda_{s-1}\|^2_{H_{s-1}}) + \sum_{s=1}^{N}\|\lambda_s-\lambda_{s-1}\|^2_{H_{s-1}}\right). \tag{8.136}$$

Adding up (8.135) and (8.136), we obtain

$$\sigma_N(d(\lambda_N)-d(\lambda)) \geq \frac{1}{2}\sum_{s=1}^{N}(-\|\lambda-\lambda_{s-1}\|^2_{H_{s-1}} + \|\lambda-\lambda_s\|^2_{H_{s-1}}) \tag{8.137}$$

$$+\sum_{s=1}^{N}(\frac{\sigma_{s-1}}{k_s}+1)\|\lambda_s-\lambda_{s-1}\|^2_{Hs-1}.$$

Therefore, keeping in mind $\|\lambda^*-\lambda_s\| \leq O(\sqrt{m})$, from (8.137) for $\lambda = \lambda^* \in L^*$, we obtain

$$\sigma_N(d(\lambda^*)-d(\lambda_N)) \leq \frac{1}{2}\sum_{s=1}^{N}(\|\lambda^*-\lambda_{s-1}\|^2_{H_{s-1}} - \|\lambda^*-\lambda_s\|^2_{H_{s-1}})$$

$$= \frac{1}{2}\sum_{s=1}^{N}(\|\lambda^*-\lambda_{s-1}\|_{H_{s-1}} - \|\lambda^*-\lambda_s\|_{H_{s-1}})(\|\lambda^*-\lambda_{s-1}\|_{H_{s-1}} + \|\lambda^*-\lambda_s\|_{H_{s-1}})$$

$$\leq O(\sqrt{m})\sum_{s=1}^{N}(\|\lambda^*-\lambda_{s-1}\|_{H_{s-1}} - \|\lambda^*-\lambda_s\|_{H_{s-1}}).$$

Using the triangle inequality

$$\|\lambda_s-\lambda_{s-1}\|_{H_{s-1}} \geq \|\lambda^*-\lambda_{s-1}\|_{H_{s-1}} - \|\lambda^*-\lambda_s\|_{H_{s-1}},$$

we have

$$\sigma_N(d(\lambda^*)-d(\lambda_N)) \leq O(\sqrt{m})\left(\sum_{s=1}^{N}\|\lambda_s-\lambda_{s-1}\|_{H_{s-1}}\right). \tag{8.138}$$

From (8.131) follows

$$d(\lambda^*)-d(\lambda_N) \leq \prod_{s=1}^{N}\left(1-\frac{\|\lambda_s-\lambda_{s-1}\|_{H_{s-1}}}{O(\sqrt{m})}\right)(d(\lambda^*)-d(\lambda_0));$$

therefore

$$\ln \frac{d(\lambda^*) - d(\lambda_N)}{d(\lambda^*) - d(\lambda_0)} \leq \sum_{s=1}^{N} \ln \left(1 - \frac{\|\lambda_s - \lambda_{s-1}\|_{H_{s-1}}}{O(\sqrt{m})}\right).$$

Keeping in mind $\ln(1+x) \leq x, \forall x > -1$, we obtain

$$\ln \frac{(d(\lambda^*) - d(\lambda_N))}{(d(\lambda^*) - d(\lambda_0))} \leq -\sum_{s=1}^{N} \frac{\|\lambda_s - \lambda_{s-1}\|_{H_{s-1}}}{O(\sqrt{m})}.$$

or

$$\Delta_N = d(\lambda^*) - d(\lambda_N) \leq (d(\lambda^*) - d(\lambda_0))e^{-\sum_{s=1}^{N} \frac{\|\lambda_s - \lambda_{s-1}\|_{H_{s-1}}}{O(\sqrt{m})}}. \tag{8.139}$$

Hence

$$\Delta_N \leq \Delta_0 e^{-\sum_{s=1}^{N} \frac{\|\lambda_s - \lambda_{s-1}\|_{H_{s-1}}}{O(\sqrt{m})}}. \tag{8.140}$$

On the other hand, from (8.138), we have

$$-O(\sqrt{m}) \sum_{s=1}^{N} \|\lambda_s - \lambda_{s-1}\|_{H_{s-1}} \leq -\sigma_N \Delta_N$$

or

$$-\frac{1}{O(\sqrt{m})} \sum_{s=1}^{N} \|\lambda_s - \lambda_{s-1}\|_{H_{s-1}} \leq \frac{-\sigma_N \Delta_N}{O(m)}. \tag{8.141}$$

From (8.139) and (8.140) follows

$$\Delta_N \leq \Delta_0 e^{\frac{-\sigma_N \Delta_N}{O(m)}}. \tag{8.142}$$

Let $1 >> \varepsilon > 0$ be the required accuracy. Therefore, if

$$\Delta_0 e^{-\frac{\sigma_N \Delta_N}{O(m)}} \leq \varepsilon, \tag{8.143}$$

then $\Delta_N \leq \varepsilon$. From (8.143) we obtain

$$-\frac{\sigma_N \Delta_N}{O(m)} \leq \ln \frac{\varepsilon}{\Delta_0},$$

or

$$\sigma_N \geq \frac{O(m)}{\Delta_N} \ln \frac{\Delta_0}{\varepsilon}. \tag{8.144}$$

Therefore, for any N

$$\sigma_N \geq O\left(\frac{m}{\varepsilon} \ln \frac{\Delta_0}{\varepsilon}\right), \tag{8.145}$$

we have $\Delta_N \leq \varepsilon$; hence the smallest integer N, for which (8.145) holds, defines the number of steps one needs to get $\Delta_N \leq \varepsilon$ □

Corollary 8.2. *Keeping in mind $\sigma_N = k_1 + \ldots + k_N$ and taking, for example, $\kappa_s = s$, we obtain $\sigma_N = \frac{N(N+1)}{2}$. From (8.145) follows*

$$\frac{N(N+1)}{2} \geq O\left(\frac{m}{\varepsilon} \ln \frac{\Delta_0}{\varepsilon}\right);$$

therefore, it requires

$$N = O\left(\frac{m}{\varepsilon} \ln \frac{\Delta_0}{\varepsilon}\right)^{\frac{1}{2}} \tag{8.146}$$

steps to find an ε-approximation for $d(\lambda^*)$.

Exercise 8.9. Find the upper bound for N if $k_{s+1} = k_s\left(1 + \frac{1}{\sqrt{m}}\right), s \geq 0$.

8.3.6 Asymptotic Convergence Rate

In this section we establish the rate of convergence for the LS multipliers method (8.94)–(8.96) and its dual equivalent (8.98) under some extra assumptions on the input data.

We say that complementarity condition is satisfied in the strict form if

$$\max\{\lambda_i^*, c_i(x^*)\} > 0, \quad i = 1, \ldots, m. \tag{8.147}$$

Theorem 8.14. *If for the problem P the complementarity condition (8.147) is satisfied, then there exists $s_0 > 0$ that for any fixed $k > 0$, the following bound*

$$d(\lambda^*) - d(\lambda_s) = o(ks)^{-1}$$

holds true for any $s \geq s_0$.

Proof. We recall that $I^* = \{i : c_i(x^*) = 0\} = \{1, \ldots, r\}$ is the active constraint set, and then from (8.147) follows $\min\{c_i(x^*) \,|\, i \notin I^*\} = \sigma > 0$. Therefore, there exists a number s_0 such that $c_i(x_s) \geq \sigma/2$, $s \geq s_0$, $i \notin I^*$. From (8.95) and A2, we have

$$\begin{aligned} &(a) \quad \lambda_{i,s+1} \leq 2\lambda_{i,s}\left(1 + e^{0.5k_{i,s}\sigma}\right)^{-1} \leq \lambda_{i,s} e^{-0.5k_{i,s}\sigma} \to 0 \text{ and} \\ &(b) \quad k_{i,s} = k\left(\lambda_{i,s}\right)^{-1} \to \infty, \quad i \notin I^*. \end{aligned} \tag{8.148}$$

For LSL we have

$$\mathscr{L}(x, \lambda_s, \mathbf{k}) = f(x) - k^{-1} \sum_{i=1}^{r} (\lambda_{i,s})^2 \psi(k_{i,s} c_i(x)) - k^{-1} \sum_{i=r+1}^{m} (\lambda_{i,s})^2 \psi(k_{i,s} c_i(x)).$$

Keeping in mind (8.148), we conclude that for $s \geq s_0$, the last term of $\mathscr{L}(x,\lambda_s,\mathbf{k})$ is eligibly small. Therefore, instead of $\mathscr{L}(x,\lambda,\mathbf{k})$, we consider the truncated LSL $\mathscr{L}(x,\lambda,\mathbf{k}) := f(x) - \sum_{i=1}^{r} \lambda_i k_i^{-1}\psi(k_i c_i(x))$ and correspondent truncated Lagrangian $L(x,\lambda) := f(x) - \sum_{i=1}^{r} \lambda_i c_i(x)$. Accordingly, instead of the original dual function and the second-order FD distance, we consider the dual function $d(\lambda) := \inf_{x\in\mathbb{R}^r} L(x,\lambda)$ and the second-order FD distance $D_2(u,\mathrm{v}) := \sum_{i=1}^{r} \mathrm{v}_i^2\varphi(u_i/\mathrm{v}_i)$ in the truncated dual space $\mathbb{R}^r$.

For simplicity, we retain the previous notations for the truncated LSL, truncated Lagrangian, correspondent dual function, and FD distance.

Below, we will assume $\{\lambda_s\}_{s\geq s_0}$ is the truncated dual sequence, that is, $\lambda_s = (\lambda_{1,s},\dots,\lambda_{r,s})$.

Let us consider the optimality criteria for the truncated interior prox method

$$\lambda_{s+1} = \arg\max\left\{ d(\lambda) - k^{-1}\sum_{i=1}^{r}(\lambda_{i,s})^2\varphi(\lambda_i/\lambda_{i,s}) \,|\, \lambda \in \mathbb{R}^r \right\}.$$

We have

$$c\,(x_{s+1}) + k^{-1}\sum_{i=1}^{r}\lambda_{i,s}\varphi'\,(\lambda_{i,s+1}/\lambda_{i,s})\,e_i = 0\,, \tag{8.149}$$

where $e_i = (0,\dots,1,\dots,0) \in \mathbb{R}^r$.

Using $\varphi'(1) = 0$, we can rewrite (8.149) as follows:

$$c\,(x_{s+1}) + k^{-1}\sum_{i=1}^{r}\lambda_{i,s}\left(\varphi'\,(\lambda_{i,s+1}/\lambda_{i,s}) - \varphi'(\lambda_{i,s}/\lambda_{i,s})\right)e_i = 0\,.$$

Using the mean value formula, we obtain

$$c\,(x_{s+1}) + k^{-1}\sum_{i=1}^{r}\varphi''\left(1 + \left(\frac{\lambda_{i,s+1}}{\lambda_{i,s}} - 1\right)\theta_i^s\right)(\lambda_{i,s+1} - \lambda_{i,s})\,e_i = 0\,, \tag{8.150}$$

where $0 < \theta_{i,s} < 1$.

We recall that $-c(x_{s+1}) \in \partial d(\lambda_{s+1})$, so (8.150) is the optimality criteria for the following problem in the truncated dual space

$$\lambda_{s+1} = \arg\max\left\{d(\lambda) - 0.5k^{-1}\|\lambda - \lambda_s\|^2_{H_s} \mid \lambda \in \mathbb{R}^r\right\}, \tag{8.151}$$

where $\|x\|_{H_s} = x^T H_s x$ and $H_s = \mathrm{diag}(h_{i,s})_{i=1}^{r}$. From (8.110a) follows

$$2 \leq h_{i,s} = \varphi''\left(1 + \left(\frac{\lambda_{i,s+1}}{\lambda_{i,s}} - 1\right)\theta_{i,s}\right) = \varphi''(\cdot) \leq 2.25, \quad i = 1,\dots,r. \tag{8.152}$$

It means that the interior prox method (8.98) is equivalent to the quadratic prox method in the rescaled truncated dual space.

We will show that the convergence analysis typical for quadratic prox method can be extended for the interior prox method (8.98) in the truncated rescaled dual space.

From (8.151) follows

$$d(\lambda_{s+1}) - (2k)^{-1}\|\lambda_{s+1} - \lambda_s\|_{H_s}^2 \geq d(\lambda_s).$$

Therefore, keeping in mind the left inequality (8.152), we obtain

$$d(\lambda^*) - d(\lambda_s) - (d(\lambda^*) - d(\lambda_{s+1})) \geq k^{-1}\|\lambda_s - \lambda_{s+1}\|^2$$

or

$$\Delta_s - \Delta_{s+1} \geq k^{-1}\|\lambda_s - \lambda_{s+1}\|^2, \tag{8.153}$$

where $\Delta_s = d(\lambda^*) - d(\lambda_s) > 0$. Using concavity of $d(\lambda)$, we obtain

$$d(\lambda) - d(\lambda_{s+1}) \leq \langle -c(x_{s+1}), \lambda - \lambda_{s+1}\rangle$$

or

$$d(\lambda_{s+1}) - d(\lambda) \geq \langle c(x_{s+1}), \lambda - \lambda_{s+1}\rangle.$$

Using (8.150), we obtain

$$d(\lambda_{s+1}) - d(\lambda) \geq -k^{-1}\langle H_s(\lambda_{s+1} - \lambda_s), \lambda - \lambda_{s+1}\rangle.$$

Thus, for $\lambda = \lambda^*$, we have

$$k^{-1}\langle H_s(\lambda_{s+1} - \lambda_s), \lambda^* - \lambda_{s+1}\rangle \geq d(\lambda^*) - d(\lambda_{s+1}) = \Delta_{s+1}.$$

Hence

$$\|H_s\| \cdot \|\lambda_{s+1} - \lambda_s\| \cdot \|\lambda^* - \lambda_s\| \geq k\Delta_{s+1}.$$

Keeping in mind the right inequality in (8.152), we obtain

$$\|\lambda_{s+1} - \lambda_s\| \geq \frac{4}{9}k\Delta_{s+1}\|\lambda_s - \lambda^*\|^{-1}. \tag{8.154}$$

From (8.153) and (8.154) follows

$$\Delta_s - \Delta_{s+1} \geq \frac{16}{81}k\Delta_{s+1}^2\|\lambda_s - \lambda^*\|^{-2}$$

or

$$\Delta_s \geq \Delta_{s+1}\left(1 + \frac{16}{81}k\Delta_{s+1}\|\lambda_s - \lambda^*\|^{-2}\right).$$

By inverting the last inequality, we obtain

$$\Delta_s^{-1} \leq \Delta_{s+1}^{-1}\left(1 + \frac{16}{81}k\Delta_{s+1}\|\lambda_s - \lambda^*\|^{-2}\right)^{-1}. \tag{8.155}$$

Further, from (8.98) follows

$$d(\lambda_{s+1}) \geq d(\lambda_{s+1}) - k^{-1}\sum_{i=1}^{r}(\lambda_{i,s})^2\,\varphi(\lambda_{i,s+1}/\lambda_{i,s})$$
$$\geq d(\lambda^*) - k^{-1}\sum_{i=1}^{r}(\lambda_{i,s})^2\,\varphi(\lambda_i^*/\lambda_{i,s})$$

or

$$k^{-1}\sum_{i=1}^{r}(\lambda_{i,s})^2\varphi(\lambda_i^*/\lambda_{i,s}) \geq \Delta_{s+1}.$$

Keeping in mind $\varphi(1) = \varphi'(1) = 0$, we obtain

$$\varphi(\lambda_i^*/\lambda_{i,s}) = \varphi''(\cdot)(\lambda_i^* - \lambda_{i,s})^2\lambda_{i,s}^{-2};$$

therefore

$$k^{-1}\sum_{i=1}^{r}\varphi''(\cdot)(\lambda_i^* - \lambda_{i,s})^2 = k^{-1}\sum_{i=1}^{r}(\lambda_{i,s})^2\,\varphi(\lambda_i^*/\lambda_{i,s}) \geq \Delta_{s+1}. \tag{8.156}$$

Taking into account $\varphi''(\cdot) \leq 2.25$ from (8.156), we obtain

$$2.25k^{-1}\|\lambda^* - \lambda_s\|^2 \geq \Delta_{s+1}$$

or

$$k\Delta_{s+1}\|\lambda^* - \lambda_s\|^{-2} \leq 2.25;$$

therefore

$$0 < \frac{16}{81}k\Delta_{s+1}\|\lambda_s - \lambda^*\|^{-2} \leq \frac{4}{9}.$$

Then, we have $(1+t)^{-1} \leq -0.6t + 1$ for all $t \in [0, 4/9]$.

Hence, for $t = \frac{16}{81}k\Delta_{s+1}\|\lambda_s - \lambda^*\|^{-2}$ from (8.155) follows

$$\Delta_s^{-1} \leq \Delta_{s+1}^{-1}(1 - 0.6k\Delta_{s+1})\|\lambda_s - \lambda^*\|^{-2})$$

or

$$\Delta_i^{-1} \leq \Delta_{i+1}^{-1} - 0.6k\,\|\lambda_{s+1} - \lambda^*\|^{-2}, \quad i = 0, \ldots, s-1. \tag{8.157}$$

Summing (8.157) for $i = 0, \ldots, s-1$, we obtain

$$\Delta_s^{-1} \geq \Delta_s^{-1} - \Delta_0^{-1} \geq 0.6k\sum_{i=0}^{s-1}\|\lambda_i - \lambda^*\|^{-2}.$$

By inverting the last inequality, we obtain

$$\Delta_s = d(\lambda^*) - d(\lambda_s) \leq \frac{5/3}{k\sum_{i=0}^{s-1}\|\lambda_i - \lambda^*\|^{-2}}$$

or

$$ks\Delta_s = \frac{5/3}{(1/s)\sum_{i=0}^{s-1}\|\lambda_i - \lambda^*\|^{-2}}.$$

From $d_H(L^*, \partial L_s) \to 0$ follows $\|\lambda_s - \lambda^*\|^{-2} \to \infty$. Therefore

$$\lim_{s\to\infty} s^{-1}\sum_{i=1}^{s}\|\lambda_i - \lambda^*\|^{-2} = \infty.$$

From Silverman–Toeplitz theorem (see Knopp (1956)) follows existence $\alpha_s \to 0$ such that

$$\Delta_s = \frac{5}{3}\alpha_s\frac{1}{ks} = o\left((ks)^{-1}\right). \quad \square \tag{8.158}$$

The estimation (8.158) can be strengthened. Under second-order sufficient optimality conditions, LS method (8.94)–(8.96) converges with Q-linear rate if $k > 0$ is fixed but large enough.

First of all, from the second-order sufficient optimality conditions follows uniqueness of the primal–dual solution; therefore, the primal $\{x_s\}_{s\in N}$ and dual $\{\lambda_s\}_{s\in N}$ sequences converge to x^* and λ^* that strict complementarity conditions (8.147) are satisfied.

From (8.95), we have $\lim_{s\to\infty} k_{i,s} = k(\lambda_i^*)^{-1}$, $i = 1,\ldots,r$, that is, the scaling parameters, which correspond to the active constraints and grow linearly with $k > 0$.

Therefore, the technique used for proving Theorem 7.2 can be applied for proving similar results for method (8.94)–(8.96).

For a given small enough $\delta > 0$, we define the extended neighborhood of λ^* as follows:

$$\Lambda(\lambda^*, \mathbf{k}, \delta) = \{(\lambda, \mathbf{k}) \in \mathbb{R}^m_+ \times \mathbb{R}^m_+ : \lambda_i \geq \delta,\ |\lambda_i - \lambda_i^*| \leq k\delta,\ i = 1,\ldots,r, k \geq k_0.$$
$$0 < \lambda_i < k\delta, \quad i = r+1,\ldots,m\}.$$

Theorem 8.15. *If $f(x)$ and all $c_i(x) \in C^2$ and sufficient second-order optimality conditions (4.73)–(4.74) hold, then there exists such small $\delta > 0$ and large $k_0 > 0$ such that for any $(\lambda, \mathbf{k}) \in \Lambda(\lambda^*, \mathbf{k}, \delta)$*

1. *there exists*

$$\hat{x} : \nabla_x \mathcal{L}(\hat{x}, \lambda, \mathbf{k}) = 0$$

and

$$\hat{\lambda}_i = \lambda_i \psi'(k_i c_i(\hat{x})),\ \hat{k}_i = k\hat{\lambda}_i^{-1}, \quad i = 1,\ldots,m;$$

2. *for the pair $(\hat{x}, \hat{\lambda})$, the bound*

$$\max\left\{\|\hat{x} - x^*\|, \left\|\hat{\lambda} - \lambda^*\right\|\right\} \leq ck^{-1}\|\lambda - \lambda^*\|. \tag{8.159}$$

holds true and $c > 0$ is independent on $k \geq k_0$;

3. *the LSL $\mathcal{L}(x, \lambda, \mathbf{k})$ is strongly convex in the neighborhood of $\hat{x}$.*

Theorem 8.15 can be proven by a slight modification of the correspondent proof of Theorem 7.2.

Corollary 8.3. *If the conditions of Theorem 8.15 are satisfied, then for the primal–dual sequence* $\{x_s, \lambda_s\}_{s\in N}$*, the following bound holds*

$$\max\{\|x_{s+1}-x^*\|, \|\lambda_{s+1}-\lambda^*\|\} \le \frac{c}{k}\|\lambda_s-\lambda^*\| \le \left(\frac{c}{k}\right)^{s+1}\|\lambda^0-\lambda^*\| \quad (8.160)$$

and $c>0$ *is independent on* $k\ge k_0$.

The numerical realization of the LS multipliers method requires replacing the unconstrained minimizer by its approximation.

It can be done using results of Theorem 7.3.

8.3.7 Generalization and Extension

The results obtained for LS remain true for any smooth approximation $\theta : (-\infty,\infty)\to(-\infty,0)$ of a non-smooth function $x_- = \min\{0,x\}$ if θ is twice continuously differentiable, increasing, concave, and satisfied

$$(a)\quad \lim_{t\to\infty}\theta(t)=0, \qquad (b)\quad \lim_{t\to-\infty}(\theta(t)-t)=0. \quad (8.161)$$

We consider $\psi:\mathbb{R}\to(-\infty,-\sigma\theta(0))$ given by formula

$$\psi(t)=\sigma(\theta(t)-\theta(0)), \quad (8.162)$$

where $\sigma=(\theta'(0))^{-1}>0$.

Along with $\psi(t)$, we consider the LF conjugate function $\psi^*(s)=\inf_t\{st-\psi(t)\}$.

To find $\psi^*(s)$, we have to solve the equation $s-\sigma\theta'(t)=0$ for t. Due to $\theta''(t)<0$, the inverse θ'^{-1} exists and $t=\theta'^{-1}(s/\sigma)=\theta^{*'}(s/\sigma)$.

By differentiating the identity $s=\sigma\theta'(\theta^{*'}(s/\sigma))$ in s, we obtain

$$\sigma\theta''\left(\theta^{*'}(s/\sigma)\right)\theta^{*''}(s/\sigma)\sigma^{-1}=1.$$

Using again $t=\theta^{*'}(s/\sigma)$, we have

$$\theta^{*''}(s/\sigma)=\left(\theta''(t)\right)^{-1}. \quad (8.163)$$

Further, for the LF conjugate $\psi^*(s)$, we obtain

$$\psi^*(s)=s\theta^{*'}(s/\sigma)-\psi\left(\theta^{*'}(s/\sigma)\right).$$

Then

$$\psi^{*'}(s) = \theta^{*'}\left(\frac{s}{\sigma}\right) + \frac{s}{\sigma}\theta^{*''}\left(\frac{s}{\sigma}\right) - \frac{1}{\sigma}\psi'\left(\theta^{*'}\left(\frac{s}{\sigma}\right)\right)\theta^{*''}\left(\frac{s}{\sigma}\right).$$

From $t = \theta^{*'}(s/\sigma)$ and $\psi'(t) = s$ follows

$$\psi^{*'}(s) = \theta^{*'}\left(\frac{s}{\sigma}\right).$$

Then, using (8.163), we have

$$\psi^{*''}(s) = \sigma^{-1}\theta^{*''}\left(\frac{s}{\sigma}\right) = \frac{1}{\sigma\theta''(t)}. \tag{8.164}$$

Let $\theta''(t_0) = \min\{\theta''(t) \mid -\infty < t < \infty\}$; such minimizer exists due to the continuity of $\theta''(t)$ and $\lim_{t\to-\infty}\theta''(t) = \lim_{t\to\infty}\theta''(t) = 0$, which follows from (8.161).

The following proposition states the basic properties of the transformation ψ.

Proposition 8.6. *Let $\theta \in C^2$ be increasing and strictly concave function, which satisfies (8.161). Then,*

C1. $\psi(0) = 0$;
C2. *(a)* $0 < \psi'(t) < \sigma$, $\forall t \in (-\infty, \infty)$ *and (b)* $\psi'(0) = 1$;
C3. $\lim_{t\to-\infty}\psi'(t) = \sigma$; $\lim_{t\to\infty}\psi'(t) = 0$;
C4. $\tau = \sigma\theta''(t_0) < \psi''(t) < 0$, $\forall t \in (-\infty, \infty)$.

Exercise 8.10. Verify properties C1–C4.

Now we consider the kernel $\varphi = -\psi^*$.

Proposition 8.7. *The kernel φ possesses the following properties:*

D1. $\varphi(s)$ *is nonnegative and strongly convex on* $(0, \sigma)$;
D2. $\varphi(1) = \varphi'(1) = 0$;
D3. $\lim_{s\to 0^+}\varphi'(s) = -\infty$;
D4. $\varphi''(s) \geq -(\sigma\theta''(t_0))^{-1}$, $\forall s \in (0, \sigma)$.

Therefore, for each smoothing function $\theta(t)$, which satisfies the conditions of Proposition 8.6, one can find a transformation with properties type A1–A5.

However, due to $\lim_{t\to-\infty}\psi'(t) = \sigma$, we have $\lim_{t\to-\infty}\psi''(t) = 0$. Therefore, $\lim_{s\to\sigma^-}\psi^{*''}(s) = -\infty$ and $\lim_{s\to 0^+}\psi^{*''}(s) = -\infty$.

To avoid the complications, discussed in Section 8.3.1, we have to modify $\psi(t)$.

We will illustrate it using the Chen–Harker–Kanzow–Smale (CHKS) smoothing function, which along with the log-sigmoid function has been used for solving complementarity problems.

For a given $\eta > 0$, the following equation $u^2 - tu - \eta = 0$ has two roots $\theta_-(t) = 0.5\left(t - \sqrt{t^2 + 4\eta}\right)$ and $\theta_+(t) = 0.5\left(t + \sqrt{t^2 + 4\eta}\right)$.

The function $\theta_- : (-\infty, \infty) \to (-\infty, 0)$ is the CHKS interior smoothing function. Then

$$\theta''(0) = \min\left\{\theta''(t) = -4\eta\left(t^2+4\eta\right)^{-3/2} \mid -\infty < t < \infty\right\} = -0.5\eta^{-0.5},$$

$\sigma = (\theta'(0))^{-1} = 2$, and $\theta(0) = -\sqrt{\eta}$. The transformation $\psi : (-\infty, \infty)$, which is given by formula

$$\psi(t) = t - \sqrt{t^2+4\eta} + 2\sqrt{\eta},$$

is called the CHKS transformation.

It is easy to see that A1–A4 from Proposition 8.6 hold for CHKS $\psi(t)$ with $\sigma = 2$ and $\tau = \min_t \psi''(t) = -\max_t 4\eta(t^2+4\eta)^{-(3/2)} = -(2\sqrt{\eta})^{-1}$.

The LF conjugate $\psi^* : (0,2) \to [0, -2\sqrt{\eta}]$ is defined by formula

$$\psi^*(s) = \inf_t\{st - \psi(t)\} = 2\sqrt{\eta}\left(\sqrt{(2-s)s} - 1\right).$$

Then, $\psi^*(0) = \psi^*(2) = -2\sqrt{\eta}$ and $\varphi : [0,2] \to [0, 2\sqrt{\eta}]$, which is given by formula $\varphi(s) = -\psi^*(s) = 2\sqrt{\eta}\left(1 - \sqrt{(2-s)s}\right)$, is called the CHKS kernel.

For the second-order CHKS φ-divergence distance $D : \mathbb{R}^m_+ \times \mathbb{R}^m_{++} \to \mathbb{R}_+$, we define by formula

$$D(u, \mathrm{v}) = \sum_{i=1}^m \mathrm{v}_i^2 \varphi(u_i/\mathrm{v}_i).$$

The properties of the CHKS kernel $\varphi(t) = 2\sqrt{\eta}\left(1 - \sqrt{(2-t)t}\right)$ are similar to those of the FD kernel φ and are given in the following proposition.

Proposition 8.8. *The kernel φ possesses the following properties:*

E1. $\varphi(t) \geq 0$, $t \in [0,2]$, *φ is strongly convex on* $(0,2)$;
E2. $\varphi(1) = \varphi'(1) = 0$;
E3. $\varphi'(t) = 2\sqrt{\eta}(t-1)(t(2-t))^{-0.5}$, $\lim_{t\to 0^+} \varphi'(t) = -\infty$, $\lim_{t \to 2^-} \varphi'(t) = \infty$;
E4. $\varphi''(t) = 2\sqrt{\eta}(t(2-t))^{-1.5} \geq 2\sqrt{\eta}$, $\forall t \in (0,2)$.

To avoid complications, which we discussed in Section 8.3.1, we modify CHKS transformation $\psi(t)$ and the corresponding kernel $\varphi(t)$ the same way it was done for LS transformation in Section 8.3.1.

The truncated CHKS transformation $\psi : (-\infty, \infty) \to (-\infty, 2\sqrt{\eta})$ is defined by formula

$$\psi(t) := \begin{cases} \psi(t) = t - \sqrt{t^2+4\eta} + 2\sqrt{\eta}, & t \geq -\sqrt{\eta} \\ q(t) \;= at^2 + bt + c, & -\infty < t \leq -\sqrt{\eta}. \end{cases} \tag{8.165}$$

We find the parameters a, b, and c from the system $\psi(-\sqrt{\eta}) = q(-\sqrt{\eta})$, $\psi'(-\sqrt{\eta}) = q'(-\sqrt{\eta})$, $\psi''(-\sqrt{\eta}) = 2a$ to ensure $\psi \in C^2$. Then, instead of C2(a), we obtain

$$0 < \psi'(t) < \infty. \tag{8.166}$$

The LF conjugate ψ^* is defined by

$$\psi^*(s) := \begin{cases} \psi^*(s) = 2\sqrt{\eta}\left(\sqrt{(2-s)s}-1\right), \; 0 < s \leq \psi'(-\eta) = 1 + \frac{1}{\sqrt{5}} \\ q^*(s) \;= (4a)^{-1}(s-b)^2 - c, \quad 1 + \frac{1}{\sqrt{5}} < s < \infty. \end{cases}$$

For the truncated CHKS kernel $\bar{\varphi} : (0,\infty) \to (0,\infty)$, we define by formula

$$\varphi(t) = \begin{cases} -\psi^*(s), \; 0 < s \leq \psi'(-\eta) = 1 + \frac{1}{\sqrt{5}} \\ -q^*(s), \;\; 1 + \frac{1}{\sqrt{5}} < s < \infty. \end{cases} \tag{8.167}$$

We collect the properties of the truncated CHKS kernel $\bar{\varphi}$ in the following Proposition.

Proposition 8.9. *The modified CHKS kernel $\bar{\varphi}$ possesses the following properties:*

F1. $\bar{\varphi}(t) \geq 0,\, t \in (0,\infty)$,
F2. $\bar{\varphi}(1) = \bar{\varphi}'(1) = 0$;
F3. $\lim_{s\to 0^+} \bar{\varphi}'(s) = -\infty,\; \lim_{s\to\infty} \bar{\varphi}'(t) = \infty$;
F4. $\bar{\varphi}''(s) \geq 2\sqrt{\eta},\, s \in (0,\infty)$.

Exercise 8.11. Consider multipliers method (8.94)–(8.96) with truncated CHKS transformation given by (8.165).

Exercise 8.12. Show that method type (8.94)–(8.96) with truncated CHKS transformation equivalent to prox method for the dual problem with second-order φ-divergence distance.

Exercise 8.13. Show that dual prox method is equivalent to interior quadratic prox in the rescaled from step to step dual space.

Let $\rho(x_s,X^*) = \min\{\|x_s - u\| | u \in X^*\}$ and $\rho(\lambda_s,L^*) = \min\{\|\lambda_s - \mathrm{v}\| | \mathrm{v} \in L^*\}$.

Theorem 8.16. *For the primal–dual sequence $\{x_s,\lambda_s\}_{s\in N}$, generated by method (8.94)–(8.96) with modified CHKS transformation given by (8.165), the following is true:*

1. *If A and B are satisfied, then*

$$f(x^*) = \lim_{s\to\infty} f(x_s) = \lim_{s\to\infty} d(\lambda_s) = d(\lambda^*),\; x^* \in X^*,\; \lambda^* \in L^*,$$

and $\lim_{s\to\infty} \rho(x_s,X^*) = 0$, $\lim_{s\to\infty} \rho(\lambda_s,L^*) = 0$.

2. *If (8.147) is satisfied, then*

$$d(\lambda^*) - d(\lambda_s) = o\left((ks)^{-1}\right).$$

3. *If sufficient second-order optimality condition (4.73)–(4.74) is satisfied, then for the primal–dual sequence generated by the NR method with truncated CHKS transformation, the bound (8.159) holds.*

For any transformation ψ given by (8.162), the corresponding truncated transformation is given by the following formula:

$$\bar{\psi}(t) := \begin{cases} \psi(t), \; t \geq t_0, \\ q(t), \; t \leq t_0 < 0, \end{cases}$$

where $q(t) = at^2 + bt + c$ and $a = 0.5\psi''(t_0)$, $b = \psi'(t_0) - t_0\psi''(t_0)$, $c = \psi(t_0) - t_0\psi'(t_0) + 0.5t_0^2\psi''(t_0)$.

Before concluding this section, we would like to make a few comments about exponential multipliers method with "dynamic" scaling parameters update, which was introduced by Tseng and Bertsekas (1993).

As we mentioned already, the authors emphasized that some aspects of convergence analysis of the exponential multiplier method have proved to be surprisingly difficult, in particular, exponential multipliers method with "dynamic" scaling parameter update.

It turns out that all results obtained for LSM method remain true for the exponential multipliers method if the exponential transformation $\psi(t) = 1 - e^{-t}$ is replaced by the truncated one.

The LF conjugate $\psi^*(s) = \inf_t\{st - \psi(t)\} = -s\ln s + s - 1$. The kernel of the correspondent second-order entropy-like distance is $\varphi(s) = -\psi^*(s) = s\ln s - s + 1$, and then $\varphi'(s) = \ln s$, $\varphi''(s) = s^{-1}$ so $\lim_{s\to\infty}\varphi''(s) = 0$. It means that 3b) from Assertion 7.2 is not satisfied.

Let us consider truncated exponential transformation

$$\bar{\psi}(t) := \begin{cases} \psi(t) = 1 - e^{-t}, & t \geq -1, \\ q(t) = at^2 + bt + c, & t \leq -1. \end{cases}$$

We find the parameters a, b, and c from $\psi(-1) = q(-1)$, $\psi'(-1) = q'(-1)$ and $\psi''(-1) = q''(-1)$ to ensure that $\bar{\psi} \in C^2$. So, $a = -0.5e$, $b = 0$, $c = 1 - 0.5e$ and

$$\bar{\psi}(t) := \begin{cases} 1 - e^{-t}, & t \geq -1, \\ q(t) = -0.5et^2 + 1 - 0.5e, & t \leq -1. \end{cases} \tag{8.168}$$

Then

$$\bar{\psi}^*(s) := \begin{cases} \psi^*(s) = -s\ln s + s - 1, & s \leq e, \\ q^*(s) = (4a)^{-1}(s-b)^2 - c = (-2e)^{-1}s^2 - 1 + 0.5e, & s \geq e. \end{cases}$$

Now the kernel $\bar{\varphi} : (0,\infty) \to (0,\infty)$

$$\bar{\varphi}(s) := \begin{cases} s\ln s - s + 1, & 0 \leq s \leq e \\ (2e)^{-1}s^2 + 1 - 0.5e, & s \geq e. \end{cases}$$

Then

$$\varphi''(s) := \begin{cases} s^{-1}, \; 0 < s \leq e \\ e^{-1}, \; s \geq e \,. \end{cases}$$

So $\min\{\varphi''(s) \mid s > 0\} = e^{-1}$, that is, for the truncated exponential transformation the property b3 holds. At the same time, $\varphi''(t) \leq 1$, $\forall t \in [1, \infty]$, that is, the property type b4 holds as well. Therefore, the results of Theorems 8.12 - 8.15 remain true for truncated exponential transformation.

Proposition 8.10. *For the primal–dual sequence $\{x_s, \lambda_s\}_{s \in N}$, which is generated by (8.94)–(8.96) with truncated exponential transformation (8.168), all statements of Theorem 8.16 hold true.*

From Remark 7.3 follows: only exponential branch of ψ given by (8.168) controls the computational process from some point on. Therefore, from Theorem 8.12 follows that the exponential multipliers method with dynamic scaling parameter update converges in value under mild assumptions on the input data. Under the strict complementarity conditions (8.147), it converges with $o((ks)^{-1})$ convergence rate, and under the standard second-order optimality condition (4.73)–(4.74), it converges with Q-linear rate if $k > 0$ is fixed but large enough.

It is worth mentioning that neither truncated exponential transformation nor its derivatives grow exponentially in case of constraints violation, which contributes to the numerical stability.

In the following section, we apply the LS method for linear programming. The convergence under very mild assumption follows from Theorem 8.12. Under the dual uniqueness, the LS method converges with quadratic convergence. The key ingredients of the proof are the A. Hoffman-type Lemma 5.8, Theorem 8.12, and the properties of the FD kernel.

8.3.8 LS Multipliers Method for Linear Programming

Let $A : \mathbb{R}^n \to \mathbb{R}^m$, $a \in \mathbb{R}^n$, $b \in \mathbb{R}^m$. We assume that the primal optimal set

$$X^* = \text{Argmin}\{\langle a, x\rangle \mid c_i(x) = (Ax - b)_i = \langle a_i, x\rangle - b_i \geq 0, \quad i = 1, \dots, m\} \quad (8.169)$$

is nonempty and bounded, and so is the dual optimal set

$$L^* = \text{Argmax}\{\langle b, \lambda\rangle \mid A^T\lambda - a = 0, \; \lambda_i \geq 0, \quad i = 1, \dots, m\} \quad (8.170)$$

The LS method (8.94)–(8.96) applied to (8.169) produces three sequences $\{x_s\}$, $\{\lambda_s\}$, and $\{\mathbf{k}_s\}$:

$$x_{s+1} : \nabla_x \mathscr{L}(x_{s+1}, \lambda_s, \mathbf{k}) = a - \sum_{i=1}^{m} \lambda_{i,s} \psi'(k_{i,s} c_i(x_{s+1})) a_i = 0 \tag{8.171}$$

$$\lambda_{s+1} : \lambda_{i,s+1} = \lambda_{i,s} \psi'(k_{i,s} c_i(x_{s+1})), \quad i = 1, \ldots, m \tag{8.172}$$

$$\mathbf{k}_{s+1} : k_{i,s+1} = k(\lambda_{s+1})^{-1}, \quad i = 1, \ldots, m. \tag{8.173}$$

From the boundedness of X^* and L^* follows Theorem 8.12. In particular

$$\lim_{s \to \infty} \langle a, x_s \rangle = \langle a, x^* \rangle = \lim_{s \to \infty} \langle b, \lambda_s \rangle = \langle b, \lambda^* \rangle .$$

Using A. Hoffman's Lemma 5.8, we can find $\alpha > 0$ such that

$$\langle b, \lambda^* \rangle - \langle b, \lambda_s \rangle \geq \alpha \rho(\lambda_s, L^*). \tag{8.174}$$

Therefore, $\lim_{s \to \infty} \rho(\lambda_s, L^*) = 0$.

If λ^* is a unique dual solution, then from (5.81) follows existence of $\alpha > 0$ such that

$$\langle b, \lambda^* \rangle - \langle b, \lambda \rangle = \alpha \|\lambda - \lambda^*\| \tag{8.175}$$

holds true $\forall \lambda \in L = \{\lambda : A^T \lambda = a, \lambda \in \mathbb{R}^m_+\}$.

Theorem 8.17. *If the dual problem (8.170) has a unique solution, then the dual sequence $\{\lambda_s\}_{s \in N}$ converges in value quadratically, that is, there is $c > 0$ independent on $k > 0$ and $s_0 > 0$ that for any $s \geq s_0$ the following estimation*

$$\langle b, \lambda^* \rangle - \langle b, \lambda_{s+1} \rangle \leq c k^{-1} \left[\langle b, \lambda^* \rangle - \langle b, \lambda_s \rangle \right]^2 \tag{8.176}$$

holds true.

Proof. It follows from (8.171)–(8.173) that

$$\nabla_x \mathscr{L}(x_{s+1}, \lambda_s, k_s) = a - \sum_{i=1}^{m} \lambda_{i,s+1} a_i = a - A^T \lambda_{s+1} = 0 \tag{8.177}$$

and $\lambda_{s+1} \in \mathbb{R}^m_{++}$. In other words, the LS method generates dual interior sequence $\{\lambda_s\}_{s \in N}$.

From (8.177), we obtain

$$0 = \langle a - A^T \lambda_{s+1}, x_{s+1} \rangle = \langle a, x_{s+1} \rangle - \sum_{i=1}^{m} \lambda_{i,s+1} c_i(x_{s+1}) - \langle b, \lambda_{s+1} \rangle$$

or

$$\langle b, \lambda_{s+1} \rangle = L(x_{s+1}, \lambda_{s+1}) .$$

From Theorem 8.11, we obtain the equivalence of the multipliers method (8.171)–(8.173) to the following interior prox for the dual problem

$$\lambda_{s+1} = \arg\max \left\{ \langle b, \lambda \rangle - k^{-1} \sum_{i=1}^{m} (\lambda_{i,s})^2 \varphi\left(\frac{\lambda_i}{\lambda_{i,s}} \right) \middle| A^T \lambda - a = 0 \right\} . \tag{8.178}$$

Keeping in mind Remark 7.3, we can assume without restricting generality that only Fermi–Dirac kernel $\varphi(t) = (2-t)\ln(2-t) + t\ln t$ is used in the method (8.178).

From (8.178), taking into account $\lambda^* \in \mathbb{R}^m_+$ and $A^T\lambda^* = a$, we obtain

$$\langle b, \lambda_{s+1}\rangle - k^{-1}\sum_{i=1}^{m}(\lambda_{i,s})^2\varphi\left(\frac{\lambda_{i,s+1}}{\lambda_{i,s}}\right) \geq \langle b, \lambda^*\rangle - k^{-1}\sum_{i=1}^{m}(\lambda_{i,s})^2\varphi\left(\frac{\lambda_i^*}{\lambda_{i,s}}\right).$$

Keeping in mind $k^{-1}\sum_{i=1}^m(\lambda_{i,s})^2\varphi((\lambda_{i,s+1})/(\lambda_{i,s})) \geq 0$, we have

$$k^{-1}\sum_{i=1}^{m}(\lambda_{i,s})^2\varphi\left(\frac{\lambda_i^*}{\lambda_{i,s}}\right) \geq \langle b, \lambda^*\rangle - \langle b, \lambda_{s+1}\rangle. \tag{8.179}$$

Let us assume that $\lambda_i^* > 0$, $i = 1, \ldots, r$; $\lambda_i^* = 0$, $i = r+1, \ldots, m$. Then

$$\varphi\left(\frac{\lambda_i^*}{\lambda_{i,s}}\right) = 2ln2, \quad i = r+1, \ldots, m.$$

Taking into account

$$\varphi(1) = \varphi'(1) = \varphi\left(\frac{\lambda_{i,s}}{\lambda_{i,s}}\right) = \varphi'\left(\frac{\lambda_{i,s}}{\lambda_{i,s}}\right) = 0,$$

we obtain

$$k^{-1}\sum_{i=1}^m(\lambda_{i,s})^2\varphi\left(\frac{\lambda_i^*}{\lambda_{i,s}}\right) =$$
$$k^{-1}\left[\sum_{i=1}^r\varphi''\left(1+\theta\frac{\lambda_i^*-\lambda_{i,s}}{\lambda_{i,s}}\right)(\lambda_i^*-\lambda_{i,s})^2 + 2\ln 2\sum_{i=r+1}^m(\lambda_i^*-\lambda_{i,s})^2\right],$$

where $0 < \theta < 1$.

From $\varphi''(\cdot) \leq 3$ follows

$$k^{-1}\sum_{i=1}^{m}(\lambda_{i,s})^2\varphi\left(\frac{\lambda_i^*}{\lambda_{i,s}}\right) \leq 3k^{-1}\|\lambda^*-\lambda_s\|^2. \tag{8.180}$$

Combining (8.179) and (8.180), we obtain

$$3k^{-1}\|\lambda^*-\lambda_s\|^2 \geq \langle b, \lambda^*\rangle - \langle b, \lambda_{s+1}\rangle. \tag{8.181}$$

From (8.175) with $\lambda = \lambda_s$ follows

$$\|\lambda_s - \lambda^*\| = \alpha^{-1}\left[\langle b, \lambda^*\rangle - \langle b, \lambda_s\rangle\right].$$

Therefore, the following bound

$$\langle b, \lambda^*\rangle - \langle b, \lambda_{s+1}\rangle \leq ck^{-1}\left[\langle b, \lambda^*\rangle - \langle b, \lambda_s\rangle\right]^2 \tag{8.182}$$

holds with $c = 3\alpha^{-2}$ for any $s \geq s_0$. □

Remark 8.4. Theorem 8.17 remains valid for the method (8.171)–(8.173), when LS transformation is replaced by the CHKS transformation given by (8.165) or the exponential transformation given by (8.168).

Notes

For MBF theory and methods see Polyak (1992) and references therein; for MBF in LP see Polyak (1992a). For MBF convergence properties for LP, see Powell (1995), Jensen and Polyak (1994) and Polyak and Teboulle (1997). For numerical results of MBF for LP, see Jensen et al. (1993).

The truncated MBF transformation was introduced in Ben-Tal et al. (1992) (see also Ben-Tal and Zibulevsky (1997)). The truncated MBF was used in the framework of NR for solving truss topology design and other engineering problems Ben-Tal and Nemirovski (1995), Ben-Tal and Zibulevsky (1997), Ben-Tal and Nemirovski (2001).

In Alber and Reemtsen (2007) a version of NR with truncated MBF transformation was used for intensity modulated radiotherapy planning.

The numerical performance of NR with truncated MBF transformation for constrained optimization problems was discussed in Breitfeld and Shanno (1996) (see also Nash et al. (1994), Griva et al. (1998), Griva and Polyak (2004, 2006) and Shanno et al. (1996)).

The truncated MBF transformation has been used in the NR framework for developing PENNON solver (see Kočvara and Stingl (2015, 2005, 2007)).

For Newton MBF method in structural optimization, see Berke et al. (1995) and Griva et al. (1998).

The NR with "dynamic" scaling parameters update for exponential transformation was introduced in Tseng and Bertsekas (1993) (see also Ben-Tal and Zibulevsky (1997)).

The interior prox method with second-order entropy-like prox with regularized MBF kernel was studied in Auslender et al. (1999) (see also Auslender and Teboulle (2005)).

The IDF and interior center methods were introduced in Huard (1970) (see also Huard (1967) and Lieu Bui-Trong and Huard (1966)).

The important notion of "analytic center" was introduced in Sonnevend (1986). The centers trajectory was used by J. Renegar in Renegar (1988) for developing a polynomial algorithm for LP. The weighted analytical centers were considered in Goffin and Vial (1993).

For modified interior distance functions and correspondent methods, see Polyak (1987, 1997).

For the exterior distance function and correspondent methods, see Polyak (2017). The smoothing technique has been studied in Chen and Mangasarian (1995) (see also Auslender et al. (1997)).

For the log-sigmoid Lagrangian and correspondent NR and prox methods with second-order Fermi–Dirac entropy distance, see Polyak (2001, 2002).

Chapter 9
Lagrangian Transformation and Interior Ellipsoid Methods

9.0 Introduction

The NR approach produced a number of multipliers methods, which are primal exterior and dual interior.

The connection between exterior primal NR and interior dual proximal point methods is well understood. Correspondent results are covered in Chaps. 7 and 8.

For many years, however, the IPMs and the EPMs were considered as fields, which has practically nothing in common.

One of the purposes of this chapter is to show a class of primal EPMs, which is equivalent to dual IPMs.

The connection is based on Lagrangian transformation (LT) scheme, which uses $\psi \in \Psi$ to rescale terms of the classical Lagrangian associated with constraints. The terms were rescaled by positive scaling parameters, one for each term or by one parameter for all terms.

The LT method finds approximation for the primal minimizer following LM and scaling parameters update.

If the parameters are updated inversely proportional to square of the Lagrange multipliers, then for nondegenerate CO the primal-dual LT method globally converges with asymptotic quadratic rate; see Polyak (2004).

In 2008, L. Matioli and C. Gonzaga used the LT scheme with truncated MBF transformation and one scaling parameter for all constraints. They called the primal exterior method MBF2 and show its equivalence to interior ellipsoid method with Dikin's ellipsoids for the dual problem.

In this chapter, we systematically study the connections between primal EPMs and dual IPMs, generated by the LT scheme.

First, the LT multiplier method is equivalent to an interior proximal point method with Bregman's or Bregman-type distance; see Bregman (1967), Bregman et al. (1999), Censor and Zenios (1992), Eckstein (1993). The interior prox, in turn, is equivalent to an interior quadratic prox (IQP) for the dual problem in the rescaled from step to step dual space.

R. A. Polyak, *Introduction to Continuous Optimization*, Springer Optimization and Its Applications 172, https://doi.org/10.1007/978-3-030-68713-7_9

Second, IQP is equivalent to an interior ellipsoid method (IEM) for the dual problem. The equivalence is used to prove convergence in value of the LT method and its dual equivalent.

Third, the LT with truncated MBF transformation leads to dual proximal point method with Bregman's distance. The MBF kernel is a self-concordant function; therefore, the correspondent dual interior prox is an interior ellipsoid method with Dikin ellipses.

Fourth, application of the LT method with truncated MBF transformation for LP leads to I. Dikin's AS type method for the dual LP, see Dikin (1967).

9.1 Lagrangian Transformation

We consider a standard convex optimization problem (P) under standard assumption A and B from Section 7.1.1.

For a given $\psi \in \Psi$ and $k > 0$, the LT $\mathscr{L} : \mathbb{R}^n \times \mathbb{R}^m_+ \times \mathbb{R}_{++} \to \mathbb{R}$ is defined as follows:

$$\mathscr{L}(x,\lambda,k) := f(x) - k^{-1} \sum_{i=1}^{m} \psi(k\lambda_i c_i(x)). \tag{9.1}$$

It follows from (2b) and (3) of $\psi \in \Psi$, convexity f, concavity c_i, $i = 1,\dots,m$, that for any given $\lambda \in \mathbb{R}^m_{++}$ and any $k > 0$, the LT $\mathscr{L}$ is convex in x.

Assertion 9.1 *For any $k > 0$ and any KKT's point (x^*,λ^*), the LT $\mathscr{L}$ possesses the following properties:*

1° $\mathscr{L}(x^*,\lambda^*,k) = f(x^*)$

2° $\nabla_x \mathscr{L}(x^*,\lambda^*,k) = \nabla_x L(x^*,\lambda^*) = \nabla f(x^*) - \sum_{i=1}^m \lambda_i^* \nabla c_i(x^*) = 0$

3° $\nabla^2_{xx} \mathscr{L}(x^*,\lambda^*,k) = \nabla^2_{xx} L(x^*,\lambda^*) - k\psi''(0) \nabla c^T_{(r)}(x^*) \Lambda^{*2}_{(r)} \nabla c_{(r)}(x^*)$,

where $\Lambda^*_{(r)} = \text{diag}[\lambda_i^*]_{i=1}^r$, $\nabla c_{(r)}(x^*) = J(c_{(r)}(x^*))$ – Jacobian of $c_{(r)}(x) = (c_1(x),\dots,c_r(x))^T$ at $x = x^*$.

Properties 1°–3° follow from formulas for $\mathscr{L}$, $\nabla_x \mathscr{L}$, $\nabla^2_{xx} \mathscr{L}$ and complementarity condition.

Under second-order sufficient optimality condition (4.73)–(4.74) from 3° and Debreu's lemma with $A = \nabla^2_{xx} L(x^*,x^*)$ and $C = \Lambda^*_{(r)} \nabla c_{(r)}(x^*)$ follows that for $k > 0$, large enough LT $\mathscr{L}(x,\lambda^*,k)$ is strongly convex in x.

With transformation, $\psi \in \Psi$ is associated kernel $\varphi \in \Phi$, that $\varphi = -\psi^*$, where ψ^* is LF transform of ψ.

Properties (1)–(4) of ψ induce particular properties of $\varphi \in \Phi$, which are collected in Assertion 7.2.

9.2 Bregman's Distance

Let $Q \subset \mathbb{R}^m$ be an open convex set, $\hat{Q}$ be the closure of Q, and $\varphi : Q \to \mathbb{R}$ be strictly convex and continuously differentiable function on Q, then Bregman's distance $\mathbb{B}_\varphi : Q \times Q \to R_+$ induced by φ is given by the following formula:

$$\mathbb{B}_\varphi(x,y) = \varphi(x) - \varphi(y) - \langle \nabla\varphi(y), x-y \rangle.$$

Let $\varphi \in \Phi$, and then function $B_\varphi : \mathbb{R}^m_+ \times \mathbb{R}^m_{++} \to \mathbb{R}_+$, which is defined by the formula:

$$B_\varphi(u,\mathrm{v}) := \sum_{i=1}^{m} \varphi(u_i/\mathrm{v}_i), \tag{9.2}$$

we call Bregman-type distance induced by kernel φ. Due to $\varphi(1) = \varphi'(1) = 0$ for any $\varphi \in \Phi$, we have:

$$\varphi(t) = \varphi(t) - \varphi(1) - \varphi'(1)(t-1), \tag{9.3}$$

which means that $\varphi(t) : \mathbb{R}_{++} \to \mathbb{R}_{++}$ is Bregman's distance between $t > 0$ and 1.

By taking $t_i = \frac{u_i}{\mathrm{v}_i}$ from (9.2), we obtain:

$$B_\varphi(u,\mathrm{v}) = B_\varphi(u,\mathrm{v}) - B_\varphi(\mathrm{v},\mathrm{v}) - \langle \nabla_u B_\varphi(\mathrm{v},\mathrm{v}), u - \mathrm{v} \rangle,$$

which we call Bregman-type distance.

Lemma 9.1. *The Bregman-type distance has the following properties:*

(1) $B_\varphi(u,\mathrm{v}) \geq 0, \forall u \in \mathbb{R}^m_+, \mathrm{v} \in \mathbb{R}^m_{++}$; $B_\varphi(u,\mathrm{v}) = 0 \Leftrightarrow u = \mathrm{v}$, $\forall \mathrm{v}, u \in \mathbb{R}^m_{++}$;
(2) $B_\varphi(u,\mathrm{v}) \geq \frac{1}{2} m_0 \sum_{i=1}^m (\frac{u_i}{\mathrm{v}_i} - 1)^2, \forall u \in \mathbb{R}^m_+, \mathrm{v} \in \mathbb{R}^m_{++}$;
(3) $B_\varphi(u,\mathrm{v}) \leq \frac{1}{2} M_0 \sum_{i=1}^m (\frac{u_i}{\mathrm{v}_i} - 1)^2, \forall u \in \mathbb{R}^m_+, \mathrm{v} \in \mathbb{R}^m_{++}$;
(4) for any fixed $\mathrm{v} \in \mathbb{R}^m_{++}$, *the gradient* $\nabla_u B_\varphi(u,\mathrm{v})$ *is a barrier function of* $u \in \mathbb{R}^m_{++}$, *that is:*

$$\lim_{u_i \to 0_+} \frac{\partial}{\partial u_i} B_\varphi(u,\mathrm{v}) = -\infty, i = 1,\ldots,m.$$

Proof. It follows from (1) of Assertion 7.2 that $\varphi(s) \geq 0$, $\forall s > 0$; therefore:

(1) $B_\varphi(u,\mathrm{v}) \geq 0, \forall u \in \mathbb{R}^m_+, \mathrm{v} \in \mathbb{R}^m_{++}$.
Then, from $\varphi(s) \geq 0$, for all $0 < s < \infty$ and $\varphi(1) = \varphi'(1) = 0$, $\varphi''(s) > 0$ follows

$$B_\varphi(u,\mathrm{v}) = 0 \Leftrightarrow u = \mathrm{v}.$$

(2) From $\varphi(1) = \varphi'(1) = 0$ follows:

$$B_\varphi(u,\mathrm{v}) = \sum_{i=1}^m \varphi\left(\frac{u_i}{\mathrm{v}_i}\right) = \sum_{i=1}^m \varphi\left(\frac{\mathrm{v}_i}{\mathrm{v}_i}\right) \tag{9.4}$$

$$+\sum_{i=1}^m \varphi'\left(\frac{\mathrm{v}_i}{\mathrm{v}_i}\right)\left(\frac{u_i}{\mathrm{v}_i}-1\right) + \frac{1}{2}\sum_{i=1}^m \varphi''\left(1+\theta_i\left(\frac{u_i}{\mathrm{v}_i}-1\right)\right)\left(\frac{u_i}{\mathrm{v}_i}-1\right)^2$$

$$= \frac{1}{2}\sum_{i=1}^m \varphi''\left(1+\theta_i\left(\frac{u_i}{\mathrm{v}_i}-1\right)\right)\left(\frac{u_i}{\mathrm{v}_i}-1\right)^2,$$

where $0 < \theta_i < 1$. From item (b) of Assertion 7.2 and 9.4 follows item (2) of the Lemma.

(3) From item (3c) of Assertion 7.2 and 9.4 follows item (3)
(4) From item (2a) of Assertion 7.2 follows:

$$\lim_{u_i \to 0_+} \frac{\partial}{\partial u_i} B_\varphi(u,\mathrm{v}) = -\infty, i = 1,\ldots,m.$$

Particular attention will be given to the MBF kernel $\varphi_2(s) = -\ln s + s - 1$.

The Bregman-type distance, given by (9.2), with MBF kernel φ_2:

$$\mathbb{B}_2(u,\mathrm{v}) = \sum_{i=1}^m \varphi_2(u_i/\mathrm{v}_i) = \sum_{i=1}^m(-\ln u_i/\mathrm{v}_i + u_i/\mathrm{v}_i - 1) \tag{9.5}$$

$$= \sum_{i=1}^m[-\ln u_i + \ln \mathrm{v}_i + (u_i - \mathrm{v}_i)/\mathrm{v}_i]$$

is, in fact, identical to Bregman's distance induced by standard log-barrier function $F(t) = -\sum_{i=1}^m \ln t_i$. □

From the definition of $\mathbb{B}_2$ follows:

$$\nabla_u \mathbb{B}_2(u,\mathrm{v}) = \nabla F(u) - \nabla F(\mathrm{v}).$$

The following three-point identity, established by Chen and Teboulle (1993), is an important element of our analysis. Let $u \in \hat{Q}$, $\mathrm{v} \in Q$, then:

$$\mathbb{B}_2(u,\mathrm{v}) - \mathbb{B}_2(u,w) - \mathbb{B}_2(w,\mathrm{v}) = \langle \nabla F(\mathrm{v}) - \nabla F(w), w - u\rangle. \tag{9.6}$$

9.3 Primal LT and Dual Interior Quadratic Prox

Let $\psi \in \Psi$, $\lambda_0 \in \mathbb{R}^m_{++}$, and $k > 0$ are given. The LT method generates a primal-dual sequence $\{x_s, \lambda_s\}_{s\in N}$:

$$x_{s+1} : \nabla_x \mathscr{L}(x_{s+1}, \lambda_s, k) = 0 \tag{9.7}$$

$$\lambda_{i,s+1} = \lambda_{i,s}\psi'(k\lambda_{i,s}c_i(x_{s+1})), i = 1,\ldots,m. \tag{9.8}$$

Theorem 9.1. *If conditions A and B hold, f is convex, c_i, $i = 1,\dots,m$ are concave, and all functions are continuously differentiable, then:*

(1) the LT method (9.7)–(9.8) is well defined, and it is equivalent to the following interior proximal point method:

$$d(\lambda_{s+1}) - k^{-1}B_\varphi(\lambda_{s+1},\lambda_s) = \max\{d(\lambda) - k^{-1}B_\varphi(\lambda,\lambda_s)|\lambda \in \mathbb{R}^m_{++}\} \tag{9.9}$$

for the dual problem, where:

$$B_\varphi(u,\mathrm{v}) := \sum_{i=1}^m \varphi(u_i/\mathrm{v}_i)$$

is Bregman-type distance, based on the kernel $\varphi = -\psi^*$.

(2) for all $i = 1,\dots,m$, we have:

$$\lim_{s\to\infty} B_\varphi(\lambda_{i,s+1}/\lambda_{i,s}) = 0. \tag{9.10}$$

(3) there exists a positive sequence $\{\varepsilon_s\}_{s\in N}$: $\lim_{s\to\infty}\varepsilon_s = 0$ that the following bound:

$$\left|\frac{\lambda_{i,s+1}}{\lambda_{i,s}} - 1\right| \le (2m_0^{-1})^{\frac{1}{2}}\varepsilon_s \tag{9.11}$$

holds for all $i = 1,\dots,m$.

Proof.

(1) From assumptions A, convexity f, concavity $c_i, i = 1,\dots,m$, property (4) of $\psi \in \Psi$ for any $\lambda \in \mathbb{R}^m_{++}$ and $k > 0$ the recession cone of Ω is empty, that is, for any nontrivial direction d and any $x \in \Omega$ and $(\lambda,k) \in \mathbb{R}^{m+1}_{++}$, we have:

$$\lim_{t\to\infty} \mathscr{L}(x+td,\lambda,k) = \infty.$$

Therefore, minimizer x_{s+1} in (9.7) exists for any $s \ge 1$. It follows from property (2a) of $\psi \in \Psi$ and (9.8) that $\lambda_s \in \mathbb{R}^m_{++} \Rightarrow \lambda_{s+1} \in \mathbb{R}^m_{++}$. Therefore, the LT method (9.7)–(9.8) is well defined. From (9.7)–(9.8) follows:

$$\begin{aligned}\nabla_x\mathscr{L}(x_{s+1},\lambda_s,k) &= \nabla f(x_{s+1}) - \sum_{i=1}^m \lambda_{i,s}\psi'(k\lambda_{i,s}c_i(x_{s+1}))\nabla c_i(x_{s+1}))\\ &= \nabla f(x_{s+1}) - \sum_{i=1}^m \lambda_{i,s+1}\nabla c_i(x_{s+1}) = \nabla_x L(x_{s+1},\lambda_{s+1}) = 0,\end{aligned} \tag{9.12}$$

therefore:

$$d(\lambda_{s+1}) = L(x_{s+1},\lambda_{s+1}) = \min\{L(x,\lambda_{s+1})|x \in \mathbb{R}^n\}.$$

From (9.8), we obtain:

$$\psi'(k\lambda_{i,s}c_i(x_{s+1})) = \lambda_{i,s+1}/\lambda_{i,s}, i = 1,\dots,m.$$

From property (3) of $\psi \in \Psi$ follows existence of the inverse ψ'^{-1}, therefore

$$c_i(x_{s+1}) = k^{-1}(\lambda_{i,s})^{-1}\psi'^{-1}(\lambda_{i,s+1}/\lambda_{i,s}), i = 1,\dots,m. \tag{9.13}$$

Using LF identity $\psi'^{-1} = \psi^{*\prime}$ we obtain

$$c_i(x_{s+1}) = k^{-1}(\lambda_{i,s})^{-1}\psi^{*\prime}(\lambda_{i,s+1}/\lambda_{i,s}),\ i = 1,\dots,m. \tag{9.14}$$

Keeping in mind:

$$-c(x_{s+1}) \in \partial d(\lambda_{s+1})$$

for $\varphi = -\psi^*$, we have:

$$0 \in \partial d(\lambda_{s+1}) - k^{-1}\sum_{i=1}^{m}(\lambda_{i,s})^{-1}\varphi'(\lambda_{i,s+1}/\lambda_{i,s})e_i.$$

The last inclusion is the optimality criteria for $\lambda_{s+1} \in \mathbb{R}^m_{++}$ to be the solution of the problem (9.9).

Thus, the LT method (9.7)–(9.8) is equivalent to the interior proximal point method (9.9).

(2) From $\varphi(1) = 0$ and (9.9) follows:

$$d(\lambda_{s+1}) \geq k^{-1}B_\varphi(\lambda_{s+1},\lambda_s) + d(\lambda_s) > d(\lambda_s),\ \forall s > 0. \tag{9.15}$$

Summing up last inequality from $s = 0$ to $s = N$, we obtain:

$$d(\lambda^*) - d(\lambda_0) \geq d(\lambda_{N+1}) - d(\lambda_0) > k^{-1}\sum_{s=0}^{N} B_\varphi(\lambda_{s+1},\lambda_s),$$

therefore:

$$\lim_{s\to\infty} B(\lambda_{s+1},\lambda_s) = \lim_{s\to\infty}\sum_{i=1}^{m}\varphi(\lambda_{i,s+1}/\lambda_{i,s}) = 0. \tag{9.16}$$

(3) From item (3) of Lemma 9.1 and (9.16) follows existence of a sequence $\{\varepsilon_s > 0\}_{s\in N}$: $\lim_{s\to\infty}\varepsilon_s = 0$ that:

$$\varepsilon_s^2 = B_\varphi(\lambda_{s+1},\lambda_s) \geq \frac{1}{2}m_0\sum_{i=1}^{m}\left(\frac{\lambda_{i,s+1}}{\lambda_{i,s}} - 1\right)^2,$$

therefore (9.11) holds. □

Corollary 9.1. *From Remark 7.1 and* (9.11) *follows existence* $s_0 > 0$ *that for* $s \geq s_0$, *only kernels* φ_i, *which correspond to the original transformations* ψ_i, $i = 1,\dots,5$

are used in the prox method (9.9), *that is, the quadratic branch of truncated* ψ_i *is irrelevant from some point on. It is true for any* $s > 1$ *if* $k_0 > 0$ *is large enough.*

Remark 9.1. The MBF kernel φ_2 is a self -concordant function in $\mathbb{R}_{++}$; hence, the Bregman's distance $\mathbb{B}_{\varphi_2}(u, \mathrm{v})$ is a self -concordant function in $u \in \mathbb{R}^m_{++}$ under any fixed $\mathrm{v} \in \mathbb{R}^m_{++}$; therefore, Newton's method for solving (9.9) in case of LP and QP calculations is efficient.

For convex optimization problems, which are not well structured, that is, the constrains and/or the objective function epigraph cannot be equipped with a self-concordant barrier, the SC theory, generally speaking, does not work, and polynomial complexity of IPM becomes problematic.

The results of the following theorem are independent of the structure of the convex optimization problem. It establishes the equivalence of the prox method (9.9) and IQP in the rescaled from step to step dual space.

In turn, the IQP is equivalent to an interior ellipsoid method (IEM) for the dual problem. The equivalence will be used later for convergence analysis of the prox method (9.9).

In the case of truncated MBF transformation, the corresponding IEM is based on the self-concordant MBF kernel φ_2; therefore, the corresponding interior ellipsoids are Dikin's ellipsoids.

Theorem 9.2. *If conditions of Theorem 9.1 are satisfied, then:*

(1) for a given $\varphi \in \Phi$, *there exists a diagonal matrix* $H_\varphi = \mathrm{diag}(h^i_\varphi \mathrm{v}_i^{-2})$ *with:*

$$m_0 \leq h^i_\varphi \leq M_0,\ \mathrm{v}_i > 0,\ \ i = 1, \dots, m \tag{9.17}$$

that $B_\varphi(u, \mathrm{v}) = \frac{1}{2}\|u - \mathrm{v}\|^2_{H_\varphi}$, *where* $\|w\|^2_{H_\varphi} = w^T H_\varphi w$;

(2) the interior prox method (9.9) *is equivalent to an IQP in the rescaled from step to step dual space, that is:*

$$\lambda_{s+1} = \arg\max\{d(\lambda) - \frac{1}{2k}\|\lambda - \lambda_s\|^2_{H^s_\varphi} | \lambda \in \mathbb{R}^m_+\}, \tag{9.18}$$

where $H^s_\varphi = \mathrm{diag}(h^{i,s}_\varphi \lambda_{i,s}^{-2}) = \mathrm{diag}(\varphi''(1 + \theta^s_i(\lambda_{i,s+1}/\lambda_{i,s} - 1))(\lambda_{i,s})^{-2})$;

$$m_0 \leq h^{i,s}_\varphi \leq M_0,$$

$0 < \theta^s_i < 1$, $i = 1, \dots, m$ *and* $s \geq 1$;

(3) the IQP (9.18) *is equivalent to an IEM for the dual problem;*

(4) there exists a converging to zero sequence $\{r_s > 0\}_{s \in N}$ *and step* $s_0 > 0$ *such that, for* $\forall s \geq s_0$, *the LT method* (9.7)–(9.8) *with truncated MBF transformation* ψ_2 *is equivalent to the following IEM for the dual problem:*

$$\lambda_{s+1} = \arg\max\{d(\lambda) | \lambda \in E(\lambda_s, r_s)\}, \tag{9.19}$$

where $H_s = diag(\lambda_{i,s})_{i=1}^m$ *and*

$$E(\lambda_s, r_s) = \{\lambda : (\lambda - \lambda_s)^T H_s^{-2} (\lambda - \lambda_s) \le r_s^2\}$$

are Dikin's ellipsoid associated with the standard log-barrier function $F(\lambda) = -\sum_{i=1}^m \ln \lambda_i$.

Proof.

(1) From $\varphi(1) = \varphi'(1) = 0$ follows:

$$B_\varphi(u, \mathrm{v}) = \frac{1}{2} \sum_{i=1}^m \varphi''\left(1 + \theta_i \left(\frac{u_i}{\mathrm{v}_i} - 1\right)\right) \left(\frac{u_i}{\mathrm{v}_i} - 1\right)^2, \tag{9.20}$$

where $0 < \theta_i < 1$, $i = 1, \ldots, m$.
From item (3b) and (3c) of Assertion 7.2 follows:

$$0 < m_0 \le \varphi''\left(1 + \theta_i \left(\frac{u_i}{\mathrm{v}_i} - 1\right)\right) \le M_0 < \infty$$

for any $\mathrm{v} \in \mathbb{R}^m_{++}$. Therefore, the diagonal elements:

$$\varphi''\left(1 + \theta_i \left(\frac{u_i}{\mathrm{v}_i} - 1\right)\right) \mathrm{v}_i^{-2} = h^i_\varphi \mathrm{v}_i^{-2}, i = 1, \ldots, m,$$

of H_φ are positive.
From (9.20) follows:

$$B_\varphi(u, \mathrm{v}) = \frac{1}{2} \|u - \mathrm{v}\|^2_{H_\varphi}. \tag{9.21}$$

(2) By taking $u = \lambda$, $\mathrm{v} = \lambda_s$, and $H_\varphi = H_\varphi^s$ from (9.20) and (9.21), we obtain (9.18).
(3) From $\lambda_{s+1} \in \mathbb{R}^m_{++}$ follows that λ_{s+1} is, in fact, an unconstrained maximizer in (9.18). Therefore, one can find $g_{s+1} \in \partial d(\lambda_{s+1})$, that:

$$g_{s+1} - k^{-1} H_\varphi^s (\lambda_{s+1} - \lambda_s) = 0. \tag{9.22}$$

Let $r_s = \|\lambda_{s+1} - \lambda_s\|_{H_\varphi^s}$, we consider ellipsoid:

$$E(\lambda_s, r_s) = \{\lambda : (\lambda - \lambda_s)^T H_\varphi^s (\lambda - \lambda_s) \le r_s^2\}$$

with center $\lambda_s \in \mathbb{R}^m_{++}$ and radius r_s. It follows from item (4) of Lemma 9.1 that $E(\lambda_s, r_s)$ is an interior ellipsoid in $\mathbb{R}^m_{++}$, that is, $E(\lambda_s, r_s) \subset \mathbb{R}^m_{++}$.
Moreover, $\lambda_{s+1} \in \partial E(\lambda_s, r_s) = \{\lambda : (\lambda - \lambda_s)^T H_\varphi^s (\lambda - \lambda_s) = r_s^2\}$; therefore, (9.22) is the optimality condition for the following optimization problem:

$$d(\lambda_{s+1}) = \max\{d(\lambda) | \lambda \in E(\lambda_s, r_s)\}. \tag{9.23}$$

Thus, the interior prox method (9.9) is equivalent to the interior ellipsoid method (9.23).

(4) Let's consider LT method (9.7)–(9.8) with truncated MBF transformation. From (9.10) follows that for $s \geq s_0$ only Bregman distance:

$$\mathbb{B}_2(\lambda,\lambda_s) = \sum_{i=1}^{m}\left(-ln\frac{\lambda_i}{\lambda_{i,s}} + \frac{\lambda_i}{\lambda_{i,s}} - 1\right)$$

is used in (9.9). Then,

$$\nabla^2_{\lambda\lambda}\mathbb{B}_2(\lambda,\lambda_s)|_{\lambda=\lambda_s} = H_s^{-2}.$$

In view of $\mathbb{B}_2(\lambda_s,\lambda_s) = 0$ and $\nabla_\lambda \mathbb{B}_2(\lambda_s,\lambda_s) = 0^m$, we obtain:

$$\begin{aligned}\mathbb{B}_2(\lambda,\lambda_s) &= \frac{1}{2}(\lambda-\lambda_s)^T H_s^{-2}(\lambda-\lambda_s) + o(\|\lambda-\lambda_s\|^2)\\ &= Q(\lambda,\lambda_s) + o(\|\lambda-\lambda_s\|^2).\end{aligned}$$

It follows from (9.10) that for large $s_0 > 0$ and any $s \geq s_0$, the term $o(\|\lambda_{s+1} - \lambda_s\|^2)$ can be ignored, and then from (9.23) follows:

$$d(\lambda_{s+1}) = \max\{d(\lambda)|\lambda \in E(\lambda_s, r_s)\},$$

where $r_s^2 = Q(\lambda_{s+1},\lambda_s)$ and

$$E(\lambda_s,r_s) = \{\lambda : (\lambda-\lambda_s)H_s^{-2}(\lambda-\lambda_s) = r_s^2\}$$

are Dikin's ellipsoid. □

9.4 Convergence Analysis

In this section, we establish convergence properties of the IQP method (9.18).

By introducing $\Lambda = (H^s_\varphi)^{\frac{1}{2}}\lambda$ one can rewrite IQP (9.18) as the following Quadratic Prox

$$\Lambda^{s+1} = \arg\max\{\mathscr{D}_s(\Lambda) - k^{-1}\|\Lambda-\Lambda_s\|^2 | \Lambda \in \mathbb{R}^m_+\},$$

where $\mathscr{D}_s(\Lambda) = d((H^s_\varphi)^{-1/2}\Lambda)$.

Note that the Euclidean distance $E(U,V) = \|U-V\|^2$ is a Bregman's distance induced by the kernel $\varphi(w) = \frac{1}{2}\|w\|^2$ because $\nabla\varphi(w) = w$ and

$$\mathbb{B}_\varphi(U,V) = \frac{1}{2}U^2 - \frac{1}{2}V^2 - \langle U-V,V\rangle = \frac{1}{2}\|U-V\|^2 = E(U,V).$$

The "three-point identity" is the basic element for convergence analysis of the classical quadratic prox. The basic ingredient of our analysis is the "three-point identity" in the rescaled dual space.

Let $H = \text{diag}(h_i)_{i=1}^m$ be a diagonal matrix with $h_i > 0, i = 1,\dots,m$ and a,b,c are three vectors in $\mathbb{R}^m$. The following three-point identity in a rescaled $\mathbb{R}^m$

$$\frac{1}{2}\|a-b\|_H^2+\frac{1}{2}\|b-c\|_H^2-\frac{1}{2}\|a-c\|_H^2=(a-b,c-b)_H=(a-b)^T H(c-b) \quad (9.24)$$

follows immediately from the standard three-point identity:

$$\frac{1}{2}\|A-B\|^2+\frac{1}{2}\|B-C\|^2-\frac{1}{2}\|A-C\|^2=\langle A-B,C-B\rangle \quad (9.25)$$

by taking $A=H^{1/2}a, B=H^{1/2}b, C=H^{1/2}c$.

From concavity of $d(\lambda)$ and boundedness of L^* follows boundedness of $L_0=\{\lambda : d(\lambda)\geq d(\lambda_0)\}$; therefore, there exists $0<L<\infty$, that:

$$\max_{i,s}\lambda_{i,s}=L. \quad (9.26)$$

We consider the dual sequence $\{\lambda_s\}_{s\in N}$ generated by IQP (9.18), the corresponding convex and bounded level sets $L_s=\{\lambda\in\mathbb{R}^m_+ : d(\lambda)\geq d(\lambda_s)\}$, and their boundaries $\partial L_s=\{\lambda\in L_s : d(\lambda)=d(\lambda_s)\}$.

In the following Theorem, we consider only well-defined kernel $\varphi : \varphi(0)<\infty$.

Theorem 9.3. *Under conditions of Theorem 9.2, the following statements are taking place:*

(1) If $\varphi\in\Phi$ is well defined, then for any $k>0$ the following bound holds:

$$d(\lambda^*)-d(\lambda_1)<k^{-1}\sum_{i=1}^m\varphi(\lambda_i^*); \quad (9.27)$$

(2) the dual sequence $\{\lambda_s\}_{s\in N}$ is monotone increasing in value, that is, $d(\lambda_s)>d(\lambda_{s-1}), s\geq 1$ and the following bound holds:

$$d(\lambda_s)>d(\lambda_{s-1})+(2k)^{-1}\frac{m_0}{L^2}\|\lambda_s-\lambda_{s-1}\|^2; \quad (9.28)$$

(3) asymptotic complementarity condition:

$$\lim_{s\to\infty}\lambda_{i,s+1}c_i(x_{s+1})=0,\ \ 1\leq i\leq m \quad (9.29)$$

is satisfied;

(4) the Hausdorff distance between the optimal set L^* and ∂L_s is monotone decreasing:

$$d_H(L^*,\partial L_s)<d_H(L^*,\partial L_{s-1}),\forall s\geq 1; \quad (9.30)$$

(5) dual sequence $\{\lambda_s\}_{s\in N}$ converges in value to the dual solution, that is:

$$\lim_{s\to\infty}d(\lambda_s)=d(\lambda^*)$$

(6)

$$\lim_{s\to\infty}d_H(L^*,\partial L_s)=0; \quad (9.31)$$

(7) primal sequence $\{x_s\}_{s\in N}$ converges in value to the primal solution:

$$\lim_{s\to\infty} f(x_s) = f(x^*).$$

Proof.

(1) For any well-defined $\varphi \in \Phi$, from (9.9) with $\lambda_0 = e \in \mathbb{R}^m_{++}$ follows

$$d(\lambda_1) - k^{-1}B_\varphi(\lambda_1, e) \geq d(\lambda^*) - k^{-1}B_\varphi(\lambda^*, e) = d(\lambda^*) - k^{-1}\sum_{i=1}^{m}\varphi(\lambda_i^*).$$

Keeping in mind $B_\varphi(\lambda_1, e) > 0$, we obtain (9.27). By taking $k > 0$ large enough, one can find an approximation for $d(\lambda^*)$ with a given accuracy. Finding the primal minimizer, however, can be very difficult for $k > 0$ large enough.

(2) From $B_\varphi(\lambda_{s-1}, \lambda_{s-1}) = 0$ and (9.9) follows:

$$d(\lambda_s) \geq d(\lambda_{s-1}) + k^{-1}B_\varphi(\lambda_s, \lambda_{s-1}).$$

From item (2) of Lemma 9.1 we have:

$$d(\lambda_s) \geq d(\lambda_{s-1}) + (2k)^{-1}m_0\sum_{i=1}^{m}(\lambda_{i,s} - \lambda_{i,s-1})^2/\lambda_{i,s-1}^2. \tag{9.32}$$

The bound (9.28) follows from (9.26) and (9.32).

(3) From the update formula (9.8) we have:

$$-k^{-1}\varphi'\left(\frac{\lambda_{i,s+1}}{\lambda_{i,s}}\right) = k^{-1}\psi'^{-1}\left(\frac{\lambda_{i,s+1}}{\lambda_{i,s}}\right) = \lambda_{i,s}c_i(x_{s+1})$$
$$= \frac{\lambda_{i,s}}{\lambda_{i,s+1}}(\lambda_{i,s+1}c_i(x_{j+1})), 1 \leq i \leq m.$$

Keeping in mind (9.11), continuity $\varphi'(t)$ in $t \geq \varepsilon$ for small enough $\varepsilon > 0$ and $\varphi'(1) = 0$, we obtain asymptotic complementarity (9.29);

(4) From concavity $d(\lambda)$ and boundedness of L^* for any $s > 0$ follows that dual level sets $L_s \subset \mathbb{R}^m_{++}$ are convex, closed, and bounded. From (9.28), we have $L^* \subset \ldots L_{s+1} \subset L_s, s \geq 0$. Therefore, from the definition of Hausdorff distance follows (9.30).

(5) From (9.28) follows that the sequence $\{d(\lambda_s)\}_{s\in N}$ is monotone increasing. It is also bounded by $f(x^*)$; therefore, there exists $\lim_{s\to\infty} d(\lambda_s) = \bar{d}$.

Let us show that $\bar{d} = d(\lambda^*) = f(x^*)$. By assuming the opposite that $\bar{d} < d(\lambda^*)$, we can find $\rho > 0$ and large enough $s_0 > 0$ that:

$$d(\lambda_s) \leq d(\lambda^*) - \rho, \ \forall s \geq s_0. \tag{9.33}$$

From the optimality condition for λ_s in (9.18) follows:

$$\langle \lambda - \lambda_s, kg_s - H_\varphi^{s-1}(\lambda_s - \lambda_{s-1}) \rangle \leq 0, \quad \forall \lambda \in \mathbb{R}^m_+.$$

Then,

$$k\langle \lambda - \lambda_s, g_s \rangle \leq (\lambda - \lambda_s)^T H_\varphi^{s-1}(\lambda_s - \lambda_{s-1}), \forall \lambda \in \mathbb{R}^m_+. \tag{9.34}$$

From concavity of d and the last inequality, we obtain:

$$k(d(\lambda_s) - d(\lambda)) \geq k\langle g_s, \lambda_s - \lambda \rangle \geq (\lambda - \lambda_s)^T H_\varphi^{s-1}(\lambda_{s-1} - \lambda_s)$$

for any $\lambda \in \mathbb{R}^m_+$.
Using the three-point identity (9.24) with $a = \lambda, b = \lambda_s, c = \lambda_{s-1}$ and $H = H_\varphi^{s-1}$ from (9.34), we have:

$$\begin{aligned} k(d(\lambda_s) - d(\lambda)) &\geq (\lambda - \lambda_s)^T H_\varphi^{s-1}(\lambda_{s-1} - \lambda_s) \qquad (9.35) \\ = \frac{1}{2}(\lambda - \lambda_s)^T H_\varphi^{s-1}&(\lambda - \lambda_s) + \frac{1}{2}(\lambda_s - \lambda_{s-1})^T H_\varphi^{s-1}(\lambda_s - \lambda_{s-1}) \\ &- \frac{1}{2}(\lambda - \lambda_{s-1})^T H_\varphi^{s-1}(\lambda - \lambda_{s-1}), \quad \forall \lambda \in \mathbb{R}^m_+. \end{aligned}$$

From (9.36) for any $\lambda^* \in L^*$ and $\forall s \geq 1$, we have:

$$\begin{aligned} 0 &\geq k(d(\lambda_s) - d(\lambda^*)) \\ &\geq \frac{1}{2}\|\lambda^* - \lambda_s\|^2_{H_\varphi^{s-1}} + \frac{1}{2}\|\lambda_s - \lambda_{s-1}\|^2_{H_\varphi^{s-1}} - \frac{1}{2}\|\lambda^* - \lambda_{s-1}\|^2_{H_\varphi^{s-1}}. \qquad (9.36) \\ &> \frac{1}{2}\left[\|\lambda^* - \lambda_s\|^2_{H_\varphi^{s-1}} - \|\lambda^* - \lambda_{s-1}\|^2_{H_\varphi^{s-1}}\right]. \end{aligned}$$

Therefore:

$$k\rho \leq k(d(\lambda^*) - d(\lambda_s)) \leq \frac{1}{2}\left[\|\lambda^* - \lambda_s\|^2_{H_\varphi^{s-1}} - \|\lambda^* - \lambda_{s-1}\|^2_{H_\varphi^{s-1}}\right],$$

which is impossible due to (9.11) and (9.30) for $s > 0$ large enough.
Hence, $\lim_{s\to\infty} d(\lambda_s) = d(\lambda^*)$.

(6) From the definition of the Hausdorff distance and $\lim_{s\to\infty} d(\lambda_s) = d(\lambda^*)$ follows (9.31).
(7) We have:

$$d(\lambda_s) = L(x_s, \lambda_s) = f(x_s) - \sum_{i=1}^m \lambda_{i,s} c_i(x_s). \tag{9.37}$$

Therefore, from item (5) and asymptotic complementarity (9.29) follows:

$$d(\lambda^*) = \lim_{s\to\infty} d(\lambda_s) = \lim_{s\to\infty} f(x_s) - \lim_{s\to\infty} \sum_{i=1}^m \lambda_{i,s} c_i(x_s) = \lim_{s\to\infty} f(x_s) = f(x^*). \ \square$$

9.5 LT with Truncated MBF and Interior Ellipsoid Method

From (7.5) and (9.11) follows that from some point on, say $s \geq s_0$, only original MBF transformation ψ_2 is used in LT method (9.7)–(9.8) and only Bregman distance $\mathbb{B}_2$ is used in the prox method (9.9).

In other words, for a given $k > 0$, the primal-dual sequence $\{x_s, \lambda_s\}_{s \in N}$ is generated by the following formulas:

$$\begin{aligned} x_{s+1} :& \nabla_k \mathcal{L}(x_{s+1}, \lambda_s, k) = \\ & \nabla f(x_{s+1}) - \sum_{i=1}^{m} \lambda_{i,s}(1 + k\lambda_{i,s}c_i(x_{s+1}))^{-1} \nabla c_i(x_{s+1}) = 0 \end{aligned} \tag{9.38}$$

$$\lambda_{s+1} : \lambda_{i,s+1} = \lambda_{i,s}(1 + k\lambda_{i,s}c_i(x_{s+1}))^{-1},\ i = 1, \ldots, m. \tag{9.39}$$

Method (9.38)–(9.39) is called *MBF*2 by Matioli and Gonzaga.

For any $\lambda^* \in L^*$ and small enough $\varepsilon > 0$, we consider an ε - optimizer $\lambda^*_\varepsilon \in \mathbb{R}^m_{++}$:

$$\lambda^*_\varepsilon = (\lambda^*_{i\varepsilon}, i = 1, \ldots, m):\ \lambda^*_{i,\varepsilon} = \begin{cases} \lambda^*_i, & \text{if } \lambda^*_i > 0 \\ \varepsilon, & \text{if } \lambda^*_i = 0. \end{cases}$$

Theorem 9.4. *Under conditions of Theorem 9.3, MBF2 methods* (9.38)–(9.39) *generate primal-dual sequence* $\{x_s, \lambda_s\}_{s \in N}$ *with the following properties:*

1. *(a)* $d(\lambda_{s+1}) > d(\lambda_s)$, $s \geq s_0$;
   (b) $\lim_{s \to \infty} \lambda_{i,s} c_i(x_s) = 0$, $i = 1, \ldots, m$
2. *(a)* $B_2(\lambda^*_\varepsilon, \lambda_s) > B_2(\lambda^*_\varepsilon, \lambda_{s+1})$ $\forall s \geq s_0$
   (b) $\lim_{s \to \infty} d(\lambda_s) = d(\lambda^*)$;
   (c) $\lim_{s \to \infty} d_H(\partial \Lambda_s, L^*) = 0$
   (d) $\lim_{s \to \infty} f(x_s) = f(x^*)$
3. *there is a subsequence* $\{s_l\}_{l \in N}$ *that for* $\bar{\lambda}_{i,s} = \lambda_{i,s}\left(\sum_{s=s_l}^{s_{l+1}} \lambda_{i,s}\right)^{-1}$ *the sequence* $\{\bar{x}_{l+1} = \sum_{s=s_l}^{s_{l+1}} \bar{\lambda}_{i,s} x_s\}_{l \in N}$ *converges and* $\lim \bar{x}_l = \bar{x} \in X^*$.

Proof.

1. (a) From (9.11) and Remark 7.1 follows existence $s_0 > 0$ that for any $s \geq s_0$ LT (9.38)–(9.39) is equivalent to the prox method (9.9) with Bregman's distance $\mathbb{B}_2$, given by (9.5).
   From (9.9) for $\lambda = \lambda_s$ follows:

$$d(\lambda_{s+1}) \geq d(\lambda_s) + k^{-1} \sum_{i=1}^{m} (-\ln(\lambda_{i,s+1}/\lambda_{i,s}) + \lambda_{i,s+1}/\lambda_{i,s} - 1). \tag{9.40}$$

   The Bregman's distance (9.5) is strictly convex in u; therefore, from (9.40) follows $d(\lambda_{s+1}) > d(\lambda_s)$ unless $\lambda_{s+1} = \lambda_s \in \mathbb{R}^m_{++}$, then $c_i(x_{s+1}) = 0$, $i = 1, .., m$ and $(x_{s+1}, \lambda_{s+1}) = (x^*, \lambda^*)$ is a KKT pair.
   (b) From $-c(x_{s+1}) \in \partial d(\lambda_{s+1})$ and concavity d follows:

$$d(\lambda) - d(\lambda_{s+1}) \leq \langle -c(x_{s+1}), \lambda - \lambda_{s+1} \rangle \tag{9.41}$$

for any $\lambda \in \mathbb{R}^m_+$. Then, for $\lambda = \lambda_s$ from (9.41) follows:

$$d(\lambda_{s+1}) - d(\lambda_s) \geq \langle c(x_{s+1}), \lambda_s - \lambda_{s+1} \rangle.$$

Using the update formula (9.39), we obtain:

$$\begin{aligned} d(\lambda_{s+1}) - d(\lambda_s) &\geq k \sum_{i=1}^{m} \lambda_{i,s} \lambda_{i,s+1} c_i^2(x_{s+1}) \qquad (9.42) \\ &= k \sum_{i=1}^{m} \lambda_{i,s}/\lambda_{i,s+1} (\lambda_{i,s+1} c_i(x_{s+1}))^2. \end{aligned}$$

Summing up (9.42) from $s = s_0$ to $s = N$, we have:

$$d(\lambda^*) - d(\lambda_{s_0}) > d(\lambda_{N+1}) - d(\lambda_{s_0}) \geq \sum_{s=s_0}^{N} \sum_{i=1}^{m} k \lambda_{i,s}/\lambda_{i,s+1} (\lambda_{i,s+1} c_i(x_{s+1}))^2.$$

Keeping in mind (9.11), we obtain asymptotic complementarity:

$$\lim_{s \to \infty} (\lambda_s, c(x_s)) = 0.$$

2. (a) Using $\nabla_{\mathrm{v}} \mathbb{B}_2(\mathrm{v}, w) = \nabla F(\mathrm{v}) - \nabla F(w)$ for $\mathrm{v} = \lambda_s$ and $w = \lambda_{s+1}$, we obtain:

$$\nabla_\lambda \mathbb{B}_2(\lambda, \lambda_{s+1})_{/\lambda=\lambda_s} = \nabla \varphi_2(\lambda_s) - \nabla \varphi_2(\lambda_{s+1}) = \left(-\sum_{i=1}^{m} \lambda_{i,s}^{-1} e_i + \sum_{i=1}^{m} \lambda_{i,s+1}^{-1} e_i \right).$$

From the three-point identity (9.6) with $u = \lambda_\varepsilon^*$, $\mathrm{v} = \lambda_s$, $w = \lambda_{s+1}$ follows:

$$\begin{aligned} &\mathbb{B}_2(\lambda_\varepsilon^*, \lambda_s) - \mathbb{B}_2(\lambda_\varepsilon^*, \lambda_{s+1}) - \mathbb{B}_2(\lambda_{s+1}, \lambda_s) = \\ &\langle \nabla \varphi(\lambda_s) - \nabla \varphi(\lambda_{s+1}), \lambda_{s+1} - \lambda_\varepsilon^* \rangle = \\ &\sum_{i=1}^{m} (-\lambda_{i,s}^{-1} + \lambda_{i,s+1}^{-1})(\lambda_{i,s+1} - \lambda_{\varepsilon,i}^*). \end{aligned} \qquad (9.43)$$

From the update formula (9.39) follows:

$$kc_i(x_{s+1}) = -\lambda_{i,s}^{-1} + \lambda_{i,s+1}^{-1}, \; i = 1, \ldots, m.$$

Keeping in mind $\mathbb{B}_2(\lambda_s, \lambda_{s+1}) \geq 0$, we can rewrite (9.43) as follows:

$$\mathbb{B}_2(\lambda_\varepsilon^*, \lambda_s) - \mathbb{B}_2(\lambda_\varepsilon^*, \lambda_{s+1}) \geq k \langle c(x_{s+1}), \lambda_{s+1} - \lambda_\varepsilon^* \rangle \qquad (9.44)$$

(b) For $\lambda = \lambda_\varepsilon^*$ from (9.41) follows:

$$\langle c(x_{s+1}), \lambda_{s+1} - \lambda_\varepsilon^* \rangle \geq d(\lambda_\varepsilon^*) - d(\lambda_{s+1}), \; s \geq s_0. \qquad (9.45)$$

The monotone increasing sequence $\{d(\lambda_s)\}_{s=s_0}^{\infty}$ is bounded from above by $f(x^*)$; therefore, there exists $\bar{d} = \lim_{s\to\infty} d(\lambda_s) \le d(\lambda^*) = f(x^*)$. Let us prove that $\bar{d} = d(\lambda^*)$.
Assuming that $d(\bar{\lambda}) < d(\lambda^*)$, we can find $\rho > 0$ that

$$d(\lambda_s) \le d(\lambda_\varepsilon^*) - \rho \text{ for } s \ge s_0 \tag{9.46}$$

Therefore, from (9.44)–(9.46), we obtain:

$$\mathbb{B}_2(\lambda_\varepsilon^*, \lambda_s) - \mathbb{B}_2(\lambda_\varepsilon^*, \lambda_{s+1}) \ge k(d(\lambda_\varepsilon^*) - d(\lambda_{s+1})) \ge k\rho. \tag{9.47}$$

Summing up (9.47) from $s = 0$ to $s = N$, we obtain:

$$B_2(\lambda_\varepsilon^*, \lambda_0) \ge B_2(\lambda_\varepsilon^*, \lambda_0) - B_2(\lambda_\varepsilon^*, \lambda_{N+1}) \ge k\rho N,$$

which is impossible for large N.
Therefore, $\lim_{s\to\infty} d(\lambda_s) = d(\lambda^*)$.

(c) From $d(\lambda_s) \to d(\lambda^*)$ follows:

$$\lim_{s\to\infty} d_H(\partial\Lambda_s, L^*) = 0.$$

(d) From (9.38) and asymptotic complementarity (9.42) follows:

$$d(\lambda^*) = \lim_{s\to\infty}(\lambda_s) = \lim_{s\to\infty} L(x_s, \lambda_s) = \lim_{s\to\infty}(f(x_s) - \sum_{i=1}^{m} \lambda_{i,s} c_i(x_s))$$

$$= \lim_{s\to\infty} f(x_s) = f(x^*). \tag{9.48}$$

3. The dual sequence $\{\lambda_s\}_{s\in N} \subset \Lambda_0$ is bounded; therefore, there is a converging subsequence $\{\lambda_{s_l}\}_{l\in N}$: $\lim_{l\to\infty} \lambda_{s_l} = \bar{\lambda}$.

Consider two subsets of indices $I_+ = \{i : \bar{\lambda}_i > 0\}$ and $I_0 = \{i : \bar{\lambda}_i = 0\}$. From the asymptotic complementarity (9.42) follows $\lim_{s\to\infty} c_i(x_s) = 0$, $i \in I_+$.

For any $i \in I_0$, we have $\lim_{l\to\infty} \lambda_{i,s_l} = 0$; therefore, without loosing the generality we can assume that:

$$\lambda_{i,s_{l+1}} \le 0.5\lambda_{i,s_l},\ i \in I_0.$$

Using the update formula (9.39), we obtain:

$$\lambda_{s_{l+1}} \prod_{s=s_l}^{s_{l+1}} (k\lambda_{i,s} c_i(x_s) + 1) = \lambda_{i,s_l} \ge 2\lambda_{i,s_{l+1}},\ i \in I_0.$$

Invoking the arithmetic-geometric means inequality for $i \in I_0$, we obtain:

$$\frac{1}{s_{l+1} - s_l} \sum_{s=s_l}^{s_{l+1}} (k\lambda_{i,s} c_i(x_s) + 1) \ge \left(\prod_{s=s_l}^{s_{l+1}} (k\lambda_{i,s} c_i(x_s) + 1)\right)^{\frac{1}{s_{l+1}-s_l}} \ge 2^{\frac{1}{s_{l+1}-s_l}}$$

or:

$$\sum_{s=s_l}^{s_{l+1}} \lambda_{i,s} c_i(x_s) > 0, \ i \in I_0.$$

From Jensen inequality and concavity c_i $1 \leq i \leq m$ follows:

$$c_i(\bar{x}_{l+1}) = c_i \left(\sum_{s=s_l}^{s_{l+1}} \bar{\lambda}_{i,s} x_s \right) \geq \sum_{s=s_l}^{s_{l+1}} \bar{\lambda}_{i,s} c_i(x_s) > 0,$$

where $\bar{\lambda}_{i,s} = \lambda_{i,s} \left(\sum_{s=s_l}^{s_{l+1}} \lambda_{i,s} \right)^{-1} \geq 0$, $\sum_{s=s_l}^{s_{l+1}} \bar{\lambda}_{i,s} = 1$, $i \in I_0$. Keeping in mind $\lim_{s \to \infty} c_i(x_s) = 0$, $i \in I_+$, we conclude that the sequence $\{\bar{x}_{l+1}\}_{l \in N}$ is asymptotically feasible; therefore, it is bounded. Without loosing generality, we can assume that $\lim_{l \to \infty} \bar{x}_l = \bar{x} \in \Omega$.

From convexity f follows:

$$f(\bar{x}_{l+1}) \leq \sum_{s=s_l}^{s_{l+1}} \bar{\lambda}_{i,s} f(x_s).$$

Therefore, from (9.48) we obtain:

$$f(\bar{x}) = \lim_{l \to \infty} f(\bar{x}_{l+1}) \leq \lim_{s \to \infty} f(x_s) = \lim_{s \to \infty} d(\lambda_s) = d(\lambda^*) = f(x^*).$$

Thus, $f(\bar{x}) = f(x^*) = d(\lambda^*) = d(\bar{\lambda})$ and $\bar{x} = x^*$, $\bar{\lambda} = \lambda^*$. □

9.6 Lagrangian Transformation and Dual Affine Scaling Method for LP

Let $a \in \mathbb{R}^n, b \in \mathbb{R}^m$, and $A : \mathbb{R}^n \to \mathbb{R}^m$ are given. We consider the following LP problem:

$$x^* \in X^* = Argmin\{\langle a, x \rangle | c(x) = Ax - b \geq 0\} \tag{9.49}$$

and the dual LP:

$$\lambda^* \in L^* = Argmin\{\langle b, \lambda \rangle | r(\lambda) = A^T \lambda - a = 0, \lambda \in \mathbb{R}^m_+\}. \tag{9.50}$$

The LT $\mathscr{L} : \mathbb{R}^n \times \mathbb{R}^m \times \mathbb{R}_{++} \to \mathbb{R}$ for LP is defined as follows:

$$\mathscr{L}(x, \lambda, k) := \langle a, x \rangle - k^{-1} \sum_{s=1}^{m} \psi(k \lambda_i c_i(x)), \tag{9.51}$$

where $c_i(x) = (Ax - b)_i = \langle a_i, x \rangle - b_i, \quad i = 1, \ldots, m.$

We assume that the optimal set $X^* \neq \phi$ is bounded and so is the dual optimal set L^*.

The LT method generates primal-dual sequence $\{x_{s+1}, \lambda_{s+1}\}_{s\in N}$ by the following formulas:

$$x_{s+1} : \nabla_x \mathscr{L}(x_{s+1}, \lambda_s, k) = 0 \tag{9.52}$$

$$\lambda_{s+1} : \lambda_{i,s+1} = \lambda_{i,s}\psi'(k\lambda_{i,s}c_i(x_{s+1})), i = 1, \ldots, m. \tag{9.53}$$

Theorem 9.5. *If the primal optimal* X^* *is bounded, then the LT method* (9.52)–(9.53) *is well defined for any transformation* $\psi \in \Psi$. *For the dual sequence* $\{\lambda_s\}_{s=0}^{\infty}$ *generated by* (9.53), *the following statements hold true:*

(1) the LT method (9.52)–(9.53) *is equivalent to the following interior prox*

$$k(b, \lambda_{s+1}) - B_\varphi(\lambda_{s+1}, \lambda_s) = \max\{k(b, \lambda) - B_\varphi(\lambda, \lambda_s) | A^T\lambda = a\},$$

where $B_\varphi(u, v) = \sum_{i=1}^m \varphi(\frac{u_i}{v_i})$ *is the Bregman-type distance;*

(2) there exists $s_0 > 0$ *that for any* $s \geq s_0$ *the LT method with truncated MBF transformation* ψ_2 *is equivalent to the affine scaling type method for the dual LP.*

Proof

(1) We use the vector form for formula (9.53) assuming that the multiplication and division are componentwise, that is, for vectors $a, b \in \mathbb{R}^n$, we have $c = ab = (c_i = a_ib_i, \quad i = 1, \ldots, n)$ and $d = a/b= (d_i = a_i/b_i, \quad i = 1, \ldots, n)$. We have:

$$\frac{\lambda_{s+1}}{\lambda_s} = \psi'(k\lambda_s c(x_{s+1})). \tag{9.54}$$

Using again the inverse function formula, we obtain:

$$k\lambda_s c(x_{s+1}) = \psi'^{-1}(\lambda_{s+1}/\lambda_s). \tag{9.55}$$

It follows from (9.52) and (9.53) that:

$$\begin{aligned}\nabla_x \mathscr{L}(x_{s+1}, \lambda_s, k) &= a - A^T\psi'(k\lambda_s c(x_{s+1}))\lambda_s = a - A^T\lambda_{s+1} \\ &= \nabla_x L(x_{s+1}, \lambda_{s+1}) = 0,\end{aligned}$$

that is,

$$\begin{aligned}L(x_{s+1}, \lambda_{s+1}) &= \langle a, x_{s+1}\rangle - \langle \lambda_{s+1}, Ax_{s+1} - b\rangle \\ &= \langle a - A^T\lambda_{s+1}, x_{s+1}\rangle + \langle b, \lambda_{s+1}\rangle = \langle b, \lambda_{s+1}\rangle.\end{aligned}$$

Using LF identity $\psi'^{-1} = \psi^{*\prime}$ and $\varphi = -\psi^*$, we can rewrite (9.55) as follows:

$$-kc(x_{s+1}) - (\lambda_s)^{-1}\varphi'(\lambda_{s+1}/\lambda_s) = 0. \tag{9.56}$$

Keeping in mind $A^T\lambda_{s+1} = a, -c(x_{s+1}) \in \partial d(\lambda_{s+1})$ and $\lambda_{s+1} \in \mathbb{R}^m_{++}$, we can view (9.56) as the optimality criteria for the following problem:

$$k(b, \lambda_{s+1}) - B_\varphi(\lambda_{s+1}, \lambda_s) = \max\{kd(\lambda) - B_\varphi(\lambda, \lambda_s) | A^T\lambda = a\}, \tag{9.57}$$

where $B_\varphi(\lambda,\lambda_s)=\sum_{i=1}^{q}\varphi(\lambda_i/\lambda_{i,s})$ is Bregman-type distance.

(2) Let us consider the LT method with truncated MBF transformation ψ_2. It follows from (9.10) that there exists s_0 that for any $s \geq s_0$ MBF kernel $\varphi_2 = -\ln s + s - 1$ and correspondent Bregman distance:

$$B_\varphi(\lambda,\lambda_s)=\sum_{i=1}^{q}\left(-ln\frac{\lambda_i}{\lambda_{i,s}}+\frac{\lambda_i}{\lambda_{i,s}}-1\right)$$

will be used in (9.57). Using considerations similar to those in item 4) Theorem 9.2, we can rewrite (9.57) as follows:

$$k\langle b,\lambda_{s+1}\rangle = \arg\max\{k\langle b,\lambda\rangle | \lambda \in E(\lambda_s,r_s), A^T\lambda = a\}, \tag{9.58}$$

where $r_s^2 = Q_{\varphi_2}(\lambda_{s+1},\lambda_s)$ and $E(\lambda_s,r_s)=\left\{\lambda : (\lambda-\lambda_s)^T H_s^{-2}(\lambda-\lambda_s) \leq r_s\right\}$ are Dikin's ellipsoid and (9.58) is affine scaling type method for the dual LP. □

Notes

The LT scheme with vector of scaling parameters one for each constraint, which are updated inversely proportional to the square of the correspondent Lagrange multipliers iterates, was introduced and studied in Polyak (2004). It was shown that under second-order sufficient optimality condition, the primal-dual LT sequence globally converges with asymptotic quadratic rate.

The LT scheme with one scaling parameter for all constraints and truncated MBF transformation was considered in Matioli and Gonzaga (2008). They called the correspondent LT method MBF2 and show that primal LT is equivalent to the dual prox with Bregman's distance.

For the equivalence of the LT method to the dual interior ellipsoid method in the rescaled from step to step dual space, see Polyak (2015).

From the SC properties of the MBF kernel follows that LT with truncated MBF transformation is equivalent to the interior ellipsoids methods with Dikin's ellipsoids.

For LP calculation, the primal LT with truncated MBF transformation is equivalent to AS-type method for the dual LP.

Chapter 10
Finding Nonlinear Equilibrium

10.0 Introduction

Optimization problems, finding a saddle point of a convex–concave function, matrix games, J. Nash equilibrium of n-person concave game, Walras–Wald equilibrium, finding nonlinear equilibrium (NE) for optimal resource allocation (ORA), and nonlinear input–output equilibrium (NIOE) are all particular cases of general nonlinear equilibrium (NE) problem.

In this chapter, we consider the basic issues associated with NE: existence, uniqueness, optimality criteria, and numerical methods.

Finding NE is equivalent to solving variational inequality (VI) with pseudo-gradient operator, which is also used for establishing optimality conditions and developing first-order method for solving VI.

Particular attention will be given to the pseudo-gradient projection (PGP) and extra pseudo-gradient (EPG) methods. As it turned out, some important equilibrium problems have such feasible sets that projection on them is a very simple operation.

In particular, application of PGP and EPG for finding NE for ORA and NIOE leads to projection on $\Omega = \mathbb{R}^n_+ \times \mathbb{R}^m_+$.

In case of J. Nash equilibrium, one has to find projection on the probability simplex, which is also a simple operation.

For all these problems, we established convergence, estimated convergence rate, and found complexity bounds under various assumptions on the input data.

Application of PGP and EPG for finding NE for ORA leads to the pricing mechanisms for establishing economic equilibrium.

When applied for finding J. Nash equilibrium, both PGP and EPG methods lead to a decomposition procedure, which requires independent projection on probability simplexes in strategic spaces of the players.

Efficient algorithm for projection on the probability simplex coupled with decomposition of the feasible sets makes both PGP and EPG efficient for finding J. Nash equilibrium.

R. A. Polyak, *Introduction to Continuous Optimization*, Springer Optimization and Its Applications 172, https://doi.org/10.1007/978-3-030-68713-7_10

10.1 General NE Problem and the Equivalent VI

We concentrate on a general problem finding nonlinear equilibrium (NE), its equivalence to VI, and its concretizations in economics, game theory, and optimization.

Let Ω be a non-empty closed convex and bounded set in $\mathbb{R}^n$. We consider a function $\Phi : \Omega \times \Omega \to \mathbb{R}$, which is continuous in $(x;y) \in \Omega \times \Omega$ and concave in $y \in \Omega$.

The general NE problem consists of finding $x^* \in \Omega$:

$$\Phi(x^*, x^*) = \max\{\Phi(x^*, y)|y \in \Omega\}. \tag{10.1}$$

Usually function Φ is differentiable, at least, in $y \in \Omega$. Vector-function $g : \Omega \to \mathbb{R}^n$, which is defined by formula

$$g(x) = \nabla_y \Phi(x,y)|_{y=x} = (g_1(x), \ldots, g_n(x))^T,$$

is called pseudo-gradient.

The pseudo-Hessian $H : \Omega \to \mathbb{R}^{n \times n}$ is the Jacobian of g, that is

$$H(x) = J(g(x)) = \nabla g(x) = \begin{bmatrix} \frac{\partial g_1}{\partial x_1}, \ldots, \frac{\partial g_1}{\partial x_n} \\ \ldots \\ \frac{\partial g_n}{\partial x_1}, \ldots, \frac{\partial g_n}{\partial x_n} \end{bmatrix}.$$

We start with existence of the NE. Let us consider a set-valued function $\omega : \Omega \to \Omega$, defined as follows:

$$\omega(x) = \text{Argmax}\{\Phi(x,y)|y \in \Omega\}. \tag{10.2}$$

From concavity $\Phi(x,y)$ in $y \in \Omega$, convexity and boundedness Ω follow, that the solution set $\omega(x)$ is closed, convex, and bounded for any given $x \in \Omega$.

The set-valued function ω is upper semicontinuous at $x_0 \in \Omega$ if for any sequence $\{x_s\}_{s \subset N} : x_s \to x_0$ and any $\{y_s \in \omega(x_s)\}_{s \subset N} : \lim_{s \to \infty} y_s = y_0$ we have $y_0 \in \omega(x_0)$.

In other words, let $L_0 = \{y \in \mathbb{R}^n : y = \lim_{s \to \infty} y_s,\ y_s \in \omega(x_s),\ x_s \to x_0\}$, then the set-valued function ω is upper semicontinuous at x_0 if $L_0 \subset \omega(x_0)$.

Let $\omega : \Omega \to 2^{\Omega}$ be a set-valued function; then $x_0 \in \Omega$ is a fixed point of ω if $x_0 \in \omega(x_0)$. The following important Kakutani's theorem establishes condition on the set-valued function ω, under which the fixed point exists.

Theorem 10.1 (Kakutani). *Let Ω be a non-empty, convex, and compact set in $\mathbb{R}^n$ and $\omega : \Omega \to 2^{\Omega}$ be an upper semicontinuous set-valued function on Ω such that for each $x \in \Omega$ the set $\omega(x)$ is a not empty convex compact; then there exists $a \in \Omega$: $a \in \omega(a)$.*

In other words, ω has a fixed point in Ω.

Existence of NE $x^* \in \Omega$ in (10.1) follows directly from Kakutani's theorem.

Theorem 10.2. *Let $\Omega \subset \mathbb{R}^n$ be a convex compact and Φ be continuous in x and $y \in \Omega$ and concave in $y \in \Omega$; then there exists $x^* \in \Omega$ defined by* (10.1).

Proof. From compactness, convexity Ω and concavity $\Phi(\cdot,y)$ in y for any given $x \in \Omega$ follow that $\omega(x) \subset \Omega$ is a convex compact.

Let us consider sequence $\{x_s\}_{s\to\infty} \subset \Omega$: $\lim_{s\to\infty} x_s = x_0$ and any sequence $\{y_s\}_{s\in N}$: $y_s \in \omega(x_s)$: $\lim_{s\to\infty} y_s = y_0$. Then, due to the continuity Φ in x and $y \in \Omega$, we have $\lim_{s\to\infty} \Phi(x_s,y_s) = \Phi(x_0,y_0)$ and $y_0 \in \omega(x_0)$. In fact, assuming $y_0 \notin \omega(x_0)$, we obtain $\Phi(x_0,y_0) < \max\{\Phi(x_0,y)|y \in \Omega\}$. Therefore, there is $\alpha > 0$ that $\Phi(x_s,y_s) \leq \max\{\Phi(x_s,y)|y \in \Omega\} - \alpha$, which is impossible for large $s > 0$ and $y_s \in \omega(x_s)$ due to the continuity of Φ. □

10.2 Problems Leading to NE

In this section, we consider a number of optimization, game theory, and mathematical economics problems, which lead to finding NE.

10.2.1 Convex Optimization

Let $f : \Omega \to \mathbb{R}$ be concave and Ω be a convex compact in $\mathbb{R}^n$. We consider the following convex optimization problem:

$$f(x^*) = \max\{f(x)|x \in \Omega\}. \tag{10.3}$$

Let

$$\Phi(x,y) = \langle \nabla f(x), y - x \rangle; \tag{10.4}$$

then for $x^* \in \Omega$, we have

$$\langle \nabla f(x^*), y - x^* \rangle \leq 0, \; \forall y \in \Omega; \tag{10.5}$$

therefore,

$$\max\{\Phi(x^*,y) = \langle \nabla f(x^*), y - x^* \rangle | y \in \Omega\} = \Phi(x^*,x^*) = 0. \tag{10.6}$$

It means if x^* is the solution of (10.3), then x^* solves NE (10.6) and vice versa.

In fact, if $x^* \in \Omega$ solves (10.6), then (10.5) is satisfied; therefore, x^* solves (10.3).

In other words, solving a convex optimization problem (10.3) is equivalent to finding equilibrium (10.1) with $\Phi(x,y)$ given by (10.4).

In the case of problem (10.3), the pseudo-gradient

$$g(x) = \nabla_y \Phi(x,y)_{|y=x} = \nabla_y[\langle \nabla f(x), y - x \rangle] = \nabla f(x)$$

is just the gradient of f.

For any $x \in \mathbb{R}^n$, the pseudo-Hessian

$$J(g(x)) = J(\nabla f(x)) = H(x)$$

is just the Hessian of f, which justifies the terminology used in the NE theory.

10.2.2 Finding a Saddle Point

Let $\Omega_1 \in \mathbb{R}^{n_1}$ and $\Omega_2 \in \mathbb{R}^{n_2}$ be two bounded, closed convex sets, and function φ : $\Omega_1 \otimes \Omega_2 \to \mathbb{R}$ is concave in $x_1 \in \Omega_1$ and convex in $x_2 \in \Omega$ function. Finding a saddle point of φ means finding $x_1^* \in \Omega_1$ and $x_2^* \in \Omega_2$:

$$\varphi(x_1, x_2^*) \leq \varphi(x_1^*, x_2^*) \leq \varphi(x_1^*, x_2), \ \forall x_1 \in \Omega_1, \ x_2 \in \Omega_2. \tag{10.7}$$

Let $x = (x_1, x_2)$ and $y = (y_1, y_2) \in \Omega = \Omega_1 \times \Omega_2$. Let us consider

$$\Phi(x,y) = \varphi(y_1, x_2) - \varphi(x_1, y_2); \tag{10.8}$$

then (10.7) is equivalent to (10.1) with $\Phi(x,y)$ given by (10.8) and

$$\Omega = \Omega_1 \otimes \Omega_2. \tag{10.9}$$

Let $x^* = (x_1^*, x_2^*) \in \Omega$; then

$$\max\{\varphi(y_1, x_2^*)|y_1 \in \Omega_1\} - \min\{\varphi(x_1^*, y_2)|y_2 \in \Omega_2\} =$$

$$\max\{\varphi(y_1, x_2^*)|y_1 \in \Omega_1\} + \max\{-\varphi(x_1^*, y_2)|y_2 \in \Omega_2\}$$

$$= \max\{\varphi(y_1, x_2^*) - \varphi(x_1^*, y_2)|y_1 \in \Omega_1, y_2 \in \Omega_2\} = \max\{\Phi(x^*, y)|y \in \Omega\} = \Phi(x^*, x^*).$$

It means that the saddle point (x_1^*, x_2^*) from (10.7) solves NE (10.1) with $\Phi(x,y)$ given by (10.8) and Ω given by (10.9).

Conversely,

$$\begin{aligned}
\Phi(x^*, x^*) &= \max\{\Phi(x^*, y)|y \in \Omega\} = \max\{\varphi(y_1, x_2^*) - \varphi(x_1^*, y_2)|y_1 \in \Omega_1, \ y_2 \in \Omega_2\} \\
&= \max\{\varphi(y_1, x_2^*)|y_1 \in \Omega_1\} + \max\{-\varphi(x_1^*, y_2)|y_2 \in \Omega_2\} \\
&= \max\{\varphi(y_1, x_2^*|y_1 \in \Omega_1\} - \min\{\varphi(x_1^*, y_2)|y_2 \in \Omega_2\}.
\end{aligned}$$

Therefore, for $x^* = (x_1^*, x_2^*) \in \Omega = \Omega_1 \times \Omega_2$ and any $x_1 \in \Omega_1, x_2 \in \Omega_2$, we have (10.7).

Thus, any solution x^* of (10.1) with $\Phi(x,y)$ given by (10.8) and Ω given by (10.9) solves (10.7) and vice versa.

The pseudo-gradient of $\Phi(x,y)$ is

$$g(x) = \nabla_y \Phi(x,y)_{|y=x} = \begin{bmatrix} \nabla_{y_1} \varphi(y_1, x_2)_{|y_1 = x_1} \\ \\ -\nabla_{y_2} \varphi(x_1, y_2)_{|y_2 = x_2} \end{bmatrix} = \begin{bmatrix} \nabla_{y_1} \varphi(x_1, x_2) \\ \\ -\nabla_{y_2} \varphi(x_1, x_2) \end{bmatrix}.$$

The correspondent pseudo-Hessian is

$$H(x) = J(g(x)) = \begin{bmatrix} \nabla^2_{x_1x_1}\varphi(x_1,x_2) & \nabla^2_{x_1x_2}\varphi(x_1,x_2) \\ -\nabla^2_{x_2x_1}\varphi(x_1,x_2) & -\nabla^2_{x_2x_2}\varphi(x_1,x_2) \end{bmatrix}.$$

10.2.3 Matrix Game

Let $A : \mathbb{R}^n \to \mathbb{R}^m$ be payoff matrix for two-person matrix game.

$$A = \begin{bmatrix} a_{11} & a_{1j} & \dots & a_{1n} \\ a_{i1} & a_{ij} & \dots & a_{in} \\ a_{m1} & a_{mj} & \dots & a_{mn} \end{bmatrix}$$

It means if the first (row) player chooses i-row and the second (column) player chooses j-column, then a_{ij} is the payoff for the first player and the loss for the second.

Each row is called pure strategy of the first player, and each column is called pure strategy of the second player.

It is easy to see that by using only pure strategies, that is, picking systematically only a particular row for the first and only a particular column for the second player, neither can win unless the matrix has a saddle point.

Let us consider the following payoff matrix:

$$B = \begin{bmatrix} 1 & 4 & 5 \\ 1 & 3 & 2 \\ 6 & 7 & 1 \end{bmatrix}.$$

If the first player will systematically pick the second row and the second player will systematically pick the second column, then the first player will win 3 and the second player will lose 3.

By picking any other row, the first player can only win less. By picking any other column, the second player can only lose more.

In other words, we have

$$\max_{1\le i\le 3}\min_{1\le j\le 3} a_{ij} = \min_{1\le j\le 3}\max_{1\le i\le 3} a_{ij} = a_{22} = 3. \qquad (10.10)$$

Therefore, matrix game B has a solution, given by pure strategies: second row for the first player and second column for the second player.

Such solution is possible only for matrices, which have a saddle point. In our case, it is a_{22}, which is max in second row and min in the second column. Violation of the optimal strategy by any player while the other picks the optimal one leads to the loss of the player, who violates the optimal strategy.

If a matrix does not have a saddle point, then solution in pure strategies is impossible. In such case, the players have to be more sophisticated to pick the rows (columns) with particular frequencies, which define the mix strategy.

Consider the following matrix game:

$$\begin{array}{cc} & \begin{array}{cc} y & 1-y \end{array} \\ \begin{array}{c} x \\ 1-x \end{array} & \begin{pmatrix} 1 & 0 \\ 0 & 1 \end{pmatrix} \end{array}$$

Obviously, there is no solution in pure strategies. Let the row player use the first row with probability $0 < x < 1$; then the second row will be used with probability $1-x$. The column player uses the first column with probability y and then the second with probability $1-y$.

Mathematical expectation of the row player's gain is

$$E(x,y) = xy + (1-x)(1-y) = 2\left(x - \frac{1}{2}\right)\left(y - \frac{1}{2}\right) + \frac{1}{2}.$$

Obviously, the optimal strategies for the row and the column players are $x^* = \left(\frac{1}{2}, \frac{1}{2}\right)$ and $y^* = \left(\frac{1}{2}, \frac{1}{2}\right)$.

Violation of the optimal row strategy $x^* = \left(\frac{1}{2}, \frac{1}{2}\right)$ leads to the loss of the row player. The same is true for the column player.

For a matrix game given by matrix A, we assume that vector $x \in S_m = \{x \in \mathbb{R}^m_+ : \sum_{i=1}^m x_i = 1\}$ defines the mixed strategy of the row player, where S_m – probability simplex in $\mathbb{R}^m$.

Let vector $y \in S_n = \{y \in \mathbb{R}^n_+ : \sum_{j=1}^n y_j = 1\}$ define the mixed strategy of the column player, where S_n – probability simplex in $\mathbb{R}^n$.

The mathematical expectation for the row player's gain is

$$E(x,y) = x^T Ay.$$

The mathematical expectation for the column player's gain is

$$-E(x,y) = -x^T Ay.$$

The solution of a matrix game is given by $x^* \in S_m$; $y^* \in S_n$:

$$E(x,y^*) \le E(x^*,y^*) \le E(x^*,y) \tag{10.11}$$

for $\forall x \in S_m$; $y \in S_n$.

It is rather remarkable that for any matrix A a solution in mixed strategies exists. It was proven by J. von Neumann (1903–1957) in 1928, see Neumann (1928). The value $E(x^*,y^*)$ is called the price of the matrix game.

From (10.11) follows that neither the row player has any incentives to replace x^* for any $x \in S_m$ nor the column player has any incentives to replace y^* for any $y \in S_n$.

In other words, the solution of the matrix game is a saddle point of the mathematical expectation of the payoff for the first player.

Let $w = (x,y) \in \mathbb{R}^m_+ \times \mathbb{R}^n_+$; then $w^* = (x^*,y^*)$ is the solution of the matrix game if and only if

$$\Phi(w^*,w^*) = \max\{\Phi(w^*,w) | w \in S\}, \tag{10.12}$$

where $\Phi(w^*,w) = E(x,y^*) - E(x^*,y)$ and $S = S_m \times S_n$.

Exercise 10.1.

(a) Show that the pair $w^* = (x^*,y^*)$, which solves (10.12), satisfies (10.11).
(b) Every saddle point (x^*,y^*) from (10.11) solves (10.12).

Thus, matrix game is a particular case of general NE problem (10.1).

10.2.4 J. Nash Equilibrium in n-Person Concave Game

The notion of equilibrium in n-person concave game was introduced by J. Nash (1928–2015) in 1950. In 1994, J. Nash shared the Nobel Prize in Economics with John Harsanyi and Reinhard Selten for their contributions to game theory.

Let each $\Omega_i \subset \mathbb{R}^{m_i}$, $1 \le i \le n$ be a convex and compact set and

$$\Omega = \Omega_1 \times \ldots \times \Omega_i \times \ldots \Omega_n.$$

The payoff function $\varphi_i(x_1,\ldots,y_i,\ldots,x_n) : \Omega \to R$ of each player $1 \le i \le n$ depends on its own strategy $y_i \in \Omega_i$ as well as strategies of other players $x_j \in \Omega_j$, $j \ne i$.

We assume that each payoff function φ_i is smooth in $(x_1,\ldots,y_i,\ldots,x_n)$ and concave in $y_i \in \Omega_i$.

Vector $x^* = (x_1^*,..,x_i^*,\ldots,x_n^*)$ defines J. Nash equilibrium if for each player $1 \le i \le n$ we have

$$\varphi_i(x_1^*,\ldots,y_i,\ldots,x_n^*) \le \varphi_i(x^*), \forall y_i \in \Omega_i \tag{10.13}$$

In other words, J. Nash equilibrium $x^* \in \Omega \times \ldots \times \Omega_i \times \ldots \Omega_n$ is such a collection of strategies $x_i^* \in \Omega_i$, $1 \le i \le n$ that no player has incentives to deviate from equilibrium strategy x_i^*, when the rest accept their equilibrium strategies.

In his paper, "Non-cooperative Games," published in 1951 in *Annals of Mathematics*, John Nash described the noncooperative n-person game as generalization of the two-person matrix game.

"For us an n-person game will be a set of n players, each with an associated finite set of pure strategies and corresponding to each player $1 \le i \le n$ payoff function p_i, which maps the set of all n_i-tuples of pure strategies into real numbers."

In a matrix game, each of the m row vectors is pure strategies of the first player, and each of the n columns is pure strategies of the second player.

In the case of n-person game, each player $1 \le i \le n$ is associated with n_i vectors $a_l^i = (a_{l1}^i,\ldots,a_{lm_i}^i)$, $1 \le l \le n_i$ from $\mathbb{R}^{m_i}$ as its pure strategies.

Even in a matrix game one cannot expect existence of equilibrium in pure strategies; therefore, in matrix game, each player uses mixed strategies, which are a convex combination of its pure strategies.

In n-person concave game, it leads to the following sets:

$$\Omega_i = \left\{ x_i \in \mathbb{R}_+^{m_i} : x_i = \sum_{l=1}^{n_i} \lambda_{il} a_l^i,\ \lambda_{il} \geq 0, \sum_{l=1}^{n_i} \lambda_{il} = 1 \right\}$$

of mixed strategies for player $1 \leq i \leq n$.

After substitution x_i into payoff functions φ_i, we obtain, instead of Ω_i probability simplex,

$$S_i = \{\lambda_i \in \mathbb{R}^{n_i} : \sum_{l=1}^{n_i} \lambda_{il} = 1,\ \lambda_{il} \geq 0,\ l = 1, \ldots, n_i, \},\ 1 \leq i \leq n.$$

Therefore, we consider J. Nash equilibrium with Ω_i replaced by a probability simplex $S_i \subset \mathbb{R}^{n_i}$ with correspondent mixed strategy $\lambda_i \in S_i$ for player $1 \leq i \leq n$, and instead of payoff functions $\varphi_i(x_1, \ldots, x_i, \ldots, x_n)$, we use payoff functions $\varphi_i(\lambda_1, \ldots, \Lambda_i, \ldots, \lambda_n)$ defined on $S = S_1 \otimes \ldots \otimes S_i \otimes \ldots \otimes S_n$.

The set $S = S_1 \otimes \ldots S_i \otimes \ldots \otimes S_n$ is a convex compact in $\mathbb{R}^N$ and $N = \sum_{i=1}^n n_i$.

The payoff function $\varphi_i(\lambda_1 \ldots \Lambda_i \ldots \lambda_n)$ for the player $1 \leq i \leq n$ is concave in $\Lambda_i \in S_i$ and continuous in $(\lambda_1, \ldots, \Lambda_i, \ldots, \lambda_n)$.

Let us consider the normalized payoff function $\Phi : S \times S \to \mathbb{R}$ defined as follows:

$$\Phi(\lambda, \Lambda) = \sum_{i=1}^{n} \varphi_i(\lambda_1, \ldots, \Lambda_i, \ldots, \lambda_n).$$

The normalized J. Nash equilibrium is defined by a vector $\lambda^* \in S$:

$$\Phi(\lambda^*, \lambda^*) = \max\{\Phi(\lambda^*, \Lambda) | \Lambda \in S\}. \tag{10.14}$$

The existence of λ^* follows from Kakutani's theorem because for any given $\lambda \in S$ the set $\omega(\lambda) = \text{Argmax}\{\Phi(\lambda, \Lambda) | \Lambda \in S\}$ is a convex compact and the map $\lambda \to \omega(\lambda)$ is upper semicontinuous.

Exercise 10.2. Show that any normalized equilibrium $\lambda^* \in S$ given by (10.14) is J. Nash equilibrium that is

$$\varphi_i(\lambda_1^*, \ldots, \lambda_i^*, \ldots, \lambda_n^*) \geq \varphi_i(\lambda_1^*, \ldots, \lambda_i, \ldots, \lambda_n^*),\ \forall \lambda_i \in S_i \text{ and } 1 \leq i \leq n. \tag{10.15}$$

The pseudo-gradient

$$g(\lambda) = \nabla_\Lambda \Phi(\lambda, \Lambda)_{|\Lambda = \lambda} =$$

$$= (\nabla_{\Lambda_1} \varphi_1(\Lambda_1, \lambda_2, \ldots, \lambda_n)|_{\Lambda_1 = \lambda_1}, \ldots, \nabla_{\Lambda_n} \varphi_n(\lambda_1, \lambda_2, \ldots, \Lambda_n))^T|_{\Lambda_n = \lambda_n}$$
$$= (g_1(\lambda), \ldots, g_i(\lambda), \ldots, g_n(\lambda))^T$$

and the pseudo-Hessian

$$H(\lambda) = J(g(\lambda)) = \begin{bmatrix} \nabla_{\lambda_1} g_1(\cdot), \ \ldots, \nabla_{\lambda_n} g_1(\cdot) \\ \\ \nabla_{\lambda_1} g_n(\cdot), \ \ldots, \nabla_{\lambda_n} g_n(\cdot) \end{bmatrix}. \tag{10.16}$$

Keeping in mind that under fixed $\lambda^* \in S$ vector λ^* solves a convex optimization problem (10.14), we have

$$\langle g(\lambda^*), \Lambda - \lambda^* \rangle \leq 0, \ \forall \Lambda \in S, \tag{10.17}$$

which is nothing but the optimality criteria for x^* to be a solution in (10.14).

In other words, finding the normalized J. Nash equilibrium (10.14) is equivalent to solving VI (10.17).

10.2.5 Walras–Wald Equilibrium

For many years, it was not clear whether J. Nash equilibrium had anything to do with economic equilibrium introduced as early as 1874 by Leon Walras (1834–1910) in his *Elements of Pure Economics*. Moreover, for a long time, it was even not clear whether the Walras equilibrium exists.

The first substantial contribution was due to Abraham Wald (1902–1950), who proved in the mid-1930s the existence of Walras equilibrium under some special assumptions on the price vector-function. The assumptions, unfortunately, were hard to justify from the economics standpoint.

In the mid-1950s, Harold Kuhn modified Walras–Wald (WW) model and proved under reasonable assumptions on the input data existence of the equilibrium; see Kuhn (1956). He used two basic tools: Kakutani's fixed point theorem (1941) and the LP duality (1947).

The interest in J. Nash equilibria was reignited by Rosen's (1965) paper, where he pointed out some connections between J. Nash equilibrium and convex optimization. Soon after, few numerical methods for finding J. Nash equilibrium were developed by Zukhovitsky et al. (1969, 1973). Also, in the late 1960s, the authors show that H. Kuhn's version of the Walras–Wald equilibrium is equivalent to J. Nash equilibrium in a concave n-person game; see Zukhovitsky et al. (1970).

We consider an economy that produces n products by consuming m resources. The resources are fixed and given by the vector $b = (b_1, \ldots, b_m)^T$.

The prices c_i for goods are not fixed; they are dependent on the production output vector $x = (x_1, \ldots, x_n)^T$.

In other words, price operator c maps the production output $x = (x_1, \ldots, x_n)^T$ into vector function $c(x) = (c_1(x), \ldots, c_n(x))^T$, where $c_j(x)$ – price per unit of product $1 \leq j \leq n$ under production output $x \in \Omega = \{x \in \mathbb{R}^n_+ : Ax \leq b\}$.

The element a_{ij} of matrix $A : \mathbb{R}^n \to \mathbb{R}^m$ defines the consumption of resource i for the production of one unit of product $1 \le j \le n$.

The WW equilibrium is defined by a pair (x^*, λ^*) :

$$\langle c(x^*), x^* \rangle = \max\{\langle c(x^*), x\rangle | Ax \le b,\ x \in \mathbb{R}^n_+\} \tag{10.18}$$

$$\langle b, \lambda^* \rangle = \min\{\langle b, \lambda\rangle | A^T\lambda \ge c(x^*),\ \lambda \in \mathbb{R}^m_+\}. \tag{10.19}$$

Let us assume that for every product $1 \le j \le n$ the price $c_j(x)$ is a continuous non-increasing function of the corresponding output x_j, when production of other goods are fixed, that is

$$t_2 > t_1 > 0 \Rightarrow c_j(x_1, \ldots, t_2, \ldots, x_n) \le c_j(x_1, \ldots, t_1, \ldots, x_n), 1 \le j \le n. \tag{10.20}$$

Then

$$\varphi_j(x_1, \ldots, x_j, \ldots, x_n) = \int_0^{x_j} c_j(x_1, \ldots, t, \ldots, x_n)dt \tag{10.21}$$

is continuous in $x = (x_1, \ldots, x_j, \ldots, x_n) \in \Omega = \{x \in \mathbb{R}^n_+ : Ax \le b\}$ and concave in x_j.

Therefore, one obtains WW equilibrium (10.18)–(10.19) by finding normalized J. Nash equilibrium with payoff functions defined by (10.21) :

$$\Phi(x^*, x^*) = \max\{\Phi(x^*, y) | y \in \Omega\}, \tag{10.22}$$

where $\Phi(x, y) = \sum_{j=1}^n \varphi_j(x_1, \ldots, y_j, \ldots, x_n)$.

For any given $x \in \Omega$, the normalized payoff function $\Phi(x, y)$ is concave in $y \in \Omega$. Hence,

$$\omega(x) = \text{Argmax}\{\Phi(x, y) | y \in \Omega\}$$

is the solution set of a convex optimization problem, when $x \in \Omega$ is fixed. Therefore, for any given $x \in \Omega$, the set $\omega(x) \subset \Omega$ is a convex compact. Moreover, it follows from continuity of Φ that $\omega(x)$ is an upper semicontinuous set-valued function on Ω. Therefore, the existence of a fixed point $x^* \in \Omega$ of the map $x \to \omega(x)$ is a direct consequence of Kakutani's theorem.

The pseudo-gradient of $\Phi(x, y)$ in (10.22) is

$$g(x) = \nabla_y \Phi(x, y)_{|y=x} = c(x) = (c_1(x), \ldots, c_n(x))^T.$$

The pseudo-Hessian is

$$H(\cdot) = J(g(x)) = J(c(x)) = \begin{bmatrix} \frac{\partial c_1(\cdot)}{\partial x_1}, & \ldots, & \frac{\partial c_1(\cdot)}{\partial x_n} \\ \\ \frac{\partial c_n(\cdot)}{\partial x_1}, & \ldots, & \frac{\partial c_n(\cdot)}{\partial x_n} \end{bmatrix}.$$

To guarantee (10.20), it is sufficient to assume that the pseudo-Hessian satisfies the dominant diagonal condition:

$$(a)\ \frac{dc_j(\cdot)}{dx_j} < 0, 1 \le j \le n \quad (b)\ \left|\frac{\partial c_j(\cdot)}{\partial x_j}\right| >> \left|\frac{\partial c_j(\cdot)}{\partial x_i}\right|, \forall i \neq j \tag{10.23}$$

Condition (10.23a) means the increase of production x_j leads to the decrease of the price c_j per unit under fixed production of the rest.

Condition (10.23b) means that for any given product $1 \le j \le n$ the impact of its production variation on its price is much stronger than the impact on its price by production variation of any other product.

Existence of the WW equilibrium (x^*, λ^*) directly follows from existence of J. Nash equilibrium and LP duality.

In fact, using the optimality criteria (10.17) with $g(x^*) = c(x^*)$, we can rewrite (10.22) as follows:

$$\max\{\langle c(x^*), x\rangle \mid Ax \le b, x \in \mathbb{R}^m_+\} = \langle c(x^*), x^*\rangle.$$

From LP duality follows existence $\lambda^* \in \mathbb{R}^m_+$:

$$\langle b, \lambda^*\rangle = \min\{\langle b, \lambda\rangle | A^T\lambda \ge c(x^*),\ \ \lambda \in \mathbb{R}^m_+\},$$

and

$$\langle c(x^*), x^*\rangle = \langle b, \lambda^*\rangle.$$

Thus, by finding J. Nash equilibrium from (10.22), one finds the WW equilibrium (10.18)–(10.19).

Finding WW equilibrium from (10.22) is equivalent to solving VI (10.17) with $g(x) = c(x)$. Solving (10.17), generally speaking, is more difficult than solving the following LP:

$$\langle c, x^*\rangle = \max\{\langle c, x\rangle | Ax \le b,\ x \in \mathbb{R}^n_+\}. \tag{10.24}$$

In the next section, we consider NE for optimal resource allocation, which is a generalization of the WW equilibrium. We will see that in a number of instances finding NE can be easier than solving LP.

10.3 NE for Optimal Resource Allocation

10.3.1 Introduction

For several decades, LP has been widely used for optimal resource allocation. In 1975, L. V. Kantorovich and T. C. Koopmans shared the Nobel Prize in Economics "for their contributions to the theory of optimal allocation limited resources."

The LP approach uses two fundamental assumptions:

(a) The price vector $c = (c_1, \ldots, c_n)^T$ for goods is fixed, given a priori and independent of the production output vector $x = (x_1, \ldots, x_n)^T$.

(b) The resource vector $b = (b_1, \dots, b_m)^T$ is also fixed, given a priori, and the resource availability is independent of the resource price vector $\lambda = (\lambda_1, \dots, \lambda_n)^T$.

Unfortunately, such assumptions do not reflect the basic market law of supply and demand. Therefore, the LP models sometimes produce solutions, which are not always practical.

Also, a small change of at least one component of the price vector c might lead to drastic changes of the primal solution. Similarly, a small variation of the resource vector b might lead to dramatic changes of the dual solution.

In the following section, we consider NE as an alternative to LP approach for optimal resource allocation (ORA).

10.3.2 Problem Formulation

The fixed price vector $c = (c_1, \dots, c_n)^T$ is replaced by a price operator $c : \mathbb{R}^n_+ \to \mathbb{R}^n_+$, which maps the production output vector $x = (x_1, \dots, x_n)^T$ into the price-vector function $c(x) = (c_1(x), \dots, c_n(x))^T$.

Similarly, the fixed resource vector $b = (b_1, \dots, b_m)^T$ is replaced by the resource operator $b : \mathbb{R}^m_+ \to \mathbb{R}^m_+$, which maps the resource price vector $\lambda = (\lambda_1, \dots, \lambda_n)^T$ into the resource availability vector – function $b(\lambda) = (b_1(\lambda), \dots, b_m(\lambda))^T$. We assume that both operators c and b are continuous on $\mathbb{R}^n_+$ and $\mathbb{R}^m_+$ accordingly.

The pair of vectors $(x^*, \lambda^*) = y^* \in \Omega = \mathbb{R}^n_+ \times \mathbb{R}^m_+$

$$x^* \in \text{Argmax}\{\langle c(x^*), x\rangle | Ax \leq b(\lambda^*), x \in \mathbb{R}^n_+\} \tag{10.25}$$

$$\lambda^* \in \text{Argmin}\{\langle b(\lambda^*), \lambda\rangle | A^T\lambda \geq c(x^*), \lambda \in \mathbb{R}^m_+\} \tag{10.26}$$

is called nonlinear equilibrium (NE).

The primal–dual LP solution, which one obtains from (10.25)–(10.26) by replacing $c(x)$ for c and $b(\lambda)$ for b, can be viewed as linear equilibrium (LE).

We will see later that strong monotonicity for both the price operator $c : \mathbb{R}^n_+ \to \mathbb{R}^n_+$ and the resource operator $b : \mathbb{R}^m_+ \to \mathbb{R}^m_+$ guarantees existence and uniqueness of NE.

Operator $b : \mathbb{R}^m_+ \to \mathbb{R}^m_+$ is strongly monotone increasing if for any pair of vectors λ_1 and λ_2 from $\mathbb{R}^m_+$ the following inequality

$$\langle b(\lambda_1) - b(\lambda_2), \lambda_1 - \lambda_2\rangle \geq \beta\|\lambda_1 - \lambda_2\|^2 \tag{10.27}$$

holds and $\beta > 0$ is independent on $\lambda_1 \in \mathbb{R}^m_+$ and $\lambda_2 \in \mathbb{R}^m_+$.

Operator $c : \mathbb{R}^n_+ \to \mathbb{R}^n_+$ is strongly monotone decreasing if for any pair of vectors x_1 and x_2 from $\mathbb{R}^n_+$ the following inequality

$$\langle c(x_1) - c(x_2), x_1 - x_2\rangle \leq -\alpha\|x_1 - x_2\|^2 \tag{10.28}$$

holds and $\alpha > 0$ is independent from $x_1 \in \mathbb{R}^n_+$ and $x_2 \in \mathbb{R}^n_+$.

It follows from (10.27): the price per unit increase for any given resource while the prices of the rest are fixed leads to availability increase of the given resource, and the margin has a positive lower bound.

It follows from (10.28): the production increase of any given product while the production of the rest is fixed leads to decrease of the price per unit for the correspondent product, and the margin has a negative upper bound.

10.3.3 NE as a VI

Our next step is to show that finding NE from (10.25)–(10.26) is equivalent to solving a particular variational inequality (VI).

Theorem 10.3. *Finding* $y^* = (x^*, \lambda^*)$ *from* (10.25)–(10.26) *is equivalent to solving the following VI:*

$$\langle g(y^*), y - y^* \rangle \leq 0, \; \forall y = (x; \lambda) \in \Omega, \tag{10.29}$$

where $g(y) \equiv g(x, \lambda) = (c(x) - A^T \lambda; Ax - b(\lambda))^T$.

Proof. Let us fix $x \in \mathbb{R}^n_+$ and $\lambda \in \mathbb{R}^m_+$ and consider the following dual pair of LP:

$$\max\{\langle c(x), X \rangle | AX \leq b(\lambda), X \in \mathbb{R}^n_+\} \tag{10.30}$$

$$\min\{\langle b(\lambda), \Lambda \rangle | A^T \Lambda \geq c(x), \Lambda \in \mathbb{R}^m_+\}. \tag{10.31}$$

Solving the dual LP pair (10.30)–(10.31) is equivalent to finding a saddle point for the correspondent Lagrangian:

$$\mathbb{L}(x, \lambda; X, \Lambda) = \langle c(x), X \rangle - \langle \Lambda, AX - b(\lambda) \rangle \tag{10.32}$$

on $\mathbb{R}^n_+ \otimes \mathbb{R}^m_+$, that is, finding

$$\max_{X \in \mathbb{R}^n_+} \min_{\Lambda \in \mathbb{R}^m_+} \mathbb{L}(x, \lambda; X, \Lambda) = \min_{\Lambda \in \mathbb{R}^m_+} \max_{X \in \mathbb{R}^n_+} \mathbb{L}(x, \lambda; X, \Lambda), \tag{10.33}$$

under fixed $x \in \mathbb{R}^n_+$, $\lambda \in \mathbb{R}^m_+$.

The problem (10.33) is equivalent to finding a J. Nash equilibrium of a concave two-person game with the following payoff functions:

$$\varphi_1(x, \lambda; X, \lambda) = \langle c(x), X \rangle - \langle \lambda, AX - b(\lambda) \rangle = \langle c(x) - A^T \lambda, X \rangle + \langle \lambda, b(\lambda) \rangle \tag{10.34}$$

and

$$\varphi_2(x, \lambda; x, \Lambda) = \langle Ax - b(\lambda), \Lambda \rangle, \tag{10.35}$$

where $X \in \mathbb{R}^n_+$ is the strategy of the first player and $\Lambda \in \mathbb{R}^m_+$ is the strategy of the second player.

Let $y = (x, \lambda)$ and $Y = (X, \Lambda)$ belong to $\Omega = \mathbb{R}^n_+ \otimes \mathbb{R}^m_+$. The corresponding normalized payoff function is defined as follows:

$$\Phi(y,Y) = \langle c(x) - A^T\lambda, X\rangle + \langle Ax - b(\lambda), \Lambda\rangle + \langle \lambda, b(\lambda)\rangle. \tag{10.36}$$

Finding a saddle point (10.33) is equivalent to finding a normalized J. Nash equilibrium $\bar{y} \in \Omega$:

$$\Phi(\bar{y},\bar{y}) = \max\{\Phi(\bar{y},Y) | Y \in \Omega\}. \tag{10.37}$$

Let us consider the corresponding pseudo-gradient:

$$\nabla_Y \Phi(y,Y)_{|Y=y} = g(y) \equiv g(x,\lambda) = (c(x) - A^T\lambda; Ax - b(\lambda))^T. \tag{10.38}$$

Then, finding $\bar{y} \in \Omega$ from (10.37) is equivalent to solving the following VI:

$$\langle g(\bar{y}), Y - \bar{y}\rangle \leq 0, \ \ \forall Y \in \Omega. \tag{10.39}$$

Conversely, if $\bar{y} \in \Omega$ solves VI (10.39), then

$$\max\{\langle g(\bar{y}), Y - \bar{y}\rangle | Y \in \Omega\} = \langle g(\bar{y}), \bar{y} - \bar{y}\rangle = 0; \tag{10.40}$$

therefore, $g(\bar{y}) \leq 0$, assuming that at least one component of $g(\bar{y})$ is positive, we obtain

$$\max\{\langle g(\bar{y}), Y - \bar{y}\rangle | Y \in \Omega\} = \infty. \tag{10.41}$$

If $\bar{y} = (\bar{x}, \bar{\lambda}) \geq 0$ solves VI (10.39), then

$$c(\bar{x}) - A^T\bar{\lambda} \leq 0 \text{ and } A\bar{x} - b(\bar{\lambda}) \leq 0. \tag{10.42}$$

Moreover, for $1 \leq j \leq n$, we have

$$(c(\bar{x}) - A^T\bar{\lambda})_j < 0 \Rightarrow \bar{x}_j = 0, \ \bar{x}_j > 0 \Rightarrow (c(\bar{x}) - A^T\bar{\lambda})_j = 0 \tag{10.43}$$

and for $1 \leq i \leq m$, we have

$$(A\bar{x} - b(\bar{\lambda}))_i < 0 \Rightarrow \bar{\lambda} = 0, \ \bar{\lambda}_i > 0 \Rightarrow (A\bar{x} - b(\bar{\lambda}))_i = 0. \tag{10.44}$$

Hence, for the primal–dual feasible pair $\bar{y} = (\bar{x}, \bar{\lambda}) \in \Omega$ complementarity condition (10.43)–(10.44) holds.

Therefore, vector $\bar{y} = (\bar{x}, \bar{\lambda})$ solves the following primal and dual LP:

$$\max\{\langle c(\bar{x}), X\rangle | AX \leq b(\bar{\lambda}), \ X \in \mathbb{R}^n_+\} = \langle c(\bar{x}), \bar{x}\rangle \tag{10.45}$$

$$\min\{\langle b(\bar{\lambda}), \Lambda\rangle | A^T\Lambda \geq c(\bar{x}), \ \Lambda \in \mathbb{R}^m_+\} = \langle b(\bar{\lambda}), \bar{\lambda}\rangle; \tag{10.46}$$

that is, $\bar{y} = y^*$. □

Now, let us show that $y^* \in \Omega$ exists.

10.3.4 Existence and Uniqueness of the NE

The arguments used for proving the existence of the normalized J. Nash equilibrium cannot be used in the case of NE problem (10.37) because the feasible set $\Omega = \mathbb{R}^n_+ \times \mathbb{R}^m_+$ is unbounded; therefore, Kakutani's theorem cannot be applied.

It turns out that if c and b are strongly monotone, then in spite of unboundedness Ω, the NE exists. We start with the following technical lemma.

Lemma 10.1. *If both operators c and b are strongly monotone, then the pseudo-gradient $g : \Omega \to \mathbb{R}^{m+n}$ is a strongly monotone operator; that is, for $\gamma = \min\{\alpha, \beta\} > 0$, the following inequality holds:*

$$\langle g(y_1) - g(y_2), y_1 - y_2 \rangle \leq -\gamma \|y_1 - y_2\|^2 \tag{10.47}$$

for any pair $(y_1; y_2) \in \Omega \otimes \Omega$.

Proof. Let $y_1 = (x_1, \lambda_1)$ and $y_2 = (x_2, \lambda_2) \in \Omega$; then

$$\begin{aligned}\langle g(y_1) - g(y_2), y_1 - y_2 \rangle &= \langle c(x_1) - A^T \lambda_1 - c(x_2) \\ &\quad + A^T \lambda_2, x_1 - x_2 \rangle + \langle A x_1 - b(\lambda_1) - A x_2 + b(\lambda_2), \lambda_1 - \lambda_2 \rangle \\ &= \langle c(x_1) - c(x_2), x_1 - x_2 \rangle - \langle A^T (\lambda_1 - \lambda_2), x_1 - x_2 \rangle \\ &\quad + \langle A(x_1 - x_2), \lambda_1 - \lambda_2 \rangle - \langle b(\lambda_1) - b(\lambda_2), \lambda_1 - \lambda_2 \rangle \\ &= \langle c(x_1) - c(x_2), x_1 - x_2 \rangle - \langle b(\lambda_1) - b(\lambda_2), \lambda_1 - \lambda_2 \rangle.\end{aligned}$$

Invoking (10.27) and (10.28), we obtain (10.47). □

We are ready to prove the existence and uniqueness of the NE.

Theorem 10.4. *If both c and b are strongly monotone, then the NE $y^* = (x^*, \lambda^*)$ exists and it is unique.*

Proof. Let us consider $y_0 \in \Omega$: $\|y_0\| \leq 1$ and a large enough number $M > 0$. Instead of (10.37), we consider the following equilibrium problem:

$$\Phi(y^*_M, y^*_M) = \max\{\Phi(y^*_M, Y) | Y \in \Omega_M\}, \tag{10.48}$$

where $\Omega_M = \{Y \in \Omega : \|Y\| \leq M\}$. The normalized payoff function $\Phi(y, Y)$, defined by (10.36), is linear in $Y \in \Omega$, and Ω_M is a convex compact set. Therefore, for a given $y \in \Omega_M$, the solution set

$$\omega_M(y) = \text{Argmax}\{\Phi(y, Y) | Y \in \Omega_M\} \tag{10.49}$$

is a convex compact. Moreover, the map $y \to \omega(y)$ is upper semicontinuous. In fact, let us consider a sequence $\{y_s\} \subset \Omega_M$: $y_s \to \bar{y}$ and any sequence of images $\{z_s \in \omega(y_s)\}$ converging to $\bar{z}$; then due to the continuity of $\Phi(y, Y)$ in y and Y, we have $\bar{z} \in \omega(\bar{y})$. Therefore, $y \to \omega(y)$ maps convex compact Ω_M into itself, and the map is upper semicontinuous; hence, Kakutani's theorem can be applied. Therefore, there exists $y^*_M \in \Omega_M$: $y^*_M \in \omega(y^*_M)$.

Let us show that the constraint $\|Y\|| \leq M$ is irrelevant in problem (10.48). Using the bound (10.47) for $y_1 = y_0$ and $y_2 = y_M^*$, one obtains

$$\gamma\|y_0 - y_M^*\|^2 \leq \langle g(y_M^*) - g(y_0), y_0 - y_M^* \rangle$$

$$= \langle g(y_M^*), y_0 - y_M^* \rangle + \langle g(y_0), y_M^* - y_0 \rangle. \tag{10.50}$$

Vector y_M^* solves VI (10.39) when Ω is replaced by Ω_M; hence,

$$\langle g(y_M^*), y_0 - y_M^* \rangle \leq 0. \tag{10.51}$$

From (10.50)–(10.51) and Cauchy–Schwarz inequality follows

$$\gamma\|y_0 - y_M^*\|^2 \leq |\langle g(y_0), y_M^* - y_0 \rangle| \leq \|g(y_0)\|\|y_M^* - y_0\| \tag{10.52}$$

or

$$\|y_0 - y_M^*\| \leq \gamma^{-1}\|g(y_0)\|. \tag{10.53}$$

Therefore,

$$\|y_M^*\| \leq \|y_0\| + \|y_M^* - y_0\| \leq 1 + \gamma^{-1}\|g(y_0)\|.$$

Hence, for $M > 0$ large enough, the inequality $\|Y\| \leq M$ is irrelevant in (10.48).

In other words, $y_M^* = y^* = (x^*, \lambda^*)$ is the NE. The strong monotonicity of operators c and b leads to uniqueness of NE.

In fact, assuming that $\bar{y} \in \Omega$, $y^* \in \Omega$ are two different solutions of the variational inequality (10.39), we obtain

$$\langle g(\bar{y}), y^* - \bar{y} \rangle \leq 0 \text{ and } \langle -g(y^*), y^* - \bar{y} \rangle \leq 0;$$

therefore, $\langle g(y^*) - g(\bar{y}), y^* - \bar{y} \rangle \geq 0$, which contradicts (10.47) for $y_1 = \bar{y}$ and $y_2 = y^*$.

The contradiction proves the uniqueness of the NE $y^* = (x^*, \lambda^*)$. □

Let $D = \{x : Ax \leq b, x \in \mathbb{R}^n_+\}$; then the classical WW equilibrium is equivalent to the following VI:

$$x^* \in D : \langle c(x^*), x - x^* \rangle \leq 0, \forall x \in D. \tag{10.54}$$

Solving (10.54), generally speaking, is more difficult than solving the corresponding primal and dual LP (10.25)–(10.26) when $b(\lambda) \equiv b$ and $c(x) \equiv c$. One can think that finding NE $y^* = (x^*, \lambda^*)$ is more difficult than solving VI (10.54).

In fact, as we will see later, finding NE $y^* = (x^*, \lambda^*)$ from (10.25)–(10.26) in a number of instances can be much easier than solving the corresponding primal and dual LP.

The fundamental difference between NE (10.25)–(10.26) and WW (10.54) follows from the geometry of their feasible sets $\Omega = \mathbb{R}^n_+ \otimes \mathbb{R}^m_+$ and D. The simplicity of Ω makes pseudo-gradient projection-type methods particularly suitable for solving VI (10.39) because it requires just matrix by vector multiplication as the main operation at each step, whereas pseudo-gradient projection methods for solving VI (10.54) require at each step solving a quadratic programming problem.

10.4 Nonlinear Input–Output Equilibrium

10.4.1 Introduction

The input–output (IO) model has been introduced before WW2. The main purpose was to better understand the interdependence of the production sectors of an economy.

Since WW2, IO model has been widely used for analysis of economic activities, planning production, economic prognoses, and international trade, just to mention a few. The applications of IO model range from a branch of a single economy to the world economy.

The main contributor to IO theory and, in particular, to the practical aspects of IO model was Wassily W. Leontief (1906–1999).

He received the Nobel Price in Economics in 1973 "for the development of the input–output method and for its application to important economic problems."

We consider nonlinear input–output equilibrium (NIOE). The main difference between NIOE and the classical Wassily Leontief's input–output (IO) model is both the production cost and the consumption are not fixed and are not given a priori.

Instead, the production cost is an operator, which maps the output into the cost per unit. The consumption is an operator as well, which maps the price for goods into the consumption.

The NIOE finds such output and such prices for goods that the following two important goals can be achieved.

First, at the NIOE, the production cost per unit is consistent with the output, and the consumption is consistent with the prices for goods.

Second, at the NIOE, the total production cost reaches its minimum, while the total consumption reaches its maximum.

10.4.2 Preliminaries

The input–output model assumes that an economy produces n products, which are also consumed in the production process. Elements $a_{ij} \geq 0$, $1 \leq i \leq n$, $1 \leq j \leq n$ of the consumption matrix $A = \|a_{ij}\|$ show the amount of product $1 \leq i \leq n$ consumed for production of one unit of product $1 \leq j \leq n$.

Let $x = (x_1, \dots, x_n)^T$ be the production vector, i.e., x_j defines how many units of product $1 \leq j \leq n$ we are planning to produce. Then, Ax is the total consumption needed to produce vector $x \in \mathbb{R}^n_+$. Therefore, the components of the vector $y =$

$x - Ax = (I - A)x$ show how much of each product is left after the production needs were covered. Vector y can be used for consumption, investments, and trades, just to mention a few possibilities.

Let $\lambda = (\lambda_1, \ldots, \lambda_n)^T$ be the price vector, i.e., λ_j defines the price of one unit of product $1 \leq j \leq n$; then $q = \lambda - A^T\lambda$ is the profit vector, i.e., q_j defines the profit out of one unit of product $1 \leq j \leq n$ under the price vector $\lambda \in \mathbb{R}^n_+$.

The IO model solves two basic problems.

1. For a fixed consumption vector $c = (c_1, \ldots, c_n)^T \in \mathbb{R}^n_+$, given a priori, the IO finds the production vector $x_c = (x_{c,1}, \ldots, x_{c,n})^T \in \mathbb{R}^n_+$, which guarantees the required consumption. Vector x_c is the solution of the following system:

$$(I - A)x = c, \; x \in \mathbb{R}^n_+. \tag{10.55}$$

2. For a fixed profit vector $q = (q_1, \ldots, q_n)^T \in \mathbb{R}^n_+$, given a priori, the IO finds such price vector $\lambda_q = (\lambda_{q,1}, .., \lambda_{q,n})^T \in \mathbb{R}^n_+$, which guarantees the given profit. Vector λ_q is the solution of the following system:

$$q = (I - A)^T\lambda, \; \lambda \in \mathbb{R}^n_+. \tag{10.56}$$

The economy is called **productive** if for any consumption vector $c \in \mathbb{R}^n_{++}$ the following system

$$x - Ax = c, \; x \geq 0 \tag{10.57}$$

has a solution.

In such case, the matrix A is called productive matrix. It is rather remarkable that if the system (10.57) has a solution for only one vector $c \in \mathbb{R}^n_{++}$, it has a solution for any given vector $c \in \mathbb{R}^n_{++}$.

It is obvious that for any consumption vector $c \in \mathbb{R}^n_+$ the system (10.57) has a positive solution if the inverse matrix $(I - A)^{-1}$ exists and it is positive.

To address the issue of matrix $B = (I - A)^{-1}$ positivity, we recall the notion of indecomposability of a non-negative matrix A.

Let S be a subset of indices $N = \{1, \ldots, n\}$, i.e., $S \subset N$ and $S' = N \setminus S$. The set of indices S is called isolated set if $a_{ij} = 0$ for all $i \in S$ and $j \in S'$. It means that the products, which belong to set S, are not used for production of any product from set S'.

In other words, if matrix A is decomposable, then there is a subset of indices that by simultaneous renumeration of rows and columns, we can find $S = \{1, \ldots, k\}$ and $S' = \{k + 1, \ldots, n\}$ such that

$$A = \begin{array}{c} \\ S \\ S' \end{array} \!\!\begin{array}{c} \begin{array}{cc} S & S' \end{array} \\ \left[\begin{array}{c|c} A_{11} & 0 \\ \hline A_{21} & A_{22} \end{array}\right] \end{array}, \tag{10.58}$$

where $A_{11} : \mathbb{R}^k \to \mathbb{R}^k$, $A_{22} : \mathbb{R}^{n-k} \to \mathbb{R}^{n-k}$.

Matrix A is indecomposable if representation (10.58) is impossible by any simultaneous rows and columns renumeration.

For indecomposable matrix, production of any product $1 \leq j \leq n$ requires product $1 \leq i \leq n$ directly $(a_{ij} > 0)$ or indirectly; that is, for any i and j such that $a_{ij} = 0$, there is a set of indices $i{=}i_1, \ldots, i_{m-1}, i_m{=}j$ such that $a_{i_s, i_{s+1}}{>}0, s{=}1, 2, \ldots, m-1$.

The following theorem plays an important role on the input–output model.

Theorem 10.5 (Frobenius–Perron). *Any non-negative indecomposable matrix A has a real positive dominant eigenvalue λ_A, i.e., for any eigenvalue λ of matrix A, the following inequality $|\lambda| \leq \lambda_A$ holds and the correspondent to λ_A eigenvector $x_A \in \mathbb{R}^n_{++}$.*

We are ready to formulate the necessary and sufficient condition for Leontief's model to be productive; see Ashmanov (1984), Gale (1960), Dorfman and Samuelson (1958), and Lancaster (1968).

Theorem 10.6 (Leontief). *For non-negative indecomposable consumption matrix A, the input–output model (10.57) is productive iff $\lambda_A < 1$.*

It follows from $\lambda_A < 1$ and Theorem 10.5 that for all eigenvalues λ of A, we have $|\lambda| < 1$; therefore, the matrix $B = \|b_{ij}\|$ has the following representation:

$$B = (I - A)^{-1} = I + A + A^2 + A^3 + \ldots$$

For non-negative indecomposable matrix A, we have

$$B - A = I + A^2 + A^3 + \ldots >> 0,$$

which means that total consumption b_{ij} of product $1 \leq i \leq n$ required to produce one unit of product $1 \leq j \leq n$ is always greater than direct consumption a_{ij}.

There is another sufficient condition for Leontief's model to be productive.

If for all $1 \leq i \leq n$ products but one we have

$$\lambda_j > (A^T\lambda)_j \text{ and } \lambda_i = (A^T\lambda)_i, i \neq j,$$

then the economy is productive; that is, for any given $c \in \mathbb{R}^n_+$, the system (10.55) has a positive solution and for any given $q \in \mathbb{R}^n_+$ the system (10.56) has a positive solution as well.

In what follows we will consider a non-negative indecomposable matrices A with $\lambda_A < 1$. Therefore, the inverse matrix

$$B = (I - A)^{-1} \tag{10.59}$$

exists and it is positive (see Leontief (1966)).

For a **productive** economy, both systems (10.55) and (10.56) have unique solutions, and both vectors x_c and λ_q are positive. We will always assume that the economy is productive.

Another possible use of IO model consists of finding "optimal" production vector $x^* \in \mathbb{R}^n_+$ and "optimal" price vector $\lambda^* \in \mathbb{R}^n_+$ by solving primal and dual LP.

Let $p = (p_1, \ldots, p_n)^T$ be fixed and a priori given production cost vector and $c = (c_1, \ldots, c_n)^T \in \mathbb{R}^n_+$ be fixed and a priori given consumption vector. Finding "optimal" production output x^* and "optimal" consumption prices λ^* leads to the following dual pair of LP problems:

$$\langle p, x^* \rangle = \min\{\langle p, x \rangle | (I - A)x \geq c, x \in \mathbb{R}^n\} \tag{10.60}$$

and

$$\langle c, \lambda^* \rangle = \max\{\langle c, \lambda \rangle | (I - A)^T \lambda \leq p, \lambda \in \mathbb{R}^n_+\} \tag{10.61}$$

For a productive economy, both the primal feasible solution

$$x_c = (I - A)^{-1} c$$

and the dual feasible solution

$$\lambda_p = ((I - A)^T)^{-1} p$$

are positive vectors.

So x_c and λ_p are primal and dual feasible solutions, for which the complementarity condition is satisfied. Therefore, $x^* = x_c$ and $\lambda^* = \lambda_p$.

In other words, the primal and dual LP (10.60) and (10.61) do not produce results different from IO model.

10.4.3 Problem Formulation

We introduce and study the nonlinear input–output equilibrium (NIOE), which extends the abilities of the classical IO model in a few directions.

We replace the fixed cost vector $p = (p_1, \ldots, p_n)^T \in \mathbb{R}^n_+$ by a cost operator $p : \mathbb{R}^n_+ \to \mathbb{R}^n_+$, which maps the production output vector $x = (x_1, \ldots, x_n)^T \in \mathbb{R}^n_+$ into cost per unit vector $p(x) = (p_1(x), \ldots, p_n(x))^T \in \mathbb{R}^n_+$.

Similarly, we replace the fixed consumption vector $c = (c_1, \ldots, c_n)^T \in \mathbb{R}^n_+$ by a consumption operator $c : \mathbb{R}^n_+ \to \mathbb{R}^n_+$, which maps the prices for goods vector $\lambda = (\lambda_1, \ldots, \lambda_n)^T \in \mathbb{R}^n_+$ into consumption vector $c(\lambda) = (c_1(\lambda), \ldots, c_n(\lambda))^T \in \mathbb{R}^n_+$.

We assume that operators p and c are continuous on $\mathbb{R}^n_+$.

We will call the nonlinear input–output equilibrium (NIOE) the following pair $(x^*, \lambda^*) = y^* \in \Omega = \mathbb{R}^n_+ \times \mathbb{R}^n_+$:

$$\langle p(x^*), x^* \rangle = \min\{\langle p(x^*), x \rangle | (I - A)x \geq c(\lambda^*), x \in \mathbb{R}^n_+\} \tag{10.62}$$

and

$$\langle c(\lambda^*), \lambda^* \rangle = \max\{\langle c(\lambda^*), \lambda \rangle | (I - A)^T \lambda \leq p(x^*), \lambda \in \mathbb{R}^n_+\} \tag{10.63}$$

The primal and dual LP (10.60) and (10.61), which one obtains in the case of $p(x) \equiv p$ and $c(\lambda) \equiv c$, can be viewed as linear input–output equilibrium, which is identical to IO model.

It follows from (10.62) that the production vector $x^* \in \mathbb{R}^n_+$ minimizes the total production cost and at the same time is such that the consumption $c(\lambda^*) \in \mathbb{R}^n_+$, defined by the price vector $\lambda^* \in \mathbb{R}^n_+$, is satisfied.

It follows from (10.63) that the price vector λ^* for goods maximizes the total consumption, and at the same time the profit per unit of goods does not exceed the production cost of the correspondent unit.

We would like to emphasize that due to (10.62) the production cost $p(x^*)$ is consistent with the optimal production vector x^* and due to (10.63) the consumption $c(\lambda^*)$ is consistent with the optimal price vector λ^* for goods.

The existence of NIOE $y^* = (x^*, \lambda^*)$ is our first concern.

We assume, at this point, that both p and c are strongly monotone operators: there are $\alpha > 0$ and $\beta > 0$ that

$$\langle p(x_1) - p(x_2), x_1 - x_2 \rangle \geq \alpha \|x_1 - x_2\|^2, \ \forall x_1, x_2 \in \mathbb{R}^n_+ \tag{10.64}$$

and

$$\langle c(\lambda_1) - c(\lambda_2), \lambda_1 - \lambda_2 \rangle \leq -\beta \|\lambda_1 - \lambda_2\|^2, \ \forall \lambda_1, \lambda_2 \in \mathbb{R}^n_+ \tag{10.65}$$

hold.

Assumption (10.64) implies that the production increase of any goods when the production of the rest is fixed leads to the cost per unit increase. Moreover, the lower bound of the margin is $\alpha > 0$.

Assumption (10.65) implies that the price per unit increase for any product when the prices for the rest are fixed leads to consumption decrease of such product. Moreover, the margin has a negative upper bound $-\beta < 0$.

10.4.4 NIOE as a VI

We first show that finding NIOE y^* is equivalent to solving a particular VI on $\Omega = \mathbb{R}^n_+ \times \mathbb{R}^n_+$.

Theorem 10.7. *For $y^* = (x^*, \lambda^*) \in \Omega$ to be a solution of (10.62) and (10.63), it is necessary and sufficient for y^* to be a solution of the following VI:*

$$\langle g(y^*), y - y^* \rangle \leq 0, \forall y \in \Omega, \tag{10.66}$$

where the operator $g : \Omega \to \mathbb{R}^{2n}$ is given by the following formula:

$$g(y) = ((I - A)^T \lambda - p(x); c(\lambda) - (I - A)x))^T \tag{10.67}$$

Proof. If $y^* = (x^*, \lambda^*)$ solves (10.62) and (10.63), then y^* is a saddle point of the Lagrangian:

$$L(y^*;X,\Lambda) = \langle p(x^*), X\rangle - \langle \Lambda, (I-A)X - c(\lambda^*)\rangle, \tag{10.68}$$

which corresponds to LP (10.62), when $x^* \in \mathbb{R}^n_+$ and $\lambda^* \in \mathbb{R}^n_+$ are fixed; that is,

$$y^* \in \operatorname*{Argmin}_{X\in\mathbb{R}^n_+} \max_{\Lambda\in\mathbb{R}^n_+} L(y^*;X,\Lambda)$$

In other words,

$$x^* \in \operatorname{Argmin}\{L(y^*;X,\lambda^*)|X \in \mathbb{R}^n_+\} =$$

$$= \operatorname{Argmin}\{\langle p(x^*) - (I-A)^T\lambda^*, X\rangle | X \in \mathbb{R}^n_+\}$$

$$= \operatorname{Argmax}\{\langle (I-A)^T\lambda^* - p(x^*), X\rangle | X \in \mathbb{R}^n_+\}. \tag{10.69}$$

Therefore,

$$(I-A)^T\lambda^* - p(x^*) \le 0. \tag{10.70}$$

On the other hand,

$$\lambda^* \in \operatorname{Argmax}\{L(y^*;x^*,\Lambda)|\Lambda \in \mathbb{R}^n_+\}$$

$$= \operatorname{Argmax}\{\langle c(\lambda^*) - (I-A)x^*, \Lambda\rangle | \Lambda \in \mathbb{R}^n_+\}. \tag{10.71}$$

Therefore,

$$c(\lambda^*) - (I-A)x^* \le 0. \tag{10.72}$$

Keeping in mind the complementarity condition $\langle g(y^*), y^*\rangle = 0$ for the dual LP pair (10.62)–(10.63), we conclude that (10.66) holds for any $y = (x,\lambda) \in \Omega$; that is, $y^* = (x^*,\lambda^*)$ solves VI (10.66).

Conversely, let $\bar{y} \in \Omega$ solve VI (10.66); then

$$\langle g(\bar{y}), y\rangle \le \langle g(\bar{y}), \bar{y}\rangle, \ \forall y \in \Omega. \tag{10.73}$$

It means that $g(\bar{y}) \le 0$; otherwise, the left-hand side can be made as large as one wants by taking the correspondent component of vector y large enough. Therefore, we have

$$(I-A)^T\bar{\lambda} - p(\bar{x}) \le 0, \ \bar{x} \ge 0 \tag{10.74}$$

and

$$c(\bar{\lambda}) - (I-A)\bar{x} \le 0, \ \bar{\lambda} \ge 0. \tag{10.75}$$

So $\bar{x}$ is a feasible solution for the primal LP:

$$\min\{\langle p(\bar{x}), x\rangle | (I-A)x \ge c(\bar{x}), x \in \mathbb{R}^n_+\} \tag{10.76}$$

and $\bar{\lambda}$ is a feasible solution for the dual LP:

$$\max\{\langle c(\bar{\lambda}), \lambda\rangle | (I-A)^T\lambda \le p(\bar{x}), \lambda \in \mathbb{R}^n_+\}. \tag{10.77}$$

From (10.73) for $y = 0 \in \mathbb{R}^{2n}$, we have

$$\langle g(\bar{y}), \bar{y}\rangle \geq 0,$$

which together with (10.74)–(10.75) leads to

$$\langle g(\bar{y}), \bar{y}\rangle = 0. \tag{10.78}$$

Therefore, $(\bar{x}, \bar{\lambda})$ is the primal and dual feasible solution for the dual LP pair (10.76)–(10.77), and the complementarity condition (10.78) is satisfied.

Thus, $\bar{x}$ solves (10.76) and $\bar{\lambda}$ solves (10.77); hence, $\bar{y} = y^*$. □

Our next step is to show that NIOE exists.

Finding a saddle point of the Lagrangian (10.68) is equivalent to finding an equilibrium in a two-person concave game with payoff function

$$\varphi_1(y; X, \lambda) = -L(y; X, \lambda) = \langle (I-A)^T\lambda - p(x), X\rangle + \langle \lambda, c(\lambda)\rangle$$

and strategy $X \in \mathbb{R}^n_+$ for the first player and payoff function

$$\varphi_2(y; x, \Lambda) = L(y; x, \Lambda) = \langle c(\lambda) - (I-A)x, \Lambda\rangle + \langle p(x), x\rangle$$

and strategy $\Lambda \in \mathbb{R}^n_+$ for the second player.

Let us consider the normalized payoff function $\Phi : \Omega \times \Omega \to R$, which is given by the following formula:

$$\Phi(y; Y) = \varphi_1(y; X, \lambda) + \varphi_2(y; x, \Lambda);$$

then finding $y^* = (x^*; \lambda^*) \in \Omega$ is equivalent to finding a fixed point of the following map:

$$y \to \omega(y) = \operatorname{argmax}\{\Phi(y; Y) | Y \in \Omega\}. \tag{10.79}$$

The normalized payoff function $\Phi(y; Y)$ is linear in $Y \in \Omega$ for any given $y \in \Omega$, and Ω is unbounded, which makes impossible to apply Kakutani's fixed point theorem to prove the existence of $y^* \in \Omega : y^* \in \omega(y^*)$.

The operator $g : \Omega \to \mathbb{R}^{2n}$ will be called pseudo-gradient because of the following formula:

$$g(y) = \nabla_Y \Phi(y; Y) = ((I-A)^T\lambda - p(x); c(\lambda) - (I-A)x) \tag{10.80}$$

10.4.5 Existence and Uniqueness of the NIOE

We start with the following lemma, which is similar to Lemma (10.1).

Lemma 10.2. *If operators p and c are strongly monotone, i.e., (10.64)–(10.65) satisfied, then the operator $g : \mathbb{R}^{2n}_+ \to \mathbb{R}^{2n}$, given by (10.80), is strongly monotone as well, i.e., there is $\gamma > 0$ such that*

$$\langle g(y_1) - g(y_2), y_1 - y_2\rangle \leq -\gamma \|y_1 - y_2\|^2 \;\; \forall y_1, y_2 \in \Omega. \tag{10.81}$$

Proof. Let us consider $y_1 = (x_1, \lambda_1)$ and $y_2 = (x_2, \lambda_2) \in \Omega$; then

$$\langle g(y_1) - g(y_2), y_1 - y_2 \rangle = \langle (I-A)^T \lambda_1 - p(x_1) - (I-A)^T \lambda_2 + p(x_2), x_1 - x_2 \rangle +$$
$$+ \langle c(\lambda_1) - (I-A)x_1 - c(\lambda_2) + (I-A)x_2, \lambda_1 - \lambda_2 \rangle =$$
$$\langle (I-A)^T(\lambda_1 - \lambda_2), x_1 - x_2 \rangle - \langle p(x_1) - p(x_2), x_1 - x_2 \rangle +$$
$$+ \langle c(\lambda_1) - c(\lambda_2), \lambda_1 - \lambda_2 \rangle - \langle (I-A)(x_1 - x_2), \lambda_1 - \lambda_2 \rangle$$
$$= -\langle p(x_1) - p(x_2), x_1 - x_2 \rangle + \langle c(\lambda_1) - c(\lambda_2), \lambda_1 - \lambda_2 \rangle$$
$$\leq -\alpha \|x_1 - x_2\|^2 - \beta \|\lambda_1 - \lambda_2\|^2.$$

Therefore, for $\gamma = \min\{\alpha, \beta\}$, we have (10.81). □

The following theorem is similar to Theorem 10.4, and we prove it for completeness.

Theorem 10.8. *If p and c are continuous operators and (10.64)–(10.65) hold, then the NIOE $y^* = (x^*, \lambda^*)$ exists and it is unique.*

Proof. We take $y_0 \in \mathbb{R}^{2n}_{++}$ such that $\|y_0\| \leq 1$, large enough number $M > 0$, and replace Ω with $\Omega_M = \{y \in \Omega : \|y - y_0\| \leq M\}$. The set Ω_M is bounded and convex, and so is the set

$$\omega_M(y) = \operatorname{argmax}\{\Phi(y, Y) | Y \in \Omega_M\}, \tag{10.82}$$

for any given $y \in \Omega$.

Due to continuity p and c, the map $y \to \omega_M(y)$ is upper semicontinuous.

Therefore, from Kakutani's theorem follows existence $y^*_M \in \Omega_M$ such that $y^*_M \in \omega_M(y^*_M)$. Now we show that the constraint $\|y - y_0\| \leq M$ in (10.82) is irrelevant. Let us consider (10.81) with $y_1 = y_0$ and $y_2 = y^*_M$; then from (10.81) follows

$$\gamma \|y_0 - y^*_M\|^2 \leq \langle g(y^*_M) - g(y_0), y_0 - y^*_M \rangle = \tag{10.83}$$
$$= \langle g(y^*_M), y_0 - y^*_M \rangle + \langle g(y_0), y^*_M - y_0 \rangle.$$

From $y^*_M \in \omega_M(y^*_M)$ follows

$$\langle g(y^*_M), Y - y^*_M \rangle \leq 0, \ \forall Y \in \Omega_M$$

That is,

$$\langle g(y^*_M), y_0 - y^*_M \rangle \leq 0.$$

Therefore, from (10.83), we have

$$\gamma \|y_0 - y^*_M\|^2 \leq \langle g(y_0), y^*_M - y_0 \rangle.$$

Using Cauchy–Schwarz inequality, we obtain $\|y_0 - y^*_M\| \leq \gamma^{-1} \|g(y_0)\|$; therefore, for M large enough, we have

$$\|y^*_M\| \leq \|y_0\| + \|y_0 - y^*_M\| \leq 1 + \gamma^{-1} \|g(y_0)\| < M.$$

So the constraint $\|y_0 - y\| \leq M$ is irrelevant in (10.82). Hence, $y^* = y^*_M$ solves VI (10.66), which is equivalent to finding NIOE from (10.62)–(10.63).

The uniqueness follows from (10.81). In fact, assuming that there are two vectors, y^* and $\bar{y}$, which solve VI (10.66), we obtain

$$\langle g(y^*), \bar{y} - y^* \rangle \leq 0 \text{ and } \langle g(\bar{y}), y^* - \bar{y} \rangle \leq 0$$

or

$$\langle g(y^*) - g(\bar{y}), y^* - \bar{y} \rangle \geq 0,$$

which contradicts (10.81) with $y_1 = y^*$ and $y_2 = \bar{y}$; that is, y^* is unique. □

Strong monotonicity assumptions (10.64) and (10.65) are just sufficient for the existence of NIOE. From now on, we assume that NIOE exists.

Finding NIOE is equivalent to solving VI (10.66) on a simple feasible set $\Omega = \mathbb{R}^n_+ \times \mathbb{R}^n_+$.

10.5 Finding NE for Optimal Resource Allocation

10.5.1 Introduction

Finding NE is equivalent to solving VI (10.29). Projection on Ω is a very simple procedure; therefore, our main focus will be two first-order PGP and EPG methods, which, as we will see later, require $O(n^2)$ operations per step.

The gradient projection method for convex optimization (see Section 5.12) was introduced by A. Goldstein (1964) and independently by Levitin and B. Polyak (1966). Some variations of this method were used by Bakushinskij and B. Polyak (1974) for solving VI. The gradient projection method for convex optimization often has mainly theoretical value because even in the case of linear constraints it requires solving at each step a QP problem. In the case of simple feasible sets, however, the gradient projection-type methods can be very efficient.

Projection on Ω is a very simple operation. The main cost per step is computing the pseudo-gradient operator, which requires two matrix by vector multiplication.

Under local strong monotonicity and Lipschitz continuity of both operators b and c, the PGP converges globally with Q-linear rate, and the ratio depends only on the condition number of the VI operator. It leads to complexity bound for the PGP method in terms of the condition number, the size of the problem, and the required accuracy.

In the absence of local strong monotonicity, the convergence of the PGP becomes problematic. Therefore, in the second part of this paragraph, we consider the EPG method for finding NE.

The EPG method was first introduced by Korpelevich (1976) for finding saddle points. In the last several years, it became an important tool for solving VI; see

Antipin (2002), Censor et al. (2011), Iusem and Svaiter (1997), Khobotov (1987), Konnov (2007), and references therein.

Application of the EPG for finding NE leads to a two-stage algorithm. At the first stage, the EPG predicts both the production output and the price vector. At the second stage, it corrects them dependent on the prices for the predicted output and resource availability for the predicted resource prices. It requires projecting the primal–dual vector on Ω twice, which is still only $O(n^2)$ operations.

The EPG method converges to the NE y^* if both the price c and resource b operators are just monotone and satisfy Lipschitz condition.

Under local strong monotonicity, the EPG method globally converges with Q-linear rate, and the ratio is defined by the condition number of the VI operator, the size of the problem, and the required accuracy $\varepsilon > 0$. For a small condition number, the EPG has a better ratio and a much better complexity bound than the PGP.

10.5.2 Basic Assumptions

We consider an economy that produces n goods by consuming m resources. There are three sets of data for problem formulation.

(1) The technological matrix $A : \mathbb{R}^n_+ \to \mathbb{R}^m_+$, which "transforms" resources into goods, i.e., a_{ij} defines the amount of factor $1 \leq i \leq m$ needed to produce one unit of good $1 \leq j \leq n$.
(2) The resource operator $b : \mathbb{R}^m_+ \to \mathbb{R}^m_+$, where $b_i(\lambda)$ is the availability of the resource $1 \leq i \leq m$ under the resource price vector $\lambda = (\lambda_1, \dots, \lambda_i, \dots, \lambda_m)^T$.
(3) The price operator $c : \mathbb{R}^n_+ \to \mathbb{R}^n_+$, where $c_j(x)$ is the price per unit of good j under the production output $x = (x_1, \dots, x_j, \dots, x_n)^T$.

We assume that the matrix A does not have zero rows or columns, which means that each resource is used for the production of at least one of the goods and each good requires at least one of the resources.

From this point on, we assume strong monotonicity of b and c only at the NE y^*:

$$\langle b(\lambda) - b(\lambda^*), \lambda - \lambda^* \rangle \geq \beta \|\lambda - \lambda^*\|^2,\ \beta > 0,\ \forall \lambda \in \mathbb{R}^m_+ \tag{10.84}$$

$$\langle c(x) - c(x^*), x - x^* \rangle \leq -\alpha \|x - x^*\|^2,\ \alpha > 0 \forall x \in \mathbb{R}^n_+ \tag{10.85}$$

In the first part, we also replace the global Lipschitz continuity of b and c by correspondent local assumptions:

$$\|b(\lambda) - b(\lambda^*)\| \leq L_b \|\lambda - \lambda^*\|,\ \forall \lambda \in \mathbb{R}^m_+ \tag{10.86}$$

and

$$\|c(x) - c(x^*\| \leq L_c \|x - x^*\|,\ \forall x \in \mathbb{R}^n_+, \tag{10.87}$$

where $\|\cdot\|$ is the Euclidean norm.

We will say that the price and resource operators are well defined if (10.84)–(10.87) hold.

The assumption (10.84) implies: an increase of the price λ_i for any resource $1 \leq i \leq m$, when the prices for the rest are fixed at the equilibrium level, leads to an increase of the resource availability $b_i(\lambda)$, and the margin for the resource increase has a positive lower bound.

The assumption (10.85) implies: any increase of production x_j, $1 \leq i \leq n$, when production of the rest is fixed at the equilibrium level, leads to a decrease of the price $c_j(x)$ per unit of good j. Moreover, the margin of the price decrease has a negative upper bound.

In other words, at the equilibrium, the resource availability has to be sensitive to the price variation of the resource, and the price for a product has to be sensitive to the production output variation.

Lipschitz conditions (10.86)–(10.87) assume that deviation from the NE cannot lead to uncontrolled changes of prices for goods and resource availability.

10.5.3 Pseudo-gradient Projection Method

We start by recalling the basic properties of the projection operator.

Let Ω be a closed convex set in $\mathbb{R}^q$; then for each $u \in \mathbb{R}^q$, there is a nearest point in Ω:

$$\mathrm{v} = P_\Omega(u) = \operatorname{argmin}\{\|w-u\| | w \in \Omega\},$$

which is called projection of u on Ω. We recall that the projection operator $P_\Omega : u \in \mathbb{R}^q \to \mathrm{v} \in \Omega$ possesses the following two properties:

(1) It is non-expansive; that is,

$$\|P_\Omega(u_1) - P_\Omega(u_2)\| \leq \|u_1 - u_2\|, \forall u_1, u_2 \in \mathbb{R}^q \tag{10.88}$$

(2) Vector $u^* \in \mathbb{R}^q$ is a solution of the following VI:

$$\langle g(u^*), u - u^* \rangle \leq 0, \forall u \in \Omega$$

if and only if for any $t > 0$ the vector u^* is a fixed point of the map $P_\Omega(I + tg) : \Omega \to \Omega$; that is,

$$u^* = P_\Omega(u^* + tg(u^*)). \tag{10.89}$$

For a vector $u \in \mathbb{R}^q$, the projection v on $\mathbb{R}^q_+$ is given by the formula

$$\mathrm{v} = P_{\mathbb{R}^q_+}(u) = [u]_+ = ([u_1]_+, \dots, [u_q]_+)^T,$$

where for $1 \leq i \leq q$ we have

$$[u_i]_+ = \begin{cases} u_i, & u_i \geq 0 \\ 0, & u_i < 0. \end{cases}$$

Therefore, projection $P_\Omega(y)$ of $y = (x, \lambda) \in \mathbb{R}^n \otimes \mathbb{R}^m$ on $\Omega = \mathbb{R}^n_+ \otimes \mathbb{R}^m_+$ is given by

$$P_\Omega(y) = [y]_+ = ([x]_+; [\lambda]_+).$$

We recall the VI operator $g : \Omega \to \mathbb{R}^{n+m}$ is defined as follows:

$$g(y) = (c(x) - A^T\lambda; Ax - b(\lambda))^T \tag{10.90}$$

We are ready to describe the PGP method for solving VI (10.29).

Let $y_0 = (x_0; \lambda_0) \in \mathbb{R}^n_{++} \otimes \mathbb{R}^m_{++}$ be a starting point, and $(x_s; \lambda_s)$ has been found already. Then

$$y_{s+1} = P_\Omega(y_s + tg(y_s)). \tag{10.91}$$

In other words, each step of the PGP method consists of updating the production vector x_s and the price vector λ_s by the following formulas:

$$x_{j,s+1} = [x_{j,s} + t(c(x_s) - A^T\lambda_s)_j]_+,\ j = 1, \ldots, n \tag{10.92}$$

$$\lambda_{i,s+1} = [\lambda_{i,s} + t(Ax_s - b(\lambda_s))_i]_+,\ i = 1, \ldots, m. \tag{10.93}$$

We will specify later the step length $t > 0$. Methods (10.92)–(10.93) can be viewed as projected explicit Euler methods for solving the following system of differential equations:

$$\frac{dx}{dt} = c(x) - A^T\lambda$$

$$\frac{d\lambda}{dt} = Ax - b(\lambda).$$

On the other hand, PGP methods (10.92)–(10.93) are pricing mechanisms for finding equilibrium.

From (10.92) follows: if the current matrix price $c_j(x_s)$ per unit of good j exceeds the expenses $(A^T\lambda_s)_j$ to produce this unit, then the production of good j has to be increased. On the other hand, if the expenses $(A^T\lambda_s)_j$ exceed the current price $c_j(x_s)$ per unit, then the production of good j has to be reduced.

From (10.93) follows: if the total consumption $(Ax_s)_i$ exceeds the current availability of resource $b_i(\lambda_s)$, then the price per unit has to be increased. If the availability $b_i(\lambda_s)$ of resource i exceeds the current consumption $(Ax_s)_i$, then the price for unit of the resource i has to be reduced.

Lemma 10.3. *If the operators b and c are strongly monotone at λ^* and x^*, i.e., (10.84)–(10.85) hold, then the operator g is strongly monotone at y^*; that is, for $\gamma = \min\{\alpha, \beta\}$, the following inequality holds:*

$$\langle g(y) - g(y^*), y - y^* \rangle \leq -\gamma \|y - y^*\|^2,\ \forall y \in \Omega. \tag{10.94}$$

Proof. We have

$$\begin{aligned}\langle g(y)-g(y^*),y-y^*\rangle &= \langle c(x)-A^T\lambda-c(x^*)+A^T\lambda^*,x-x^*\rangle \\ &\quad+\langle Ax-b(\lambda)-Ax^*+b(\lambda^*),\lambda-\lambda^*\rangle \\ &= \langle c(x)-c(x^*),x-x^*\rangle-\langle A^T(\lambda-\lambda^*),x-x^*\rangle \\ &\quad+\langle A(x-x^*),\lambda-\lambda^*\rangle-\langle b(\lambda)-b(\lambda^*),\lambda-\lambda^*\rangle.\end{aligned}$$

Using (10.84) and (10.85) for $\gamma=\min\{\alpha,\beta\}$, we obtain (10.94). □

Lemma 10.4. *If b and c satisfy local Lipschitz conditions (10.86)–(10.87), then the operator* $g:\Omega\to\mathbb{R}^{n+m}$ *given by (10.90) satisfies local Lipschitz condition at* y^*, *i.e., there is an* $L>0$ *such that*

$$\|g(y)-g(y^*)\|\le L\|y-y^*\|,\ \forall y\in\Omega \tag{10.95}$$

The proof of Lemma 10.4 and the upper bound for L will be discussed later.

Remark 10.1. We assume that for a given $x\in\mathbb{R}^n$ finding $c(x)$ does not exceed $O(n^2)$ operations and for a given $\lambda\in\mathbb{R}^m$ finding $b(\lambda)$ does not exceed $O(m^2)$ operations. We also assume that $n\ge m$. From (10.92)–(10.93) follows that each step of the PGP method (10.91) does not require more than $O(n^2)$ operations.

Example 10.1 *Let* $c(x)=\nabla(\frac{1}{2}x^TCx+c^Tx)$ *and* $b(\lambda)=\nabla(\frac{1}{2}\lambda^TB\lambda+b^T\lambda)$, *where* $C:\mathbb{R}^n\to\mathbb{R}^n$ *is symmetric negative definite and* $B:\mathbb{R}^m\to\mathbb{R}^m$ *is symmetric positive definite; then each step of the PGP method (10.91) requires* $O(n^2)$ *operations.*

Let $\varkappa=\gamma L^{-1}$ be the condition number of the VI operator g. The following theorem establishes the global Q-linear convergence rate and the complexity of the PGP method (10.91).

Theorem 10.9. *If operators b and c are well defined, that is, (10.84)–(10.87) hold, then:*

(1) For any $0<t<2\gamma L^{-2}$, *the PGP method (10.91) globally converges to NE* $y^*=(x^*,\lambda^*)$ *with Q-linear rate and the ratio* $0<q(t)=(1-2t\gamma+t^2L^2)^{1/2}<1$, *i.e.,*

$$\|y_{s+1}-y^*\|\le q(t)\|y_s-y^*\|. \tag{10.96}$$

(2) For $t=\gamma L^{-2}=\min\{q(t)|t>0\}$, *the following bound holds:*

$$\|y_{s+1}-y^*\|\le(1-\varkappa^2)^{1/2}\|y_s-y^*\|. \tag{10.97}$$

(3) For the PGP complexity, we have the following bound:

$$Comp(PGP)=O(n^2\varkappa^{-2}\ln\varepsilon^{-1}), \tag{10.98}$$

where $\varepsilon>0$ *is the required accuracy.*

Proof.

(1) From (10.91), non-expansive property of operator P_Ω (10.88) and optimality criteria (10.89) follow:

$$\begin{aligned}\|y_{s+1}-y^*\|^2 &= \|P_\Omega(y_s+tg(y_s))-P_\Omega(y^*+tg(y^*))\|^2\\ &\le \|y_s+tg(y_s)-y^*-tg(y^*)\|^2\\ &= \langle y_s-y^*+t(g(y_s)-g(y^*)),y_s-y^*+t(g(y_s)-g(y^*))\rangle\\ &= \|y-y^*\|^2+2t\langle y_s-y^*,g(y_s)-g(y^*)\rangle\\ &+t^2\|g(y_s)-g(y^*)\|^2 \end{aligned} \tag{10.99}$$

For well-defined b and c from (10.94), (10.95), and (10.99), we obtain

$$\|y_{s+1}-y^*\|^2 \le \|y_s-y^*\|^2(1-2t\gamma+t^2L^2)$$

Hence, for $0<t<2\gamma L^{-2}$, we have $0<q(t)=(1-2t\gamma+t^2L^2)^{\frac{1}{2}}<1$. In other words, the projection operator (10.91) is contractive, which means that for any given $t\in(0,2\gamma L^{-2})$ the PGP method globally converges with Q-linear rate; that is, (10.96) holds.

(2) For $t=\gamma L^{-2}=\operatorname{argmin}\{q(t)|t>0\}$, we have

$$q=q(\gamma L^{-2})=(1-(\gamma L^{-1})^2)^{\frac{1}{2}}=(1-\varkappa^2)^{\frac{1}{2}}$$

that is, (10.97) holds.

(3) Let $0<\varepsilon\ll 1$ be the required accuracy; then in view of (10.97), it takes $O((\ln q)^{-1}\ln\varepsilon)$ steps to find an ε-approximation for the NE $y^*=(x^*,\lambda^*)$. It follows from Remark 10.1 that each PGP step (10.91) does not require more than $O(n^2)$ operations. Therefore, finding the ε-approximation to NE $y^*=(x^*,\lambda^*)$ requires

$$N=O\left(n^2\frac{\ln\varepsilon}{\ln q}\right)=O\left(n^2\frac{\ln\varepsilon^{-1}}{\ln q^{-1}}\right)$$

operations.

In view of $(\ln q^{-1})^{-1}=(-\frac{1}{2}\ln(1-\varkappa^2))^{-1}$ and keeping in mind $\ln(1+x)\le x$, $\forall x>-1$, we have $\ln(1-\varkappa^2)\le-\varkappa^2$ that is $-\frac{1}{2}\ln(1-\varkappa^2)\ge\frac{1}{2}\varkappa^2$ or $(\ln q^{-1})^{-1}=(-\frac{1}{2}\ln(1-\varkappa^2))^{-1}\le 2\varkappa^{-2}$, so for the overall complexity of the PGP method, we obtain (10.98). □

If $\gamma=\min\{\alpha,\beta\}=0$, then pseudo-gradient $g:\Omega\to\mathbb{R}^{m+n}$ defined by (10.90) is not even locally strongly monotone; therefore, (10.96) cannot guarantee convergence of the PGP method (10.91). In the following section, we consider the EPG method for finding NE (10.25)–(10.26), which converges in the absence of local strong monotonicity of both operators b and c.

First, we show that EPG converges to the NE for any monotone operators b and c, which satisfy Lipschitz condition on $\Omega=\mathbb{R}^n_+\otimes\mathbb{R}^m_+$, i.e.,

$$\|g(y_1) - g(y_2)\| \leq L\|y_1 - y_2\|, \forall y_1, y_2 \in \Omega \tag{10.100}$$

10.5.4 Extra Pseudo-gradient Method for Finding NE

Application of G. Korpelevich extragradient method for solving VI (10.29) leads to the EPG method for finding NE $y^* = (x^*, \lambda^*)$. Each step of EPG method consists of two phases: the predictor phase and the corrector phase. We start with initial approximation $y_0 = (x_0; \lambda_0)) \in \mathbb{R}^n_{++} \otimes \mathbb{R}^m_{++}$. Let us assume that vector $y_s = (x_s, \lambda_s)$ has been found already.

The predictor phase consists of finding

$$\hat{y}_s = P_\Omega(y_s + tg(y_s)) = [y_s + tg(y_s)]_+. \tag{10.101}$$

The corrector phase finds the new approximation:

$$y_{s+1} = P_\Omega(y_s + tg(\hat{y}_s)) = [y_s + tg(\hat{y}_s)]_+. \tag{10.102}$$

The step length $t > 0$ will be specified later.

In other words, the first phase predicts new production vector:

$$\hat{x}_s = [x_s + t(c(x_s) - A^T\lambda_s)]_+ \tag{10.103}$$

and new price vector:

$$\hat{\lambda}_s = [\lambda_s + t(Ax_s - b(\lambda_s))]_+. \tag{10.104}$$

The pair $(\hat{x}_s; \hat{\lambda}_s)$, in turn, predicts the price vector $c(\hat{x}_s) = (c_1(\hat{x}_s), \ldots, c_n(\hat{x}_s))$ and the resource vector $b(\hat{\lambda}_s) = (b_1(\hat{\lambda}_s), \ldots, b_m(\hat{\lambda}_s))$.

Then, the second phase finds new production vector:

$$x_{s+1} = [x_s + t(c(\hat{x}_s) - A^T\hat{\lambda}_s)]_+ \tag{10.105}$$

and new price vector:

$$\lambda_{s+1} = [\lambda_s + t(A\hat{x}_s - b(\hat{\lambda}_s))]_+. \tag{10.106}$$

The meaning of formulas (10.103)–(10.104) and (10.105)–(10.106) is similar to the meaning of formulas (10.92)–(10.93).

Formulas (10.101)–(10.102) can be viewed as pricing mechanisms for finding the NE $y^* = (x^*; \lambda^*)$.

Theorem 10.10. *If c and b are monotone operators and Lipschitz condition (10.100) is satisfied, then for any $t \in (0, (\sqrt{2}L)^{-1})$, EPG methods (10.101)–(10.102) generate a sequence $\{y_s\}_{s\in N}$ convergent to NE:* $\lim_{s\to\infty} y_s = y^*$.

Proof. Let us consider vector $h_s = y_s + tg(y_s) - \hat{y}_s$; then from (10.101), we have

$$\langle h_s, y - \hat{y}_s \rangle \leq 0, \quad \forall y \in \Omega = \mathbb{R}^n_+ \otimes \mathbb{R}^m_+;$$

that is, for a given $t > 0$ and $\forall y \in \Omega$, we have

$$\langle tg(y_s) + (y_s - \hat{y}_s), y - \hat{y}_s \rangle \leq 0. \tag{10.107}$$

For $h_{s+1} = y_s + tg(\hat{y}_s) - y_{s+1}$ from (10.102) follows $\langle h_{s+1}, y - y_{s+1} \rangle \leq 0$, $\forall y \in \Omega$.
Therefore, for a given $t > 0$ and $\forall y \in \Omega$, we have

$$\langle tg(\hat{y}_s) + (y_s - y_{s+1}), y - y_{s+1} \rangle \leq 0. \tag{10.108}$$

From (10.101), (10.102) and non-expansive property of the operator P_Ω, which is defined by (10.88), as well as Lipschitz condition (10.100) follows

$$\begin{aligned} ||y_{s+1} - \hat{y}_s|| &= ||P_\Omega(y_s + tg(\hat{y}_s)) - P_\Omega(y_s + tg(y_s))|| \\ &\leq t\,||g(\hat{y}_s) - g(y_s)|| \\ &\leq tL\,||\hat{y}_s - y_s||\,. \end{aligned} \tag{10.109}$$

From (10.108) for $y = y^*$, we have

$$\langle y_s - y_{s+1} + tg(\hat{y}_s), y^* - y_{s+1} \rangle \leq 0. \tag{10.110}$$

By taking $y = y_{s+1}$ in (10.107), we obtain

$$\langle y_s - \hat{y}_s, y_{s+1} - \hat{y}_s \rangle + t\langle g(y_s), y_{s+1} - \hat{y}_s \rangle \leq 0,$$

or

$$\langle y_s - \hat{y}_s, y_{s+1} - \hat{y}_s \rangle + t\langle g(\hat{y}_s), y_{s+1} - \hat{y}_s \rangle - t\langle g(\hat{y}_s) - g(y_s), y_{s+1} - \hat{y}_s \rangle \leq 0. \tag{10.111}$$

Then, from (10.109) follows

$$\begin{aligned} \langle g(\hat{y}_s) - g(y_s), y_{s+1} - \hat{y}_s \rangle &\leq ||g(\hat{y}_s) - g(y_s)||\,||y_{s+1} - \hat{y}_s|| \\ &\leq tL^2\,||\hat{y}_s - y_s||^2\,. \end{aligned}$$

Therefore, from (10.111), we have

$$\langle y_s - \hat{y}_s, y_{s+1} - \hat{y}_s \rangle + t\langle g(\hat{y}_s), y_{s+1} - \hat{y}_s \rangle - (tL)^2\,||\hat{y}_s - y_s||^2 \leq 0. \tag{10.112}$$

By adding (10.110) and (10.112), we obtain

$$\begin{aligned}&\langle y_s - y_{s+1}, y^* - y_{s+1}\rangle + t\langle g(\hat{y}_s), y^* - y_{s+1}\rangle + \langle y_s - \hat{y}_s, y_{s+1} - \hat{y}_s\rangle \\ &\quad + t\langle g(\hat{y}_s), y_{s+1} - \hat{y}_s\rangle - (tL)^2 ||\hat{y}_s - y_s||^2 \\ &\quad = \langle y_s - y_{s+1}, y^* - y_{s+1}\rangle + t\langle g(\hat{y}_s), y^* - \hat{y}_s\rangle \\ &\quad + \langle y_s - \hat{y}_s, y_{s+1} - \hat{y}_s\rangle - (tL)^2 ||y_s - \hat{y}_s||^2 \le 0.\end{aligned} \tag{10.113}$$

From $\langle g(y^*), y - y^*\rangle \le 0$, $\forall y \in \Omega$, we have $\langle g(y^*), \hat{y}_s - y^*\rangle \le 0$ or $t\langle -g(y^*), y^* - \hat{y}_s\rangle \le 0$.

Adding the last inequality to the left-hand side of (10.113) and using the monotonicity inequality

$$\langle g(\hat{y}_s) - g(y^*), y^* - \hat{y}_s\rangle \ge 0$$

from (10.113) follows

$$2\langle y_s - y_{s+1}, y^* - y_{s+1}\rangle + 2\langle y_s - \hat{y}_s, y_{s+1} - \hat{y}_s\rangle - 2(tL)^2 ||\hat{y}_s - y_s||^2 \le 0. \tag{10.114}$$

Using the three-point identity

$$2\langle u - \mathrm{v}, w - \mathrm{v}\rangle = ||u - \mathrm{v}||^2 + ||\mathrm{v} - w||^2 - ||u - w||^2 \tag{10.115}$$

with $u = y_s$, $\mathrm{v} = y_{s+1}$, and $w = y^*$, we obtain

$$2\langle y_s - y_{s+1}, y^* - y_{s+1}\rangle = ||y_s - y_{s+1}||^2 + ||y_{s+1} - y^*||^2 - ||y_s - y^*||^2.$$

Using the same identity with $u = y_s$, $\mathrm{v} = \hat{y}_s$, and $w = y_{s+1}$, we obtain

$$2\langle y_s - \hat{y}_s, y_{s+1} - \hat{y}_s\rangle = ||y_s - \hat{y}_s||^2 + ||\hat{y}_s - y_{s+1}||^2 - ||y_s - y_{s+1}||^2.$$

Therefore, we can rewrite (10.114) as follows:

$$||y_{s+1} - y^*||^2 + (1 - 2(tL)^2)\,||y_s - \hat{y}_s||^2 + ||\hat{y}_s - y_{s+1}||^2 \le ||y_s - y^*||^2. \tag{10.116}$$

By adding up the last inequality from $s = 0$ to $s = N$, we obtain

$$||y_{N+1} - y^*||^2 + (1 - 2(tL)^2) \sum_{s=0}^{N} ||y_s - \hat{y}_s||^2 + \sum_{s=0}^{N} ||\hat{y}_s - y_{s+1}||^2 \le ||y_0 - y^*||^2,$$

which means that for $0 < t < \frac{1}{\sqrt{2}L}$ we obtain

$$\sum_{s=0}^{N} ||y_s - \hat{y}_s||^2 < \infty \quad , \quad \sum_{s=0}^{N} ||\hat{y}_s - y_{s+1}||^2 < \infty.$$

In other words, we have

(a) $||y_s - \hat{y}_s|| \to 0$ and (b)$||\hat{y}_s - y_{s+1}|| \to 0$.

It follows from (10.116) that $\{||y_s - y^*||\}_{s \in N}$ is a monotone decreasing sequence; hence, the sequence $\{y_s\}_{s \in N}$ is bounded.

Therefore, there exists a convergent subsequence $\{y_{s_i}\}_{s_i\geq 1}^{\infty}$; that is, $\lim_{s_i\to\infty} y_{s_i} = \bar{y}$. Due to (a), we have $\lim_{s_i\to\infty} \hat{y}_{s_i} = \bar{y}$, and due to (b), we have $\lim_{s_i\to\infty} y_{s_i+1} = \bar{y}$. Keeping in mind the continuity of the operator g, we obtain

$$\bar{y} = \lim_{s_i\to\infty} y_{s_i+1} = \lim_{s_i\to\infty} [y_{s_i} + tg(\hat{y}_{s_i})]_+$$
$$= [\bar{y} + tg(\bar{y})]_+;$$

that is, $\bar{y} = P_\Omega(\bar{y} + tg(\bar{y}))$ for $t > 0$. Therefore, from (10.89) follows $\bar{y} = y^*$, which together with $||y_{s+1} - y^*|| < ||y_s - y^*||$ for $s \geq 1$ leads to $\lim_{s\to\infty} y_s = y^*$. □

10.5.5 Convergence Rate

It follows from (10.84), (10.85), and Lemma 10.3 that for $\gamma = \min\{\alpha, \beta\}$, we have

$$\langle g(y) - g(y^*), y - y^* \rangle \leq -\gamma ||y - y^*||^2 \quad , \forall y \in \Omega \tag{10.117}$$

or

$$\langle g(y), y - y^* \rangle - \langle g(y^*), y - y^* \rangle \leq -\gamma ||y - y^*||^2 , \forall y \in \Omega.$$

Keeping in mind that $\langle g(y^*), y - y^* \rangle \leq 0$, $\forall y \in \Omega$ from (10.117), we obtain

$$\langle g(y), y - y^* \rangle \leq -\gamma ||y - y^*||^2 \quad , \forall y \in \Omega. \tag{10.118}$$

Theorem 10.11. *If (10.84) and (10.85) are satisfied and Lipschitz condition (10.100) holds, then for* $\mathrm{v}(t) = 1 + 2\gamma t - 2(tL)^2$ *and the ratio* $q(t) = 1 - 2\gamma t + 4(\gamma t)^2(\mathrm{v}(t))^{-1}$*, the following bounds hold:*

(1) $||y_{s+1} - y^*||^2 \leq q(t)\, ||y_s - y^*||^2$, $0 < q(t) < 1$, $\forall t \in (0, (\sqrt{2}L)^{-1})$,
(2) For $t = \frac{1}{2L}$*, we have*

$$q\left(\frac{1}{2L}\right) = \frac{1+\varkappa}{1+2\varkappa},$$

(3) For any $\varkappa \in (0, 0.5]$*, we have*

$$||y_{s+1} - y^*|| \leq \sqrt{1 - 0.5\varkappa}\, ||y_s - y^*||, \tag{10.119}$$

(4)

$$Comp(EPG) \leq O(n^2 \varkappa^{-1} \ln \varepsilon^{-1}). \tag{10.120}$$

Proof.

(1) It follows from (10.101)–(10.102), the non-expansive property of the projection operator P_Ω, and Lipschitz condition (10.100) that

$$\begin{aligned}||\hat{y}_s - y_{s+1}|| &= ||P_\Omega(y_s + tg(y_s)) - P_\Omega(y_s + tg(\hat{y}_s))|| \\ &\leq t\,||g(y_s) - g(\hat{y}_s)|| \\ &\leq tL\,||y_s - \hat{y}_s||\,.\end{aligned}$$

Using arguments from the proof of Theorem 10.10, we obtain

$$\begin{aligned}&\langle y_s - y_{s+1}, y^* - y_{s+1}\rangle + \langle y_s - \hat{y}_s, y_{s+1} - \hat{y}_s\rangle + \\ &t\langle g(\hat{y}_s), y^* - \hat{y}_s\rangle - (tL)^2\,||\hat{y}_s - y_s||^2 \leq 0.\end{aligned} \tag{10.121}$$

From (10.118) with $y = \hat{y}_s$, we obtain

$$\langle g(\hat{y}_s), y^* - \hat{y}_s\rangle \geq \gamma ||\hat{y}_s - y^*||^2\,.$$

Therefore, we can rewrite (10.121) as follows:

$$\begin{aligned}&2\langle y_s - y_{s+1}, y^* - y_{s+1}\rangle + 2\langle y_s - \hat{y}_s, y_{s+1} - \hat{y}_s\rangle + \\ &2\gamma t\,||\hat{y}_s - y^*||^2 - 2(tL)^2\,||\hat{y}_s - y_s||^2 \leq 0.\end{aligned} \tag{10.122}$$

Applying identity (10.115) to the scalar products in (10.122), we obtain

$$\begin{aligned}&||y_s - y_{s+1}||^2 + ||y_{s+1} - y^*||^2 - ||y_s - y^*||^2 + ||y_s - \hat{y}_s||^2 + \\ &||\hat{y}_s - y_{s+1}||^2 - ||y_s - y_{s+1}||^2 + 2\gamma t\,||\hat{y}_s - y^*|| - 2(tL)^2\,||y_s - \hat{y}s||^2 \leq 0,\end{aligned}$$

or

$$\begin{aligned}&||y_{s+1} - y^*||^2 + ||\hat{y}_s - y_{s+1}||^2 + (1 - 2(tL)^2)\,||y_s - \hat{y}_s||^2 + \\ &2\gamma t\,||\hat{y}_s - y^*||^2 \leq ||y_s - y^*||^2\,.\end{aligned} \tag{10.123}$$

Using

$$\begin{aligned}||\hat{y}_s - y^*||^2 &= \langle \hat{y}_s - y_s + y_s - y^*, \hat{y}_s - y_s + y_s - y^*\rangle \\ &= ||\hat{y}_s - y_s||^2 + 2\langle \hat{y}_s - y_s, y_s - y^*\rangle + ||y_s - y^*||^2\end{aligned}$$

we can rewrite (10.123) as follows:

$$\begin{aligned}&||y_{s+1} - y^*||^2 + ||\hat{y}_s - y_{s+1}||^2 + (1 - 2(tL)^2)\,||\hat{y}_s - y_s||^2 + \\ &\quad 2\gamma t\,||\hat{y}_s - y_s||^2 + 4\gamma t\langle \hat{y}_s - y_s, y_s - y^*\rangle + 2\gamma t\,||y_s - y^*||^2 \leq ||y_s - y^*||^2,\end{aligned}$$

or

$$\begin{aligned}&||y_{s+1} - y^*||^2 + ||\hat{y}_s - y_{s+1}||^2 + (1 + 2\gamma t - 2(tL)^2)\,||\hat{y}_s - y_s||^2 + \\ &\quad 4\gamma t\langle \hat{y}_s - y_s, y_s - y^*\rangle \leq (1 - 2\gamma t)\,||y_s - y^*||^2\,.\end{aligned} \tag{10.124}$$

By introducing $\mathrm{v}(t) = 1 + 2\gamma t - 2(tL)^2$, we can rewrite the third and the fourth term of the left-hand side as follows:

$$\left\| \sqrt{\mathrm{v}(t)}(\hat{y}_s - y_s) + 2(y_s - y^*)\frac{\gamma t}{\sqrt{\mathrm{v}(t)}} \right\|^2 - \frac{4(\gamma t)^2 \, ||y_s - y^*||^2}{\mathrm{v}(t)}.$$

Therefore, from (10.124), we have

$$||y_{s+1} - y^*||^2 + ||\hat{y}_s - y_{s+1}||^2 +$$
$$\left\| \sqrt{\mathrm{v}(t)}(\hat{y}_s - y_s) + 2(y_s - y^*)\frac{\gamma t}{\sqrt{\mathrm{v}(t)}} \right\|^2 \le \left(1 - 2\gamma t + \frac{4(\gamma t)^2}{\mathrm{v}(t)}\right) ||y_s - y^*||^2$$

Hence, for $q(t) = 1 - 2\gamma t + 4(\gamma t)^2(\mathrm{v}(t))^{-1}$, we obtain

$$||y_{s+1} - y^*||^2 \le q(t) \, ||y_s - y^*||^2.$$

(2) For $t = \frac{1}{2L}$ and $\varkappa = \gamma L^{-1}$, we have

$$q\left(\frac{1}{2L}\right) = 1 - \varkappa + \frac{\varkappa^2}{0.5 + \varkappa} = \frac{1 + \varkappa}{1 + 2\varkappa} \tag{10.125}$$

It is easy to see that for every $t \in (0, (\sqrt{2}L)^{-1})$ we have $0 < q(t) < 1$.

(3) From (10.125) for any $0 \le \varkappa \le 0.5$ follows

$$q\left(\frac{1}{2L}\right) \le 1 - 0.5\varkappa$$

Therefore, the bound (10.119) holds.

(4) It follows from (10.119) that for a given accuracy $0 < \varepsilon \ll 1$ and $q = \sqrt{1 - 0.5\varkappa}$, the EPG method requires

$$s = O\left(\frac{\ln \varepsilon^{-1}}{\ln q^{-1}}\right)$$

step to get $y_s : \|y_s - y*\| \le \varepsilon$.

It follows from (10.101)–(10.102) and Remark 10.1 that each step of EPG requires $O(n^2)$ operations; therefore, the overall complexity of the EPG method is bounded by $O(n^2(\ln \varepsilon^{-1})(\ln q^{-1})^{-1})$.

Then, $(\ln q^{-1})^{-1} = (-\frac{1}{2}\ln(1 - 0.5\varkappa))^{-1}$. Due to $\ln(1 + x) \le x$, $\forall x > -1$, we obtain $\ln(1 - 0.5\varkappa) \le -0.5\varkappa$; hence, $-\frac{1}{2}\ln(1 - 0.5\varkappa) \ge 0.25\varkappa$ and $(\ln q^{-1})^{-1} \le 4\varkappa^{-1}$.

Therefore, the overall EPG complexity is

$$Comp(EPG) \le O(n^2 \varkappa^{-1} \ln \varepsilon^{-1}),$$

i.e., the bound (10.120) holds true. □

Remark 10.2. For small $\varkappa > 0$, the complexity bound (10.120) is much better than the PGP bound (10.98). On the other hand, the EPG requires two projections at each step instead of one, as in the case of PGP. Keeping in mind the relatively low cost to project on Ω, one can still expect the EPG to be more efficient. However, in the case when $1 > \varkappa > 0.5$ and n is large enough, the PGP could be more efficient.

10.5.6 Bound for the Lipschitz Constant

The important part of both PGP and EPG methods is the Lipschitz constant $L > 0$ in (10.100).

Let us find an upper bound for $L > 0$.

To simplify our considerations, we assume that the matrix A is rescaled, so

$$||A||_I = \max_{1\leq j\leq n} \sum_{i=1}^{m} |a_{ij}| \leq 1 \text{ and } ||A||_{II} = \max_{1\leq i\leq m} \sum_{j=1}^{n} |a_{ij}| \leq 1. \tag{10.126}$$

We assume as always that the components of vector functions $c(x)$ and $b(\lambda)$ satisfy Lipschitz condition; that is, for any $1 \leq j \leq n$, there is $L_{c,j}$ such that

$$|c_j(x_1) - c_j(x_2)| \leq L_{c,j}\,||x_1 - x_2||, \quad \forall (x_1, x_2) \in \mathbb{R}^n_+ \otimes \mathbb{R}^n_+ \tag{10.127}$$

and for any $1 \leq i \leq m$, there is $L_{b,i}$ such that

$$|b_i(\lambda_1) - b_i(\lambda_2)| \leq L_{b,i}\,||\lambda_1 - \lambda_2||, \quad \forall (\lambda_1, \lambda_2) \in \mathbb{R}^m_+ \otimes \mathbb{R}^m_+ \tag{10.128}$$

Using (10.127), we obtain

$$\begin{aligned} ||c(x_1) - c(x_2)|| &= \sqrt{\sum_{j=1}^{n} (c_j(x_1) - c_j(x_2))^2} \leq \sqrt{\sum_{j=1}^{n} L^2_{c,j}\,||x_1 - x_2||^2} \\ &\leq L_c \sqrt{n||x_1 - x_2||^2} = L_c\sqrt{n}\,||x_1 - x_2||, \end{aligned}$$

where $L_c = \max_{1\leq j\leq n} L_{c,j}$.

Using (10.128), we obtain

$$\begin{aligned} ||b(\lambda_1) - b(\lambda_2)|| &= \sqrt{\sum_{i=1}^{m} (b_i(\lambda_1) - b_i(\lambda_2))^2} \leq \sqrt{\sum_{i=1}^{m} L^2_{b,i}||\lambda_1 - \lambda_2||^2} \\ &\leq L_b \sqrt{m||\lambda_1 - \lambda_2||^2} = L_b\sqrt{m}\,||\lambda_1 - \lambda_2||, \end{aligned}$$

where $L_b = \max_{1\leq i\leq m} L_{b,i}$. Therefore,

$$\begin{aligned}
||g(y_1) - g(y_2)|| &\le \left|\left|c(x_1) - A^T\lambda_1 - c(x_2) + A^T\lambda_2\right|\right| + ||Ax_1 - b(\lambda_1) - Ax_2 + b(\lambda_2)|| \\
&\le ||c(x_1) - c(x_2)|| + \left|\left|A^T\right|\right| ||\lambda_1 - \lambda_2|| + ||A|| \, ||x_1 - x_2|| + ||b(\lambda_1) - b(\lambda_2)|| \\
&\le L_c\sqrt{n}\,||x_1 - x_2|| + \left|\left|A^T\right|\right| ||\lambda_1 - \lambda_2|| + ||A|| \, ||x_1 - x_2|| + L_b\sqrt{m}\,||\lambda_1 - \lambda_2|| \\
&= (L_c\sqrt{n} + ||A||)\,||x_1 - x_2|| + (L_b\sqrt{m} + \left|\left|A^T\right|\right|)\,||\lambda_1 - \lambda_2|| \qquad (10.129)
\end{aligned}$$

For $||A|| = \sqrt{\lambda_{\max}(A^TA)}$ and $\left|\left|A^T\right|\right| = \sqrt{\lambda_{\max}(AA^T)}$, in view of (10.126), we have (see Gantmacher 1959)

$$||A|| \le \sqrt{n}\,||A||_I \le \sqrt{n}$$

and

$$\left|\left|A^T\right|\right| \le \sqrt{m}\,\left|\left|A^T\right|\right|_I \le \sqrt{m}$$

Hence, from (10.129) follows

$$||g(y_1) - g(y_2)|| \le \sqrt{n}(L_c + 1)\,||x_1 - x_2|| + \sqrt{m}(L_b + 1)\,||\lambda_1 - \lambda_2||$$

Assuming $n > m$ and taking $\hat{L} = \max\{L_c, L_b\}$, we obtain

$$\begin{aligned}
||g(y_1) - g(y_2)|| &\le \hat{L}(\sqrt{n} + 1)\,[||x_1 - x_2|| + ||\lambda_1 - \lambda_2||] \\
&\le \sqrt{2}\hat{L}(\sqrt{n} + 1)\,||y_1 - y_2||
\end{aligned}$$

In other words, $L \le \sqrt{2}\hat{L}(\sqrt{n} + 1) = O(\sqrt{n})$.

10.5.7 Finding NE as a Pricing Mechanizm

The "symmetrization" of the classical WW equilibrium was achieved by replacing the fixed resource vector b by the resource operator $b : \mathbb{R}^m_+ \to \mathbb{R}^m_+$. This is not only justifiable from the market standpoint, but it also leads to new methods for solving correspondent VI, which are based on projection-type techniques. At each step, the production vector x_s and the price vector λ_s are updated by simple formulas, and it can be done in parallel. In other words, one can view both PGP and EPG as primal–dual decomposition methods.

The complexity bounds (10.98) and (10.120) show that in a number of instances finding NE by PGP or EPG can be cheaper than solving a correspondent LP by interior point methods.

Both PGP and EPG can be used for large-scale resource allocation problems when simplex or interior point methods for solving LP are difficult to use due to the necessity of solving large linear systems of equations at each step.

The "symmetrization" also helps to avoid the combinatorial nature of LP. On the other hand, finding NE drastically reduces the complexity as compared with using

PGP or EPG for finding WW equilibrium, which requires at each step solving one or two quadratic programming problems:

$$P_\Omega(x+tg(x)) = \operatorname{argmin}\{||y-(x+tg(x))||^2 \,|y \in \Omega\},$$

where $\Omega = \{x : Ax \leq b, x \geq 0\}$.

Both the PGP and the EPG can be viewed as pricing mechanisms for finding NE, which make the prices $c(x^*)$ consistent with the output x^* and the resource availability $b(\lambda^*)$ consistent with prices λ^* for the resource.

Moreover, we have

$$(c(x^*) - A^T\lambda^*)_j < 0 \Rightarrow x_j^* = 0 \tag{10.130}$$

$$x_j^* > 0 \Rightarrow (c(x^*) - A^T\lambda^*)_j = 0 \tag{10.131}$$

$$(Ax^* - b(\lambda^*))_i < 0 \Rightarrow \lambda_i^* = 0 \tag{10.132}$$

$$\lambda_i^* > 0 \Rightarrow (Ax^* - b(\lambda^*))_i = 0 \tag{10.133}$$

It follows from (10.130) that at the equilibrium the market is cleared from goods, the prices for which cannot cover their production expenses. It follows from (10.132) that a resource has no value if its supply at the equilibrium is greater than its demand. It follows from (10.131) that at the equilibrium for each product on the market, the price is equal to its production expenses. It follows from (10.133) that for every resource, which has at the equilibrium a positive price, the supply is equal to the demand.

Finally, at the equilibrium, the total cost of the goods on the market is equal to the total production cost, i.e.,

$$\langle c(x^*), x^* \rangle = \langle b(\lambda^*), \lambda^* \rangle.$$

10.6 Finding Nonlinear Input–Output Equilibrium

10.6.1 Introduction

Finding NIOE $y^* = (x^*, \lambda^*)$ is equivalent to solving a particular VI on $\Omega = \mathbb{R}^n_+ \times \mathbb{R}^n_+$.

For solving the correspondent VI, we use PGP and EPG, for which projection on Ω is the main operation per step. Numerically, it leads to two matrix-by-vector multiplications, which require $O(n^2)$ arithmetic operations per step.

Both methods decompose the NIOE problem in the primal and in the dual spaces allowing the computation of the production and the price vectors simultaneously. Both PGP and EPG can be viewed as pricing mechanisms for establishing NIOE.

The main distinction of the EPG is its ability to handle NIOE problems when both the production and the consumption operators are not strongly monotone but just monotone.

This is one of the finest properties of the EPG method introduced by G. Korpelevich in the 1970s.

In the case of NIOE, the application of EPG leads to two-stage algorithm. At the first stage, EPG predicts the production vector $\hat{x} \in \mathbb{R}^n_+$ and the price vector $\hat{\lambda} \in \mathbb{R}^n_+$. At the second stage, EPG corrects them dependent on the production cost per unit vector $p(\hat{x})$ and consumption vector $c(\hat{\lambda})$.

10.6.2 Basic Assumptions

In what is following, we replace global strong monotonicity assumptions (10.64)–(10.65) for both operators p and c by less restrictive assumptions of local strong monotonicity, only at the NIOE $y^* = (x^*, \lambda^*)$.

We assume the existence of such $\alpha > 0$ and $\beta > 0$ that

$$\langle p(x) - p(x^*), x - x^* \rangle \geq \alpha \|x - x^*\|^2, \ \forall x \in \mathbb{R}^n_+ \tag{10.134}$$

$$\langle c(\lambda) - c(x^*), \lambda - \lambda^* \rangle \leq -\beta \|\lambda - \lambda^*\|^2, \forall \lambda \in \mathbb{R}^n_+ \tag{10.135}$$

holds.

In the next section, we also replace the global Lipschitz continuity of p and c by corresponding assumptions at the NIOE $y^* = (x^*, \lambda^*)$.

$$\|p(x) - p(x^*)\| \leq L_p \|x - x^*\|, \ \forall x \in \mathbb{R}^n_+ \tag{10.136}$$

and

$$\|c(\lambda) - c(\lambda^*)\| \leq L_c \|\lambda - \lambda^*\|, \forall \lambda \in \mathbb{R}^n_+. \tag{10.137}$$

We will say that both production p and consumption c operators are well defined if (10.134)–(10.137) hold.

Assumption (10.134) means that the production cost operator p is sensitive to the production change only at the equilibrium. Assumption (10.135) means that the consumption operator c is sensitive to the price change only at the equilibrium. Lipschitz conditions (10.136)–(10.137) mean that the production and consumption are under control in the neighborhood of the NIOE $y^* = (x^*, \lambda^*)$.

10.6.3 PGP Method for Finding NIOE

Let $y_0 = (x_0; \lambda_0) \in \mathbb{R}^n_{++} \times \mathbb{R}^n_{++}$ be the starting point, and $y_s = (x_s, \lambda_s)$ has been found already.

The PGP method finds the next approximation by the following formula:

$$y_{s+1} = P_\Omega(y_s + tg(y_s)). \tag{10.138}$$

The step length $t > 0$ will be specified later.

In other words, the PGP method simultaneously updates production vector x_s and the price vector λ_s by the following formulas:

$$x_{j,s+1} = [x_{j,s} + t((I-A)^T\lambda_s - p(x_s))_j]_+, \quad 1 \le j \le n \tag{10.139}$$

$$\lambda_{i,s+1} = [\lambda_{i,s} + t(c(\lambda_s) - (I-A)x_s)_i]_+, \quad 1 \le i \le n. \tag{10.140}$$

The PGP can be viewed as the primal–dual decomposition method, which computes the primal and dual vectors independently and simultaneously.

Formulas (10.139)–(10.140) can be viewed as pricing mechanisms for establishing NIOE.

From (10.139) follows: if the profit $q_{j,s} = ((I-A)^T\lambda_s)_j$ per unit of product $1 \le j \le n$ is greater than the production cost $p_j(x_s)$, then the production of $x_{j,s}$ has to be increased.

On the other hand, if production cost $p_j(x_s)$ is greater than the profit $q_{j,s} = ((I-A)^T\lambda_s)_j$ per unit of product $1 \le j \le n$, then the production $x_{j,s}$ has to be reduced.

From (10.140) follows: if consumption $c_i(\lambda_s)$ of product $1 \le i \le n$ is greater than production $((I-A)x_s)_i$, then the price per unit $\lambda_{i,s}$ has to be increased. If the consumption $c_i(\lambda_s)$ is less than the production $((I-A)x_s)_i$, then the price $\lambda_{i,s}$ has to be reduced.

PGP methods (10.140)–(10.141) can be viewed as projected explicit Euler methods for solving the following system of ordinary differential equations:

$$\frac{dx}{dt} = (I-A)^T\lambda - p(x)$$

$$\frac{d\lambda}{dt} = c(\lambda) - (I-A)x$$

with $x(0) = x_0$ and $\lambda(0) = \lambda_0$.

To prove convergence of the PGP method (10.138), we need the following lemma similar to Lemma 10.3.

Lemma 10.5. *If operators p and c are strongly monotone at x^* and λ^*, i.e., (10.134)–(10.135) hold, then the operator $g : \Omega \to \mathbb{R}^{2n}$ given by (10.68) is strongly monotone at y^*; that is, for $\gamma = \min\{\alpha, \beta\} > 0$, the following bound holds:*

$$\langle g(y) - g(y^*), y - y^* \rangle \le -\gamma\|y - y^*\|^2, \forall y \in \Omega. \tag{10.141}$$

The proof is similar to the proof of Lemma 10.3.

Lemma 10.6. *If operators p and c satisfy local Lipschitz conditions (10.136)–(10.137), then pseudo-gradient operator $g : \Omega \to \mathbb{R}^{2n}$ given by (10.68) satisfies local*

Lipschitz condition at y^*; that is, there is $L > 0$ such that

$$\|g(y) - g(y^*)\| \le L\|y - y^*\|, \ \forall y \in \Omega. \tag{10.142}$$

The upper bound for L will be established later.

Remark 10.3. Let us assume that for a given $x \in \mathbb{R}^n_+$ and for a given $\lambda \in \mathbb{R}^n_+$, computing $p(x)$ and computing $c(\lambda)$ do not require more than $O(n^2)$ operations. It is true, for example, if $c(\lambda) = \nabla(\frac{1}{2}\lambda^T C\lambda + d^T\lambda)$ and $p(x) = \nabla(\frac{1}{2}x^T Px + q^T x)$, where $C : \mathbb{R}^n \to \mathbb{R}^n$ is a symmetric negative semidefinite matrix and $P : \mathbb{R}^n \to \mathbb{R}^n$ is a symmetric positive semidefinite matrix. Then, each PGP step does not require more than $O(n^2)$ operations.

Let $\varkappa = \gamma L^{-1}$ be the condition number of the VI operator $g : \Omega \to \mathbb{R}^{2n}$. The following theorem establishes global Q-linear convergence rate and complexity of the PGP method (10.138).

Theorem 10.12. *If operators p and c are well defined, then:*

1. *For any $0 < t < 2\gamma L^{-2}$, the PGP method (10.138) globally converges to NIOE $y^* = (x^*, \lambda^*)$ with Q-linear rate and the ratio $0 < q(t) = (1 - 2t\gamma + t^2L^2)^{\frac{1}{2}} < 1$, i.e.,*

$$\|y_{s+1} - y^*\| \le q(t)\|y_s - y^*\| \tag{10.143}$$

2. *For $t = \gamma L^{-2} = \min\{q(t)|t > 0\}$, the following bound holds:*

$$\|y_{s+1} - y^*\| \le (1 - \varkappa^2)^{\frac{1}{2}}\|y_s - y^*\| \tag{10.144}$$

3. *The PGP complexity is given by the following bound:*

$$comp(PGP) = O(n^2\varkappa^{-2}\ln\varepsilon^{-1}), \tag{10.145}$$

where $\varepsilon > 0$ is the given accuracy.

The proof is similar to the proof of Theorem 10.9.

Exercise 10.3. If operators p and c are monotone and satisfy Lipschitz condition, then the operator g is monotone; that is,

$$\langle g(y_1) - g(y_2), y_1 - y_2\rangle \le 0 \tag{10.146}$$

holds and Lipschitz condition

$$\|g(y_1) - g(y_2)\| \le L\|y_1 - y_2\| \tag{10.147}$$

holds for any $y_1, y_2 \in \Omega$.

If $\gamma = 0$, then PGP method (10.139)–(10.140) might not converge to the equilibrium (x^*, λ^*). Therefore, in the following section, we apply extra pseudo-gradient (EPG) method for solving VI (10.66).

10.6.4 EPG Method for Finding NIOE

Application of the EPG method for solving VI (10.66) leads to the following method for finding NIOE.

Each step of the EPG method consists of two phases: in the prediction phase, we predict the production vector $\hat{x} \in \mathbb{R}^n_+$ and the price vector $\hat{\lambda} \in \mathbb{R}^n_+$; in the corrector phase, we correct the production and the price vectors by using the predicted cost per unit vector $p(\hat{x})$ and the predicted consumption vectors $c(\hat{\lambda})$.

We start with $y_0 = (x_0, \lambda_0) \in \mathbb{R}^n_{++} \times \mathbb{R}^n_{++}$ and assume that the approximation $y_s = (x_s, \lambda_s)$ has been found already. We find the next approximation y_{s+1} using the two-phase EPS method.

In the prediction phase, we find

$$\hat{y}_s = P_\Omega(y_s + tg(y_s)) = [y_s + tg(y_s)]_+. \tag{10.148}$$

In the correction phase, we find the new approximation:

$$y_{s+1} = P_\Omega(y_s + tg(\hat{y}_s)) = [y_s + tg(\hat{y}_s)]_+. \tag{10.149}$$

The step length $t > 0$ will be specified later.

In other words, the EPG method first predicts the production vector:

$$\hat{x}_s = [x_s + t((I-A)^T \lambda_s - p(x_s))]_+ \tag{10.150}$$

and the price vector:

$$\hat{\lambda}_s = [\lambda_s + t(c(\lambda_s) - (I-A)x_s)]_+. \tag{10.151}$$

Then, EPG finds the new production vector:

$$x_{s+1} = [x_s + t((I-A)^T \hat{\lambda}_s - p(\hat{x}_s))]_+ \tag{10.152}$$

and the new price vector:

$$\lambda_{s+1} = [\lambda_s + t(c(\hat{\lambda}_s) - (I-A)\hat{x}_s)]_+. \tag{10.153}$$

The meaning of formulas (10.150)–(10.153) is similar to the meaning of formulas (10.139)–(10.140).

EPG methods (10.148)–(10.149), in fact, are pricing mechanisms for establishing NIOE. The following theorem establishes the convergence of EPG.

The proof is similar to the proof of Theorem 10.10; we provide it for completeness.

Theorem 10.13. *If p and c are monotone operators and Lipschitz condition (10.147) is satisfied, then for any $t \in (0, (\sqrt{2}L)^{-1})$, EPG methods (10.148)–(10.149) generate a sequence $\{y_s\}_{s\in N}$ converging to NIOE:* $\lim_{s\to\infty} y_s = y^*$.

Proof. It follows from (10.148)–(10.149), non-expansiveness of P_Q, and Lipschitz condition (10.147) that

$$\|y_{s+1} - \hat{y}_s\| = \|P_\Omega(y_s + tg(\hat{y}_s)) - P_\Omega(y_s + tg(y_s))\|$$

$$\leq t\|g(\hat{y}_s) - g(y_s)\| \leq tL\|\hat{y}_s - y_s\|. \tag{10.154}$$

From (10.148) follows

$$\langle tg(y_s) + y_s - \hat{y}_s, y - \hat{y}_s \rangle \leq 0, \ \forall y \in \Omega;$$

therefore, by taking $y = y_{s+1}$, we obtain

$$\langle tg(y_s) + y_s - \hat{y}_s, y_{s+1} - \hat{y}_s \rangle \leq 0$$

or

$$\langle y_s - \hat{y}_s, y_{s+1} - \hat{y}_s \rangle + t\langle g(\hat{y}_s), y_{s+1} - \hat{y}_s \rangle -$$

$$- t\langle g(\hat{y}_s) - g(y_s), y_{s+1} - \hat{y}_s \rangle \leq 0. \tag{10.155}$$

From (10.154) and Lipschitz condition (10.147), we have

$$\langle g(\hat{y}_s) - g(y_s), y_{s+1} - \hat{y}_s \rangle \leq \|g(\hat{y}_s) - g(y_s)\|\|y_{s+1} - \hat{y}_s\| \leq tL^2\|\hat{y}_s - y_s\|^2. \tag{10.156}$$

From (10.155) and (10.156), we obtain

$$\langle y_s - \hat{y}_s, y_{s+1} - \hat{y}_s \rangle + t\langle g(\hat{y}_s), y_{s+1} - \hat{y}_s \rangle - (tL)^2\|\hat{y}_s - y_s\|^2 \leq 0. \tag{10.157}$$

From (10.149) follows

$$\langle tg(\hat{y}_s) + y_s - y_{s+1}, y - y_{s+1} \rangle \leq 0, \ \forall y \in \Omega. \tag{10.158}$$

Therefore, for $y = y^*$, we have

$$\langle y_s - y_{s+1} + tg(\hat{y}_s), y^* - y_{s+1} \rangle \leq 0. \tag{10.159}$$

Also from $\langle g(y^*), y - y^* \rangle \leq 0$, $\forall y \in \Omega$, we obtain $\langle g(y^*), \hat{y}_s - y^* \rangle \leq 0$, so for $\forall t > 0$, we have

$$t\langle -g(y^*), y^* - \hat{y}_s \rangle \leq 0. \tag{10.160}$$

By adding (10.157), (10.159), and (10.160) and using the monotonicity of g, i.e.,

$$\langle g(\hat{y}_s) - g(y^*), y^* - \hat{y}_s \rangle \geq 0,$$

we obtain

$$2\langle y_s - y_{s+1}, y^* - y_{s+1} \rangle + 2\langle y_s - \hat{y}_s, y_{s+1} - \hat{y}_s \rangle - 2(tL)^2\|\hat{y}_s - y_s\|^2 \leq 0. \tag{10.161}$$

Using the three-point identity

$$2\langle u - \text{v}, w - \text{v}\rangle = \|u - \text{v}\|^2 + \|\text{v} - w\|^2 - \|u - w\|^2 \tag{10.162}$$

twice first with $u = y_s, \text{v} = y_{s+1}$, and $w = y^*$ and second with $u = y_s, \text{v} = \hat{y}_s$, and $w = y_{s+1}$, we obtain

$$2\langle y_s - y_{s+1}, y^* - y_{s+1}\rangle = \|y_s - y_{s+1}\|^2 + \|y_{s+1} - y^*\|^2 - \|y_s - y^*\|^2 \tag{10.163}$$

and

$$2\langle y_s - \hat{y}_s, y_{s+1} - \hat{y}_s\rangle = \|y_s - \hat{y}_s\|^2 + \|\hat{y}_s - y_{s+1}\|^2 - \|y_s - y_{s+1}\|^2. \tag{10.164}$$

From (10.161), (10.163), and (10.164) follows

$$\|y_{s+1} - y^*\|^2 + (1 - 2(tL)^2)\|y_s - \hat{y}_s\|^2 + \|\hat{y}_s - y_{s+1}\|^2 \leq \|y_s - y^*\|^2. \tag{10.165}$$

Summing up (10.165) from $s = 0$ to $s = N$, we obtain

$$\|y_{N+1} - y^*\|^2 + (1 - 2(tL)^2)\sum_{s=0}^{N} \|y_s - \hat{y}_s\|^2 + \tag{10.166}$$

$$+\sum_{s=0}^{N} \|\hat{y}_s - y_{s+1}\|^2 \leq \|y_0 - y^*\|^2.$$

It follows from (10.166) that for $0 < t \leq \frac{1}{\sqrt{2}L}$, we have $\sum_{s=0}^{\infty} \|y_s - \hat{y}_s\|^2 < \infty$ and $\sum_{s=0}^{\infty} \|\hat{y}_s - y_{s+1}\|^2 < \infty$, which means that

$$(a) \lim_{s\to\infty} \|y_s - \hat{y}_s\| \to 0 \text{ and } (b) \lim_{s\to\infty} \|\hat{y}_s - y_{s+1}\| \to 0. \tag{10.167}$$

Also from (10.165) follows

$$\|y_{s+1} - y^*\| \leq \|y_s - y^*\| \ \forall s \geq 0.$$

Thus, $\{y_s\}_{s\in N}$ is a bounded sequence; therefore, there is a converging sub-sequence $\{y_{s_k}\}_{s_k>1}^{\infty} : \lim_{k\to\infty} y_{s_k} = \bar{y}$. It follows from (10.167a) that $\lim_{k\to\infty} \hat{y}_{s_k} = \bar{y}$, and from (10.167b) follows $\lim_{k\to\infty} y_{s_k+1} = \bar{y}$. □

From the continuity of the operator g, we have

$$\bar{y} = \lim_{k\to\infty} y_{s_k+1} = \lim_{k\to\infty} [y_{s_k} + tg(\hat{y}_{s_k})]_+ = [\bar{y} + tg(\bar{y})]_+. \tag{10.168}$$

From (10.168) follows $\bar{y} = y^*$, which together with (10.168) leads to $\lim_{s\to\infty} y_s = y^*$.
□

10.6.5 Convergence Rate and Complexity of the EPG Method

Under Lipschitz continuity and local strong monotonicity of both operators $p : \mathbb{R}^n_+ \to \mathbb{R}^n_+$ and $c : \mathbb{R}^n_+ \to \mathbb{R}^n_+$, the EPG method converges with linear rate.

The EPG requires two projections on Ω, while PGP requires only one such projection, but the main operation per step is four matrix-by-vector multiplications, which require $O(n^2)$ operations.

To establish the convergence rate and complexity of EPG, we will need two inequalities, which follow from the local strong monotonicity of g.

First, by adding (10.157) and (10.159), we obtain

$$\langle y_s - y_{s+1}, y^* - y_{s+1}\rangle + t\langle g(\hat{y}_s), y^* - \hat{y}_s\rangle + \langle y_s - \hat{y}_s, y_{s+1} - \hat{y}_s\rangle - \tag{10.169}$$
$$-(tL)^2\|y_s - \hat{y}_s\|^2 \leq 0.$$

Second, it follows from (10.141) that

$$\langle g(y), y - y^*\rangle - \langle g(y^*), y - y^*\rangle \leq -\gamma\|y - y^*\|^2. \tag{10.170}$$

Keeping in mind $\langle g(y^*), y - y^*\rangle \leq 0, \ \forall y \in \Omega$ from (10.170), we obtain

$$\langle g(y), y^* - y\rangle \geq \gamma\|y - y^*\|^2, \ \forall y \in \Omega. \tag{10.171}$$

The following theorem is similar to Theorem 10.11; we provide it for completeness.

Theorem 10.14. *If for both p and c operators the local strong monotonicity (10.134)–(10.135) hold and Lipschitz condition (10.147) is satisfied, then*

1. *There exists $0 < q(t) < 1, \ \forall t \in (0, (\sqrt{2}L)^{-1})$ such that*
$$\|y_{s+1} - y^*\| \leq q(t)\|y_s - y^*\| \tag{10.172}$$
2. $$\|y_{s+1} - y^*\| \leq \sqrt{1 - 0.5\varkappa}\|y_s - y^*\|, \ \forall \varkappa \in [0, 0.5] \tag{10.173}$$
3. $$Comp(EPG) \leq O(n^2\varkappa^{-1}\ln\varepsilon^{-1}) \tag{10.174}$$

where $\varepsilon > 0$ is the required accuracy.

Proof.

1. From (10.171) with $y = \hat{y}_s$, we have
$$\langle g(\hat{y}_s), y^* - \hat{y}_s\rangle \geq \gamma\|\hat{y}_s - y^*\|^2.$$
Therefore, we can rewrite (10.169) as follows:
$$2\langle y_s - y_{s+1}, y^* - y_{s+1}\rangle + 2\langle y_s - \hat{y}_s, y_{s+1} - \hat{y}_s\rangle \tag{10.175}$$

$$+2\gamma t\|\hat{y}_s - y^*\|^2 - 2(tL)^2\|\hat{y}_s - y_s\|^2 \leq 0.$$

Using identity (10.162), first with $u = y_s, v = y_{s+1}, w = y^*$ and second with $u = y_s, v = y_{s+1}, w = \hat{y}_s$ from (10.175), we obtain

$$\|y_{s+1} - y^*\|^2 + \|\hat{y}_s - y_{s+1}\|^2 + (1 - 2(tL)^2)\|y_s - \hat{y}_s\|^2 \tag{10.176}$$
$$+2\gamma t\|\hat{y}_s - y^*\|^2 \leq \|y_s - y^*\|^2.$$

Using

$$\|\hat{y}_s - y^*\|^2 = \|\hat{y}_s - y_s\|^2 + 2(\hat{y} - y_s, y_s - y^*) + \|y_s - y^*\|^2$$

from (10.176), we obtain

$$\|y_{s+1} - y^*\|^2 + \|\hat{y}_s - y_{s+1}\|^2 + (1 - 2(tL)^2)\|\hat{y}_s - y_s\|^2 +$$
$$+2\gamma t\|\hat{y}_s - y_s\|^2 + 4\gamma t\langle\hat{y}_s - y_s, y_s - y^*\rangle + 2\gamma t\|y_s - y^*\|^2 \leq \|y_s - y^*\|^2$$

or

$$\|y_{s+1} - y^*\|^2 + \|\hat{y}_s - y_{s+1}\|^2 + (1 + 2\gamma t - 2(tL)^2)\|\hat{y}_s - y_s\|^2 +$$
$$+4\gamma t\langle\hat{y}_s - y_s, y_s - y^*\rangle \leq (1 - 2\gamma t)\|y_s - y^*\|^2. \tag{10.177}$$

Let $\mu(t) = 1 + 2\gamma t - 2(tL)^2$; then the third and the fourth term of the left-hand side can be rewritten as follows:

$$\|\sqrt{\mu(t)}(\hat{y} - y_s) + 2(y_s - y^*)\frac{\gamma t}{\sqrt{\mu(t)}}\|^2 - \frac{4(\gamma t)^2}{\mu(t)}\|y_s - y^*\|^2.$$

Therefore, from (10.177), we obtain bound (10.172):

$$\|y_{s+1} - y^*\| \leq q(t)\|y_s - y^*\|$$

with $q(t) = (1 - 2\gamma t + 4(\gamma t)^2(\mu(t))^{-1})^{\frac{1}{2}}$. It is easy to see that for $t \in (0, (\sqrt{2}L)^{-1})$, we have $0 < q(t) < 1$.

2. For $\varkappa = \gamma L^{-1}$ and $t = (2L)^{-1}$, we obtain

$$q((2L)^{-1}) = [(1+\varkappa)(1+2\varkappa)^{-1}]^{\frac{1}{2}}.$$

For $0 \leq \varkappa \leq 0.5$, we have

$$q = q((2L)^{-1}) \leq \sqrt{1 - 0.5\varkappa};$$

therefore, bound (10.173) holds.

3. It follows from (10.173) that for a given accuracy $0 < \varepsilon << 1$, the EPG method requires

$$s = O\left(\frac{\ln \varepsilon^{-1}}{\ln q^{-1}}\right)$$

steps to get $y_s : \|y_s - y^*\| \leq \varepsilon$. Then, we have

$$(\ln q^{-1})^{-1} = \left(-\frac{1}{2}\ln(1-0.5\varkappa))^{-1}\right).$$

Using $\ln(1+x) \leq x$, $\forall x > -1$, we obtain $\ln(1-0.5\varkappa) \leq -0.5\varkappa$ or

$$-0.5\ln(1-0.5\varkappa) \geq 0.25\varkappa.$$

Therefore, $(\ln q^{-1})^{-1} \leq 4\varkappa^{-1}$. Keeping in mind that each EPG step requires $O(n^2)$ operation, we obtain bound (10.174) $\square$

Remark 10.4. For small $0 < \varkappa < 1$, EPG complexity bound (10.174) is much better than PGP bound (10.145); therefore, the necessity to project on Ω twice per step is easily compensated by faster convergence of the EPG.

In the case of $1 > \varkappa > 0.5$ and large n, however, the PGP could be more efficient.

10.6.6 Lipschitz Constant

Let us estimate the Lipschitz constant $L > 0$ in (10.147), which plays an important role in both PGP and EPG methods.

First of all, due to the productivity of the matrix A, we have $\max_{1\leq j\leq n}\sum_{i=1}^{n} a_{ij} \leq 1$ and $\max_{1\leq i\leq n}\sum_{j=1}^{n} a_{ij} \leq 1$.

Therefore, for matrix

$$D = I - A = \begin{pmatrix} 1-a_{11} & \cdots & -a_{1n} \\ \vdots & \vdots & \ddots & \vdots \\ -a_{n1} & \cdots & 1-a_{nn} \end{pmatrix} = \begin{pmatrix} d_{11} & \cdots & d_{1n} \\ \vdots & \vdots & \ddots & \vdots \\ d_{n1} & \cdots & d_{nn} \end{pmatrix}$$

we obtain

$$\|D\|_I = \max_{1\leq j\leq n}\sum_{i=1}^{n}|d_{ij}| \leq 2, \quad \|D\|_{II} = \max_{1\leq i\leq n}\sum_{j=1}^{n}|d_{ij}| \leq 2. \tag{10.178}$$

Let us consider operators $p : \mathbb{R}^n_+ \to \mathbb{R}^n_+$ and $c : \mathbb{R}^n_+ \to \mathbb{R}^n_+$. We assume that for both operators Lipschitz condition is satisfied for each component of correspondent vector function; that is, for $x_1, x_2 \in \mathbb{R}^n_+$, we have

$$|p_j(x_1) - p_j(x_2)| \leq L_{p,j}\|x_1 - x_2\|, 1 \leq j \leq n \tag{10.179}$$

and for $\lambda_1, \lambda_2 \in \mathbb{R}^n_+$, we have

$$|c_i(\lambda_1) - c_i(\lambda_2)| \leq L_{c,i}\|\lambda_1 - \lambda_2\|, \ 1 \leq i \leq n. \tag{10.180}$$

It follows from (10.179) that for any $x_1, x_2 \in \mathbb{R}^n_+$, we have

$$\|p(x_1)-p(x_2)\| = \sqrt{\sum_{j=1}^{n}(p_j(x_1)-p_j(x_2))^2} \le \sqrt{\sum_{i=1}^{n} L_{p,j}^2\|x_1-x_2\|^2} \le$$

$$L_p\sqrt{n\|x_1-x_2\|^2} = L_p\sqrt{n}\|x_1-x_2\|, \tag{10.181}$$

where $L_p = \max_{1\le j\le n} L_{p,j}$.

Similarly, we obtain

$$\|c(\lambda_1)-c(\lambda_2)\| = \sqrt{\sum_{i=1}^{n}(c_i(\lambda_1)-c_i(\lambda_2))^2} \le$$

$$\sqrt{\sum_{i=1}^{n} L_{c,i}^2\|\lambda_1-\lambda_2\|^2} \le L_c\sqrt{n}\|\lambda_1-\lambda_2\|, \tag{10.182}$$

where $L_c = \max_{1\le i\le n} L_{c,i}$.

We are ready to find the upper bound for L in (10.147). Using (10.80) for any pair $y_1, y_2 \in \Omega = \mathbb{R}^n_+ \times \mathbb{R}^n_+$, we have

$$\|g(y_1)-g(y_2)\| \le \|(I-A)^T\lambda_1 - p(x_1) - (I-A)^T\lambda_2 + p(x_2)\|$$

$$+\|c(\lambda_1)-(I-A)x_1 - c(\lambda_2) + (I-A)x_2\|$$

$$\le \|p(x_1)-p(x_2)\| + \|(I-A)^T\|\|\lambda_1-\lambda_2\| +$$

$$+\|c(\lambda_1)-c(\lambda_2)\| + \|(I-A)^T\|\|x_1-x_2\| \le$$

$$(L_p\sqrt{n}+\|I-A\|)\|x_1-x_2\| + (L_c\sqrt{n}+\|(I-A)^T\|)\|\lambda_1-\lambda_2\|. \tag{10.183}$$

For $D = I-A$ and $D^T = (I-A)^T$, we have

$$\|D\| = \sqrt{\lambda_{max}(D^TD)} \text{ and } \|D^T\| = \sqrt{\lambda_{max}(DD^T)}$$

Keeping in mind (10.178), we obtain

$$\|D\| \le \sqrt{n}\|D\|_I = 2\sqrt{n}, \ \ \|D^T\| \le \sqrt{n}\|D^T\|_{II} = 2\sqrt{n}. \tag{10.184}$$

Using (10.181), (10.182), and (10.184) from (10.183), we obtain

$$\|g(y_1)-g(y_2)\| \le \sqrt{n}(L_p+2)\|x_1-x_2\| +$$

$$+\sqrt{n}(L_c+2)\|\lambda_1-\lambda_2\| \le \sqrt{2}\hat{L}(\sqrt{n}+2)\|y_1-y_2\|,$$

where $\hat{L} = \max\{L_p, L_c\}$. Therefore, for the Lipschitz constant in (10.147), we have the following bound:

$$L \le \sqrt{2}\hat{L}(\sqrt{n}+2) = O(\sqrt{n}). \tag{10.185}$$

The NIOE adds some new features to the classical IO model. In particular, it finds the output consistent with production cost and consumption consistent with prices for goods.

At the NIOE, the output minimizes the total production cost, while the price vector maximizes the total consumption.

It looks like solving problem VI (10.67) is much more difficult than solving IO systems (10.56)–(10.57). In reality, both PGP and EPG are very simple methods, which require neither solving subproblems nor systems of equations at each step. Instead, the most expensive operation at each step is computing the pseudo-gradient g. It requires for PGP two and for EPG four matrix-by-vector multiplications or $O(n^2)$ operations.

Therefore, in some instances, finding NIOE can be even less expensive than solving large systems of linear equations.

In particular, it follows from bounds (10.174) and (10.185) that EPG complexity is $O(n^{2.5} \ln \varepsilon^{-1})$ for a given fixed $\gamma > 0$.

Therefore, for very large n, the EPG method can be used when solving linear systems (10.56)–(10.57) is very difficult or even impossible due to their size.

Primal–dual decomposition for both methods allows computing the primal and dual vectors in parallel.

By applying parallel technique for matrix-by-vector multiplication, complexity of both PGP and EPG can be substantially improved; see Gusev and Evans (1993).

10.7 Finding J. Nash Equilibrium in n-Person Concave Game

From our remarks in Section 10.2.4 follows that the data for finding J. Nash equilibrium is given by n payoff functions $\varphi_i(x_1,\ldots, y_i, \ldots,x_n)$ continuous in $(x_1,\ldots,y_i,..,x_n)$ and concave in $y_i \in S_i$, where $S_i = \{x_i \in |\mathbb{R}^{n_i}_+ : \sum_{j=1}^{n_i} x_{ij} = 1\}$ is the probability simplex (PS) in $\mathbb{R}^{n_i}$.

J. Nash equilibrium is such a vector $x^* \in S = S_1 \otimes \ldots \otimes S_i \ldots \otimes S_n$ that

$$\varphi_i(x_1^*,\ldots,y_i,\ldots,x_n^*) \leq \varphi_i(x_1^*,\ldots,x_i^*,\ldots,x_n^*) \tag{10.186}$$

for any $1 \leq i \leq n$.

It means: if all but one player accepts the equilibrium strategy, then any player who violates it, can only decrease its payoff function.

The normalized payoff function $\Phi : S \otimes S \to R_1$ is given by

$$\Phi(x,y) = \sum_{i=1}^{n} \varphi_i(x_1,\ldots,y_i,\ldots,x_n). \tag{10.187}$$

The normalized equilibrium $x^* \in S$, which solves the following problem

$$\Phi(x^*;x^*) = \max\{\Phi(x^*,y)|y \in S\}, \tag{10.188}$$

also solves (10.186).

Keeping in mind that for a fixed $x^* \in S$ problem (10.188) is a convex optimization problem, we obtain

$$\langle g(x^*), y - x^* \rangle \leq 0, \ \ \forall y \in S, \tag{10.189}$$

where the pseudo-gradient $g : S \to \mathbb{R}^N$, $N = \sum_{i=1}^m n_i$ is given by the following formula:

$$\begin{aligned} g(x) = \nabla_y \Phi(x,y)_{|y=x} = \\ (\nabla_{y_1} \varphi_1(y_1, x_2, \ldots, x_n)_{|y_1=x_1}, \ldots, \nabla_{y_n} \varphi_n(x_1, \ldots, y_n)_{|y_n=x_n})^T \\ = (g_1(x), \ldots, g_n(x))^T. \end{aligned} \tag{10.190}$$

So finding J. Nash equilibrium (10.186) is equivalent to solving VI (10.189) with operator g given by (10.190).

Application of the PGP method for solving VI (10.189) leads to the following sequence $\{x_s\}_{s \in N}$:

$$x_{s+1} = P_S(x_s + tg(x_s)), \tag{10.191}$$

where $t > 0$ will be specified later.

Projection $y \in R^N$ on S leads to the following QP problem:

$$\hat{x} = (\hat{x}_1, \ldots, \hat{x}_n) = P_S(y) = \min\{\frac{1}{2}\|y - x\|^2 | x \in S\} = \tag{10.192}$$

$$= \min\{\frac{1}{2}[\|y_1 - x_1\|^2 + \ldots + \|y_n - x_n\|^2] | x \in S\}.$$

The structure of the feasible set S naturally decomposes problem (10.192) into n QP problems:

$$\hat{x}_i = \min\{\frac{1}{2}\|y_i - x_i\|^2 | x_i \in S_i\}, \ \ i = 1, \ldots, n. \tag{10.193}$$

It means each step of PGP method (10.191) consists of finding

$$\hat{x}_{i,s+1} = \min\{\frac{1}{2}\|(x_s + tg_s)_i - x_i\|^2 | x_i \in S_i\}, \ i = 1, \ldots, n. \tag{10.194}$$

Application of the EPG method for solving VI (10.189) requires two projections on S per step.

The first finds the predictor strategy:

$$\hat{x}_s = (x_{1,s}, \ldots, x_{n,s}) = P_S(x_s + tg(x_s)). \tag{10.195}$$

The second finds the new strategies for each player:

$$x_{s+1} = (x_{1,s+1}, \ldots, x_{n,s+1}) = P_S(x_s + tg(\hat{x}_s)). \tag{10.196}$$

Again, one finds both predictor $\hat{x}_s$ and corrector x_{s+1} by solving (10.192) type problems.

In other words, the main numerical tool for finding J. Nash equilibrium of the n- person concave game (10.186) is projection into PS.

Likely, the projection into PS is a relatively simple procedure. For PS in $\mathbb{R}^n$, it requires $O(n \ln n)$ operations. Pseudo-gradient $g : S \to \mathbb{R}^N$ is well defined if:

1. It is locally strongly monotone, so there exists $\gamma > 0$ such that

$$\langle g(x) - g(x^*), x - x^* \rangle \leq -\gamma \|x - x^*\|, \ \forall x \in S \tag{10.197}$$

2. Local Lipschitz condition

$$\|g(x) - g(x^*)\| \leq L\|x - x^*\|, \ \forall x \in S \tag{10.198}$$

holds.

Let $\varkappa = \gamma L^{-1}$ be the condition number of the operator g.

Theorem 10.15. *If the pseudo-gradient g is well defined, then for any $0 < t < 2\gamma L^{-1}$, the PGP method (10.191) globally converges to J. Nash equilibrium $x^* \in S$ with Q-linear rate*

$$\|x_{s+1} - x^*\| \leq q(t)\|x_s - x^*\|, \tag{10.199}$$

where $q(t) = (1 - 2t\gamma + t^2L^2)^{\frac{1}{2}}$, and for $t = \min\{q(t)|t > 0\} = \gamma L^{-2}$, the following bound

$$\|x_{s+1} - x^*\| \leq (1 - \varkappa^2)^{\frac{1}{2}} \|x_s - x^*\| \tag{10.200}$$

holds.

The proof is similar to the proof of Theorem 10.9.

If the pseudo-gradient is just monotone

$$\langle g(x) - g(y), x - y \rangle \leq 0, \forall x, y \in S, \tag{10.201}$$

then Theorem 10.15, generally speaking, does not guarantee convergence of the PGP sequence $\{x_s\}_{s \in N}$ given by (10.191) to J. Nash equilibrium.

If g is just monotone and Lipschitz condition

$$\|g(x) - g(y)\| \leq L\|x - y\|, \forall x, y \in S \tag{10.202}$$

holds, then EPG methods (10.195)–(10.196) can be used.

EPG methods (10.195)–(10.196) generate a sequence $\{x_s\}_{s \in N}$, which converges to the J. Nash equilibrium.

Theorem 10.16. *If for the operator $g : S \to \mathbb{R}^N$ conditions* (10.201)–(10.202) *hold, then for any $t \in (0, (\sqrt{2}L)^{-1})$, EPG methods* (10.195)–(10.196) *generate a sequence $\{x_s\}_{s \in N}$: $\lim_{s \to \infty} x_s = x^*$.*

The proof is similar to the proof of Theorem 10.10.

If local strong monotonicity assumption (10.197) is satisfied and local Lipschitz condition (10.198) holds, then EPG methods (10.195)–(10.196) generate a sequence $\{x_s\}_{s \in N}$ that converges to J. Nash equilibrium with linear rate.

The following theorem is similar to Theorem 10.11 and can be proven using similar arguments.

Theorem 10.17. *If local strong monotonicity (10.197) and local Lipschitz condition (10.198) hold, then for* $\mathrm{v}(t) = 1 + 2\gamma t - 2(tL)^2$ *and the ratio* $q(t) = 1 - 2\gamma t + 4(\gamma t)^2(\mathrm{v}(t))^{-1}$*, the following bound*

1.
$$\|x_{s+1} - x^*\| \leq (q(t))^{\frac{1}{2}} \|x_s - x^*\|,\ 0 < q(t) < 1,\ \forall t \in (0, (\sqrt{2}L)^{-1})$$
holds.

2. *For* $t = (2L)^{-1}$*, we have*
$$q((2L)^{-1}) = (1 + \varkappa)(1 + 2\varkappa)^{-1}.$$

3. *For any* $\varkappa \in [0, 0.5]$*, the following bound* $\|x_{s+1} - x^*\| \leq \sqrt{1 - 0.5\varkappa}\|x_s - x^*\|$ *holds.*

The main operation at each step in both PGP and EPG methods is projection on PS.

In the following section, we consider the projection on PS from the numerical standpoint; see Wang and Carreira (2013) and references therein.

10.7.1 Projection Onto Probability Simplex

Let $S_q = \{x \in \mathbb{R}^q_+ : \sum_{i=1}^q x_i = 1\}$ be the probability simplex in $\mathbb{R}^q$.

Finding the Euclidean projection of a vector $y \in \mathbb{R}^q$ onto S_q leads to the following QP:
$$\hat{x} = P_{S_q}(y) = \mathrm{argmin}\{\frac{1}{2}\|y - x\|^2 | x \in S_q\}. \tag{10.203}$$

The projection $\hat{x}$ exists, and it is unique because $\frac{1}{2}\|y - x\|^2$ is strongly convex in $x \in \mathbb{R}^q$.

Let us consider Lagrangian, which corresponds to (10.203).
$$L_y(x, \lambda, \alpha) = \frac{1}{2}\|y - x\|^2 - \lambda[(x, e) - 1] - (\alpha, x), \tag{10.204}$$

where $e = (1, \ldots, 1)^T \in \mathbb{R}^q$ and $\alpha \in \mathbb{R}^q_+$.

From the KKT's conditions follows
$$\nabla_{x_i} L_y(x, \lambda, \alpha) = x_i - y_i - \lambda - \alpha_i = 0,\ \ i = 1, \ldots, q. \tag{10.205}$$
$$x_i \geq 0 \tag{10.206}$$
$$\alpha_i \geq 0 \tag{10.207}$$
$$x_i \alpha_i = 0 \tag{10.208}$$

$$\sum_{i=1}^{q} x_i = 1. \tag{10.209}$$

For $x_i > 0$ from (10.208) follows $\alpha_i = 0$; therefore, from (10.205), we obtain

$$x_i > 0 \Rightarrow \alpha_i = 0 \Rightarrow x_i = y_i + \lambda. \tag{10.210}$$

For $x_i = 0$ from (10.205) follows

$$y_i + \lambda = -\alpha_i \leq 0. \tag{10.211}$$

It means that active constraints of the system (10.206) correspond to smallest components of vector $y = (y_1, \ldots, y_q)$.

Without loss of generality, we can assume

$$y_1 \geq y_2 \geq \ldots y_r \geq y_{r+1} \ldots \geq y_q.$$

Let $r = \max\{1 \leq j \leq q : y_j + r^{-1}(1 - \sum_{i=1}^{j} y_i) > 0\}$, so $y_j + r^{-1}(1 - \sum_{i=1}^{j} y_i) \leq 0,\ j \geq r+1$.

Let $\hat{x}_i = 0$, $i \geq r+1$; then from (10.209) and (10.210), we obtain

$$1 = \sum_{i=1}^{r} x_i = \sum_{i=1}^{r} (y_i + \lambda);$$

therefore,

$$\lambda = r^{-1}(1 - \sum_{i=1}^{r} y_i).$$

Then from (10.210) follows

$$\hat{x}_i = y_i + r^{-1}(1 - \sum_{i=1}^{r} y_i),\ \ i = 1, \ldots, r. \tag{10.212}$$

Exercise 10.4. Let r be the number of possible components in the solution $\hat{x}$ of problem (10.186).

Show that

$$r = \max\left\{1 \leq j \leq q : y_j + j^{-1}\left(1 - \sum_{i=1}^{j} y_i\right) > 0\right\}.$$

10.7.2 Algorithm for Projection onto PS

Algorithm for finding $\hat{x} = P_q(y)$ consists of the following steps:

1. Sort $y := \bar{y} : \bar{y}_1 \geq \bar{y}_2 \geq \ldots \geq \bar{y}_q$.

2. Find $r = \max\{1 \leq j \leq n : y_j + j^{-1}(1 - \Sigma_{j=1}^{r} y_j) > 0\}$.
3. Find $\lambda = r^{-1}(1 - \Sigma_{j=1}^{r} y_j)$.
4.
$$\hat{x} = (\hat{x}_1, \dots, \hat{x}_q) : \hat{x}_j = \max\{y_j + \lambda, 0\}, \quad j = 1, \dots, q.$$

The main numerical operation for finding projection onto PS is sorting components of vector $y = (y_1, \dots, y_q)$.

It requires $O(q \ln q)$ operations.

The main advantage of both PGP (10.191) and EPG (10.195)–(10.196) is the possibility to decompose projections on Ω in (10.191) and (10.195)–(10.196) into projections on the player's probability simplexes S_i, $i = 1, \dots, n$.

In other words, to find the projection in (10.191) or (10.195) and (10.196), each player can independently project onto correspondent PS.

Moreover, it can be done simultaneously.

Each step of PGP requires total $\Sigma_{i=1}^{n} O(m_i \ln m_i) = M$ operations.

In the case of simultaneous projection onto S_i by players $1 \leq i \leq q$, the time for one step is proportional to

$$C = \max_{1 \leq i \leq q} O(m_i \ln m_i).$$

Notes

The general NE has been considered in Polyak (1978). The controlling sequence methods for finding equilibrium were introduced in Polyak and Primak (1977a, 1977b).

For the original formulation of J. Nash's equilibrium in n-person concave game, see Nash (1951). Existence and uniqueness of the equilibrium in n-person concave game were considered in Rosen (1965). For existence of Walras–Wald (WW) equilibrium, see Kuhn (1956).

The equivalence of WW and Nash's equilibrium was established in Zukhovitsky et al. (1970). Methods of finding equilibrium in n-person concave game got their origin in convex optimization; see Zukhovitsky et al. (1969, 1973). One can find the methods for solving correspondent VI in Antipin (2002), Bakushinskij and B. Polyak 1974, Farcchinei and Pang (2003), Iusem and Svaiter (1997), Khobotov (1987), Konnov (2007), Korpelevich (1976), and references therein.

The IO model and its applications were introduced and studied in Leontief (1966); see also Ashmanov (1984), Dorfman and Samuelson (1958), Gale (1960), Lancaster (1968), and references therein.

For LP for optimal allocation limited resources, see Dantzig (1963), Dorfman and Samuelson (1958), Kantorovich (1939), Kantorovich (1959), and references therein.

The generalized Walras–Wald equilibrium as an alternative to LP for optimal resource allocation was considered in Polyak (2008), where PGP for finding generalized Walras–Wald equilibrium was used and convergence rate was established.

The PGP and EPG for finding NE for ORA have been used in Polyak (2015), where convergence, convergence rate, and complexity were established under natural assumptions on the input data.

The nonlinear input–output equilibrium (NIOE) was introduced in Polyak (2016), where PGP and EPG have been applied for finding NIOE.

Chapter 11
Applications and Numerical Results

11.0 Introduction

In this chapter we describe several real-life applications and provide results obtained by solving truss topology design (TTD), intensity-modulated radiation therapy planning, support vector machine, non-negative least squares, and economic equilibrium. We also provide numerical results of testing both nonlinear and linear optimization problems. The results obtained strongly corroborate the theory.

In particular, for NR methods, we systematically observed the "hot" start phenomenon for both nonlinear and linear optimization problems.

11.1 Truss Topology Design

Truss topology design is an important structural design problem aimed at making reliable constructions such as bridges, cantilevers, etc. One famous example of a structure made of trusses is the Eiffel Tower.

A good design of structures made of trusses implies that certain characteristics such as stiffness, stability, or cost-related quantities are met. Often such characteristics can be conflicting. The stability of a bridge can be increased by increasing the cost. Therefore, the question arises how to create a structure that meets certain engineering characteristics while minimizing the cost.

Conceptually a solution to the TTD problems provides coordinates to the end points for elastic bars that form trusses (see Figure 11.1). Those end points are called nodes. To design a structure of trusses, one has to decide on the shape of the bars and where those bars are connected to one another. Of course, that should take into account characteristics of the material used, not to mention some technological and cost constraints.

Suppose we have n nodes in 3D space with unknown positions. That defines $M = 3n$ unknown variables (x_i, y_i, z_i), $i = 1, \ldots, n$, which are coordinates of the nodes. In

R. A. Polyak, *Introduction to Continuous Optimization*, Springer Optimization and Its Applications 172, https://doi.org/10.1007/978-3-030-68713-7_11

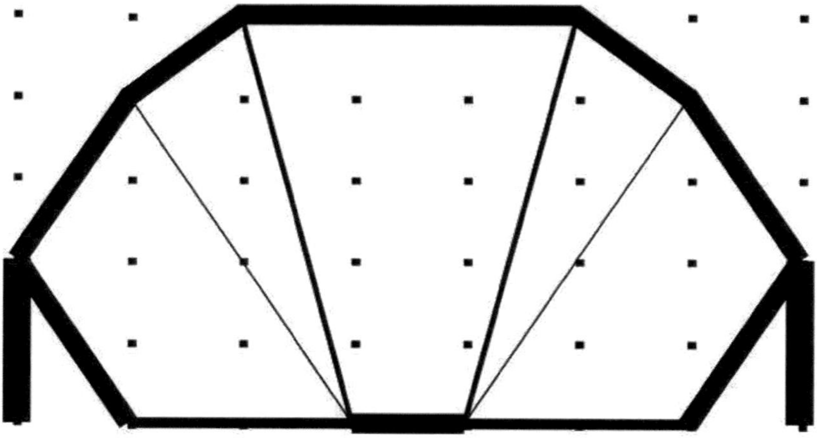

Fig. 11.1 Truss – Topology design

more general terms, we define variables (x_i, y_i, z_i), $i = 1, \ldots, n$, as displacements with respect to some fixed initial position.

That initial position of the nodes may come from engineering or visual considerations. For notational simplicity, we treat $M = 3n$ displacement coordinates as one vector $x \in \mathbb{R}^M$.

Any pair of nodes can be potentially linked with a bar. The solution to the problem should provide an optimal displacement of each node and determine which pairs of nodes should be linked with bars of certain volumes. Assuming that any pair of nodes can be potentially connected, the maximum number of bars is $N = C(n, 2) = n(n-1)/2$.

In our model, the bars are line segments or 1D objects of finite lengths that connect two points in space. Each segment has a volume v_j associated with it given length and thickness of a bar. In particular, when $\mathrm{v}_j = 0$, there is no bar between the corresponding pair of nodes. The vector of volumes $\mathrm{v} = (\mathrm{v}_1, \ldots, \mathrm{v}_N)$ is another vector of variables in the model. Minimization of the total cost leads to minimization of the total volume of all bars while providing the structure capability to resist certain external forces.

The external forces can be potentially applied to every node, which are 3D vectors $F_i = (F_i^x, F_i^y, F_i^z)$, $i = 1, \ldots, n$. We combine the components of the load in a M dimensional vector f that matches vector x. Usually f is taken from a set $\mathscr{F} \subset \mathbb{R}^M$ of possible loading scenarios.

The tension or compression $A(\mathrm{v})x$ caused by displacement x depends on the technological matrix $A(\mathrm{v})$, which, in turn, depends on the vector of volumes v.

At the equilibrium, the tension is equal to the external forces:

$$A(\mathrm{v})x = f.$$

The compliance of the truss under the load is $c = f^T x$, which describes the total energy of the deformation and often has to be minimized.

In case of the *linear elastic model* of the material, the technological matrix $A(\mathrm{v})$ is linear in v, i.e.:

$$A(\mathrm{v}) = \sum_{j=1}^{N} \mathrm{v}_j A_j,$$

where A_i is a positive semidefinite matrix that contains the information about the Young modulus of the bars.

Therefore, the resulting optimization problem is as follows:

$$\min f^T x$$

subject to:

$$\sum_{j=1}^{N} \mathrm{v}_j A_j x = f, \quad \sum_{j=1}^{N} \mathrm{v}_j = V, \quad \mathrm{v}_j \geq 0, j = 1, \ldots, N \tag{TP}$$

The nonlinear constraints make (TP) problem difficult. Using duality consideration (see Ben-Tal and Nemirovski (1995)), in case of a single loading scenario, the problem can be rewritten as the following convex minimax problem:

$$\min_{x \in \mathbb{R}^M} \max_{1 \leq j \leq N} \{V x^T A_j x - 2 f^T x\} \tag{$P1$}$$

In order to use the NR methodology, an additional variable y is introduced to convert $(P1)$ into the following convex optimization problem:

$$\min y$$

subject to

$$y - \frac{V}{2} x^T A_j x + f^T x \geq 0, \; j = 1, \ldots, N. \tag{$C1$}$$

The TTD problems were solved with NR method using truncated MBF transformation. The result below are taken from Ben-Tal and Nemirovski (1995) and shown in Table 11.1.

Table 11.1 Numerical results

Variables	Constraints	Newton steps
98	150	12
126	1234	22
242	4492	23
342	8958	22
450	15,556	41
656	32,679	47

It follows from Table 11.1 that the number of Newton steps required to solve a truss topology design problem stays within a few dozens even for large TTD problems.

The TTD problem usually has many more constraints than variables. This is the type of problems for which NR technique is particularly efficiency. After very few Lagrange multipliers updates with a reasonably large scaling parameter, we have the following situation:

(1) The Lagrange multipliers for the passive constraints become very small. It defines bars that are not in the optimal solution. Practically it leads to substantial reduction of the number of constraints.
(2) Then, the approximation for the primal minimizer is in the Newton area after each updated LM, which leads to at most $O(\ln\ln\varepsilon^{-1})$ Newton steps for finding a new approximation for the minimizer with accuracy given $\varepsilon > 0$. This is the so-called "hot" start, which has been observed practically for all problems solved.
(3) The condition number of the LEP's Hessian remains stable and so is the Newton area. This allows finding solution with a high accuracy.

There are a number of other engineering applications addressed by Kočvara and Stingl (2005, 2007, 2015) using PENNON solver. The applications include free material optimization, optimal sensor localization, and optimal control, just to mention a few; see Jarre et al. (1996).

11.2 Intensity-Modulated Radiation Therapy Planning

Planning radiation therapy for cancer treatment is another important area, where the NR methodology has been successfully applied by Alber and Reemtsen (2007). Intensity-modulated radiation therapy (IMRT) is a widely used procedure for cancer treatment. The goal is to destroy tumorous cells in a body using a clever distribution of radiation beams. Each radiation beam is not powerful enough to destroy cells along the direction of the beam and therefore can penetrate harmlessly through the healthy tissue. However, regions of intersection of several radiation beams undergo more intense radiation capable to destroy cells in those regions.

The goal of planning radiation therapy is finding the location, direction, and intensity of each beam, that the result is desirable distribution of the radiation intensity to particular parts of the body. Needless to say that finding such a distribution of radiation beams could be very challenging.

Usually positions and directions of the radiation beams are fixed and characterized by a particular IMRT device. The goal of solving an optimization problem is finding the intensity of beams, which meet the desirable spatial distribution. Assuming that the total number of the radiation beams is N, we need to find the weights ϕ_i, $i = 1,\ldots,N$ that range between 0 and 1, which correspond to the fraction of the maximum intensity of a beam. For example, if $\phi_1 = 1$, then the first beam is used to its maximum intensity, while if $\phi_1 = 0$, then the first beam is not used at all.

To describe the optimization problem, we divide the region that must receive the radiation into three-dimensional grid of *voxels*. We assume that each voxel is the smallest unit of volume to receive a constant dose of radiation within its boundaries.

Therefore, the more refined the grid of voxels, the more complex distribution of the radiation intensities can be achieved. Let M be the total number of voxels.

We assume that a beam i can deliver the intensity a_{ij} into the voxel j, when used to its maximum capacity. The value a_{ij} is the function of tissue type, position in the body, and whether it is healthy or not. It is given as data for the model. The actual radiation intensity contributed by the beam i for the voxel j is $\phi_i a_{ij}$, when a choice of ϕ_i is made.

Therefore, $d_j = \sum_{i=1}^{N} \phi_i a_{ij}$ is the total dose deposited in voxel j by all radiation beams. We want the dose to be as close as possible to the desired level $\hat{d}_j$ for each voxel. The natural objective then is to minimize:

$$\sum_{j=1}^{M} (d_j - \hat{d}_j)^2 \qquad (IMRT)$$

subject to:

$$\sum_{i=1}^{N} \phi_i a_{ij} = d_j,\ j = 1,\ldots,M,$$

$$0 \le \phi_i \le 1,\ i = 1,\ldots,N.$$

The IMRT problem is QP and can be further modified to achieve more complex tasks. For example, deviation of the radiation dose from the target values could be less desirable for some group of voxels than that for the others. Therefore, the objective function can be modified to include the weights $w_j > 0$, which quantify a relative tolerance of the deviations from the target values:

$$\sum_{j=1}^{M} w_j (d_j - \hat{d}_j)^2$$

Moreover, some deviations can be simply unacceptable as they can destroy healthy tissue. So the additional set of constraints:

$$l_j \le d_j \le u_j,\ j = 1,\ldots,M$$

can be added to restrict the radiation doses to certain ranges.

The effect of deviation of the radiation dose from the target value is not symmetric. For example, for healthy tissue, there should be no problem if the actual radiation dose is significantly below the target value, while exceeding the target value could destroy healthy cells. On the other hand, for tumorous cells, if the radiation dose falls significantly below the target value, then the radiation may not destroy the tumorous cells, while significantly exceeding the target value could be acceptable.

Such an asymmetric treatment can be modeled with the following objective function:

$$\sum_{j=1}^{M} w_j^l (\max(0, l_j - d_j))^2 + \sum_{j=1}^{M} w_j^u (\max(0, d_j - u_j))^2,$$

or

$$\sum_{j=1}^{M} w_j^l(\max(0,\hat{d}_j - d_j))^2 + \sum_{j=1}^{M} w_j^u(\max(0,d_j - \hat{d}_j))^2,$$

where the parameters w_j^l and w_j^u penalize differently for the deviation below and above the bounds or the target values.

These objective functions are only once continuously differentiable, which precludes using Newton's method. Therefore, often the following objective function can be used:

$$\sum_{j=1}^{M} \exp(\alpha_j(d_j - \hat{d}_j)).$$

Parameters α_j can be used to attain a nonsymmetric penalization for the deviation from the target values $\hat{d}_j$.

The IMRT problem can be further modified to address more complex scenarios. We refer the reader to Alber and Reemtsen (2007) for more details and numerical results obtained by a version of NR method applied for 127 clinical examples (Table 11.2).

The authors used the Polak–Ribiere version of conjugate gradient method for the unconstrained minimization within NR framework. They report the average number of Lagrange multipliers update called the "outer iterations" after running multiple cases and confirm (although not shown in the table) that the number of inner iterations (i.e., iterations needed to find an approximation for unconstrained minimum) strongly decreases when the number of outer iterations are increasing. The number of LM update is small, which is typical for NR methods.

The ability of the NR methodology to find the solution with a high accuracy is critical for the IMRT problems. The described methodology is implemented in IMRT devices and successfully used in hospitals for cancer treatment.

Table 11.2 Typical mean values for the size of clinical optimization problems and the associated number of iterations required to solve them

Case class	Cases	Constr	Variables	Outer iter	Inner iter
Prostate	43	6	2600	3.7	140
Prostate/lymph nodes	13	11	9600	6.3	330
Head and neck + nodes	59	26	10,200	7.4	510
Breast	6	11	13,100	5.1	190
Paranasal sinus	6	21	3200	7.8	460

11.3 QP and Its Applications

Let $Q = Q^T$ be positive semidefinite matrix and vectors x, a, a_i, $i = 1, \dots, m$ belong to $\mathbb{R}^n$,$b \in \mathbb{R}^m$ and $f(x) = 0.5x^T Qx + a^T x$.

In this section, along with QP:

$$f(x^*) = \min\{f(x)|a_i^T x \leq b_i,\ i = 1, \dots, m\},$$

we consider two applications of (QP): *non-negative least squares problems* and *soft-margin support vector machines*. Both applications fall in the category of finding unknown dependencies from available empirical data.

In many cases, a good model for an unknown function can be selected from a predefined family of parametrized functions. The goal is finding unknown parameters that specify a particular representative from the family of functions minimizing the least squares criterion.

There is a vast literature dedicated to the least squares method and more generally *regression analysis*. Most of the literature is dedicated to the linear regression, when the unknown parameters linearly enter a modeling function. For example, if a polynomial function is used to model the dependencies, we need to find coefficients. Therefore, finding those polynomial coefficients would be the goal of linear regression.

Suppose one has m pairs of functional argument – value dependencies (x_1, y_1), $(x_2, y_2), \dots, (x_m, y_m)$, which constitute examples or *training data*. We assume that row vectors x_i, $i = 1 \dots, m$ are elements of the n-dimensional Eucledean space. The goal is to find a function f that accurately relates an argument x to $y : y = f(x)$.

Usually instead of finding f, we settle for an estimator $\hat{f}$ taken from some class of functions $\mathscr{F}$.

There are several challenges related to finding an estimator $\hat{f}$. First, we have to define the class of functions $\mathscr{F}$. As we will see later having, $\mathscr{F}$ too small or too large may result in a poor choice of the estimator $\hat{f}$. Second, $\hat{f}$ must be selected based on a finite and often limited number of examples.

If $\hat{f}$ does not reflect good enough dependencies among the training data, then even for small $\max_i |\hat{f}(x_i) - y_i|$, it does not guarantee good prediction dependencies in the future.

By adopting the least squares (LS) criteria, we obtain the following problem:

$$\min_{a \in \mathbb{R}^n} \sum_{i=1}^{m} \left(a^T x_i - y_i\right)^2 = \min_{a \in \mathbb{R}^n} \|Xa - y\|^2, \qquad (LS)$$

where X is an m by n matrix made up by row vectors x_i, $i = 1, \dots, m$, vector $y \in \mathbb{R}^m$ is given, and $m > n$. The Euclidean norm $\|x\| = \sqrt{x^T x}$ is used.

The solution of LS one finds by solving the following normal system of linear equations:

$$X^T X a = X^T y, \qquad (NS).$$

If X is a full rank matrix, then $(X^T X)^{-1}$ exists and:

$$a = (X^T X)^{-1} X^T y.$$

11.3.1 Non-negative Least Squares

Often X is not a full rank matrix, and there is extra condition on a, for example, non-negativity, then we are dealing with non-negative LS problem (NNLS). Unknown parameters can represent some weights showing relative importance of the factors. In one of the examples, we consider later such weights required for *support vector machines* methodology.

We describe NNLS problem using unknown variables $x \in \mathbb{R}^n$, and then the NNLS finds

$$f(x^*) = \min\left\{\sum_{i=1}^{m} (\langle a_i, x\rangle - y_i)^2 \,|x \in \mathbb{R}^n_+\right\} = \min\{\|Ax - y\|^2 |x \in \mathbb{R}^n_+\}, \quad (NNLS)$$

where matrix $A : \mathbb{R}^n \to \mathbb{R}^m$, $m > n$ with row vectors a_i, $i = 1, \ldots, m$ from $\mathbb{R}^n$.

The NNLS is an important linear algebra problem, which has been studied for a long time. The research on NNLS was summarized in the classical monograph by Lawson and Hanson (1995). Since the 1970s, their active set method and its modifications were one of the main tools for solving NNLS.

The active set approach requires at each step solving a standard LS sub-problem, which is equivalent to solving a linear system of equations. The combinatorial nature of the active set methods does not allow establishing meaningful bounds for the number of steps.

On the other hand, NNLS is a QP problem and can be solved by IPMs in polynomial time.

In fact, it takes $O(m^3 \ln \varepsilon^{-1})$ operations to find an $\varepsilon > 0$ approximation for $f(x^*)$. The IPMs require, however, solving linear systems of equations at each step, which for large-scale NNLS can be difficult.

In this section for solving NNLS, the *gradient projection (GP) methods* from Section 5.13.1 are applied.

Instead of solving a linear system of equations, the GP at each step requires matrix by vector multiplication. What is even more important, the GP methods have no combinatorial features. It allows establishing both convergence rate and complexity bounds under various assumptions on the input data.

Particular attention will be given to the fast gradient projection (FGP) method from Section 5.13.2, which has significant practical importance.

The FGP requires $O(\lambda^{\frac{1}{2}} \|x_0 - x^*\| n^2 \varepsilon^{-\frac{1}{2}})$ operations for finding $f(x_k) : \Delta_k = f(x_k) - f^* \leq \varepsilon$, where $\lambda = \text{maxeigeval} A^T A$, small enough $\varepsilon > 0$ is the required accuracy, and x_0 is the starting point.

Therefore, for large n, FGP has the potential to be an efficient alternative to IPMs for solving the NNLS problems. Moreover, matrix by vector multiplication is much cheaper than solving the same size system of linear equations, and it admits fast parallel computations, which can substantially speed up the process and improve the complexity bound (see, for example, Gusev and Evans (1993), Quinn (2004), Migdalas et al. (1997), and Benthem and Keenan (2004)).

Projection on $\mathbb{R}^n_+$ is a very simple operation.

Let:

$$[a_i]_+ = \begin{cases} a_i, & \text{if } a_i > 0 \\ 0, & \text{if } a_i \le 0, \end{cases}$$

then $P_{\mathbb{R}^n_+} a = [a]_+ = ([a_1]_+, \ldots, [a_n]_+)^T$.

We are back to NNLS.

The gradient:

$$\nabla f(x) = A^T(Ax - b) = Qx - q,$$

where $Q = A^T A$ and $q = A^T b \in \mathbb{R}^n$, satisfies the Lipschitz condition:

$$\|\nabla f(x) - \nabla f(y)\| \le \|Q\| \|x - y\|,$$

therefore, for any $L \ge \text{maxeigval}\, Q = \lambda$, we obtain:

$$\|\nabla f(x) - \nabla f(y)\| \le L\|x - y\| \tag{11.1}$$

Let $\Psi(x, X) = f(x) + \langle X - x, \nabla f(x)\rangle + \frac{L}{2}\|X - x\|^2$, then:

$$\text{argmin}\{\Psi(x, X) | X \in \mathbb{R}^n\} = x - L^{-1}\nabla f(x).$$

First, we consider the following GP method:

$$x_{s+1} = \left[x_s - L^{-1}\nabla f(x_s)\right]_+. \tag{11.2}$$

The bound below follows from Theorem 5.17:

$$\Delta_k = f(x_k) - f^* \le \frac{L}{2k}\|x_0 - x^*\|^2. \tag{11.3}$$

The bound (11.3) can be improved by applying FGP method from Section 5.12.2.

At each step, FGP generates a predictor vector x_{k+1} and a corrector vector X_{k+1}. The predictor x_{k+1} is computed as an extrapolation of two successive correctors. The corrector X_{k+1} one obtains as a result of one GP step with x_{k+1} as a starting point.

Let $x_0 = x_1 \in \mathbb{R}^n$, the upper bound L for the Lipschitz constant is given, and $t_1 = 1$. We assume that X_k has been found already. A step of the FGP method consists of three parts:

(a) find step size $t_{k+1} = \frac{1+\sqrt{1+4t_k^2}}{2}$;
(b) find the new predictor $x_{k+1} = X_k + \frac{t_k - 1}{t_{k+1}}(X_k - X_{k-1})$;
(c) find new approximation:

$$X_{k+1} = \text{argmin}\{\langle X - x_k, \nabla f(x_k)\rangle + \frac{L}{2}\|X - x_k\|^2 | X \in \mathbb{R}^n_+\} = \left[x_{k+1} - \frac{1}{L}\nabla f(x_{k+1})\right]_+$$

The bound below was established in Theorem 5.16:

$$\Delta_k \leq \frac{2L\|x_0 - x^*\|^2}{(k+2)^2}, \tag{11.4}$$

where $\Delta_k = f(x_k) - f^*$.

The most costly operation per step is finding $\nabla f(x_{s+1}) = Qx_{s+1} - q$, which requires matrix by vector multiplication or $O(n^2)$ operations.

If A is a full rank matrix, then convergence rate for the GP method can be improved.

Let A be a full rank matrix, i.e., rank $A = n$, and then $f : \mathbb{R}^n \to \mathbb{R}$ is strongly convex or the gradient $\nabla f : \mathbb{R}^n \to \mathbb{R}^n$ is a strongly monotone operator, i.e., there exists $l > 0$:

$$\langle \nabla f(x) - \nabla f(y), x - y \rangle \geq l\|x - y\|^2, \forall x, y \in \mathbb{R}^n \tag{11.5}$$

and for $Q : \mathbb{R}^n \to \mathbb{R}^n$, we have:

$$\langle Qx, x \rangle \geq l\|x\|^2, \forall x \in \mathbb{R}^n. \tag{11.6}$$

We remind that the gradient ∇f satisfies Lipschitz condition (11.1).

The GP method is defined by formula:

$$x_{s+1} = [x_s - t\nabla f(x_s)]_+ . \tag{11.7}$$

From Theorem 5.17 follows:

1. for $0 < t < 2/(l+L)$, the following bound holds:

$$\|x_{s+1} - x^*\|^2 \leq \left(1 - t\frac{2lL}{l+L}\right)\|x_s - x^*\|^2; \tag{11.8}$$

2. for $t = 2/(l+L)$, we have:

$$\|x_{s+1} - x^*\| \leq \left(\frac{1-\varkappa}{1+\varkappa}\right)\|x_s - x^*\|, \tag{11.9}$$

where $0 < \varkappa = l/L < 1$ is the condition number of the matrix $Q = A^T A$;

3. let $\varepsilon > 0$ be the given accuracy, and then the complexity of the GP method (11.7) is:

$$Comp(GP) = O(n^2\varkappa^{-1}\ln\varepsilon^{-1}). \tag{11.10}$$

FGP does not require much extra work as compared to GP (11.2), but FGP has much better convergence rate and better complexity bound.

11.3.2 Support Vector Machines

For using the NNLS approach, one has to know a priori a class of functions $\mathscr{F}_a$, from which the best estimator is selected. Often, however, it is reasonable to ask what can be done if there is no good justification for choosing a particular class $\mathscr{F}_a$. Is it possible to build an accurate function estimator if the only available information is the limited quantity of training data $(x_1,y_1),(x_2,y_2),\ldots,(x_m,y_m)$ and nothing is a priori known about functional dependency between x and y? Another interesting questions: is it possible to design a system with the class $\mathscr{F}_a$ of nonlinear functions with the capacity controlled by the value of one of the parameters? That would allow to have a general enough class of functions $\mathscr{F}_a$ that could accommodate nonlinear data dependencies with a controlled complexity to prevent a phenomenon called *overfitting*.

Answers to those challenging questions were provided by V. Vapnik and C. Chervonenskis in the 1970s. They suggested a way of building a function estimator using the methodology known as *support vector machines (SVM)*. The SVM can be used to estimate a discrete-valued or a real-valued function simply from a set of training data $(x_1,y_1),(x_2,y_2),\ldots,(x_m,y_m)$. Moreover, a value of one SVM parameter controls the complexity of $\mathscr{F}_a$. The parameter is called the *Vapnik–Chervonenskis dimension* or simply *VC dimension*.

The SVM is also known as a method of *machine learning* (ML). ML is an active area of research that studies algorithms that allow computers to learn a general concept from the training data with an ultimate goal of learning the true concept that the training data represent; see Vapnik (2000).

Since machine learning algorithms generalize beyond the training examples, the success of ML algorithms is characterized by their generalization ability, i.e., how accurate ML algorithms learn the true concept.

To link machine learning to our previous discussion, let us point out that in our case the function f is the true concept and the estimator $\hat{f}$ is the output produced by SVM leaning algorithm. If $\hat{f}$ approximates f well after processing a small number m of training examples, then we say the SVM possesses a good generalization ability.

The *VC theory* is an important component of machine learning. Detailed discussion of the VC theory goes beyond the scope of this book. But we briefly mention one of its important aspects. A key discovery of the VC theory is that two and only two factors are responsible for generalization ability of a ML algorithm: one is the empirical loss, which defines how well the estimator approximates training data. Another is the capacity or VC dimension which defines the diversity of the set of functions from which an estimator is selected. If VC dimension is finite, then one can achieve a good generalization provided the empirical risk is small.

Thus, the ability of the SVM explicitly minimizes both the VC dimension and the empirical loss which makes it a very powerful tool for machine learning. In the rest of the subsection, we will formulate optimization problems one has to solve by using SVM methodology and explain how the optimization techniques described earlier in the book can be used for it.

We restrict ourselves to the case of building *classifier* for only two classes, i.e., the f can assume only binary values. For classifiers with more than two classes, see Vapnik (1998, 2000).

We consider m pairs of labeled training data $(x_1,y_1),(x_2,y_2),\ldots,(x_m,y_m)$, where $x \in \mathbb{R}^n$ and $y_i \in \{-1,1\}$ represent the label indicating the class to which x_i belongs.

The main idea of the SVM is to build a separating hypersurface that splits the training data into two classes (class "+1" and class "−1") using functions taken from the class with a controlled capacity or VC dimension. The separation of data points achieves the following important goal. The separating hypersurface or the decision boundary allows to build a classifier that assigns values "+1" and "−1 by looking to which side of the decision boundary a data point, that needed to be classified, belongs. Note that if the training data points are completely separated, then the task of minimizing the empirical risk is perfectly done if classifier makes no mistake on the training data.

However, according to the VC theory, minimizing the empirical risk is only a part of the story. The second part is controlling the VC dimension.

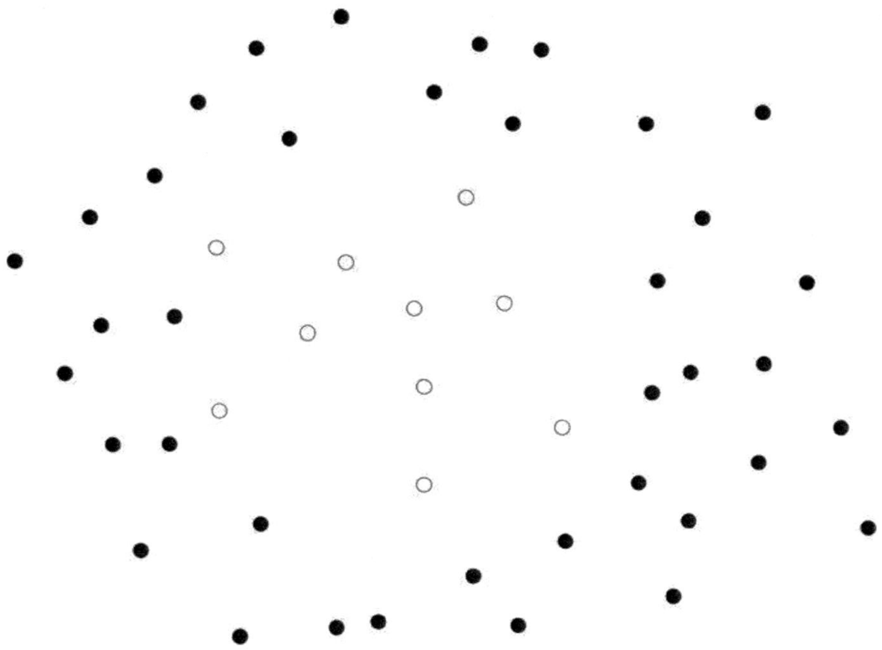

Fig. 11.2 Linearly nonseparable case in $\mathbb{R}^2$

One important observation is that for the SVM, it is enough to consider the class of linear separators or hyperplanes. If the data points cannot be separated using a hyperplane in $\mathbb{R}^n$, they can be mapped in a higher-dimensional space $\mathbb{R}^N$, $N > n$ using a transformation $\Phi : \mathbb{R}^n \to \mathbb{R}^N$, where the transformed data points $z_i = \Phi(x_i)$, $i = 1, \dots, m$ can be separated with a hyperplane.

For example, imagine that one would like to separate the white points from the black ones in $\mathbb{R}^2$ (see Figure 11.2). The separation cannot be done with a line; however, it can be done with an appropriate ellipse. Selection of an ellipse in $\mathbb{R}^2$ means choosing of a maximum of six coefficients, that linearly enter an equation of a second-order curve. Therefore, the selection of an ellipse in $\mathbb{R}^2$ is equivalent to the selection of a hyperplane in $\mathbb{R}^6$.

Of course, a mapping to a higher-dimensional space increases the VC dimension. However, as long the latter stays finite and can be controlled, the SVM algorithm can achieve a good generalization.

Thus, using hyperplanes for separation is not a limitation. On a contrary, it allows to separate very complicated patterns. At the same time, separating hyperplanes can be found by solving convex optimization problems, as we will see later.

There is another important aspect addressed by the VC theory. It turns out that the VC dimension can be decreased with the possibility of separating two sets of points with a hyperplane surrounded by a region with a positive thickness called *margin.* In other words, if instead of separating data with a hyperplane, one can separate with "a thick tube" (see Figure 11.3); then the thicker the tube; the smaller the VC dimension of the class of estimator functions.

Thus the goal of the SVM is to separate training data by finding a hyperplane with a largest possible margin. Suppose we would like to separate white and black points in $\mathbb{R}^n$ (see Figure 11.3). We assume that while points are labeled with $+1$, i.e., $y_i = +1,\ i \in white$, the black points are labeled with -1, i.e., $y_i = -1,\ i \in black$. So the SVM's goal is to find a vector $w \in \mathbb{R}^n$ and a scalar b that define a hyperplane $w^T x + b = 0$ that separates white and black points with a largest possible margin. Since one can define the same hyperplane by multiplying both w and b by any nonzero scalar, we will search for a hyperplane such that $\|w\| = 1$ to eliminate that ambiguity.

A separating hyperplane satisfies the following conditions:

$$w^T x_i + b > 0,\ i \in white,$$

$$w^T x_i + b < 0,\ i \in black.$$

In case the hyperplane separates the white and black points with a margin, the pair (w,b) satisfied the following conditions:

$$w^T x_i + b \geq \Delta,\ i \in white,$$

$$w^T x_i + b \leq -\Delta,\ i \in black$$

for some $\Delta > 0$.

Keeping in mind that $y_i = +1,\ i \in white$ and $y_i = -1,\ i \in black$, the above two inequalities can be combined into the following one:

$$y_i[w^T x_i + b] \geq \Delta,\ i = 1,\ldots,m.$$

Therefore, to find a hyperplane with the largest possible margin, we must solve the following optimization problem:

$$\max \Delta$$

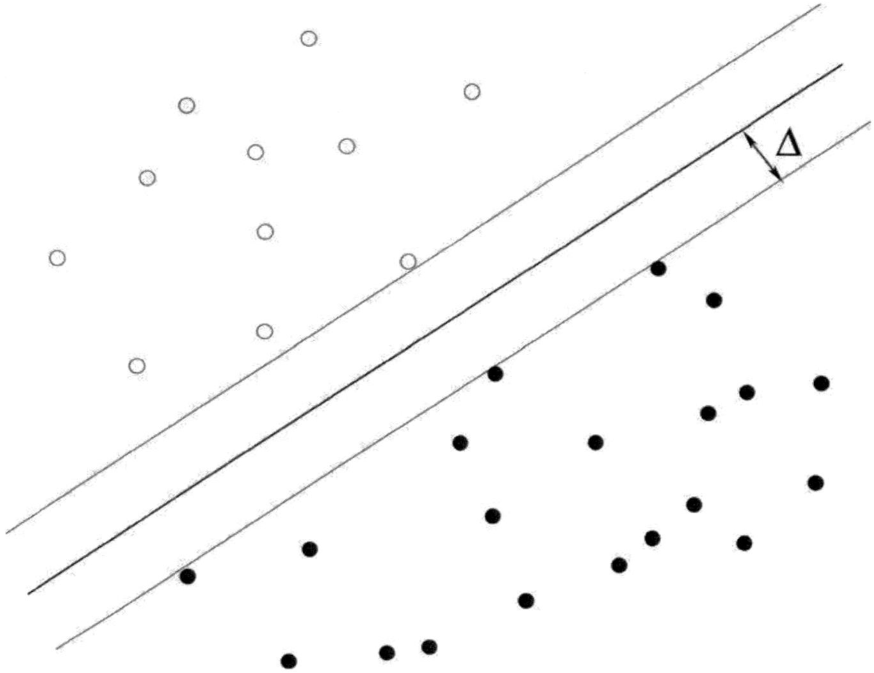

Fig. 11.3 Separation with a positive margin

subject to:

$$y_i[w^T x_i + b] \geq \Delta, i = 1, \ldots, m,$$

$$\|w\| = 1.$$

The last condition can be replaced by an equivalent one $w^T w = 1$. We obtain the following problem:

$$\max \Delta$$

subject to:

$$y_i[w^T x_i + b] \geq \Delta, i = 1, \ldots, m, \qquad (SVM1)$$

$$w^T w = 1.$$

The (SVM1) can be converted to an equivalent convex optimization problem by introducing the variables $\bar{w} = w/\Delta$ and $\bar{b} = b/\Delta$. Keeping in mind that we are only interested in a positive margin, $\Delta > 0$, the constrains:

$$y_i[w^T x_i + b] \geq \Delta, i = 1, \ldots, m$$

will be replaced by:

$$y_i[\bar{w}^T x_i + \bar{b}] \geq 1, i = 1, \ldots, m,$$

while the constraint:

$$w^T w = 1$$

is replaced by:

$$\Delta^2 \bar{w}^T \bar{w} = 1.$$

Therefore, the margin is related to the norm of $\bar{w}$ as:

$$\Delta = \frac{1}{\|\bar{w}\|},$$

thus, maximizing the margin Δ is equivalent to minimizing the norm $\|\bar{w}\|$.

After dropping the bars for notational simplicity, we obtain the following QP:

$$\min 0.5 w^T w$$

subject to:

$$y_i[w^T x_i + b] \geq 1, i = 1, \ldots, m, \qquad (SVM2)$$

which is equivalent to (SVM1).

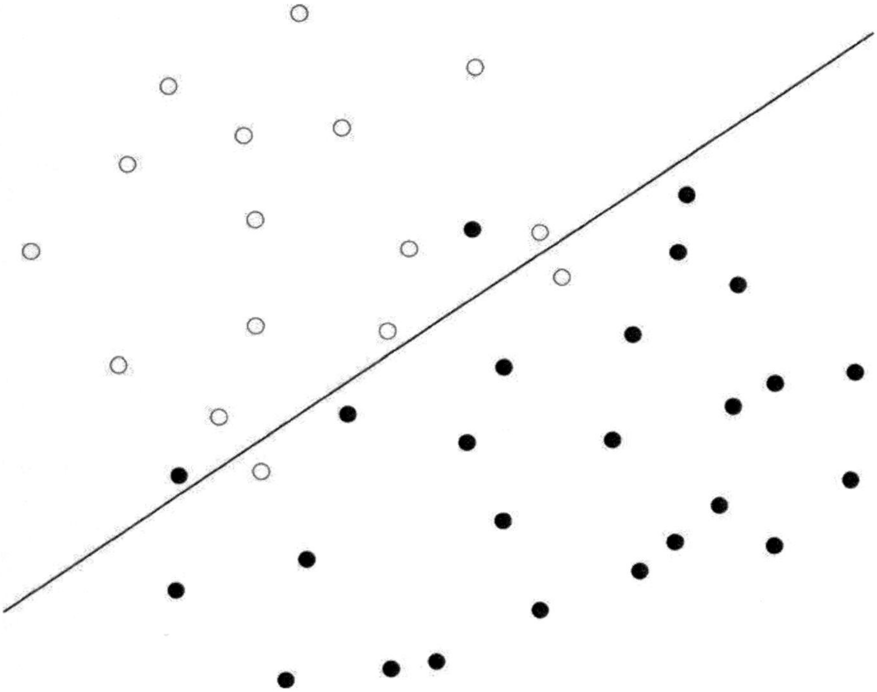

Fig. 11.4 Almost separable case

According to the VC theory, it means: by maximizing the margin, we minimize the VC dimension while keeping the empirical risk at zero.

If the data points x_i cannot be separated with a hyperplane (see Figure 11.4), then the problems (SVM1) and (SVM2) are infeasible. If a hyperplane separates with a "good" margin, a large number of data points except a few of them, then the empirical risk is still low, while a large margin makes the VC still small. Such a desirable hyperplane cannot be found by solving the problems (SVM1) and (SVM2). Therefore, to address the not separable case, we need to modify further the SVM problems.

We allow the constraints to be violated, but to make the violations $\xi_i \geq 0$, $i = 1,\ldots,m$ undesirable, we add to the objective function a penalty term, which penalize the violation. Then, we obtain the following QP problem:

$$\min\{0.5w^T w + C\sum_{i=1}^{m}\xi_i\}$$

subject to:

$$y_i[w^T x_i + b] \geq 1 - \xi_i,\ i = 1,\ldots,m, \qquad (SVM3)$$

$$\xi_i \geq 0, i = 1,\ldots,m,$$

where the parameter $C > 0$ specifies a relative trade-off between minimizing the empirical risk versus minimizing the VC dimension.

Which C should be taken? We do not know the answer *a priori*. The answer depends on a particular data set. Practically C is selected through the process of validation, when some portion of training data is put aside for testing. The "good" value of C is selected by checking how accurate the SVM is for various values of C on the validation set.

We are not planning to discuss validation methods in detail and refer the reader to the literature on machine learning; see Vapnik (2000).

Instead, we focus on how to solve SVM optimization problems. It is important to remember that by choosing various values of C, we can control the capacity of the class of functions, from which the estimator is selected.

Too large C may result in the increase of the capacity of the class and *overfitting* the training data. Too small C may result in a poor estimator, which can poorly classify even the training data. We need the value of C to be in a "sweet spot," and one has to find it by the validation.

Thus, problem (SVM3) with a right choice of C is more practical than (SVM1) and (SVM2).

Now let us focus on how to solve (SVM3), keeping in mind that the number of unknown variables (SVM3) can be large. The good news, however, is we can consider the dual problem to (SVM3), which will have just m variables and simpler constraints.

The Lagrangian for (SVM3) is:

$$L(w,b,\xi,\alpha,\mu) = 0.5w^T w + C\sum_{i=1}^{m}\xi_i - \sum_{i=1}^{m}\alpha_i\left(y_i[w^T x_i + b] - 1 + \xi_i\right) - \sum_{i=1}^{m}\mu_i\xi_i,$$

where $\alpha \in \mathbb{R}_+^m$ and $\mu \in \mathbb{R}_+^m$ are the vectors of the Lagrange multipliers. Differentiating the Lagrangian with respect to the w, b, ξ yields:

$$\frac{\partial L}{\partial w} = w - \sum_{i=1}^{m}\alpha_i y_i x_i = 0 \quad \rightarrow \quad w = \sum_{i=1}^{m}\alpha_i y_i x_i,$$

$$\frac{\partial L}{\partial b} = -\sum_{i=1}^{m}\alpha_i y_i = 0, \tag{11.11}$$

$$\frac{\partial L}{\partial \xi_i} = C - \alpha_i - \mu_i = 0,\ i = 1,\dots,m. \tag{11.12}$$

Keeping in mind $\mu_i \geq 0$, $i = 1,\dots,m$, from (11.12), we obtain $\alpha_i \leq C$, $i = 1,\dots,m$.

After substitution $w = \sum_{i=1}^{m}\alpha_i y_i x_i$ into $L(w,b,\xi,\alpha,\mu)$ and keeping in mind (11.11) and (11.12), we have:

$$\{L(w,b,\xi,\alpha,\mu) : w = \sum_{i=1}^{m}\alpha_i y_i x_i, (11.11), (11.12)\} \equiv$$

$$\sum_{i=1}^{m}\alpha_i - 0.5\sum_{i=1}^{m}\sum_{j=1}^{m} y_i y_j x_i^T x_j \alpha_i \alpha_j$$

Therefore, we end up dealing with the following dual support vector machine problem:

$$\max\left\{\sum_{i=1}^{m}\alpha_i - 0.5\sum_{i=1}^{m}\sum_{j=1}^{m} y_i y_j x_i^T x_j \alpha_i \alpha_j\right\}$$

subject to:

$$\sum_{i=1}^{m}\alpha_i y_i = 0,$$

$$0 \leq \alpha_i \leq C,\ i = 1,\dots,m. \tag{DSVM1}$$

Now we briefly come back to the issue of mapping the data points from the original space of features $\mathbb{R}^n$ to a higher-dimensional space $\mathbb{R}^N$, $N > n$, using the transformation $\Phi : \mathbb{R}^n \to \mathbb{R}^N$.

The data points x_i, $i = 1,\dots,m$ enter the (DSVM1) only by forming the dot products $x_i^T x_j$. Therefore, the dot products between all pairs is what we need to know in order to solve (DSVM1). One can think about dot products as a measure of correlation or similarity between the data points.

The similarity measure between some objects is defined in such a way that it forms a dot product in $\mathbb{R}^N$. Then the (DSVM1) could be used without explicitly defining the transformation Φ as long as the dot product in $\mathbb{R}^N$ is defined. In other words, if one can define a function $K : \mathbb{R}^n \times \mathbb{R}^n \to \mathbb{R}$ in such a way that it is a dot product in $\mathbb{R}^N$, i.e., $K(x_i,x_j) = \Phi(x_i)^T\Phi(x_j)$, then the explicit form of Φ is not required to solve (DSVM1).

There is a necessary and sufficient condition (Mercer's theorem), which can be imposed on K to guarantee that it represents the dot product in $\mathbb{R}^N$. For the kernel $K : \mathbb{R}^n \times \mathbb{R}^n \to \mathbb{R}$ to be a dot product in $\mathbb{R}^N : K(x_i,x_j) = \Phi(x_i)^T\Phi(x_j)$, it is necessary and sufficient that the following matrix:

$$G = \begin{pmatrix} K(x_1,x_1) & K(x_1,x_2) & \dots & K(x_1,x_m) \\ K(x_2,x_1) & K(x_2,x_2) & \dots & K(x_2,x_m) \\ \vdots & \vdots & \ddots & \vdots \\ K(x_m,x_1) & K(x_m,x_2) & \dots & K(x_m,x_m) \end{pmatrix}$$

is symmetric and positive semidefinite. The most commonly used kernels are the exponential kernel: $K(x_i,x_j) = \exp(\gamma\|x_i - x_j\|_2^2)$ and the polynomial kernels $K(x_i,x_j) = \langle x_i,x_j \rangle^d$.

For a given kernel, we obtain the following dual SVM:

$$\max \left\{ \sum_{i=1}^m \alpha_i - 0.5 \sum_{i=1}^m \sum_{j=1}^m y_i y_j K(x_i,x_j)\alpha_i\alpha_j \right\}$$

subject to:

$$\sum_{i=1}^m \alpha_i y_i = 0,$$

$$0 \le \alpha_i \le C,\, i = 1,\dots,m. \qquad (DSVM2)$$

Before we turn our attention on how to solve (DSVM2), let us mention how to construct an estimator once (DSVM1) or (DSVM2) are solved.

In the case of (DSVM1), we have $w = \sum_{i=1}^m \alpha_i y_i x_i$, and the value of b is given by the optimal Lagrange multiplier for the only equality constraint in (DSVM1).

From (DSVM1), solution follows the estimator:

$$y = \hat{f}(x) = \begin{cases} +1, \text{ if } \sum_{i=1}^m \alpha_i y_i x_i^T x + b \ge 0 \\ -1, \text{ if } \sum_{i=1}^m \alpha_i y_i x_i^T x + b < 0 \end{cases}$$

From the solution of (DSVM2), we obtain the following estimator:

$$y = \hat{f}(x) = \begin{cases} +1, \text{ if } \sum_{i=1}^m \alpha_i y_i K(x_i,x) + b \ge 0 \\ -1, \text{ if } \sum_{i=1}^m \alpha_i y_i K(x_i,x) + b < 0 \end{cases}$$

Finally, the dimension of (DSVM2) is defined by the number of training examples m. The dimension of the feature space n is irrelevant.

The data x_i, $i = 1,\dots,m$ could be elements of Hilbert space, or it could be even not mathematical objects. For example, it can be animals, which one wants to put into classes. As long as one can define a similarity measure between any pair x_i and

x_j, that we call a kernel function, which satisfies the Mercer's conditions, we can use the SVM.

Now let us discuss how (DSVM2) can be solved. First, we reformulate (DSVM2) as a minimization problem. Let:

$$f(\alpha) = 0.5 \sum_{i=1}^{m} \sum_{j=1}^{m} y_i y_j K(x_i, x_j) \alpha_i \alpha_j - \sum_{i=1}^{m} \alpha_i,$$

then the (DSVM2) is equivalent to:

$$\min f(\alpha)$$

subject to:

$$g(\alpha) = \sum_{i=1}^{m} \alpha_i y_i = 0,$$

$$0 \leq \alpha_i \leq C, i = 1, \ldots, m. \qquad (DSVM3)$$

The Hessian $\nabla^2 f$ is, generally speaking, a dense matrix made up of mostly nonzero elements. Therefore, the IPMs would require solving at each step a dense linear system of equations with the size equal to the number of data points m.

Instead we adopt FGP method from Section 5.12.2 to solve (DSVM3).

The FGP requires at each step a matrix by vector multiplications as the most expensive operation.

Let us introduce:

$$B = \{\alpha \in \mathbb{R}^m : 0 \leq \alpha_i \leq C, i = 1, \ldots m\}.$$

Note, the projection operator $P : \mathbb{R}^m \to B$ is computationally inexpensive and requires at most $O(m)$ arithmetic operations (see Figure 11.5):

1. Loop over all $i = 1, \ldots, m$.
2. If $\alpha_i < 0$ then Set $\alpha_i = 0$
3. If $\alpha_i > C$ then Set $\alpha_i = C$
4. Return α.

Fig. 11.5 Operator P : Projection of $\alpha \in \mathbb{R}^m$ onto the set B

Then the (DSVM3) can be rewritten as follows:

$$f(\alpha^*) = \min\{f(\alpha) | g(\alpha) = 0, \ \alpha \in B\}. \qquad (11.13)$$

To accommodate the only linear equation $g(\alpha) = 0$, we employ augmented Lagrangian method.

For the augmented Lagrangian for (11.13), we have:

$$\mathcal{L}_k(\alpha, \lambda) = f(\alpha) - \lambda g(\alpha) + 0.5k g(\alpha)^2,$$

where $\lambda \in \mathbb{R}$ is the Lagrange multiplier, which corresponds to the only equality constraint $g(\alpha) = 0$ and $k > 0$ is the scaling parameter.

Augmented Lagrangian method consists of sequential minimizations of $\mathscr{L}_k(\alpha,\lambda)$ in α on the set B:

$$\hat{\alpha} \approx \alpha(\lambda) = \underset{\alpha \in B}{\operatorname{argmin}} \; \mathscr{L}_k(\alpha,\lambda). \tag{11.14}$$

followed by the Lagrange multiplier λ update (see Figure 11.6):

1. Set $\alpha = 0$, $\lambda = 0$. Select $k > 0$, $\delta > 1$, $\varepsilon > 0$ required accuracy
2. Find $\hat{\alpha} \approx \underset{\alpha \in B}{\operatorname{argmin}} \; \mathscr{L}_k(\alpha,\lambda)$ with required accuracy
3. Find $\hat{\lambda} = \lambda - kg(\hat{\alpha})$.
4. Set $\alpha := \hat{\alpha}$, $\lambda := \hat{\lambda}$, $k := \delta k$.
5. If $|g(\alpha)| > \varepsilon$, then Goto 2.

Fig. 11.6 Augmented Lagrangian method

For minimization on B in Step 2 of Fig. 11.6, we are using the FGP algorithm from Section 5.12.2.

The FGP method requires estimation of the Lipschitz constant L of $\nabla_\alpha \mathscr{L}_k(\alpha,\lambda)$, that is, such $L > 0$ that the inequality:

$$\|\nabla_\alpha \mathscr{L}_k(\alpha_1,\lambda) - \nabla_\alpha \mathscr{L}_k(\alpha_2,\lambda)\| \le L\|\alpha_1 - \alpha_2\|.$$

hold for any $\alpha_1, \alpha_2 \in \mathbb{R}^m$.

For the gradient and the Hessian of $\mathscr{L}_k(\alpha,\lambda)$, we have:

$$\nabla_\alpha \mathscr{L}_k(\alpha,\lambda) = M\alpha - e - (\lambda - k(\sum_{i=1}^{m} y_i \alpha_i))y \tag{11.15}$$

$$\nabla^2_{\alpha\alpha} \mathscr{L}_k(\alpha,\lambda) = M + ky^T y,$$

where $M \in \mathbb{R}^{m\times m}$ is the Kernel matrix with the elements $M_{ij} = y_i y_j K(x_i,x_j)$, $i = 1,\dots,m$, $j = 1,\dots,m$; $y = (y_1,\dots,y_m)^T$; and $e = (1,\dots,1)^T \in \mathbb{R}^m$.

Since $\mathscr{L}$ is a quadratic function with respect to α, for the Lipschitz constant L, we have $L = \|\nabla^2_{\alpha\alpha} \mathscr{L}_k(\alpha,\lambda)\| = maxeigenvalue\,(\nabla^2_{\alpha\alpha} \mathscr{L}_k(\alpha,\lambda))$. The constant L depends only on the Kernel matrix M and the parameter k.

In case of Gaussian Kernels, we have:

$$L \le \operatorname{trace}(\nabla^2_{\alpha\alpha} \mathscr{L}_k(\alpha,\lambda)) = \operatorname{trace}(M + ky^T y)$$

$$= m + km = m(k+1).$$

Therefore, we used $L = (k+1)m$ for the FGP.

Note that the matrix-vector product $M\alpha$ is the most computationally expensive part for $\nabla_\alpha \mathscr{L}_k(\alpha,\lambda)$ calculation, which takes $\mathscr{O}(m^2)$ arithmetic operations. The projection operator $P : \mathbb{R}^m \to B$ is computationally inexpensive (see Figure 11.5) and requires only $\mathcal{O}(m)$ arithmetic operations. Therefore, one iteration of FGP requires $\mathcal{O}(m^2)$ operations.

For the stopping criteria, we use the following merit function that measures the violation of the first-order optimality conditions for problem (11.14):

$$\mu(\alpha,\lambda) = \max_{1\le i\le m} \mu_i(\alpha,\lambda),$$

where:

$$\mu_i(\alpha,\lambda) = \begin{cases} |(\nabla_\alpha \mathscr{L}(\alpha,\lambda))_i|, & \text{if } 0<\alpha_i<C, \\ \max\{0, -(\nabla_\alpha \mathscr{L}(\alpha,\lambda))_i\}, & \text{if } \alpha_i = 0, \\ \max\{0, (\nabla_\alpha \mathscr{L}(\alpha,\lambda))_i\}, & \text{if } \alpha_i = C, \end{cases}$$

Note that $\mu(\alpha,\lambda)$ measures the violation of the optimality conditions for the problem (11.14).

The augmented Lagrangian method with the FGP is a subroutine (Figure 11.7) we call AL-FGP method:

1. Input (α,λ).
2. Set $\bar{\alpha} = \alpha$, $t = 1$. Select $L > 0$.
3. Set $\hat{\alpha} = P_{Box}(\alpha - \frac{1}{L}\nabla_\alpha \mathscr{L}_k(\alpha,\lambda))$
4. Set $\bar{t} = 0.5(1+\sqrt{1+4t^2})$
5. Set $\alpha = \hat{\alpha} + \frac{t-1}{\bar{t}}(\hat{\alpha}-\bar{\alpha})$
6. Set $\bar{\alpha} = \hat{\alpha}$, $t = \bar{t}$
7. If $\mu(\hat{\alpha},\lambda) > RequiredAccuracy$, Goto 3.
8. Output $\hat{\alpha}$.

Fig. 11.7 Fast gradient projection algorithm

The results of AL-FGP method testing are shown in Table 11.3. It shows the number of training instances and attributes and the total number of FGP iterations required to train the SVM for each training data set. One iteration is a step of FGP within the augmented Lagrangian iteration.

Keeping in mind that the AL-FGP does not require solving linear systems of equations, we can expect that the FGPM runs faster than methods that require solving such systems. In Table 11.4 we present results of testing the primal-dual NR method, which solves the same set of problems.

The improvement in time for AL-FGP over PDNR is summarized in Table 11.5.

It follows from the tables that AL-FGP requires on the order of $10^2 - 10^3$ more iterations to solve the problems than PDNR does. AL-FGP, however, does not require solving linear systems, which takes $\mathcal{O}(m^3)$ operations per iteration. Instead, each FGP iteration requires $\mathcal{O}(m^2)$ of arithmetic operations because of the matrix-vector multiplication being the most computationally expensive part. Therefore, we can expect that starting with training data of about a few thousand data points, the AL-FGP will outperform the methods that require solving linear systems. The results in Tables 11.3, 11.4, and 11.5 corroborate such a point of view. Starting with the problem of recognizing handwritten digits of the size 3498 data points, AL-FGP

Table 11.3 Numerical results for AL-FGP

Data set name	Instances	Attributes	Iterations	Solution time (s)
Arcene	100	10,000	289	0.38
Seeds	210	7	11,245	0.62
Dexter	300	20,000	503	2.92
Haberman's survival	306	3	23,611	2.63
Breast cancer wisconsin	569	30	806	0.38
Balance scale	625	4	49,321	22.62
CNAE-9	1080	856	14,239	20.22
Contraceptive method choice	1473	9	33,081	80.99
Madelon	2000	500	1,315	8.58
Recog. of handwritten digits	3498	16	2,008	29.35
Statlog (Landsat satellite)	4435	36	1,957	46.21
Page block classification	5473	10	20,518	692.99
EEG eye state	14,980	14	3,869	1007.14
MAGIC gamma Tel. data	19,020	10	35,959	14,714.7

Table 11.4 Numerical results for PDNR

Data set name	Instances	Attributes	Iterations	Solution time (s)
Arcene	100	10,000	5	0.5
Seeds	210	7	87	0.39
Dexter	300	20,000	5	3.06
Haberman's survival	306	3	42	0.58
Breast cancer wisconsin	569	30	5	0.91
Balance scale	625	4	53	2.69
CNAE-9	1080	856	28	8.14
Contraceptive method choice	1473	9	91	52.12
Madelon	2000	500	5	3.7
Recog. of handwritten digits	3498	16	6	444.54
Statlog (Landsat satellite)	4435	36	5	613.04
Page block classification	5473	10	47	1736.13
EEG eye state	14,980	14	5	6010.57
MAGIC gamma Tel. data	19,020	10	27	45,058.57

Table 11.5 Numerical results for solving large problems

Data Set Name	Instances	Attr.	PDNR Time (s)	AL-FGP Time (s)	Impr. factor
Recog. of Handwritten Digits	3,498	16	444.54	29.35	15.15
Statlog (Landsat Satellite)	4,435	36	613.04	46,21	13.27
Page Blocks Classification	5,473	10	1736.13	692.99	2.51
EEG Eye State	14,980	14	6010.6	1007.1	5.97
MAGIC Gamma Tel. Data	19,020	10	45,059	14,711	3.06

consistently outperforms PDNR. The numerical results indicate that for large problems, the first-order methods become preferable over the second-order methods as computational and memory requirements become better. In the next subsection, we

Table 11.6 DFGP vs. DPG methods

Variables	Constraints	DFPG Method		DPG Method	
n	m	Iteration	Time(sec)	Iteration	Time(sec)
100	50	329	0.0235656	4337	0.302921
		280	0.02037611	9184	0.596667
		271	0.01836622	8582	0.5620883
		278	0.02176196	2762	0.1793624
200	100	534	0.06932351	31,477	3.465273
		391	0.0462308	24,132	2.6858285
		402	0.04651364	5447	0.602366
		356	0.05332621	7689	0.8508355
400	200	424	0.11949022	8734	2.2083445
		526	0.1362238	13,603	3.414913
		813	0.20793851	73,766	18.565112
		457	0.1187053	13,549	3.3947103
800	400	681	0.60041203	32,783	27.770106
		514	0.43933119	76,704	63.624228
		758	0.6550702	26,355	22.271063
		1037	0.87214279	70,217	58.25214
1600	800	726	4.73007466	110,836	715.59573
		553	3.48821862	94,567	606.56422
3200	1600	698	17.6589915	120,000	3008.1289
		1851	45.0301598	120,000	2924.3177

show some numerical results obtained by using first-order methods for QP problems.

11.3.3 Fast Gradient Projection for Dual QP. Numerical Results

In the previous two subsections, we discussed how the FGP method can be used for solving NNLS and SVM problems.

In the current subsection, we provide numerical results for a general quadratic programming problem with linear inequality constraints:

$$f(x^*) = \max\{0.5x^T Qx + a_0^T x - b_0 |\ Ax \le b\}, \tag{P}$$

where $A : \mathbb{R}^n \to \mathbb{R}^m$ and $b \in \mathbb{R}^m$. We will assume that the matrix Q is negative definite. Then the dual to problem (P) is:

$$d(\lambda^*) = \min\{0.5\lambda^T H\lambda - g^T\lambda + f |\ \lambda \in \mathbb{R}^m_+\}, \tag{D}$$

where $H = AQ^{-1}A^T$, $g = AQ^{-1}a_0 + b$, $f = 0.5a_0^T Q^{-1}a_0 + b_0$.

The dual problem (D) is a convex quadratic programming problem with only non-negativity constraints. Therefore, both GP and FGP are appropriate for solving such a problem.

Table 11.7 DFGP method for large QP

Variables	Constraints	L Constant	DFPG Method	
n	m	L	Iteration	Time(sec)
3200	1600	5.7241106	785	25.03843
		3.9762428	751	23.32618
		2.7892649	791	24.06845
5000	2000	7.7319259	962	56.92798
		2.5401688	609	35.77332
		3.3512473	768	45.09735
6400	3200	99.819553	2542	301.9402
		4.4427132	815	96.38931
		5.0307601	782	92.15438
8000	4000	4.3058827	878	160.3511
		4.0467058	898	164.8367
		4.3819769	895	163.0681
10000	4000	7.727307	1120	255.0111
		9.0908216	1287	293.8916
		7.134779	1166	265.7037

When applying GP and FGP to the dual problem (D), the correspondent methods are called dual gradient projection (DGP) and dual fast gradient projection (DFGP) methods; see Polyak et al. (2013).

Table 11.6 shows numerical results obtained by DGP and DFGP methods for randomly generated QP. The iteration limit was set to 120000 iterations, and the requested accuracy for the duality gap was set to $\varepsilon = 10^{-6}$. Both algorithms were implemented in MATLAB and tested on a regular laptop. As we can see from the table, the DFGP systematically outperforms the DGP. Figure 11.8 shows the dynamics of the duality gap reduction while solving a problem with $n = 1000$ and $m = 500$.

Table 11.7 shows the performance of DFGP for larger problems.

The numerical results suggest that DFGP is an efficient algorithm for large-scale quadratic problems. The most time-consuming part of the algorithm is a matrix-vector multiplication that can be accelerated using a multi-core parallel environment. Therefore, the FGP method could be used for solving large QP.

11.4 Finding Nonlinear Equilibrium

In this section, we show numerical results obtained by applying PGP and EGP methods for finding nonlinear equilibrium for resource allocation problems (see Sections 10.3 and 10.5). The stopping criteria measures the primal-dual gap and primal and dual infeasibility.

We have generated a number of problems of different sizes with condition number (CN) $\varkappa \approx 0.05$ and solved them with accuracy 10^{-6}. Table 11.8 shows the number of rows (m) and columns (n) of matrix A, the CN ($\varkappa = \gamma/L$) of the problem,

Table 11.8 Nonlinear equilibrium problems

m	n	CN	# of PGP iter	# of EPG iter
100	300	0.052	9409	989
150	550	0.054	8538	943
500	700	0.058	7481	884
1000	2000	0.052	6910	784
2000	4000	0.054	8586	945

and the number iterations needed for both EPG and PGP methods to reach the given accuracy.

Note that independently of the size of the problems, the number of iterations does not change much.

Figure 11.8 shows rates of convergence for both EPG and PGP methods for NE with $m = 2000$ and $n = 4000$. The horizontal axis shows the number of iterations, while the vertical axis shows the accuracy of the solution obtained expressed in terms of the number of decimal digits after the decimal point in the match between the primal and the dual objective functions and primal and dual infeasibility.

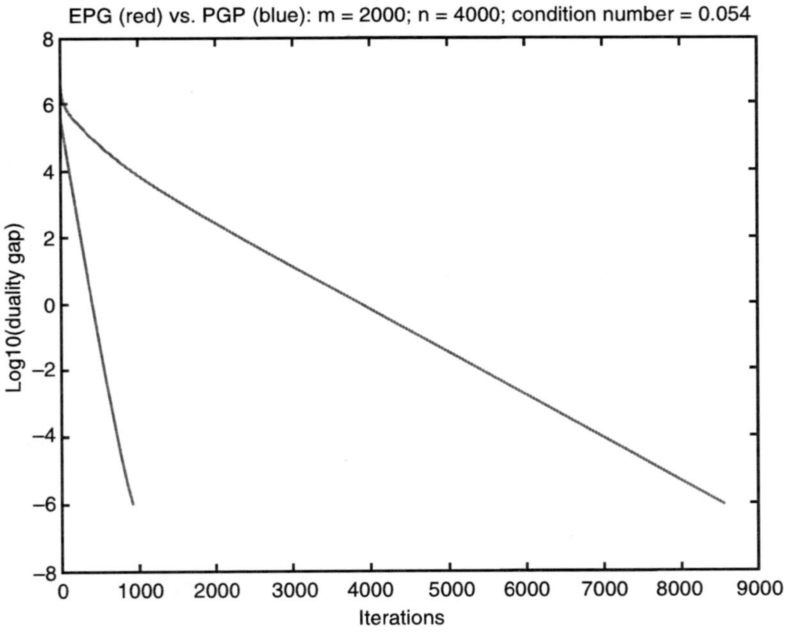

Fig. 11.8 Convergence rates of EPG and PGP methods for finding nonlinear equilibrium with $m = 2000$ and $n = 4000$

For a rather dense matrix A with $m = 2000$ and $n = 4000$, it took approximately 1 min. The rest of the NE problems were solved in few seconds on a regular laptop.

11.5 The "Hot" Start Phenomenon

In this section, we provide numerical results obtained by NR methods described in Chaps. 7 and 8. In particular, we systematically observed "hot" start phenomenon.

It is a particular pattern typical for NR methods. The number of Newton steps, required for finding an approximation for an unconstrained primal minimizer, decreases substantially with the increase of LM update. The decrease is not necessarily monotonic, but the number of Newton steps per update after few updates becomes small, often equal to one or two.

A typical behavior of NR algorithm is shown in Tables 11.9, 11.10, and 11.11. The tables show the primal-dual gap and the primal infeasibility after every LM update and the number of Newton steps per update.

Table 11.9 Name, markowitz2; $n = 1200$, $q = 1201$; objective, convex quadratic; constraints, linear

it	gap	inf	# of steps
0	7.032438e+01	1.495739e+00	0
1	9.001130e-02	5.904234e-05	10
2	4.205607e-03	3.767383e-06	12
3	6.292277e-05	2.654451e-05	13
4	1.709659e-06	1.310097e-05	8
5	1.074959e-07	1.381697e-06	5
6	7.174959e-09	3.368086e-07	4
7	4.104959e-10	3.958086e-08	3
8	1.749759e-11	2.868086e-09	2
9	4.493538e-13	1.338086e-10	2
Total number of Newton steps			59

Table 11.10 Name, moonshot; $n = 786$, $q = 592$; objective, linear; constraints, nonconvex quadratic

it	gap	inf	# of steps
1	3.21e+00	4.98e-05	10
2	8.50e-06	5.44e-10	1
3	1.19e-08	1.24e-11	1
4	3.96e-09	4.16e-12	1
Total number of Newton steps			13

The "hot" start was systematically observed on both linear and nonlinear optimization problems. The results for linear programming one can find in Jensen et al. (1993).

Table 11.11 COPS: journal bearing: $n = 5000$, $m = 5000$, nonlinear objective, bounds

it	f	$\|\nabla L(x,\lambda)\|$	gap	constr violat	# of steps
0	−4.504e+02	5.1364e+01	1.6229e+03	0.0000e+00	0
1	−8.002e−02	9.1922e−08	9.2602e−03	7.0564e−03	20
2	−1.550e−01	9.3995e−13	1.6896e−05	3.4093e−05	7
3	−1.550e−01	1.4592e−15	9.1966e−09	6.0043e−07	4
4	−1.550e−01	7.7398e−17	1.1702e−10	1.3002e−08	2
5	−1.550e−01	6.2450e−17	1.5082e−11	2.7984e−09	1
6	−1.550e−01	6.5919e−17	1.1229e−12	4.4897e−10	1
7	−1.550e−01	6.5919e−17	6.5135e−14	5.9776e−11	1
8	−1.550e−01	6.2450e−17	3.9621e−15	6.6580e−12	1
Total number of Newton steps					37

The "hot" start phenomenon is typical for nonlinear rescaling-augmented Lagrangian method as well (see Table 11.12). Isometrization of α-pinene is taken from the COPS set. The problem has linear and nonlinear equality constraints and bounds. The tables shows the iteration number, the norm of the Lagrangian gradient, the primal-dual gap, the constrained violation, and the number of Newton iterations required for LM update.

Table 11.12 COPS: isometrization of α -pinene $n = 4000$, $m = 4000$, nonlinear objective, nonlinear constraints

it	f	$\|\nabla L(\cdot)\|$	gap	constr violat	# of steps
0	1.096e+10	9.6378e+10	0.0000e+00	2.3600e+01	0
1	2.095e+01	6.1850e−04	2.6426e+00	4.1064e−05	17
2	1.989e+01	1.0998e−01	2.5621e−01	2.8963e−06	4
3	1.987e+01	2.9277e+00	1.6708e−02	2.4864e−07	2
4	1.987e+01	3.0175e−04	9.4867e−04	2.1777e−08	2
5	1.987e+01	1.2649e−03	2.1393e−06	1.3047e−09	1
6	1.987e+01	4.4104e−06	1.1108e−07	7.1941e−11	1
7	1.987e+01	1.4076e−08	5.5255e−09	3.6255e−12	1
8	1.987e+01	1.5019e−09	2.5360e−10	1.6685e−13	1
Total number of Newton steps					29

In this chapter, we considered several applications that are successfully addressed by the optimization methods described in the book. The realm of applications of nonlinear optimization methods spans much beyond those we have observed. Our goal was to develop some intuition about the choices of optimization algorithms. For example, for optimization problem with hundred thousands of variables, using Newton's type methods does not make sense. For such problems, the first-order methods are preferable, especially if the projection of a feasible set is not expensive. However, when the problem has a reasonable size, up to a few thousands of variables and constraints, but involves nonlinear functions, second-order methods, in particular, Newton NR or PDNR, could be a better choice.

Thus, there is no "one-fits-all" nonlinear optimization algorithm that can address all optimization problems simply because the realm of nonlinear problems is incredibly rich.

Therefore, there will be a permanent search for new methods to address new important applications that will appear in the future.

Notes

The TTD problems have been considered in Ben-Tal et al. (1992) – Ben-Tal and Nemirovski (2001) and references therein; see also Jarre et al. (1996), Kočvara and Stingl (2005), and Kočvara and Stingl (2007). For the PENNON solver, see Kočvara and Stingl (2015). Other numerical results obtained by NR with truncated MBF transformation for structural optimization can be found in Berke et al. (1995), Griva et al. (1998).

Application of NR for IMRT was considered in Alber and Reemtsen (2007). For SVM via NR see Griva et al. (2007). For numerical results obtained by NR with truncated MBF transformation, see Breitfeld and Shanno (1996), Nash et al. (1994), and Griva et al. (1998) – Griva and Polyak (2006).

For PET image reconstruction, see Bailey et al. (2005) and Shepp and Vardi (1982) and references therein. For the "hot" start phenomenon, see Griva et al. (1998) – Griva and Polyak (2006), Griva and Polyak (2008), and Melman and Polyak (1996). For the "hot" start in LP, see Jensen et al. (1993). For the numerical performance of the dual fast gradient projection method for QP, see Polyak et al. (2013). For application of the fast projection gradient for NNLS, see Polyak (2015).

Concluding Remarks

1. During my almost 60 years "tenure" in Continuous Optimization (CO), the field went from its infancy to great maturity. The important sign of maturity of any field is when seemingly unrelated facts turned out to be intrinsically connected. The duality is critical in this regard. The power of duality, however, is not only in establishing such connections but also in helping to understand the convergence mechanisms in a number of CO methods.
2. Over the years the standards for research in CO have been fundamentally changed. The complexity of a new method became the main criteria.
   Unfortunately, for many methods, including IPMs, there is a big gap between theoretical complexity and real performance, which is often much better.
3. For a number of CO methods, the duality helps to select parameters responsible for convergence, rate of convergence, and complexity, which are critical for the gap reduction between theoretical complexity and real numerical performance.
4. The progress in CO multiplied by dramatically improved computational capability of modern computers allowed solving very difficult, large-scale, real-life problems, which just recently seemed to be intractable.
   For example, the TTD problems with few thousand variables and couple of hundred thousand quadratic constraints can be solved in a timely manner with high accuracy.
5. In spite of substantial attention in the book to the first-order methods for solving optimization and equilibrium problem, we have not covered, for obvious reasons, some very recent trends and results in this field.
   We hope that a correspondent volume will appear in the nearest future.
6. It is rather amazing that in the twenty-first century, we still fundamentally relay on ideas and methods developed more than 200–300 years ago by Fermat, Newton, Lagrange, Legendre, Fourier, and Cauchy.
   It means only one thing: good ideas are never dying.
7. Let me finish by wishing the new generation of optimizers bright ideas to open new directions in CO, which has so many important applications.

R. A. Polyak, *Introduction to Continuous Optimization*, Springer Optimization and Its Applications 172, https://doi.org/10.1007/978-3-030-68713-7

Appendix

Some Matrix Inequalities

We consider several matrix inequalities, which we used, starting with Debreu's Lemma Debreu (1952). For some other matrix inequalities, Lyapunov theorem and simple iteration method, see B. Polyak (1987) and references therein.

Lemma 1. *(Debreu) Let* $A = A^T : \mathbb{R}^n \to \mathbb{R}^n$, $C : \mathbb{R}^n \to \mathbb{R}^r (n > r)$, $\operatorname{rank} C = r$ *and*

$$\langle Ax, x\rangle \geq m\|x\|^2, m > 0, \forall x : Cx = 0; \tag{1}$$

then there is $0 < \alpha < m$ *and large* $k_0 > 0$ *that for any* $k \geq k_0$ *the following inequality holds*

$$\langle (A + kC^T C)x, x\rangle \geq \alpha\|x\|^2, \forall x \in \mathbb{R}^n. \tag{2}$$

Proof. First of all from $\operatorname{rank} C = r$ follows existence $\mu > 0$ that $\|C^T y\| \geq \mu\|y\|, \forall y \in \mathbb{R}^r$.

In fact,

$$\|C^T y\| = \sqrt{\langle C^T y, C^T y\rangle} = \sqrt{\langle CC^T y, y\rangle} \geq \sqrt{\mu^2 \langle y, y\rangle} = \mu\|y\|, \tag{3}$$

where $\mu^2 > 0$ -mineigval of CC^T. Therefore $\|(CC^T)^{-1}\| \leq \mu^{-2}$ or $\|(CC^T)^{-1}\|^{-1} \geq \mu^2$.

For $x \in \mathbb{R}^n$ we have $x = x_1 + x_2$, where $Cx_1 = 0$ and $(x_1, x_2) = 0$; then $x_2 = C^T (CC^T)^{-1} Cx$. Using Cauchy–Schwarz inequality, we obtain

$$\|Cx\| \geq \frac{\mu^2 \|x_2\|}{\|C^T\|}. \tag{4}$$

For the left-hand side of (2), we obtain

R. A. Polyak, *Introduction to Continuous Optimization*, Springer Optimization and Its Applications 172, https://doi.org/10.1007/978-3-030-68713-7

$$
\begin{aligned}
\langle (A+kC^TC)x,x\rangle &= \langle Ax_1,x_1\rangle + 2\langle Ax_1,x_2\rangle + \langle Ax_2,x_2\rangle + k\|Cx\|^2 \\
&\geq m\|x_1\|^2 - \|A\|\|x_2\|^2 - 2\|A\|\|x_1\|\|x_2\| + \frac{k\mu^4}{\|C^T\|^2}\|x_2\|^2 \\
&= \alpha(\|x_1\|^2 + \|x_2\|^2) + (m-\alpha)\left(\|x_1\| - \frac{\|A\|}{m-\alpha}\|x_2\|\right)^2 \\
&\quad + \left(\frac{k\mu^4}{\|C^T\|^2} - \alpha - \|A\| - \frac{\|A\|^2}{(m-\alpha)}\right)\|x_2\|^2 \\
&\geq \alpha(x,x) + \|x_2\|^2\left(\frac{k\mu^4}{\|C^T\|^2} - \alpha - \|A\| - \frac{\|A\|^2}{m-\alpha}\right).
\end{aligned}
$$

So for $k_0 = \frac{\|A\|\|C^T\|^2}{\mu^4}(1 + \frac{\|A\|}{m-\alpha} + \frac{\alpha}{\|A\|})$ and any $k \geq k_0$, we obtain (2).

Corollary 1. *It follows from (2) that the inverse $(A+kC^TC)^{-1}$ exists for any $k \geq k_0$, also $\|(A+kC^TC)^{-1}\| \leq \alpha^{-1}$ and $\alpha > 0$ is independent on $k \geq k_0$.*

Lemma 2. *Let $C : \mathbb{R}^n \to \mathbb{R}^r$ be a full rank matrix and (1) holds; then there is $k_0 > 0$ large enough and $\beta > 0$ that for any $k \geq k_0$ the following bound holds*

$$\|(A+kC^TC)^{-1}C^T\| \leq \beta k^{-1}, \tag{5}$$

$\beta > 0$ is independent on $k \geq k_0$.

Proof. Let $t \in \mathbb{R}^r$ and $x = (A+kC^TC)^{-1}C^Ty \in \mathbb{R}^n$, then we have

$$(A+kC^TC)x = C^Ty,\ \forall y \in \mathbb{R}^r. \tag{6}$$

Also let $x = x_1 + x_2$ be such that $Cx_1 = 0, (x_1,x_2) = 0$; then (4) holds.

By multiplying both sides of (6) by x_1 and keeping in mind $Cx_1 = 0$, we obtain

$$\langle (A+kC^TC)x,x_1\rangle = \langle x_1,C^Ty\rangle = \langle Cx_1,y\rangle = 0;$$

therefore,

$$\langle Ax,x_1\rangle + k\langle C^TCx,x_1\rangle = \langle Ax,x_1\rangle + k\langle Cx,Cx_1\rangle = \langle Ax,x_1\rangle = 0.$$

Then from (1) follows

$$0 = \langle Ax,x_1\rangle = \langle Ax_1,x_1\rangle + \langle Ax_2,x_1\rangle \geq m\langle x_1,x_1\rangle + \langle Ax_2,x_1\rangle,$$

or $-\langle Ax_2,x_1\rangle \geq m\langle x_1,x_1\rangle$.

Using Cauchy–Schwarz inequality, we obtain

$$\|A\|\|x_1\|\|x_2\| \geq -\langle Ax_2,x_1\rangle \geq m\|x_1\|^2,$$

or

$$\|x_1\| \leq \frac{\|A\|}{m}\|x_2\|. \tag{7}$$

So

$$\|x\|^2 = \|x_1\|^2 + \|x_2\|^2 \leq \left(1 + \frac{\|A\|^2}{m^2}\right) \|x_2\|^2. \tag{8}$$

From (4) and (6), we obtain

$$\begin{aligned} \|C^T\|\|y\| &\geq \|C^T y\| = \|(A + kC^T C)x\| \geq -\|A\|\|x\| + k\|C^T(Cx)\| \\ &\geq -\|A\|\|x\| + k\mu\|Cx\| \geq -\|A\|\|x\| + k\mu^3\|C^T\|^{-1}\|x_2\| \end{aligned} \tag{9}$$

From (8) follows

$$\|x_2\| \geq \left(1 + \frac{\|A\|^2}{m^2}\right)^{-\frac{1}{2}} \|x\|$$

From (7)–(9), we have

$$\begin{aligned} \|C^T\|\|y\| &\geq \left(-\|A\| + \frac{k\mu^3}{\|C^T\|}\left(1 + \frac{\|A\|^2}{m^2}\right)^{-\frac{1}{2}}\right)\|x\| \\ &= k\left(-\frac{\|A\|}{k} + \frac{\mu^3}{\|C^T\|}\left(1 + \frac{\|A\|^2}{m^2}\right)^{-\frac{1}{2}}\right)\|x\| = k\beta_0\|x\|. \end{aligned}$$

Therefore for $k_0 > 0$ large enough, any $k \geq k_0$, and $\beta \geq \|C^T\|\beta_0^{-1}$, we have

$$\|x\| \leq \frac{\beta}{k}\|y\|$$

and $\beta > 0$ is independent on $k \geq k_0$.

Hence

$$\|(A + kC^T C)^{-1} C^T y\| \leq \frac{\beta}{k}\|y\|, \ \forall y \in \mathbb{R}^r,$$

which is equivalent (5).

Corollary 2. *Let $B = (A + kC^T C)^{-1} C^T$; then from $\|B\| = \|B^T\|$ follows $\|C(A + kC^T C)^{-1}\| \leq \frac{\beta}{k}$.*

Lemma 3. *Under conditions of Lemma 1 for $k_0 > 0$ large enough and any $k \geq k_0$, there is $\gamma > 0$ independent on k that*

$$\|I - kC(A + kC^T C)^{-1} C^T\| \leq \gamma/k.$$

Proof. Let us consider $D = I - kC(A + kC^T C)^{-1} C^T$; then

$$\begin{aligned} CC^T D &= CC^T - C\left[(A + kC^T C) - A\right](A + kC^T C)^{-1} C^T \\ &= CC^T - CC^T + CA(A + kC^T C)^{-1} C^T = CA(A + kC^T C)^{-1} C^T. \end{aligned}$$

From Cauchy–Schwarz inequality, bound (4), and $\|(CC^T)^{-1}\| \leq \mu^{-2}$, we obtain

$$\|D\| = \|(CC^T)^{-1}CC^T D\| \le \|(CC^T)^{-1}\|\|C\|\|A\|\|(A+kC^T C)^{-1}C^T\| \le \frac{\|C\|\|A\|}{\mu^2}\frac{\beta}{k} = \frac{\gamma}{k},$$

where $\gamma = \beta\mu^{-2}\|C\|\|A\| > 0$ is independent on $k > 0$ for any $k \ge k_0$ and $k_0 > 0$ large enough.

Lemma 4. *Let* $A = A^T : \mathbb{R}^n \to \mathbb{R}^n, C : \mathbb{R}^n \to \mathbb{R}^m (n > m)$, $\operatorname{rank} C = m$, *and*

$$\langle Ax, x\rangle > 0, \forall x : Cx = 0; \tag{10}$$

then there exists

$$B^{-1} = \begin{pmatrix} A & C^T \\ C & 0 \end{pmatrix}^{-1} \tag{11}$$

and $\beta_0 > 0$ *that* $\|B^{-1}\| \le \beta_0$.

Proof. Let us show that B is a full rank matrix, that is, $rankB = m+n$. Let $y = \begin{pmatrix} u \\ \mathrm{v} \end{pmatrix}$, $u \in \mathbb{R}^n, \mathrm{v} = \mathbb{R}^m$; then from $By = 0$ follows

$$Au + C^T \mathrm{v} = 0 \tag{12}$$

$$Cu = 0. \tag{13}$$

By multiplying both sites (12) by u, we obtain

$$\langle Au, u\rangle + \langle u, C^T \mathrm{v}\rangle = \langle Au, u\rangle + \langle Cu, \mathrm{v}\rangle = \langle Au, u\rangle = 0. \tag{14}$$

Therefore from (10) follows $u = 0$. From (12) follows $C^T \mathrm{v} = 0$ and from that $\operatorname{rank} C = m$ we obtain $\mathrm{v} = 0$. In other words $By = 0 \Rightarrow y = 0$. Therefore $\operatorname{rank} B = m+n$ and B^{-1} exists and there is $\beta_0 > 0$ such that $\|B^{-1}\| \le \beta_0$

Lemma 5. *Under conditions of Lemma 1, there exists* $k_0 > 0$ *large enough that for any* $k \ge k_0$ *matrix*

$$B_k = \begin{pmatrix} A & C^T \\ C & -k^{-1}I \end{pmatrix}$$

has an inverse that $\|B_k^{-1}\| \le \rho$ *and* $\rho > 0$ *is independent on k for any* $k \ge k_0$, *when* $k_0 > 0$ *is large enough.*

Proof. Let us consider $y = \begin{pmatrix} u \\ \mathrm{v} \end{pmatrix} \in \mathbb{R}^{n+m}$. From $B_k y = 0$ follows $Au + C^T \mathrm{v} = 0$ and $Cu = k^{-1}\mathrm{v}$; therefore

$$Au + kC^T Cu = 0.$$

Keeping in mind (2) for any $k \ge k_0$, we obtain

$$0 = \langle (A + kC^T C)u, u\rangle \ge \alpha\|u\|^2, \ \alpha > 0$$

Therefore $u = 0$; then $C^T \mathrm{v} = 0$ and $\operatorname{rank} C = m$ follows $\mathrm{v} = 0$. Hence B_k is not singular. It has an inverse B_k^{-1}, and there is $\rho > 0$ independent on $k \ge k_0$ that $\|B_k^{-1}\| \le \rho$.

Corollary 3. *Under conditions of Lemma 1, there is large enough k_0 that for any $k \geq k_0$ the following inequality holds.*

$$\|B^{-1} - B_k^{-1}\| \leq \frac{1}{k}\beta_0\rho. \tag{15}$$

In fact,

$$\begin{aligned}\|B_k^{-1} - B^{-1}\| &= \|B^{-1}(B - B_k)B_k^{-1}\| \\ &\leq \|B^{-1}\|\|B_k^{-1}\|\|B - B_k\| = \frac{1}{k}\|B\|^{-1}\|B_k^{-1}\| \leq \frac{1}{k}\beta_0\rho.\end{aligned}$$

Simple Iteration Methods (SIM)

Let $A : \mathbb{R}^n \to \mathbb{R}^n$ and $\lambda_1, \ldots, \lambda_n$ be eigenvalues of A. The spectral radius of A is the number

$$\rho(A) = \max_{1 \leq i \leq n} |\lambda_i|. \tag{16}$$

Another important characteristic of A is the norm

$$\|A\| = \max_{\|x\|=1} \|Ax\|, \tag{17}$$

where $\|x\| = \langle x, x \rangle^{\frac{1}{2}}$.

For non-symmetric matrices $\rho(A) \leq \|A\|$, for example, let $A = \left(\begin{smallmatrix} 0 & 0 \\ a & 0 \end{smallmatrix}\right)$, $a \neq 0$; then $\rho(A) = 0 < \|A\| = 1$. However, for symmetric matrices $\rho(A) = \|A\|$, because for a symmetric matrix all eigenvalues $\lambda_i, i = 1, \ldots, n$ are real and there is a full orthogonal system of eigenvectors.

The following important formula establishes connection between (16) and (17)

$$\rho(A) = \lim_{k \to \infty} \|A^k\|^{\frac{1}{k}}. \tag{18}$$

It follows from (18) that for $\lim_{k \to \infty} A^k = 0$ it is necessary and sufficient $\rho(A) < 1$. Moreover, for a given $\varepsilon > 0$, there is $c(\varepsilon)$ that

$$\|A^k\| \leq c(\varepsilon)(\rho(A) + \varepsilon)^k \text{ for } k = 1, 2, \ldots \tag{19}$$

and $\rho(A) + \varepsilon < 1$.

If follows from (19) that $\rho(A) < 1$ is necessary and sufficient condition for a sequence $\{x^s\}_{s \in N} : x_{s+1} = Ax_s$ converge to zero.

Exercise 1. If $\rho(A) < 1$, then matrix equation

$$A^T U A = U - C$$

has a solution U, which is a symmetric matrix if C is symmetric and $U \succeq C$ if $C \succeq 0$.

Hint: From (19) follows existence $0 < q < 1$ that $\|A^s\| \le cq^s$; therefore there is $U = \sum_{s=0}^{\infty}(A^T)^s C A^s$.

Matrix A is stable if for its eigenvalues $\lambda_1, \dots, \lambda_n$, the following holds

$$\gamma = \max_{1 \le l \le n} Re\lambda_i < 0. \tag{20}$$

Exercise 2. Show that stability of A is necessary and sufficient for

$$\lim_{t \to \infty} e^{At} = 0$$

and for any given $\varepsilon > 0$ there is $c = c(\varepsilon)$ that

$$\|e^{At}\| \le c(\varepsilon) e^{(\gamma+\varepsilon)t}, \forall t \ge 0,$$

where $\gamma = \max Re\lambda_i$.

Hint: If λ is an eigenvalue of A, then $f(\lambda)$ is an eigenvalue of $B = f(A)$, i.e.,

$$\rho(B) = \max_{1 \le i \le n} e^{Re\lambda_i} = e^{\gamma}.$$

Theorem 1. *(Lyapunov) Let matrix A be stable and $C = C^T$; then the system*

$$AU + UA^T = -C \tag{21}$$

has a solution U, that $U \succ 0 (U \succeq 0)$ if $C \succ 0 (C \succeq 0)$

Proof. It follows from Exercise 2 that $U = \int_0^\infty e^{At} C e^{A^T t} dt$ exists; also $Z(t) = e^{At} C e^{A^T t}$ is solution of the following differential equation:

$$\dot{Z}(t) = AZ + ZA^T, Z(0) = C.$$

On the other hand

$$U = \int_0^\infty Z(t) dt;$$

therefore

$$AU + UA^T = \int_0^\infty (AZ + ZA^T) dt = \int_0^\infty \dot{Z}(t) dt = -Z(0) = C.$$

Hence $U = \int_0^\infty e^{At} C e^{A^T t} dt$ is the solution of system (21), and if $C \succ 0 (C \succeq 0)$, then $U \succ 0 (U \succeq 0)$.

The connection between stable matrices and matrices with $\rho(A) < 1$ establishes the following Theorem.

Theorem 2. *Let A be a stable matrix, with eigenvalues $\lambda_i, i = 1,2,\dots,n$, that is, (20) holds; then for any $0 < \gamma < \min_{1\le i\le n}\{-2Re\lambda_i|\lambda_i|^{-2}\}$, the matrix $B = I + \gamma A$ has a spectral radius $\rho(B) < 1$.*

For eigenvalues B we have $\mu_i = 1 + \gamma\lambda_i$; therefore

$$\begin{aligned}|\mu_i|^2 &= (1+\gamma Re\lambda_i)^2 + \gamma^2(Im\lambda_i)^2 = 1 + 2\gamma Re\lambda_i + \gamma^2[(Re\lambda_i)^2 + (Im\lambda_i)^2]\\ &= 1 + 2\gamma Re\lambda_i + \gamma^2|\lambda_i|^2.\end{aligned}$$

So for

$$\gamma < \min\{-2Re\lambda_i|\lambda_i|^{-2}\},$$

we have $|\mu_i| < 1$, that is $\rho(B) < 1$.

SIM for Nonlinear System

The SIM generates a sequence $\{x_s\}_{s\in N}$ by formula

$$x_{s+1} = C(x_s), \tag{22}$$

where $C : \mathbb{R}^n \to \mathbb{R}^n$ is a given map.

Let us assume that the map C has a fixed point $x^* \in \mathbb{R}^n$, that is,

$$x^* = C(x^*). \tag{23}$$

Let us consider conditions on the map C, which guarantee convergence $\{x_s\}_{s\in N}$ to x^*.

Theorem 3. *Let x^* be a fixed point in (22), $C(x)$ be continuous differentiable at the neighborhood x^*, and spectral radius $\rho = \rho(\nabla C(x^*))$ of the Jacobian matrix*

$$\nabla C(x^*) = \begin{pmatrix}\nabla c_1(x^*)\\ \nabla c_n(x^*)\end{pmatrix}$$

be less than 1; then the sequence $\{x_s\}_{s\in N}$ generated by (22) converges locally to x^* and the following bound holds.

For any $0 < \varepsilon < 1-\rho$, there is $\delta > 0$ and $C > 0$ that

$$\|x_s - x^*\| \le C(\rho+\varepsilon)^s$$

if $\|x_0 - x^*\| \le \delta$

Proof. Let $A = \nabla C(x^*)$; then keeping in mind smoothness of $C(x)$ in the neighborhood of x^*, we obtain

$$C(x) = C(x^*) + A(x-x^*) + o(x-x^*).$$

Therefore we can rewrite (22) as follows:

$$u_{s+1} = Au_s + \mathrm{v}_s, \tag{24}$$

where $u_s = x_s - x^*, \mathrm{v}_s = o(x_s - x^*)$.

Therefore

$$u_{s+1} = A^{s+1}u_0 + \sum_{i=0}^{s} A^{s-i}\mathrm{v}_i$$

or

$$\|u_{s+1}\| \leq \|A^{s+1}\|\|u_0\| + \sum_{i=0}^{s} \|A^{s-i}\|\|\mathrm{v}_i\|. \tag{25}$$

From (19) follows $\|A^s\| \leq C(\varepsilon)(\rho+\varepsilon)^s$. Using $\|\mathrm{v}_s\| = o(\|x_s - x^*\|)$ we can find $C > C(\varepsilon)$ independent on s that

$$\|x_s - x^*\| \leq C(\rho+\varepsilon)^s.$$

In case $\nabla C(x^*) = 0$ the sequence (22) converges to the fixed point $x^* = C(x^*)$ with quadratic rate.

Implicit Function Theorem

Theorem 4. *Let $f : \mathbb{R}^{n+m} \to \mathbb{R}^m$ be vector function of $x \in \mathbb{R}^n$ and $y \in \mathbb{R}^m$ such that*
(1) $f(\bar{x},\bar{y}) = 0 \in \mathbb{R}^m$,
(2) f continuous and has continuous nonsingular matrix $\nabla_y f(x,y)$ in an open set

$$S(\bar{u},\varepsilon) = \{u = (x,y) \in \mathbb{R}^{n+m} : \|u - \bar{u}\| < \varepsilon\}.$$

Then there exist $\varepsilon_x > 0$ and $\varepsilon_y > 0$, open sets $S_{\bar{x}} = \{x \in \mathbb{R}^n : \|x - \bar{x}\| < \varepsilon_x\}$ and $S_{\bar{y}} = \{y \in \mathbb{R}^m : \|y - \bar{y}\| < \varepsilon_y\}$, and continuous function $y : \mathbb{R}^n \to \mathbb{R}^m$ that $y(\bar{x}) = \bar{y}$ and

$$f(x,y(x)) \equiv 0, \forall x \in S(\bar{x},\varepsilon_x)$$

The vector function $y(x)$ is unique in a sense that if $x \in S_{\bar{x}}, y \in S_{\bar{y}}$ and $C(x,y) = 0$, then

$$y = y(x). \tag{26}$$

Moreover, if C is k times continuously differentiable, the same is true for y. In particular,

$$\nabla_x y(x) = -(\nabla_y f(x,y(x)))^{-1}\nabla_x f(x,y(x)), x \in S_{\bar{x}} \tag{27}$$

Corollary 4. *Let us consider a map $C : \mathbb{R}^n \to \mathbb{R}^n$ such that $C(x^*) = 0$. We assume that there is $\varepsilon > 0$ that the Jacobian $\nabla C(x)$ is continuous in $S(x^*,\varepsilon) = \{x \in \mathbb{R}^n : \|x - x^*\| \leq \varepsilon\}$ and $det\nabla C(x^*) \neq 0$, then the system*

$$C(x) = u \tag{28}$$

has a solution $x(u) : x(0) = x^*$ *and there is small enough* $\delta > 0$ *that for* $\|u\| \leq \delta$

$$x(u) = x^* - (\nabla c(x^*))^{-1} u + o(\|u\|). \tag{29}$$

Hausdorff Distance Between Two Compact Sets

Let X and Y be two bounded and closed sets in $\mathbb{R}^n$ and $d(x,y) = \|x - y\|$ be the Euclidean distance between $x \in X, y \in Y$. Then the Hausdorff distance between X and Y is defined as follows:

$$\begin{aligned} d_H(X,Y) &:= \max\{\max_{x \in X} \min_{y \in Y} d(x,y), \max_{y \in Y} \min_{x \in X} d(x,y)\} \\ &= \max\{\max_{x \in X} d(x,Y), \max_{y \in Y} d(y,X)\}. \end{aligned}$$

For any pair of compact sets X and $Y \subset \mathbb{R}^n$,

$$d_H(X,Y) = 0 \Leftrightarrow X = Y.$$

Let $Q \subset \mathbb{R}^m_{++}$ be a compact set, $\hat{Q} = \mathbb{R}^m_{++} \setminus Q$, $S(u,\varepsilon) = \{\mathrm{v} \in \mathbb{R}^m_+ : \|u - \mathrm{v}\| \leq \varepsilon\}$ and

$$\partial Q = \{u \in Q | \exists \mathrm{v} \in Q : \mathrm{v} \in S(u,\varepsilon), \exists \hat{\mathrm{v}} \in \hat{Q} : \hat{\mathrm{v}} \in S(u,\varepsilon)\}, \forall \varepsilon > 0$$

be the boundary of Q.

Let $A \subset B \subset C$ be convex and compact sets in $\mathbb{R}^m_+$; then the following inequality follows from the definition of Hausdorff distance:

$$d_H(A, \partial B) < d_H(A, \partial C). \tag{30}$$

References

Adler, I., Monteiro, R.: Limiting behavior of the affine scaling continuous trajectories for linear programming problems. Math. Program. **50**, 29–51 (1991)

Adler, I., Karmarkar, N., Resende, M., Veiga, G.: An implementation of Karmarkar's algorithm for linear programming. Math. Program. **44**, 297–335 (1989). Errata in Math. Program. **50**, 415 (1991)

Alber, M., Reemtsen, R.: Intensity modulated radiotherapy treatment planning by use of a barrier-penalty multiplier method. Optim. Methods Software **22**(3), 391–411 (2007)

Anstreicher, K.M.: Potential reduction algorithms, In: Terlaky, T. (ed.) Interior Point Methods of Mathematical Programming, pp. 125–158. Kluwer Academic Publishers, Dordrecht (1996)

Antipin, A.: A gradient-type method for finding the saddle point of the augmented lagrangian (In Russian). Ekonomika i Matemat. Metody **13**(3), 560–65 (1977)

Antipin, A.: Methods of Nonlinear Programming Based on the Direct and Dual Augmentation of the Lagrangian. Moscow VNIISI (1979)

Antipin, A.: The Gradient and Exstragradient Approaches in Bilinear Equilibrium Programming A. In: Dorodnizin Computing Center RAS (in Russian) (2002)

Arnold, V.: Small denominators and problem of stability in classical and celestial mechanics. Uspehi Matematicheskih Nauk **18**(6), 91–192 (1963)

Ashmanov, S.: Introduction into Mathematical Economics. Moscow, Nauka (1984)

Auslender, A., Teboulle, M.: Asymptotic Cones and Functions in Optimization and Variational Inequalities. Springer, Berlin (2003)

Auslender, A., Teboulle, M.: Interior projection-like methods for monotone variational inequalities. Math. Program. **104**(1), 39–68 (2005)

Auslender, R., Cominetti, R., Haddou, M.: Asymptotic analysis for penalty and barrier methods in convex and linear programming. Math. Oper. Res. **22**(1), 43–62 (1997)

Auslender, A., Teboulle, M., Ben-Tiba S.: Interior proximal and multipliers methods based on second-order homogeneous kernels. Math. Oper. Res. **24**(3), 645–668 (1999)

R. A. Polyak, *Introduction to Continuous Optimization*, Springer Optimization and Its Applications 172, https://doi.org/10.1007/978-3-030-68713-7

Bailey, D., Townsend, D., Valk, P., Maisey, M.: Positron Emmission Tomography. In: Basic Sciences Secaucus. Springer, London (2005)

Bakushinskij, A., Polyak, B.: On the solution of variational inequalities. Sov. Math. Doklady **14**, 1705–1710 (1974)

Barnes, E.R.: A variation on Karmarkar's algorithm for solving linear programming problem. Math. Program. **36**, 174–182 (1986)

Barnes, E.R.: Some results concerning convergence of the affine scaling algorithm. Contemp. Math. **114**, 131–139 (1990)

Bauschke, H., Matouskova, E., Reich, S.: Projection and proximal point methods, convergence results and counterexamples. Nonlinear Anal. **56**(5), 715–738 (2004)

Beck, A., Teboulle, M.: Fast gradient based algorithm for constrained total variation image denoising and deblurring problems. Trans. Imag. Proc. **18**, 2419–2434 (2009)

Beck, A., Teboulle, M.: A fast iterative shrinkage—thresholding algorithm for linear inverse problem. SIAM J. Imag. Sci. **2**(1), 183–202 (2009a)

Ben-Tal, A., Nemirovski, A.: Optimal design of engineering structures. In: Optima, vol. 47 (1995)

Ben-Tal, A., Nemirovski, A.: Lectures on Modern Convex Optimization Analysis, Algorithms and Engineering Applications. SIAM, Philadelphia (2001)

Ben-Tal, A., Zibulevsky, M.: Penalty-barrier methods for convex programming problems. SIAM J. Optim. **7**, 347–366 (1997)

Ben-Tal, A., Yuzefovich, B., Zibulevsky, M.: Penalty-Barrier Multipliers Method for Minimax and Constrained Smooth Convex Optimization. Technion, Research report, pp. 9–92 (1992)

Benthem, M., Keenan, M.: Fast algorithm for the solution of large-scale non-negativity constrained least squares problems, J. Chemom. **18**, 441–450 (2004)

Berke, L., Khot, N., Polyak, R., Schneur, R.: Structural optimization using Newton modified barrier method. Struct. Optim. **10**(3), 209–216 (1995)

Bertsekas, D.P.: Multiplier methods: a survey. Automatica **12**, 45–143 (1976)

Bertsekas, D.P.: Constrained Optimization and Lagrange Multiplier Methods. Academic Press, New York (1982)

Bertsekas, D.P.: Nonlinear Programming. Athena Scientific, Belmont (1999)

Bixby, R.E.: A brief history of linear and mixed—integer programming computation. In: Documenta Mathematica Optimization Stories, Berlin (2012)

Boyd, S., Vanderberghe, L.: Convex Optimization. Cambridge University, United Kingdom (2004)

Bregman, L.: The relaxation method for finding the common point of convex sets and its application to the solution of problems in convex programming. USSR Comput. Math. Math. Phys. **7**, 200–217 (1967)

Bregman, L., Censor, Y., Reich, S.: Dykstra algorithm as the nonlinear extention of Bregman's optimization method. J. Convex Anal. **6**(2), 319–333 (1999)

Breitfeld, M., Shanno, D.: Computational experience with modified log-barrier methods for nonlinear programming. Annals Oper. Res. **62**, 439–464 (1996)

Carroll, C.: The created response surface technique for optimizing nonlinear-restrained systems. Oper. Res. **9**(2), 169–184 (1961)
Censor, Y., Zenios, S.: The Proximal minimization algorithm with d–functions. J. Optim. Theory Appl. **73**, 451–464 (1992)
Censor, Y., Gibali, A., Reich, S.: The subgradient extragradient method for solving variational inequalities in hilbert space. J. Optim. Theory Appl. **148**, 318–335 (2011)
Chen, C., Mangasarian, O.L.: Smoothing methods for convex inequalities and linear complementarity problems. Math. Program. **71**, 51–69 (1995)
Chen, G., Teboulle, M.: Convergence analysis of a proximal–like minimization algorithm using Bregman functions. SIAM J. Optim. **3**(4), 538–543 (1993)
Cheney, W., Goldstein, A.: Note on paper by Zuhovickii concerning the Tchebycgeff problem for linear equations. J. Soc. Indust. Appl. Math. **6**(3), 233–239 (1958)
Courant, R.: Variational methods for the solution of problems of equilibrium and vibrations. Bull. Am. Math. Soc. **49**, 1–23 (1943)
Dantzig, G.: Linear Programming and Extensions. Princeton University, Princeton (1963)
Dantzig, G., Wolfe, P.: The decomposition principle for linear programs. Oper. Res. **8**(1), 101–111 (1960)
Dantzig, G., Fulkerson, D., Johnson, S.: Solution of a large—scale traveling—salesman problem. J. Oper. Res. Soc. Am. **2**(4), 393–410 (1954)
Dantzig, G., Ford, L., Fulkerson, D.: In: Kuhn, H., Tucker, A. (eds.) Algorithm for Simultaneous Solution the Primal and Dual Linear Programming Problem in Linear Inequalities and Related System (1956)
Daube-Witherspoon, M., Muehllehner: An iterative space reconstruction algorithm suitable for volume ECT. IEEE Trans. Med. Imaging **5**, 61–66 (1986)
Debreu, G.: Definite and Semidefinite Quadratic forms. Economica **20**, 295–300 (1952)
Demyanov, V., Rubinov, A.: Approximate Methods in Optimization Problems. American Elsevier Publishing, New York (1970)
Den Hertog, D., Roos, C., Vial, J.: A complexity reduction for the long-step path-following algorithm for linear programming. SIAM J. Optim. **2**, 71–87 (1992)
Dennis, J.E., Schnabel, R.B.: Numerical Methods for Unconstrained Optimization and Nonlinear Equations. SIAM, Philadelphia (1996)
Dikin, I.: Iterative solution of problems of linear and quadratic programming. Doklady Akademii Nauk SSSR **174**, 747–748 (1967). Translated in Soviet Mathematics Doklady **8**, 674–675 (1967)
Dikin, I.: On the convergence of an iterative process. Upravlyaemye Sistemi **12**, 54–60 (1974, in Russian)
DiPillo, G., Grippo, L.: A new class of augmented lagrangian in nonlinear programming. SIAM J. Control Optim. **77**(1), 618–628 (1979)
Dorfman, R., Samuelson, R.: Solow Linear Programming and Economic Analysis. Mc Graw-Hill, New York (1958)
Eckstein, J.: Nonlinear proximal point algorithms using bregman functions with applications to convex programming. Y. Math. Oper. Res. **18**(1), 202–226 (1993)

Eggermont, P.: Multiplicative iterative algorithm for convex programming. Linear Algebra Appl. **130**, 25–32 (1990)

Ermoliev, Y., Shor, N.: Minimization of nondifferetiable functions. Kibernetica **1**, 101–102 (1967)

Farcchinei, F., Pang, J.: Finite—Dimentional Variational Inequalities and Complementarity Problems, vol 1 and 2. Springer, Berlin (2003)

Farkas, J.: Teorie der einfachen Ungleivhungen. J. Reine und Angewandte Mathematik **124**, 1–27 (1902)

Fiacco, A.V., McCormick, G.P.: Nonlinear Programming: Sequential Unconstrained Minimization Techniques. SIAM, Philadelphia (1990)

Frank, M., Wolfe P.: An algorithm for quadratic programming. Nav. Res. Logist. Q. **3**, 95–110 (1956)

Freund, R.: Theoretical efficiency of a shifted barrier function algorithm for linear programming. Linear Algebra Appl. **152**, 19–41 (1991)

Freund, R., Grigas, P.: New analysis and results for the Frank–Wolfe method. Math. Prog. **155**(1–2), 199–230 (2016)

Frisch, K.: The Logarithmic Potential Method for Solving Linear Programming Problems, Memorandum. University Institute of Economics, Oslo (1955)

Gale, D.: The Theory of Linear Economic Models, NY (1960)

Gantmacher, D.: The Theory of Matrices, ANS (1959)

Garber, D., Hazan, E.: Faster Rates for the Frank–Wolfe Method over Strongly Convex Sets (2014, 2015). arxiv: 1406.1305v2[mathOC]

Gill, P., Murray, W., Saunders, J., Tomlin, J., Wright, M.: On projected Newton Barrier methods for linear programming and an equivalence to Karmarkar's projective method. Math. Program. **36**, 183–209 (1986)

Goffin, J.: On the convergence rates of subgradient optimization models. Math. Program. **13**(3), 329–347 (1977)

Goffin, J., Vial, J.: On the computation of weighted analytic centers and dual ellipsoids with the projective AlgoRithm. Math. Program. **60**, 81–92 (1993)

Goldfarb, D., Todd, M.: A relaxed version of Karmarkar's method. Meth. Program. **40**, 289–315 (1988)

Goldfarb, D., Mints, K., Polyak, R., Yuzefovich, I.: Modified Barrier—augmented lagrangian method for constrained minimization. Comput. Optim. Appl. **14**, 55–74 (1999)

Goldshtein, E., Tretiakov, N.: Modified Lagrangian Functions. Nauka, Moscow (1989)

Goldstein, A.: Convex programming in gilbert space. Bull. Am. Math Soc. **70**, 709–710 (1964)

Gonzaga, C.: An algorithm for solving linear programming problems in $O(n^3L)$ operations. In: Megiddo, N. (ed.) Progress in Mathematical Programming: Interior Point and Related Methods, pp. 1–28. Springer, New York (1989)

Gonzaga, C.: Path-following methods for linear programming. SIAM Rev. **34**(2), 167–227 (1992)

Griva, I., Polyak, R.: Primal-dual nonlinear rescaling methods for convex optimization. J. Optim. Theory Appl. (JOTA) **122**(1), 111–156 (2004)

Griva, I., Polyak, R.: Primal-dual nonlinear rescaling method with dynamic scaling parameter update. Math. Program. Ser. A **106**, 237–259 (2006)

Griva, I., Polyak, R.: 1.5-Q- superlinear convergence of an exterior point method for constrained optimization. J. Global Optim. **40**, 679–695 (2008)

Griva, I., Polyak, R.: Proximal point nonlinear rescaling method for convex optimization. Numer. Algebra Control Optim. **1**(3), 283–299 (2013)

Griva, I., Polyak, R., Sobieski, J.: The Newton log-sigmoid method in constrained optimization, a collection of technical papers. In: Proceedings of the 7th AIAA/USAF/NASA/ISSMO Symposium on Multidisciplinary Analysis and Optimization, vol. 3, pp. 2193–2201 (1998)

Griva, I., Polyak, R., Shen-Shyang, H.: Support vector machine via nonlinear rescaling method. Optim. Lett. **1**, 367–378 (2007)

Griva, I., Shanno, D., Vanderbei, R., Benson, H.: Global convergence of a primal-dual interior-point method for nonlinear programming. Algorithmic Oper. Res. **3**(1), 12–29 (2008)

Grossman, K., Kaplan, A.A.: Nonlinear programming based on unconstrained minimization. Novosibirsk, Nauka (1981), in Russian

Guler, O.: On the convergence of the proximal point algorithm for convex minimization. SIAM J. Control Optim. **29**, 403–419 (1991)

Guler, O.: New proximal point algorithms for convex minimizations. SIAM J. Optim. **2**(4), 649–664 (1992)

Guler, O., den Hertog, D., Roos, C., Terlaky, T., Tsuvhiya, T.: Degeneracy in interior point methods for linear programming: a survey. Ann. Oper. Res. **46**, 107–138 (1993)

Gusev, M., Evans, D.: The fastest matrix vector multiplication. Parallel Algorithms Appl. **1**(1), 57–67 (1993)

Haarhoff, P., Buys, J.: A new method for the optimization of a nonlinear function subject to nonlinear constraints. Comput. J. **13**(2), 178–84 (1970)

Hestenes, M.R.: Multipliers and gradient methods. J. Optim. Theory Appl. **4**, 303–320 (1969)

Hoffman, A.: On approximate solution of system of linear inequalities. J. Res. Nat. Bureau Stand. **49**, 263–265 (1952)

Huard, P.: A method of centres by upper-bounding function with applications. In: Non-linear Programming. North Holland, Amsterdam, pp. 1–30 (1967)

Huard, P.: Resolution of mathematical programming with nonlinear constraints by the method of centers. In: Abadie, J. (ed.) Nonlinear Programming, pp. 207–219. North Holland, Amsterdam (1970)

Ioffe, A., Tichomirov, V.: Duality of convex functions and extremum problems. Uspexi Mat. Nauk **23**(6)(144), 51–116 (1968)

Ioffe, A.D., Tihomirov, V.M.: Theory of Extremal Problems. North-Holland, Amsterdam (2009)

Iusem, A., Svaiter, B.A.: Variant of Korpelevich's method for the variational inequalities with a new search strategy. Optimization **42**(4), 309–321 (1997)

Iusem, A., Svaiter, B., Teboulle, M.: Entropy-like proximal methods in convex programming. Math. Oper. Res. **19**, 790–814 (1994)

Jarre, F., Sonnevend, G., Stoer, G.: An Implementation of the method of analytic centers. In: Benoussan, A., Lions, J.L. (eds.) Lecture Notes in Control and Information Sciences, vol. 111. Springer, Berlin (1988)

Jarre, F., Kocvara, M., Zowe, J.: Interior point methods for mechanical design problems. Preprint No. 173, Institut fur Angewandte Matematik, Universitat Erlangen Nurnber, Martensstr. 3, D-91058 Erlangen, Germany (1996)

Jensen, D., Polyak, R.: The convergence of MBF method for convex programming. IBM J. Res. Dev. **38**, 307–321 (1994)

Jensen, D., Polyak, R., Schneur, R.: Experience with Modified Barrier Function Methods for Linear Programming, Research Report Department of Mathematical Sciences, vol. 10598. IBM T.J. Watson Research Center, New York, pp. 1–35 (1993)

Jensen, B., Roos, C., Terlaky, T., Vial, J.: Primal-dual algorithms for linear programming based on th logarithmic barrier method. J. Optim. Theory Appl. **83**, 1–26 (1994)

Jensen, B., Roos, C., Terlaky, T.: A polynomial dikin-type primal-dual algorithm for linear programming. Math. Oper. Res. **21**, 341–353 (1996)

Kantorovich, L.: Mathematical methods of organizing and planning production. Manage. Sci. **6**(4), 366–422 (1939). JSTOR 2627082

Kantorovich, L.:The Best Use of Economic Resources. Pergamon Press, New York (1959)

Kantorovich, L., Akilow, G.P.: Functional Analysis in Normed Spaces. MacMillan, New York (1964)

Kanzow, C.: Global convergence properties of some iterative methods for linear complementarity problems. SIAM Optim. **6**, 326–341 (1996)

Karmarkar, N.: A new polynomial-time algorithm for linear programming. Combinatorica **4**, 373–395 (1984)

Khachiyan, L.: A polynomial algorithm in linear programming. Doklady Akademiia Nauk SSSR **244**, 1093–1096 (1979). (Translated into English in Soviet Mathematics Doklady **20**, 191–194)

Khobotov, E.: A modification of the extragradient method for solving variational inequalities and some optimization problems (Russian). Zh. Vychisl. Mat. i Mat Fiz. **27**(10), 1462–1473, 1957 (1987)

Knopp, K.: Infinite Sequence and Series. Dover Publication Inc., New York (1956)

Kočvara, M., Stingl, M.: Resent Progress in the NLP-SDP Code PENNON, Workshop "Optimization and Applications", Oberwalfach (2005)

Kočvara, M., Stingl, M.: On the solution of large-scale SDP problems by the modified barrier method using iterative solvers. Math. Program. Series B **109**(2–3), 413–444 (2007)

Kočvara, M., Stingl, M.: PENNON: Software for Linear and Nonlinear Matrix Inequalities (2015). arXiv:1504.07212v2 [mat. OC]

Kojima, M., Mizuno, S., Yoshise, A.: A polynomial-time algorithm for a class of linear complementarity problems. Math. Program. **44**, 1–26 (1989)

Kojima, M., Mizuno, S., Yoshise, A.: A primal-dual interior point algorithm for linear programming, In: Megiddo, N. (ed.) Progress in mathematical programming: interior point and related methods, pp 29–47. Springer, New York (1989a)

Kojima, M., Megiddo, N., Noma, T., Yoshise, A.: A unified approach to interior point algorithms for linear complementarity problems. Lecture Notes in Computer Science, vol. 538. Springer, Berlin (1991)

Konnov, I.: Equilibrium Models and Variational Inequalities, vol 210. Elsevier, Amsterdam (2007)

Koopmans, T.C.: Optimum utilization of the transportation system. Econometrica **17**, 136–146 (1949)

Korpelevich, G.: Extragradient method for finding saddle points and other problems. Matecon **12**(4), 747–756 (1976)

Kort, B., Bertsekas, D.: Multiplier methods for convex programming. In: Proceedings 1973 IEEE Conference on Decision and Control, San Diego, California, pp 428–432 (1973)

Kuhn, H.: On a theorem of wald, linear inequalities and related systems. In: Annals of Mathematics Studies, vol. 38. Princeton University, Princeton, pp. 265–273 (1956)

Kuhn, H., Tucker, A.: Nonlinear programming. In: Neyman, J. (ed.) Proceedings of the Second Berkeley Symposium on Mathematical Statistics and Probability. University of California Press, Berkeley, pp. 481–492 (1951)

Lancaster, K.: Mathematical Economics. The Macmillan Company, New York (1968)

Lawson, C., Hanson, R.: Solving Least Squares Problems. In: SIAM Classic in Apply Mathematics, Philadelphia (1995)

Lemarechal, C. In: Dixon, L.C.W., Spedilato, E., Szego, G.P. (eds.) Nonlinear Optimization: Theory and Algorithms, Boston (1980)

Leontief, W.: Input-Output Economics. Oxford University Press, Oxford (1966)

Levitin, E., Polyak, B.: Minimization methods in the presence of constraints. Z. Vycisl. Mat. i Mat Fizz **6**, 787–823 (1966)

Lieu, B.-T., Huard, P.: La methods des centres dans un espace topologique. Num. Mat. **8**(1), 56–67 (1966)

Lusternik: On conditional extrema of functionals. Mathematicheskij Sbornik **41**(3), 390–401 (1934)

Lustig, I., Marsten, R., Shanno, D.: Computational experience with a primal-dual interior point method for linear programming. Linear Algebra Appl. **152**, 191–222 (1991)

Lustig, I., Marsten, R., Shanno, D.: On implementing Mehrotra's predictor-corrector interior point method for linear programming. SIAM J. Optim. **2**, 435–449 (1992)

Lustig, I., Marsten, R., Shanno, D.: Interior point methods for linear programming: computational state of the art. ORSA J. Comput. **6**(1), 1–14 (1994)

Mangasarian, O.: Mathematical programming in neural networks. ORSA J. Comput. **5**(4), 349–360 (1993)

Martinet, B.: Regularization d' inequations variationelles par approximations successive. Revue Francaise d'Automatique et Informatique Recherche Operationelle **4**, 154–159 (1970)

Martinet, B.: Perturbation des Methodes d'optimization, Application, R.A.I.R.O. In: Analyse numerique/Numerical Analysis (1978)

Matioli, L., Gonzaga, C.: A new family of penalties for augmented lagrangian methods. Numer. Linear Algebra Appl. **15**, 925–944 (2008)

Megiddo, N.: Pathways to the optimal set in linear programming. In: Progress in Mathematical Programming—Interior Point and Related Methods. Springer, Berlin, pp. 131–158 (1989)

Megiddo, N., Shub, M.: Boundary behavior of interior point methods in linear programming. Math. Oper. Res. **14**(1), 97–146 (1989)

Mehrotra, S.: Higher order methods and their performance. Technical Report 90-16R1, Department of Industrial Engineering and Management Science, Northwestern University, Evanston, IL 60208, USA (1990). Revised July 1991

Mehrotra, S.: On the implementation of a (primal-dual) interior point method. SIAM J. Optim. **2**(4), 575–601 (1992)

Melman, A., Polyak, R.: The Newton modified barrier method for QP problems. Ann. Oper. Res. **62**, 465–519 (1996)

Mifflin, R.: Rates of convergence for a method of centers algorithm. JOTA **18**(2), 199–228 (1976)

Migdalas, A., Pardalos, P., Storoy, S.: Parallel Computing in Optimization. Kluwer Academic Publishers, Dordrecht (1997)

Mizuno, S., Todd, M.: An $O(n^3 L)$ adaptive path following algorithm for a linear complementarity problem. Math. Program. **52**, 587–595 (1991)

Monteiro, R., Adler, I., Resende, M.: A polynomial - time primal-dual affine scaling algorithm for linear and convex quadratic programming and its power series extension. Math. Oper. Res. **15**, 191–214 (1990)

Mordukhovich, B.: Variational Analysis and Applications. Springer, Berlin (2018)

Moreau, J.: Proximite' et Dualite' Dans un Espace Hilbertien. Bull. Soc. Math. France **93**, 273–299 (1965)

Motzkin, T.: New techniques for linear inequalities and optimization. In: Project SCOOP, Symposium on Linear Inequalities and Programming, Planning Research Division, Director of Management Analysis Service, U.S. Air Force, Washington, vol. 10 (1952)

Nash, J.: Non-cooperative games. Ann. of Math. 54(2), MR0043432(13, 261g) (1951)

Nash, S., Polyak, R., Sofer, A.: A numerical comparison of barrier and modified – barrier methods for large–scale bound–constrained optimization. In: *Large scale Optimization: State of Art*. Kluwer Academic, Dordrecht (1994)

Nemirovski, A., Yudin, D.: Informational Complexity and Efficient Methods for Solution of Convex Extremal Problems. Wiley, New York (1983)

Nesterov, Y.: A method for solving convex programming problems with convergence rate $O(k^{-2})$. Dokl. Akad. Navk. SSSR **269**(3), 543–547 (1983, in Russian)

Nesterov, Y.: Minimization methods for nonsmooth convex and quasiconvex functions. Economika i Mat. Metody **11**(3), 519–531 (1984) (In Russian, translated as Mat. Econ)

Nesterov, Y.: Introductory Lectures on Convex Optimization. Kluwer Academic Publishers, Norwell, (2004)

Nesterov, Y., Nemirovski, A.: Interior Point Polynomial Algorithms in Convex Programming. SIAM, Philadelphia (1994)

Neumann, J.: Zur Theorie der Gesellschaftsspiele. Math. Ann. **100**, 295–320 (1928)

Pardalos, P., Resende, M.: Interior point methods for global optimization. In: Terlaky, T. (ed.) IPM of Mathematical Programming Applied Optimization, vol. 5. Kluwer, Dordrecht, pp. 467–500 (1996)

Pardalos, P., Wolkowicz, H.: Topics in Semidefinite and Interior - Point Methods. Fields Institute Communications Series 18, American Mathematical Society, New York (1998)

Pardalos, P., Ye, Y., Han, G.: Computational Aspects of an Interior Point Algorithm for Quadratic Problems with box Constraints. In: Large-Scale Numerical Optimization. SIAM, Philadelphia, pp. 92–112 (1990)

Polak E.: Computational Methods in Optimization: A Unified Approach. Academic Press, New York (1971)

Polyak, B.: Gradient methods for the minimization of functionals USSR. Comput. Maths. Math. Phys. **3**(4), 864–878 (1963)

Polyak, B.: A General Method for Solving Extremum Problems. Soviet Math. Dokl. **8**(3), 593–597 (1967)

Polyak, B.: Iterative methods using lagrange multipliers for solving extremal problems with constraints of the equation type. Comput. Math. Math Phys. **10**(5), 42–52 (1970)

Polyak, B.: Introduction to Optimization. Optimization Software, New York (1987)

Polyak, B.: Newton's methods and its use in optimization. Eur. J. OR **181**, 1086–1096 (2007)

Polyak, B., Tret'yakov, N.: The method of penalty estimates for conditional extremum problems. Comput. Math. Math. Phys. **13**(1), 42–58 (1973)

Polyak, R.: Algorithm for Simultaneous Solution Primal and Dual Convex Programming Problems, Transaction of the Scientific Seminar. Cibernetic Institute, Kiev (1966)

Polyak, R.: On acceleration of the convergence of methods of convex programming. Dokl. Akad. Nauk SSSR **212**(5), 1063–1066 (1973)

Polyak, R.: On finding a fixed point of a class of set-valued mappings. Dokl. Akad. Nauk SSSR **242**(6), 1284–1288 (1978)

Polyak, R.: Modified Barrier and center methods, reposts from the Moscow Refusnik seminar. In: Annals of the New York Academy of Sciences, New York, vol. 491, pp. 194–196 (1987)

Polyak, R.: Smooth optimization methods for minimax problems. SIAM J. Control Optim. **26**(6), 1274–1286 (1988)

Polyak, R.: Modified Barrier functions (theory and methods). Math. Program. **54**(2), 177–222 (1992)

Polyak, R.: Modified Barrier Functions in Linear Programming. IBM Research Report, RC 17790 (No. 78331), pp. 1–55 (1992)

Polyak, R.: Modified Interior Distance Functions in Optimization Methods in Partial Differential Equations. In: Series: Contemporary Mathematics, vol. 209. AMS, New York (1997)

Polyak, R.: Log-sigmoid multipliers method in constrained optimization. Ann. Oper. Res. **101**, 427–460 (2001)

Polyak, R.: Nonlinear rescaling vs. smoothing technique in constrained optimization. Math. Program. **92**, 197–235 (2002)

Polyak, R.: Lagrangian transformation in convex optimization, Research Report-072004. Department of SEOR and Mathematical Science Department, GMU, Fairfax, pp 1–23 (2004)

Polyak, R.: Nonlinear rescaling multipliers methods as interior quadratic prox. In: *Computational Optimization and Applications*, vol 35, pp. 347–373 (2006)

Polyak, R.: Primal-dual exterior point method for convex optimization. Optim. Methods Software **23**(1), 141–160 (2008)

Polyak, R.: Finding generalized Warlas-Wald equilibrium. Methods Funct. Anal. Topology **14**(3), 242–254 (2008)

Polyak, R.: Regularized Newton method for unconstrained convex optimization. Math. Program. Ser. B **120**, 125–145 (2009)

Polyak, R.: On the local Quadratic convergence of the primal-dual augmented lagrangian method. Optim. Methods Software **24**, 369–379 (2009)

Polyak, R.: Lagrangian transformation and interior ellipsoid methods. J. Optim. Theory Appl. (JOTA) **164**(3), 966–992 (2015)

Polyak, R.: Nonlinear equilibrium for optimal resources allocation. In: Contemporary Mathematics, vol. 636. AMS, New York, pp. 1–17 (2015)

Polyak, R.: The projected gradient method for non-negative least squares. In: Contemporary Mathematics, vol. 636. AMS, New York, pp. 167–179 (2015)

Polyak, R.: Nonlinear input-output equilibrium. In: Contemporary Mathematics, vol. 659. AMS, New York (2016)

Polyak, R.: The Legendre Transformation in Modern Optimization in "Optimization and its Applications in Control and Data Sciences". Springer, Berlin (2016)

Polyak, R.: Exterior distance function. Pure Appl. Funct. Anal. **2**, 369–394 (2017)

Polyak, R.: Complexity of the regularized Newton method. Pure Appl. Funct. Anal. **3**(2), 327–347 (2018)

Polyak, R., Costa, J., Neyshabouri, S.: Dual fast projected gradient method for quadratic programming. Optim. Lett. **7**(4), 631–645 (2013)

Polyak, R., Primak, M.: Methods of controlling sequences for solving equilibrium problems, Part 1. Cibernetics **13**(2), 92–101 (1977a). (English translation available)

Polyak, R., Primak, M.: Methods of controlling sequences for solving equilibrium problems, Part 2. Cibernetics **13**(4), 56–62 (1977b). (English translation available)

Polyak, R., Teboulle, M.: Nonlinear rescaling and proximal–like methods in convex programming. Math. Program. **76**, 265–284 (1997)

Potra, F.: A quadratically convergent predictor - corrector method for solving linear programs from infeasible starting points. Math. Program. **67**(3), 383–406 (1994)

Potra, F., Wright, S.: Interior-point methods. J. Comput. Appl. Math. **124**, 281–302 (2000)

Powell, M.: A method for nonlinear constraints in minimization problems. In: Fletcher (ed.) Optimization. London Academic Press, London, pp. 283–298 (1969)

Powell, M.: Some convergence properties of the modified log barrier methods for linear programming. SIAM J. Optim. **50**(4), 695–739 (1995)

Pshenichnyj, B.: The Linearization Method for Constrained Optimization. Springer, Berlin (1994)

Quinn, M.: Parallel programming in C with MPI and Open MP. McGraw-Hill, New York (2004)

Ray, A., Majumder, S.: Derivation of some new distributions in statistical mechanics using maximum entropy approach. Yugoslav J. Oper. Res. **24**(1), 145–155 (2014)

Reich, S., Sabach, S.: Two strong convergence theorems for a proximal method in reflexive Banach spaces. Numer. Funct. Anal. Optim. **31**, 22–44 (2010)

Renegar, J.: A polynomial -time algorithm, based on Newton's method, for linear programming. Math. Program. **40**, 59–93 (1988)

Renegar, J.: A mathematical view of interior - point methods in convex optimization. In: MPS-SIAM Series on Optimization. SIAM, New York (2001)

Rockafellar, R.T.: Convex Analysis. Princeton University, Princeton (1970)

Rockafellar, R.T.: A dual approach to solving nonlinear programming problems by unconstrained minimization. Math. Program. **5**, 354–373 (1973)

Rockafellar, R.T.: Monotone operators and the proximal point algorithm. SIAM J. Control Optim. **14**, 877–898 (1976)

Rockafellar, R.T.: Augmented lagrangians and applications of the proximal points algorithms in convex programming. Math. Oper. Res. **1**, 97–116 (1976)

Rockafellar, R.T., Wets R.J.-B: Variational Analysis. Springer, Berlin (2009)

Roos, C., Terlaky, T.: Advances in linear optimization. In: Dell'Amico, M., Maffioli, F., Martello, S. (eds.) Annotated Bibliography in Combinatorical Optimization, Chapter 7. Wiley, New York (1997)

Roos, C., Vial, J.-Ph.: Analytic centers in linear programming. Technical Report 88–74, Faculty of Mathematics and Computer Science, TU Delft, NL-2628 BL Delft, The Netherlands (1988)

Roos, C., Terlaky, T., Vial, J.-Ph.: Theory and Algorithms for Linear Optimization: An Interior Point Approach. Wiley, Chichester (1997)

Rozen, J.: The gradient projection method for nonlinear programming part 1: linear constraints. SIAM Journal Appl. Math. **8**, 181–218 (1960)

Rozen, J.: The gradient projection method for nonlinear programming part 2: nonlinear constraints. SIAM J. Appl. Math. **9**, 514–553 (1961)

Rosen, J.: Existence and uniqueness of equilibrium points for concave n-persone games. Econometrica **33**, 520–534 (1965)

Shanno, D., Breitfeld, M., Simantiraki, E.: Implementing barrier methods for nonlinear programming. In: Terlaky, T. (ed.) Interior Point Methods of Mathematical Programming, pp 369–398, Kluwer Academic Publishers, Dordrecht (1996)

Shepp, L., Vardi, Y.: Maximin likelihood reconstruction in emission tomography. IEEE Trans. Med. Imaging **1**(2), 113–122 (1982)

Shor, N.: Nondifferentiable Optimization and Polynomial Problems. Kluwer Academic, Boston (1998)

Sonnevend, Gy.: An "analytic center" for polyhedrons and new classes of global algorithms for linear (smooth, convex) programming. In: Prekopa, A., Szelezsan, J., Strazicky, B. (eds.) System modeling and optimization: proceedings of the 12th IFIP-conference held in budapest, Hungary, September 1985. Lecture Notes in Control and Information Sciences, vol. 84, pp 866–876. Springer, Berlin (1986)

Stiefel, E.: Note on Jordan elimination, linear programming and Tchebycheff approximation. Numerische Matematik **2**(1), 1–17 (1960)

Teboulle, M.: Entropic proximal mappings with application to nonlinear programming. Math. Oper. Res. **17**, 670–690 (1992)

Tikhonov, A.: Solution of incorrectly formulated problems and the regularization method. Soviet Math. Dokl. **4**, 1035–1038 (1963)

Tikhonov, A., Arsenin, V.: Solution of ILL – Posed Problems V.H. Winston & Sons, Washington (1977)

Todd, M.: Recent development and new directions in linear programming. In: Iri, M., Tanabe, K. (eds.) Mathematical programming: recent developments and applications, pp. 109–157. Kluwer Academic Press, Dordrecht (1989)

Todd, M.: A lower bound on the number of iterations of primal-dual interior-point methods for linear programming. In: Griffits, G.F. (ed.) Numerical Analysis (1993). Pitman Research, vol. 303

Todd, M., Ye, Y.: A lower bound on the number of iterations of long-step and polynomial interior-point linear programming algorithms. Ann. Oper. Res. **62**, 233–252 (1996)

Tseng, P., Bertsekas, D.: On the convergence of the exponential multipliers method for convex programming. Math. Program. **60**, 1–19 (1993)

Tsuchiya, T.: Global convergence of the affine scaling methods for degenerate linear programming problems. Math. Program. **52**, 377–404 (1991)

Tsuchiya, T.: Affine scaling algorithm. In: Terlaky, T. (ed.) Interior point methods of mathematical programming, pp. 35–82. Kluwer Academic Publishers, Dordrecht (1996)

Tsuchiya, T., Muramatsu, M.: Global convergence of the long-step affine scaling algorithm for degenerate linear programming problems. SIAM J. Optim. **5**(3), 525–551 (1995)

Vaidya, P.: An algorithm for linear programming which requires $O((m+n)n^2 + (m+n)^{1.5}nL)$ arithmetic operations. Math. Program. **47**, 175–201 (1990)

Vanderbei, R.: Linear Programming: Foundations and Extensions. Kluwer Academic, Boston (1996)

Vanderbei, R., Meketon, M., Freedman, B.: A modification of Karmarkar's linear programming algorithm. Algorithmica **1**(4), 395–407 (1986)

Vapnik, V.: Statistical Learning Theory, Wiley, New York (1998)
Vapnik, V.: The Nature of Statistical Learning Theory, 2nd edn. Springer, Berlin (2000)
Vardi, Y., Shepp, L., Kaufman, L.: A statistical model for position emission tomography. J. Am. Statist. Assoc. **80**, 8–38 (1985)
Vial, J.: A projective algorithm for linear programming with no regularity condition. Oper. Res. Lett. **12**(1), 1–2 (1992)
Wang, W., Carreira–Perpinan, M.: Projection onto the Probability Simplex: An Efficient Algorithm with a Simple Proof and an Application (2013). arXiv:1309.1541 [cs LG]
Wolkowicz, H., Saigal, R., Vandenberghe, L. (eds.): Handbook of Semidefinite Programming: Theory, Algorithms, and Applications, vol. 27. Kluwer Academic Publisher, Dordrecht (2000)
Wright, S.: Primal-Dual Interior-Point Methods. SIAM, Philadelphia (1997)
Ye, Y.: Karmarkar's Algorithm and the Ellipsoid Method. Oper. Res. Lett. **6**, 177–182 (1987)
Ye, Y.: An $O(n^3 L)$ potential reduction algorithm for linear programming. Math. Program. **50**, 239–258 (1991)
Ye, Y.: Interior Point Algorithms: Theory and Analysis. Wiley, New York (1997)
Ye, Y., Pardalos, P.: A class of linear complementarity problems solvable in polynomial time. Linear Algebra Appl. **152**, 3–17 (1991)
Ye, Y., Todd, M.: Containing and shrinking ellipsoids in the path-following algorithm. Math. Program. **47**, 1–10 (1990)
Ye, Y., Guler, O., Tapia, R., Zhang, Y.: A quadratically convergent $O(\sqrt{n}L)$-iteration algorithm for linear programming. Math. Program. **59**, 151–162 (1993)
Ye, Y., Todd, M., Mizuno, S.: An $O(\sqrt{n}L)$ - iteration homogeneous and self-dual linear programming algorithm. Math. Oper. Res. **19**, 53–67 (1994)
Zhang, Y., Tapia, R.: Superlinear and quadratic convergence of primal-dual interior-point methods for linear programming revisited. J. Optim. Theory Appl. **73**(2), 229–242 (1992)
Zoutendijk, G.: Methods of Feasible Directions. Elsevier, Amsterdam (1960)
Zuchovitsky, S.: Algorithm for finding chebushev approximation of overdetermined system of linear equations. Dokl. Acad. Nauk SSSR **79**(4), 561–564 (1951)
Zukhovitsky, S., Polyak, R., Primak, M.: An algorithm for solving convex Chebyshev approximation problem. Dokl. Acad. Nauk SSSR **151**(1), 27–30 (1963)
Zukhovitsky, S., Polyak, R., Primak, M.: An algorithm for solving convex programming problem. Dokl. Acad. Nauk SSSR **153**(3), 991–994 (1963a)
Zukhovitsky, S., Polyak, R., Primak, M.: Numerical method for solving convex programming problem in Hilbert space. Dokl. Acad. Nauk SSSR **163**(2) (1965)
Zukhovitsky, S., Polyak, R., Primak, M.: Two methods for finding equilibrium of N-person concave games. Soviet Math. Dokl. **10**(2) (1969)
Zukhovitsky, S., Polyak, R., Primak, M.: N-person concave games and one production model. Soviet Math. Dokl. **11**(2), (1970)
Zukhovitsky, S., Polyak, R., Primak, M.: Concave N-Person Games: Numerical Methods, MAKETON (1973). (English translation available)